WILD MUSHROOMS

WILD MUSHROOMS
Characteristics, Nutrition, and Processing

Edited by
Sanju Bala Dhull, Aarti Bains, Prince Chawla, and
Pardeep Kumar Sadh

CRC Press
Taylor & Francis Group
Boca Raton London New York

CRC Press is an imprint of the
Taylor & Francis Group, an **informa** business

First edition published 2022
by CRC Press
6000 Broken Sound Parkway NW, Suite 300, Boca Raton, FL 33487-2742

and by CRC Press
4 Park Square, Milton Park, Abingdon, Oxon, OX14 4RN

CRC Press is an imprint of Taylor & Francis Group, LLC

© 2022 Sanju Bala Dhull, Aarti Bains, Prince Chawla, and Pardeep Kumar Sadh

Library of Congress Cataloguing-in-Publication Data
Names: Dhull, Sanju Bala, editor.
Title: Wild mushrooms : characteristics, nutrition, and processing/edited by Sanju Bala Dhull, Aarti Bains, Prince Chawla, Pardeep Kumar Sadh.
Description: First edition. | Boca Raton : Taylor and Francis, 2022. | Includes bibliographical references and index.
Identifiers: LCCN 2021061511 (print) | LCCN 2021061512 (ebook) | ISBN   9780367692513 (hardback) | ISBN 9780367715564 (paperback) | ISBN   9781003152583 (ebook)
Subjects: LCSH: Mushrooms. | Mushroom culture. | Mushrooms--Nutrition. | Mushrooms--Therapeutic use.
Classification: LCC QK617 .W49 2022 (print) | LCC QK617 (ebook) | DDC 579.6/163--dc23/eng/20220118
LC record available at https://lccn.loc.gov/2021061511
LC ebook record available at https://lccn.loc.gov/2021061512

ISBN: 978-0-367-69251-3 (hbk)
ISBN: 978-0-367-71556-4 (pbk)
ISBN: 978-1-003-15258-3 (ebk)

DOI: 10.1201/9781003152583

Typeset in Times
by MPS Limited, Dehradun

Contents

Part I
Introduction to wild mushrooms

Chapter 1
An introduction to wild mushrooms and their exploitation for human well-being:

Shweta Sharma, V. P. Sharma, Satish Kumar, Abhishek Sharma, Kumari Manorma, and P. K. Chauhan

Chapter 2

Anita Klaus and Wan Abd Al Qadr Imad Wan-Mohtar

Chapter 3

Rohit Biswas

Chapter 4

Jovana Vunduk, Daniel Tura, and Alona Yu. Biketova

Part II
Health aspects of wild mushrooms

Chapter 5

Somanjana Khatua and Krishnendu Acharya

Chapter 6

Jaspreet Kaur, Jyoti Singh, Vishesh Bhadariya, Simran Gogna, Sapna Jarial, Prasad Rasane, and Kartik Sharma

Chapter 7

Melinda Fogarasi, Anca Fărcaș, Sonia Socaci, Maria-Ioana Socaciu, and Cristina Anamaria Semeniuc

Preface

Over the past years, wild mushrooms have attained remarkable interest in the field of medicine and food processing due to their proficient nutritional and therapeutic properties. Globally, about 14,000 species of mushrooms are known and, among them, about 2,000 species are considered edible mushrooms. As well, almost 200 species of mushrooms have been commercially cultivated for the formulation of ayurvedic medicine and human consumption. Moreover, wild mushrooms, due to potential nutritional and health attributes, can be compared with various meat, fish, egg, and dairy products. Besides nutritional importance, mushrooms are well known for their bioactive compounds (lectins, polysaccharides such as β-glucans, polysaccharide-peptides and polysaccharide-protein complexes, lanostanoids, other terpenoids, alkaloids, sterols, and phenolic structured compounds), which are responsible for different biological and therapeutic activities including antimicrobial, antioxidant, anti-inflammatory, anti-diabetic, anti-cancerous, antiviral, and anti-immunomodulatory activities. There have been several books published covering specific applications of mushrooms; however, this book aims to provide a holistic and comprehensive view of the subject starting with the problem and then discussing classical to modern approaches of the utilization of wild varieties of mushrooms for different food and therapeutic applications. The main areas of interest include various aspects of the cultivation and commercialization of mushroom, its health attributes and utilization in waste management. The concept of this book was developed in India, shared among all authors and editors of the book, and finally with the CRC Taylor and Francis publishing team. The editors would like to thank all those who contributed to the discussion, planning, writing, and publishing of this book. We hope that this compilation will provide information on most of the important aspects of wild edible mushrooms. The book chapters are divided into four sections. **Part I: Introduction to wild mushrooms** covers the historical background, cultivation strategy, processing, preservation, commercialization, nutraceuticals, and dietary supplement formulation of mushrooms. **Part II: Health aspects of wild mushrooms** discuss mushrooms as a protein source, mineral source, its therapeutic importance, nature of bioactive components of mushrooms, and *in vitro* and *in vivo* bioactivity of edible mushrooms. **Part III: Analysis of mushroom** reveals oxidative stress, cellular longevity, techniques of analysis of mushrooms, toxicology of mushroom, and influence of food processing and functional aspects of mushrooms. **Part IV: Specific applications of wild mushrooms** discusses extracellular enzymes, degradation of xenobiotic components, reduction of pesticides, utilization of wild mushrooms in waste management, and cultivation of wild mushrooms using lignocellulosic biomass-based residue as a substrate. We hope that students, teachers, researchers, and companies involved in mushroom utilization and discovery will find *Wild Mushrooms: Characteristics, Nutrition, and Processing* to be a useful resource. With great pleasure, we extend our sincere thanks to all the contributors for their timely response, excellent contributions, and consistent support and cooperation. We welcome any suggestions and comments from readers for future improvement of the book in the new edition.

Editors

Editors

Sanju Bala Dhull, Ph.D., is an associate professor in the Department of Food Science and Technology, Chaudhary Devi Lal University, Sirsa, with more than 14 years of teaching and research experience. Her areas of interest include characterization and modification of biomolecules, edible films, nanoparticles, nanoemulsions, and their applications in foods . She has published more than 40 research papers, 2 books, and 15 book chapters. She has also presented more than 20 research papers at various national and international conferences. She is life member of the Association of Food Scientists and Technologists (India) and the Association of Microbiologists of India. She also serves as editorial board member and reviewer of several national and international journals.

Aarti Bains, Ph.D., is an assistant professor in microbiology, (Department of Biotechnology), CT Institute of Pharmaceutical Sciences, South Campus, Jalandhar, Punjab. She has a M.Sc. in microbiology from Himachal Pradesh University and completed her Mphil, Ph.D. in microbiology from Shoolini University, Solan. Her areas of interest include the extraction and identification of bioactive compounds from wild edible mushrooms and their biological activity. She has published 20 research and review articles and 4 international book chapters. She has eight years of research and teaching experience and has guided seven master's students. She is a reviewer of several national and international journals.

Prince Chawla, Ph.D., joined Lovely Professional University, Phagwara, Punjab as an assistant professor in food technology and nutrition (School of Agriculture). Dr. Chawla is an alumnus of Chaudhary Devi Lal University, Sirsa, and Shoolini University, Solan. Dr. Chawla's chief interests are in mineral fortification, functional foods, protein modification, and the detection of adulterants from foods using nanotechnology. He has worked in the Department of Biotechnology and the Department of Science and Technology's funded research projects and has eight years of research experience. He has 3 patents, 2 books, 52 international research papers, and 7 international book chapters. Dr. Chawla is guiding 4 Ph.D. students and has guided 11 M.Sc. students. He is a recognized reviewer of more than 30 international journals and reviewed several research and review articles.

Pardeep Kumar Sadh, Ph.D., is an assistant professor of biotechnology, Chaudhary Devi Lal University, Sirsa, Haryana, India. He earned his M.Sc., M.Phil, and Ph.D., from the Department of Biotechnology, Chaudhary Devi Lal University, Sirsa, Haryana, India. He has published more than 25 national and international research articles and 3 book chapters with international and national publishers. His areas of interest for research include protein engineering, structural analysis using various techniques, bioresource technology, functional foods, food biotechnology, and agricultural waste management. He was a recipient of the Rajiv Gandhi National Fellowship during his Ph.D. from University Grant Commission under the Government of India. He has been awarded best poster awards in various conferences and certificates of excellence for reviewing international reputed journals. He is the referee and recognized reviewer of several international and national journals.

Contributors

Krishnendu Acharya
Molecular and Applied Mycology and Plant
 Pathology Laboratory, Centre of Advanced
 Study, Department of Botany
University of Calcutta
Kolkata, West Bengal, India

Liliana Aguilar-Marcelino
Centro Nacional de
 InvestigaciónDisciplinariaenSalud
 Animal e Inocuidad, INIFAP
Morelos, México

Laith Khalil Tawfeeq Al-Ani
A. Department of plant protection,
 College of Agriculture
University of Baghdad, Baghdad, Iraq. B.
 School of Biology Science, UniversitiSains
 Malaysia
Minden, Pulau Pinang, Malaysia

Vishesh Bhadariya
Department of Petroleum and Chemical
 Engineering, School of Chemical
 Engineering and Physical Sciences
Lovely Professional University
Phagwara, Punjab, India

Alona Yu Biketova
Jodrell Laboratory, Royal Botanic
 Gardens, Kew
Surrey, UK

Rohit Biswas
Department of Agricultural and Food
 Engineering, Indian Institute of Technology
Kharagpur, West Bengal

Fabio Ribeiro Braga
Universidade Vila Velha
Brazil

Parveen Chauhan
Faculty of Applied Sciences and
 Biotechnology
Shoolini University
Solan, Himachal Pradesh, India

P.K. Chauhan
Faculty of Applied Sciences and
 Biotechnology
Shoolini University
Solan, Himachal Pradesh, India

Maria Simona Chis
Department of Food Engineering, Faculty of
 Food Science and Technology
University of Agricultural Science and
 Veterinary Medicine
Cluj-Napoca, Romania

Anamika Das
Department of Paramedical Sciences
Guru Kashi University
Bhatinda, India

Sanju Bala Dhull
Department of Food Science
 and Technology
Chaudhary Devi Lal University
Sirsa, Haryana, India

Kanika Dulta
Faculty of Applied Sciences and
 Biotechnology
Shoolini University
Solan, Himachal Pradesh, India

Anca Fărcas
Department of Food Science
University of Agricultural Sciences and
 Veterinary Medicine of Cluj-Napoca
Cluj-Napoca, Romania

Melinda Fogarasi
Department of Food Engineering
University of Agricultural Sciences and
    Veterinary Medicine of Cluj-Napoca
Cluj-Napoca, Romania

Simran Gogna
Department of Food Technology and Nutrition,
    School of Agriculture
Lovely Professional University
Punjab, India

Sapna Jarial
Department of Agricultural Economics &
    Extension, School of Agriculture
Lovely Professional University
Punjab, India

Gaurav Joshi
School of Pharmacy
Graphic Era Hill University
Dehradun, Uttarakhand, India

Karishma Joshi
Centre for Environmental Science &
    Technology
Banaras Hindu University
Varanasi Uttar Pradesh, India

Jaspreet Kaur
Department of Food Technology and Nutrition,
    School of Agriculture
Lovely Professional University
Punjab, India

Somanjana Khatua
Molecular and Applied Mycology and Plant
    Pathology Laboratory, Centre of Advanced
    Study Department of Botany
University of Calcutta
Kolkata, West Bengal, India

and

Department of Botany, Krishnagar
    Government College
Krishnagar, West Bengal, India

Mohd. Kashif Kidwai
Department of Energy and Environmental
    Sciences
Chaudhary Devi Lal University
Sirsa, Haryana, India

Anita Klaus
University of Belgrade, Faculty of Agriculture
Belgrade, Republic of Serbia

Maja Kozarski
University of Belgrade, Faculty of Agriculture
Republic of Serbia

Satish Kumar
ICAR-Directorate of Mushroom Research
Chambaghat, Solan, Himachal Pradesh

Vinod Kumar
Peoples' Friendship University of Russia
(RUDN University)
Moscow, Russian Federation

Priyanka Kundu
Department of Food Technology and Nutrition,
    School of Agriculture
Lovely Professional University
Punjab, India

Simona Maria Man
Department of Food Engineering,
    Faculty of Food Science
    and Technology
University of Agricultural Science and
    Veterinary Medicine
Cluj-Napoca, Romania

Kumari Manorma
Dr YSP & UHF
Nauni, Solan, Himachal Pradesh

Mihaela Mihai
Department of Food Engineering, Faculty of
    Food Science and Technology
University of Agricultural Science and
    Veterinary Medicine
Cluj-Napoca, Romania

Sevastita Muste
Department of Food Engineering, Faculty
    of Food Science and Technology
University of Agricultural Science and
    Veterinary Medicine
Cluj-Napoca, Romania

AnaVictoria Valdivia Padilla
Tecnologico de Monterrey, School of
    Engineering and Sciences, Campus
    Queretaro, Av. Epigmenio Gonzalez
San Pablo, Querétaro, Mexico

Adriana Păucean
Department of Food Engineering, Faculty of
    Food Science and Technology
University of Agricultural Science and
    Veterinary Medicine
Cluj-Napoca, Romania

Predrag Petrović
Innovation Center of the Faculty of
    Technology and Metallurgy
University of Belgrade
Belgrade, Serbia

Anamaria Pop
Department of Food Engineering, Faculty of
    Food Science and Technology
University of Agricultural Science and
    Veterinary Medicine
Cluj-Napoca, Romania

Prasad Rasane
Department of Food Technology and Nutrition,
    School of Agriculture
Lovely Professional University
Punjab, India

Pawan Kumar Rose
Department of Energy and Environmental
    Sciences
Chaudhary Devi Lal University
Sirsa, Haryana, India

Bibekananda Sarkar
Department of Zoology, B.S.S. College
Supaul, Bihar, India

Cristina Anamaria Semeniuc
Department of Food Engineering
University of Agricultural Sciences and
    Veterinary Medicine of Cluj-Napoca
Cluj-Napoca, Romania

Abhishek Sharma
CSIR-Imtech Chandigarh
Punjab, India

Ashutosh Sharma
Tecnologico de Monterrey, School of
    Engineering and Sciences, Campus
    Queretaro, Av. Epigmenio Gonzalez
Querétaro, Mexico

Kartik Sharma
Department of Biotechnology, Council of
    Scientific and Industrial Research-Institute
    of Himalayan Bioresource Technology
    (CSIR-IHBT)
Palampur, Himachal Pradesh, India

Shweta Sharma
ICAR-Directorate of Mushroom Research
Chambaghat, Solan, Himachal Pradesh

VP Sharma
ICAR-Directorate of Mushroom Research
Chambaghat, Solan, Himachal Pradesh

Jyoti Singh
Department of Food Technology and Nutrition,
    School of Agriculture
Lovely Professional University
Punjab, India

Neha Singh
Rajendra Mishra School of Engineering
    Entrepreneurship, Indian Institue of
    Technology
Kharagpur, West Bengal

Somvir Singh
Faculty of Applied Sciences and
    Biotechnology
Shoolini University
Solan, Himachal Pradesh, India

Filippe Elias de Freitas Soares
Department of Chemistry
Universidade Federal de Lavras
Minas Gerais, Brazil

Sonia Socaci
Department of Food Science
University of Agricultural Sciences and
    Veterinary Medicine of Cluj-Napoca
Cluj-Napoca, Romania

Maria-Ioana Socaciu
Department of Food Engineering
University of Agricultural Sciences and
    Veterinary Medicine of Cluj-Napoca
Cluj-Napoca, Romania

Chitra Sonkar
Department of Food Process Engineering
Sam Higginbottom University of Agriculture,
    Technology and Sciences
Allahabad, Uttar Pradesh

Arti Thakur
Faculty of Applied Sciences and
    Biotechnology
Shoolini University
Solan, Himachal Pradesh, India

Daniel Tura
National Institute of Research and
    Development for Biological
    Sciences, Institute for Biological
    Research Cluj-Napoca
Romania

Leo J. L. D. van Griensven
Mushrooms4Health B.V., Horst
The Netherlands

Jo vana Vunduk
Ekofungi Ltd., Padinska Skela bb
Belgrade, Serbia

Jovana Vunduk
Institute of General and Physical
    Chemistry
Belgrade, Serbia

Wan Abd Al QadrImad Wan-Mohtar
Institute of Biological Sciences,
    Faculty of Science
Universiti Malaya
Kuala Lumpur, Malaysia

PART I

Introduction to wild mushrooms

An introduction to wild mushrooms and their exploitation for human well-being: An overview

Shweta Sharma[1], V. P. Sharma[1], Satish Kumar[1], Abhishek Sharma[2], Kumari Manorma[3], and P. K. Chauhan[4]

[1]ICAR-Directorate of Mushroom Research, Chambaghat, Solan
[2]CSIR-Imtech Punjab, India
[3]Dr YSP&UHF Nauni
[4]Shoolini University, Solan

CONTENTS

1.1 INTRODUCTION

For thousands of years, wild mushrooms have been taken into account for culinary and medicinal uses. The archaeological record reveals edible species associated with people living 13,000 years ago in Chile, but it is in China where the eating of wild fungi is first reliably noted, several hundred years before the birth of Christ (Aaronson, 2000). Edible fungi were collected from forests in ancient Greek and Roman times and were highly valued, though more by high-ranking people than by peasants (Buller, 1914). Caesar's mushroom (*Amanita caesarea*) is a re-minder of an ancient tradition that still exists in many parts of Italy, embracing a diversity of edible species dominated today by truffles (*Tuber sp.*) and porcini (*Boletus edulis*). The current chapter summarizes the following main points related to wild mushrooms and their impact on humans.

1.2 GENERAL OVERVIEW ON MUSHROOMS

A mushroom is a macrofungi part of the fungi kingdom that grows and is similar to plants that lack roots, stems, leaves, seeds, and flowers, etc.; mushrooms are devoid of chlorophyll green pigments that are the major property of plants and, due to absence of chlorophyll, they are generally saprophytic (Chang & Miles, 1987, 1992). The mushroom body is made up of fine threads i.e. hyphae and clusters of these threads acknowledged as mycelium. A compact mass of hyphae grows vegetative beneath the soil or on the waste material of surrounding until the fruiting season comes. A mushroom, macrofungi, consists of cap and stalk. In ancient times a particular word i.e. "toadstool". Being saprophytic in nature mushrooms survive on dead and decaying matter, these macrofungi have the potential to convert decaying matter into their food (Hadar et al., 1992; Jaradat, 2010). Under favorable seasonal conditions like temperature, moisture, nourishment source, and light, the vegetative mycelia grow into a small pinhead and are finally converted into a mature fruit body. The fully grown mature fruit body of mushrooms consists of three main parts: stipe, pileus, and gills. The gills portion of a mushroom bear spores that spread by air into the atmosphere. These spores subsequently germinate into a new mycelium to reiterate the new life cycle under appropriate substrate (Figure 1.1).

Mature mushroom fruit bodies are generally utilized as food delicacies and medicine; however, there are many species in the world identified as unpalatable or poisonous (Boa, 2004). Most frequently collected wild mushrooms i.e. Chanterelle (*Cantharellus cibarius*), black morel (*Morchella conica*), yellow morel (*Morchella esculenta*), pine mushroom (*Armillaria* ponderosa),

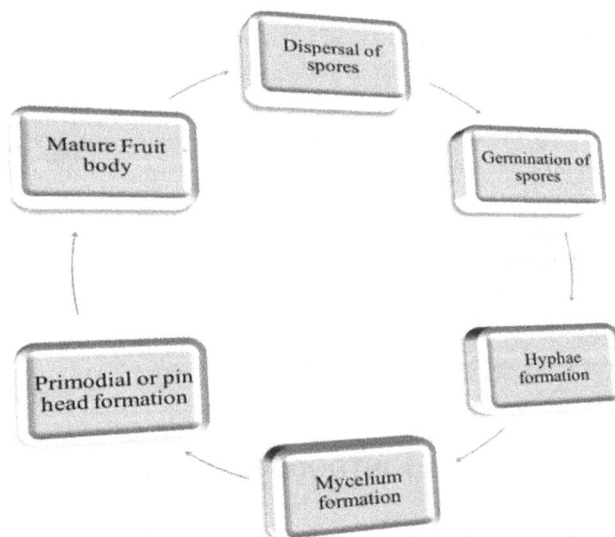

Figure 1.1 A life cycle of macrofungi (mushrooms).

matsutake (*Tricholoma matsutake*), boletus (*Boletus edulis*), and sweet tooth or hedgehog (*Dentinum repandum*) by the tribal communities and mushroom hunters have commercial perspective. Other species of wild mushrooms, i.e. *Agaricus campestris, Lyophyllum multiceps, Caprinus commatus, Laetiporus (Polyporus) sulphurcus,* and *Morchella augusticeps* are foraged on a leisure basis. Although mushrooms are an excellent source of nutrients, proteins, and vitamins, people are uncertain about their consumption. Local folks or tribal communities distributed the knowledge on their edibility to the public domain in local markets and fairs. Therefore, these wild mushrooms must be domesticated for commercial use; otherwise, they will remain unknown in the jungle and will become extinct (Purkayastha & Chandra, 1985; Srivastava & Soreng, 2012).

1.3 STATUS OF WILD MUSHROOMS

Both microfungi and macrofungi are found abundantly on Earth. According to the latest data, there are more than 15 lakh species of macrofungi (mushrooms) recorded on Earth, out of which only 1.0 lakhs are studied and described so far. The latest reports all around the world showed that consumption of mushrooms recorded from China, Spain, and Egypt are more than 6,000, 18 000, and 4,000 years back, respectively (Chang, 2006; Power et al., 2015; Straus et al., 2015; Zhang et al., 2015).

The main practice to identify edible mushrooms from wild habitats by a human is not much clear; however, there are some approaches like cautious tasting, smell, texture, and lack of adverse reaction decided that either they can be eaten or not. Another mushroom species can be avoided due to its lack of taste and difficulty to digest, they can be toxic/poisonous. U.S. Army Survival Manual FM 2020 published a document i.e. "universal edibility test" to recognize the edible mushroom from the wild (Survival Use of Plants, 2020). Chinese researchers work hard on the different types of mushrooms and also on their valuable medicinal properties (Wang et al., 2009). Mexico is another country with remarkable information on edible mushrooms (Garibay-Orijel et al., 2020).

Wild mushrooms are also an important non-timber forest source utilized by the tribal and local community and their use has been well maintained in several countries all over the world (Harkonen et al., 1993; Jones & Whalley, 1994; Chang & Lee, 2004; Roberto et al., 2005). These mushrooms are available for sale in local markets (Moreno-Black et al., 1996; Roberto et al., 2005) or commercially used as food or therapeutic use (Oso, 1977; Vaidya & Rabba, 1993; Chamberlain, 1996). The mushroom species collected from a different region, with their scientific name, local name, family, ethnomedicinal or medicinal uses, and seasonal occurrence are listed in Table 1.1 (Semwal et al., 2014; Thawthong et al., 2014).

1.4 WHY IS DOMESTICATION/ARTIFICIAL CULTIVATION OF WILD MUSHROOMS IMPORTANT?

Mushrooms are delicate, fleshy, spore-bearing fruiting members of the fungi kingdom. Wild, edible mushrooms have played a significant role as a human delicacy from ancient times. Though, practical knowledge for their cultivation and farming is comparatively new (Martínez-Carrera et al., 2000). China was the first country that autonomously developed the cultivation methods of black ear mushroom (*Auricularia* sp.) and shiitake [*Lentinula edodes* (Berk.) Pelger] about a thousand years ago. About 350 years ago France developed the cultivation method of white button mushroom [*Agaricus bisporus* (Lange)] Imbach. However, in the last 50 decades the methods of cultivation and domestication of important wild mushrooms have been considerably improved (Chang & Miles, 1989; Chawla et al., 2019). Industries for commercial production of wild edible mushrooms have also been rising (Chang & Miles, 1987). Statistics data showed that mainly 11 main genera of mushrooms

Table 1.1 List of wild mushrooms with their applications and seasonal collection

S. no.	Scientific name	Family	Medicinal uses and mode of consumption	Medicinal properties	Time of collection
1.	*Agaricusaugustus* Fries	Agaricaceae	Most commonly consumed as food, general tonic and used as supplementary diet to patients suffering from asthma, stroke, heart ailments, and diabetes	Antioxidant	July–September
2.	*Agaricusbisporus* (Lange) Imbach	Agaricaceae	Culinary use with nutritional properties	Antioxidant	July–October
3.	*Agaricuscampestris* L.: Fr.	Agaricaceae	Culinary use with nutritional properties	Antihyperglycemic, insulin releasing	July–October
4.	*Boletus edulis* Bull.	Boletaceae	Iron-rich, recommend to pregnant women after deliveries also used as expectorant to treat cough and antidepressant	Antioxidant	May–June, and October–November
5.	*Bovistaplumbea* Pers.	Agaricaceae	Used against respiratory tract and also against frostbite	To treat tonsillitis, sore throats, and skin ulcers	May–October
6.	*Cantharell-uscibarius* Fr.	Cantharellales	Culinary use and also used to cure wounds, bone ailments, and general weakness	Good source of essential amino acids and vitamin A. Effective against night blindness, dry skin, respiratory tract infection, soreness of eye, and sarcoma	July–August
7.	*Coprinuscomatus* (O.F. Mull.: Fr.) Pers.	Agaricaceae	Fresh decoction of juvenile mushrooms is exploited against respiratory ailments and diabetes	Regulate blood sugar, also act as antitumor, antibacterial, and immunomodulator	October–November
8.	*Disciotisven-osa* (Pers.) Arnould	Morchellaceae	Culinary use and some tribal communities use against the common cold	No details	March–April
9.	*Flammulinavelutipes* (Curt.)Singer	Physalacriaceae	Culinary use and also consume as an immunomodulator tonic	Used to cure liver and gastroenteric ulcers diseases	November–December
10.	*Fomesfomentarius* (L.) Fr.	Polyporaceae	Medicinal use against skin ailments	Diuretic, antimicrobial, wound healing	Whole year
11.	*Ganodermaapplanatum* (Pers.) Pat.	Ganodermataceae	Medicinal use Decoction recommended to patients with weak immune systems. Folk herbalists used it to cure cough and the common cold	Immuno-stimulator, anti-cancerous, antibiotic, and antiviral properties	May–August

No.	Species	Family	Use	Properties/Details	Season
12.	*Ganodermalucidum* (Leysser) Karsten	Ganodermataceae	Consume as a tea against various ailments, prescribed to the patients recovering from prolonged illness	Used against hepatitis, chronic bronchitis; coronary heart disease, allergies, and altitude sickness. For aging-related conditions, anticancer and as an immunomodulator	May–August
13.	*Geoporasumneriana* (Cooke) M. Torre	Pyronemataceae	Culinary use and its tonic used to cure common cold andconstipation by some tribal communities	No details	March–April
14.	*Gyromitra* sp.	Discinacare	Culinary use	No record	August–September
15.	*Helvellaacetabulum* (L.) Quel	Helvellaceae	Medicinal use against chronic cough	No record	April–May
16.	*Helvellacrispa* Bull	Helvellaceae	Medicinal use against asthma, cough. intestinal inflammation, and mouth ulceration	Antioxidant, anti-allergic	May–August
17.	*Hericiumcoralloides* (Scop.) Pers.	Hericiaceae	Medicinal use as anti-cold also consume as a brain tonic, useful for diabetic patients	Works against dementia, other brain disorders, and ulcers and is also a useful remedy against pancreatitis, Crohn's disease,and hemorrhoids	June–December
18.	*Hygrophorus* sp.	Hygrophoraceae	Culinary use	–	July–September
19.	*Lactariusdeliciosus* (L.: Fr.) Gray	Russulaceae	Medicinal use as sex stimulant	Anti-tumor, antioxidant, and immunostimulant	July–August
20.	*Lentinustigrinus* (Bull.) Fr.	Polyporaceae	Culinary as well as medicinal use as a brain tonic and helpful against dry cough and asthma	Antibacterial, hypoglycemic, and antioxidant	May–October
21.	*Lycoperdonperlatum* Pers.	Agaricaceae	Medicinal use as healing of wounds and frostbite	Antimicrobial and antioxidant	June–August
22.	*Lycoperdonpyriforme* Schaeff.	Agaricaceae	Medicinal use as Healing of wounds and frostbites	Anticoagulant	June–August
23.	*Morchellaconica* (Pers.) Fr.	Morchellaceae	Culinary as well as medicinal use against heart ailments arthritis	Vitamin D supplement, antioxidant, antimicrobial, and immunostimulant	March–May
24.	*Morchellaesculenta* Fr.	Morchellaceae	Culinary as well as medicinal use against heart ailments, arthritis	Vitamin D supplement, antioxidant, immunostimulantNephroprotective against drug-induced kidney damages	May–June

(Continued)

Table 1.1 (Continued) List of wild mushrooms with their applications and seasonal collection

S. no.	Scientific name	Family	Medicinal uses and mode of consumption	Medicinal properties	Time of collection
25.	*Morchellahybrid* (Fr.) Cetto	Morchellaceae	Do	Do	March–May
26.	*MorchellaVulgaris* (Pers.) Boud.	Morchellaceae	Do	Do	March–April
27.	*Neolentinuslepideus* (Fr.) Redhead	Gloeophyllaceae	Culinary use	Antibacterial. Inhibitory effect on free radicals,antioxidant activity	June–October
28.	*Phallus impudicus* L.	Phallaceae	Culinary as well as medicinal use	Immunomodulator, anti-inflammatory, anti-stress, anti-tumor, wound healing, and anticancer	May–August
29.	*Pleurotusostreatus* (Jacq:. Fr.) P. Kumm.	Pleurotaceae	Culinary as well as medicinal use againstprolonged illness	Effective in thetreatment of lumbago, numbed limbs, and tendon and blood vessel discomfort and reduces cholesterol	June–November
30.	*Ramariastricta* (Pers.) Quel.	Gomphaceae	Culinary as well as medicinal use against asthma and other respiratory ailments also reduce lung inflammation	No reports	June–October
31.	*Rhizopogonvillosulus* Zeller	Rhizopogonaceae	Culinary as well as medicinal use against kidney stones and urinary tract infections	Squirrel diet	March–May
32.	*Rhizopogonroseolus* Corda	Rhizopogonaceae	Medicinal use as treatment of urinary tract infections	Antioxidant,antimicrobial	March–November
33.	*Sparassiscrispa* Wulfen	Sparassidaceae	Medicinal use as a general tonic, blood purifier, and anti-cold by folk herbalist	Enhances hemoglobin level and anti-tumor properties	June–October
34.	*Sparassisspathulata* (Schwein.) Fr.	Sparassidaceae	Culinary as well medicinal use	Not reported	June–October
35.	*Trametesversicolor* (L.) Lloyd.	Polyporaceae	Medicinal use works against skin diseases such as rashes, itching, dryness, and healing of wounds when used with butter or oil	Secondary prevention strategy as immune therapy against breast cancer	Round the year

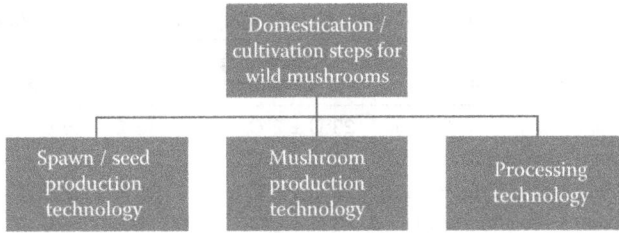

Figure 1.2 Flowchart illustrating steps of domestication of wild mushrooms.

i.e. *Agaricus, Lentinula, Volvariella, Pleurotus, Auricularia, Flammulina, Tremella, Hypsizygus, Ganoderma, Grifola,* and *Hericium* are widely cultivated.

There are mainly three fundamental steps involved in the domestication and commercial production of wild mushrooms: 1) spawn or seed production technology, 2) mushroom production technology, and 3) processing technology (Martínez-Carrera et al., 2000; Sharma & Kumar, 2011; Figure 1.2). A variety of methods and techniques have been developed and described in detail for each technology (Flegg et al., 1985). Spawn or seed production technology involves the isolation of pure strains from wild mushrooms present in their natural habitat, either by spore culture technique or by tissue culturing method (Royse et al., 1985; Kozak & Krawczyk, 1999; Sharma et al., 2019).

Wild strains are generally not suitable for commercial production; hence, to overcome this problem intensive selection and breeding through traditional and molecular methods. A selected strain with valuable gastronomic and salutiferous properties is used for spawn preparation on cereal grains (e.g. rye, wheat, rice, maize, and sorghum, etc.) or sometimes on the organic substrate (i.e. wheat straw, cotton waste, coffee pulp, sawdust, and mustard straw) in autoclavable glass bottle or bags. The substrate for the spawn is inoculated with mushroom mycelium and kept at a temperature (25°C) for complete mycelial colonization on the substrate (Chang & Miles, 1989). Large-scale production technology for mushrooms is initiated with the construction of infrastructure equipped with all cropping conditions, i.e. temperature, light, relative humidity, and ventilation, etc. Appropriate raw substrates are used according to the availability of mushroom cultivation. These raw materials are selective and different for each species of mushroom. Production of fruit bodies varies according to each species, spawn quality, substrate quality, and environmental conditions (Sharma et al., 2020) Table 1.2.

In addition to standardizing cultivation protocols for valuable wild mushrooms, production of high-yielding strains supplemented with characteristics such as diseases resistance, temperature tolerance, and good-quality fruit bodies (regarding shape, size color, and taste) is also a matter of concern. Nowadays, several techniques, such as molecular breeding and single spore isolation, are used for improvement in the quality of mushrooms.

1.5 WILD MUSHROOMS EXPLOITED FOR HUMAN WELL-BEING

Apart from numerous excellent nutritional properties, several edible wild mushrooms have also been used as a remedial purpose in many countries. Currently, around 300 species of mushrooms are identified for their remedial and healing qualities against various diseases (Ying et al., 1987). These medicinal mushrooms are employed as customary medicinal components for the cure of health-related problems. Thus, due to their tremendous health benefits, researchers and scientists all over the world keenly worked on the identification, domestication, commercialization, and evaluation of nutraceuticals properties of wild mushrooms. Extensive scientific investigations,

Table 1.2   Temperature requirement for different growth stages of medicinal mushrooms

S.no.	Medicinal mushroom	Spawning condition	Pinhead condition	Fruiting condition
1.	Shiitake mushroom (*Lentinula edodes*)	23°C ± 2°C	18°C – 20°C	18°C – 20°C
2.	Reishi mushroom (*Ganodermalucidum*)	23°C ± 2°C	30°C – 32°C	30°C – 32°C
3.	Dhingri/oyster Mushroom (*Pleurotus* sp.)	23°C ± 2°C	18°C – 22°C	18°C – 22°C
4.	Maitake mushroom (*Grifolafrondosa*)	23°C ± 2°C	16°C – 18°C	16°C – 18°C
5.	Monkey head mushroom (*Hericiumerinaceus*)	23°C ± 2°C	16°C – 18°C	16°C – 18°C
6.	Coral tooth fungi (*Hericiumcoralloides*)	23°C ± 2°C	16°C – 18°C	20°C – 25°C
7.	Black ear mushroom (*Auriculariapolytricha*)	23°C ± 2°C	18°C – 20°C	18°C – 20°C
8.	Turkey tail mushroom (*Trametesvesicolor*)	25°C ± 2°C	18°C – 24°C	18°C – 24°C
9.	*Cordycepsmilitaris*	18°C – 22°C	18°C – 22°C	18°C – 22°C
10.	*Tremellafuciformis*	25°C – 28°C	18°C – 23°C	23°C – 24°C
11.	*Agaricusblazei*	23°C ± 2°C		
12.	Milky mushroom (*Calocybeindica*)	23°C ± 2°C	28°C – 30°C	28°C – 30°C
13.	Paddy straw mushroom (*Volvariella* sp.)	23°C ± 2°C	28°C – 32°C	28°C – 32°C
14.	Black poplar mushroom (*Agrocybeaegerita*)	23°C ± 2°C	14°C – 16°C	14°C – 16°C
15.	*Schizophyllumcommune*	23°C ± 2°C	28°C – 30°C	28°C – 30°C
16.	*Flammulinasp*	22°C – 25°C	10°C – 14°C	10°C – 14°C
17.	*Lentinussajor-caju*	22°C – 25°C	28°C – 30°C	28°C – 30°C
18.	*Lentinus conatus*	23°C – 26°C	20°C – 24°C	22°C – 24°C
19.	*Panusvelupitus*	22°C – 25°C	28°C – 30°C	28°C – 30°C
20.	*Panuslecometi*	22°C – 26°C	22°C – 25°C	22°C – 25°C

particularly from eastern countries like China, Korea, and Japan have confirmed the traditional curative uses of mushrooms. The Mediterranean region, especially Italy and Greece, established many production lines based on the utilization of wild edible fungi. These production units are based on the precise recognition of wild mushrooms with important organoleptic traits. Several imperative, wild, medicinal mushrooms are discussed as follows (Figure 1.3).

1.5.1 *Ganoderma* sp.

Ganoderma lucidum, acknowledged as an "herb of mortality and longevity", has been used since 3–4 A.D. in China. *Ganoderma* has been used in Chinese medicine as a tonic since ancient periods due to its health benefits and non-poisonous traits (Liu, 1999; Chawla et al., 2020). *Ganoderma* is a notable medicinal mushroom throughout the world due to its significant curative characteristics (Hawksworth, 2001; Kingston & Newman, 2005). This mushroom is known with different names such as Reishi, Lingzhi, and Mannentake in Japan, Korea, China, and other East Asian countries. *Ganoderma* sp. usually grows on dead deciduous forest trees. Its fruit body is generally dark red with a shiny woody texture. From the last few decades, demand for *Ganoderma sp.* increases all across the world due to its extensive medicinal properties such as anti-microbial, anticancer, anti-diabetic, anti-aging, immuno-modulating agents, etc. Researchers reported that polysaccharides extracted from *Ganoderma lucidum,* i.e. polysaccharides "Gl-PS" when admini-strated on mice lowered their blood sugar level. "Gl-PS" polysaccharide has a hypoglycemic effect

Figure 1.3   Pictures of different domesticated valuable medicinal and gastronomic mushrooms (a. *Ganoderma* sp., b. *Lentinula* sp., c. *P. ostreatus*, d. *Tremella* sp., e. *Agaricus* sp., f. *Agrocybe* sp., g. *Cordyceps sinensis*, h. *Cordyceps militaris*, i. *Hericium erinaceus*, j. *Hericium coralloides*, k. *Grifola frondosa*, l. *Auricularia* sp.).

on normal mice and also possesses the insulin-releasing activity that facilitates the Ca2+ inflow to the pancreatic beta cells (Zhang et al., 2004).

Currently, various *Ganoderma* products (*Ganoderma* powder, tea, nutritional supplements, and several medicines) derived from different parts, such as fruiting body, mycelium, and spores, are available in markets for public use (Wachtel-galor et al., 2011). Several bioactive compounds have been synthesized by the cell wall of *Ganoderma* sp., which has significant medicinal characteristics useful for the development of novel drugs.

1.5.2 Shiitake (*Lentinula edodes*)

Lentinula edodes (Berk.) Pegler, commonly known variously as the black oak mushroom, shiitake, or shiang-gu, is a gilled mushroom with polyporaceous affinities (Pegler, 1983). More than 800 years back, shiitake was first cultivated in China (Chang & Miles, 1987) and, at present, China accounts for about 70% of world production. *L. edodes* is the second most widely cultivated mushroom species among the five most cultivated edible mushrooms all over the world. *Lentinus edodes* is currently the world's top cultivated edible mushroom with about 22% of the world's supply. These mushrooms grow in the wild in China, Japan, Korea, the Himalayas, the Philippines, Papua New Guinea, and in the north of Thailand (Mori et al., 1974; Chang & Miles, 1992; Campbell & Slee, 2004; Sharma et al., 2016). *L. edodes* is valued for its unique flavor, which derives mainly from its content of the modified amino acid lanthionine and the nucleotide guanine-5-monophosphate (Mizuno, 1995; Yang et al., 1998). The fruit bodies are rich in minerals, vitamins, and essential amino acids (especially lysine and leucine) and are high in fiber content but contain less than 10% crude fat (Ho et al., 1994; Mizuno, 1995). Apart from its importance as a mushroom crop, it has also been found that *L. edodes* contains medicinal compounds, including lentinan, which has antitumor activity, the hypocholesterolemic eritadenine, and cortinellin, an antibacterial agent (Sugiyama et al., 1997; Przybylowicz & Donoghue, 1988). Pegler (1983) interrelated phenotypical dissimilarity with geographic allocation to identify three species of shiitake, i.e. *Lentinus lateritia* in Southeast Asia and Australasia and *Lentinus novaezelandieae* in New Zealand. Other species of *Lentinus* include *L. crinitus* and *L. tigrinus* (Bisen et al., 2010).

1.5.3 *Pleurotus ostreatus*

Pleurotus ostreatus, also known as the oyster mushroom, is a common edible mushroom. Its large-scale production for food was first done by Germany during the World War (Stamets, 2000). This mushroom is predominately found in the wild among all other mushrooms, although this mushroom can be easily cultivated on the waste straw of wheat and rice nowadays. *P. ostreatus* also have some cultivation similarities with P. *eryngii*. *P. ostreatus* has the peculiar aroma of the chemical benzaldehyde (Stamets, 2000). This mushroom can be easily identified in the wild as it grows on decaying wood in large clusters with a short stem with white gills containing whitish to mauve or pale purplish-colored spores (Philips & Roger, 2006). Seasonal appearance of this mushroom is generally between October and April. *P. pulmonarius* (pale-colored), *P. populinus* (appears on the wood of aspen), and other similar species are commonly found in between April and September. *P. ostreatus* is generally saprophytic but tends to be a facultative parasite during stress conditions (Stamets, 2005).

Due to its unique taste, essence, and high nutritional and medicinal value, *P. ostreatus* mushroom is consumed all around the world. *P. ostreatus* has numerous medicinal characteristics, i.e. antibacterial, anticancer, antidiabetic, anticholesterol, antioxidants, antiarthritic, and antiviral. Thus, having high nutritional values along with potent medicinal qualities, *P. ostreatus* is considered an important nutraceutical functional food (Deepalakshmi & Mirunalini, 2014). Besides many health benefits, these mushrooms can also be used for mycoremediation (degrade hazardous waste material) and biocontrol purposes as they attack and kill nematodes and bacteria (Barh et al., 2019).

1.5.4 *Tremella* species

Tremella sp., also known as "jelly fungi" due to its gelatinous appearance, belongs to the *Tremellaceae* family and is parasitic to other fungi. Here, *tremere* means "to temble" in Latin; initially Linnaeus kept *Tremella* with cyanobacteria, myxomycetes (slime molds), algae, and seaweeds along with fungi. But, Persoon revised it in 1794 and 1801 and repositioned *Tremella* as a new genus under the ICBN (International Code of Botanical Nomenclature), i.e. *Tremella mesenterica* (Chen et al., 2006). Presently, almost 100 species are identified around the world, out of which two species, i.e. *T. aurantialba* and *T. fuciformis,* are commercially produced for food (Zhang et al., 2004). As *Tremella* sp. is parasitic, hence it produces a particular type of hyphal and haustorial cell that infiltrates the host plant. The basidia are globose or ellipsoids with vertical and diagonal septa. Conidia are present on conidiophores as similar patterns present in yeast cells (Chen et al., 2006). *Tremella mesenterica* species is found extensively in forests with broadleaf, conifer, and mixed tree vegetation. Though the *Tremella* mushroom is insipid and flavorless, it is still considered edible. *Tremella mesenterica* synthesizes various bioactive compounds with various useful biological activities (Deng et al., 2009)

1.5.5 *Agaricus* species

This is an imperative genus of mushrooms including both edible and non-edible species with more than 300 members globally. The genus generally comprises two widely used mushrooms, i.e. common white button mushroom (*Agaricus bisporus*) and field mushroom (*A. campestris*). The fruit bodies of *Agaricus* genus are mainly characterized with fleshy white/brown colored pileus with spore-bearing gills inside them (Masuda et al., 2008). These mushrooms have a stipe that lifts the pileus above the compost/substrate and also possess a partial mask/veil and protect the growing gills and afterward forms a ring on the stalk (Regulo et al., 2013). *Agaricus* sp. (common mushroom) is also known with many different names, i.e. button mushroom, white mushroom, champignon mushroom, table mushroom, crimini mushroom, Roman brown mushroom, Italian mushroom, and Swiss brown mushroom, all around the world.

White button mushrooms rank first globally on their consumption basis. It is an edible mushroom with lots of nutritional (protein and vitamin D) and medicinal traits. Species other than *Agaricus bisporus* also have numerous valuable health benefits along with gastronomic properties. For example, *Agaricus subrufescens* is an optional culinary mushroom with a slightly sweet taste with an almond-like aroma (Chuchin & Wu, 1984). This mushroom is widely used as a substitute for cancer medicine in Japan (Hyodo et al., 2005). It is considered a health beneficial food. *Agaricus blazei Agaricus brasiliensis* KA21), another species of *Agaricus* genus, is reported to show immune-enhancing properties like leukocyte increase, antitumor/anticancer, endotoxin shock – alleviating and hepatopathy alleviating effects in experimental mice (Liu et al., 2008). While an experiment was conducted on humans, it has been reported that with the intake of this particular mushroom, blood cholesterol, sugar level, percent body, and visceral fat level declines, and NK (natural killer) cell action rises (Liu et al., 2008). *Agaricus blazei* secretes oligosaccharides and beta-glucans that play novel activities as antihypertriglyceridemic, antihyperglycemic, anti-arteriosclerotic, and anti-hypercholesterolemic in diabetic mice (Kim et al., 2005). The metabolite extracted from *Agaricus blazei* showed an improvement in insulin resistance against type 2 diabetes. Consumption of *Agaricus blazei* enhances the concentration of adiponectin, which provides many health benefits in human beings (Hsu et al., 2007).

1.5.6 *Agrocybe aegerita*

Agrocybe aegerita (Brig.) Sing. is commonly known as "black poplar mushroom". It has a unique flavor and nutritive value. This mushroom generally is cultivated in Japan, Korea, Australia, and China. *Agrocybe aegerita* synthesizes important valuable bioactive metabolites such as agrocybenine, cylindan, and indole derivatives with antifungal, anticancer, and free radical scavenging activity, respectively (Zhong & Xiao, 2009). Two compounds, glucan and heteroglycan, are extracted from the fruit body of the *Agrocybe* mushroom. Kiho et al. (1994) reported that glucan showed significant hypoglycemic activity in normal and diabetic mice.

1.5.7 *Cordyceps* sp.

Cordyceps, or caterpillar mushroom, is an entomopathogenic fungus that is related to the *Ascomycetes* family. Genus *Cordyceps* is a natural parasite on arthropods, especially lepidopteron larvae and pupae (Guo et al., 2016). Generally, two species, *C. sinensis* and *C. militaris,* are popular among all 450 reported species. This mushroom is widely used in traditional Chinese medicinal practices (Georges, 2007). Chinese folk utilize the fruiting body with insect carcass for numerous medicinal purposes. *C. sinensis* is more expensive and complicated to cultivate artificially than the *C. militaris* (Lin et al., 2016). *Cordyceps* mushrooms secrete several novel bioactive constituents like cordycepin (substitute of adenosine nucleosides) and cordycpic acid (D-mannitol), polysaccharides, sterols, and macrolides (Chatterjee et al., 1957; Bok et al., 1999; Yang et al., 2009; Wu et al., 2014).

1.5.8 *Hericium* sp.

Hericium erinaceus (Bull. Fr.) Pers. is known as monkey head mushroom in China and "cen-dawanbungakobis" in Malaysia. *Hericium* sp. also has some other names such as lion's mane, old man's beard, bearded tooth fungus, hedgehog mushroom, satyr's beard, pompom, and yamabush-itake (Stamets, 2005). This belongs to the *Basidiomycetes* class and *Hericiaceae* family. It can be identified by its long spines (greater than 1 cm length), short stalks, and form a whitish cluster of downward cascading spines that are commonly found native to North America, Europe, and Asia. *H. erinaceous* is an edible mushroom with many medicinal qualities including anticancer, neuro-protection, antioxidant, and anti-inflammatory. *H. erinaceous* also consists of other nutrients such as proteins, unsaturated fatty acids, carbohydrates, and micronutrients including phosphorus, sulfur, calcium, magnesium, zinc, iron, and copper, which play an important role in the multiple physio-logical systems (nervous, digestive, circulatory, and immune systems) of the organism. Fruiting bodies of *H. erinaceus* may have beneficial effects against stomach, esophageal, and skin cancers (Kim, 2012). *H. erinaceus* (its fruiting body, mycelium, and products in the medium) also contain some lower MW pharmaceutical constituents, such as the novel phenols (hercenones A and B) and Y-A-2, which may have chemotherapeutic effects on cancer (Lee et al., 2000). Another species of *Hericium* genus is *H. coralloides,* a saprophytic fungus, commonly known as coral tooth fungus. It grows on dead hardwood trees. When young, the fungus is soft and edible, but as it ages the branches and hanging spines become brittle and turn into a light shade of yellowish-brown in color. This is also an edible mushroom with therapeutic value (Sharma et al., 2020)

1.5.9 *Grifola frondosa*

Maitake (*Grifola frondosa*) is a well-accepted mushroom on the Asian continent for its deli-cious flavor and immune-stimulating property. Maitake is also known as dancing mushroom as researchers and mushroom hunters say it was so rare that anyone who found it danced with joy. Hen of the woods or sheep's head are other names of *Grifola frondosa*. Recently, researchers all across the country are working on its large-scale cultivation (Sharma et al., 2020). The *Grifola frondosa* mushroom possess many healthy beneficial traits as it helps in the reduction of blood sugar. Alpha-glucosidase inhibitor component is produced by the maitake mushroom fruit body and helps in maintaining the blood sugar level (Matsuur et al., 2002; Hong et al., 2007).

1.5.10 *Auricularia polytricha*

The black ear mushroom has the oldest record of cultivation by the Chinese, dating back to 600 AD. It is commercially cultivated in some of the Southeast Asian countries. This mushroom ranks fourth among all cultivated edible mushroom. This mushroom is believed to cure sore throat, anemia, and certain digestive disorders, especially piles on regular consumption. In India, this mushroom is collected and consumed in the northeastern states (Sharma et al., 2020).

1.6 THE ROLE OF WILD MUSHROOMS IN THE ENVIRONMENT

Recently, many studies showed that wild mushrooms are exploited for decomposition of forest litter, cellulose, hemicelluloses, and lignin compounds, and other biodegradation of environmentally hazardous materials (Aggelis et al., 2002). Several studies reported wild mushrooms are effective agents for waste material biodegradation (Wang et al., 2009). There are several wood-decaying basidiomycetes, i.e. *Ganoderma* sp, *Lentinus tigrinus, Panellus stipticus, Phanerochaete chrysosporium, Abortiporus biennis, Inonotus hispidus, Pleurotus* sp., *Bjerkandera adusta, Trametes hirsuta, Irpex lacteus*, and *Dichomitus squalens* that are potent agents to degrade/detoxify a broad range of pollutants, e.g. chlorinated compounds, textile or industrial effluents, aromatic hydrocarbons, preservatives, agro-industrial waste (fungicides, herbicides, and insecticides, etc.) into non-toxic simple organic compounds (Anastasi et al., 2010; Da Silva Coelho et al., 2010; Inoue et al., 2010; Ntougias et al., 2012). These macrofungi can break down the lignin and other related constituents like hemicelluloses and cellulose into a simpler form by secreting laccases, phenoloxidases, and peroxidases enzymes (Martinez et al., 2005; Baldrian, 2006). The aforesaid mushroom species could be eventually exploited by many significant biotechnological processes meant for providing eco-friendly and economical approaches related to waste management. *Lentinula* sp. has been utilized in bio-pulping practice and this species also been used to degrade some of the toxic waste natter present in municipal dumps. Rhodes (2012) concluded that macrofungi are a remarkable degrading agent of waste by-products and also a significant element of the food web. These macrofungi cannot destroy/degrade forest litter directly; first of all, they cover the entire forest floor with their mycelial mat and then disintegrate the waste material. Degradation speed generally depends upon the nutrients present in the soil. Several macrofungi species, e.g. *Pleurotus* sp., *Lentinula* sp., *Ganoderma* sp., *Agaricus* sp., *Lentinus Squarrosulus, Irex lacteus, Stropharia coronilla, Lentinus tigirinus,* and *Nematolana prowardii,* interact with obstinate waste litter such as chitin, keratin, lignin, and fats, and disintegrate them into simple molecules, i.e. cellulose, sugars (Barh et al., 2019).

In addition to this, some findings show that native diversity of mushroom species, e.g. *Cordyceps* sp. and *Pleurotus* sp., are used as biological control agents against phytopathogenic pests, i.e. insects, nematodes, respectively (Chitwood, 2002; Shah & Pell, 2003). Fungal inoculum especially ecto-mycorrhizal mushroom species are applied for the establishment of plants/trees under reforestation initiative schemes (Kropp & Langlois, 1990; Bonet et al., 2006; Rincón et al., 2007).

1.7 ETHNOBIOLOGY AND WILD MUSHROOMS

Many aboriginal tribal communities all across the globe utilize wild varieties of mushrooms as their natural source of diet and income (Semwal et al., 2014; Ao et al., 2016; Borah et al., 2018). Macrofungi like mushrooms were traditionally used by the human civilization since the primeval era. Customary or native information of the wild is generally employed by the local folk for their cultural practices. People from ancient times of many continents and sub-continents like Greece, Iran, China, Japan, and India exploit mushrooms for various rituals (Lowy, 1971; Sharma, 2003). A 3500 BC old Charaka Samhita of Indian treatise reported the utilization of mushrooms as delicacies and medicine. In the previous era, wild mushrooms were considered an exceptional kind of food with several nutraceuticals qualities and regarded as "food of Gods" by the Romans. Mushrooms were considered a source of strength and nutrition and hence given to soldiers before battle by the Greeks. Various records have confirmed the utilization of wild mushrooms as medicines since the Neolithic and Paleolithic eras (Samorini, 2001). Boa (2004) revealed that wild mushrooms are not simply a source of food but also a revenue-generating source for developed and developing countries.

In India, more than 280 edible species of wild mushrooms are recorded, out of which a few are domesticated (Purkayastha & Chandra, 1985). The domestication of wild varieties of mushrooms drew attention all over the globe as cultivated varieties are accessible throughout the year for people to use. India has an extraordinarily diverse climate and seasonal conditions that lead to rich mushroom diversity. Several mushroom species are sold in local markets of India by tribal communities (Tanti et al., 2011). Tribal communities from different parts of the world forage various kinds of wildly grown mushrooms for food or remedial purposes.

1.8 CONCLUSION

Mushrooms, due to their stupendous unique flavor, aroma, and texture, are acknowledged as nourishing food and a significant source of nutraceutical health-beneficial components. Usually mushrooms contain high dietary fiber content and are low in energy. They are an exceptional source of antioxidants along with several secondary metabolites and phenolic compounds. Around the world, there are more than 1,100 species of mushrooms utilized by different continents and subcontinental countries as delicacies and medicine. Edible, wild mushrooms with high nutritional and medicinal value are used as remedies for many dreadful diseases like cancer, tumor, diabetes, etc. Several species, i.e. *Lentinula* sp., *Ganoderma* sp., *Pleurotus* sp., and *Cordyceps* sp., are considered curative foods due to their anti-cancerous, anti-cholesterolaemic, and anti-viral behavior. Along with tremendous therapeutic traits, wild, edible mushrooms also help in the maintenance of bone density, skin health, blood sugar level, cholesterol level, improve our immune system, and enhance the energy level.

With so many outstanding properties, the wild, edible mushrooms are still not fully explored and many of them are not properly domesticated for the public domain. The vast diversity of wild mushroom species that have been investigated is still less in number and slight information is available. There is a necessity to do more and more research and exploration of wildly present mushrooms, mainly for their gastronomic and medicinal benefits. Domestication of wild mushrooms is a challenge, as is commercialization of wild mushrooms. Commercial cultivation of wild mushrooms leads to opening the way to national and international trade of beneficial wild mushrooms throughout the world. In addition to the gastronomic and medicinal utilization, the wild mushrooms should be well documented because such information and facts would not only lead to sustainable exploitation but also would give a view to attaining significant strains with potential bioactive metabolites.

ACKNOWLEDGMENT

The authors wish to thank the director, ICAR–Directorate of Mushroom Research Chambaghat Solan, for providing laboratory facilities and financial assistance under the project.

On behalf of all the authors, the corresponding author states that there is no conflict of interest.

REFERENCES

Aaronson, S. (2000). Fungi. In: K. F. Kiple & K. C. Ornelas (Eds.), *The Cambridge World History of Food.* Cambridge: Cambridge University Press, pp 313–336.

Aggelis, G., Ehaliotis, C., Nerud, F., Stoychev, I., Lyberatos, G., & Zervakis, G. I. (2002). Evaluation of white-rot fungi for detoxification and decolorization of effluents from the green olives debiterring process. *Appl Microbiol Biotechnol*, *59*, 353–360.

Anastasi, A., Spina, F., Prigione, V., Tigini, V., Giansanti, P., & Varese, G. C. (2010). Scale-up of a bioprocess for textile wastewater treatment using *Bjerkandera adusta*. *Biores Technol*, *101*, 3067–3075.

Ao, T., Deb, C. R., & Khruomo, N. (2016). Wild edible mushrooms of Nagaland, India: A potential food resource. *J Exper BioAgri Sci*, *4*(1), 59–65.

Baldrian, P. (2006). Fungal laccases – Occurrence and properties. *FEMS Microbiol Rev*, *30*, 215–242.

Barh, A., Kumari, B., Sharma, S., Aneepu, S. K., Kumar, A., Kamal, S., & Sharma, V. P. (2019). Mushroom mycoremediation: Kinetics and mechanism. *Smart Bioremediation Technologies: Microbial Enzymes (ed.)*, *1*, 1–22.

Bisen, P. S., Baghel, R. K., Sanodiya, B. S., Thakur, G. S., & Prasad, G. B. K. S. (2010). *Lentinus edodes*: A macrofungus with pharmacological activities. *Cur Med Chem*, *17*, 2419–2430.

Boa, E. (2004). Wild edible fungi: A global overview of their use and importance to people. *Non-Wood Forest Products*, No. 17. Rome, Italy: FAO, Forestry Department.

Bok, J. W., Lermer, L., Chilton, J., Klingeman, H. G., & Towers, G. H. (1999). Antitumor sterols from the mycelia of *Cordyceps sinensis*. *Phytochemi*, *51*(7), 891–898.

Bonet, J. A., Fischer, C., & Colinas, C. (2006). Cultivation of black truffle to promote reforestation and land-use stability. *Agron Sustain Dev*, 26, 69–76.

Borah, N., Semwal, R. L., & Garkoti, S. C. (2018). Ethnomycological knowledge of three indigenous communities of Assam, India. *Ind J Tradi Knowleg*, *17*(2), 327–335.

Buller, A. H. R. (1914). The fungus lore of the Greeks and Romans. *Trans Br Mycol Soc*, *5*, 21–66.

Campbell, A. C., & Slee, R. W. (2004). The introduction of *Lentinus edodes*, the Japanese wood mushroom, to the UK. *Nat in Dev*, *10*, 25–33.

Chamberlain, M. (1996). Ethnomycological experiences in South West China. *The Mycolog*, *10*, 173–176.

Chang, S. T. (2006). The world mushroom industry: Trends and technological development. *Inter J Med Mush*, *8*(4), 297–314.

Chang, Y. S., & Lee, S. S. (2004). Utilization of macrofungi species in Malaysia. *Fung Dives*, *15*, 15–22.

Chang, S. T., & Miles, P. G. (1987). Historical record of the early cultivation of *Lentinus* in China. *Mush J Trop*, *7*, 31–37.

Chang, S. T., & Miles, P. G. (1992). Mushroom biology—A new discipline. *The Mycolog*, *6*, 64–65.

Chang, S. T., & Miles, P. G. (1989). *Edible Mushrooms and Their Cultivations*. Boca Raton, Florida: CRC Press, 79.

Chatterjee, R., Srinivasan, K. S., & Maiti, P. C. (1957). *Cordyceps sinensis* (Berkeley) Saccardo: Structure of cordycepic acid. *J Am Pharm Assoc*, *46*(2), 114–118.

Chawla, P., Chawla, V., Bains, A., Singh, R., Kaushik, R., & Kumar, N. (2020). Improvement of mineral absorption and nutritional properties of *Citrullus vulgaris* seeds using solid-state fermentation. *J Am College Nutrition*, *39*(7), 628–635. 10.1080/07315724.2020.1718031

Chawla, P., Kumar, N., Kaushik, R., & Dhull, S. B. (2019). Synthesis, characterization and cellular mineral absorption of gum arabic stabilized nanoemulsion of *Rhododendron arboreum* flower extract. *J Food Sci Technol*, *56*(12), 5194–5203.

Chen, S. O., Phung, S. R. S., Hur, G., Ye, J. J., Kwok, S. L., & Shrode, G. E. (2006). Antiaromatase activity of phytochemicals in white button mushrooms (*Agaricus bisporus*). *Cancer Res*, *66*(24), 1026–1034.

Chitwood, D. J. (2002). Phytochemical based strategies for nematode control. *Annu Rev Phytopathol*, *40*, 221–249.

Chuchin, C. C., & Wu, C. M. (1984). Volatile components of mushroom (*Agaricus subrufecens*). *J Food Sci*, *49*, 1208.

Da Silva Coelho, J., de Oliveira, A. L., de Souza, C. G. M., Bracht, A., & Peralta, R. M. (2010). Effect of the herbicides bentazon and diuron on the production of ligninolytic enzymes by *Ganoderma lucidum*. *Int Biodeter Biodegr*, *64*, 156–161.

Deepalakshmi, K., & Mirunalini, S. (2014). Pleurotus *ostreatus*: An oyster mushroom with nutritional and medicinal properties. *J Biochem Tech*, *5*(2), 718–726.

Deng, G., Lin, H., & Seidman, A. (2009). A phaseI/II trial of polysaccharide extracts from *Grifola frondosa* (Maitake mushroom) in breast cancer patients: Immunological effects. *J Canc Res Clini Onco, 135*(9), 1215–1221.

Flegg, P. B., Spencer, D. M., Wood, D. A., & Mushroom. (1985). *The Biology and Technology of Cultivated Mushrooms.* New York, NY, USA: John Wiley and Sons, pp. 279–293.

Garibay-Orijel, R., Argüelles-Moyao, A., Álvarez-Manjarrez, J., Ángeles-Argáiz, R. E., García-Guzmán, O. M., & Hernández- Yáñez, H. (2020). Diversity and importance of edible mush-rooms in ectomycorrhizal communities in Mexican neotropics. In: J. Pérez-Moreno, A. Guerin-Laguette, R. Flores Arzú, & F. Q. Yu (Eds.), *Mushrooms, Humans and Nature in a Changing World.* Cham, Switzerland: Springer International Publishing, pp. 407–424.

Georges, M. H. (2007). *Healing Mushrooms.* New York, USA: Square One Publisher.

Guo, M., Guo, S., Huaijun, Y., Bu, N., & Dong, C. H. (2016). Comparison of major bioactive compounds of the Caterpillar medicinal mushroom, *Cordyceps militaris* (ascomycetes), fruiting bodies cultured on wheat substrate and pupae. *Int J Med Mush, 18*(4), 327–336.

Hadar, Y., Keren, Z., Gorodecki, B., & Ardon, O. (1992). Utilization of lignocellulosic waste by the edible mushroom, *Pleurotus. Biodegradation, 3*, 189–205.

Harkonen, M., Buyck, B., Saarimäki, T., & Mwasumbi, L. (1993). Tanzanian mushrooms and their uses. *Russula Karstenia, 33*, 11–50.

Hawksworth, D. L. (2001). Mushrooms: The extent of the unexplored potential. *Int J Med Mush, 3*, 333–337.

Ho, K., Hun, Z., & Yei, J. (1994). *The Chinese Shiitake.* Shanghai, China: Shanghai Science & Technology Press.

Hong, L., Xun, M., & Wutong, W. (2007). Anti-diabetic effect of an alpha-glucan from fruit body of maitake (*Grifola frondosa*) on KK-Ay mice. *J Pharm Pharmacol, 59*, 575–582.

Hsu, C. H., Liao, Y. L., Lin, S. C., Hwang, K. C., & Chou, P. (2007). The mushroom *Agaricus Blazei* Murill in combination with metformin and gliclazide improves insulin resistance in type 2 diabetes: A randomized, double-blinded, and placebo-controlled clinical trial. *J Altern Complement Med, 13*, 97–102.

Hyodo, I., Amano, N., Eguchi, K., Narabayashi, M., Imanishi, J., & Hirai, M. (2005). Nationwide survey on complementary and alternative medicine in cancer patients in Japan. *J Clin Oncol, 23*, 2645–2654.

Inoue, Y., Hata, T., Kawai, S., Okamura, H., & Nishida, T. (2010). Elimination and detoxification of triclosan by manganese peroxidase from white rot fungus. *J Hazard Mater, 180*, 764–767.

Jaradat, A. A. (2010). Genetic resources of energy crops: Biological systems to combat climate change. *Aust J Crop Sci, 4*, 309–323.

Jones, E. B., & Whalley, J. (1994). A fungus foray to Chiang Mai market in northern Thailand. *Mycologist, 8*, 87–90.

Kiho, T., Sobue, S., & Ukai, S. (1994). Structural features and hypoglycemic activities of two polysaccharides from a hot-water extract of *Agrocybe cylindracea. Carbohydr Res, 251*, 81–87.

Kim, J. H. (2012). Biological activities of water extract and solvent fractions of an edible mushroom, *Hericium Erinaceus. Kor J Mycol, 40*, 159–163.

Kim, Y. W., Kim, K. H., Choi, H. J., & Lee, D. S. (2005). Anti-diabetic activity of beta-glucans and their enzymatically hydrolyzed oligosaccharides from *Agaricus blazei. Biotechnol Lett, 27*, 483–487.

Kingston, D. G., & Newman, D. J. (2005). The search for novel drug lead for predominately antitumor therapies by utilizing Mother Nature's pharmacophoric libraries. *Curr Opin Drug Discov Devel, 8*, 207–227.

Kozak, M., & Krawczyk, J. (1999). *Growing Shiitake Mushrooms in a Continental Climate.* Peshtiga, Wisconsin: Field and Forest Production.

Kropp, B. R., & Langlois, C. G. (1990). Ectomycorrhizae in reforestation. *Can J Forest Res, 20*(4), 438–451.

Lee, E. W., Shizuki, K., Hosokawa, S., Suzuki, M., Suganuma, H., Inakuma, T., Li, J., Ohnishi-Kameyama, M., Nagata, T., Furukawa, S., & Kawagish, H. (2000). Two novel diterpenoids, erinacines H and I from the mycelia of *Hericium erinaceum. Biosci Biotechnol Biochem, 64*, 2402–2405.

Lin, S., Liu, Z. Q., Xue, Y. P., Baker, P. J., Wu, H., Xu, F., Teng, Y., Brathwaite, M. E., & Zheng, Y. G. (2016). Biosynthetic pathway analysis for improving the cordycepin and cordycepic acid production in *Hirsutella sinensis*. *Appl Biochem Biotechnol*, *179*(4), 633–649.

Liu, G. T. (1999). Recent advances in research of pharmacology and clinical application of *Ganoderma* (P. Karst) species (*Aphyllophoromycetideae*) in China. *Int J Med Mush*, *1*, 63–67.

Liu, Y., Fukuwatari, Y., Okumura, K., Okumura, K., Takeda, K., & Ishibashi, K. (2008). Immunomodulating activity of *Agaricus brasiliensis* KA21 in Mice and in human volunteers. Evid based complement. *Alternat Med*, *5*, 205–219.

Lowy, B. (1971). New records of mushroom stories in Guatemala. *Mycologia*, *63*(5), 983–993.

Martinez, A. T., Speranza, M., Ruiz-Duenas, F. J., Ferreira, P., Camarero, S., & Guillen, F. (2005). Biodegradation of lignocellulosics: Microbiological, chemical and enzymatic aspects of fungal attack to lignin. *Int Microbiol*, *8*, 195–204.

Martínez-Carrera, D., Aguilar, A., Martínez, W., Bonilla, M., Morales, P., & Sobal, M. (2000). Commercial production and marketing of edible mushrooms cultivated on coffee pulp in Mexico. In: T. Sera, C. Soccol, A. Pandey, & S. Roussos (Eds.), *Coffee Biotechnology and Quality*. Dordrecht, The Neterhlands: Kluwer Academic Publishers, pp. 471–488.

Masuda, Y., Murata, Y., Hayashi, M., & Nanba, H. (2008). Inhibitory effect of MD fraction on tumor metastasis: Involvement of NK cell activation and suppression of intercellular adhesion molecule (ICAM)-1 expression in lung vascular endothelial cells. *Bioloi Pharm Bulletin*, *31*(6), 1104–1124.

Matsuur, H., Asakawa, C., Kurimoto, M., & Mizutani, J. (2002). Alpha-glucosidase inhibitor from the seeds of balsam pear (Momordica charantia) and the fruit bodies of *Grifola frondosa*. *Biosci Biotechnol Biochem*, *66*, 1576–1578.

Mizuno, T. (1995). Shiitake, *Lentinus edodes*: Functional properties for medicinal and food purposes. *Food Rev Inter*, *11*, 111–128.

Moreno-Black, G., Akanan, W., Somnasang, P., Thamathawan, S., & Brozvosky, P. (1996). Non domesticated food resources in the marketplace and marketing system of northeastern Thailand. *J Ethnobiology*, *16*, 99–117.

Mori, K., Fukai, S., & Zennyoji, A. (1974). Hybridization of shiitake (*Lentinula edodes*) between cultivated strains of Japan and wild strains grown in Taiwan and New Guinea. *Mush Sci*, *9*, 391–403.

Ntougias, S., Baldrian, P., Ehaliotis, C., Nerud, F., Merhautova, V., & Zervakis, G. I. (2012). Biodegradation and detoxification of olive mill wastewater by selected strains of the mushroom genera *Ganoderma* and *Pleurotus*. *Chemos*, *88*(5), 620–626.

Oso, B. A. (1977). Mushrooms in Yoruba mythology and medicinal practices. *Econo Bot*, *31*, 367–371.

Pegler, D. N. (1983). The genus *Lentinula* (*Tricholomataceae* tribe *Collybieae*). *Sydo*, *36*, 227–239.

Philips, N., & Roger, F. (2006). *Mushrooms*. 6th ed. London, United Kingdom: Macmillan Publishers, pp. 266.

Power, R. C., Salazar-García, D. C., Straus, L. G., Morales, M. R. G., & Henry, G. (2015). Micro remains from El Mirón Cave human dental calculus suggest a mixed plant–animal subsistence economy during the Magdalenian in Northern Iberia. *J Archae Sci*, *60*, 39–46.

Przybylowicz, P., & Donoghue, J. (1988). Nutritional and health aspects of shiitake. In: P. Przybylowicz & J. Donoghue (Eds.), *Shiitake Grower's Handbook the Art and Science of Mushroom Cultivation*. Dubuque, IA: Kendall-Hunt Publishing, pp. 183–188.

Purkayastha, R. P., & Chandra, A. (1985). *Manual of Indian Edible Mushrooms*. New Delhi, India: Jagendra Book Agency.

Regulo, C. L. H., Michele, L. L., Anne- Marie, F., Marie, F. O., Nathalie, F., & Catherine, R. R. (2013). Potential of European wild strains of *Agaricus subrufescens* for productivity and quality on wheat straw based compost. *W J Microbio Biotech*, *29*(7), 1243–1253.

Rhodes, J. (2012). Feeding and healing the world: Through regenerative agriculture and permaculture. *Sci Prog*, *95*(4), 345–446.

Rincón, A., de Felipe, M. R., & Ferna´ndez-Pascual, M. (2007). Inoculation of *Pinus halepensis* Mill. with selected ectomycorrhizal fungi improves seedling establishment 2 years after planting in a degraded gypsum soil. *Mycorrhiza*, *18*, 23–32.

Roberto, G.-O., Cifuentes, J., Estrada-Torres, A., & Caballero, J. (2005). Fungal biodiversity people using macro-fungal diversity in Oaxaca, Mexico. *Fungal Diversity*, *21*, 41–67.

Royse, D. J., Schizler, L. C., & Dichle, D. A. (1985). Shiitake mushrooms – Consumption, production and cultivation. *Interdisciplinary Sci Rev*, *10*, 329–340.

Samorini, G. (2001). New data on the ethnomycology of psychoactive mushrooms. *Inter J Med Mush*, *3*(2–3), 257–278.

Semwal, K. C., Stephenson, S. L., Bhatt, V. K., & Bhatt, R. P. (2014). Edible mushrooms of the Northwestern Himalaya, India: A study of indigenous knowledge, distribution and diversity. *Mycosphere*, *5*(3), 440–461.

Shah, P. A., & Pell, J. K. (2003). Entomopathogenic fungi as biological control agents. *Appl Microbiol Biotechnol*, *61*, 413–423.

Sharma, N. (2003). Medicinal uses of macrofungi. *Ethnobotany*, *15*, 97–99.

Sharma, V. P., Gupta, M., Kamal, S., Sharma, S., Aneepu, S. K., Kumar, S., & Sanyal, S. K. (2019). Genetic diversity and physiological relatedness among 30 strains of *Lentinula edodes* (Berk) Pegler. *Sydowia*, *71*, 91–101.

Sharma, V. P., & Kumar, S. (2011). Spawn production technology. In: M. Singh, B. Vijay, S. Kamal, & G. C. Wakchaure (Eds.), *Mushrooms Cultivation, Marketing and Consumption*. Chambaghat, Solan: Directorate of Mushroom Research (ICAR).

Sharma, V. P., Kumar, S., & Sharma, S. (2020). Technologies developed by ICAR-DMR for commercial uses. ICAR – Directorate of Mushroom Research Solan (HP), p. 82.

Sharma, V. P., Sharma, S., Kumar, S., Gupta, M., & Kamal, S. (2016). Cob web and dry bubble diseases in Lentinula edodes cultivation – A new report. In: *Science and Cultivation of Edible Fungi*. Baar & Sonnenberg (Eds.), International Society for Mushroom Science, pp. 130–134.

Srivastava, A. J., & Soreng, P. K. (2012). An effort to domesticate wild edible mushrooms growing in the forest of Jharkhand. *Inter J Rec Tred Sci Technol*, *3*(3), 88–92.

Stamets, P. (2000). Growth parameters for gourmet and medicinal mushroom species. In: B. Bodine Ford, & P. Stamets (Eds.), *Growing Gourmet and Medicinal Mushrooms*. 3rd ed. Berkeley, California, USA: Ten Speed Press, pp. 308–315.

Stamets, P. (2005). The role of mushrooms in nature. In: *Growing Gourmet and Medicinal Mushrooms*. 3rd ed. Berkeley, California, USA: Ten Speed Press, pp. 10–11.

Straus, L. G., Morales, M. R. G., Carretero, J. M., & Marín-Arroyo, A. B. (2015). "The Red Lady of El Mirón". Lower Magdalenian life and death in oldest Dryas Cantabrian Spain: An overview. *J Archaeol Sci*, *60*, 134–137.

Sugiyama, K., Yamakawa, A., Kawagishi, H., & Saeki, S. (1997). Dietary eritadenine modifies plasma-phosphatidylcholine molecular species profile in rats fed different types of fat. *J Nutri*, *127*, 593–599.

Survival Use of Plants. (2020). Retrieved from http://www.survivalebooks.com/09.htm#par1

Tanti, B., Gurung, L., & Sarma, G. C. (2011). Wild edible fungal resource used by the ethnic tribes of Nagaland, India. *Ind J Tradi Knowleg*, *10*(3), 512–515.

Thawthong, A., Karunarathna, S. C., Thongklang, N., Chukeatirote, E., Kakumyan, P., Chamyuang, S., Rizal, L. M., Mortimer, P. E., Xu, J., Callac, P., & Hyde, K. D. (2014). Discovering and domesticating wild tropical cultivatable mushrooms. *Chiang Mai J Sci*, *41*(4), 731–764.

Vaidya, J. G., & Rabba, A. S. (1993). Fungi in folk medicine. *Mycology*, *7*, 131–133.

Wachtel-Galor, S., Yuen, J., John, A., Bushwell, & Benzie, F. F. I. (2011). Ganoderma lucidum (Lingzhi or Reishi): A medicinal mushroom. In: F.F.I. Benzie, & S. Wachtel-Galor (Eds.) *Herbal Medicine: Bimolecular and Clinical Aspects*. 2nd ed. Boca Raton: CRC Press.

Wang, C., Sun, H., Li, J., Li, Y., & Zhang, Q. (2009). Enzyme activities during degradation of polycyclic aromatic hydrocarbons by white rot fungus *Phanerochaete chrysosporium* in soils. *Chemosphere*, *77*, 733–738.

Wu, D. T., Xie, J., Wang, L. Y., Ju, Y. J., Lv, G. P., Leong, F., Zhao, J., & Li, S. P. (2014). Characterization of bioactive polysaccharides from Cordyceps *militari*s produced in China using saccharide mapping. *Journal of Functional Foods*, *9*, 315–323.

Yang, M. S., Chyau, C. C., Horng, D. T., & Yang, J. S. (1998). Effects of irradiation and drying on volatile components of fresh shiitake (*Lentinus edodes* Sing.). *J Sci Food Agric*, *76*, 72–76.

Yang, F. Q., Feng, K., Zhao, J., & Li, S. P. (2009). Analysis of sterols and fatty acids in natural and cultured *Cordyceps* by one-step derivation followed with gas chromatography–mass spectrometry. *J Pharm Biomed Anal*, 9(5), 1172–1178.

Ying, J. Z., Mao, X. L., Ma, Q. M., Zong, S. C., & Wen, H. A. (1987). *Illustrations of Chinese Medicinal Fungi*. Beijing: Science Press, pp. 579–582.

Zhang, N., Chen, H., Zhang, Y., Xing, L., Li, S., Wang, X., & Sun, Z. (2015). Chemical composition and antioxidant properties of five edible *Hymenomycetes* mushrooms. *Inter J Food Sci Techno*, 50, 465–471.

Zhang, G. Z., Du, J., & Fu, X. Y. (2004). Study on the characters of PPO and anti-browning inhibitors in mushrooms. *J Xinjiang Agri Univ*, 27(1), 75–78.

Zhong, J. J., & Xiao, J. H. (2009). Secondary metabolites from higher fungi: Discovery, bioactivity and biopro-duction. *Adv Biochem Eng Biotechnol*, 113, 79–150.

CHAPTER 2

Cultivation strategies of edible and medicinal mushrooms

Anita Klaus[1] and Wan Abd Al Qadr Imad Wan-Mohtar[2]
[1]University of Belgrade, Faculty of Agriculture, Belgrade, Republic of Serbia
[2]Institute of Biological Sciences, Faculty of Science, Universiti Malaya, Kuala Lumpur, Malaysia

CONTENTS

DOI: 10.1201/9781003152583-3

2.1 INTRODUCTION

Mushrooms have always attracted the attention of various civilizations that have used them as food, medicine, or means in spiritual rituals. So far, the oldest evidence of mushrooms use is that isolated from the dental plaque of a woman named The Red Lady, fossil remains excavated at El Mirón Cave, in the mountainous region of Cantabria in Spain, whose age is estimated at 18,700 years. The application of modern techniques has determined the presence of different types of mushrooms spores, embedded in the teeth of this fossil, which testifies to the use of mushrooms as food or because of some special powers, even in that ancient time (Power et al., 2015). For centuries, people, for their various needs, simply collected mushrooms that were available in the environment. And then they started growing *Auricularia auricula* (600 AD), *Flammulina velutipes* (800 AD), *Lentinula edodes* (1000 AD), *Agaricus bisporus* (1600 AD), *Volvariella volvacea* (1700 AD), *Tremella fuciformis* (1800 AD), and *Pleurotus ostreatus* (1900 AD) (Chang & Miles, 1987). Due to the very pronounced adaptability to different environments, the ability to grow on different substrates, and a relatively short life cycle, these species of saprophytic mushrooms have been successfully grown in artificial conditions. Thus, these human communities provided larger quantities of mushrooms, for hundreds of years growing them only on wooden logs, so they were less dependent on naturally available strains. However, the greatest progress in mushroom cultivation was achieved by the technique of growing *A. bisporus* on a composted substrate, first applied in France around 1600. Over the last hundred years, this technique has been significantly improved, enabling the production of large quantities of *A. bisporus*, which has become both the most popular and sought after in the Western world. Taking into account the world production of fungi, Royse et al. (2017) reported that in the first place is *L. edodes* (22%), followed by *P. ostreatus* (19%), *A. auricula* (18%), while *A. bisporus* is in fourth place (15%); the production of *F. velutipes* (11%) and *V. volvacea* (5%) is also significant, while several other species are produced in smaller amounts (10%).

Numerous studies have highlighted mushrooms as a natural source of very important bioactive components, so in the last 40 years, there has been an increase in world mushroom production by as much as 2,500%, while population growth in that period was about 700% (Royse, 2014). Such an intensive growth of production was achieved by the development of various techniques for indoor growing, which overcame the shortcomings of outdoor cultivation, i.e. enabled control of parameters that directly contribute to the successful production of mushrooms. Therefore, this chapter will be dedicated to different mushroom-growing techniques, especially in light of the most efficient use of agro-industrial waste.

2.2 WHY CULTIVATE EDIBLE AND MEDICINAL MUSHROOMS?

Mushrooms are considered a delicacy that have exceptional nutritional value. These organisms contain a high level of protein, dietary fiber, vitamins (especially B, C, and D), and minerals, while on the other hand they have a low content of total fat (with a significant proportion of unsaturated fatty acids) and have no cholesterol (Kalač, 2016). Valued for their specific and unique taste, mushrooms are also used as additions to various dishes, improving their nutritional and sensory characteristics, but also contributing to health aspects (Gargano et al., 2017; Novakovic et al., 2020). According to Royse et al. (2017), the average consumption per person was 5 kg of mushrooms per year, and it is thought to increase over time due to the growing awareness of people about the benefits of a mushroom-rich diet. The most commonly grown edible species are *A. bisporus*, *Pleurotus* spp., *L. edodes*, *V. volvacea*, *F. velutipes*, *Pholiota nameko*, *Agrocybe aegerita*, etc. However, a strict division into edible and medicinal mushrooms cannot be made because many edible species also have a variety of beneficial health effects (Stamets, 2000; Miles & Chang, 2004; Klaus et al., 2013; Kozarski et al., 2015). Several medically important mushrooms can be used for food (Guillamón et al., 2010; Klaus, 2011). On the other hand, the cultivation of species that are inedible due to their hard texture and/or unpleasant smell and taste, but are a rich source of biologically active substances, is also intensive. The most famous and most cultivated among these mushrooms are *Ganoderma lucidum*, *G. applanatum*, *Phellinus linteus*, *Inonotus obliquus*, *Schizophyllum commune*, and *Coriolus versicolor* (Klaus et al., 2011; Kozarski et al., 2012; Shrestha et al., 2012; Chen et al., 2019; Peng & Shahidi, 2020). Very extensive research, conducted in recent decades, indicates there is still untapped potential of mushrooms. They are rich in many molecules showing high biological activity, such as polysaccharides (β-glucans are especially important), phenolic components, triterpenes, steroids, alkaloids, antibiotics, etc. (De Silva et al., 2013). So far, more than 130 medical effects of mushrooms are known, among others antioxidant, antimicrobial, antihypertensive, antitumor, immunomodulating, cholesterol-lowering, cardiovascular, neuroprotective, antiviral, etc. (Wasser, 2014; Klaus et al., 2015; Petrović et al., 2016; Linnakoski et al., 2018). Throughout the long history of mushroom use, they have played a significant role as a food, as well as a means of treating various health disorders. As recent findings have revealed, the presence of many significant, biologically active compounds in their fruiting body, mycelium, and cultivation broth, thus increasing interest in these very valuable organisms, it is obvious that the demand for mushrooms has also increased.

2.3 MUSHROOM CULTIVATION TECHNIQUES

To date, the exact number of fungi is still unknown, but it is estimated that up to 5,100,000 species live on our planet, of which over 100,000 have been described in the literature (Blackwell, 2011), and at least 14,000 species reproduce by macroscopic fruiting bodies (Miles & Chang, 2004). Mushrooms, organisms belonging to the kingdom Fungi, are divided into divisions of Ascomycota and Basidiomycota, and grow all over the world, in habitats rich in lignocellulosic material, such as forests, meadows, pastures, or parks. According to current knowledge, over 3,000 species of mushrooms are edible, about 2,000 species have some medicinal properties, more than 60 species are grown commercially, while on an industrial scale it is possible to grow just over ten species (Miles & Chang, 2004). The number of industrially grown species is so low due to the very specific nutritional and ecological requirements of mushrooms, which science and technology at the current level of development still cannot meet. The development of mushroom growing techniques or fungi culture appeared later and developed much more slowly than other human activities such as agriculture or animal husbandry. The reasons for this are a very specific mushroom life cycle, diet, complex metabolism, and complicated requirements in terms of environmental parameters. In addition, mushrooms generally have long incubation periods and are susceptible to many parasites, predators, and competing organisms (Stamets, 2000). Significant progress in the field of mushroom growing was achieved due to discoveries in microbiology and the development of industrial machines in the early 19th century. New findings have made it possible to isolate pure mushroom cultures as well as to select the best strains of species which cultivation was already known (Siniscalco, 2013). This has greatly contributed to the safer production of mushrooms, and thus to the provision of increasing quantities of high-quality protein food. Various cultivation techniques that rely on the saprophytic nature of mushrooms have been applied to the cultivation of not only edible but also several medically important species.

To achieve successful cultivation of any type of mushroom, it is necessary to meet all its requirements, which are in line with the ecological role of that species in nature. Mushrooms develop as saprophytes on organic matter, and in relation to the substrate requiring for development, they are divided into primary, secondary, and tertiary decomposers (Rahi et al., 2009).

Primary decomposers, including *L. edodes* and *P. ostreatus*, can provide all energy needs, using several organic materials for their diet, so they do not depend on the products of other organisms. These species synthesize various enzymes that break down large molecules of lipids, proteins, and glycides. As sources of energy and carbon, necessary for the growth of mycelium, they can use hemicellulose, cellulose, lignin, or starch (Raimbault, 1998). Mushrooms that cause brown and white rot have an irreplaceable role in nature due to their ability to decompose lignocellulosic materials such as wood and straw, thus contributing to their recycling. This property is widely used for the transformation of various lignocellulosic waste to produce edible/medicinal mushrooms (Kumla et al., 2020) or to obtain diverse biologically active components, e.g. enzymes (Nguyen et al., 2018).

Secondary decomposers, such as *A. bisporus*, require a pre-prepared, partially decomposed organic substrate to develop successfully. Metabolic processes of naturally present microbiota, in the process of composting, cause changes in the substrate, necessary for further development of mycelium (Miles & Chang, 2004). Thus, the proper growth of secondary decomposers is conditioned by the previous activities of other microorganisms. Tertiary decomposers mainly live in the soil, such as *Agrocybe* spp., and play a significant role in completing the decomposition process. Mushrooms from this group, together with primary and secondary decomposers, participate in the complete decomposition of organic lignocellulosic material in nature and thus enable the circulation of matter (Stamets, 2000; Bains & Chawla, 2020). Relying on the natural characteristics of mushrooms, their life cycle, and diet, different cultivation systems have been developed, and adapted to the specific needs of each species.

2.3.1 Outdoor log culture

This simple method of cultivation involves growing mushrooms outdoors, in completely natural conditions. It was developed in China and Japan more than a thousand years ago, but even today it has its significance among small producers who supply the local market and thus provide additional income for their families. Although it requires a lot of work and is slow, compared to cultivation techniques that use a sterilized substrate, it is still important for growing several species of mushrooms, such as *L. edodes*, *Auricularia auricula* – judae, *Pholiota nameko*, *Hericium erinaceus*, *Pleurotus* spp., *Hypholoma capnoides*, *H. sublateritium*, *Ganoderma lucidum*, and *Hypsizygus* spp. (Stamets, 2000).

The trunks of broad-leaf hardwoods (poplar, alder, cotton wood, oak, etc.), with a firm outer bark, are used. Species with thin bark are not suitable because they are easily damaged by changes in weather conditions, especially moisture. If the bark falls from the trunk, the mycelium can no longer form fruit bodies effectively. It is common for trees to be cut down during the winter or early spring before the leaves begin to grow (Stamets, 2000). This period of the year is suitable because the logs then have the highest content of carbohydrates and other important organic components (which will be necessary for the growth of mycelium and fruiting). Then the bark is firmly attached to the wooden part, so special care should be taken not to create cracks, damage the bark and the trunk during cutting. The highest yields and mushrooms of the best quality are obtained when using mixed logs of older and younger trees, aged 15 to 20 years and 18 cm in diameter (Miles & Chang, 2004; Bains & Chawla, 2020). Logs should have 40%–45% moisture and a pH of 4.5–5.5. Lower moisture content can slow down or completely prevent the growth of mushrooms, and a higher pH value is also suitable for the development of other, harmful species of fungi. The trees are cut into logs 1–1.5 m long, and then they are dried in sunlight until optimal humidity is achieved. At the same time, the undesirable microbiota on the bark is destroyed, and the bark itself adheres better to the central woody part, which reduces the possibility of peeling, which is extremely undesirable for mushroom growth. Increasingly, inoculation is performed immediately after cutting and log formation because drying begins at the site of inoculation, and mycelial growth is in the direction of the drying process (Miles & Chang, 2004). The traditional method involved placing newly cut logs close to those already producing fungi, so spores were transmitted (Stamets, 2000). In this way, the spores begin to germinate on the new logs as well, the mycelium grows through them, and the growth of the fruiting bodies continues; this method, successful for centuries, is still used today as quite acceptable. A significant improvement in this method was the introduction of the plug and sawdust spawn.

2.3.1.1 Sawdust spawn and wood plug spawn preparation

Mushrooms are naturally adapted to the climatic conditions under which they grow. Therefore, when choosing the type of mushroom for inoculation of the appropriate trunk, it is necessary to pay attention to the natural requirements of that strain, to obtain maximum yields. Within one type of mushroom, several strains have different optimal growth temperatures, so it is best to use the mycelium of different strains to allow fruiting throughout the year. Pure mushroom culture grown on an agar plate is not suitable for direct use, so it is transferred to a suitable medium that is easily inoculated into the trunk. Common media are sawdust spawn and wood plug spawn (Miles & Chang, 2004).

Sawdust spawn – completely clean sawdust is mixed with all additives and as much water is added as is needed to drip only slightly when the sawdust is squeezed by hand. This mixture is sterilized in an autoclave (121°C, 2 hours) in bottles or polypropylene bags (preferably with filters that allow gas exchange). The cooled substrate is inoculated with an appropriate mother culture and incubated at 24°C–25°C until complete colonization.

Wood plug spawn –wedge-shaped or cylindrical pieces of wood are used instead of sawdust (Stamets, 2000). The procedure for preparation of the pieces and inoculation with the mother culture is the same as described previously. When the surface of the plug is completely covered with mycelium, it can be used for inoculation.

Preparation of both types of spawn is done in the cleanest possible environment, in the shade, and on days without precipitation.

2.3.1.2 Log inoculation

Whether sawdust spawn or wood plug spawn is used, the logs must first be properly prepared (Miles & Chang, 2004). Holes 1.0–1.5 cm in diameter and 1.5–2.0 cm deep are drilled in the logs (the depth of the hole should be at least 1 cm below the bark). The holes are arranged at 20–30 cm along the length of the log. There should be a distance of 5–6 cm between the rows of holes, i.e. 12 cm between every other row (Figure 2.1).

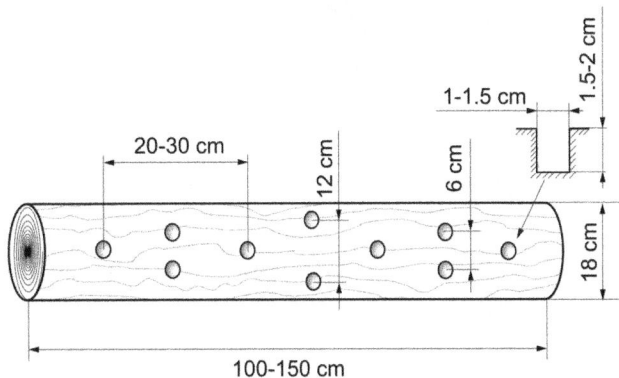

Figure 2.1 Arrangement of holes on the log.

When the holes on the logs are drilled, the prepared spawn is chopped to the size of a thumb, it should weigh about 1 g, be left loose, and carefully inserted into each hole. It is then lightly pressed, and the holes are closed with plugs made of bark (or other material), slightly larger than the hole itself and 3–4 mm thick. The plugs are tapped with a hammer and then covered with a layer of wax or paraffin to maintain a constant level of moisture and prevent contamination (Stamets, 2000).

2.3.1.2.1 Stacking of inoculated logs

For successful fruiting, it is necessary to choose a suitable place for stacking inoculated logs (Miles & Chang, 2004). It is best to be a forest, located south or southeast, a place with low humidity and good ventilation, which is reached by about 30% of sunlight through the branches. It is recommended that the logs be lined up under the treetops. When growing mushrooms in this way, watering is not necessary, but the logs are arranged in such a way as to prevent the loss of moisture by evaporation. One of the common ways of arranging inoculated logs is shown in Figure 2.2. The mycelium grows in logs under the weather conditions that prevail in the environment, and the formation of fruiting bodies is initiated by seasonal changes. In the case of *L. edodes* cultivation, when the incubation is completed (after 12–15 months), the logs are soaked with water for 24 hours, and the first fruiting bodies appear 7–10 days after soaking (Leatham, 1982).

Since outdoor cultivation is applied to various types of fungi, it is necessary to take into account the specific requirements of each species in terms of required humidity, temperature, ventilation, amount of sunlight, arrangement of inoculated logs, etc., to obtain optimal yields of quality fruiting bodies.

2.3.1.3 Stump inoculation

Another possibility for the production of high-quality mushrooms under completely natural conditions is growing on stumps. All over the world, stumps are forest, lignocellulosic waste that have no special economic value. Sooner or later they are decomposed by forest fungi. Stumps that have just been cut, healthy, and without competitive microbiota should be chosen. They should be located in the shaded part of the forest, where there is enough water that will provide additional support to the growth of mycelium through the roots and capillary system. Special attention should be paid to the choice of tree type, i.e. stump that is most suitable for growing a particular type of

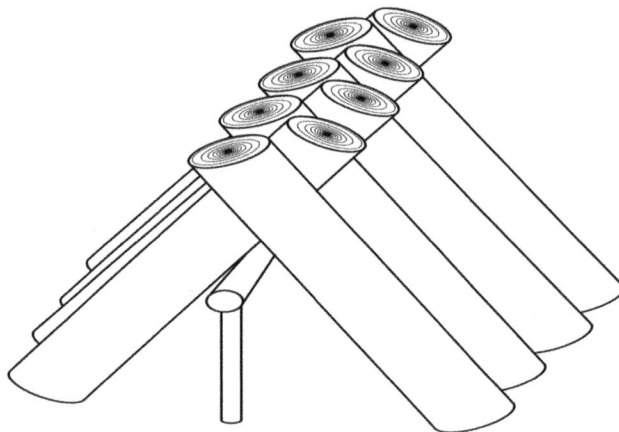

Figure 2.2 Arrangement of inoculated logs in the forest.

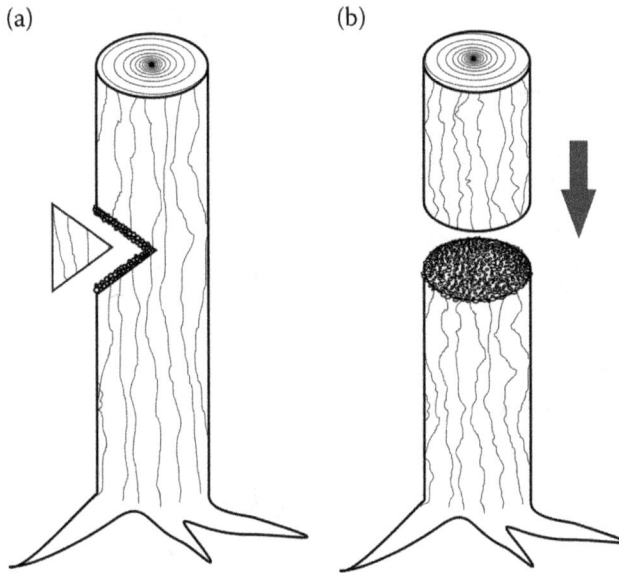

Figure 2.3 Stump inoculation: a) wedge formation; b) disk formation.

mushroom. It is also important to use mushroom strains obtained from native species because they are already well adapted to the specific environment. Edible as well as medicinal mushrooms can be grown in this way, e.g. *P. ostreaus, Grifola frondosa, Laetiporus sulphureus, L. edodes*, etc.

Stump inoculation can be done in several ways. If there are cracks on the stumps, the plug spawn can be inserted directly into them. Inoculation of the stump can be done by cutting a wedge into which sawdust is then inserted, and the cut piece is put back in place (Figure 2.3a). Similarly, the disc can be excised, inoculated, and restored (Figure 2.3b). In both cases, nailing can ensure firm contact between the cut parts (Stamets, 2000).

Considering the ecological parameters in the forest community and the correct choice of species of mushrooms and stumps, a rapid return of nutrients to the chain of circulation in nature can be achieved, from which other forest creatures also benefit, and the ecosystem is improved.

2.3.2 Indoor cultivation

Due to the possibility of controlling numerous parameters that affect the growth of mycelium and the formation of fruiting bodies, indoor cultivation offers many advantages over outdoor cultivation, which is completely dependent on weather conditions and seasonal changes. Although the mushroom texture is poorer, the level of bioactivity is lower, the shelf life is shorter, production is more expensive, and requires more work, the majority of mushrooms are produced by indoor cultivation, e.g. as much as 95% of *L. edodes* sold in stores (Hill, 2010). When growing *L. edodes*, a biological efficiency (according to Liang et al. (2019) defined as the percentage ratio of the mushroom fresh weight per gram of dry substrate) – BE, of up to 80% can be achieved with this method (Royse, 1985).

However, much higher yields in a shorter time, the ability to control microbiological contamination and thus incomparably higher earnings have contributed to the rapid development of indoor mushroom growing techniques. Anyway, all indoor mushroom growing systems rely on the use of an optimal substrate obtained from sawdust of the appropriate type of wood, enriched with additives specific for the mushroom being grown.

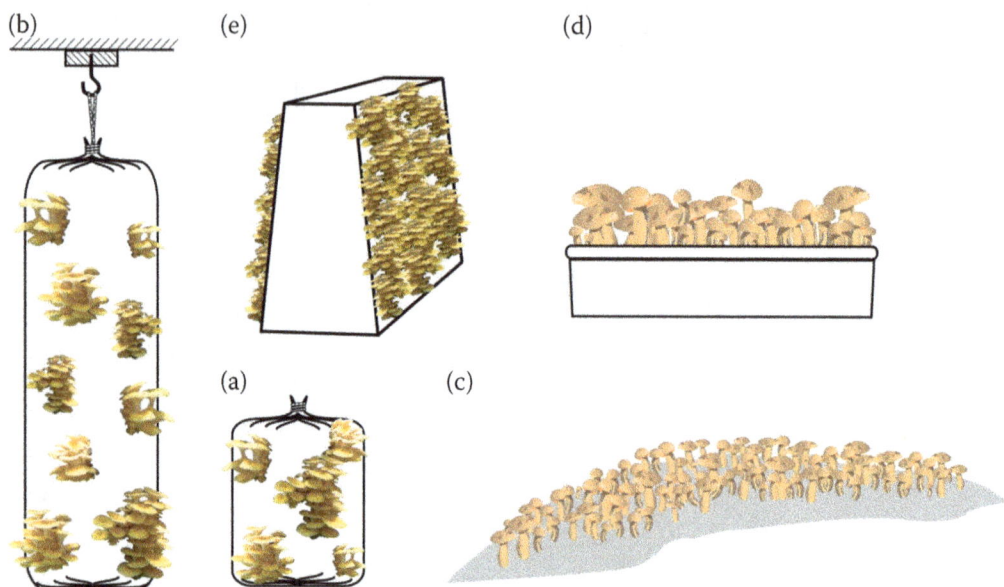

Figure 2.4 Indoor mushroom growing systems: a) artificial logs; b) column culture; c) mound culture; d) tray culture; e) wall culture.

According to Stamets (2000), indoor mushroom growing involves several cultivation systems, such as artificial logs, column culture, mound culture, tray culture, and wall culture (Figure 2.4).

a. Artificial logs are plastic bags filled with a suitable substrate made of sawdust with the addition of various types of agro-industrial waste which exist in every part of the world, such as corncobs, wheat straw, rice straw, sunflower seed hulls, coffee wastes, cotton seed hulls, etc. Polypropylene plastic bags should be used due to the heat resistance of this polymer to the autoclave sterilization temperature (121°C, 2 hours). It is recommended that the sawdust used for this purpose be of uniform size, without hard and sharp pieces that could damage the bag and thus open the way for harmful microorganisms. The bags are filled with a substrate of suitable composition, shaped into a small log with light pressure, and autoclaved. After cooling, the substrate is inoculated with pre- viously prepared mushroom spawn, mixed, and incubated in the dark, at 20°C–25°C. The incubation time depends on the type of mushroom. Incubation lasts until the mycelium has fully grown through the substrate. Then the bags are opened, or holes are drilled on their surfaces, and moistening and intensive ventilation are applied, to initiate the formation of fruiting bodies.

b. Column culture is a growing system in which the pasteurized and the cooled substrate is inoculated with mushroom mycelium, and then packed in long bags. When the inoculated substrate settles, after approximately 1 hour, holes about 3 mm in diameter are drilled in the surface of the bag, at an appropriate distance. Each hole is made with four slits about 5 cm long so that these lids open as the mushrooms grow. This method of cultivation is especially suitable for growing *P. ostreatus* because it allows the smooth formation of "bouquets" of mushrooms, just as it happens in natural conditions.

c. Mound culture is a way of growing mushrooms on piles of a certain shape. The substrate of the appropriate composition is sterilized and inoculated with mushroom mycelium, and then poured on the floor, forming piles between which people can move, for easier manipulation.

d. Tray culture is a growing system that makes very good use of space because it allows a very dense arrangement of trays. The previously prepared and inoculated substrate is packed on shelves, and the manipulation is simple. Also, there is the possibility of adding a casing layer for species that require it. A special advantage is the simple placement of perforated plastic films, which aims to provide the specific microclimate necessary for the formation of the primordium, as well as for the formation of a "bouquet" of mushrooms that will have a uniform size and weight.

e. Wall culture is an improvement of the tray system performed by vertical placement of shelves, which allows fruiting on both sides. It is also possible to cover both sides with perforated plastic foils that maintain the required humidity and enable the even formation of properly formed mushrooms. Depending on the possibilities, needs, and types of mushrooms, various variations of this cultivation system have been developed.

2.4 SOLID-STATE FERMENTATION (SSF)

Solid-state fermentation (SSF) is a cultivation technique based on the natural process of fermentation that takes place on a solid substrate, performed by different types of microorganisms that occurs in conditions with very low content of available water, just as is the case in nature (Manan & Webb, 2017). However, a solid substrate must contain the amount of water necessary for biochemical processes and cell growth, while being a source of nutrients and physical support to the growing microbiota (Pandey, 2003). As fungi are naturally adapted to survive in conditions with a very low content of available water, this technique is particularly suitable for their cultivation, providing an artificially created environment similar to that which exists in natural conditions. Such conditions favor the development of fungi because the majority of other microorganisms cannot survive on solid substrates without enough water; thus, the possibility of contamination is significantly lower (Soccol et al., 2017).

The fermentation, which is the core of these processes, has been known to various human communities since time immemorial, and it is still used today without any special changes. These techniques are relatively simple, requiring appropriate raw materials, pre-treatment of the substrate, and an inoculum to ferment the substrate (Manan & Webb, 2017).

It is likely that they first began to grow yeasts and molds that normally develop in spoiled food and organic residues. Over time, people have noticed that these spontaneous processes have created products such as bread, fermented meat, or alcoholic beverages, which suit their needs; historical data prove that these production technologies were developed in ancient Egypt (3000–3800 BC) (Patrick & Durham, 1970; Leroy et al., 2013; Barakat, 2019). Solid-state fermentation, as one of the oldest biotechnological processes, has been used in Asian countries, too. Data aged around 3000 BC indicate probably the oldest known fermentation used for food production, i.e. Japanese "koji", for which steamed rice as a solid base is inoculated with molds *Aspergillus oryzae*. The same mold, grown on a variety of solid substrates, has also been used in Japan to produce a range of other products such as soy sauce, miso, or sake, the unavoidable ingredients of traditional cuisine. Among the more famous products are certainly Indonesian tempeh or Indian ragi; the solid substrate for their production are steamed and cracked legume seeds that are completely penetrated and covered with mold mycelium. Boiled and fermented soybeans completely overgrown with *Rhizopus oligosporus* mycelium are used for the production of tempeh (Raimbault, 1998; Hölker et al., 2004; Mani & Ming, 2016). The technology of growing microorganisms on a solid substrate has a very long history in the Western world as well, where probably the oldest product (2000 BC) is cheese matured due to the activity of the mold *Penicillium roquefortii* (Vallone et al., 2014). Further improvement of these techniques was intuitively developed by various cultures, according to their abilities (Letti et al., 2018).

Certainly, one of the most important roles of using solid substrates is that for mushroom production, starting with outdoor cultivation, which is completely dependent on weather conditions and biological factors, through various indoor cultivation techniques that have been developed and improved over many years. In the light of modern requirements that imply a high yield of mushrooms and low energy consumption, the use of SSF as an environmentally friendly process is justified (Soccol et al., 2017).

2.4.1 Substrate properties

Mushrooms reproduce by spores which, when ripe, are released from ascus (Ascomycota) or basidium (Basidiomycota). When they reach a favorable nutrient substrate, and if all other ecological parameters are met, they begin to germinate forming the first hypha. This hypha grows, branches, and forms a mycelium – a branched network of hyphae. The mycelium feeds by secreting enzymes that catalyze the extracellular digestion of large, complex molecules, and then absorb the newly formed smaller and simpler molecules (Boddy et al., 2008). All of these processes can take place properly only if the mycelium has enough available moisture. As the mushroom fruiting body is largely made up of water, which also has an irreplaceable role in their physiology, the solid substrate must contain sufficient water for proper metabolism and fruiting. On the other hand, if water is present in excess, it will settle to the bottom of the bioreactor, interfere with the normal growth of mycelium and contribute to the creation of favorable conditions for the development of undesirable microbiota. When choosing a substrate, the amount of water that can be absorbed should be taken into account, i.e. the amount of water that will be available to the mycelium for metabolic processes (a_w). It is also related to the requirements of the type of mushroom being grown on that substrate. Some species can grow at relatively low a_w, so this is their advantage over other species, but most require high a_w (Boddy et al., 2008). Hence, to achieve proper mycelial growth and the formation of fruiting bodies, it is crucial to ensure optimal hydration of the solid substrate.

Substrate structure, i.e. the size and shape of the granules directly affect the transfer of water, gases, and heat through the solid matrix (Chen, 2015). The substrate must be formed to allow unimpeded transport of oxygen and carbon dioxide, which is necessary for proper respiration. The mycelium can usually adapt to slightly higher concentrations of carbon dioxide. In the absence of oxygen, the growth of mycelium slows down and even completely stops, and the anaerobic microbiota takes over and changes the environment so that it no longer supports the growth of mycelium (Li et al., 2016). Too small granules make the substrate non-porous, especially it becomes compact after hydration, preventing the supply of oxygen and removal of excess heat, which allows the mycelium to grow only on the surface.

If the granules are too large and non-porous, hydration is difficult, and thus the growth of mycelium through the substrate. Too loose substrate prevents uniform mycelial growth and the formation of a compact block (Gianotti et al., 2009). A very important parameter that directly affects the proper development of the mycelium is the ratio of carbon and nitrogen (C/N). Nitrogen is necessary for the synthesis of proteins, essential molecules that participate in the construction of mycelium and fruiting bodies. If there is not enough nitrogen, the growth of mycelium stops. In the presence of a slight excess of nitrogen, the mycelium grows quickly, but the fruiting bodies are formed later (Yang et al., 2013). Excess nitrogen is undesirable because its metabolism produces by-products that reduce the amount of carbon available, which completely stops the growth of mycelium. There is no universally desirable C/N, it is the specificity of each species. For the normal development of the mycelium, oxygen and hydrogen are also needed, which, together with carbon and nitrogen come from the substrate, but also water and the atmosphere. According to the special requirements of mushrooms, it is necessary to add mineral salts or natural sources in which they are already contained, to provide sufficient elements such as Mg, Ca, P, K, Na, etc. (Manzi et al., 1999). The optimal pH of the substrate that favors most mushrooms is 5.0–7.0. If the mycelium releases acids during growth and metabolic processes, the pH of the substrate decreases due to the accumulation of acids. Such conditions favor the development of bacteria that are adapted to the acidic environment. During development, bacteria produce an additional amount of acid, which causes the pH of the substrate to drop, even more, resulting in the cessation of mycelial growth. To prevent harmful acidification of the substrate, it is common to include an appropriate buffer component in the substrate formulation. Good results are achieved by adding calcium carbonate which can maintain the pH of the substrate in an acceptable range throughout the growing period, and at the same time is a decisive source of calcium (Karunarathna et al., 2014).

The substrate prepared according to the requirements of a certain type of mushroom must also be microbiologically stable, i.e. it must not contain a microbiota that could endanger the development of the mycelium. For some mushroom species, it is sufficient to apply pasteurization, which destroys vegetative microbial cells, because the remaining viable forms, primarily spores, do not endanger the further growth of mycelium. On the other hand, the development of some mycelium requires the elimination of all sporogenic forms, so such substrates must be sterilized. In addition to the resistance of microorganisms to be eliminated, the sensitivity of the substrate to heat transfer as well as the volume of the bioreactor must be taken into account when choosing heat treatment (Oseni et al., 2012).

2.4.2 External factors

Properly prepared and inoculated substrate should be placed in an atmosphere that will provide optimal growth conditions for a respective type of mushroom. Under controlled environmental conditions, starting from spawning, the mycelium grows and colonizes the substrate. It is solid-state fermentation, i.e. the initial phase of mushroom cultivation (Zervakis & Koutrotsios, 2017). To initiate primordium formation, a change of external factors should occur, imitating the natural environment. The main factors that, according to the requirements of the mushroom, are manipulated in the environment are humidity, temperature, air change, and light. Even spraying of water in the room should provide very high humidity (95%–100%), i.e. fog-like conditions. After the formation of primodium, the humidity gradually drops to 90%–95%, respectively. In addition to moisture, temperature plays a key role in the development of mycelium, primordia, and fruiting body formation. Although there is an optimal temperature for each mushroom species, most edible and medicinal mushrooms grow quite well in the temperature range of 20°C–35°C. Some strains like *A. bisporus* and *P. ostreatus* require a drop in temperature below a certain threshold (to be around 15°C–16°C), as a trigger for primordium formation. Other species, such as *V. volvacea*, will fructification in the temperature range 27°C–32°C (Stamets, 2000). Air exchange is one of the necessary parameters that are applied with the appropriate intensity starting from the moment when the formation of the primordium should begin. A drastic decrease in the amount of carbon dioxide (preferably below 500 ppm for most species) and an increase in the amount of oxygen initiates the mycelium to begin primordium formation. The appropriate intensity of air changes during growth and maturation allows constant evaporation of moisture from the surface of fruiting bodies, directly affecting their proper formation (Stamets, 2000). Light affects the elongation of the stipe and the development of the cap, but it is also a signal that spores can be released in an open environment. Light properties, wavelength, and intensity are specific parameters for each mushroom species. If mushrooms grow in nature in the shade of a tree canopy, then it is necessary to provide an adequately smaller amount of light in artificial growing conditions. Some species, e.g. those belonging to the genus *Pleurotus*, require certain lighting for fruiting. But, it is extremely important to provide optimal lighting because bright light causes a change in the color of the fruiting body, so some strains become pale while others darken. Sensitivity to environmental conditions is used to favor some desired mushroom traits. Thus, *F. velutipes* is grown in the absence of light to obtain elongated stipes and white carpophores (Sharma et al., 2010), while in an atmosphere with an increased amount of carbon dioxide, elongated antler-shaped fruiting bodies of *G. lucidum* can be obtained (Sudheer et al., 2018).

2.5 AGRO-INDUSTRIAL WASTE

As a result of the intensive development of the population around the world, increasing amounts of waste from agricultural production, forestry, and the food industry are generated annually. Disposal of this diverse waste on land without prior appropriate treatment is a global problem because its presence inevitably leads to environmental pollution, and thus negative health consequences.

The environmental impact of high-income countries, which produce much higher amounts of food waste than low-income countries, is particularly negative (Leite et al., 2021). In light of these problems, waste management legislation has become more restrictive in recent years. The European Commission's Circular Economy Action Plan has included several actions aimed at promoting resource efficiency and living in a sustainable world where waste generation would be significantly lower (European Commission, 2019). By-products obtained from agricultural production and industrial processing of agricultural products constitute agro-industrial waste (Sadh et al., 2018). Waste generated by agricultural production consists of residue from the field and those derived from the process. Stems, leaves, straw, roots, seeds, and pods remain after the crop is harvested from field residues. During the processing of crops, peels, pulp, shells, and husks remain as process residues. Industrial residues are peels, pomace, bagasse, and bran that are formed as by-products of the fruit and vegetable processing industry as well as the food industry (Kumla et al., 2020). Current agricultural production implies an unequal distribution of agricultural waste across continents (Figure 2.5).

Uncontrolled disposal and accumulation of agro-industrial waste greatly endanger ecosystems. However, the consequences of such human activities can be mitigated by the purposeful use of these materials, which still have huge potential due to a significant amount of unused sugars, proteins, fibers, and minerals (Mussatto et al., 2012). These residues can be used as raw materials for various research and industry needs, but also for the production of biofuels, fungi, or products obtained by the activity of appropriate types of microorganisms such as tempeh (Sadh et al., 2018). The possibilities of using agro-industrial waste for the production of mushrooms and their lignocellulosic enzymes via solid-state fermentation are especially promising, considering that this waste is still a rich source of energy, but also of nutrients and carbon necessary for the mushrooms metabolic processes (Grimm & Wösten, 2018). To date, numerous studies have been conducted regarding the cultivation of mushrooms on a variety of agro-industrial waste, and some mushroom species cultivation on the selected waste is shown in Table 2.1.

Some research conducted in the last ten years confirming the possibility of cultivating edible and medicinal mushrooms species on selected agro-industrial waste. As saprotrophs, mushrooms secrete enzymes into the external environment to break down complex macromolecules, then absorb the newly formed smaller and simpler compounds to be used for various metabolic processes. Thus, mushrooms decompose natural biopolymers, but also those created by human activities. Due to lignocellulosic enzymes, they play a key role in the decomposition of wood and other plants in nature, and a significant number of these mushrooms belong to the group of white-rot fungi. These organisms have promising potential for bioconversion of lignocellulosic waste into several valuable products. Understanding the essence of these biotechnological processes has resulted in a significant increase in mushroom production on agro-industrial waste as a basic substrate (Gargano et al., 2017). Raising awareness about healthy food and the multiple benefits of mushrooms, the popularity of vegan food, cost-effective production, and the possibility of cultivation on agro-industrial waste contributed to the global mushroom market growth in recent years. According to the report Mushroom Cultivation – Global Market Outlook (2019–2027), the global

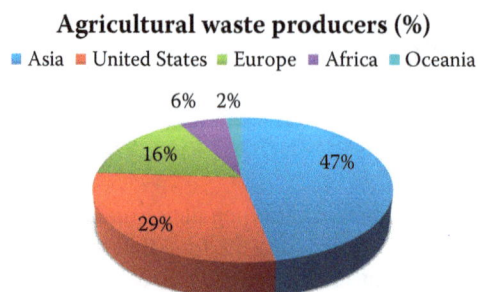

Agricultural waste producers (%)

■ Asia ■ United States ■ Europe ■ Africa ■ Oceania

6%   2%

16%

47%

29%

Figure 2.5 Distribution of agricultural waste production by continent (according to Cherubin et al., 2018).

Table 2.1 Some research conducted in the last ten years confirming the possibility of cultivating edible and medicinal mushrooms species on selected agro-industrial waste

Substrate	Mushroom	Reference
Bajra straw	*Pleurotus sapidus*	Telang et al., 2010
Banana leaves	*Volvariella volvacea*	Zikriyani et al., 2018
Barley straw	*Pleurotus ostreatus*	Dahmardeh 2013
Cassava peel	*Pleurotus ostreatus*	Kortei et al., 2014
Coffee pulp	*Lentinula edodes*	Mata et al., 2016
Corn cob	*Grifola frondosa* *Oudemansiella canarii* *Pleurotus eryngii* *Pleurotus ostreatus*	Song et al., 2018 Xu et al., 2016 Sardar et al., 2017 Han et al., 2020
Corn straw	*Grifola frondosa* *Pleurotus florida*	Song et al., 2018 Salami et al., 2017
Cotton seed hull	*Oudemansiella canarii* *Pleurotus ostreatus*	Xu et al., 2016 Yang et al., 2013
Corn stalks	*Auricularia polytricha*	Liang et al., 2019
Date-palm leaves	*Pleurotus ostreatus*	Alananbeh et al., 2014
Grape pomace	*Pleurotus ostreatus*	Doroški et al., 2020
Jowar straw	*Pleurotus sapidus*	Telang et al., 2010
Millet straw	*Agaricus bisporus*	Zhang, Wai, et al., 2019
Olive cake	*Fomes fomentarius*	Neifar et al., 2013
Olive cultivation residues	*Hericium erinaceus*	Koutrotsios et al., 2016
Palm oil wastes	*Auricularia polytricha*	Abd Razak et al., 2012
Pineapple leaf	*Ganoderma lucidum*	Hariharan & Nambisan, 2012
Rice straw	*Ganoderma lucidum* *Grifola frondosa* *Volvariella volvacea* *Pleurotus sapidus*	Postemsky et al., 2014 Song et al., 2018 Thuc et al., 2020 Telang et al., 2010
Rice husks	*Ganoderma lucidum*	Postemsky et al., 2014
Soybean straw	*Grifola frondosa* *Pleurotus sapidus*	Song et al., 2018 Telang et al., 2010
Sugarcane bagasse	*Pleurotus eryngii*	Sardar et al., 2017
Sugarcane leaves	*Pleurotus ostreatus*	Hossain, 2018
Sugar cane waste	*Pleurotus ostreatus*	Aguilar-Rivera & De Jesús-Merales, 2010
Sunflower stalk	*Pleurotus sapidus*	Telang et al., 2010
Tur straw	*Pleurotus sapidus*	Telang et al., 2010
Wheat bran	*Fomitopsis pinicola*	Tu et al., 2020
Wheat straw	*Ganoderma lucidum* *Pleurotus sapidus*	Ćilerdžić et al., 2018 Telang et al., 2010

mushroom growing market was worth \$16.37 billion in 2019 and is projected to reach \$26.09 billion by 2027, with an annual growth rate (CAGR) of 6.0% over this period. The predominant component of agro-industrial waste is lignocellulosic material composed of cellulose as the most abundant component, followed by hemicellulose and lignin. Cellulose is a relatively stable homopolymer made up of several hundred to many thousands of β-anhydro-glucose units interconnected by β-1,4 glycosidic bonds. Hemicellulose is a heteropolymer based on polysaccharide backbones which structure depends on the sugar units, the chain length, and the way the side chains are connected. Typical binding sugars involved in the construction of hemicellulose are pentose (arabinose and xylose), hexose (glucose, galactose, and mannose), hexuronic acids (glucuronic, galacturonic, and 4-O-methyl-d-glucuronic), as well as small amounts of fucose and rhamnose and an acetyl group. Lignin is a highly branched, amorphous, hydrophobic, and rigid polymer containing various functional groups such as carboxyl, methoxy, aliphatic, phenol hydroxyl, and carbonyl. These functional groups contribute to the very complex structure characteristic of lignin. In complex polymerization reactions, a unique, three-dimensional, highly branched lignin configuration is formed. The composition of lignin depends on the respective plant species and the environment in which it is formed (Zhou et al., 2016). Complex lignocellulosic material in nature is broken down by organisms that possess lignocellulosic enzymes such as bacteria, fungi, earthworms, and woodlice, which ensure the circulation of carbon in nature through these activities (Bredon et al., 2018). Due to the extremely complex binding between lignocellulose polymers, degradation is performed by the synergistic action of several carbohydrate-active enzymes (Lombard et al., 2013). Complete degradation of lignocellulose is achieved by the combined action of hydrolytic enzymes involved in the degradation of cellulose and hemicellulose as well as oxidative enzymes responsible for the degradation of lignin. Mushrooms are especially efficient decomposers of lignocellulosic waste. Thus, by controlled solid-state fermentation, agro-industrial waste is converted into high-grade protein food, fruiting bodies, or mushrooms. Spent substrate remaining after mushroom cultivation can be used as fodder or fertilizer but is also a valuable substrate for a new commercially viable mushroom production (de Mattos-Shipley et al., 2016; Grimm & Wösten, 2018). Most agro-industrial waste has a low nitrogen content. An important element in the selection of components for the cultivation of respective mushrooms is the ratio of carbon and nitrogen (C/N) in the waste because it directly affects the growth of mycelium and weight of fruiting bodies, as well as protein content in mushrooms (Grimm & Wösten, 2018). Thus, it is desirable to appropriately combine the organic part (soybean meal, cereal husk, cereal bran) with inorganic supplements (urea, ammonium chloride) to obtain the most favorable C/N ratio (Cueva et al., 2017). The growth of mycelium and the formation of fruiting bodies are significantly affected by the enrichment of waste with the addition of calcium carbonate, gypsum, and epsom salts (Grimm & Wösten, 2018). Another important factor to consider when selecting agro-industrial waste is biological efficiency (BE), a parameter that assesses substrate conversion efficiency. According to Liang et al. (2019), this value is calculated as a percentage of the fresh weight of harvested mushrooms to the dry weight of waste used for cultivation. A higher BE value indicates a higher potential for waste utilization. The BE value must be greater than 50% for mushroom cultivation to be profitable. The chemical composition of agro-industrial waste differs in different parts of the world, as well as the methods of cultivation and environmental conditions, and accordingly, the differences in the chemical composition of mushrooms are confirmed (Grimm & Wösten, 2018; Sadh et al., 2018).

2.6 CULTIVATION OF THE EDIBLE AND MEDICINAL MUSHROOM *LENTINULA EDODES*

The white-rot basidiomycete *Lentinula edodes* (Berk.) Pegler (1975), also known as shiitake, is native to East Asia. This edible mushroom has been popular for centuries because of its unique taste, flavor, and dietary significance, and is also appreciated due to its therapeutic properties (Ahn et al., 2017). Recent scientific studies have confirmed numerous beneficial effects on human health, including anti-inflammatory, immunostimulatory, and antioxidant, and polysaccharides from the group of β-glucans have been identified as components that contribute the most to biological activity, while lentinan has been singled out as particularly important (Kozarski et al., 2012; Finimundy et al., 2014; Muszyńska et al., 2018). Due to pronounced biologically active properties, and especially its anticancer effect, lentinan derived from *L. edodes* has been approved in Japan and China since the

1980s as an adjuvant in the treatment of some types of cancer (Zhang, Zhang, et al., 2019). In addition to traditional cultivation on cut logs, several indoor cultivation systems which include controlled conditions, have been developed. This enabled the cultivation of this precious mushroom in other parts of the world where climate conditions are not suitable for outdoor growing, which led to the multiplication of its production. Thus, people on almost all continents today can use shiitake for food as well as afford its medicinal benefits. A synthetic log system that implies indoor cultivation in polypropylene bags containing a substrate compacted in log form, is widely used in Europe and America. A substrate is based on 45%–79% of sawdust (Gong et al., 2014), but current trends include the use of different waste from agricultural production (Gao et al., 2020), to protect forests that are increasingly endangered. Although rice straw, as a suitable substrate for growing *L. edodes*, is available worldwide, the largest producer of rice, and thus rice straw, is certainly China. In this country, a total amount of straw of 870 million tons was recorded in 2019, while 730 million tons were used for various purposes (Xinxin et al., 2020). It is obvious that the use of rice straw for growing mushrooms, including *L. edodes*, could significantly contribute to the protection of forests. After preparation, sterilization, and inoculation of the substrate (as already explained in section 2.3.2.), the mycelium grows and permeates the substrate. During growth, hyphae secrete enzymes to break down cellulose, hemicellulose, and lignin, complex substrate molecules, which form smaller soluble molecules that are absorbed by the mycelium and used for metabolic processes. When the substrate is completely permeated with mycelium, the mycelial coat begins to harden, and the hyphae intensively absorb and accumulate the nutrients necessary in larger quantities to support the formation of fruiting bodies. In some places of the mycelial coat, bumps of unequal shapes and sizes begin to form, representing clumps of mycelium; most of them fail, and only some develop future primordiums. Due to the presence of bumps, air spaces are formed between the plastic bag and the mycelial coat, thus contributing to increased air circulation and the formation of brown pigment. Bumps no longer form 60–75 days after inoculation. More intense pigmentation is achieved by higher air flow, i.e. by slightly loosening the plug. During this period, the coat hardens, and it takes 20–30 days to be completely dry and hard. The moisture content inside the compost increases up to 78%. Thus, a formed artificial log with an exterior is hard and dry and the interior is very moist is most suitable for fruiting. Proper and complete formation of the mycelial coat is necessary because it has the same function as bark in wood logs – it prevents water loss, retains CO_2 in the interior, and thus improves mycelial growth, and also prevents the penetration of contaminants (Miles & Chang, 2004). As Rossi et al. (2003) proposed, at this stage, the bag should be removed and the ambient humidity maintained at 90%–95% to prevent moisture loss from inside the logs. Fruiting bodies are formed more evenly if the logs are subjected to induction, which means that they are immersed in cold water (16°C–26°C, 24 hours) and then left in a room where the relative humidity is maintained in the range of 90%–90% and temperature 21°C–23°C. The first flush occurs 8 days after induction and, usually, four flushes are obtained. Compared to outdoor cultivation, this method is much more efficient, and handling plastic bags is easier. The first harvest is possible about 5 months earlier, the end of the harvest is shorter up to 2 years (Miles & Chang, 2004), and biological efficiency (BE) is achieved up to 80% (Royse, 1985). Although the growth of mycelium is faster and the yield is higher, the quality of mushrooms is still lower than that obtained by outdoor cultivation. It takes a lot of work as well as energy consumption to sterilize the substrate in heat-resistant bags and subsequent inoculation of the cooling medium in each bag (Miles & Chang, 2004).

2.7 CULTIVATION OF THE EDIBLE AND MEDICINAL MUSHROOM *PLEUROTUS OSTREATUS*

Pleurotus ostreatus (Jacq.) P. Kummer (1871), also known as oyster mushroom, belongs to the group of white-rot fungi based on its participation in very important natural processes of hardwood decomposition by secreting ligninolytic and cellulolytic enzymes (Tsujiyama & Ueno, 2013). As a rare other species, the highly aggressive and productive *P. ostreatus* is well adapted to life in the temperate climate zone where it is commonly and wildly propagated in forests (Shnyreva et al., 2017). Due to its pleasant taste and texture, many people have used it for centuries as a highly valued food. According to Bellettini et al. (2019), today it is the second most industrially grown mushroom globally, as well as one of the most commonly consumed in China. However, the value of this mushroom is not only in its edibleness but also in its possession of biologically active components with numerous health benefits. Among others, antioxidant and cytotoxic potential against colorectal cancer cells have been demonstrated (Doroški et al., 2020), as well as its hypoglycemic properties (Meetoo et al., 2007). Lovastatin, approved by the FDA in 1987 in the treatment of elevated blood cholesterol (Stamets, 2000), which can contribute to reducing the risk of cardiovascular disease, has been derived from the fruiting bodies of *P. ostreatus*, as a secondary metabolite (Atlı et al., 2019). Cultivation of *P. ostreatus* is simpler compared to most other fungi, an exceptional advantage is the possibility of development on a wide array of lignocellulosic materials, it is not demanding, and is very adaptable to environmental conditions. A special advantage is the very efficient conversion of the substrate into mushrooms, so that the biological efficiency (BE), according to Stamets (2000) often even exceeds 100%. All these benefits have contributed to intensive production which, especially in poor areas, promotes the development of the rural economy and helps to overcome the lack of protein-rich food. Various cultivation systems, according to the possibilities, needs, and available lignocellulosic substrate, can be used for the production of *P. ostreatus*. In addition to traditional outdoor cultivation, indoor cultivation is very popular because higher yields are achieved due to the possibility of controlling conditions. Column culture (already mentioned in section 2.3.2) is especially suitable because "bouquets" of mushrooms can grow freely on the entire surface of the bags and the manipulation during fruiting is simple. It is recommended that the mother culture be prepared in sterile, laboratory conditions. According to Owaid et al. (2015), pasteurization gives quite satisfactory results in the elimination of potentially dangerous microorganisms from straw, a frequently used substrate for oyster cultivation. However, if possibilities allow, it is always safer to apply sterilization (Ryu et al., 2015). After inoculation, the bags are hung on the ceiling or lean against the beams. During the permeation of mycelium through the substrate, it is necessary to maintain a temperature of about 24°C–25°C, while the concentration of CO_2 should not exceed 15%–20% (Miles & Chang, 2004). To ensure the proper formation of fruiting bodies and avoid elongation of the stipes, it is necessary to maintain a concentration of CO_2 < 1,000 ppm during the fruiting period (Stamets, 2000).

A special role in the life cycle of *P. ostreatus* has light because it belongs to the group of mushrooms that require periods of illumination to achieve normal fruiting. Although the photoperiod has no significance for the development of the mycelium, it is very necessary for the growth of fruiting bodies. A recent study in the field of fungal photobiology, conducted by Colavolpe and Albertó (2014), indicates that mushrooms which fruit is dependent on a certain amount of light use common regulatory pathways to develop basidium and form viable spores. According to Jaramillo Mejía and Albertó (2013), the photoperiod required for mycelia stimulation of that mushroom group to start fruit body formation should be 200–640 lux, 8–12 hours per day, while Stamets, (2000) recommends 200 lux, 12 hours per day for *P. ostreatus* primordia formation and 50–500 lux illumination for fruit body formation.

2.8 CULTIVATION OF THE EDIBLE AND MEDICINAL MUSHROOM *AURICULARIA AURICULA – JUDAE*

Auricularia auricula – judae (Fr.) J. Schröt., well known as black ear or wood ear, can be found in temperate and subtropical regions worldwide, growing as a wood rot mushroomon in living and dead trees (Priya et al., 2016). It was previously thought to be a single species; however, a recent study conducted by Wu et al. (2014) revealed that it is a species complex, so they introduced species name *A. heimuer*. It is one of the earliest cultivated mushrooms in China where it is still grown in large quantities, so in 2017, as much as 6.3 billion kg were produced in this country (Chinamushroombusinessnetwork, 2018). Although it is not often consumed in the Western world, in Asia *A. auricula* is very popular as an edible and medicinal mushroom. Recent studies have confirmed several medicinal properties of this mushroom, among others antitumor (Reza et al., 2014), antioxidant (Ma et al., 2013), antiviral (Nguyen et al., 2012), and antimicrobial (Oli et al., 2020). The first written trace of the simple cultivation of *A. heimuer* on logs was left by Gong Su in Tang materia Medica 1300 years ago (Huang et al., 2010). Due to increasing demand and relatively small investments, this method (described in section 2.3.1) is still used in Asian countries. Stamets (2000) recommends soaking the logs in water for 24 hours to start the formation of mushrooms. However, as outdoor cultivation is not able to meet the growing needs for this mushroom, other methods of cultivation are widely used, too. One of them is the artificial log method, similar to that used for growing *L. edodes* (already explained in section 2.3.2.). According to the instructions of Stamets (2000), the basis of the substrate is sawdust enriched with appropriate additives, this mixture is filled into polypropylene bags, sterilized, cooled, and inoculated with *A. auricula* mycelium. When the substrate is completely permeated with mycelium, the bags open and intense wetting begins to initiate the formation of the primordium. Cultivation on compost obtained from straw, rice bran and appropriate supplements also provides good yields of this mushroom. After thoroughly mixing all the necessary ingredients, water is added and the humidity is maintained at 65%–70%. The composting process usually takes five days, then the substrate is sterilized, cooled, and inoculated (Quimio, 1982). Excellent yields are also achieved by using column culture (explained in section 2.2). For this purpose, wheat straw to which 5% of wheat bran is added is the basis for *A. auricula* cultivation. After 16–18 hours of soaking with water, the straw is filled into bags, sterilized for about 1.5 hours. It is then inoculated and incubated at 25°C–26°C for 20–25 days. When the substrate is completely permeated with mycelium, holes are made and bags are hung on the ceiling. It is necessary to maintain a relative humidity of 85%–90%, a temperature of 25°C–26°C, to provide 1–2 hours of diffuse lighting and adequate aeration. The first fruiting bodies appear after 10–12 days, and after 4–5 days they are mature enough to be harvested. With this method, a very high BE of as much as 140% can be achieved, in 3–4 waves (Stamets, 2000).

2.9 CULTIVATION OF THE EDIBLE AND MEDICINAL MUSHROOM *AGARICUS BISPORUS*

Agaricus bisporus (J.E.Lange) Imbach (1946), or white button mushroom, is a saprophyte widespread on grasslands in Europe and North America (does not grow on trees). It has a significant ecological role as a secondary decomposer in terrestrial ecosystems as it contributes to carbon cycling by degrading plant biomass in a humic-rich environment (Morin et al., 2012). This mushroom has become very popular all over the world due to its exceptional nutritional value which comes from its high protein levels, low-fat content, and high fiber content (Atila et al., 2017). Also, there is a growing interest in this fungus as a health promoter, because numerous studies have proven that the fruiting body of *A. bisporus* contains bioactive components such as ergothioneine, ergosterol, β-glucans, flavonoids, and vitamin D (Blumfield et al., 2020). The presence of lectins, triterpenoids, nucleosides, essential peptides, glycoproteins, peptides, and fatty acids, as well as their derivatives contributes to the potential use of this mushroom as an antioxidant, antimicrobial, antidiabetic, hepatoprotective anticancer, antihypertensive, and antihypercholesterolemic agent (Kozarski et al., 2011; Atila et al., 2017; Kozarski et al., 2020). Initial cultivation of *A. bisporus* on a composted substrate, a primitive method used around 1600 in

France, has undergone many changes and refinements over the centuries, so today it is a very complex technique that requires sophisticated equipment and large capital investments. However, given the growing demands of consumers, *A. bisporus* is now produced in more than 100 countries around the world (Miles & Chang, 2004). The appropriate substrate is obtained by composting, a complex biological process involving numerous microorganisms. According to Owaid et al. (2017), various lignocellulosic components are used, with the addition of chicken, pigeon or horse manure, mineral salts, and other additives, in accordance with the respectively applied technique to obtain a quality substrate suitable for composting. As Miles & Chang (2004) reported, there are several approaches and modifications of substrate preparation, but all techniques have in common that the substrate is composted and then pasteurized. The composting process begins by mixing and moistening the appropriate components as much as is necessary for the naturally occurring microorganisms, which perform composting, to have enough water available for metabolic processes. At the beginning of composting, cellulose and hemicellulose are converted into sugars, straw and other raw materials soften due to the microbial activity, soluble sugars decompose, and the C/N ratio decreases. Lignin is a resistant component that impregnates plant fibers, and bacteria generally cannot break it down. Nitrogenous substances such as nitrates and ammonia, present in the substrate, enable the growth of bacteria and fungi that convert them into proteins; these microbiological processes lead to an increase in the protein content of the compost (Miles & Chang, 2004). At the end of this phase, the temperature in the compost heap reaches 60°C because heat is released due to microbiological activity (Owaid, 2009). Mesophilic microorganisms that have started composting can no longer survive at this temperature, so thermophilic species of the genus *Actinomyces* become dominant. In the center of the compost heap, an anaerobic zone is formed in which the desired microorganisms cannot survive. On the other hand, a temperature of about 35°C in the outer zone is also not sufficient for the growth of microorganisms involved in composting. Therefore, it is necessary to mix a pile of compost to achieve equal conditions in all parts of the volume for a time, to provide ventilation, i.e. the amount of oxygen needed for microorganisms and the optimal temperature (Miles & Chang, 2004). In the further process, pasteurization, it is necessary to destroy insects and pests that come with the substrate, spores of contaminating microorganisms, as well as to equalize the temperature of the substrate to 50°C–55°C, thus providing optimal conditions for the development of thermophilic microbiota. All applied measures enable the formation of a substrate that will favor the growth of mushrooms. In the next phase, the cooled substrate is inoculated with mycelium, and after complete permeation of the mycelium through the substrate, a casing layer is placed on the surface. The casing layer is commonly a peat moss which pH is adjusted on 6.5–8.0 by adding lime. It should be sterilized before application. The purpose of adding a cover is to maintain moisture and evenly fruiting (Miles & Chang, 2004).

2.10 CULTIVATION OF THE MEDICINAL MUSHROOM *GANODERMA LUCIDUM*

Ganoderma lucidum (W. Curt.:Fr.) P. Karst., in Asia known as Ling Zhi or Reishi, is widely distributed in temperate and subtropical climate areas of Asia, Europe, America, and Africa (Siwulski et al., 2015). As a wood-decaying mushroom that causes white rot, this species plays a very important role in the breakdown of different types of hardwood trees due to the spectrum of lignocellulosic enzymes it possesses (Zhou et al., 2018). Since time immemorial, *G. lucidum* has symbolized longevity, success, holiness, goodness, and happiness in the life of people in Asian countries, so it has been called by many symbolic names such as "celestial herb", "herb of spiritual potency", or "mushroom of immortality" (De Silva et al., 2012). In recent decades, there has been a growing interest in this mushroom in the Western world too, so many extensive studies which have been conducted revealed that *G. lucidum* contains over 400 biologically active compounds, including polysaccharides, triterpenes, steroids, proteins, amino acids, alkaloids, and nucleosides

(Cör et al., 2018). These extremely valuable components show several health benefits, e.g. anticancer (Rossi et al., 2018), antioxidative and immunomodulating (Kozarski et al., 2011), antidiabetic (Ma et al., 2015), antimicrobial (Celal, 2019), antiviral (Linnakoski et al., 2018), and antiacetylcholinesterase (Cör et al. 2018). Given the role it played in the tradition of the Asian people (especially China, Japan, Korea), *G. lucidum* was an economically important mushroom even 4,000 years ago (Wasser, 2005). Due to the discoveries of modern science about extremely beneficial effects on health, there is a constant growth in demand for this mushroom, so the global market of *G. lucidum* in 2019 was $3.09 billion, and is expected to reach $5.06 billion by 2027, recording a CAGR of 8.1% from 2021 to 2027 (Allied Market Research, 2021). Various cultivation techniques are being applied to meet the expectations of such a growing market. Traditional methods of growing on logs (explained in section 2.3.1.2.) or tree stumps (explained in section 2.3.1.3.), which are inoculated with spawn directly under natural conditions (Stamets, 2000), are still used to produce quality fruiting bodies, although it takes longer and yields are lower. The cultivation method, which gives high yields, involves the use of sterilized wood segments, best obtained from oak, with a diameter of about 12 cm and a length of up to 15 cm, with flat sections (Chang & Buswell, 1999). The formation of segments should be done in the dormant period when the moisture content in the wood is 45%–55%. The prepared wooden segments are buried in the ground inside a greenhouse or other suitable facility. The soil should not be too moist, but it must provide drainage, air permeability, and retention of the required amount of water. This method of cultivation is more economically viable because it allows higher BE, faster growth, and development of quality fruiting bodies, but requires more investment than cut logs and stumps methods (Miles & Chang, 2004). As *G. lucidum* is relatively rare in nature, quantities are very limited and it is very difficult to monitor the quality of fruiting bodies in the forest, cultivation in a controlled environment is the recommended solution for commercial production. Cultivation in polypropylene bags or bottles is commonly used for this purpose (Zhou et al., 2011). As Ćilerdžić et al. (2018) proposed the substrate should be obtained by mixing and wetting the appropriate components and additives so that the moisture is about 65%. The substrate was then filled into polypropylene bags and sterilized at 121°C for 2 hours. After cooling, the substrate is inoculated with mycelium and left at 25°C to be completely permeated. In the next step, bags are opened and fructification is performed in the growth chambers, under controlled conditions that are in line with the needs of the growth phase. According to Stamets (2000), it is necessary to provide a day-night light regime for the primordium formation (12 hours on/off, 500–1,000 lux), as well as for the fruiting bodies formation (750–1,500 lux). In some growing chambers, bottles or bags are placed horizontally, one on top of the other, forming a wall-like shape, and fruiting bodies are formed on top of the bag or bottle so that one has the impression of growing out of the wall (mentioned in section 2.3.2.). An exceptional advantage of this cultivation method is the possibility of growing *G. lucidum* on wheat straw, which is a widely available and inexpensive substrate, a byproduct in agricultural production. The period of complete permeation of straw with mycelium is short and the formation of primordium and fruiting bodies is rapid (Ćilerdžić et al., 2018). However, for the cultivation of this mushroom, very different lignocellulosic substrates obtained by a combination of suitable components can be used, to achieve the most efficient production. Thakur and Sharma (2015) reported a significantly high yield and BE of 27.9% by using 80% *Jacaranda mimosifolia* sawdust supplemented with 20% of wheat bran.

2.11 SUBMERGED-LIQUID FERMENTATION (SLF)

For decades, people have been growing edible and medicinal mushrooms. The development of these mushrooms through fruiting body (cell) production is a lengthy process that takes several months before the first product appears, resulting in a wide range of product quality. Regulated cultivation in heterotrophic bioreactors has been designed to reduce cultivation time and improve quality. In comparison to the average 6-month duration needed for conventional mushroom cultivation through SSF, technology for quick cultivation (10 days or less) using *G. lucidum* as the prime model has recently been established as a promising cultivation strategy. For these mushrooms, the invention of a quick cultivation system known as submerged-liquid fermentation (SLF) methods has sped up mycelium production (cells). SLF on *G. lucidum* is a better option because the mycelium can be processed in large quantities, using a cheap and safe method, and to a consistent consistency. Such efficient SLF systems based on bioreactors can be integrated into the food and biomass chain for landless food and vital bioactive compound production, particularly in white fungal biotechnology applications.

2.11.1 Isolation and morphological identification of pure mycelium of mushrooms

Edible and medicinal mushrooms can be cultivated in submerged-liquid fermentation (SLF) once they are successfully isolated and identified from the wild (Hassan et al., 2019). These mushrooms must be isolated and replicated in a standard sawdust mushroom spawn bag before molecular identification. The chosen mushroom's fruiting body is freshly picked and morphologically analyzed based on its shape, color, hardness, and spore type. Typically, it was traditionally grown on a farm via solid-substrate fermentation (SSF); however, SLF provides an alternative way using the landless food production concept. These liquid-based mushrooms are commonly cultivated due to the mycelium's identified health-preserving compounds, such as potent anti-inflammatory, antioxidant, antibacterial, antiviral, antifungal, and anticancer properties (Jaros et al., 2018; Klaus et al., 2021), in addition to biofilm formation (Vunduk et al., 2019) and recent bioremediation wastewater treatment (Mooralitharan et al., 2021). A great mushroom mycelium application is a mycoprotein generated by Kim et al. (2011) from *Agaricus bisporus* commercial submerged fermentation. While fungal mycelium could provide a unique chewing sensation, the negative consumer perception of fungal mycelium necessitates the development of meat analogues made from true mushroom mycelium. When compared to soy protein, the development of an industrial bioprocess for *A. bisporus* mycelium permitted the production of a highly appropriate meat analogue with superior textural properties and umami characteristics. Novel protein sources are desperately required to meet the ever-increasing protein demand of the world's population. Using *Pleurotus sapidus* on apple pomace, a similar study is focusing on the development of protein-rich mushroom mycelia on industrial side streams (Ahlborn et al., 2019). As a result of this technique, nutritionally important mushroom mycelia enriched in vitamin D2 has emerged. Meanwhile, the submerged fermentation of edible mushroom *Pleurotus ostreatus* biomass was extracted for 19 pharmaceutical metabolites (Papaspyridi et al., 2012). It has provided a promising alternative for efficient biomass and useful metabolite production, with faster production in a shorter period, less space, and fewer contamination risks (Tang et al., 2007). Furthermore, the SLF system's recent application in white fungal biotechnology has encouraged stakeholders to invest in it. Many edible mushrooms used in traditional folk medicine, such as *L. edodes* (shiitake mushroom), *G. frondosa* (maitake mushroom), *H. erinaceus*, *F. velutipes*, *Tremella mesenterica*, and *P. ostreatus*, are considered a good source of bioactive compounds (Sullivan et al., 2006), and attempts to cultivate them in SLF can be referred to using the Figure 2.6 blueprint, with *Ganoderma* as the model.

Figure 2.6 provides a clear blueprint for the identification of wild mushrooms before SLF cultivation. As a pilot example, a wild medicinal *G. lucidum* was isolated from Mount Avala, Serbia, and morphologically identified based on its brown-liquorish cap and woody stipe (Hassan et al., 2019) at the base of wild oak (*Quercus robur* L.). The carpophores and 100x basidiospores structure (exosporium and endosporium), as well as the hymenium, which resembles *G. lucidum*'s

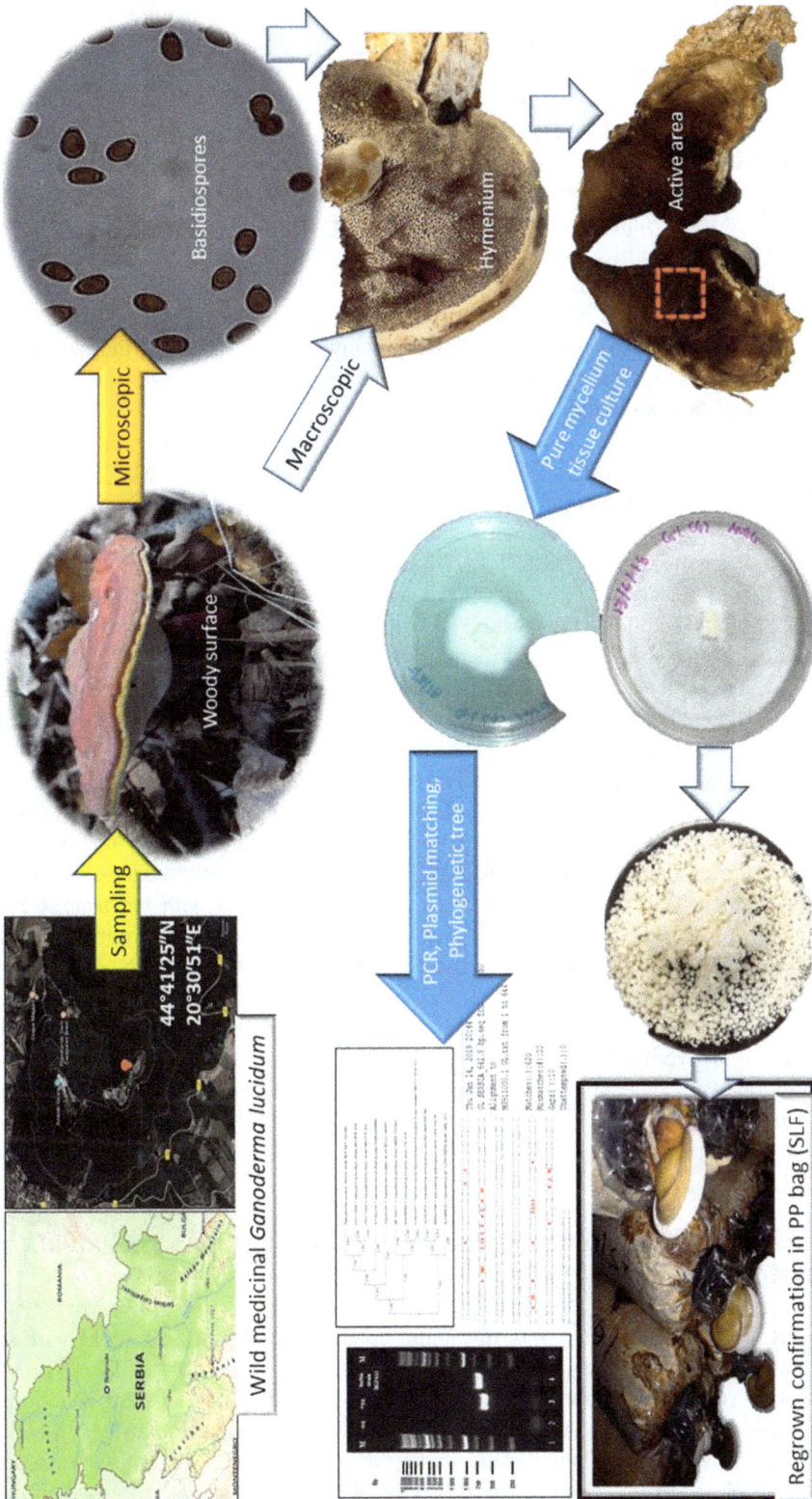

Figure 2.6 Isolation and identification blueprint of wild edible and medicinal mushrooms.

hymenium pores (Hennicke et al., 2016), were examined under microscopic and macroscopic conditions to confirm the right mushroom species (Figure 2.6). A tissue culture procedure is performed on the *Ganoderma* fruiting body to create a pure mycelium culture. The fruiting body was first washed in 99.9% ethanol for 10 seconds before being dried in laminar flow. It was broken open to reveal the active area (red dotted square in Figure 2.6) and finely cut at the marked area using a scalpel. Finally, the tissue was put on malt extract agar (MEA) and allowed to develop at room temperature before mycelium appeared. To obtain pure mycelium, the mycelium was sub-culture onto fresh MEA (Supramani et al., 2019). This practice is repeated several times (some-times up to eight times) until a singular mycelium mat is obtained on the agar plate (PDA or MEA). They successfully isolated mycelium then underwent triple molecular identification, which is PCR, phylogenetic tree, and plasmid matching comparison. To execute this, the *G. lucidum* mycelium on square cut MEA is finely ground in liquid nitrogen using a set of mortar and pestle to extract the fungal DNA. It follows with gDNA extraction and PCR procedure with slight mod-ification in the amplification (Liu et al., 2000; Zhou et al., 2007; Tamura et al., 2013). The PCR product was amplified using two standard fungal internal transcribed spacer primers (ITS1 and ITS4). Each primer (0.5 pmol), dNTP mix (200 M, Promega), PCR buffer (ThermoFisher Scientific, Waltham, USA), DNA polymerase (0.5 U, Promega, Madison, USA), and water were included in the PCR reactions. The following conditions were used for amplification of the targeted fragments: initial denaturation at 98°C for 120 seconds (1 cycle); 25 cycles of annealing and extension (98°C for 15 seconds; 60°C for 30 seconds; 72°C for 30 seconds); and 1 cycle of final extension at 72°C for 10 minutes. Purification and sequencing of PCR-amplified fungal PCR yields are subjected to 1% agarose gel PCR purification tool (Tiangen Biotech Co., China) analysis for 1 hour at 80 V, and sequenced using a BigDye® Terminator v3.1 sequencer (Applied Biosystems Co., USA). For data analysis, Clustal Omega by Sievers and Higgins (2018) was used to align the sequences. For essential identification and information on previous taxonomic and phylogenetic studies, the sequences were compared to similar sequences of *Ganoderma* species using BLAST software (NCBI) (adapted from https://blast.ncbi.nlm.nih.gov/Blast.cgi). NCBI GeneBank was used to store fungal sequences. The evolutionary distance (K_{nuc}) of identical fungal species was calculated using the neighboring-joining (NJ) method in Molecular Evolutionary Genetic Analysis (MEGA-X), and then a phylogenetic tree was created. The species with the nearest Knuc are thought to be the same. To confirm the species, A plasmid Editor (ApE) program was used to compare the sequences of the closest K_{nuc} species and the sequence of gDNA for mismatches which gave a factual verification of the *Ganoderma* genus and species (Hassan et al., 2019; Supramani et al., 2019). In the final verification, the isolated pure mycelium is pre-grown in liquid culture, generating active hyphal tips (Figure 2.6) and transferred to a spawn culture for fruiting body regeneration. The newly formed fruiting body will be verified microscopically and macro-scopically with its isolated origin as a "regrown confirmation in PP bag" before SLF. This strategy is efficient and can be adapted to most of the wild and edible medicinal mushroom species worldwide.

2.11.2 Conversion of solid-state fermentation (SSF) to submerged-liquid fermentation (SLF)

To convert purified SSF *Ganoderma* mycelium culture to SLF, cut agar squares of mycelial SSF plate culture are transferred in 100 mL of 250 mL Erlenmeyer flask (SLF). The standard practice is, four mycelial agar squares (5 mm × 5 mm × 5 mm) are prepared using a sterilized cork borer or stainless-steel scalpel (Wan-Mohtar, Young, et al., 2016; Wan-Mohtar, Abd Malek, et al., 2016) (Figure 2.7). When using small-scale shake flasks or agar cultures, most mushroom sci-entists use small pieces of mycelium that are still attached to the agar plates on which the fungi were produced, and inoculate these directly into the fermentation broth. This seed culture

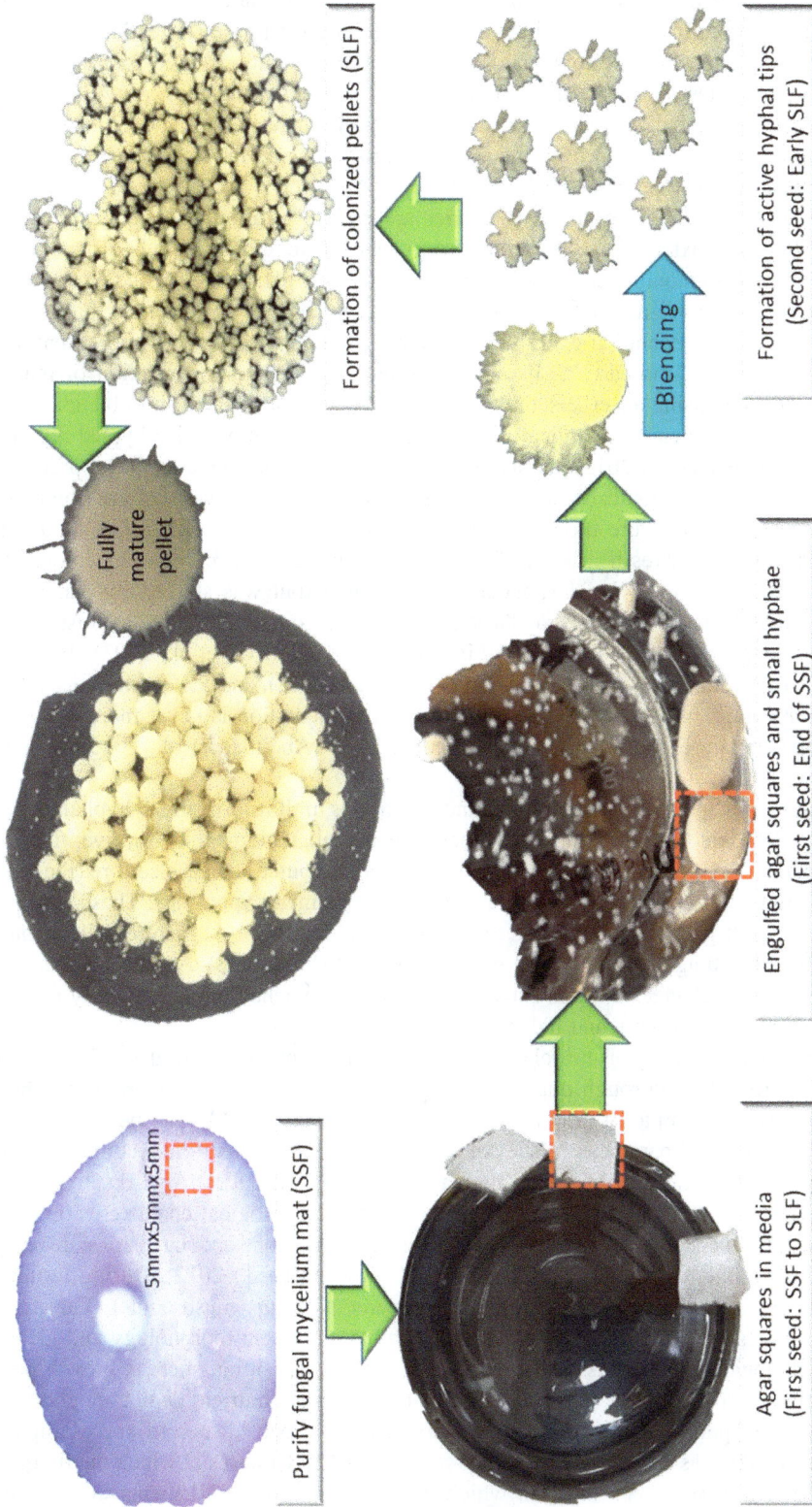

Figure 2.7 Conversion of SSF to SLF using the medicinal mushroom *Ganoderma lucidum*.

technique is used to inoculate a bioreactor so that the mycelium can adapt from a solid to a liquid environment, lowering inoculum densities and reducing the lag phase (Fazenda et al., 2010). To avoid a long lag time, the environmental and nutritional conditions used to prepare the seed culture, as well as those in the fermentation, must remain constant. *G. lucidum* inoculum standardization is also essential, as is the removal of cut mycelia-mats at the same radial distance from the agar colony center. This approach guarantees that all inocula have the same amount of mycelium at the same developmental time.

At the end of the SSF stage, the agar squares are grown in formulated chemical factors containing glucose, yeast extract, potassium dihydrogen phosphate, dipotassium hydrogen phosphate, magnesium sulfate, ammonium chloride, and physical factors; pH, agitation, and temperature. It is called "first seed culture" as the fungal mycelium colonizes the agar squares until it reached threshold growth (first generation pellet) and produces protuberance in the liquid system reaching "second seed culture". This early SLF stage requires inoculum homogenization of mature first seed. For *G. lucidum*, using a sterile stainless steel Warring Blender and a short blending time, the mycelium can be homogenized aseptically to maximize the amount of active hyphal tips (Fazenda et al., 2010; Stanbury et al., 2013). To avoid potential mycelial damage from the procedure, the inoculum blending time should be no more than 20 seconds in a low-speed environment. Such treatment has helped to ensure process reproducibility. An active inoculum can reduce the lag period in subsequent cultures if it is transferred at the right time (i.e. correct physiological state). Years ago, the impact of *G. lucidum* inoculum size on output was studied by Yang and Liau (1998), who found that a 7-day shake flask inoculum was optimal. Many scientists have used inoculum percentages (v/v) for *G. lucidum*, including 17% (Berovic et al., 2003), 10% (Wagner et al., 2003), 10% (Sanodiya et al., 2009), 12% (Liu et al., 2012), and 5% (Kim et al., 2006). The maximum percentage (v/v) used in *Ganoderma* liquid fermentation is never more than 20%, according to the literature and extensively verified in 2020 (Abdullah et al., 2020; Wan-Mohtar, Taufek, et al., 2020).

In the late SLF stage, the fungal protuberance protrudes until it forms fresh hyphal tips (second generation pellet) that re-engulf the first-generation mycelial biomass to survive. The second-generation pellet evolved and produced protuberances at the outer layer. Once it reaches maturity, these protuberances are shaved off from the surface and formed hyphal tips. These active hyphal tips will form small pellets and continue to mature until day 10. Fully mature pellets shaved off their outer layer reaching maximum third-generation pellet (Figure 2.8). Autolysis starts to initiate with yellowish color changes. At this stage, the fungus has reached the late to early death phase, which is extracted for biomass and bioactive metabolites.

If a better understanding of the relationship between morphology and desired bioproducts is desired in Figure 2.8, a thorough quantitative structural analysis of *Ganoderma* morphology is required. This is an example of how methods like image analysis (IA) can be used to calculate important morphological parameters (Riley et al., 2000), and IA is therefore recommended for any work involving the production of fungal hyphae (Tucker et al., 1992). As a result, IA has been developed as a fast and precise method for quantitative morphological characterization, as parameters of an observed object such as size, total, shape, position, and intensity can be quickly evaluated (Treskatis et al., 2000; Park et al., 2002; Tepwong et al., 2012). During fermentation, light microscopy (IA) is used to characterize filamentous fungi (mushrooms) (Fazenda et al., 2010). IA attempts to qualitatively characterize the most prominent morphology and identify any related morphological changes associated with any process transition, such as growth phase, nitrogen limitation, repeated-batch cycle, harvesting time, famine, nutrient depletion, and catabolite repression. The dispersed type can be further divided into freely dispersed and clumps in suspension culture (Thomas & Paul, 1996; Tepwong et al., 2012). Cultures in which the fungus grows as pellets are less viscous than those in which it grows as scattered filaments in terms of morphological structure (Tang et al., 2007; Fraga et al., 2014).

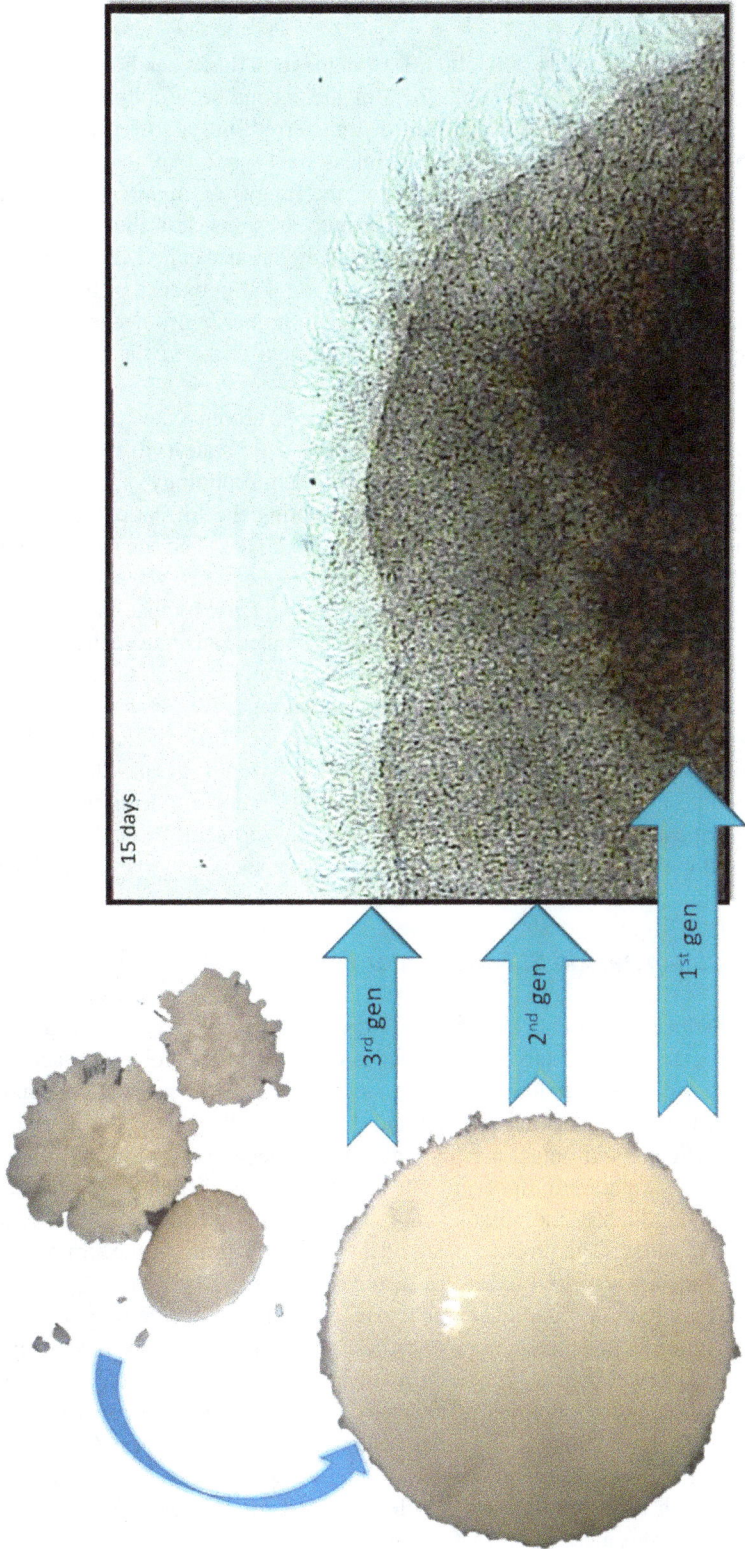

Figure 2.8 Fungal pellet behavior in submerged liquid fermentation of the medicinal mushroom *Ganoderma lucidum*.

2.11.3 SLF in the bioreactor system

Submerged-liquid fermentation in 200–500 mL Erlenmeyer's flasks can be scaled up to a larger bioreactor device (2,000–5,000 mL). At 10%–20% of the second seed culture, the active fungal pellets from the flask fermentation are inoculated or transferred into a stirred-tank reactor vessel (Figure 2.9). With heated flames at the top plate stainless steel screw cap, the inoculum is poured aseptically. In both the shake flask and the bioreactor, the fungus *G. lucidum* displayed different morphological characteristics during the various SLF processes. As described by Žnidaršič-Plazl and Pavko (2001), this fungus forms small, loose mycelial aggregates called clumps that later grow into spherical aggregates (pellets) consisting of highly entangled networks of hyphae. The shake flask (low shear) had morphological variations, likely due to higher hydromechanical stress within the bioreactor (high shear), which disrupted its life cycle and changed the formation of clamp connections.

The clamp connections are intended to ensure that each basidiomycete "cell" or compartment has a compatible pair of nuclei, thus maintaining the dikaryotic state (Fazenda et al., 2008). In Figure 2.9, raising the shear rate in the bioreactor changes the morphology of *G. lucidum*, splitting the pellets into more dispersed hyphae, and thereby promoting the formation of the filamentous type, results in an exponential active phase (Fazenda et al., 2010). The means of moving nuclei from one hyphal fragment to another, also known as clamp connections, were discovered based on these events in Figure 2.9, which depicts the specifics of *G. lucidum* hyphae in the bioreactor and shake flask during SLF. The red arrow indicates the clamp connections, the septum is indicated by the yellow arrow, and the hyphal tip is indicated by the blue arrow. Overall, the discovery of clamp connections in SLF cultures was both surprising and encouraging, given that submerged hyphae in liquid cultures may fail to form clamp connections (Carlisle et al., 2001), which could be due to the current stirred-tank bioreactor's use of extreme agitation (Fazenda et al., 2010). The primary fungal mycelium would not divide synchronously without the clamp connections (Deacon, 2013), and the cells' dikaryotic stage would be disrupted, halting development. When fully mature fungal pellets have achieved maximum clean pellet structure, they are usually harvested and subjected to downstream steps.

2.11.3.1 Critical parameters in SLF bioreactor system

Chemical, biochemical, physical, and morphological factors all play a role in growing *G. lucidum* in liquid media. The biocatalyst's behavior is determined by biological factors, whereas physical and chemical elements define the environment. The best fermentation conditions are determined by the design of the desired product and the fungus strain used. Furthermore, culture conditions affect growth rate and fungal morphology. The fermentation macro-environment can be influenced by aeration, temperature, agitation rate or shear force, fermenter design, and culture time. These factors may influence the fungus's morphological and physiological behavior, as well as the bioprocess's efficiency. The physical state of the fungus will affect the strain's growth and productivity. Table 2.2 shows several examples of physical causes.

Chemical influences have a significant impact on biotechnology process evolution. Metals, ions, additives, medium pH, medium composition, carbon source, complex media, by-products, and nitrogen source are mentioned in Table 2.3 as metals, ions, additives, medium pH, medium composition, carbon source, complex media, derivatives, and nitrogen source.

Biological factors such as inoculum age and volume significantly impact *G. lucidum* growth and product development. As a result, proper inoculum characterization is critical for future industrial applications such as culture isolation, selection, and maintenance (Seviour et al., 2011; Stanbury et al., 2013). Furthermore, these tests must be conducted on the chosen inoculum because it is susceptible to infection, spontaneous mutation, degradation, and death (Fazenda et al., 2008).

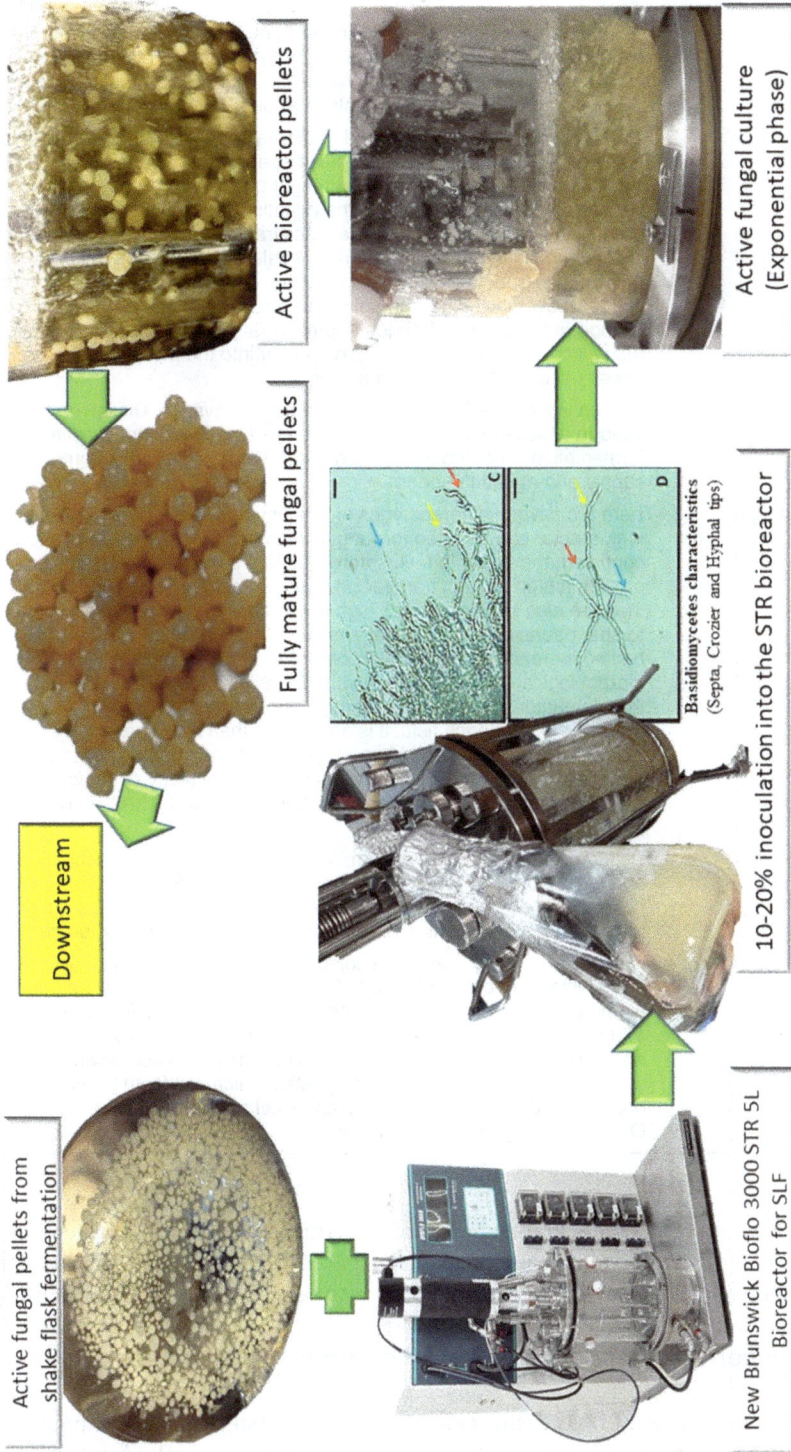

Figure 2.9 Submerged-liquid fermentation of the medicinal mushroom *Ganoderma lucidum* in the bioreactor system.

Table 2.2   Physical factors involved in submerged-liquid fermentation of *G. lucidum*

Physical factors	Explanation
1. **Temperature**	In *Ganoderma* fungal cultures, temperature influences growth rate, medium evaporation, dissolved oxygen (DO), pellet production, and product creation. *G. lucidum* growth has been observed in SLF at temperatures ranging from 25°C to 36°C, with most studies taking place at 30°C.
2. **Agitation**	Rushton turbines are widely used in bioreactor systems to provide continuous agitation for *G. lucidum*. The impeller speed of a bioreactor, which plays a significant role in determining the fungal growth rate through mixing, heat transfer, and mass, is equal to the agitation rate during growth and bioproduct production. There can be a visible mass transfer gradient in the bioreactor vessel in such fungal cultures, particularly at scale, triggering both mycelial morphology and product spectrum changes. Agitation's shear force can harm mycelium in various ways, including damaging cell structure, stimulating morphological change, and causing changes in growth rate and bioproduct formation. As a result, achieving sufficient oxygen transfer into the medium while avoiding shear stress necessitates optimal agitation.
3. **Aeration**	Aeration is a critical factor in Basidiomycete cultivation. Dissolved oxygen (DO) is used to assess aeration, and it is the most vital variable to optimise in aerobic fermentation. In any operated fermenter setup, aeration is controlled by air supply and agitation speed.
4. **Fermenter design**	There are several different types of fermenters on the market, but only the stirred-tank reactor can be used for fungal culture (STR). There are two forms of STR: confused and unbaffled. Rushton impellers are widely used because they provide a high oxygen transfer rate and promote aerobic fungi growth; however, they are also known for their high shear force, which reduces product yield due to shear-sensitive cells. Under similar culture conditions, a low shear oscillatory baffled bioreactor (OBR) was reported to produce high EPS, but no morphological differences were noted. Baffles improve the efficiency of mixing in the bioreactor. Even though the cells are subjected to a hydrodynamic force in the baffle's turbulence field, it is not greater than the impeller zone's force. When fungal cells engulf and build up at the bioreactor's baffle, mixing and cell growth are disrupted for long continuous cultures, particularly *G. lucidum*. As a result, further research into this species' culture is needed, particularly in terms of fermenter design.
5. **Culture time**	The fungal harvesting time determines the efficiency of polysaccharide production and is variable due to carbon source depletion in SSF. Aside from polysaccharide consistency, the age of the fungal culture has a significant impact on polysaccharide development. As a result, it must be considered.
6. **Foaming**	When a fermenter medium is subjected to aeration and agitation, the foam may form, which can be problematic. The foam could clog the exit air filters, causing an increase in pressure in the bioreactor's headspace. When the foam grows in a fungal culture, the mycelium follows it and gradually develops on the wall and headspace. As a result, chemical and mechanical antifoam can be used in large-scale cultivation to combat foaming. Mechanical antifoam is favoured because it does not reduce OTRs, whereas chemical antifoam has been shown to reduce OTRs, affecting cell development.

As a result, inoculum preservation and maintenance are critical, and frozen stock cultures should be prepared as soon as possible after isolation (Walser et al., 2001). As discussed in section 2.11.2, the ideal inoculum as the primary biological factor is between 10% and 20% blended mycelium.

2.11.3.2 Extraction of compounds from SLF of a bioreactor culture

Extraction or downstream of bioreactor cultures starts with fungal pellets' filtration to obtain dry cell weight (DCW) or dried biomass. In Figure 2.5, filtering of a 200-mL sample through a pre-dried and weighed GF/C filter (Whatman Ltd., UK) using a Buchner funnel filter set connected to a water pump, followed by repeated washing (three times) of the mycelial biomass with distilled

Table 2.3 Chemical parameters involved in submerged-liquid fermentation of G. lucidum

Chemical factors	Explanation
1. **Medium pH**	The initial pH of the medium may have an impact on *G. lucidum* growth and polysaccharide development. Besides salt solubility, the ionic state of a substrate, cell membrane structure, cell morphology, substrate absorption rate (glucose consumption), and bioproduct formation are all affected by pH medium. The effect of initial pH on *G, lucidum* was investigated in this review. The *G. lucidum* favoured and performed best at pH 4 in general.
2. **Medium composition**	*G. lucidum* requires a specific medium composition to thrive. Fazenda et al. (2008) used a wide variety of complex, synthetic media, or chemically defined media, as well as waste substrate, in their research. The optimised media must result in an increase in biomass and the formation of bioproducts. Biomass concentration is higher in complex media than in chemically based media. The medium composition can be investigated by adjusting the carbon: nitrogen (C:N) ratio source.
3. **Carbon source**	Disaccharides (lactose, maltose, and sucrose) and monosaccharides (glucose, fructose, and sucrose) are two types of carbon sources (fructose, galactose, and glucose). The production of EPS is affected by the different types of carbon sources. Simple carbon sources and molasses (complex carbon source) aided in higher EPS production and mycelial development, according to a study by Xiao et al. (2006).
4. **Complex media**	In the large-scale cultivation of *Ganoderma* sp. using waste materials, complex media were used. Complex media can reduce the cost of mushroom mycelium cultivation as compared to synthetic media. Many researchers believe that complex media is the best liquid culture medium for *G. lucidum*.
5. **Nitrogen source**	Nitrate, nitrite, ammonium salts, casein, peptone, amino acids, yeast extract, beet, and cane molasses are all popular nitrogen sources for *G. lucidum* growth. Such species are N-limited in the wild because their natural habitat is deficient in nutrients except for carbon and has very low N content in forests. New strategies for cultivating fungus can be established by understanding this mechanism for surviving in an N-limited climate. The type of nitrogen source has a significant impact on *G. lucidum* cell growth and EPS development.
6. **Metal and ions**	*G. lucidum* needs adequate amounts of critical metals and trace elements for growth and product creation. These salts (e.g. magnesium sulphate and potassium phosphate) are essential for fungal growth. Simultaneously, trace elements (manganese, zinc, iron, copper, and molybdenum) serve as enzyme co-factors. Other ions essential for Basidiomycete growth include magnesium (Mg^{2+}) and potassium (K^+); Mg^{2+} is needed for enzymatic reactions, while K^+ promotes hyphal tip extension (Fazenda et al., 2008).

water yields the DCW from the collected *G. lucidum* fermentation method. If more than 200 mL of fungal culture was used, the filtration system would be blocked by tiny mycelium and reduce sucking pressure. Until measuring, the mycelial or pellet filter cake is dried overnight in the dryer (low-temperature food dehydrator (Wan-Mohtar et al., 2020) and cooled 24 hours in a desiccator. DCW in g/L is calculated by subtracting the pre-weighed filter mass from the mass with the filtrate and multiplying by the dilution factor (gravimetric method). The dried fungal biomass contains lipids, protein, phenols, and enzymes beneficial for biotechnology applications (Yadav, 2019). A recent application by Wan-Mohtar, Halim-Lim, et al. (2020) has reported that dried *Ganoderma* biomass contains high protein content (32%), 13.8% of fiber, 48.38% carbohydrate, and 4.45% of lipid in a proximate composition analysis. The dried biomass also possesses a vital functional compound called endo-polysaccharide (ENS) (Abdullah et al., 2020); however, it needs to be extracted. To do so, combine the dried biomass with 1 g of distilled water in 20 mL of distilled water (1:20 ratio). Then it's sterilized for 30 minutes at 121°C (hot-water extraction). The mixture is extracted after sterilization to obtain the supernatant. The crude ENS is then precipitated by adding 1:4 of 99.9% (v/v) ethanol to one volume of cell-free filtrate and leaving it overnight at 4°C. After that, the precipitate is separated by centrifugation at 10,000 g for 10 minutes, which is done twice. The residue is then filtered twice with 5 mL of 95% (v/v) ethanol and washed through a pre-dried and weighted GF/ C filter pad. It was then placed in desiccators and allowed to dry to a constant weight. The weight of ENS was calculated (Ubaidillah et al., 2015), a similar gravimetric procedure to measure DCW. ENS morphologically resembles burnt sugar and has the same solubility as EPS (Figure 2.10).

The filtrate of initial fungal pellets is washed away into the bottom of the Buchner flask using a triple-wash of distilled water as EPS is attached to the surface of fungal pellets. The critical step to obtain high crude EPS extraction is using cold-ethanol shock treatment (Wan-Mohtar, Young, et al., 2016; Wan-Mohtar, Abd Malek, et al., 2016). Normally, by doing this, EPS quickly develops and formed a white-gel structure with tiny bubbles attached, indicating the appearance of an insoluble protein-bound β-glucan (Rop et al., 2009). The EPS is extracted from the harvested fermentation broth's filtered supernatants. The crude EPS is then precipitated by adding four volumes of 99.9% (v/v) ethanol to one volume of cell-free filtrate and leaving it overnight at 4°C. After that, the precipitate is separated by centrifugation at 10,000 g for 10 minutes, which is done twice. The residue is then washed through a pre-dried and weighted GF/C filter pad. It is then placed in desiccators and allowed to dry to a constant weight, after which the weight of EPS is calculated. The crude EPS of *G. lucidum* resembles white powder when thoroughly dried at a low temperature. It contains glucose as a monomer unit and is grouped as glucans. Filamentous fungi like mushrooms produce related polymers known as -glucans or -1,3-D-glucans and -1,6-D-glucans. The relative molecular weights of -glucans vary greatly, ranging from tens to thousands of kilodaltons (depending on origin). The solubility of fungal glucans in water is primarily determined by their structure, linked to their origin. Their solubility rises as the temperature rises (Rop et al., 2009). Glucans that are bound to proteins are insoluble. Their molecules can form gels after partial hydrolysis. Since native molecules lack this capacity, -glucans can be classified as partially soluble and partially insoluble food ingredients. Combination bonds (1 → 3) and (1 → 4) form the fundamental structure of glucans. This glucan is responsible for the use of white fungal biotechnology in recent years.

2.11.4 A recent 6-year application of SLF bioreactor extracts in white fungal biotechnology

Based on the medicinal mushroom SLF via a bioreactor, with *G. lucidum* as the core example, the fungal culture produces four essential biocompounds, which are dried biomass, live biomass, exopolysaccharide (EPS), and sulfated EPS (Wan-Mohtar, Young, et al., 2016; Supramani et al., 2019). Each of these compounds has been proven its value in the new industrial mushroom application concepts. White biotechnology uses living cells (fungi) to create industrial goods that are more easily degradable, use fewer resources, and produce less waste. The addition of fungus to this definition results in "white fungal biotechnology", which involves biodiversity fungi from various environments and industrial applications in several industries (Yadav, 2019). Figure 2.11 depicts the future beneficial function of fungi using a *G. lucidum* bioreactor culture, which could be used in sustainable systems. The first two essential compounds are in live and dried biomass, which was produced via a landless food concept. A circular agricultural system is established using stirred-tank (STR) bioreactors (Wan-Mohtar, Taufek, et al., 2020).

Live bioreactor-produced *Ganoderma* biomass was firstly introduced in 2019 by (Hanafiah et al., 2019) to treat synthetic sewage loading using a batch reactor, which is suitable for Malaysian wastewater and discharges (Mumtaz et al., 2010). Such live biomass achieved 96% of chemical oxygen demand (COD) and 93.2% of ammonia (NH_3-N) removal from synthetic sewage pollutants in an acidic environment at pH 4. The live biomass absorbs NH3-N in an increasing COD/nitrogen

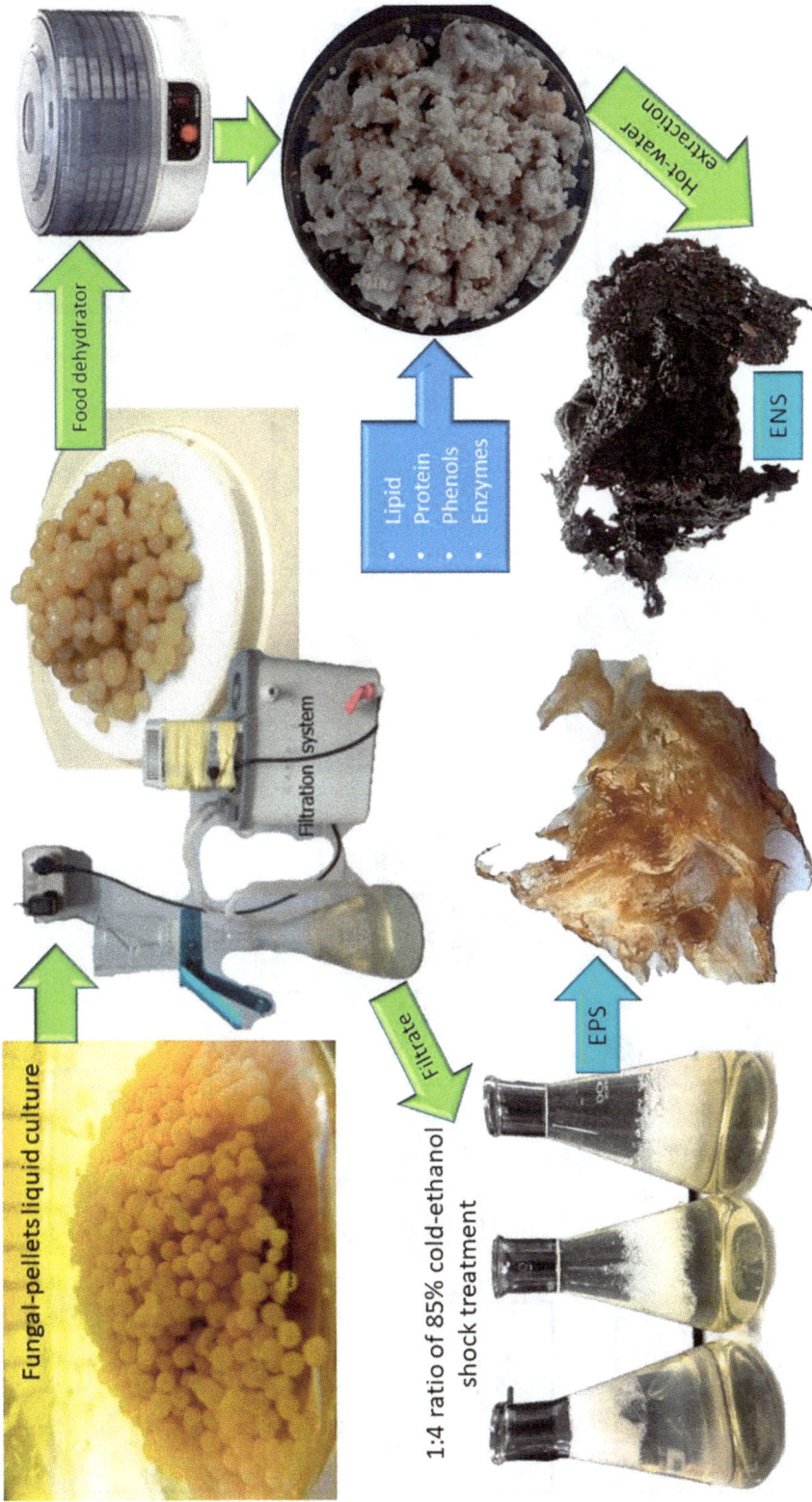

Figure 2.10 Downstream extraction of bioreactor cultures of *Ganoderma lucidum*.

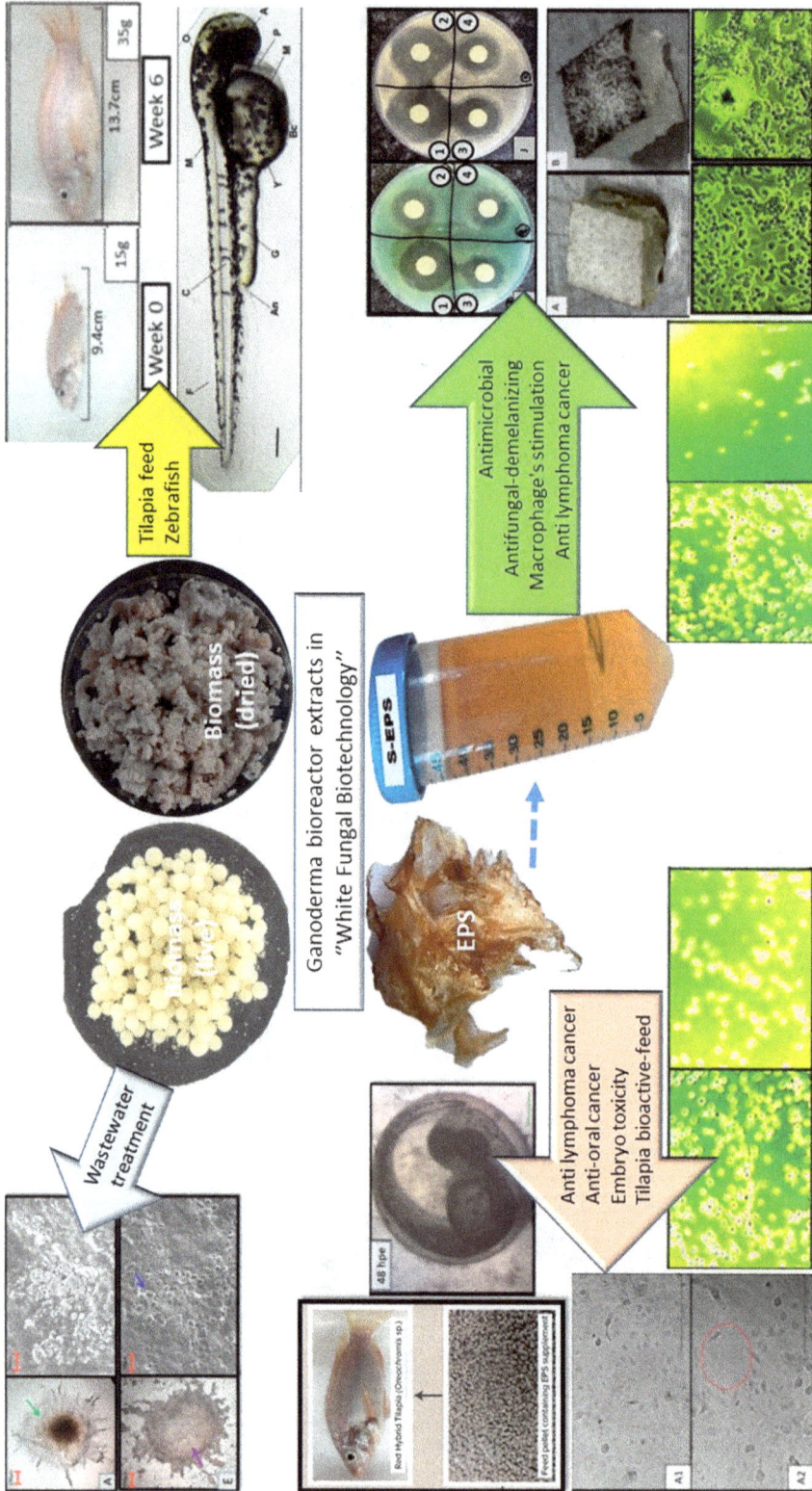

Figure 2.11 A recent 6-year application of *Ganoderma* bioreactor extracts.

gradient morphologically using a scanning electron microscope (VPSEM) and successfully retains its spherical shapes and oval masses. This experiment demonstrated that the live biomass performed effectively in an acidic environment with a pH range of 2.5 to 3.5, consistent with the literature research. This living biomass can also secrete a collection of extracellular enzymes that break down the organic substrate and use it as an energy and nutrient source (Harms et al., 2011). Furthermore, the outer layer of live biomass acts as a buffer against inhibitory compounds and has advantages over traditional live bacteria wastewater treatment (Sankaran et al., 2010). To completely adapt the bioreactor culture of medicinal mushrooms to real-world applications, further research on the impact of this fungus on real domestic wastewater treatment is needed. Once live *Ganoderma* biomass is carefully dried, the resulting product contains a high amount of protein (32.2%), lipid (6.3%), carbohydrate (48.4%), and fiber (13.8%) (Wan-Mohtar, Halim-Lim, et al., 2020), which is used as a cost-effective source of protein in tilapia feed while boosting its immunity. Wan-Mohtar, Taufek, et al. (2020) fed 15 g/kg of dried *G. lucidum* biomass to 15 tilapia per tank in 70-L tanks, which lasted for 42 days in a controlled environment. Compared to the control, this treatment resulted in 100% tilapia survival, longer body length, and higher body weight gain. Furthermore, the dried biomass treatment increased hemoglobin (HGB), red blood cells (RBC), white blood cells (WBC), and antioxidative responses, implying that the diets can improve tilapia survival and development. According to Rahmann et al., (2019), the landless food concept is a circular and self-sustaining strategy that not only benefits tilapia growth rate and internal health but may also occupy less land than traditional SSF cultivation and may be advantageous in high-population, low-income countries with small-scale farming practices as economic biotechnology. This strategy combined to replace 1 ha of mushroom cropland with 1 m^2 of bioreactor space. During live *Ganoderma* biomass extraction, the prominent compound solubilizes into liquid, forming exopolysaccharide (EPS). There are four main applications of bioreactor EPS: anti-lymphoma cancer (Wan-Mohtar, Young, et al., 2016), anti-oral cancer (Abdullah et al., 2020), embryotoxicity (Taufek et al., 2020), and tilapia bioactive-feed (Wan-Mohtar et al., 2021). For the anti-lymphoma effect, 60 μg/mL of EPS gave 10% cytotoxic effects against cancerous human Caucasian histiocytic lymphoma cell line (U937) cells from a 37-year-old-male patient (Wan-Mohtar, Young, et al., 2016). Meanwhile, the same EPS showed antiproliferative activity against oral cancer cell line (ORL-48) cells, with a reduction in cell growth at IC$_{50}$ of 0.23 mg/mL (Abdullah et al., 2020). With these successes, this EPS was proven safe when tested in zebrafish embryos, comparable with mammals' toxicity at LC$_{50}$ 2648 μg/mL (Taufek et al., 2020). As a precommercialization attempt, 3 g/kg of EPS has been proven as immune booster fish-feed on farmed-red hybrid tilapia. Such EPS increased 66.1% of tilapia weight, boosted antioxidant activity, hemoglobin, red blood cells, and white blood cell. This highlights the potential use of EPS white fungal biotechnology. Since the EPS derived from the bioreactor fermentation process are water-insoluble, it can be enhanced using sulfation (Wan-Mohtar, Young, et al., 2016). As a result, an unavoidable process to increase its solubility in water and bioactivity must be introduced. A recent review by Arlov et al. (2021) stated that sulfated polysaccharides exhibit novel and augmented biological properties by facilitating interactions with other molecules. In 2016, sulfated EPS provided antimicrobial responses against ten species of bacteria, with the most significant effect against gram-positive *Staphylococcus aureus*, gram-negative *Escherichia coli*, and antibiotic-resistant *Klebsiella pneumoniae* British strain (Wan-Mohtar, Young, et al., 2016). At 60 g/mL, the same compound had the most potent antiproliferative effect with the fewest cancerous lymphoma (U937) cell development, demonstrating that the EPS sulfation process enriched the antiproliferative intensity. In 2017, researchers discovered that similar sulfated EPS stimulated RAW264.7 macrophages and had antifungal activity against *Aspergillus niger* A60 at a fungicidal concentration of 100 mg/mL. These findings point to a novel bifunctional property of sulfated EPS that is important to biomedical applications in 2021 (Arlov et al., 2021).

2.12 CONCLUSION

People have been cultivating edible and medicinal mushrooms for hundreds of years. The growth of these mushrooms through fruiting body production (SSF) and mycelium biomass (SLF) are complicated processes that produce a wide range of product quality. More than 130 medical effects and superfoods are developed using traditional fruiting body cultivation techniques. Due to their ability to degrade lignocellulosic materials, this method is also common, useful, and of great importance in the decomposition of lignocellulosic agro-industrial waste. Outdoor log SSF systems have a lower yield than indoor SSF systems, which have greater control over fruiting body formation and are not dependent on weather changes. With the world's population growing, controlled cultivation in heterotrophic SLF bioreactors has been established to shorten cultivation times and increase mushroom biomass yield. This analysis provides stakeholders and mushroom growers with up-to-date knowledge on how to cultivate edible and medicinal mushrooms based on their settings and preferences.

REFERENCES

Abd Razak, D. L., Abdullah, N., Khir Johari, N. M., & Sabaratnam, V. (2012). Comparative study of mycelia growth and sporophore yield of *Auricularia polytricha* (Mont.) Sacc on selected palm oil wastes as fruiting substrate. *Applied Microbiology and Biotechnology*, *97*(7), 3207–3213.

Abdullah, N. R., Sharif, F., Azizan, N. H., Hafidz, I. F. M., Supramani, S., Usuldin, S. R. A., Ahmad, R., & Wan, W. A. A. Q. I. (2020). Pellet diameter of *Ganoderma lucidum* in a repeated-batch fermentation for the trio total production of biomass-exopolysaccharide-endopolysaccharide and its anti-oral cancer beta-glucan response. *AIMS Microbiology*, *6*(4), 379.

Aguilar-Rivera, N., & De Jesús-Merales, J. (2010). Edible mushroom *Pleurotus ostreatus* production on cellulosic biomass of sugar cane. *Sugar Tech*, *12*(2), 176–178.

Ahlborn, J., Stephan, A., Meckel, T., Maheshwari, G., Rühl, M., & Zorn, H. (2019). Upcycling of food industry side streams by basidiomycetes for production of a vegan protein source. *International Journal of Recycling of Organic Waste in Agriculture*, *8*(1), 447–455.

Ahn, H., Jeon, E., Kim, J.-C., Kang, S. G., Yoon, S., Ko, H.-J., Kim, P.-H., & Lee, G.-S. (2017). Lentinan from shiitake selectively attenuates AIM2 and non-canonical inflammasome activation while inducing pro-inflammatory cytokine production. *Scientific Reports*, *7*, 1314.

Alananbeh, K. M., Bouqellah, N. A., & Al Kaff, N. S. (2014). Cultivation of oyster mushroom *Pleurotus ostreatus* on date-palm leaves mixed with other agro-wastes in Saudi Arabia. *Saudi Journal of Biological Sciences*, *21*(6), 616–625.

Allied Market Research. (2021, March 8). https://finance.yahoo.com/

Arlov, Ø., Rütsche, D., Asadi Korayem, M., Öztürk, E., & Zenobi-Wong, M. (2021). Engineered sulfated polysaccharides for biomedical applications. *Advanced Functional Materials*, *31*(19), 2010732.

Atila, F., Owaid, M. N., & Shariati, M. A. (2017). The nutritional and medical benefits of *Agaricus bisporus*: A review. *Journal of Microbiology,Biotechnology and Food Sciences*, *7*(3), 281–286.

Atlı, B., Yamaç, M., Yıldız, Z., & Şölener, M. (2019). Solid state fermentation optimization of *Pleurotus ostreatus* for lovastatin production. *Pharmaceutical Chemistry Journal*, *53*(9), 858–864.

Bains, A., & Chawla, P. (2020). In vitro bioactivity, antimicrobial and anti-inflammatory efficacy of modified solvent evaporation assisted Trametes versicolor extract. *3 Biotech*, *10*(9), 1–11.

Barakat, H. (2019). *Giving Life: A History of Bread in Egypt*. Rawi Magazine.

Bellettini, M. B., Fiorda, F. A., Maieves, H. A., Teixeira, G. L., Ávila, S., Hornung, P. S., Júnior, A. M., & Ribani, R. H. (2019). Factors affecting mushroom *Pleurotus* spp. *Saudi Journal of Biological Sciences*, *26*(4), 633–646.

Berovic, M., Habijanic, J., Zore, I., Wraber, B., Hodzar, D., Boh, B., & Pohleven, F. (2003). Submerged cultivation of *Ganoderma lucidum* biomass and immunostimulatory effects of fungal polysaccharides. *Journal of Biotechnology*, *103*, 77–86.

Blackwell, M. (2011). The fungi: 1, 2, 3... 5.1 million species? *American Journal of Botany, 98*(3), 426–438.

Blumfield, M., Abbott, K., Duve, E., Cassettari, T., Marshall, S., & Fayet-Moore, F. (2020). Examining the health effects and bioactive components in *Agaricus bisporus* mushrooms: A scoping review. *The Journal of Nutritional Biochemistry, 84*, 108453.

Boddy, L., Frankland, J. C., Van West, P., & British Mycological Society. (2008). *Ecology of Saprotrophic Basidiomycetes*. Elsevier Academic Press.

Bredon, M., Dittmer, J., Noël, C., Moumen, B., & Bouchon, D. (2018). Lignocellulose degradation at the Holobiont level: Teamwork in a keystone soil invertebrate. *Microbiome, 6*(1), 162.

Carlisle, M. J., Watkinson, S. C., & Gooday, G. W. (2001). *The Fungi*. Academic Press.

Celal, B. (2019). Antioxidant and antimicrobial capacities of Ganoderma lucidum.*Journal of Bacteriology & Mycology: Open Access, 7*(1), 5–7.

Chang, S. T., & Buswell, J. A. (1999). *Ganoderma lucidum* (Curt.: Fr.) P. Karst. (Aphyllophoromycetideae)–A mushrooming medicinal mushroom. *International Journal of Medicinal Mushrooms, 1*(2), 139–146.

Chang, S. T., & Miles, P. G. (1987). Historical record of the early cultivation of *Lentinus* in China. *Mushroom Journal for the Tropics, 7*, 31–37.

Chen, H. (2015). *Modern Solid State Fermentation: Theory and Practice*. Springer.

Chen, W., Tan, H., Liu, Q., Zheng, X., Zhang, H., Liu, Y., & Xu, L. (2019). A review: The bioactivities and pharmacological applications of *Phellinus linteus. Molecules, 24*(10), 1888.

Cherubin, M. R., Oliveira, D. M. da S., Feigl, B. J., Pimentel, L. G., Lisboa, I. P., Gmach, M. R., Varanda, L. L., Morais, M. C., Satiro, L. S., Popin, G. V., Paiva, S. R. de, Santos, A. K. B. dos, Vasconcelos, A. L. S. de, Melo, P. L. A. de, Cerri, C. E. P., & Cerri, C. C. (2018). Crop residue harvest for bioenergy production and its implications on soil functioning and plant growth: A review. *Scientia Agricola, 75*(3), 255–272.

Chinamushroom business network. (2018). Investigation Report on *Auricularia auricula* Industry in 2017. http://zixun.mushroommarket.net/201808/16/185667.html.

Ćilerdžić, J. Lj., Vukojević, J. B., Klaus, A. S., Ivanović, Ž. S., Blagojević, J. D., & Stajić, M. M. (2018). Wheat straw – A promising substrate for *Ganoderma lucidum* cultivation. *Acta Scientiarum Polonorum Hortorum Cultus, 17*(1), 13–22.

Colavolpe, M. B., & Albertó, E. (2014). Cultivation requirements and substrate degradation of the edible mushroom *Gymnopilus pampeanus*—A novel species for mushroom cultivation. *Scientia Horticulturae, 180*, 161–166.

Cör, D., Knez, Ž., & Knez Hrnčič, M. (2018). Antitumour, antimicrobial, antioxidant and anti-acetylcholinesterase effect of *Ganoderma lucidum* terpenoids and polysaccharides: A review. *Molecules, 23*(3), 649.

Cueva, M. B. R., Hernández, A., & Niño-Ruiz, Z. (2017). Influence of C/N ratio on productivity and the protein contents of *Pleurotus ostreatus* grown in differents residue mixtures. *Ciencias Agrarias, 49*(2), 331–344.

Dahmardeh, M. (2013). Use of oyster mushroom (*Pleurotus ostreatus*) grown on different substrates (wheat and barley straw) and supplemented at various levels of spawn to change the nutritional quality forage. *International Journal of Agriculture and Forestry, 3*(4), 138–140.

Deacon, J. W. (2013). *Fungal Biology*. John Wiley & Sons.

De Silva, D. D., Rapior, S., Hyde, K. D., & Bahkali, A. H. (2012). Medicinal mushrooms in prevention and control of diabetes mellitus. *Fungal Diversity, 56*(1), 1–29.

De Silva, D. D., Rapior, S., Sudarman, E., Stadler, M., Xu, J., Aisyah Alias, S., & Hyde, K. D. (2013). Bioactive metabolites from macrofungi: Ethnopharmacology, biological activities and chemistry. *Fungal Diversity, 62*(1), 1–40.

Doroški, A., Klaus, A., Kozarski, M., Cvetković, S., Nikolić, B., Jakovljević, D., Tomasevic, I., Vunduk, J., Lazić, V., & Djekic, I. (2020). The influence of grape pomace substrate on quality characterization of *Pleurotus ostreatus* – Total quality index approach. *Journal of Food Processing and Preservation, 45*(1), e15096.

European Commission. (2019). European Commission – Implementation of the circular economy action plan.

Fazenda, M. L., Harvey, L. M., & McNeil, B. (2010). Effects of dissolved oxygen on fungal morphology and process rheology during fed-batch processing of *Ganoderma lucidum. Journal of Microbiology and Biotechnology, 20*(4), 844–851.

Fazenda, M. L., Seviour, R., McNeil, B., & Harvey, L. M. (2008). Submerged culture fermentation of "higher fungi": The macrofungi. *Advances in Applied Microbiology*, *63*, 33–103. Academic Press.

Finimundy, T. C., Dillon, A. J. P., Henriques, J. A. P., & Ely, M. R. (2014). A review on general nutritional compounds and pharmacological properties of the *Lentinula edodes* mushroom. *Food and Nutrition Sciences*, *5*(12), 1095–1105.

Fraga, I., Coutinho, J., Bezerra, R. M., Dias, A. A., Marques, G., & Nunes, F. M. (2014). Influence of culture medium growth variables on *Ganoderma lucidum* exopolysaccharides structural features. *Carbohydrate Polymers*, *111*, 936–946.

Gao, S., Huang, Z., Feng, X., Bian, Y., Huang, W., & Liu, Y. (2020). Bioconversion of rice straw agro-residues by *Lentinula edodes* and evaluation of non-volatile taste compounds in mushrooms. *Scientific Reports*, *10*(1).

Gargano, M. L., van Griensven, L. J. L. D., Isikhuemhen, O. S., Lindequist, U., Venturella, G., Wasser, S. P., & Zervakis, G. I. (2017). Medicinal mushrooms: Valuable biological resources of high exploitation potential. *Plant Biosystems –An International Journal Dealing with All Aspects of Plant Biology*, *151*(3), 548–565.

Gianotti, B. M., Cleaver, M. P., Cleaver, P. D., Bailey, C., & Holliday, J. C. (2009). *21st Century Diversified Agriculture Project for Ghana –Aloha Ecowas Development Corporation LTD – Diversified Agriculture Part 1: Simplified and Lower Cost Methods for Mushroom Cultivation inAfrica.* http://www.alohaecowas.com/diversified-agriculture-part1.html

Gong, W., Xu, R., Xiao, Y., Zhou, Y., & Bian, Y. (2014). Phenotypic evaluation and analysis of important agronomic traits in the hybrid and natural populations of *Lentinula edodes*. *Scientia Horticulturae*, *179*, 271–276.

Grimm, D., & Wösten, H. A. B. (2018). Mushroom cultivation in the circular economy. *Applied Microbiology and Biotechnology*, *102*(18), 7795–7803.

Guillamón, E., García-Lafuente, A., Lozano, M., D´Arrigo, M., Rostagno, M. A., Villares, A., & Martínez, J. A. (2010). Edible mushrooms: Role in the prevention of cardiovascular diseases. *Fitoterapia*, *81*(7), 715–723.

Han, M.-L., An, Q., He, S.-F., Zhang, X.-L., Zhang, M.-H., Gao, X.-H., Wu, Q., & Bian, L.-S. (2020). Solid-state fermentation on poplar sawdust and corncob wastes for lignocellulolytic enzymes by different *Pleurotus ostreatus* strains. *BioResources*, *15*(3), 4982–4995.

Hanafiah, Z. M., Mohtar, W. H. M. W., Hasan, H. A., Jensen, H. S., Klaus, A., & Wan, W. A. A. Q. I. (2019). Performance of wild-Serbian *Ganoderma lucidum* mycelium in treating synthetic sewage loading using batch bioreactor. *Scientific Reports*, *9*(1), 1–12.

Hariharan, S., & Nambisan, P. (2012). Optimization of lignin peroxidase, manganese peroxidase, and lac production from *Ganoderma lucidum* under solid state fermentation of pineapple leaf. *BioResources*, *8*(1), 250–271.

Harms, H., Schlosser, D., & Wick, L. Y. (2011). Untapped potential: Exploiting fungi in bioremediation of hazardous chemicals. *Nature Reviews Microbiology*, *9*(3), 177–192.

Hassan, N. A., Supramani, S., Sohedein, M. N. A., Usuldin, S. R. A., Klaus, A., Ilham, Z., Chen, W. H., & Wan, W. A. A. Q. I. (2019). Efficient biomass-exopolysaccharide production from an identified wild-Serbian *Ganoderma lucidum* strain BGF4A1 mycelium in a controlled submerged fermentation. *Biocatalysis and Agricultural Biotechnology*, *21*, 101305.

Hennicke, F., Cheikh-Ali, Z., Liebisch, T., Maciá-Vicente, J. G., Bode, H. B., & Piepenbring, M. (2016). Distinguishing commercially grown *Ganoderma lucidum* from *Ganoderma lingzhi* from Europe and East Asia on the basis of morphology, molecular phylogeny, and triterpenic acid profiles. *Phytochemistry*, *127*, 29–37.

Hill, D. B. (2010). Introduction to shiitake: The "forest" mushroom. *Kentucky Shiitake Production Workbook*.

Hölker, U., Höfer, M., & Lenz, J. (2004). Biotechnological advantages of laboratory-scale solid-state fermentation with fungi. *Applied Microbiology and Biotechnology*, *64*(2), 175–186.

Hossain, M. M. (2018). Effect of different substrates on yield of *Pleurotus ostreatus* mushroom. *Environment and Ecology*, *36*(1), 312–315.

Huang, N. L., Lin, Z. B., & Chen, G. L. (2010). *The Chinese Medicinal and Edible Fungi*. Science Press, pp. 1834.

Imbach, E. J. (1946). Pilzflora des Kantons Luzern und der angrenzen Innerschweiz. *Mitteilungen der naturforschenden Gesellschaft Luzern (in German). 15*, 5–85.

Jaramillo Mejía, S., & Albertó, E. (2013). Heat treatment of wheat straw by immersion in hot water decreases mushroom yield in *Pleurotus ostreatus. RevistaIberoamericana de Micología, 30*(2), 125–129.

Jaros, D., Köbsch, J., & Rohm, H. (2018). Exopolysaccharides from Basidiomycota: Formation, isolation and techno-functional properties. *Engineering in Life Sciences, 18*(10), 743–752.

Kalač, P. (2016). *Edible Mushrooms: Chemical Composition and Nutritional Value*, Academic Press.

Karunarathna, S. C., Hyde, K. D., Chukeatirote, E., & Klomklung, N. (2014). Optimal conditions of mycelial growth of three wild edible mushrooms from northern Thailand. *Acta BiologicaSzegediensis, 58*(1), 39–43.

Kim, K., Choi, B., Lee, I., Lee, H., Kwon, S., Oh, K., & Kim, A. Y. (2011). Bioproduction of mushroom mycelium of *Agaricus bisporus* by commercial submerged fermentation for the production of meat analogue. *Journal of the Science of Food and Agriculture, 91*(9), 1561–1568.

Kim, H. M., Park, M. K., & Yun, J. W. (2006). Culture pH affects exopolysaccharide production in submerged mycelial culture of *Ganoderma lucidum. Applied Biochemistry and Biotechnology, 134*(3), 249–262.

Klaus, A. (2011). Chemical characterization, antimicrobial and antioxidant properties of polysaccharide of lignicolous fungi *Ganoderma* spp. *Laetiporus sulphureus* and *Schizophyllum commune* [PhD Thesis].University of Belgrade, Faculty of Agriculture, Institute for Food Technology and Biochemistry.

Klaus, A., Kozarski, M., Niksic, M., Jakovljevic, D., Todorovic, N., Stefanoska, I., & Van Griensven, L. J. L. D. (2013). The edible mushroom *Laetiporus sulphureus* as potential source of natural antioxidants. *International Journal of Food Sciences and Nutrition, 64*(5), 599–610.

Klaus, A., Kozarski, M., Niksic, M., Jakovljevic, D., Todorovic, N., & Van Griensven, L. J. L. D. (2011). Antioxidative activities and chemical characterization of polysaccharides extracted from the basidiomycete *Schizophyllum commune. LWT – Food Science and Technology, 44*(10), 2005–2011.

Klaus, A., Kozarski, M., Vunduk, J., Todorovic, N., Jakovljevic, D., Zizak, Z., Pavlovic, V., Levic, S., Niksic, M., & Van Griensven, L. J. L. D. (2015). Biological potential of extracts of the wild edible Basidiomycete mushroom *Grifola frondosa. Food Research International, 67*, 272–283.

Klaus, A., Wan, W. A. A. Q. I., Nikolić, B., Cvetković, S., & Vunduk, J. (2021). Pink oyster mushroom *Pleurotus flabellatus* mycelium produced by an airlift bioreactor-the evidence of potent in vitro biological activities. *World Journal of Microbiology and Biotechnology, 37*(1), 1–14.

Kortei, N. K., Dzogbefia, V. P., & Obodai, M. (2014). Assessing the effect of composting cassava peel based substrates on the yield, nutritional quality, and physical characteristics of *Grifola frondosa* (Jacq. ex Fr.) Kummer. *Biotechnology Research International, 2014*, 1–9.

Koutrotsios, G., Larou, E., Mountzouris, K. C., & Zervakis, G. I. (2016). Detoxification of olive mill wastewater and bioconversion of olive crop residues into high-value-added biomass by the choice edible mushroom *Hericium erinaceus. Applied Biochemistry and Biotechnology, 180*(2), 195–209.

Kozarski, M., Klaus, A., Niksic, M., Jakovljevic, D., Helsper, J. P. F. G., & Van Griensven, L. J. L. D. (2011). Antioxidative and immunomodulating activities of polysaccharide extracts of the medicinal mushrooms *Agaricus bisporus, Agaricus brasiliensis, Ganoderma lucidum* and *Phellinus linteus. Food Chemistry, 129*(4), 1667–1675.

Kozarski, M., Klaus, A., Nikšić, M., Vrvić, M. M., Todorović, N., Jakovljević, D., & Van Griensven, L. J. L. D. (2012). Antioxidative activities and chemical characterization of polysaccharide extracts from the widely used mushrooms *Ganoderma applanatum, Ganoderma lucidum, Lentinus edodes* and *Trametes versicolor. Journal of Food Composition and Analysis, 26*(1-2), 144–153.

Kozarski, M., Klaus, A., Vunduk, J., Jakovljevic, D., Jadranin, M., & Niksic, M. (2020). Health impact of the commercially cultivated mushroom *Agaricus bisporus* and wild-growing mushroom *Ganoderma resinaceum* – A comparative overview. *Journal of the Serbian Chemical Society, 85*(6), 721–735.

Kozarski, M., Klaus, A., Vunduk, J., Zizak, Z., Niksic, M., Jakovljevic, D., Vrvic, M. M., & Van Griensven, L. J. L. D. (2015). Nutraceutical properties of the methanolic extract of edible mushroom *Cantharellus cibarius* (Fries): Primary mechanisms. *Food & Function, 6*(6), 1875–1886.

Kumla, J., Suwannarach, N., Sujarit, K., Penkhrue, W., Kakumyan, P., Jatuwong, K., Vadthanarat, S., & Lumyong, S. (2020). Cultivation of mushrooms and their lignocellulolytic enzyme production through the utilization of agro-industrial waste. *Molecules, 25*(12), 2811.

Kummer, P. (1871). *Der Fhrer in die Pilzkunde*(1st ed.).

Leatham, G. F. (1982). *Cultivation of Shiitake, the Japanese Forest Mushroom, on Logs: A Potential Industry for the United States*. University Of Wisconsin-Extension.

Leite, P., Sousa, D., Fernandes, H., Ferreira, M., Costa, A. R., Filipe, D., Gonçalves, M., Peres, H., Belo, I., & Salgado, J. M. (2021). Recent advances in production of lignocellulolytic enzymes by solid-state fermentation of agro-industrial wastes. *Current Opinion in Green and Sustainable Chemistry*, *27*, 100407.

Leroy, F., Geyzen, A., Janssens, M., De Vuyst, L., & Scholliers, P. (2013). Meat fermentation at the crossroads of innovation and tradition: A historical outlook. *Trends in Food Science & Technology*, *31*(2), 130–137.

Letti, L. A. J., Vitola, F. M. D., de Melo Pereira, G. V., Karp, S. G., Medeiros, A. B. P., da Costa, E. S. F., Bissoqui, L., & Soccol, C. R. (2018). Solid-state fermentation for the production of mushrooms. In: A. Pandey, C. Larroche, & C. R. Soccol (Eds.), *Current Developments in Biotechnology and Bioengineering, Current Advances in Solid-State Fermentation*. Elsevier, pp. 285–318.

Li, H., Tan, F., Ke, L., Xia, D., Wang, Y., He, N., Zheng, Y., & Li, Q. (2016). Mass balances and distributions of C, N, and P in the anaerobic digestion of different substrates and relationships between products and substrates. *Chemical Engineering Journal*, *287*, 329–336.

Liang, C.-H., Wu, C.-Y., Lu, P.-L., Kuo, Y.-C., & Liang, Z.-C. (2019). Biological efficiency and nutritional value of the culinary-medicinal mushroom *Auricularia* cultivated on a sawdust basal substrate supplement with different proportions of grass plants. *Saudi Journal of Biological Sciences*, *26*(2), 263–269.

Linnakoski, R., Reshamwala, D., Veteli, P., Cortina-Escribano, M., Vanhanen, H., & Marjomäki, V. (2018). Antiviral agents from fungi: Diversity, mechanisms and potential applications. *Frontiers in Microbiology*, *9*, 2325.

Liu, D., Coloe, S., Baird, R., & Pederson, J. (2000). Rapid mini-preparation of fungal DNA for PCR. *Journal of Clinical Microbiology*, *38*(1), 471.

Liu, G. Q., Wang, X. L., Han, W. J., & Lin, Q. L. (2012). Improving the fermentation production of the individual key triterpene ganoderic acid me by the medicinal fungus *Ganoderma lucidum* in submerged culture. *Molecules*, *17*(11), 12575–12586.

Lombard, V., Golaconda Ramulu, H., Drula, E., Coutinho, P. M., & Henrissat, B. (2013). The carbohydrate-active enzymes database (CAZy) in 2013. *Nucleic Acids Research*, *42*(D1), D490–D495.

Ma, H.-T., Hsieh, J.-F., & Chen, S.-T. (2015). Anti-diabetic effects of *Ganoderma lucidum*. *Phytochemistry*, *114*, 109–113.

Ma, H., Xu, X., & Feng, L. (2013). Responses of antioxidant defenses and membrane damage to drought stress in fruit bodies of *Auricularia auricula*-judae. *World Journal of Microbiology and Biotechnology*, *30*(1), 119–124.

Manan, M. A., & Webb, C. (2017). Design aspects of solid state fermentation as applied to microbial bioprocessing. *Journal of Applied Biotechnology & Bioengineering*, *4*(1).

Mani, V., & Ming, L. C. (2016). Tempeh and other fermented soybean products rich in isoflavones. In: J. Frias, C. Martinez-Villaluenga, & E. Peñas (Eds.), *Fermented Foods in Health and Disease Prevention*. Elsevier, pp. 453–474.

Manzi, P., Gambelli, L., Marconi, S., Vivanti, V., & Pizzoferrato, L. (1999). Nutrients in edible mushrooms: An inter-species comparative study. *Food Chemistry*, *65*(4), 477–482.

Mata, G., Salmones, D., & Pérez-Merlo, R. (2016). Hydrolytic enzyme activities in shiitake mushroom (*Lentinula edodes*) strains cultivated on coffee pulp. *Revista Argentina de Microbiología*, *48*(3), 191–195.

de Mattos-Shipley, K. M. J., Ford, K. L., Alberti, F., Banks, A. M., Bailey, A. M., & Foster, G. D. (2016). The good, the bad and the tasty: The many roles of mushrooms. *Studies in Mycology*, *85*, 125–157.

Meetoo, D., McGovern, P., & Safadi, R. (2007). An epidemiological overview of diabetes across the world. *British Journal of Nursing*, *16*(16), 1002–1007.

Miles, P. G., & Chang, S. T. (2004). *Mushrooms: Cultivation, Nutritional Value, Medicinal Effect, and Environmental Impact*. CRC Press.

Mooralitharan, S., Hanafiah, Z. M., Abd Manan, T. S. B., Hasan, H. A., Jensen, H. S., Wan, W. A. A. Q. I., & Mohtar, W. H. M. W. (2021). Optimization of mycoremediation treatment for the chemical oxygen demand (COD) and ammonia nitrogen (AN) removal from domestic effluent using wild-Serbian *Ganoderma lucidum* (WSGL). *Environmental Science and Pollution Research*, *28*, 32528–32544.

Morin, E., Kohler, A., Baker, A. R., Foulongne-Oriol, M., Lombard, V., Nagye, L. G., Ohm, R. A., Patyshakuliyeva, A., Brun, A., Aerts, A. L., Bailey, A. M., Billette, C., Coutinho, P. M., Deakin, G., Doddapaneni, H., Floudas, D., Grimwood, J., Hilden, K., Kues, U., & LaButti, K. M. (2012). Genome sequence of the button mushroom *Agaricus bisporus* reveals mechanisms governing adaptation to a humic-rich ecological niche. *Proceedings of the National Academy of Sciences, 109*(43), 17501–17506.

Mumtaz, T., Yahaya, N. A., Abd-Aziz, S., Yee, P. L., Shirai, Y., & Hassan, M. A. (2010). Turning waste to wealth-biodegradable plastics polyhydroxyalkanoates from palm oil mill effluent – A Malaysian perspective. *Journal of cleaner production, 18*(14), 1393–1402.

Mushroom Cultivation – *Global Market Outlook (2019–2027)*, ID: 5221377, Report, (2021, January 5). https://www.businesswire.com/news/home/20210105005608/en/Mushroom-Cultivation-Industry---Global-Market-Outlook-2019-2027-Market-is-Expected-to-Reach-26.09-Billion---ResearchAndMarkets.com

Mussatto, S. I., Ballesteros, L. F., Martins, S., & Teixeira, J. A. (2012). Use of agro-industrial wastes in solid-state fermentation processes. In: Industrial Waste. InTech Open Access Publisher, pp. 121–140.

Muszyńska, B., Grzywacz-Kisielewska, A., Kała, K., & Gdula-Argasińska, J. (2018). Anti-inflammatory properties of edible mushrooms: A review. *Food Chemistry, 243*, 373–381.

Neifar, M., Jaouani, A., Ayari, A., Abid, O., Salem, H. B., Boudabous, A., Najar, T., & Ghorbel, R. E. (2013). Improving the nutritive value of Olive Cake by solid state cultivation of the medicinal mushroom *Fomes fomentarius*. *Chemosphere, 91*(1), 110–114.

Nguyen, T. L., Chen, J., Hu, Y., Wang, D., Fan, Y., Wang, J., Abula, S., Zhang, J., Qin, T., Chen, X., Chen, X., khakame, S. K., & Dang, B. K. (2012). In vitro antiviral activity of sulfated *Auricularia auricula* polysaccharides. *Carbohydrate Polymers, 90*(3), 1254–1258.

Nguyen, K., Wikee, S., & Lumyong, S. (2018). Brief review: Lignocellulolytic enzymes from polypores for efficient utilization of biomass. *Mycosphere, 9*(6), 1073–1088.

Novakovic, S., Djekic, I., Klaus, A., Vunduk, J., Đorđević, V., Tomovic, V., Koćić-Tanackov, S., Lorenzo, J. M., Barba, F. J., & Tomasevic, I. (2020). Application of porcini mushroom (*Boletus edulis*) to improve the quality of frankfurters. *Journal of Food Processing and Preservation, 44*(8).

Oli, A. N., Edeh, P. A., Al-Mosawi, R. M., Mbachu, N. A., Al-Dahmoshi, H. O. M., Al-Khafaji, N. S. K., Ekuma, U. O., Okezie, U. M., & Saki, M. (2020). Evaluation of the phytoconstituents of *Auricularia auricula*-judae mushroom and antimicrobial activity of its protein extract. *European Journal of Integrative Medicine, 38*, 101176.

Oseni, T., Dlamini, S., Earnshaw, D., & Masarirambi, M. (2012). Effect of substrate pre-treatment methods on oyster mushroom (*Pleurotus ostreatus*) production. *International Journal of Agriculture and Biology, 14*, 251–255.

Owaid, M. N. (2009). *Biotechnology for Local Compost Preparation Used to Produce Mushroom Agaricus Bisporus* (p. 129) [M.Sc. Thesis]. Department of Life Sciences in, Iraq the College of Science – University of Al-Anbar.

Owaid, M. N., Abed, A. M., & Nassar, B. M. (2015). Recycling cardboard wastes to produce blue oyster mushroom *Pleurotus ostreatus* in Iraq. *Emirates Journal of Food and Agriculture, 27*(7), 537–541.

Owaid, M. N., Barish, A., & Ali Shariati, M. (2017). Cultivation of *Agaricus bisporus* (button mushroom) and its usages in the biosynthesis of nanoparticles. *Open Agriculture, 2*(1), 537–543.

Pandey, A. (2003). Solid-state fermentation. *Biochemical Engineering Journal*, 13, 81–84. doi:10.1016/S13 69-703X(02)00121-3

Papaspyridi, L.-M., Aligiannis, N., Topakas, E., Christakopoulos, P., Skaltsounis, A.-L., & Fokialakis, N. (2012). Submerged fermentation of the edible mushroom *Pleurotus ostreatus* in a batch stirred tank bioreactor as a promising alternative for the effective production of bioactive metabolites. *Molecules, 17*(3), 2714–2724.

Park, J. P., Kim, Y. M., Kim, S. W., Hwang, H. J., Cho, Y. J., Lee, Y. S., ... Yun, J. W. (2002). Effect of aeration rate on the mycelial morphology and exo-biopolymer production in *Cordyceps militaris*. *Process Biochemistry, 37*(11), 1257–1262.

Patrick, C. H., & Durham, N. C. (1970). *Alcohol, Culture, and Society*. (pp. 26–27). Duke University Press (reprint edition by AMS Press).

Pegler, D. (1975). The classification of the genus *Lentinus* Fr. (Basidiomycota). *Kavaka*, 3, 11–20.

Peng, H., & Shahidi, F. (2020). Bioactive compounds and bioactive properties of chaga (*Inonotus obliquus*) mushroom: A review. *Journal of Food Bioactives, 12*, 9–75.

Petrović, P., Vunduk, J., Klaus, A., Kozarski, M., Nikšić, M., Žižak, Ž., Vuković, N., Šekularac, G., Drmanić, S., & Bugarski, B. (2016). Biological potential of puffballs: A comparative analysis. *Journal of Functional Foods, 21*, 36–49.

Postemsky, P. D., Delmastro, S. E., & Curvetto, N. R. (2014). Effect of edible oils and Cu (II) on the biodegradation of rice by-products by *Ganoderma lucidum* mushroom. *International Biodeterioration & Biodegradation, 93*, 25–32.

Power, R. C., Salazar-García, D. C., Straus, L. G., González Morales, M. R., & Henry, A. G. (2015). Microremains from El Mirón Cave human dental calculus suggest a mixed plant–animal subsistence economy during the Magdalenian in Northern Iberia. *Journal of Archaeological Science, 60*, 39–46.

Priya, R. U., Geetha, D., & Darshan, S. (2016). Biology and cultivation of black ear mushroom – *Auricularia* spp. *Advances in Life Sciences, 5*(22), 10252–10254.

Quimio, T. H. (1982). Physiological considerations of *Auricularia* spp. In: S. T. Chang & T. H. Quimio (Eds.), *Tropical Mushrooms: Biological Nature and Cultivation Methods*. Chinese University Press, pp. 397–408.

Rahi, D. K., Rahi, S., Pandey, A. K., & Rajak, R. C. (2009). Enzymes from mushrooms and their industrial application. In M. Rai(Ed.), *Advances in Fungal Biotechnology*. I K International Publishing House, pp. 136–184.

Rahmann, G., Grimm, D., Kuenz, A., & Hessel, E. (2019). Combining land-based organic and landless food production: A concept for a circular and sustainable food chain for Africa in 2100. *Organic agriculture, 10*, 9–21.

Raimbault, M. (1998). General and microbiological aspects of solid substrate fermentation. *Electronic Journal of Biotechnology, 1*(2), 174–188.

Reza, Md. A., Hossain, Md. A., Lee, S.-J., Yohannes, S. B., Damte, D., Rhee, M., Jo, W.-S., Suh, J.-W., & Park, S.-C. (2014). Dichlormethane extract of the jelly ear mushroom *Auricularia auricula - judae* (higher Basidiomycetes) inhibits tumor cell growth in vitro. *International Journal of Medicinal Mushrooms, 16*(1), 37–47.

Riley, G. L., Tucker, K. G., Paul, G. C., & Thomas, C. R. (2000). Effect of biomass concentration and mycelial morphology on fermentation broth rheology. *Biotechnology and Bioengineering, 68*(2), 160–172.

Rop, O., Mlcek, J., & Jurikova, T. (2009). Beta-glucans in higher fungi and their health effects. *Nutrition Reviews, 67*(11), 624–631.

Rossi, P., Difrancia, R., Quagliariello, V., Savino, E., Tralongo, P., Randazzo, C. L., & Berretta, M. (2018). B-glucans from *Grifola frondosa* and *Ganoderma* lucidum in breast cancer: An example of complementary and integrative medicine. *Oncotarget, 9*(37), 24837–24856.

Rossi, I. H., Monteiro, A. C., Machado, J. O., Andrioli, J. L., & Barbosa, J. C. (2003). Shiitake (*Lentinula edodes*) production on a sterilized bagasse substrate enriched with rice bran and sugarcane molasses. *Brazilian Journal of Microbiology, 34*(1), 66–71.

Royse, D. J. (1985). Effect of spawn run time and substrate nutrition on yield and size of the shiitake mushroom. *Mycologia, 77*(5), 756–762.

Royse, D. J. (2014). A global perspective on the high five: *Agaricus, Pleurotus, Lentinula, Auricularia & Flaminulina. Proceedings of the 8th International Conference on Mushroom Biology and Mushroom Products (ICMBMP8)*, I & II, 1–6. ICAR-DMR, Solan HP & WSMBMP, New Delhi.

Royse, D. J., Baars, J., & Tan, Q. (2017). Current overview of mushroom production in the world 5. In: D. C. Zied & A. Pardo-Giménez (Eds.), *Edible and Medicinal Mushrooms: Technology and Applications*. Wiley-Blackwell.

Ryu, J.-S., Kim, M. K., Im, C. H., & Shin, P.-G. (2015). Development of cultivation media for extending the shelf-life and improving yield of king oyster mushrooms (*Pleurotus eryngii*). *Scientia Horticulturae, 193*, 121–126.

Sadh, P. K., Duhan, S., & Duhan, J. S. (2018). Agro-industrial wastes and their utilization using solid state fermentation: A review. *Bioresources and Bioprocessing, 5*(1).

Salami, A., Bankole, F., & Salako, Y. (2017). Nutrient and mineral content of oyster mushroom (*Pleurotus florida*) grown on selected lignocellulosic substrates. *Journal of Advances in Biology & Biotechnology, 15*(1), 1–7.

Sankaran, S., Khanal, S. K., Jasti, N., Jin, B., Pometto III, A. L., & Van Leeuwen, J. H. (2010). Use of filamentous fungi for wastewater treatment and production of high value fungal byproducts: A review. *Critical Reviews in Environmental Science and Technology*, *40*(5), 400–449.

Sanodiya, B. S., Thakur, G. S., Baghel, R. K., Prasad, G. B., & Bisen, P. S. (2009). *Ganoderma lucidum*: A potent pharmacological macrofungus. *Current Pharmaceutical Biotechnology*, *10*(8), 717–742.

Sardar, H., Ali, M. A., Anjum, M. A., Nawaz, F., Hussain, S., Naz, S., & Karimi, S. M. (2017). Agro-industrial residues influence mineral elements accumulation and nutritional composition of king oyster mushroom (*Pleurotus eryngii*). *Scientia Horticulturae*, *225*, 327–334.

Seviour, R. J., McNeil, B., Fazenda, M. L., & Harvey, L. M. (2011). Operating bioreactors for microbial exopolysaccharide production. *Critical Reviews in Biotechnology*, *31*(2), 170–185.

Sharma, V., Kumar, S., & Tewari, R. (2010). *Flammulina velutipes*, the culinary medicinal winter mushroom. NRCM Silver Jubilee (1983–2008), Technical Bulletin, National Research Centre for Mushroom (Indian Council of Agricultural Research) Chambaghat, Solan- 173213 (HP).

Shnyreva, A. A., Kozhevnikova, E. Y., Barkov, A. V., & Shnyreva, A. V. (2017). Solid-state cultivation of edible oyster mushrooms, *Pleurotus* spp. under laboratory conditions. *Advances in Microbiology*, *7*(02), 125–136.

Shrestha, B., Zhang, W., Zhang, Y., & Liu, X. (2012). The medicinal fungus *Cordyceps militaris*: Research and development. *Mycological Progress*, *11*(3), 599–614.

Sievers, F., & Higgins, D. G. (2018). Clustal Omega for making accurate alignments of many protein sequences. *Protein Science*, *27*(1), 135–145.

Siniscalco, C. (2013). *History of Italian Mycology and First Contribution to the Correct Nomenclature of Mushrooms*. IstitutoSuperiore per La Protezione E La RicercaAmbientale; ISPRA Handbooks and guidelines.

Siwulski, M., Sobieralski, K., Golak-Siwulska, I., Sokół, S., & Sękara, A. (2015). *Ganoderma lucidum* (Curt.: Fr.) Karst. – Health-promoting properties. A review. *Herba Polonica*, *61*(3), 105–118.

Soccol, C. R., Costa, E. S. F. da, Letti, L. A. J., Karp, S. G., Woiciechowski, A. L., & Vandenberghe, L. P. de S. (2017). Recent developments and innovations in solid state fermentation. *Biotechnology Research and Innovation*, *1*(1), 52–71.

Song, B., Ye, J., Sossah, F. L., Li, C., Li, D., Meng, L., Xu, S., Fu, Y., & Li, Y. (2018). Assessing the effects of different agro-residue as substrates on growth cycle and yield of *Grifola frondosa* and statistical optimization of substrate components using simplex-lattice design. *AMB Express*, *8*(1), 46.

Stamets, P. (2000). *Growing Gourmet & Medicinal Mushrooms: Shokuyō oyobiyakuyō kinoko no saibai]: A Companion Guide to The Mushroom cultivator*. Ten Speed Press.

Stanbury, P. F., Whitaker, A., & Hall, S. J. (2013). *Principles of Fermentation Technology* (Second ed.). Elsevier Science Ltd.

Sudheer, S., Taha, Z., Manickam, S., Ali, A., & Cheng, P. G. (2018). Development of antler-type fruiting bodies of *Ganoderma lucidum* and determination of its biochemical properties. *Fungal Biology*, *122*(5), 293–301.

Sullivan, R., Smith, J. E., & Rowan, N. J. (2006). Medicinal mushrooms and cancer therapy: Translating a traditional practice into Western medicine. *Perspectives in Biology and Medicine*, *49*(2), 159–170.

Supramani, S., Rahayu Ahmad, Z. I., Annuar, M. S. M., Klaus, A., & Wan, W. A. A. Q. I. (2019). Optimisation of biomass, exopolysaccharide and intracellular polysaccharide production from the mycelium of an identified *Ganoderma lucidum* strain QRS 5120 using response surface methodology. *AIMS Microbiology*, *5*(1), 19.

Tamura, K., Stecher, G., Peterson, D., Filipski, A., & Kumar, S. (2013). MEGA6: Molecular evolutionary genetics analysis version 6.0. *Molecular Biology and Evolution*, *30*(12), 2725–2729.

Tang, Y. J., Zhu, L. W., Li, H. M., & Li, D. S. (2007). Submerged culture of mushrooms in bioreactors – Challenges, current state-of-the-art, and future prospects. *Food Technology and Biotechnology*, *45*(3), 221–229.

Taufek, N. M., Harith, H. H., Abd Rahim, M. H., Ilham, Z., Rowan, N., & Wan, W. A. A. Q. I. (2020). Performance of mycelial biomass and exopolysaccharide from Malaysian *Ganoderma lucidum* for the fungivore red hybrid Tilapia (*Oreochromis* sp.) in Zebrafish embryo. *Aquaculture Reports*, *17*, 100322.

Telang, S., Patil, S., & Baig, M. (2010). Biological efficiency and nutritional value of *Pleurotus sapidus* cultivated on different substrates. *Food Science Research Journal*, *1*(2), 127–129.

Tepwong, P., Giri, A., & Ohshima, T. (2012). Effect of mycelial morphology on ergothioneine production during liquid fermentation of *Lentinula edodes*. *Mycoscience*, *53*(2), 102–112.

Thakur, R., & Sharma, B.M. (2015). Deployment of indigenous wild *Ganoderma lucidum* for better yield on different substrates. *Journal of Agricultural Research*, *10*(33), 3338–3341.

Thomas, C. R., & Paul, G. C. (1996). Applications of image analysis in cell technology. *Current Opinion in Biotechnology*, *7*(1), 35–45.

Thuc, L. V., Corales, R. G., Sajor, J. T., Truc, N. T. T., Hien, P. H., Ramos, R. E., Bautista, E., Tado, C. J. M., Ompad, V., Son, D. T., & Hung, N. V. (2020). Rice-straw mushroom production. In: M. Gummert, N. Hung, P. Chivenge, & B. Douthwaite (Eds.), *Sustainable Rice Straw Management*. Springer, pp. 93–109.

Treskatis, S. K., Orgeldinger, V., Wolf, H., & Gilles, E. (2000). Morphological characterization of filamentous microorganisms in submerged cultures by on-line digital image analysis and pattern recognition. *Biotechnology and Bioengineering*, *53*(2), 191–201.

Tsujiyama, S., & Ueno, H. (2013). Performance of wood-rotting fungi-based enzymes on enzymic saccharification of rice straw. *Journal of the Science of Food and Agriculture*, *93*(11), 2841–2848.

Tu, J., Zhao, J., Liu, G., Tang, C., Han, Y., Cao, X., Jia, J., Ji, G., & Xiao, H. (2020). Solid state fermentation by *Fomitopsis pinicola* improves physicochemical and functional properties of wheat bran and the bran-containing products. *Food Chemistry*, *328*, 127046.

Tucker, K. G., Kelly, T., Delgrazia, P., & Thomas, C. R. (1992). Fully-automatic measurement of mycelial morphology by image analysis. *Biotechnology Progress*, *8*(4), 353–359.

Ubaidillah, N., Hafizah, N., Abdullah, N., & Sabaratnam, V. (2015). Isolation of the intracellular and extracellular polysaccharides of *Ganoderma neojaponicum* (Imazeki) and characterization of their immunomodulatory properties. *Electronic Journal of Biotechnology*, *18*(3), 188–195.

Vallone, L., Giardini, A., & Soncini, G. (2014). Secondary metabolites from *Penicillium roqueforti*, a starter for the production of Gorgonzola cheese. *Italian Journal of Food Safety*, *3*(3), 2118.

Vunduk, J., Wan, W. A. A. Q. I., Mohamad, S. A., Abd Halim, N. H., Dzomir, A. Z. M., Žižak, Ž., & Klaus, A. (2019). Polysaccharides of *Pleurotus flabellatus* strain Mynuk produced by submerged fermentation as a promising novel tool against adhesion and biofilm formation of foodborne pathogens. *LWT – Food Science and Technology*, *112*, 108221.

Wagner, R., Mitchell, D., & Berovic, M. (2003). Current techniques for the cultivation of *Ganoderma lucidum* for the production of biomass, ganoderic acid and polysaccharides. *Food Technology and Biotechnology*, *41*, 371–382.

Walser, P. J., Hollenstein, M., Klaus, M. J., & Kes, U. (2001). Genetic analysis of basidiomycete fungi. In: N. Talbot (Ed.), *Molecular and Cellular Biology of Filamentous Fungi: A Practical Approach*. UK: Oxford University Press, pp. 59–90.

Wan-Mohtar, W. A. A. Q. I., Abd Malek, R., Harvey, L. M., & McNeil, B. (2016). Exopolysaccharide production by *Ganoderma lucidum* immobilised on polyurethane foam in a repeated-batch fermentation. *Biocatalysis and Agricultural Biotechnology*, *8*, 24–31.

Wan-Mohtar, W. A. A. Q. I., Halim-Lim, S. A., Kamarudin, N. Z., Rukayadi, Y., Abd Rahim, M. H., Jamaludin, A. A., & Ilham, Z. (2020). Fruiting-body-base flour from an Oyster mushroom waste in the development of antioxidative chicken patty. *Journal of Food Science*, *85*(10), 3124–3133.

Wan-Mohtar, W. A. A. Q. I., Taufek, N. M., Thiran, J. P., Rahman, J. F. P., Yerima, G., Subramaniam, K., & Rowan, N. (2021). Investigations on the use of exopolysaccharide derived from mycelial extract of *Ganoderma lucidum* as functional feed ingredient for aquaculture-farmed red hybrid Tilapia (*Oreochromis* sp.). *Future Foods*, *3*, 100018.

Wan-Mohtar, W. A. A. Q. I., Taufek, N. M., Yerima, G., Rahman, J., Thiran, J. P., Subramaniam, K., & Sabaratnam, V. (2020). Effect of bioreactor-grown biomass from *Ganoderma lucidum* mycelium on growth performance and physiological response of red hybrid tilapia (*Oreochromis* sp.) for sustainable aquaculture. *Organic Agriculture*, *2020*, 1–9.

Wan-Mohtar, W. A. A. Q. I., Young, L., Abbott, G. M., Clements, C., Harvey, L. M., & McNeil, B. (2016). Antimicrobial properties and cytotoxicity of sulfated (1,3)-beta-D-glucan from the mycelium of the mushroom *Ganoderma lucidum*. *Journal of Microbiology and Biotechnology*, *26*(6), 999–1010.

Wasser, S. (2005). Reishi or Ling Zhi (*Ganoderma lucidum*). *Encyclopedia of Dietary Supplements*, *1*, 603–622.

Wasser, S. (2014). Medicinal mushroom science: Current perspectives, advances, evidences, and challenges. *Biomedical Journal*, *37*(6), 345.

Wu, F., Yuan, Y., Malysheva, V. F., Du, P., & Dai, Y. C. (2014). Species clarification of the most important and cultivated *Auricularia* mushroom "Heimuer": Evidence from morphological and molecular data. *Phytotaxa*, *186*(5), 241.

Xiao, J. H., Chen, D. X., Wan, W. H., Hu, X. J., Qi, Y., & Liang, Z. Q. (2006). Enhanced simultaneous production of mycelia and intracellular polysaccharide in submerged cultivation of *Cordyceps jiangxiensis* using desirability functions. *Process Biochemistry*, *41*(8), 1887–1893.

Xinxin, L., Zuliang, S., Jiuchen, W., & Rongfeng, J. (2020). Review on the crop straw utilization technology of China. *American Journal of Environmental Science and Engineering*, *4*(4), 61–64.

Xu, F., Li, Z., Liu, Y., Rong, C., & Wang, S. (2016). Evaluation of edible mushroom *Oudemansiella canarii* cultivation on different lignocellulosic substrates. *Saudi Journal of Biological Sciences*, *23*(5), 607–613.

Yadav, A. N. (2019). Fungal white biotechnology: Conclusion and future prospects. In: A.N. Yadav, S. Singh, S. Mishra, & A. Gupta (Eds.), *Recent Advancement in White Biotechnology through Fungi*. Springer, pp. 491–498.

Yang, W., Guo, F., & Wan, Z. (2013). Yield and size of oyster mushroom grown on rice/wheat straw basal substrate supplemented with cotton seed hull. *Saudi Journal of Biological Sciences*, *20*(4), 333–338.

Yang, F. C., & Liau, C. B. (1998). The influence of environmental conditions on polysaccharide formation by *Ganoderma lucidum* in submerged cultures. *Process Biochemistry*, *33*(5), 547–553.

Zervakis, G. I., & Koutrotsios, G. (2017). *Medicinal Plants and Fungi: Recent Advances in Research and Development* (Vol. 4, pp. 365–396). Springer.

Zervakis, G. I., & Koutrotsios, G. (2017). Solid-state fermentation of plant residues and agro-industrial wastes for the production of medicinal mushrooms. In: D. Agrawal, H. S. Tsay, L. F. Shyur, Y. C. Wu, & S. Y. Wang (Eds.), *Medicinal Plants and Fungi: Recent Advances in Research and Development. Medicinal and Aromatic Plants of the World*. Springer, pp. 365–396.

Zhang, H.-L., Wei, J.-K., Wang, Q.-H., Yang, R., Gao, X.-J., Sang, Y.-X., Cai, P.-P., Zhang, G.-Q., & Chen, Q.-J. (2019). Lignocellulose utilization and bacterial communities of millet straw based mushroom (*Agaricus bisporus*) production. *Scientific Reports*, *9*(1), 1151.

Zhang, M., Zhang, Y., Zhang, L., & Tian, Q. (2019). Mushroom polysaccharide lentinan for treating different types of cancers: A review of 12 years clinical studies in China. *Progress in Molecular Biology and Translational Science*, 297–328.

Zhou, X., Broadbelt, L. J., & Vinu, R. (2016). Mechanistic understanding of thermochemical conversion of polymers and lignocellulosic biomass. In: Kevin M. Van Geem (Ed.), *Thermochemical Process Engineering* . Academic Press, Vol. 49, pp. 95–198.

Zhou, X., Lin, J., Yin, Y., Zhao, J., Sun, X., & Tang, K. (2007). Ganodermataceae: Natural products and their related pharmacological functions. *American Journal of Chinese Medicine*, *35*(4), 559–574.

Zhou, X.-W., Su, K.-Q., & Zhang, Y.-M. (2011). Applied modern biotechnology for cultivation of *Ganoderma* and development of their products. *Applied Microbiology and Biotechnology*, *93*(3), 941–963.

Zhou, S., Zhang, J., Ma, F., Tang, C., Tang, Q., & Zhang, X. (2018). Investigation of lignocellulolytic enzymes during different growth phases of *Ganoderma lucidum* strain G0119 using genomic, transcriptomic and secretomic analyses. *PLOS ONE*, *13*(5), e0198404.

Zikriyani, H., Saskiawan, I., & Mangunwardoyo, W. (2018). Utilization of agricultural waste for cultivation of paddy straw mushrooms (*Volvariella volvacea* (Bull.) Singer 1951). *International Journal of Agricultural Technology*, *14*(5), 805–814.

Žnidaršič-Plazl, P., & Pavko, A. (2001). The morphology of filamentous fungi in submerged cultivations as a bioprocess parameter. *Food Technology and Biotechnology*, *39*(3), 237–252.

CHAPTER 3

Preservation and processing technology of wild mushrooms

Rohit Biswas
Research Scholar, Department of Agricultural and Food Engineering, Indian Institute of Technology, Kharagpur, India

CONTENTS

DOI: 10.1201/9781003152583-4

3.1 INTRODUCTION

Mushrooms are fast-blooming species of fungi that are sources of various nutritional components. The mushroom grows in the fleshy body over a substrate (Nketia et al., 2020). A wide variety of mushrooms encompassing variety are segregated as edible or cultivated mushrooms and wild mushrooms (Lee et al., 2012; Sharma, 2018). As per Varma et al. (2011), there are more than 2,000 varieties of mushrooms that are edible and non-poisonous. Soltaninejad (2018) reported that 2,700 species of mushrooms are edible, whereas 100 species were found to be toxic to humans, with a total of 140,000 species of mushrooms found in the world. Most of the mushrooms are macrofungi that are extensively cultivated in today's world due to their organoleptic properties (Ma et al., 2018). Mushrooms have been prevailing in various tribal and ethnic groups for centuries, and it has been a prime source of various nutritional requirements, especially protein (Karun & Sridhar, 2017). Primarily the poisonous mushroom can be segregated into three major categories such as (Govorushko et al., 2019):

1. Local acting poison – consisting mainly of yellow-stained, dark-scaled, and some others. The toxicity emerges after 1–2 hours of consumptions (Morel et al., 2018).
2. Nervous system affecting – consisting of panther cap, fly agaric, and destroying angel. Their toxic symptoms mainly include nausea, diarrhoea, and sweating emerging after 2 hours of consumption; there also exist a condition of hallucination and uncontrolled laughter/crying and the rare condition of loss of consciousness (Kosentka et al., 2013).
3. Lethal group – consisting mainly of death cap and sulphur tuft. This group of mushroom toxicity targets the body's vital organs, leading to irreversible changes to the human body. The toxic onset is comparatively late, ranging from 8–48 hours from consumption of half or even a third of the mushroom (Cai et al., 2016; Barman et al., 2018).

Despite the mushroom's toxicity, especially with the wild mushroom, mushroom use proliferates since ancient times; as Hippocrates described in the fourth century on mushroom's therapeutic use (Muszyńska et al., 2020). Also, China has a documented history of mushroom cultivation and medicinal use (Ma et al., 2018). Mushrooms have an untapped nutritional component apart from these various other bioactive compounds and act as dietary supplements and have therapeutic uses (Borthakur & Joshi, 2019; Krüzselyi et al., 2020).

A mushroom is a rich source of some vital nutrients, containing 95% moisture, less than 10% fat, and more than 16% protein; apart from this, it consists of a healthy level of essential amino acids such as aspartic, glutamic, alanine, and arginine (Nketia et al., 2020). Various mushroom therapeutic effects have been identified due to the presence of certain specific chemicals such as phenolic compounds functioning as potent antioxidants, β-glucans have an immunomodulatory effect, triterpenoids and some proteins are cited as immune system modulators, and reactive oxygen species acting as proinflammatory medicinal usage (Soković et al., 2018). Peroxidase, enzymes-glucose oxidase, laccases, and superoxide dismutase produced as primary metabolites may prevent oxidative stress. Alkaloids, antibiotics, lecithin, lactones, and terpenoids produced as secondary metabolites have therapeutic effects (Wasser, 2010).

Mushrooms have been cultivated for decades, and some of the commercially cultivated ones include button mushrooms, oyster mushrooms, and shitake mushrooms. The button mushroom conquers most of the mushroom segment's market share (Lee et al., 2012; Borthakur & Joshi, 2019;

Nketia et al., 2020). Despite being into cultivation and commercial availability, wild mushrooms hold significant potential due to their high protein content and medicinal uses. Edible wild mushrooms are harvested mainly from the wild woodlands as field availability of the mushroom is highly dependent on seasonal changes (Zhang et al., 2014; Soković et al., 2018).

This chapter's content mainly focuses on various processing techniques employed in the commercially available mushrooms and the wild mushrooms. Specific focus will be on processing flows, and some light will be curtailed onto particular parametric calculations that are essential to the understanding of the processing techniques. The processing of mushrooms is a prime concern due to the possible nature of the mushroom. As a mushroom consists of a high amount of moisture ranging between 90%–95% of the total weight (Nketia et al., 2020), the deterioration tends to start when harvested. This processing plays a crucial role in maintaining the quality of the mushroom for future use. Various processing techniques are currently available within the market that are employed for the processing of the mushroom.

3.2 SPECIES

The family of mushrooms encompasses a variety of species that are known to mankind. Some of the notable species that are widely used, either cultivated or found wild and used as edible mushrooms, are discussed below. Six species are currently widely used for edible purposes due to their nutritional or medicinal usage.

3.2.1 *Agaricus* spp.

Agaricus is the most common and essential mushroom species that is widely cultivated – commonly known as the button mushroom (Soković et al., 2018). *Agaricus* are characterized by a fleshy cap and underside gill that produces naked spores (Ogidi et al., 2020). *Agaricus* spp. is commercially available in the market due to its health and medicinal usage; it is cultivated widely. The *Agaricus* spp. has some specific medicinal usage, especially *A. brasiliensis* Waser, which has antimicrobial, anti-QS, immunomodulatory, anti-inflammatory, and antioxidant properties due to which it is cultivated (Llarena-Hernández et al., 2013).

3.2.2 *Ganoderma* Species

Ganoderma is a species that mainly grows in the wild in the woods (Ogidi et al., 2020). The species is characterized by thick, smooth, shiny skin with perennial and woody bracket portions forming the mushroom's flesh known as "conks", which render it an inedible form mushroom (Soković et al., 2018; Ogidi et al., 2020). This species is the most tested mushroom for its various medicinal purposes in Chinese traditional medicine (Baby et al., 2015). The species is widely used as a tonic to treat various diseases, encompassing neurasthenia, diabetics, gastric ulcers, cancer, and some in China, Japan and other Asian countries (Lee et al., 2012).

3.2.3 *Lentinus* spp.

Lentinus spp. is one of the most cultivated species of shiitake mushroom. They have been a medicinal and healthy food for years in Japan (Lee et al., 2012). *Lentinus* spp. possess extracellular enzymes that feed saprobically on dead or fallen woods of broad-leaf trees (Soković et al., 2018; Ogidi et al., 2020). Lentinan is a compound found in *Lentinus* spp. with antibacterial, antiviral, and anti-parasitic capabilities (Ngai & Ng, 2003).

3.2.4 *Grifolafrondosa*

Grifolafrondosa has a sizeable fruiting body measuring 60 cm in diameter and arranged in a cluster of wavy spoon-shaped caps. The mushroom is perennial, growing at the same place in succession for several years (Ogidi et al., 2020). *G. frondosa* is an edible fungus mainly in Japan and China, encompassing nutritional and medical functionings. The major compound in the fungus is β-glucans, which has anti-inflammatory, antivirus, immunomodulation, and antidiabetic use (Wu et al., 2021).

3.2.5 *Rigidoporus* Species

Rigidoporus species are root decaying mushrooms that mainly grow onto the root bark with flattened mycelia strand adhered onto it of thickness 1–2 mm. The fungus's fruiting body has a spread of 20 cm and has a leathery, faintly velvety texture (Ogidi et al., 2020). *Rigidoporus* species has a wood-decaying property due to extracellular hydrolytic and oxidative secretion than tend to decay the wood growing (Oghenekaro et al., 2015).

3.2.6 *Pleurotusostreatus*

Pleurotusostreatus, more commonly known as oyster mushrooms, is a widely available mushroom cultivated and harvested from the wild (Ogidi et al., 2020). The fungus is rich in non-starch polysaccharides, including β-glucans and other phenolic compounds (Deepalakshmi & Sankaran, 2014).

3.3 CULTIVATION

Mushroom farming is a skilled labor-extensive process involving various processes to be taken care of to obtain a profitable yield. Proper manure or compost, spawns added with the correct environmental consideration, such as temperature and humidity, play a crucial role in mushroom farming (Sharma, 2018). Mushroom farming requires a series of processes with individual duration and specific conditions to be fulfilled to proliferate the mushroom growth discussed below.

3.3.1 Compost Preparations

Compost preparation is the first objective that needs to fulfilled correctly as the mushroom and the quality of compost determine the overall yield. Compost can be composed of various substances such as straw, gypsum, rice bran, and manure that are commonly used. Apart from this, compost is vulnerable to microbial growth that may eventually contaminate the compost hindering the mushroom growth. Proper moisture level and regular turnover should be done to avoid the compost's drying due to heat generation during mushroom growth. Gypsum plays a crucial role in maintaining the aeration, due to which it is added at regular intervals (Mohan, 2009; Royse & Beelman, 2016; Sharma, 2018).

3.3.2 Spawning

Spawning is done due to the uncertainty in germination and growth of the mushroom. Mushroom spawns are germinated onto grains until the mycelium grows into a white cotton-like structure. Furthermore, the spawns are mixed with the prepared compost and are covered with newspaper and watered regularly to maintain the growth's moisture. Spawning extends for 30–35 days depending on the type of compost and environmental conditions (Mohan, 2009; Dg, 2013; Sharma, 2018; Mahari et al., 2020).

3.3.3 Casing

The casing covering the spawn has been mixed with sterilized manure or soil. The casing is done after approximately 15–20 days or until the mushroom's pinhead appears onto the compost. It facilitates mycelium growth and prevents microbial infiltration, thus providing a sterile environment for the mushroom to grow (Dg, 2013; Sharma, 2018; Mahari et al., 2020).

3.3.4 Harvesting

Harvesting is done before the mushroom sheds the spore for future propagation. And, significantly, the harvesting is done either by handpicking or with the help of a knife (Sharma, 2018; Mahari et al., 2020).

To the above consideration, some primary environmental consideration needs to be maintained. Depending on variety of mushrooms, temperature (25°C during spawning and 17°C–30°C during fructification), moisture content (68%–75%), and specific nutrient components are required as per specific strain of mushroom. Particular mushrooms, such as *Volvariella volvaeaea*, *Lentinusedodes*, *Tremellafuciformis*, and *Auriculariapolytricha*, require higher than 30°C to grow (Dg, 2013; Royse & Beelman, 2016; Mahari et al., 2020).

3.4 SUBSTRATE FOR CULTIVATION

The substrate is a significant component of mushroom cultivation. Various substrate types are made depending upon the requirement and type of mushroom to be cultivated (Mahari et al., 2020). Agricultural waste plays a crucial role in mushroom substrate formation, as most of the substrate uses an agricultural waste base (Assan & Mpofu, 2014; Bains & Chawla, 2020). In general, rice husk, wheat hay, sawdust, waste paper, and manure are used widely (Baysal et al., 2003; Assan & Mpofu, 2014).

Wendiro et al. (2019) researched some of the indigenous substrates for mushroom cultivation, especially for the saprophytic category that is also the need of an hour in terms of sustainable utilization of natural resources. Some of the substrates studied are listed below.

Grass waste or forest litter were used for the cultivation of *Obunegyere* mushrooms that are second-level saprophytic mushrooms. They grow in close harmony with the terminates that also grow in piles of grass and banana leaves. As elders believed terminates break down the piles of grass, making suitable arrangements for the mushroom mycelium to proliferate. *Obunegyere* generally grows in the rainy season, which corresponds to their reproductive phase during the sighting of a pinhead of mushroom in light piles of grass or banana, they are also watered to maintain the moisture level and are harvested upon suitable size (Hsieh et al., 2017; Kamthan & Tiwari, 2017).

Cattle manure, which is widely available in a rural areas compared to more modern cultivation sites, is used directly as a mushroom substrate. As discussed, piles of cattle manure are placed in an open area and left to decompose, on which mushrooms proliferate. The manure is checked regularly for mushroom growth and harvested accordingly. Cow manure is generally required in the substrate but not as a whole base, but the minimum amount required for optimal growth has not been quantified (Mshandete et al., 2013; Jeznabadi et al., 2016).

The banana residue used as the substrate is common in the Buikwe district, where the banana residue is available as a by-product of juice extraction mixed with spear grass. The by-product is piled under the tress and left checked for fruiting of mushrooms (Carvalho et al., 2012).

Sorghum waste is another substrate used more widely even in commercial spaces. After being fermented and collected as waste, the sorghum seeds are stoked and left for decomposition, after which they are mixed with other agricultural waste such as maize stalks and are used as mushroom substrate (Narh et al., 2011).

Maize waste is the maize cob left after the milling process. During rainy seasons, the maize cob is stocked in piles for the mushroom to proliferate onto the cobs and then are harvested commercially (Adjapong et al., 2015).

Deadwood mushroom is a dead or fallen tree trunk that is used as the substrate. Many mushrooms, also know as the wood mushroom, grow onto the dead trees due to the extracellular secretion that helps them penetrate the woods and proliferate (Teke et al., 2019).

The substrate described shows the use of various agricultural wastes to be used as sustainable alternatives to natural resource harvesting. As some wild mushrooms require more specific environmental conditions and substrates to grow, these sustainable techniques provide a solution (Wendiro et al., 2019).

The substrate can be wastage after mushroom production that may pile up as agricultural waste. Mahari et al. (2020) suggested three significant ways of the utilization of spent substrate as compost with the help of various microbes to breakdown essential nutrients that are required for soil conditioning, plant growing media, and bio-fertilizer due to rich sources of lignocellulose decomposed by mushrooms and high level of organic matter, nitrogen, potassium, and other nitrogen and valorization of the substrate as alternative energy sources to produce biofuel energy.

3.5 BLANCHING

Blanching is the foremost process in most preservation techniques. The primary objective of blanching is to inactivate the enzyme along with partial destruction of microflora or surface disinfection. Various techniques are available for blanching to achieve the required effect of a blanching suitable temperature and time combination (Jabłonska-Rys et al., 2019). Following in Figure 3.1 are the basic steps of the blanching process that could be understood and applied to any technique used.

Blanching's first step involves sorting and grading mushrooms based on several physical conditions such as bruises, cuts, damaged parts, size, and color, followed by cleaning the

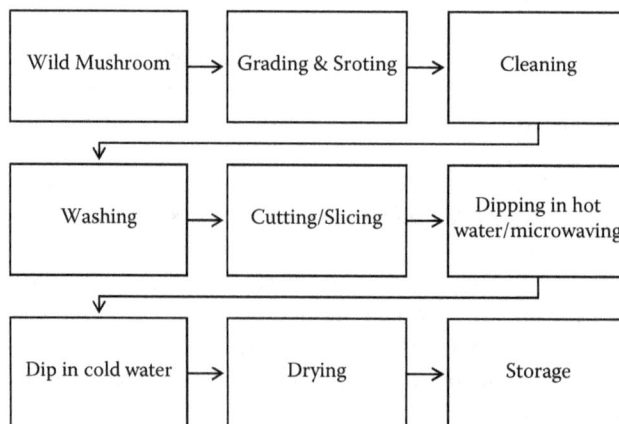

Figure 3.1 Blanching.

mushrooms of any debris, twigs, and unwanted materials. Washing is a prime concern as wild mushrooms are grown in various conditions that need to be removed. For most lactic acid fermentation, cold water cleaning is carried out, and handwashing can also be done to remove the unwanted (Liu et al., 2016). Various other solutions are also used for washing to fulfill the specific purpose for higher brightness value (5% H_2O_2 + 4.5% Na isoascorbate + 0.2% cysteine HCL + 0.1% Na2EDTA), decrease in soluble polyphenol content (2 to 5 ppm chlorine), and increase in polyphenol content (5% aq. H_2O_2 + 4% Na erythorbate) is done (Sapers et al., 1999; Czapski, 2002). Further cleaning is also a step that needs to be done as a precaution, as minor damage or bruise onto the delicate mushroom skin would cause darkening of the mushroom (Bernaś et al., 2006). Blanching is accomplished with various means, such as dipping in hot water, microwaving, or hot solutions. The water ratio to mushroom is 5:1 to dip the mushroom and provide the required blanching temperature (Drewnowska et al., 2017).

Typically, in lactic acid fermentation, it is recommended to blanch in plain or distilled water, whereas Khaskheli et al. (2015) recommended a 0.05% KMS solution. A specific purpose, such as prevention of darkening for the mushroom that may be subjected to further processing, is recommended to be blanched in 0.05%–0.5% citric acid (Martin-Belloso & Llanos-Barriobero, 2001).

Although blanching has various benefits in storage and processing, numerous research also highlighted the drawback of specific blanching techniques onto nutritional properties, some of which are reduced mineral content (A. *bisporus* blanching for 15 minutes at 95°C–100°C) and reduction in protein, fat, and carbohydrate content (P. *ostreatus* blanching for 1 minutes at 88°C) is observed (Coşkuner & Özdemir, 2000; Muyanja et al., 2014).

3.6 DRYING

Drying has been used for ages as a preservation method, and mushroom processing also involves various drying techniques. Drying is the cheapest and widely used preservation technique to extend its shelf life. As mainly the mushroom flourishes toward the most humid part of the season, especially wild mushrooms, special care should be taken to avoid microbial infestation during the whole drying process (Ruan-Soto et al., 2017).

Drying could be done in a variety of ways, such as traditional, mechanical, and modern. Some of the drying techniques are listed in Figure 3.2 (Balan & Radhakrishnan, 2014; Drewnowska et al., 2017; Ruan-Soto et al., 2017; Zhang et al., 2018; Piskov et al., 2020).

Drying involves a series of steps that need to be performed to obtain the final dried product. Depending on the product species and the final product requirement, drying parameters are adjusted accordingly, and drying is performed. The typical steps that are involved in drying irrespective of drying techniques are shown in Figure 3.3.

3.6.1 Traditional Drying

Traditional drying involves three different methods of drying: home, kitchen or fireplace, and solar drying. Solar drying has been prevailing for ages in our history due to simple processes and availability.

Traditional home drying is generally done in a remote place, whereas solar drying is either completely unavailable or partially available. In the place with partial availability of solar final drying of the mushroom is done in solar or mixed type drying is performed (Ruan-Soto et al., 2017).

Solar drying is the cheapest form of drying that is virtually available for free. Despite being the most popular drying mode, the solar-dried product is darker in color due to its long drying time compared to modern techniques (Sharma, 2018; Zhang et al., 2018). Due to the higher drying time, the final product has lower quality and least acceptability. The solar drying is mainly performed on stainless steel plates or trays in the open with an ambient temperature of 25°C–30°C. The drying

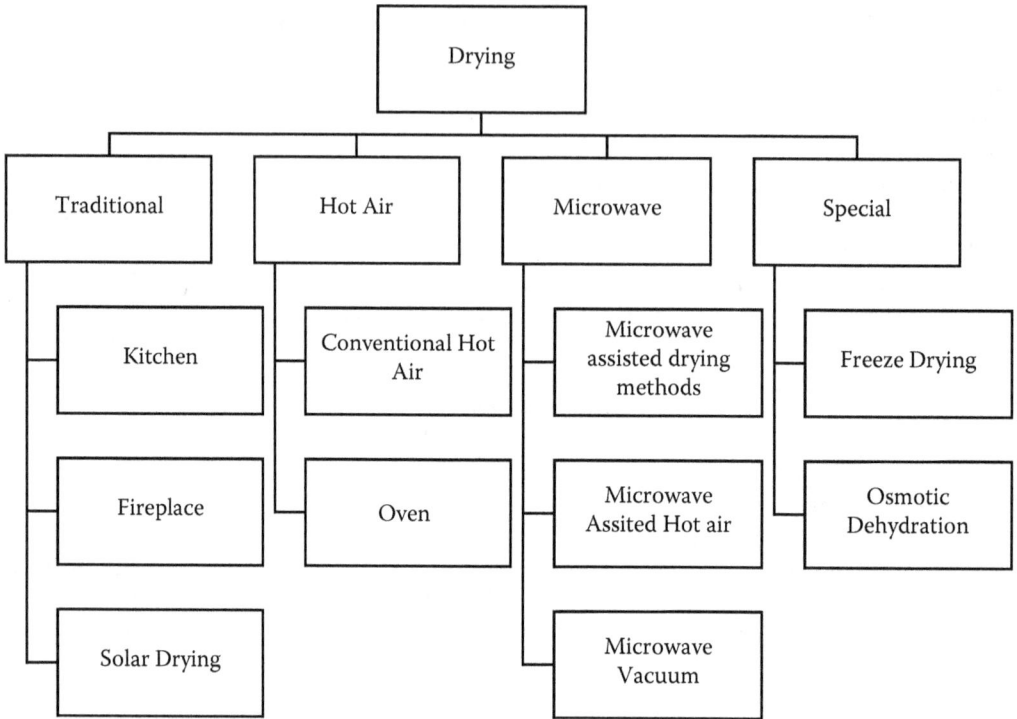

Figure 3.2 Drying methods.

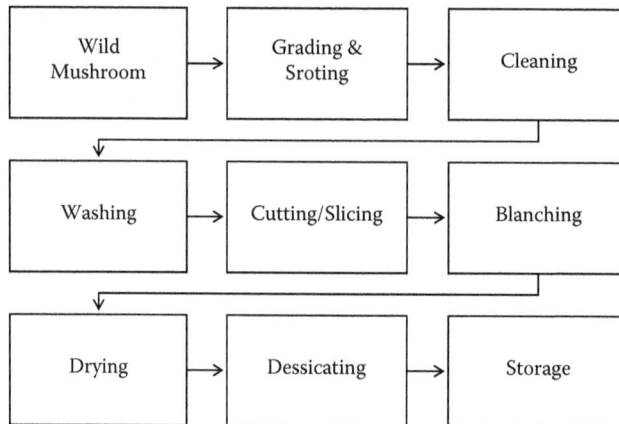

Figure 3.3 Drying steps.

time usually ranges from 24–48 hours, depending upon the solar intensity and product moisture (Zhang et al., 2018; Piskov et al., 2020).

3.6.2 Hot Air Drying

Hot air drying is the most primitive type of mechanical drying with the most straightforward procedure that can be performed with a bit of skill. A variety of mechanical equipment is present for hot air drying. Some of the standard equipment available are cabinet dryer, tray dryer, vacuum oven dryer, and hot air oven (Piskov et al., 2020).

Most of the hot air drying involves forced convention of air that is blown over the product as well as through the perforated tray onto which the product is placed. The drying temperature depends on the final product requirement; the higher the temperature, the darker the final product and vice versa. Generally, the temperature of the hot air oven ranges from 45°C–80°C. The same applies to drying time commercially. The drying time ranges from 6–24 hours. Higher temperature drying is mainly done for analytical purposes (Yuen et al., 2014; Drewnowska et al., 2017; Kic, 2018; Piskov et al., 2020).

Vacuum oven drying is a method involving drying mushrooms at a lower pressure, which is in the range of 60–80 Pa, which drastically reduces the drying time and at the same time maintaining the final drier quality of the product. Vacuum drying also helps retain various heat-sensitive compounds that are otherwise destroyed in conventional drying (Balan & Radhakrishnan, 2014).

3.6.3 Microwave Drying

Microwave drying is a modern-day drying technique involving microwave radiation to remove the product's moisture. The significant benefit of microwave drying is reduced drying time, resulting in higher product quality (Sun, 2012). Microwave frequency in the home appliance is 2,450 Mhz; in commercial equipment, two different frequencies are available, that is 915 and 2,450 Mhz (Balan & Radhakrishnan, 2014).

Microwave-assisted drying involves the application of a microwave in association with conventional drying at different stages of drying. Majorly, microwave is applied at three different stages of the drying process:

- The first method is to apply the microwave at the start of drying to quickly heat the center of the product. The microwave heating helps heat the center faster, expelling moisture to the product's surface, which can be removed later with conventional drying. The heating also creates a puffed product due to pores' creation for the moisture movement (Drouzas et al., 1999).
- The second method is the application of a microwave at the start of the drying curve's falling rate, which helps create the required vapor pressure, eventually assisting faster removal of moisture from the product (Andrés et al., 2004).
- The third method involves the application of the microwave at the last of the drying. At this point, of least moisture content and maximum shrinkage diffusion of moisture are difficult with conventional air drying. Thus, microwaves help eliminate the moisture at the last of the drying process to the maximum extent (Maskan, 2000; Balan & Radhakrishnan, 2014).

The microwave-assisted hot air drying method combines both microwave power and conventional hot air drying techniques to obtain a final product with a better overall structure and bulk volume. The method significantly reduces the drying time while maintaining the required product characteristics. A minimum of 1 m/s of air velocity was identified for avoiding product darkening during drying (Balan & Radhakrishnan, 2014).

Microwave-vacuum drying is the combination of microwave and low-pressure environments to facilitate faster drying. As discussed in vacuum air drying, lower pressure reduces the boiling, eventually minimizing overall drying time and maintaining the final product quality. This method is costly due to the extended operation of vacuum required throughout the drying time, impacting its commercial usage (Gunasekaran, 1999; Balan & Radhakrishnan, 2014).

The above-discussed microwave drying method focuses primarily on the microwave, which is a form of electromagnetic radiation. A microwave works on dielectric heating, which depends on the raw material's various dielectric components, especially moisture. An increase in moisture content would have increased dielectric properties, but the dielectric properties decrease with a further increase in moisture after a certain level. A microwave is a bulk heating method from the inside out due to which the penetration depth of the energy is an essential factor to consider. Penetration depth is dependent on wavelength, the dielectric constant of the product, and loss factor, as shown below

$$D = \frac{\lambda_0 \sqrt{2}}{2\pi} \left[\varepsilon' \sqrt{1 + \left(\frac{\varepsilon''}{\varepsilon'}\right)^2} - 1 \right]^{-\frac{1}{2}} \qquad (3.1)$$

where
D is the penetration depth or the distance to the center of the product from the surface
λ_0 is the free space wavelength
ε'' is the loss factor
ε' is the dielectric constant

Equation 3.1 shows the relation of the microwave properties with that of the product thickness. The product that is to be dried needs to be double the size of the penetration depth to achieve even drying. Wild mushrooms, either harvested in the wild or cultivated, have different sizes and thicknesses for which proper cutting needs to be done to achieve maximum efficiency (Schiffmann, 1995).

A microwave is an energy-intensive process and requires a higher amount of energy to perform drying. The energy utilization and drying efficiency can be obtained from (Gunasekaran, 1999)

$$EUF = \frac{Total\ energy\ absorbed}{Total\ energy\ input}$$

3.6.4 Freeze Drying

Freeze drying is the most advanced technique of drying in terms of final product quality retention. The significant benefit of freeze drying is lower temperature due to which heat-sensitive components such as vitamins could be retained (Pei et al., 2014). The basic concept in freeze drying is the ice's sublimation directly to water vapor at reduced pressure or vacuum (Liapis & Bruttini, 1987). The freeze drying is performed in three steps as listed below:

- Freezing involves cooling the product to a temperature where it freezes completely. Quick freezing is performed to obtain a smaller ice crystal size as during sublimation; smaller pores are created rather than large extrudates of water which may occur in slow freezing. Freezing is generally done depending upon the composition of the product and the moisture content level. Some of the temperature and time combinations are −22°C for 24 hours (Tarafdar et al., 2017) and −40°C for 72 hours (Piskov et al., 2020). This temperature and time combination is highly dependent upon the product characteristics.
- Primary drying is the follow-up process after freezing involving the placement of the frozen product into a vacuum chamber with pressure ranging from 0.001–80 mmHg. Mushrooms are dried at 0.3–1 mmHg with the surrounding medium like air (Piskov et al., 2020). The process can be performed either with hot air or a microwave. Microwave drying at the primary stage has an added benefit of reduced overall drying time to 2–4 hours compared to 24–36 hours for hot air drying; thus maintaining the maximum product quality (Pei et al., 2014). Strict precautions are taken to maintain the product's temperature well below 30°C (Piskov et al., 2020). The primary stage of drying continues until the product temperature exceeds the triple point, and further sublimation of ice could not occur, which marks the onset of secondary drying (Tarafdar et al., 2017).

- Secondary drying is the removal of moisture that is left over and needs to be removed. It targets primarily the water that does not freeze (bound or sorbed water). In this stage, the temperature is raised above the triple point under vacuum to evaporate the leftover moisture present in the product. The second stage could also be done with a microwave to maintain the optimal final product quality. Microwave drying is preferred if the moisture level during the second stage of drying is high, leading to structural collapse (Liapis & Bruttini, 1987; Tarafdar et al., 2017).

Freeze drying is highly dependent upon the diffusivity of the overall moisture content during both stages of drying as due to direct sublimation of ice into water vapor requires travel of moisture from the center of the food to the surface for the removal purpose. Understanding the diffusivity of moisture specific mathematical models are helpful to visualize the drying.

The total mass flow rate of the system can be given as

$$N_t = N_w + N_g$$

N_t is the system's total mass flow rate, including the mass flow rate of water vapor and gases as N_w and Ng, respectively, to obtain the mass flow rate of water vapor and gases following. Equation 3.2 is employed

$$N_w = -D_{wg}\frac{\partial C_{pw}}{\partial x} + \left(\frac{C_{pw}}{C_{pw} + C_{pg}}\right)N_t \tag{3.2}$$

$$N_g = -D_{wg}\frac{\partial C_{pg}}{\partial x} + \left(\frac{C_{pg}}{C_{pw} + C_{pg}}\right)N_t \tag{3.3}$$

Equation 3.3 gives the relation of total mass flow rate N_t with diffusivity of the water vapor in inert gas and specific heat capacity. The effective diffusivity of water vapor to inert gas can be obtained from Equation 3.4 (Geankoplis, 2003):

$$D_{wg} = \frac{1.8583 * 10^{-7}T_i^{\frac{3}{2}}}{P\sigma_{gw}^2\Omega_{gw}}\left(\frac{1}{M_g} + \frac{1}{M_w}\right)^{\frac{1}{2}} \tag{3.4}$$

where
σ_{gw} is the average collision diameter
Ω_{gw} is the collision integrals
M_g is the molecular weight of inert gas
M_w is the molecular weight of water vapor

The mathematical models help observe various changes that may occur during the various phases of drying, and a specific understanding of drying with regards to its product characteristics could be understood further. The overall drying time and power requirement can also be formulated based on the basic understanding of the moisture removal rate.

3.6.5 Osmotic Dehydration

Osmotic dehydration is not specifically a drying method for significant mushroom processing. It is an add-on to microwave drying, which helps in various functionaries of the microwave drying techniques. Osmotic dehydration includes mushroom treatment with a hypertonic salt solution for

1–3 hours of various concentrations, such as 10%–25%, depending upon the requirement. The salt concentration helps to draw moisture from the product and also enhance the taste. The solute that is settled inside the mushroom during osmotic dehydration helps to retain the structural integrity during microwave drying (Balan & Radhakrishnan, 2014).

Drying involves reducing moisture to a safe level where microbial contamination could be avoided to extend the product shelf life. However, the dried mushroom is not edible in most scenarios, and it is needed to be rehydrated to make it palatable. Various drying techniques affect the rehydration ratio as freeze drying has the highest rehydration ratio compared to solar dried. The rehydration ratio is obtained from the ratio of difference of rehydrated and dried weight of mushrooms to that of dried weight of mushrooms (Piskov et al., 2020):

$$RR = \frac{W_1 - W_0}{W_0}$$

3.7 FREEZING

Freezing is another vital method of preservation for wild mushrooms. Freezing of wild mushrooms requires some technical understanding related to various changes that may occur during freezing. Apart from freezing, the follow-up process is thawing, which is carried out for making the mushroom edible. Though thawing is a relatably simple process to do, various changes and after-effects are observed depending upon the method used for thawing. Freezing is mainly carried out using either a blast freezer or individual quick freezing (Li et al., 2018; Sharma, 2018). A blast freezer is done at generally −30°C to −20°C at an air velocity of 8.23 (m/s), whereas IQF is carried out at −62.5°C (Li et al., 2018; Sharma, 2018; Reid et al., 2017; Drewnowska et al., 2017). Cryogenic freezing is also an advanced technique used for IQF in which Lq. N_2 or solid CO_2 is employed for freezing at -80°C to -100°C (Sharma, 2018). A different storage shelf life could be achieved for different freezing methods ranging from 14 days to 6 months for blast freezing or IQF (air) and up to 1 year for cryogenic freezing (Sharma, 2018; Rathore et al., 2017; Reid et al., 2017; Drewnowska et al., 2017). The basic freezing process involves the same steps, as shown in Figure 3.3, except in place or drying freezing is performed using the available method.

Cooling is another preservation process that involves lowering a temperature above the freezing point but near 0°C. The freezing can be done with any available method to reach the required temperature. Cooling is relatively cheap compared to freezing but with a shorter shelf life. Cooling the fungus to 0°C can lower the mushroom respiration rate by three times than mushroom at 10°C, which extends the shelf life up to 9 days (Zhang et al., 2018).

Thawing is a process that involves melting back the mushroom to its original state to make it edible. Li et al. (2018) discussed that mushrooms could be thawed using three different methods, i.e. thawing at room temperature, thawing with water at 4°C, and microwave thawing till the internal temperature of mushroom reaches 4°C. Thawing of IQF frozen mushrooms showed the highest water holding capacity was when thawed with a microwave and had a lower thawing loss.

Despite being the method that did not involve preservation by heat, freezing still affects the mushroom's nutritional aspect. As reported by Liu et al. (2014), there is a reduction of 24.3% in proteins and free amino acid reduction when freezing at -25°C was done and stored for 6 months. As per Jaworska et al. (2011), mushrooms, when frozen and stored for 12 months, showed decreased amino acids levels. Studies done by Czapski and Szudyga (2000) found that blanching before freezing leads to lower mushroom whiteness and brightness value.

Freezing requires several technological approaches to select the proper freezing techniques and understand the change in the mushroom's physiochemical properties. Some basic calculations that need to be done to evaluate the freezing process are water holding capacity, thawing loss, and moisture sorption.

Water holding capacity is the measure of mushroom ability to hold water of the frozen mushroom, as, during freezing, the moisture tends to leave the mushroom and freeze at the surface, which decreases the water content (Li et al., 2018) (Equation 3.5).

$$Water\ holding\ capacity\ (\%) = \left[\frac{w_1 - (w_3 - w_2)}{w_1} * 100 \right] \tag{3.5}$$

W_1 is the mushroom's initial weight before freezing, W_2 is the weight of the mushroom after thawing and dewatering, and W_3 is the weight after removingfrozen surface water with the help of a centrifuge.

Thawing loss is the amount of weight loss in the form of water that extrudes from the mushroom cell during thawing. The various freezing methods have different thawing loss, dependent upon the size of the ice crystal formed (Li et al., 2018) (Equation 3.6).

$$Thawing\ loss\ (\%) = \frac{w_1 - w_2}{w_1} * 100 \tag{3.6}$$

To understand the basic freezing process, change in water activity with temperature change is also critical. To understand this relation, a water sorption isotherm is essential to consider the BET and GAB model and are two widely accepted equations to quantify the relation.

The BET equation is (Brunauer et al., 1938)

$$M_w = \frac{M_b B a_w}{(1 - a_w)[1 + (B - 1)a_w]} \tag{3.7}$$

Equation 3.7 involves BET monolayer moisture content (g/g dry solid) (M_b), constant related to heat sorption (B), and water activity (a_w). The BET isotherm is appropriate for a range of water activity of 0.5 and 0.45.

Another model GAB equation also has been widely used for food (Anderson, 1946; Guizani et al., 2013):

$$M_w = \frac{M_g C K a_w}{(1 - K a_w)[(1 - K a_w)(1 - K a_w + C K a_w)]} \tag{3.8}$$

Equation 3.8 involves GAB monolayer moisture content (g/g dry solid) (M_g), constant related to monolayer heat of sorption (C), and a factor related to the heat of sorption of multilayer water (K).

3.8 CANNING

Canning is another preservation process that includes packaging of mushrooms into tin or aluminium cans followed by sterilization of the same, eventually partially cooking the mushroom. Mushroom canning is a relatively cheap process with a remarkable shelf life of about more than 2 years (Martínez-Carrera et al., 1998). Canning includes two different processes that differ in their filling. The canning can be done with a brine solution as filler or can be pickled using various vegetables, either partially cooked or uncooked (Martínez-Carrera et al., 1998; Pogoń et al., 2017; Pursito et al., 2020; Nketia et al., 2020). The basic process of canning is depicted in Figure 3.4.

```
┌──────────────┐    ┌──────────┐    ┌──────────┐    ┌──────────┐
│ Fresh wild   │───▶│ Sorting  │───▶│ Grading  │───▶│ Cleaning │
│ mushroom     │    │          │    │          │    │          │
└──────────────┘    └──────────┘    └──────────┘    └──────────┘
        ▼
┌──────────────┐    ┌──────────┐    ┌──────────┐    ┌──────────────────┐
│ Washing      │───▶│ Blanching│───▶│ Sorting  │───▶│ Filling mushroom │
│              │    │          │    │          │    │ (Glass/Can)      │
└──────────────┘    └──────────┘    └──────────┘    └──────────────────┘
        ▼
┌──────────────────┐ ┌──────────┐    ┌──────────┐    ┌──────────┐
│ Filling media    │▶│ Sealing  │───▶│ Retorting│───▶│ Cooling  │
│ (Brine/Vegetables)│ │          │    │          │    │          │
└──────────────────┘ └──────────┘    └──────────┘    └──────────┘
        ▼
┌──────────────┐
│ Storage      │
│              │
└──────────────┘
```

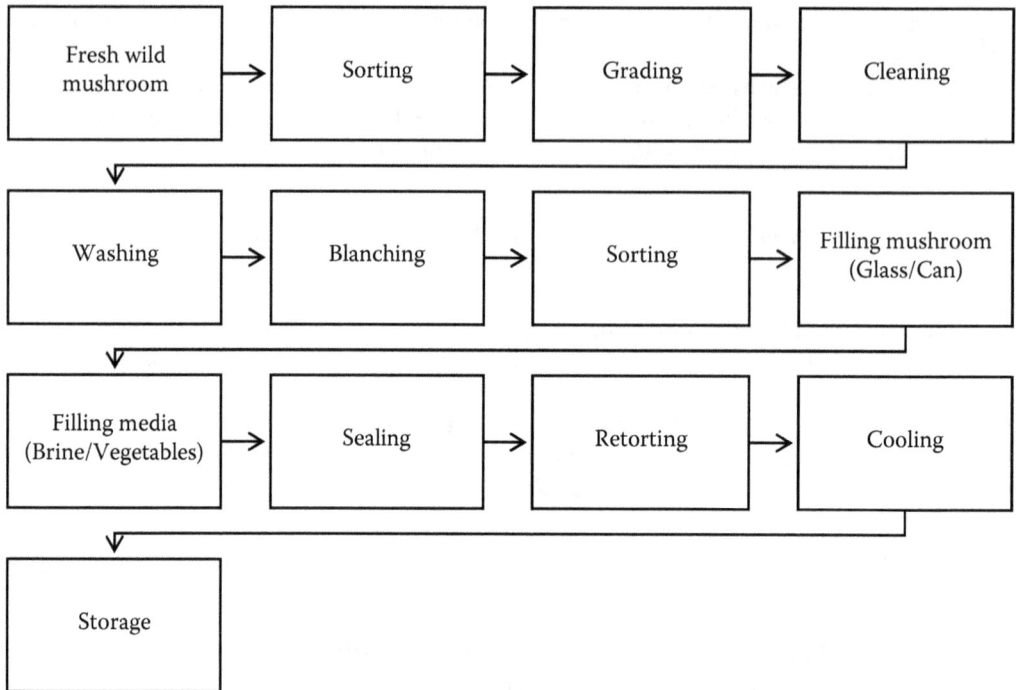

Figure 3.4   Canning (Pursito et al., 2020; Martínez-Carrera et al., 1998; Soliman et al., 2017).

As shown in Figure 3.4, canning depicts the filling of a simple brine solution or vegetable that is partially cooked or uncooked (Pursito et al., 2020). The canning is presumably done in both can and glass bottles. A glass bottle is a more traditional way of canning the mushroom, including vegetable filler composed of carrots, onions, peppers, potatoes, and certain spices such as pepper, allspice, mustard, and bay leaf. The mushroom is either cooked with these vegetables or pickled uncooked (Martínez-Carrera et al., 1998; Pogoń et al., 2017; Soliman et al., 2017).

Mushroom canning is being performed with various filler liquids that impart various properties or tastes to the canned mushroom. Primary filler liquids are listed below:

- Salt brine solution is the most common and widely used solution for canning with a variety of compositions, most commonly 2% brine solution (Nketia et al., 2020), mild brine solution (3.5% salt), and strong brine solution (4.5% salt) (Pogoń et al., 2017).
- Acetic acid brine solution is the second most widely used solution for canning and prominently used in pickling. The concentration of 0.7% (mild) and 1.5% (strong) is used for the brine purpose (Pogoń et al., 2017).
- Vegetable mixture blend containing 2% brine (salt), carrot juice (maintained at 5 pH), and tomato juice are mixed in defined proportions and added with citric acid, sodium benzoate, and ascorbic acid before boiling and then poured in as filler (Soliman et al., 2017).

The canning is mainly performed as either steam-operated or direct steam generated with continuous or batch line process depending upon production size. The retort is mainly operated at 15 psi for 30 minutes correlating to 121°C required to destroypathogenic microbes. The retort eventually partially cooks the mushroom and the vegetable if filled (Pursito et al., 2020; Nketia et al., 2020). The canned mushroom can also be pasteurized depending upon shelf life's requirement at 95°C–100°C for 15–30 minutes (Pogoń et al., 2017; Soliman et al., 2017).

Canning sterility is a term for eradicating microbes essential for the stated longer shelf life and can be obtained as F_0. The F_0 value is calculated using the general method, which expresses the heat adequacy required for the sterilization process equalized to the heating time required at 121°C to inactivate *C. botulinum* spores with T as the temperature at any given time and T_{ref} as reference processing temperature and z-value as 10°C (Pursito et al., 2020) (Equation 3.9):

$$F_0 = \int_0^t 10^{\frac{T - T_{ref}}{z}} dt \qquad (3.9)$$

3.9 PICKLING

Pickling is a traditional way of preservation of wild mushrooms. Pickling involves fermenting the mushroom into acetic acid or salt brine solution to enhance the mushroom's organoleptic properties and extend the shelf life (Jabłońska-Ryś et al., 2019). Mushrooms are preserved more traditionally with added vegetables that ferment along the mushroom, giving a wide variety of texture and note to the final product (Temesgen & Workneh, 2015). Pickling of mushrooms involves several steps, as shown in Figure 3.5, depending on the final product's requirement.

Pickling starts with sorting and grading wild mushrooms for optimum size and grade and removing unwanted materials. Sorting mushroom is followed by cleaning and washing with clean tap water (Khaskheli et al., 2015; Singh et al., 2018). Blanching is done with KMS solution to remove any extra microbial load from the mushroom due to wild mushrooms being collected from woods (Khaskheli et al., 2015; Singh et al., 2016). The fungus can be subjected to drying dependent upon the final product requirement with any of the previous drying methods discussed (Temesgen & Workneh, 2015). After

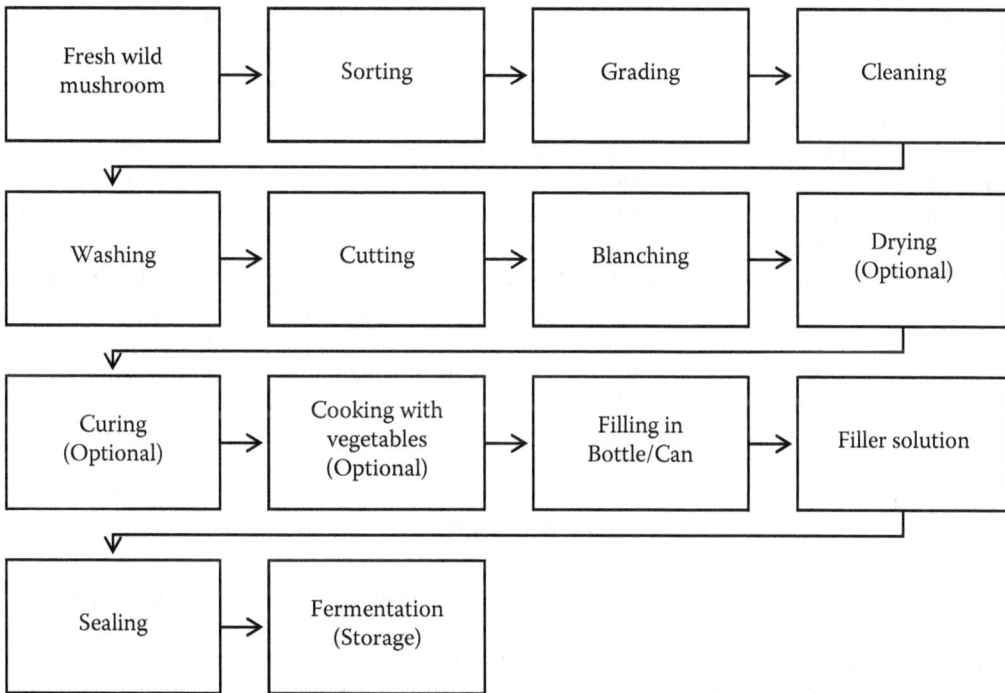

Figure 3.5 Pickling.

drying, rehydration is essential to increase the fungus's water activity for the fermentation to occur, carried in warm water for 2–5 minutes (Singh et al., 2016). Curing adds spices onto the mushroom to remove excess moisture and incorporates specific taste parameters (Khaskheli et al., 2015). Martin-Belloso and Llanos-Barriobero (2001) reported that blanching before drying might cause mushroom darkening during drying, so blanching can be optional if drying occurs. Same as curing, marination is also a process of coating the mushroom with spices intended for the pickling without the aim of removal of water using salt, followed by the cooking of mushroom with the vegetable until a soft texture is obtained (Khaskheli et al., 2015; Temesgen & Workneh, 2015; Singh et al., 2018). The cooked mushroom is poured into the bottles/cans and topped with the filler solution. A variety of different filler solutions are used as fermentation media for the pickling process, such as white vinegar (Temesgen & Workneh, 2015), vinegar and oil (Khaskheli et al., 2015), spirit vinegar of 10% acidity (Drewnowska et al., 2017), and oil (Sharma, 2018). The mushroom is kept at a low temperature in the range of 10°C–18°C for 3 weeks to a month, dependent upon the fungus requirement's final characteristics (Jabłońska-Ryś et al., 2019).

Pickling is a traditional method of preservation that has various physiochemical changes during pickling. The significant mushroom component is a carbohydrate followed by a protein that is generally used for lactic acid fermentation (Guillamón et al., 2010). Jabłońska-Ryś et al. (2019) reported a decrease in fructose and mannitol content upon fermentation for 12 days at 12°C and an increase in tryptamine and histamine content depicted the deterioration after a threshold level.

3.10 IRRADIATION

Irradiation is a relatively advanced type of preservation method involving the fungus to be subjected to radiation, either UV (Sławińska et al., 2016), gamma (Fernandes et al., 2013b), or e-beam irradiator (Fernandes et al., 2015).

For the UV type of irradiation system, the mushrooms are subjected to UV beams for 30 minutes on to the grill side facing the UVB, or the fungi are sliced for more even radiation, followed by either drying of mushrooms or directly packaging in a plastic bag and stored at low temperature (Sławińska et al., 2016).

Gamma radiation involves radiation produced by Co-60 in a radiation chamber into which the mushrooms are placed for the irradiation process (Fernandes et al., 2012; Fernandes et al., 2013a). Radiation of 1–3 kGy is recommended for fresh mushroom shelf life extension and 10–50 kGy for dried mushroom seasoning as per the FDA (Fernandes et al., 2014a).

e-beam irradiator is done on fungi with the help of 10 MeV of energy with a pulse duration of 5.5 µs at a pulse frequency of 440 Hz and an average current of 1.1 mA. The dosage of irradiation absorbed depends on the 20% uncertainty of dosing (Fernandes et al., 2014b, 2015).

Gamma radiation is considered the safest option for irradiation with several benefits, such as diminishing enzymatic browning. UV beam also helped decrease the microbial load for 21 days of shelf life, and e-beam radiation reduces the aerobic and psychotropic population (Zhang et al., 2018).

3.11 PACKAGING

Packaging is often considered the final step of all the processing techniques, i.e. storing processed products in a container or plastic bag. In modern times, various packaging methods and materials have in existence to extend shelf life. PLA (poly(L-lactic acid) films are used for the packaging of mushrooms, which is a biodegradable plasticized film of PLA enhancing the texture, sensory attributes, and diminishing PPO effect; on the other hand, it has an effect on weight loss and TSS, which is higher compared to processed mushrooms (Han et al., 2015). Bilayer active packaging with modified atmosphere packaging has a drastic effect on extending the mushroom's shelf life up to 11 days depending on the mushroom's coating. Gelatin coating also played a crucial role in maintaining the final product's quality and physical attributes (Zhang et al., 2018; Lyn et al., 2020).

3.12 OTHER PROCESSING TECHNIQUES

3.12.1 Pulsed electric Field and Ultrasound

Pulsed electric field and ultrasound are two specific methods of preservation by inactivation of microbes with the help of short pulses of electricity, creating transient pores in microbes and sound pulse for creation of a cavity by the generation of high energy shock waves and intensive shear force, respectively (Roselló-Soto et al., 2018; Zhang et al., 2018).

3.12.2 Ozone

Ozone treatment is relatively safe compared to other water treatments due to the treatment media being a gas. Ozone is a potent antimicrobial agent that helps extend its shelf life due to its strong oxidative property. However, being gas, it is unlikely to have any effect on food matrics. Exposure to ozone for 60 minutes at different concentrations has shown a 2–4 log of microbial loads reduction in mushrooms (Prabha et al., 2015; Zhang et al., 2018).

3.12.3 Electrolyzed Water

Electrolyzed water is a disinfecting mushroom technique using a salt solution electrolyzed using electricity to produce free chlorine, which forms hypochlorous acid, a strong oxidative-reduction potential and combined effect help to disinfect mushrooms. The U.S. FDA approves the process with the limitation of a maximum of 200 ppm of free available chlorine (Lee et al., 2014; Zhang et al., 2018).

3.13 CONCLUSION

Mushrooms being procured in the wild have various applications, such as food, medicinal, and therapeutic uses. A mushroom is rich in protein and several other phytochemicals that have a specific purpose. Wild mushrooms have been in use for ages for all sorts of purposes, where wild farming was the majority source of food. In today's context, the wild mushroom can also be harvested, as discussed in this chapter, to have a continuous supply throughout the year as wild mushrooms are highly dependent upon seasonal changes. Commercial cultivation has improved with various cultivation techniques available depending upon the growing condition of the specific mushroom. As discussed, a mushroom has a high moisture content due to which it is prone to microbial and chemical deterioration shortly after the harvest. Due to the perishable mushroom nature, various processing techniques are essential for mushroom availability throughout the year. Various processes standardized to different properties of mushrooms depend on the final requirement of the product. As discussed in the chapter, drying is the most common and widely utilized processing technique due to availability and ease. Some specific techniques such as freeze drying, ozone treatment, sterilization, and electrolytic water are more advanced in nature with promising results but, due to the operations' complexity and cost, these processes are not widely used. Canning and pickling are the more traditional approaches toward the processing techniques with more traditional touches and tailored changes according to the marketplace.

REFERENCES

Adjapong, A. O., Ansah, K. D., Angfaarabung, F., & Sintim, H. O. (2015). Maize residue as a viable substrate for farm-scale cultivation of oyster mushroom (Pleurotusostreatus). *Advances in Agriculture*, *2015*, 1–6.

Anderson, R. B. (1946). Modifications of the Brunauer, Emmett and Teller equation1. *Journal of the American Chemical Society*, *68*(4), 686–691.

Andrés, A., Bilbao, C., & Fito, P. (2004). Drying kinetics of apple cylinders under combined hot air–microwave dehydration. *Journal of Food Engineering*, *63*(1), 71–78.

Assan, N., & Mpofu, T. (2014). The influence of substrate on mushroom productivity. *Scientific Journal of Crop Science*, *3*(7), 86–91.

Baby, S., Johnson, A. J., & Govindan, B. (2015). Secondary metabolites from Ganoderma. *Phytochemistry*, *114*, 66–101.

Bains, A., & Chawla, P. (2020). In vitro bioactivity, antimicrobial and anti-inflammatory efficacy of modified solvent evaporation assisted Trametes versicolor extract. *3 Biotech*, *10*(9), 1–11.

Balan, N., & Radhakrishnan, M. (2014). Microwave drying of edible mushroom. *Mushroom Research*, *23*, 81–87.

Barman, B., Warjri, S., Lynrah, K. G., Phukan, P., & Mitchell, S. T. (2018). Amanita nephrotoxic syndrome: Presumptive first case report on the Indian subcontinent. *Indian Journal of Nephrology*, *28*(2), 170.

Baysal, E., Peker, H., Yalinkiliç, M. K., & Temiz, A. (2003). Cultivation of oyster mushroom on waste paper with some added supplementary materials. *Bioresource Technology*, *89*(1), 95–97.

Bernaś, E., Jaworska, G., & Lisiewska, Z. (2006). Edible mushrooms as a source of valuable nutritive constituents. *Acta Scientiarum Polonorum Technologia Alimentaria*, *5*(1), 5–20.

Borthakur, M., & Joshi, S. R. (2019). Wild mushrooms as functional foods: The significance of inherent perilous metabolites. In: J. Singh, & P. Gehlot (Eds.). *New and Future Developments in Microbial Biotechnology and Bioengineering*. Elsevier, pp. 1–12.

Brunauer, S., Emmett, P. H., & Teller, E. (1938). Adsorption of gases in multimolecular layers. *Journal of the American Chemical Society*, *60*(2), 309–319.

Cai, Q., Cui, Y. Y., & Yang, Z. L. (2016). Lethal amanita species in China. *Mycologia*, *108*(5), 993–1009.

Carvalho, C. S. M. D., Aguiar, L. V. B. D., Sales-Campos, C., Minhoni, M. T. D. A., & Andrade, M. C. N. D. (2012). Applicability of the use of waste from different banana cultivars for the cultivation of the oyster mushroom. *Brazilian Journal of Microbiology*, *43*(2), 819–826.

Coşkuner, Y., & Özdemir, Y. (2000). Acid and EDTA blanching effects on the essential element content of mushrooms (Agaricus bisporus). *Journal of the Science of Food and Agriculture*, *80*(14), 2074–2076.

Czapski, J. (2002). Mushrooms quality as affected by washing and storage conditions. *Vegetable Crops Research Bulletin*, *56*, 722–725.

Czapski, J., & Szudyga, K. (2000). Frozen mushrooms quality as affected by strain, flush, treatment before freezing, and time of storage. *Journal of Food Science*, *65*(4), 722–725.

Deepalakshmi, K., & Sankaran, M. (2014). Pleurotusostreatus: An oyster mushroom with nutritional and medicinal properties. *Journal of Biochemical Technology*, *5*(2), 718–726.

Dg. (2013). RR mushroom sales & marketing, Types of edible mushroom in India & their medicinal facts [cited 2013 August 18]: Available from: https://mushroomsales.wordpress.com/2013/08/18/types-ofedible-mushroom-in-india-there-medicinal-facts/

Drewnowska, M., Hanć, A., Barałkiewicz, D., & Falandysz, J. (2017). Pickling of chanterelle Cantharellus cibarius mushrooms highly reduce cadmium contamination. *Environmental Science and Pollution Research*, *24*(27), 21733–21738.

Drouzas, A. E., Tsami, E., & Saravacos, G. D. (1999). Microwave/vacuum drying of model fruit gels. *Journal of Food Engineering*, *39*(2), 117–122.

Fernandes, Â., Antonio, A. L., Barreira, J. C., Oliveira, M. B. P., Martins, A., & Ferreira, I. C. (2012). Effects of gamma irradiation on physical parameters of Lactariusdeliciosus wild edible mushrooms. *Postharvest Biology and Technology*, *74*, 79–84.

Fernandes, Â., Barreira, J. C., Antonio, A. L., Oliveira, M. B. P., Martins, A., & Ferreira, I. C. (2014a). Effects of gamma irradiation on chemical composition and antioxidant potential of processed samples of the wild mushroom Macrolepiota procera. *Food Chemistry*, *149*, 91–98.

Fernandes, Â., Barreira, J. C., Antonio, A. L., Oliveira, M. B. P., Martins, A., & Ferreira, I. C. (2014b). Feasibility of electron-beam irradiation to preserve wild dried mushrooms: Effects on chemical composition and antioxidant activity. *Innovative Food Science & Emerging Technologies, 22*, 158–166.

Fernandes, Â., Barreira, J. C., Antonio, A. L., Rafalski, A., Oliveira, M. B. P., Martins, A., & Ferreira, I. C. (2015). How does electron beam irradiation dose affect the chemical and antioxidant profiles of wild dried Amanita mushrooms?. *Food Chemistry, 182*, 309–315.

Fernandes, Â., Barreira, J. C., Antonio, A. L., Santos, P. M., Martins, A., Oliveira, M. B. P., & Ferreira, I. C. (2013a). Study of chemical changes and antioxidant activity variation induced by gamma-irradiation on wild mushrooms: Comparative study through principal component analysis. *Food Research International, 54*(1), 18–25.

Fernandes, Â., Barros, L., Barreira, J. C., Antonio, A. L., Oliveira, M. B. P., Martins, A., & Ferreira, I. C. (2013b). Effects of different processing technologies on chemical and antioxidant parameters of Macrolepiota procera wild mushroom. *LWT-Food Science and Technology, 54*(2), 493–499.

Geankoplis, C. J. (2003). Transport processes and separation process principles (includes unit operations). *Prentice Hall Professional Technical Reference.* Third Edition. 381–425.

Govorushko, S., Rezaee, R., Dumanov, J., & Tsatsakis, A. (2019). Poisoning associated with the use of mushrooms: A review of the global pattern and main characteristics. *Food and Chemical Toxicology, 128*, 267–279.

Guillamón, E., García-Lafuente, A., Lozano, M., Rostagno, M. A., Villares, A., & Martínez, J. A. (2010). Edible mushrooms: Role in the prevention of cardiovascular diseases. *Fitoterapia, 81*(7), 715–723.

Guizani, N., Rahman, M. S., Klibi, M., Al-Rawahi, A., & Bornaz, S. (2013). Thermal characteristics of Agaricus bisporus mushroom: Freezing point, glass transition, and maximal-freeze-concentration condition. *International Food Research Journal, 20*(4), 1945–1952.

Gunasekaran, S. (1999). Pulsed microwave-vacuum drying of food materials. *Drying Technology, 17*(3), 395–412.

Han, L., Qin, Y., Liu, D., Chen, H., Li, H., & Yuan, M. (2015). Evaluation of biodegradable film packaging to improve the shelf-life of Boletus edulis wild edible mushrooms. *Innovative Food Science & Emerging Technologies, 29*, 288–294.

Hsieh, H. M., Chung, M. C., Chen, P. Y., Hsu, F. M., Liao, W. W., Sung, A. N., …, & Ju, Y. M. (2017). A termite symbiotic mushroom maximizing sexual activity at growing tips of vegetative hyphae. *Botanical Studies, 58*(1), 1–14.

Jabłońska-Ryś, E., Skrzypczak, K., Sławińska, A., Radzki, W., & Gustaw, W. (2019). Lactic acid fermentation of edible mushrooms: Tradition, technology, current state of research: A Review. *Comprehensive Reviews in Food Science and Food Safety, 18*(3), 655–669.

Jaworska, G., Bernaś, E., & Mickowska, B. (2011). Effect of production process on the amino acid content of frozen and canned Pleurotusostreatus mushrooms. *Food Chemistry, 125*(3), 936–943.

Jeznabadi, E. K., Jafarpour, M., & Eghbalsaied, S. (2016). King oyster mushroom production using various sources of agricultural wastes in Iran. *International Journal of Recycling of Organic Waste in Agriculture, 5*(1), 17–24.

Kamthan, R., & Tiwari, I. (2017). Agricultural wastes-potential substrates for mushroom cultivation. *European Journal of Experimental Biology, 7*(5), 31.

Karun, N. C., & Sridhar, K. R. (2017). Edible wild mushrooms of the Western Ghats: Data on the ethnic knowledge. *Data in Brief, 14*, 320–328.

Khaskheli, S. G., Zheng, W., Sheikh, S. A., Khaskheli, A. A., Liu, Y., Wang, Y., & Huang, W. (2015). Effect of processing techniques on the quality and acceptability of Auricularia auricula mushroom pickle. *Journal of Food and Nutrition Research, 3*(1), 46–51.

Kic, P. (2018). Mushroom drying characteristics and changes of colour. In *17th International Scientific Conference Engineering for Rural Development.* Latvia University of Agriculture, Jelgava (pp. 432–438).

Kosentka, P., Sprague, S. L., Ryberg, M., Gartz, J., May, A. L., Campagna, S. R., & Matheny, P. B. (2013). Evolution of the toxins muscarine and psilocybin in a family of mushroom-forming fungi. *PloS One, 8*(5), e64646.

Krüzselyi, D., Móricz, Á. M., & Vetter, J. (2020). Comparison of different morphological mushroom parts based on the antioxidant activity. *LWT-Food Science and Technology, 127*, 109436.

Lee, N. Y., Kim, N. H., Jang, I. S., Jang, S. H., Lee, S. H., Hwang, I. G., & Rhee, M. S. (2014). Decontamination efficacy of neutral electrolyzed water to eliminate indigenous flora on a large-scale of cabbage and carrot both in the laboratory and on a real processing line. *Food Research International*, *64*, 234–240.

Lee, K. H., Morris-Natschke, S. L., Yang, X., Huang, R., Zhou, T., Wu, S. F., …, & Itokawa, H. (2012). Recent progress of research on medicinal mushrooms, foods, and other herbal products used in traditional Chinese medicine. *Journal of Traditional and Complementary Medicine*, *2*(2), 1–12.

Li, T., Lee, J. W., Luo, L., Kim, J., & Moon, B. (2018). Evaluation of the effects of different freezing and thawing methods on the quality preservation of Pleurotuseryngii. *Applied Biological Chemistry*, *61*(3), 257–265.

Liapis, A. I., & Bruttini, R. (1987). Freeze drying. *Handbook of Industrial Drying*, *2*, 309–343.

Liu, Y., Huang, F., Yang, H., Ibrahim, S. A., Wang, Y. F., & Huang, W. (2014). Effects of preservation methods on amino acids and 5′-nucleotides of Agaricus bisporus mushrooms. *Food Chemistry*, *149*, 221–225.

Liu, Y., Xie, X. X., Ibrahim, S. A., Khaskheli, S. G., Yang, H., Wang, Y. F., & Huang, W. (2016). Characterization of Lactobacillus pentosus as a starter culture for the fermentation of edible oyster mushrooms (Pleurotus spp.). *LWT-Food Science and Technology*, *68*, 21–26.

Llarena-Hernández, R. C., Largeteau, M. L., Farnet, A. M., Foulongne-Oriol, M., Ferrer, N., Regnault-Roger, C., & Savoie, J. M. (2013). Potential of European wild strains of Agaricus subrufescens for productivity and quality on wheat straw based compost. *World Journal of Microbiology and Biotechnology*, *29*(7), 1243–1253.

Lyn, F. H., Adilah, Z. M., Nor-Khaizura, M. A. R., Jamilah, B., & Hanani, Z. N. (2020). Application of modified atmosphere and active packaging for oyster mushroom (Pleurotusostreatus). *Food Packaging and Shelf Life*, *23*, 100451.

Ma, G., Yang, W., Zhao, L., Pei, F., Fang, D., & Hu, Q. (2018). A critical review on the health promoting effects of mushrooms nutraceuticals. *Food Science and Human Wellness*, *7*(2), 125–133.

Mahari, W. A. W., Peng, W., Nam, W. L., Yang, H., Lee, X. Y., Lee, Y. K., …, & Lam, S. S. (2020). A review on valorization of oyster mushroom and waste generated in the mushroom cultivation industry. *Journal of Hazardous Materials*, *400*, 123156.

Martin-Belloso, O., & Llanos-Barriobero, E. (2001). Proximate composition, minerals and vitamins in selected canned vegetables. *European Food Research and Technology*, *212*(2), 182–187.

Martínez-Carrera, D., Sobal, M., Aguilar, A., Navarro, M., Bonilla, M., & Larqué-Saavedra, A. (1998). Canning technology as an alternative for management and conservation of wild edible mushrooms in Mexico. *Micol. Neotrop. Apl*, *11*, 35–51.

Maskan, M. (2000). Microwave/air and microwave finish drying of banana. *Journal of Food Engineering*, *44*(2), 71–78.

Mohan, D. (2009). Cultivate mushrooms. Dathu Mohan, Mushroom cultivating methods, Technology for Mushroom cultivation [cited 2009 April 21]: Available from: http://cultivatemushrooms.blogspot.com/2009/04/technologyfor-mushroom-cultivation.html

Morel, S., Arnould, S., Vitou, M., Boudard, F., Guzman, C., Poucheret, P., …, & Rapior, S. (2018). Antiproliferative and antioxidant activities of wild Boletales mushrooms from France. *International Journal of Medicinal Mushrooms*, *20*(1), 13–19.

Muszyńska, B., Kała, K., Lazur, J., & Włodarczyk, A. (2020). Imleriabadia culinary-medicinal mushroom with interesting biological properties. *Food Bioscience*, *37*, 100663.

Mshandete, A. M., Raymond, P., & Kivaisi, A. K. (2013). Cultivation of oyster mushroom (Pleurotus HK-37) on solid sisal waste fractions supplemented with cow dung manure. Journal of Biology and Life Science, *4*(1), 273–286.

Muyanja, C., Kyambadde, D., & Namugumya, B. (2014). Effect of pretreatments and drying methods on chemical composition and sensory evaluation of oyster mushroom (Pluerotus Oestreatus) Powder and Soup. *Journal of Food Processing and Preservation*, *38*(1), 457–465.

Narh, D. L., Obodai, M., Baka, D., & Dzomeku, M. (2011). The efficacy of sorghum and millet grains in spawn production and carpophore formation of Pleurotusostreatus (Jacq. Ex. Fr) Kummer. *International Food Research Journal*, *18*(3), 1092–1097.

Ngai, P. H., & Ng, T. B. (2003). Lentin, a novel and potent antifungal protein from shitake mushroom with inhibitory effects on activity of human immunodeficiency virus-1 reverse transcriptase and proliferation of leukemia cells. *Life Sciences*, *73*(26), 3363–3374.

Nketia, S., Buckman, E. S., Dzomeku, M., &Akonor, P. T. (2020). Effect of processing and storage on physical and texture qualities of oyster mushrooms canned in different media. *Scientific African*, *9*, e00501.

Oghenekaro, A. O., Daniel, G., & Asiegbu, F. O. (2015). The saprotrophic wood-degrading abilities of Rigidoporus microporus. *Silva Fennica*, *49*(4), 1–10.

Ogidi, C. O., Oyetayo, V. O., & Akinyele, B. J. (2020). Wild medicinal mushrooms: Potential applications in phytomedicine and functional foods. In: A. K. Passari , & S. Sánchez (Eds.). *An Introduction to Mushroom*. IntechOpen.

Pei, F., Shi, Y., Gao, X., Wu, F., Mariga, A. M., Yang, W., ..., & Hu, Q. (2014). Changes in non-volatile taste components of button mushroom (Agaricus bisporus) during different stages of freeze drying and freeze drying combined with microwave vacuum drying. *Food Chemistry*, *165*, 547–554.

Piskov, S., Timchenko, L., Grimm, W. D., Rzhepakovsky, I., Avanesyan, S., Sizonenko, M., & Kurchenko, V. (2020). Effects of various drying methods on some physico-chemical properties and the antioxidant profile and ACE inhibition activity of oyster mushrooms (Pleurotus Ostreatus). *Foods*, *9*(2), 160.

Pogoń, K., Gabor, A., Jaworska, G., & Bernaś, E. (2017). Effect of traditional canning in acetic brine on the antioxidants and vitamins in Boletus edulis and Suillus luteus mushrooms. *Journal of Food Processing and Preservation*, *41*(2), e12826.

Prabha, V., Barma, R. D., Singh, R., & Madan, A. (2015). Ozone technology in food processing: A review. *Trends in Biosciences*, *8*(16), 4031–4047.

Pursito, D. J., Purnomo, E. H., Fardiaz, D., & Hariyadi, P. (2020). Optimizing steam consumption of mushroom canning process by selecting higher temperatures and shorter time of retorting. *International Journal of Food Science*, *2020*, 1–8.

Rathore, H., Prasad, S., & Sharma, S. (2017). Mushroom nutraceuticals for improved nutrition and better human health: A review. *Pharma Nutrition*, *5*(2), 35–46.

Reid, T., Munyanyi, M., & Mduluza, T. (2017). Effect of cooking and preservation on nutritional and phytochemical composition of the mushroom Amanita zambiana. *Food Science & Nutrition*, *5*(3), 538–544.

Roselló-Soto, E., Poojary, M. M., Barba, F. J., Koubaa, M., Lorenzo, J. M., Mañes, J., & Moltó, J. C. (2018). Thermal and non-thermal preservation techniques of tiger nuts' beverage "horchata de chufa". Implications for Food Safety, Nutritional and Quality Properties. *Food Research International*, *105*, 945–951.

Royse, D. J., & Beelman, R. B. (2016). Pennstate extension, six steps to mushroom farming [Updated 2016 June 27]: Available from: https://extension.psu.edu/six-steps-to-mushroomfarming

Ruan-Soto, F., Ordaz-Velázquez, M., García-Santiago, W., & Pérez-Ovando, C. (2017). Traditional processing and preservation of wild edible mushrooms in Mexico. *Annals of Food Processing and Preservation*, *2*(1), 1013.

Sapers, G. M., Miller, R. L., Choi, S. W., & Cooke, P. H. (1999). Structure and composition of mushrooms as affected by hydrogen peroxide wash. *Journal of Food Science*, *64*(5), 889–892.

Schiffmann, R. F. (1995). Microwave and dielectric drying. *Handbook of Industrial Drying*, *1*, 345–372.

Sharma, K. (2018). Mushroom: Cultivation and processing. *International Journal of Food Processing Technology*, *5*(2), 9–12.

Singh, J., Sindhu, S. C., & Sindhu, A. (2016). Development and evaluation of value added pickle from dehydrated shiitake (Lentinus edodes) mushroom. *International Journal of Food Science and Nutrition*, *1*(1), 24–26.

Singh, M. P., Sodhi, H. S., & Ranote, P. S. (2018). Optimization and storage studies of A. bisporus vinegar-oil pickle to utilize stipe as a value added product. *International Journal of Current Microbiology and Applied Sciences*, *7*(12), 1676–1689.

Sławińska, A., Fornal, E., Radzki, W., Skrzypczak, K., Zalewska-Korona, M., Michalak-Majewska, M.,..., & Stachniuk, A. (2016). Study on vitamin D2 stability in dried mushrooms during drying and storage. *Food Chemistry*, *199*, 203–209.

Soković, M., Glamočlija, J., Ćirić, A., Petrović, J., & Stojković, D. (2018). Mushrooms as sources of therapeutic foods. In: A. Grumezescu , & A. M. Holban (Eds.). *Therapeutic Foods*. Academic Press, pp. 141–178.

Soliman, A., Abbas, M., & Ahmed, S. (2017). Preparation, canning and evaluation process of vegetable mixture diets (ready-to-eat) supplemented with mushroom. *Suez Canal University Journal of Food Sciences*, 4(1), 19–28.

Soltaninejad, K. (2018). Outbreak of mushroom poisoning in Iran: April–May, 2018. *The International Journal of Occupational and Environmental Medicine*, 9(3), 152.

Sun, D. W. (Ed.). (2012). *Thermal Food Processing: New Technologies and Quality Issues*. Crc Press.

Tarafdar, A., Shahi, N. C., Singh, A., & Sirohi, R. (2017). Optimization of freeze-drying process parameters for qualitative evaluation of button mushroom (Agaricus bisporus) using response surface methodology. *Journal of Food Quality, 2017*.

Teke, A. N., Kinge, T. R., Bechem, E. E. T., Ndam, L. M., & Mih, A. M. (2019). Mushroom species richness, distribution and substrate specificity in the Kilum-Ijim forest reserve of Cameroon. *Journal of Applied Biosciences*, 133, 13592–13617.

Temesgen, M., & Workneh, T. S. (2015). Effect of osmotic and pickling pretreatments on nutritional quality and acceptance of traditional fermented oyster mushrooms. *Food Science and Quality Management*, 37, 64–73.

Varma, A., Gaur, K. J., & Bhatia, P. (2011). Mushrooms and poisoning. *Journal of the Indian Medical Association*, 109(11), 826–828.

Wasser, S. P. (2010). Medicinal mushroom science: History, current status, future trends, and unsolved problems. *International Journal of Medicinal Mushrooms*, 12(1), 1–16.

Wendiro, D., Wacoo, A. P., & Wise, G. (2019). Identifying indigenous practices for cultivation of wild saprophytic mushroom s: Responding to the need for sustainable utilization of natural resources. *Journal of Ethnobiology and Ethnomedicine*, 15(1), 1–15.

Wu, J. Y., Siu, K. C., & Geng, P. (2021). Bioactive ingredients and medicinal values of grifolafrondosa (Maitake). *Foods*, 10(1), 95.

Yuen, S. K., Kalianon, K., & Atong, M. (2014). Effect of different drying temperatures on the nutritional quality of edible wild mushroom, Volvariellavolvacea obtained nearby forest areas. *International Journal of Advanced Research*, 2(5), 859–864.

Zhang, K., Pu, Y. Y., & Sun, D. W. (2018). Recent advances in quality preservation of post-harvest mushrooms (Agaricus bisporus): A review. *Trends in Food Science & Technology*, 78, 72–82.

Zhang, Y., Geng, W., Shen, Y., Wang, Y., & Dai, Y. C. (2014). Edible mushroom cultivation for food security and rural development in China: Bio-innovation, technological dissemination and marketing. *Sustainability*, 6(5), 2961–2973.

CHAPTER 4

Medicinal mushroom nutraceutical commercialization: Two sides of a coin

Jovana Vunduk[1,2]**, Daniel Tura**[3]**, and Alona Yu. Biketova**[4]

[1]Institute of General and Physical Chemistry, Belgrade, Serbia
[2]Ekofungi Ltd., Padinska Skela bb, Belgrade, Serbia
[3]National Institute of Research and Development for Biological Sciences, Institute for Biological Research Cluj-Napoca, Romania
[4]Jodrell Laboratory, Royal Botanic Gardens, Kew, UK

CONTENTS

4.1 INTRODUCTION

The elusive nature of macrofungi counteracts traditional classification of pharmaceuticals. They are not just regular food either. Mushrooms can be considered as food and medicine at the same time. In both cases, the traditional knowledge is extensive, goes several thousand years back

DOI: 10.1201/9781003152583-5

in China, and it is supported by ancient texts and cultivation techniques (Mengesha & Chaithanya, 2020). Similarly, the rest of the world has its mushroom-oriented traditions, *Psilocybe* species and their spiritual use in pre-Spanish South America, various species used as food and medicine in Africa, truffles as food and medicine in the Middle East, and delicacy food in ancient Greece and Rome, as well as Ötzie, a 5,000-year-old mummy discovered in Europe who carried in his bag two medicinal mushroom species (Trutmann, 2012; El Enshasy et al., 2013; Vunduk et al., 2015). Chinese, Japanese, Korean, and Slavic people utilized mushrooms as medicine for millennia (Moradali et al., 2007). However, the knowledge about different uses of edible and medicinal mushrooms seemed to succumb to the contemporary use of fast food and potent medications like antibiotics – the scenario strongly present in Western countries. Fungi were not just forgotten but even fell into the state of being afraid of, known as mycophobia – especially true when mushrooms are regarded as food. As reported by Brandenburg and Ward (2018), 7,428 cases of mushroom poisoning per year and around three fatalities occurred in the USA. Ironically, in the same country, 128,000 people get hospitalized and 3,000 die from foodborne illnesses each year (official records of Center for Disease Control; "Burden of Foodborne Illness: Overview", 2019). Recent scientific discoveries triggered a sort of renaissance for the kingdom of fungi, especially mushrooms. A search through the PubMed database showed a tremendous increase in the number of published research papers focused on mushroom's biological activity (from less than 50 papers per year published before 2000 to more than 100 starting from 2003 (https://pubmed.ncbi.nlm.nih.gov/?term=Mushrooms%20biological%20activity&timeline=expanded). A total of more than 50,000 research papers covering the topic of pharmacological activities of mushrooms were previously reported by Morris et al. (2016). With 130 confirmed medicinal activities and ever more appreciated nutritive value no wonder that mushrooms became a prized ingredient in the food industry, especially as a functional food or superfood (Morris et al., 2016; Ma et al., 2018; Badalyan et al., 2019). Inevitably, such a gem with dual scope market possibilities was spotted and entered the rapidly developing field of nutraceutical/dietary supplements.

It appeared as perfectly fitted for this industry due to several reasons: evidence of traditional use, broad scope biological activities, *in vitro* and *in vivo* confirmed effects as a basis for health claims, diverse species – about 300 (even 700 according to some authors), possibility to be cultivated using biotechnology, fit for industrial size scale-up as well as to be processed using novel technologies like nanotechnology, and no or scarce evidence of side effects (Lakhanpal & Monika, 2005; Jarić et al., 2011; Morris et al., 2016; Wasser, 2017; Vunduk et al., 2019; Klaus et al., 2020; Klaus et al., 2021).

But how is it possible that medicinal mushrooms are not considered pharmaceuticals? Is the pharma industry avoiding this potent source of biologically active compounds? The answers are not quite straightforward. Unlike synthetic medicines, the majority of mushrooms' active constituents are complex, diverse, and often presented as crude extracts – essentially, a combination of several active compounds (Klaus et al., 2015; Ma et al., 2018). That is why these products are called preparations and not drugs. The mentioned characteristics of mushrooms render them as unpractical raw material for pharma. Although macrofungi possess diverse pharmacologically active constituents, polysaccharides are often emphasized as key components as well as the most explored ones (Moradali et al., 2007). The main biological activity expressed by these macromolecules is immunomodulation or host defense potentiation (Lakhanpal & Monika, 2005; Zhang, Li, et al., 2017; Zhao et al., 2020). This makes medicinal mushrooms seen mainly as preventive, thus nutraceuticals. On the other hand, low molecular weight compounds attract pharma and researchers' attention due to their ability to surpass cell membrane and exhibit direct influence in diverse conditions, like metastasis (Petrova et al., 2006). The clinical trials and improvement of purification technology enabled drug-like production process and quality control, and the mushroom-based prescription drugs are in use in some countries like Japan (Wasser, 2017).

Although mushroom nutraceuticals were defined at the end of the 20th century as a mushroom sourced compound with dietetic supplement potential, used as preventive or disease treatment without side effects, the full-range expansion of the nutraceutical niche is yet expected. In the meantime, other more expanded definitions of mushroom nutraceuticals appeared, confirming the growing scientific and market interest in this type of product (Lakhanpal & Monika, 2005). From the commercialization perspective, mushrooms were traditionally sold as food, the price depending on the species, quality, local culture, and season. The increasing demand affected the rise in the exploration of cultivation techniques, genetic manipulation, and disease resistance, resulting in a significant increase in annual mushroom production worldwide (from 0.17 million tons in 1960 to 34.8 million tons in 2014) (Chakravarty, 2011; Singh et al., 2017; Shwet et al., 2019; Doroški et al., 2020). But the encounter with mortality (especially in the *baby boomer* generation) brought attention toward new commercialization possibilities – food should provide more than energy and nutrients (Childs, 2000; Hilton, 2017). It should serve as a sort of medicine or nutraceutical. Evidently, mushrooms fit into this category. However, the opportunity for mushroom-derived supplements' commercial expansion has its issues. The current chapter provides evidence of the state of different and, in some cases, contradictory aspects concerning mushroom nutraceutical formulations: consumer familiarity with the ingredients present in formulations, strategies for new product development, and clinically relevant dosage as well as *fairy dusting* (the practice of adding the active component in a dose too small to be effective). The aspects like drug delivery methods as well as raw material and final product safety are also addressed. In addition, the growing trend of *organic raw materials* and their significance for this industry were evaluated. Finally, new commercialization possibilities with an emphasis on existing evidence of potential mushroom-based supplements as support of the immune system during the COVID-19 pandemic are discussed.

4.2 MUSHROOM NUTRACEUTICALS: THE ELEPHANT IN THE ROOM

Is there such a thing as a mushroom nutraceutical? Yes and no. The answer to this question depends on the acceptance of this terminology in various countries by their regulatory organizations and scientific staff. The expression was defined by Chang and Buswell in 1996, pointing out that this sort of product belongs to dietary supplements rather than food. Other researchers readily adopted it and even emphasized its superiority among other products from the same category (Wasser & Akavia, 2008). Since pharmacologically active preparations from mushrooms come in form of more or less crude extracts they cannot be categorized as commercial drugs. Furthermore, the predominant preventive mode of action (immunomodulation and antioxidant activity) undoubtedly puts it in the supplement group (Kozarski et al., 2015). Sometimes the term *mushroom nutraceutical* is used as a synonym for functional food. Although such food has a role in disease prevention, it is not consumed in the form of a pill or capsule. In some cases, the distinction between a nutraceutical and functional food is hard. Mushroom powder with high vitamin D content (synthesized through UV stimulation during mushroom cultivation) can be consumed in the form of a capsule or pure powder (Cardwell et al., 2018). It has not been purified, meaning that it is food-derived and consumed as food. At the same time, it can be considered as a dietary supplement, improving vitamin D deficiency and cognitive performance (Zajac et al., 2020). Another example of this duality is mushrooms enriched with selenium (Witkowska, 2014). If we consult the scientific literature, the expression *mushroom nutraceutical* is well defined and often used (Morris et al., 2016; Badalyan et al., 2019). Thus, a mushroom nutraceutical is a preparation based on one or all parts of the fungus (mycelium, fruit body, or spores), with a different degree of purification/extraction, consumed as pharmaceutical (pill or capsule) with no intent to replace food or to cure but to be preventive of choice. If we follow the European regulative and definition for dietary supplements with rules for maximal allowed daily intake, mushroom nutraceuticals cannot be claimed as supplements (Reis et al., 2017; Santini et al., 2018). So, do they exist at all? The actual market recognizes mushroom nutraceuticals, which is evident from reports and predictions offered by numerous market analysis companies. In these reports, mushroom nutraceuticals are often presented with promising headlines, like *Medicinal Mushrooms Gain Foothold in Nutraceutical Applications* ("Global Medicinal Mushroom Extract Market: Information by type, form, function, and region – Forecast till 2027, https://www.marketresearchfuture.com/reports/medicinal-mushroom-extract-market-4737). The numbers in these reports also speak in favor of mushroom-derived nutraceuticals. The size and predicted worth of this specific market is estimated to be more than 15 million US$ by 2026 with a compound annual growth rate of 6.10%–9.85%. There is even a type of consumer known as the *nutraceutical consumer*. With such a bright prospect and the growing individual health-awareness materialized through the shift toward prevention instead of curation, the existing predictions seem sensible (Childs, 2000). On the other hand, there is a problem with the regulation of mushroom nutraceuticals, because the term *nutraceutical* is not uniformly accepted worldwide, thus creating confusion. Each country has its own opinion on nutraceuticals as well as legislation, and the topic has been thoroughly discussed by several authors (Shinde et al., 2014; Aronson, 2016; Dailu et al., 2018; Santini et al., 2018). So the burden of the more general term *nutraceutical* affects mushroom-derived prevention-intended preparations.

Other terms appearing in connection with mushroom nutraceuticals/dietary supplements are connected with their specific activity, like immunoceuticals since immune enhancement is one of the most examined activity of mushrooms (Zhang, Li, et al., 2017). Similarly, compounds like mushroom polysaccharides are marked as immune response modifiers or chemopreventive agents (Gao & Zhou, 2004). Finally, all mushroom compounds (derived from or produced by) are sometimes named *mychochemicals,* pointing to their mycological origin.

4.3 MUSHROOM IDENTIFICATION AND NOMENCLATURE

A correct taxonomic identification of fungi (fungal ID) is essential for production of mushroom dietary supplements. Unfortunately, this aspect is superficially overlooked especially by start-up companies, when people without experience and without a clear standard procedure deal with fungal taxonomy and nomenclature. This leads to establishing wrong fungal IDs which, in a chain reaction, may cause concerns regarding poisoning, toxicity, and adverse therapeutic effects in people and animals when using the finished product. There were cases where, besides gastro-intestinal disturbance, wrong fungal ID resulted in fever, nephropathy, acute hepatitis, and even coma (Wasser, 2011; Chang & Wasser, 2012). Sometimes misidentifications happen on purpose when rare and expensive species are exchanged with similar looking cheaper ones, like in case with *Ophiocordyceps sinensis* (Wasser, 2011; Zhang et al., 2020). To avoid these problems and obtain standardized high-quality products, proper taxonomical identification is indispensable (Raja, Miller, et al., 2017).

There are a few methods of fungal ID: morpho-anatomical, chemotaxonomic, mating compatibility test with reference strains, and molecular identification (DNA-based). The oldest method is classical morpho-anatomical, which requires macro- and microscopical analysis of fungal structures, mainly of fruit bodies. Also, there are a few monographs dedicated to fungal ID based on cultural characteristics of mycelial colonies, grown on agar plates (Stalpers, 1978; Buchalo et al., 2009). Chemotaxonomy can be applied in the case of detection of genus- and species-specific medicinal compounds, e.g. in *Ganoderma* spp. (Richter et al., 2014). Both morpho-anatomical and chemotaxonomic methods analyze phenotypical characteristics and reflect phenetic species concept. Mating compatibility test represents biological species concept: if two species are intercompatible, they are grouped as one biological species (Boidin, 1986; Petersen & Hughes, 1993). Given the broad acceptance of this idea, mating compatibility tests have been used to evaluate the taxonomic identity of fungal species classified by morphological characters (Bao et al., 2004). As for culinary-medicinal mushrooms, such tests were performed on cultures of

Pleurotus spp. (Petersen & Hughes, 1993; Vilgalys et al., 1993; Zervakis, 1998; Bao et al., 2004), *Armillaria mellea* (Anderson & Ullrich, 1979), and *Gymnopus dryophilus* (Vilgalys, 1991) complexes of species.

The most recent methods of fungal ID are molecular ones, which reflect genetic species concept and shaped the path of fungal taxonomy. These methods are based on studies of fungal DNA and include: DNA-DNA hybridization, restriction enzyme analysis, electrophoretic karyotyping, fingerprinting (e.g. arbitrary primed PCR), sequence analysis (e.g. DNA barcoding), etc. (Bruns et al., 1991; Bin et al., 2008; Xu, 2016). DNA barcoding is one of the most rapid, effective, and widely used methods of fungal ID. It is based on production of short DNA sequences from a specific gene/non-coding region or several genes and their comparison with sequences deposited in DNA barcode databases, such as User-friendly NordicBarcode of Life Data System ITS Ectomycorrhiza Database (UNITE) (https://unite.ut.ee/), the Barcode of Life Data System (BOLD) (Barcode of Life Data System V4, 2021), and NCBI GenBank (GenBank Overview, 2013). Genes and non-coding regions of nuclear ribosomal DNA cluster (rDNA): internal transcribed spacer (ITS), large subunit (nrLSU), and small subunit (nrSSU) are considered the most universal and most commonly used markers for fungal DNA barcoding (Raja, Baker, et al., 2017). Especially the ITS region is considered as the official DNA barcoding marker for species-level identification of fungi (Schoch et al., 2012; Kõljalg et al., 2013; Xu, 2016). While ITS sequences are effective at distinguishing most of medicinal mushrooms, there could be limitations for using it as a marker for some macrofungi such as members of the genus *Cantharellus* with exceptionally long ITS, which amplification may bias PCR reactions (Buyck et al., 2014). For such cases, it is necessary to use other marker genes and their combinations, e.g. nrLSU, nrSSU, translation elongation factor 1-α (TEF1-α), DNA-directed RNA polymerase II subunit 1 (RPB1), and subunit 2 (RPB2).

Unfortunately, there is a lack of information regarding the identification of fungi used for food supplements and nutraceuticals production. Raja, Miller, et al. (2017) provided an interesting outlook of this aspect. In short, the authors reported 31% of the studies used morphological ID, 27% – molecular ID, and only 14% of research papers published in *Journal of Natural Products* performed fungal identification based on a combination of morphology and molecular data. Moreover, 28% of papers published in the same journal did not report any fungal ID, whose secondary metabolites were used in formulating natural products. On how important is the fungal ID reflects the fact that even strains of the same species originating from different localities exhibit different chemical profiles, which directs it toward different purposes. For example, *G. lucidum* originating from Serbia has a higher content of proteins and has better performance as the mycoremediation agent (water purification), while the same species from Asia has higher medicinal potential and serves better as a source of medicinal compounds (Abdullah et al., 2020; Mooralitharan et al., 2021). Optimal cultivation parameters, like temperature, are also different for these two strains.

Some of the most common misidentifications are among phenotypically similar species belonging to the same genus and even to different ones. A striking example of problematic fungal ID based on morphological analysis is *Pleurotus "florida"*, which has a large commercial distribution and appears frequently in literature on cultivated mushrooms (Buchanan, 1993). This name was introduced by Eger (1965), but was never published nor intended as a binomial (Eger et al., 1979). Different strains of *"florida"* were found to be genetically compatible with two species: *P. pulmonarius* (Bresinsky et al., 1977; Hilber, 1982) and *P. ostreatus* (Eger et al., 1979; Hilber, 1982, 1997). Also, Hilber (1989) noted that *P. floridanus* Singer is unrelated to the two species represented by *Pleurotus "florida"*.

DNA-based identification is very common today as the most reliable way for fungal ID, which has its merit of providing resolution of the most complicated taxonomies. In the scientific circles, it is widely promoted in comparison with classical morpho-anatomical methodology due to somewhat inconsistent morphology of higher fungi influenced by the environment and subjectiveness of a taxonomist. However, DNA-based identification has some limitations, which may lead to fungal species misidentification. Based on authors' experience and some publications (Xu, 2016; Raja, Baker, et al., 2017; Zamora et al., 2018), there are issues, which can cause misleading fungal ID originating from DNA barcoding analysis:

1. Contamination and mix-up of specimens.
2. Wrong choice of the marker gene for a particular group of fungi.
3. Wrong choice of primers; it is important to take in account taxonomic specificity of primers.
4. Two different taxa may share identical DNA sequences at a given locus, even for already tested barcoding markers. Conversely, not all members of a species can be assumed to share the same DNA sequence at a specific locus.
5. Intraspecific (or even intraindividual) differences in the DNA sequence of a marker may be comparable to or exceed interspecific differences.
6. Insufficient sequence length.
7. Poor quality of sequence due to contamination, problems with sequencing, or presence of a few copies of the gene with different length.
8. Some sequences generated through different sequencing techniques may be artifacts (e.g. chimeric sequences) and consequently not represent reality.
9. Unintentionally or intentionally altered DNA sequences using editing software.
10. Problems related to online DNA sequence databases such as NCBI GenBank:
    - The lack of knowledge on how to use BLAST analysis and poor result interpretation by non-familiar users;
    - Wrong taxonomic annotation of sequences or lack of annotation;
    - Poor sequence availability for some species; most of described fungal species have not yet been sequenced;
    - Type strains/samples are obvious or not clearly indicated;
    - Lack of published sequence diversity (for some species we may see a long list of sequences coming from the same study of non-specialists in fungal taxonomy);
    - Taxonomic names are not up to date; the NCBI GenBank is a rather static database, which is rarely updated by its curators and very few authors update their nomenclature on published sequences;
    - Sequence quality issues.

UNITE (Kõljalg et al., 2005) and BOLD (Ratnasingham & Hebert, 2007) databases are curated by experts, have more reliable data than NCBI GenBank, and recommended for ITS sequence-based fungal ID analysis. They include their own well-annotated fungal ITS sequences as well as reliable sequences from NCBI GenBank. Both UNITE and BOLD also contain detailed specimen metadata to offer easy tracking of specimen and taxa information. Moreover, UNITE provides *species hypotheses* (which correspond to *operational taxonomic units*) based on sequence similarity with their reference sequences (Kõljalg et al., 2013).

Some mushroom raw material producers work with third-party companies in establishing fungal ID. Unfortunately, here, situations can happen when sequence-based ID laboratories provide their clients with wrong ID results, and if they are not verified in-house by the knowledgeable raw material producer, a mushroom product gets marketed under the wrong name.

As we see, identification methods when performed separately could lead to misidentifications, but taken together they support each other and provide a robust and reliable ID. The most precise fungal ID can be achieved by combining methods, which represent genetic identification supported by phenetic or biological data. For practical reasons, medicinal mushrooms should be first identified through morpho-anatomical analysis (since majority of them can be easily identified by these classical methods) and then verified with a genetic method (e.g. DNA barcoding), especially if it concerns their commercial use. Such an approach can eliminate bias in fungal ID and facilitate client credibility in mushroom nutraceuticals.

Currently, there are many attempts to identify the fungus in its blended powder form and currently several reports (Xiang et al., 2013; Zhang, Zhang, et al., 2017; Raja, Baker, et al., 2017) discuss methods on how to identify such fungi which practically shows the major lack of client credibility in raw material manufacturers and regulatory organizations. The latter should use taxonomists or staff familiar with medicinal mushroom fungal taxonomy. However, since there is still obscurity in regards to this matter, some mushroom-based products available on the market today remain a mystery.

Consistency of nomenclature and taxonomy for medicinal mushrooms is also a very important aspect. It is true that fungal taxonomy changes rapidly, but raw material and nutraceutical producers should follow it and check the correct fungal names in Index Fungorum and Mycobank websites (*www.indexfungorum.org* (2021), *www.mycobank.org* (2021). It is true that sometimes even these two sites lack data on the synonymy of some taxa. For example, based on the study of Kerrigan (2005), *Agaricus subrufescens* Peck should be treated as a priority name for its synonyms: *Agaricus blazei* Murrill *sensu* Heinemann, *Agaricus brasiliensis* Wasser et al. nom illeg., and *Agaricus rufotegulis* Nauta. Among them, only the last name is mentioned as a synonym of *A. subrufescens* in Index Fungorum. This study also proved that two commercial samples of "medicinal agaricus" from Japan and China belong to *A. subrufescens*.

There are some important recommendations to mushroom nutraceutical producers. Essential is to have on hand is a file comprising taxonomical analyses of every mushroom used in the production line ready to be shown to possible clients responsible for further processing and formulation of the raw materials. This will also offer regulatory organizations transparency in regards to the fungal ID used in production. When medicinal mushroom raw material producers will understand this, they will not only cut costs on periodical ID analyses, but their product will step out of the doubt that wraps some of their customers and competitors. Since the ID process is misunderstood at this level, this calls for a fungal taxonomist able to clearly state the identity of existing fungi into production or new fungi to be launched into production for obtaining raw materials should send the identification data to a third party for validation. This third-party should be a university be a university representative familiar with the identification procedure. This last step is of uttermost necessity to avoid intentional in-house ID manipulation by companies willing to cross the line in what regards ID-related ethics.

4.4 RAW MATERIAL: HOW TO MAKE A FAIR CHOICE?

Many mushroom nutraceuticals on the market originate from mycelium colonized grains, mycelium biomass obtained via fermentation processes, fruit bodies, or spores. A very popular debate among the raw material producers remains the methods through which the product is made, the form that it is marketed as, and its potency. In regards to this, everyone follows their philosophy and highlights their products as being superior to some other products derived from other procedures and formulations. In this scenario, start-up companies struggle more than their competitors and we may easily observe an evolutionary scale that divides them into companies with settled and non-settled standard operating procedures (SOP). Unfortunately, this is also reflected in the quality of their products and consistency. For example, a mushroom farmer growing edible mushrooms for several years deciding to tap into the production of medicinal mushroom fruit bodies may quickly adjust and settle their SOPs while companies growing mycelium on grain are more likely to have non-settled SOPs even after years of existence because there will always be an issue that requires them to adjust their procedures to get their final product in good shape.

The number of companies supplying raw ingredients for mushroom nutraceuticals is great and it grows (observation based on the number of exhibitors on expos specialized in nutraceutical ingredients). There are plenty of choices: whole fruit bodies, slices of fruit bodies, crude extracts, customized extracts, compounds of choice (polysaccharides, terpenes, spore oil, erinacine, cordycepin, etc.), mycelium extracts, fermented products, concentrated liquids, myceliated grains, powdered mycelium, or fruit bodies. Moreover, the material can be of wild origin or cultivated.

With so many variables to consider in still underdeveloped fields when it comes to regulation of quality and safety an interesting combination of analytical data and *in silico* methods was proposed by VanderMolen et al. (2017). The authors applied a multifaceted approach based on five key principles: morphological and molecular identification, dereplication, chemical analysis-based comparison with known culinary mushrooms, literature review, and in-market data. The first step proves to be highly critical since species misidentification often happens (Wasser, 2011). Thus, when searching for literature data research papers including molecular and morphological identification of the examined mushroom-material provides a higher level of confidence (Vunduk et al., 2019). Just recently, a group of authors published a new-evidence based classification system upon which known mushroom species are assigned with a final edibility status (Li et al., 2021). The authors provided the most comprehensive list of edible mushrooms, which is good to be consulted when determining to use or not some mushroom species when formulating a nutraceutical.

Often, a mushroom nutraceutical represents a blend of mycelium and grain termed as mushroom spawn – generally used by the mushroom cultivation industry. This practice is a sort of mischief since a significant part of the whole product is grain (about 70%). The chemical structure is different from that of a mushroom, although both contain glucans. Particularly, grain-derived glucans are different than those isolated from macrofungi. Polysaccharides from mushrooms are polymers with high molecular mass with many side chains (Gong et al., 2020). The polysaccharides existing in grain are linear and with lower molecular mass, they are also water-soluble (Wang & Ellis, 2014). Having in mind that the biological-immunomodulating response of mushroom polysaccharides is highly dependent on its physicochemical characteristics and it is of paramount importance to distinguish if the material is pure fungi or a mixture of a fungus with grain (Meng et al., 2016). However, mycelium is easier to obtain (due to shorter cultivation period), so it is much cheaper to produce compared to growing of fruit bodies. It can still be a good raw material depending on the cultivation method. Simple raw materials available as powder may be called *full spectrum* termed for medicinal mushroom dietary supplement products containing not only mycelium grown on grain, but also fruit bodies and spores. This blend is sought to be a superior product (compared to plain dried and ground mushroom spawn), but still controversial among competitors that use fruit body–derived raw materials only. By some others it is understood as a mixture more likely in equal parts, but the reality may be different, because the fruit body amount in mycelium grown on grain raw materials is probably under 3% and represents fruit bodies that start to grow during the long incubation periods.

The technology applied for mycelium production with a liquid medium is known as submerged fermentation and it has several benefits in comparison to traditional or solid-state cultivation (Hassan et al., 2019; Klaus et al., 2021). Instead of several months to years, the final product is obtained in a matter of days. The process is highly controlled so the mycelium quality is easy to monitor and standardize. By applying submerged fermentation additional products like exopolysaccharides are also produced (Zhang et al., 2019). They can be collected and further processed. Additionally, the health-supporting claims can be substantiated with the real data, since research papers mainly consider mycelium obtained by submerged fermentation instead of the grain-based cultivation process. Obodai et al. (2017) demonstrated that the chemical constituents (amount, type, and structure) are influenced by origin of the mushroom part, which is used as raw material. Natural conditions, substrate, and temperature cause the variations in a mushroom's composition; thus, controlled and standardized cultivation is of most importance.

The extraction which follows the cultivation step is also a critical point when it comes to final product effectiveness. Solvent choice (polarity) affects the compounds, which will be extracted from raw material (Cör et al., 2017). Traditionally, mushroom biomolecules are obtained by water and ethanol extraction. Solvent temperature is also important due to the presence of thermo-labile compounds in crude extracts (Gong et al., 2020). Sometimes additional treatments, like active molecule modification, are necessary or recommended. Oshiman et al. (2002) demonstrated the increase in the antitumor activity of β-glucan from *Agaricus subrufescens* (as *A. blazei*). Acid treatment lowered its molecular mass and orally ingested β-1,6-glucan expressed antitumor activity in an animal model. Relatively recently, the industry adopted supercritical fluid extraction, which proved efficient in a laboratory setting (Mazzutti et al., 2012). There are two options for solvent: CO_2 and H_2O. And while supercritical water extraction is more environmentally friendly, CO_2 is the fluid of choice in the industry (Castro-Puyana et al., 2017; Barbosa et al., 2020). CO_2 appears to be safer when pressurized and the equipment for its application, so it is less prone to deterioration. Additional extraction techniques, known as non-traditional, are ultrasound-assisted, microwave-assisted, and enzyme-assisted extraction. They are often superior to traditional water or alcohol extraction since there is an optimal contact-time between solvent and raw material and reduced destruction of active compounds (Gong et al., 2020). The outcome is a higher yield of a product and improved quality. The final step in extract production is drying, which also affects biological activity, as proven by Ma et al. (2013). The authors demonstrated higher antioxidant ability when polysaccharides from the medicinal mushroom *Inonotus obliquus* were freeze dried. Spray drying is another technique used in the final stage of extract preparation (Assadpour & Jafari, 2019). The technique is practical since it also enables improvement of product stability when it includes microencapsulation (Shao et al., 2019). However, spray drying is performed with dextrin since the pure extract tends to agglomerate. Adding dextrin in the final product messes with the purity of extract because it is a low-molecular-weight sugar from wheat. Not just that it adds to the total fiber content in the product, but it can express side effects like constipation (McRorie, 2015).

A particularly interesting raw material is mushroom spores. They serve the reproductive purpose, so their chemical structure differs significantly from mycelium and fruit body (Petrović et al., 2019). Sliva et al. (2003) tested different commercial *Ganoderma lucidum* supplements (containing spores and powder) and did not find any correlation between composition, purity of preparation, and its effectiveness toward inhibition of highly invasive human breast and prostate cancer cells. The authors explained that commercial products were not clearly labeled, so it stayed unknown which concentration of active ingredient each of the products contained. Companies have the freedom to make their proprietary blends and formulations, and not specify the percentage of active ingredient or concentration on the label, further convoluting this already complicated market (Childs, 2000). In the end, non-standardized production, collection procedures, and lack of formulation transparency affect consumers' trust.

Spores are rich in lipids, especially unsaturated fatty acids, known as effective in different health conditions like hypercholesterolemia (Goodfellow et al., 2000). Recently, a highly branched water-soluble β-glucan was isolated from *G. lucidum* spores and it has been proved to possess immune-promoting activity *in vivo* (Wang, Liu et al., 2017). Spores are rich in active molecules but they are poorly absorbed due to a hard cell wall protecting the genetic material. The solution is to perform pre-treatment, to break the spore cell wall. A patent from 1997, based on a combination of biochemical and mechanical processes, claim to be effective in breaking of the spore cell wall, while maintaining active ingredients biological activity (https://patents.google.com/patent/CN1165 032A/en). Supercritical fluid extraction is also used for the extraction of sensitive components from spores (lipids and triterpenes) (Lin, 2009) (Figure 4.1).

As we may conclude, when choosing a medicinal mushroom source for raw material, one should pay special attention to the following criteria:

- Fungus ID–related documentation should be provided by the producer;
- Raw material origin: strains/specimens collected in nature, hybrids, or genetically modified strains;
- The location of the production facility. It should be located away from major sources of air pollution, such as atomic power plants, mining sites, or heavily polluted urban places. Mushrooms are known to accumulate heavy metals, such as lead, arsenic, or mercury, and therefore a heavy metal test is necessary in this case;

(a)   (b)

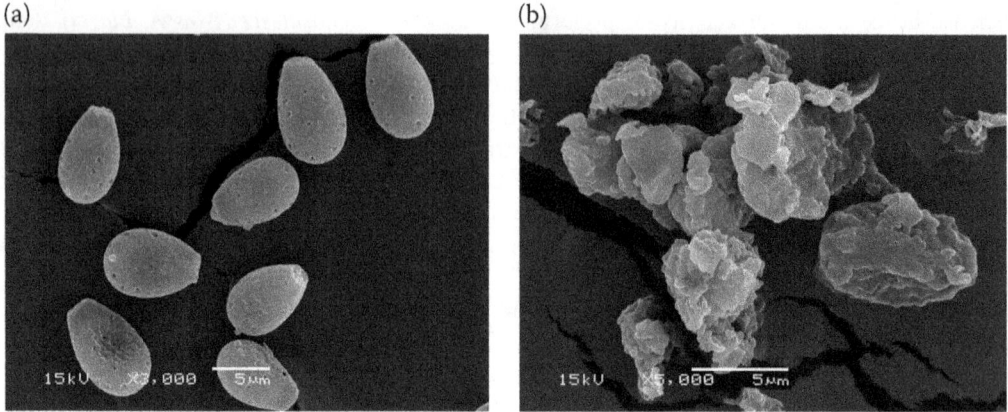

Figure 4.1   (a) *G. lucidum* spores with cell wall (b) and after the cell wall being broken. By courtesy of GAN-OHERB GROUP.

- The raw material must be produced in a contaminant-free facility (including dust, fiberglass, or other particles);
- Microbiological control by specialized labs for bacteria and mold presence. Certificates of analyses have to be provided by the producer or in-house tests need to be performed periodically to ensure a high product quality free of toxins;
- *Organic* certification: whether the raw material has *organic* origin or not;
- The type of the raw material: spawn or fruit body originated, simple powder or extract. It is another important characteristic, because this reflects the way it was produced, the amount of active ingredients together with potency, and effects on human and animals;
- Proper storage of medicinal mushroom raw material must be away from moisture and rodents.
- Composition of raw materials for purchase and launch into production: one mushroom species or a blend;
- What type of grain was used to be colonized with mycelium and what is the percentage of grain left in the final product;
- Analysis certificates showing content of β-glucan and other active ingredients;
- Whether the product contains D-vitamin or other added minerals and vitamins. Some raw materials are forcefully pumped with UV light originating vitamin D generally considered a good source but still, different than the natural vitamin D obtained from the sun;
- Scientific studies about efficiency of medicinal mushroom(s) used as a raw material, performed by third-party laboratories, represent essential points in highlighting product quality;
- Active ingredient response on the human or animal body in regards to dosage and any possible side effects.

The previous list defines the quality of the raw material, but close attention should be addressed to the company – the producer of the raw material and the philosophy of the staff. Moreover, visiting the production site is essential in establishing a close relationship with the producer and gaining trust in the product.

4.5 ORGANIC RAW MATERIALS FOR MUSHROOM NUTRACEUTICALS

Intensive indoor and outdoor mushroom production was always sensitive and prone to diseases and pests affecting the yield and quality. It is a constant challenge approached with a combination of hygiene management and the use of chemicals in the form of pesticides and fungicides. In large facilities and monoculture areas of production (like in the USA or China), pest-related disease outbreaks are common and spread like wildfire. Namely, Kennet Square, Pennsylvania, is known as the mushroom disease capital of the world (Ellor, 2021). Once present, phorid and sciarid flies move from tunnel to tunnel transmitting spores of mycoparasitic fungi, which, for example, cause dry (*Lecanicillium fungicola*) and wet bubble (*Hypomyces perniciosus*) diseases. The use of chemicals to prevent and treat already existing diseases is tempting to some growers and unfortunately they are often used in an inappropriate manner. Another challenge in mushroom production is the lack of pest management products that are effective enough, so the producers apply products intended for vegetables and other crops (Tian et al., 2020). No wonder several studies reviewed by Tian et al. (2020) reported on the presence of pyridaben, pyriproxyfen, cypermethrin, permethrin, and malathion in cultivated mushrooms. Moreover, the pesticide market is expanding without being followed by risk assessment in mushroom production. The previous list leads to the products that are not only unsuitable for nutraceutical formulation, but even pose a more or less serious health risk. Besides, mushrooms accumulate not only chemicals used for pest management but metals and metalloids present in the substrate. Melgar et al. (2014) studied mushrooms and mushroom supplements from Spain for the presence of arsenic and found that the concentration of this toxic metalloid was slightly higher in mushroom supplements than in wild or cultivated mushrooms, probably due to extraction and concentration steps in supplements preparation. The values they found in supplements posed no health risk.

There are two options in developing fair and safe mushroom dietary supplements. In case conventionally produced mushrooms are raw material, it is of paramount importance to perform as much as it is possible thorough analysis for the presence of pesticides, fungicides, heavy metals, and radionuclides. Another option is a sustainable strategy for the mushroom industry with better management of chemical pest control, high-quality crops, and lower negative impact on the environment (Jess et al., 2017). That is why certified *organic mushroom cultivation* is more and more popular in recent years. Currently, *organic mushroom farming* allows no chemical pesticides. Control of pests and diseases is obtained using physical barriers and biological agents. Also, when the production cycle comes to an end, meaning that production space is empty, proper hygiene and disinfection are a must. For a grow-room, cleanup steam is generally recommended, but other chemical agents are allowed as well with prior permission. Since mushroom cultivation includes two types of raw materials, spawn and substrate, each has its regulation. Thus, spawn used in *organic cultivation* does not need to be certified as *organic*, which is not the case with raw materials for substrate preparation. However, 25% of raw material is allowed to be from *non-organic* sources (Organic Mushroom Production, 2021). Just in the USA, this sector increased by 11% from the growing seasons 2005–2006 to 2006–2007 (Akavia et al., 2009). The main species was *A. bisporus*, while other specialty mushrooms accounted for 22% of the total produced amount. Recently, throughout growing season 2019–2020 the USA growers produced 127 million pounds of specialty mushrooms certified organic throughout. Out of the total amount, 62% of mushrooms were sold as certified organic and *Agaricus* mushrooms accounted the most of it (88%) (https://www.nass.usda.gov/Publications/Todays_Reports/reports/mush0820.pdf). Statistics on *organic mushroom farming* in other parts of the world are scarce, including Europe. Current estimates might be significantly larger than the real data due to the legislation gray zone. Namely, European law does not differentiate between *organic* certified produced mushrooms and those collected from the wild, which can also bear organic certificates. Besides, the production of *Agaricus* species requires compost based on horse manure or chicken litter, but only the first material can fulfill requirements for *organic cultivation*. Having in mind that chicken litter used by intensive production systems can not originate from households raising chickens (due to small quantities). Therefore, the material containing antibiotics and hormones, heavily used in chicken production, cannot account for organic mushroom cultivation. On the other hand, how many horses does Europe have? And how many are bred in an antibiotic-free way? Optimistically, we can estimate that 10,000 tons of *organic* compost can be produced annually which equals approximately 2,000 tons of *Agaricus* species.

Organic raw materials coming from the United States are the most prized, while China, although being a significant *organic producer*, still works on building trust. China has wide areas of good-quality land and resources, which in combination with guidelines and certificates for organic production, provide high-quality materials. For example, Heilongjiang Province accounts for 1/9 of China's arable land, with the lowest use of pesticides and fertilizers in the country. In the year 2019, this province had 1999 organic product certificates, which represent around 10% of the country's total. The same province had 604 companies registered as organic producers (Ministry of Agriculture and Rural Affairs of the People's Republic of China, n.d., http://www.moa.gov.cn/gk/tzgg_1/tz/201912/t20191210_6333030.htm). Benefits from organic production were recognized by the Chinese government, which actively supports the development of national organic product bases and also issued *The Reform Plan for Establishing A Green Ecologically-Oriented Agricultural Subsidy System.* When it comes to mushroom solid-state-based production, China has an elaborated system meaning that fruit bodies are cultivated outdoor, in ecological zones, like Qingyuan, with strict rules for crop rotation/movement. Mushrooms are cultivated under tents with pure running water, which is regularly tested for fulfilling requirements of organic production. China also has the national certification and accreditation public platform (National Certification and Accreditation Information Public Service Platform, 2021), which provides different types of information in connection with organic production as well as the online search of various products including mushrooms. For example, if the platform is searched for "*reishi organic products*", the system provides a count of 186 producers that are registered in China (Figure 4.2).

According to China's government, a public announcement 5.63% of sampled products did not meet organic certification requirements in 2020 (Announcement of the State Administration for Market Regulation on the Issuance of the Third Phase of Certification Risk Warning for 2020). They even introduced a 5-year restriction to apply for an organic certificate for those companies found to be seriously inconsistent with requirements. Interestingly, the majority of non-compliant companies were mushroom producers. It has been found that products contained chlorpyrifos (an organophosphate pesticide) and carbendazim (broad-spectrum fungicide). Some Chinese organic products, like *G. lucidum* and *L. edodes* blended with millet and fresh ginger, were banned in the United States as

Figure 4.2 Inside of the organic certified *G. lucidum* production tent. By courtesy of GANOHERB GROUP.

Figure 4.3 Xianzhilou Organic Ganoderma Expo Park (GANOHERB was previously named Xianzhilou). By courtesy of GANOHERB GROUP.

non-compliant with organic certification (The Crisis Facing Organic Food Certification, 2015). As commented, Chinese standards for organic products are stricter than in the EU and USA, but lacking strict supervision. Knowing that China is the largest producer of mushrooms and usually the main supplier of raw materials, it is expected that most nutraceutical producers will consider it as its supplier. Thus, it is wise to do a personal check of the facility for cultivation, or site, as well as to constantly monitor if the material fulfills the stated quality (Figure 4.3).

4.6 SAFETY CONCERNS AND QUALITY ASSURANCE/CONTROL

Numerous studies and reviews on mushroom bioactivity, therapeutic compounds, and the potential for nutraceutical development, followed by a small number of clinical studies, have been published. On top of that, macrofungi are known for their high nutritive value and prospects in functional food application (Prasad et al., 2015). Altogether, these assets create a super-therapeutic image followed by increased sales of mushroom-based nutraceuticals. Like with any other compound and formulation being industrially produced, principles of good manufacturing practice as well as different standards should be applied to provide efficient, reliable, and effective nutraceuticals. With no standardized identification methods, raw material quality and cultivation standards, allowed daily doses, and mandatory clinical studies, how do you expect a safe product? Some of these aspects are addressed depending on the country of production, but the wholesome principle still does not exist. The governments rather transfer it to the conscience of the industry (like in the case of the USA and Food and Drug Administration Agency). Some, more obvious safety aspects, which are usually part of hazard analysis and critical control points standard HACCP, like metal particle detecting, microbiological and heavy metal analysis, are conducted by the nutraceutical producers. However, pesticides, radioactivity, and polycyclic aromatic hydrocarbons might occur in raw material as residues. Mushrooms accumulate heavy metals and radioactive elements from their surroundings and serve as organisms suitable for biomonitoring

(Duff & Ramsey, 2008). This can pose a problem if mushrooms are collected from nature, so the standardized cultivation was supposed to overcome bio-concentration of toxic elements, such as arsenic, cadmium, chromium, lead, nickel, and mercury (Zou et al., 2019). In practice, the presence of metals is monitored by inductively coupled plasma mass spectrometry or inductively coupled plasma-atomic emission spectrometry. Sample preparation is very important since different parts of the mushroom fruit body accumulate different amounts of metals (Dowlati et al., 2021). It is equally important to consider the growing phase and the age of mycelium if the producer uses them instead of the fruit body.

The process of macrofungi cultivation is marked as sustainable and eco-friendly since the residues and waste from agroindustry and agriculture can be converted into food and medicine. Like in every monoculture production, cultivated mushrooms are prone to diseases that are controlled by the application of fungicides, bactericides, and insecticides (Tian et al., 2020). As already mentioned, macrofungi accumulate molecules from the substrate and, besides heavy metals, pesticide residues can be found as well. The residues of pesticides are monitored by gas chromatography or liquid chromatography coupled with mass spectrometry. Furthermore, the inoculated substrate is packed in plastic bags containing and releasing phthalate esters and metalloids. Ma et al. (2021) screened eight mushroom species cultivated in central China for the presence of contaminants from the cultivation process. The analysis revealed the presence of six phthalate esters and nine potentially toxic metalloids, exceeding the maximal weekly intake defined for China. An alternative is to perform submerged fermentation and minimize the possibility of contamination with unwanted substances or microorganisms. However, this type of cultivation cannot be used in the production of fruit bodies.

Besides being safe, mushroom nutraceuticals are expected to have the label-specified amount of active compounds, usually polysaccharides or terpenes. However, methods for their measurement are not standardized, which leaves the space for manipulations, especially when declaring polysaccharide content. As already mentioned, mushrooms can be grown on grains without being transferred to substrate for fructification. The material (mycelium colonized grain aka spawn) is dried, milled, and presented as the final product. In this case, the total amount of polysaccharides is higher than in pure mycelium or fruit body since grains contain this compound as well, but having a different structure. Without the characterization or the necessity to report the nature and source of polysaccharides, manufacturers declare a significantly higher amount of active ingredient. This cannot be in agreement with the statements connected with its biological activity since up to 70% of the final product is grain. Methods for polysaccharide qualitative and quantitative analysis used in practice are high-performance liquid chromatography (HPLC), size exclusion chromatography, high-performance liquid chromatography with an ultraviolet detector, gas chromatography, and UV-visible spectrophotometry. HPLC is the most commonly used due to the consistency and stability of obtained fingerprints (Zheng et al., 2020). Improvements in resolution, separation efficiency, and sensitivity are expected through the wider application of ultra-performance liquid chromatography. This technique has been accessed for qualitative and quantitative analysis of polysaccharides from traditional Chinese medicines (Kuang et al., 2011). The researchers managed to detect a higher number of chemical constituents in less time than with conventional HPLC. The method has been described as sensitive, precise, and accurate.

Besides different nomenclature and legislation in each country, the major problem with analyzing the quality of dietary supplements/nutraceuticals is the diversity of ingredients. Exotic plants, not common for Western consumers or countries, are constantly bioprospected for new commercial product development (Pandey et al., 2020). Fungi are especially interesting, and from 2019 we entered *the age of fungiceuticals* (Pandey et al., 2020). With so many known fungal species – around 148,000 (Cheek et al., 2020), which represent less than 10% of the estimated existing ones (2.2–3.8 million, based on Hawksworth & Lücking, 2017), having their own primary and secondary metabolites, it is unrealistic to expect the existence of reliable analytical methods capable of precise identification of each of them. The situation is further complicated with the existence of highly sought products, like those that improve sexual performance or products for bodybuilders. This is a specific niche mostly driven by sentiment, so the consumer is expected to be less careful when buying the product. Under such circumstances, deliberate *spiking* often occurs. A typical example is *O. sinensis*. This mushroom became popular after the 1991 athlete meeting in China, with market-present amounts significantly overcoming the quantity of raw material: a few thousand kg collected just in 2001 (Li et al., 2006). Also, the price per kg reached more than 10,000 US$. All mentioned open prospects for adulteration. Recent experiments showed that natural *O. sinensis* can be used like the cultivated one from a nutritive aspect (Zhou et al., 2019). Still, the natural one contained significantly higher amounts of adenosine and cordycepin, crucial for the medicinal properties of this mushroom. It has been demonstrated that the majority of products claiming to contain *O. sinensis* were adulterated with flour, starch, dye, and soybean (Liu et al., 2011). However, adulterants are not always so benign. Sometimes, synthetic drugs are added to obtain efficacy (Dwyer et al., 2018). Even toxic compounds like aluminum sulfate solution can be used in order to increase product's weight (Liu et al., 2011). Another problem is the lack of standard materials, so even when an analytical method is available there is no substance with which to compare examined material. Also, standards can be incorrectly identified by the supplier (Betz et al., 2011).

In order to purchase the best raw material, a nutraceutical producer needs to know and understand the importance of the aspects of quality and safety analysis. Not paying attention to this can lead to serious problems, like poisoning, with a product that otherwise should be preventive or therapeutic.

4.7 SIDE EFFECTS, CONTRAINDICATIONS, AND CONCERNS

The field of nutraceuticals/dietary supplements is rising, but it is highly study-unsubstantiated concerning the pharmacodynamics, pharmacokinetics, and possible adverse events. These are over-the-counter products, meaning that they are not necessarily safe and thoroughly accessed for safety (Dailu et al., 2018). Problems can occur due to product content, low quality of ingredients, contamination during production, or interactions with other compounds and medications. Badalyan et al. (2019) found about 600 research papers dealing with the health benefits of mushrooms and a small number of clinical studies. But what is with adverse events, especially in the case of people with impaired immunity receiving specific therapy? Mushrooms are heavily advertised as important adjuvants in cancer therapy, making it ever more important to study interactions with pharmaceuticals as well as to describe side effects (Siddiqui & Moghadasian, 2020). The same applies to the stability of active compounds; they are prone to changes during shelf life, and once safe can become the source of problems after chemical changes due to inappropriate storage conditions. Also, the more complex the nutraceutical the harder is to determine its shelf life and safety. The drug delivery system can also pose a problem. As Helal et al. (2019) stated, nanosystems have adverse effects like respiratory disorder, cardiovascular diseases, carcinogenicity, and lowering of life expectancy, inflammation, and oxidative stress. Existing studies usually examine one compound or extract while commercial products represent a combination of several ingredients, often of different origin (like mixtures of a mushroom extract with botanicals or vitamins and minerals). These protocols are established for dietary supplements like vitamins, but not for botanicals since analytical methods for their analysis are still developing (Mehta, 2010).

The existing literature on this topic is scarce and it seems that the scientific community is stubbornly refusing to deal with this hot topic. Sweet et al. (2013) found only one case study reporting adverse events in connection with mushroom dietary supplements when used during cancer therapy. The study (Mukai et al., 2006) involved three patients (with ovary and breast cancer) all of whom have taken unidentified supplements based on *A. subrufescens* (as *A. blazei*). Reported symptoms were fever, fatigue, increased liver enzymes followed by liver dysfunction, and death (in two cases). It has been observed that these results are contradictory with animal studies that all confirm the hepatoprotective activity of medicinal mushrooms, especially in the case of the chemical-injured liver. Although a modification of action of cytochrome P450 isoforms was proposed there is no clear explanation of these events. Interestingly, the clarification of this study never occurred although up to 80% (by some authors it goes as high as 91%) of patients with cancer take some complementary and alternative medicine product (Bernstein & Grasso, 2001, Yates et al., 2005). A recent review on herb-drug interaction focused on two medicinal mushrooms, *G. lucidum* and *Trametes versicolor*, found no reported adverse events in the case of clinical, animal, and *in vitro* studies (Lam et al., 2020). On the contrary, both species exhibited synergistic effects stimulating the immune system, decreased immunosuppression in patients using chemotherapy, and alleviated general and specific side effects of standard cytotoxic drugs. Still, the authors warned about the low quality of the existing clinical studies and lack of following of the world's recommendations when performing clinical studies with traditional Chinese medicine preparations. Another study estimated the in vivo (animal model) long-term actions of low levels of dietary lectins from peanuts and mushrooms on rats' intestinal and pancreatic growth (Kelsall et al., 2002). Pancreatic growth was observed pointing to a potential carcinogenesis-promoter activity of dietary lectins.

Species from the *Ophiocordyceps* and *Cordyceps* genera are part of traditional Chinese medicine and are well known for numerous health benefits although with several exceptions. As reviewed by Tuli et al. (2014), *Cordyceps militaris* can provoke an allergic response, adverse gastrointestinal behaviors (dry mouth, diarrhea, nausea), with a suggestion to be avoided by patients with autoimmune diseases. An interesting report by Wu et al. (1996) discussed lead poisoning caused by *O. sinensis* powder, which only demonstrates the importance of the source of raw material as well as its quality check. Furthermore, this parasitic mushroom can interact with some drugs due to its hypoglycemic and anti-viral compounds. Similar information about *O. sinensis* is rarely publicly listed like on "Cordyceps: Side Effects, Dosages, Treatment, Interactions, Warnings" (https://www.rxlist.com/consumer_cordyceps/drugs-condition.htm), where a list of some 72 compounds/herbs/drugs known to mildly interact with *O. sinensis* is given.

Studies concerning adverse effects in pregnant and lactating women do not exist so it is generally suggested to avoid mushroom supplement use in these periods. Recently, Sun and Niu (2020) conducted a study to access the impact of mushroom diet on pregnancy-related complications, macrosomia, and pregnancy-induced hypertension. In both cases, a positive impact of the mushroom diet was observed.

Surprisingly, consumers in the USA do not pay much attention to label statements. Dodge (2015) summed the consumer believe labels and dietary supplements. As reported, erroneous beliefs include understanding that the government approved the supplement, that before being marketed supplements were tested for safety and effectiveness, that content of the product is analyzed, and if known to the producer adverse effects would be stated on the product. More striking is the fact that even health professionals (65%) do not have an adequate source of information about dietary supplements, and although they ask their patients about supplement use they rarely report adverse effects (only 18%) or do not know where to report them (73%) (Cellini et al., 2013). To minimize the consequences of unregulated supplements intake, the Food and Drug Administration (FDA) published a list of actions that may be harmful to the consumer. The list includes: combining supplements, using supplements with medications, taking too much, and substituting supplements with prescription medicines (What You Need to Know about Dietary Supplements, 2019). Karbownik et al. (2019) discovered a negative correlation between knowledge about dietary supplements and trust in advertising them.

4.8 BIOAVAILABILITY OF MUSHROOMS' ACTIVE COMPOUNDS AND DRUG DELIVERY SYSTEMS

Phytochemicals and mycochemicals are natural compounds with complex structures. Contrary to the pharmaceuticals, the basic molecule of a nutraceutical has not been designed. The natural origin of the product, although preferred by 80% of the global population, is not always bioavailable or efficient in its original form (Helal et al., 2019). Low level of solubility, as well as permeability, low stability, fast metabolism, and ineffective targeting, are the problems often occurring with substances of natural origin (Helal et al., 2019). Mushroom-derived active compounds are not excepted from these problems.

There is a long-established opinion that humans and primates are not able to digest chitin, an integral compound of mushrooms, lichens, and insects. For some primates, mushrooms are a significant part of their diet. Moreover, mushroom compounds are likely more bioavailable to humans than we assumed before, because most studies show that acidic chitinase (hCHIA) is functional in the stomach of the majority of participants (Janiak et al., 2017). However, more studies are needed.

Generally, mushrooms' bioactive molecules can be categorized as high- and low-molecular-mass compounds. Polysaccharides, such as β-glucans, heteroglycans, peptidoglycans, lectins, dietetic fibers, non-digestible polysaccharides, and chitin substances, as well as RNA compounds have high molecular mass. Terpenes, steroids, and phenolic compounds are all characterized by low molecular mass (Wasser & Weis, 1999). This characteristic is very important for industrial applications since it affects the bioavailability and actual biological effect expressed *in vivo*. The highest therapeutic potential of all mushroom constituents is expressed by polysaccharides, in particular glucans, due to their natural structural variability (Rop et al., 2009). Based on the assumption of Rowan and Smith (2002), only four monomer sugars can produce 35,560 unique tetrasaccharides. Different mushroom species were found to be sources of anticancer polysaccharides, established as anticancer pharmaceutical agents: *Schizophyllum commune*, *G. lucidum*, *Lentinula edodes*, *Grifola frondosa*, *T. versicolor*, *Tropicoporus linteus* (Maity et al., 2021). However, the anticancer effect of the original molecule was better expressed when the injection route (instead of oral ingestion) was applied (Rowan & Smith, 2002). This is due to glucan's low solubility in water (Han et al., 2008). Characteristics, like linkages in the main chain, the number, and structure of side branches, type of monosaccharide units, functional groups, and bound with proteins, also influence the biological activity (immunomodulatory, antioxidant) of glucans as well as solubility (Rop et al., 2009). Different opinions exist concerning the structure-activity connection. For example, it was believed that the triple helix structure guarantees the highest biological response. Further, explorations proved that loss of helix form, like in alkali hydrolysis, does not necessarily mean the loss of biological activity. The current opinion is that high molecular mass and highly branched structure is of the greatest importance for β-glucans activity (Synytsya & Novák, 2013).

In many cases, modification of an original molecule had a positive effect on its immunomodulation and antitumor activity through the increase of solubility (Wasser & Weis, 1999). Chemical methods like carboxymethylation, sulfation, and phosphorylation are successfully used to increase the solubility of polysaccharides (Synytsya & Novák, 2013). After being carboxymethylated (1→3)-β-glucan from the sclerotia of *Pleurotus tuber-regium* had higher water solubility and improved *in vitro* and *in vivo* anticancer activity (Zhang et al., 2004). There are other methods to modify glucans: physical and biological (Wang, Sheng, et al., 2017). The outcomes are different, as Li et al. (2015) explained since different methods affect:

- the change in spatial structure without affecting the primary structure, like with physical modification; the outcome is the change in physicochemical properties and bioactivity;
- the introduction of functional groups changes both conformation and primary structure, so it is a method of choice when it comes to an increase in solubility.

Among low-molecular-mass compounds from mushrooms, polyphenols are considered very important. They express antioxidant, anti-inflammatory, and antitumor activity (Palacios et al., 2011). Usually, total phenolic compounds are measured in mushroom species (Li, Li, et al., 2020). However, specific phenolic compounds can have different bioactivity and to evaluate some species, the researcher needs to make a profile of the compounds present as well as measure their concentration (Islam et al., 2016). Like with polysaccharides, the bioavailability of phenolics from mushrooms depends on several factors: cultivation technique, source of origin, processing of raw material, the presence of other compounds in a complex matrix, and physicochemical characteristics. Dietetic fibers increase flavonoid absorption, so when present with polysaccharides, these fungal compounds are more bioavailable (D'Archivio et al., 2010). Even the same species harvested during the different stages of development contain different amounts of phenolics and also different phenolic compounds profile, as demonstrated by Petrović, et al. (2019) when mycelium, immature, and mature fruit bodies of puffballs were examined. The chemical structure is among the most important factors for phenolics bioavailability. They are mostly present in the bound form, bounded to sugar derivatives, which have to be hydrolyzed in the intestinal tract (Barros et al., 2009).

The mushroom crude extract is usually a combination of several compounds present in different amounts. Most of them have strong and unpleasant flavors (Ribeiro et al., 2015). Mostly, the unpleasant taste turns off healthy consumers from long-term use of mushroom nutraceuticals, also shown in the case of other functional foods, like beverages. Nguyen et al. (2019) used extracts of *G. lucidum* in the formulation of a novel Shiraz wine product. The increased amounts of mushroom extract (more than 2 g/L) appeared as less acceptable for the consumer due to the bitter taste originating from triterpene acids (Nguyen et al., 2020).

Traditionally, mushroom nutraceuticals started their market life in the form of pills and capsules consisted of crude or refined extracts. As the market developed the need for resolving the problems like low stability, bioavailability, and sensory attributes became an important and integral part of new supplement formulations. The industry approached these issues with the implementation of technology through improved drug delivery systems. Put simply, the active compound should be ingested in the form that enables a safe route to the place of its action. The two most promising options are encapsulation and nanotechnology, also known as green technologies.

4.8.1 Encapsulation of mushroom nutraceuticals

Encapsulation enables the protection of an active compound from its surroundings by creating a physical barrier-wall material. In the case of a nutraceuticals environment, this is the package as well as the digestion route. Some compounds, like polyphenols, present in the mushroom extract are prone to oxidation when exposed to light and oxygen, or degradation under high temperature (e.g. spore oil), and have to be preserved until entering the digestive tract (Pettinato et al., 2020). The encapsulation methods are diverse: spray drying, spray chilling, extrusion coating, coacervation, fluidized bed drying, supercritical antisolvent precipitation, ionic gelation, as well as liposome entrapment (Dias et al., 2017; Santana & Macedo, 2019). In an industrial setting, factors like price, simplicity, and efficacy come as even more important than the preservation of active compounds, so the method of choice does not have to be the most efficient in maintaining the bioavailability. It should rather enable the competitive price with compounds that are efficient enough. Usually, a compromise has to be made, and in this case, spray drying appears as the most common technique in practice (Ribeiro et al., 2015; Shao et al., 2019). The encapsulation materials have to fulfill physicochemical, safety, technological, and commercial requirements. The materials can be of natural or synthetic origin, like alginate, carrageenan, chitosan, gelatin, cellulose acetate phthalate, starches, whey proteins, and cashew gum (Dias et al., 2017). In the case of nutraceuticals, mushroom derived as well, maltodextrin is among the most common encapsulating materials. Lately, promising materials, by-products of the food industry, are emerging as encapsulating agents (Comunian et al., 2021). Sometimes, a combination of encapsulating materials can be used, like in the case of a study conducted by Shao et al. (2019). Here, maltodextrin was combined with whey protein, soybean protein isolate, and sodium caseinate. These biomaterials were chosen due to their good emulsification and film-forming properties. It was shown that microcapsules of *G. lucidum* polysaccharide had better-controlled release ability and better stability after 10 weeks of storage. Petrovic et al. (2019) used calcium-alginate for the entrapment of chaga (*I. obliquus*) extract rich in phenolic compounds. The outcome was pH-responsive formulation with a prospect in the therapy of gastrointestinal conditions (Figure 4.4).

4.8.2 Nanotechnology-mycosynthesis

One of the most exploited and promising fields to apply nanotechnology is the improvement of drug delivery (Pandey et al., 2020). The minute size of particles (less than 100 nm) change materials attributes with the prevalence of quantum effects. When it comes to nutraceuticals, the main benefit of nanotechnology is improved bioavailability. Additionally, nanoparticles (NPs) enable targeted drug delivery thus eliminating acute toxicity. Active compounds are more easily absorbed by microvilli in the intestine while the small size of particles enables passage of the location-specific barriers (Delcassian & Patel, 2020). Classical NPs synthesis has its drawbacks and more efficient and environment-friendly green nanotechnology establishes in many fields, especially medicine and pharmacy (Nabipour & Hu, 2020; Pandey et al., 2020). If fungi, viruses, and bacteria, or extracts are used as reduction and stabilizing agents for metal NP synthesis the process is

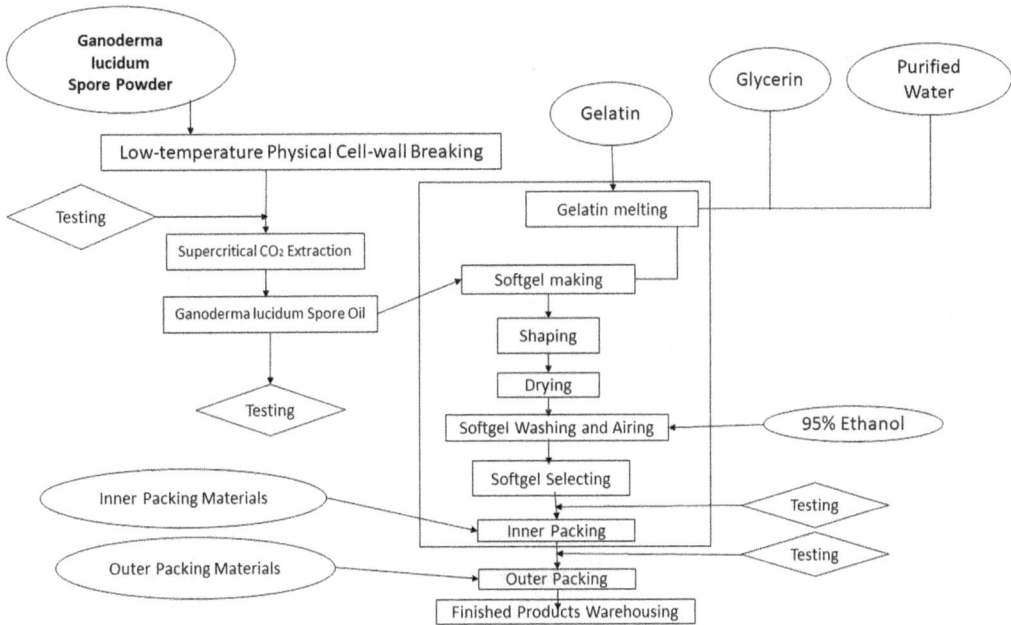

Figure 4.4 Flowchart of the production process of *G. lucidum* spore oil . By courtesy of GANOHERB GROUP.

called biosynthesis (Pandey et al., 2020). Specifically, mycosyntheis involves mushroom-derived extracts (Owaid, 2019). The benefits are eco-friendliness, reliability, and non-toxicity. The research community readily accepted the prospects of this technology manifesting through a significant number of original articles dealing with mushroom extract- or mycelium-based NPs. A thorough review of *Pleurotus* spp. exploitation for mycosynthesis was prepared by Owaid (2019), demonstrating the prevalence of Ag NPs. Aygün et al. (2020) used *G. lucidum* extract to prepare AgNPs. These NPs demonstrated improved biological activity accessed in terms of antioxidant, DNA cleavage, and antimicrobial activities. The authors suggested a synergistic effect valuable for the pharmaceutical and medicinal field. The popularity of silver NPs is due to their antibacterial properties expressed when destabilized AgNP releases silver ions. Besides, physicochemical characteristics contributed to the greatest marketing value of AgNPs in comparison with other metal NPs (Calderón-Jiménez et al., 2017). Jaloot et al. (2020) were the first to apply *Inonotus hispidus* mushroom's extract for AgNPs which showed an antibacterial and antifungal effect. Another study, by Klaus et al. (2020), demonstrated improved antimicrobial properties of AgNPs obtained with extracts of three mushroom species (*Agaricus bisporus, A. subrufescens* (as *A. brasiliensis*), and *P. linteus*). In a study by Nagajyothi et al. (2014), green technology–based silver chaga NPs expressed cytotoxicity against several cancer cell lines. Other metal NPs included Au (Narayanan et al., 2015; Owaid et al., 2019; Dheyab et al., 2020), bimetallic Ag@Au (Bhanja et al., 2020), and zinc oxide (Rafeeq et al., 2021) in combination with different mushroom species and purposes. Despite an increasing number of mushroom-NPs, there is no data on toxicity. On the other hand, *in vitro* and *in vivo* studies demonstrated readily interaction with molecules in biological media. Some of the effects of high exposure to AgNP are apoptosis, oxidative stress, reduction in cell viability, DNA damage, and carcinogenicity (Calderón-Jiménez et al., 2017). It has been proven that nano-silver accumulates in different organs including the liver, kidneys, and spleen. However, the critical oral dose is still unknown. Without adequate studies accessing the toxicity, risks, and accumulation in prolonged use, this mycosynthesis cannot expect significant growth in the nutraceutical market (Figure 4.5).

4.9 DOSING

In creating a fair nutraceutical product, besides safety, efficacy comes as a prerogative. The compound(s) present in the formulation has to be added in a sufficient amount to express beneficial effect. So, how much is enough? The terminology, in this case, is adopted from pharmacology and includes *efficient dose* and *efficient concentration*. In both cases, the dose is the quantity of drug (substance) which produces a biological response. But there is a difference: if the testing was performed *in vivo* the correct term is *effective dose*. Otherwise, as in the case of *in vitro* experiments, the term *effective concentration* is applied. In the field of mushroom nutraceutical, the difference between these two terms might point to the type of experiments used to support label claims. The clinically relevant dose will be the quantity of active ingredient which will express the specific effect supported by the scientific findings. Only in this case, the commercial product will be in line with the nutraceutical definition (Reis et al., 2017). In order to determine efficient/ clinically relevant dose there are two options: to perform a clinical study, following the protocols for pharmaceutical drugs, or to consult literature and search for the existing evidence *in vitro* or *in vivo*. The latter approach is a cheaper option but less reliable. The issues occur due to variations in the mushroom source, way of cultivation, and extraction technology. Many studies performed *in vitro* on cell cultures are not implemented at the product level by the industry. They are small-scale, highly specific, applying preparation techniques not adopted by raw material suppliers. In

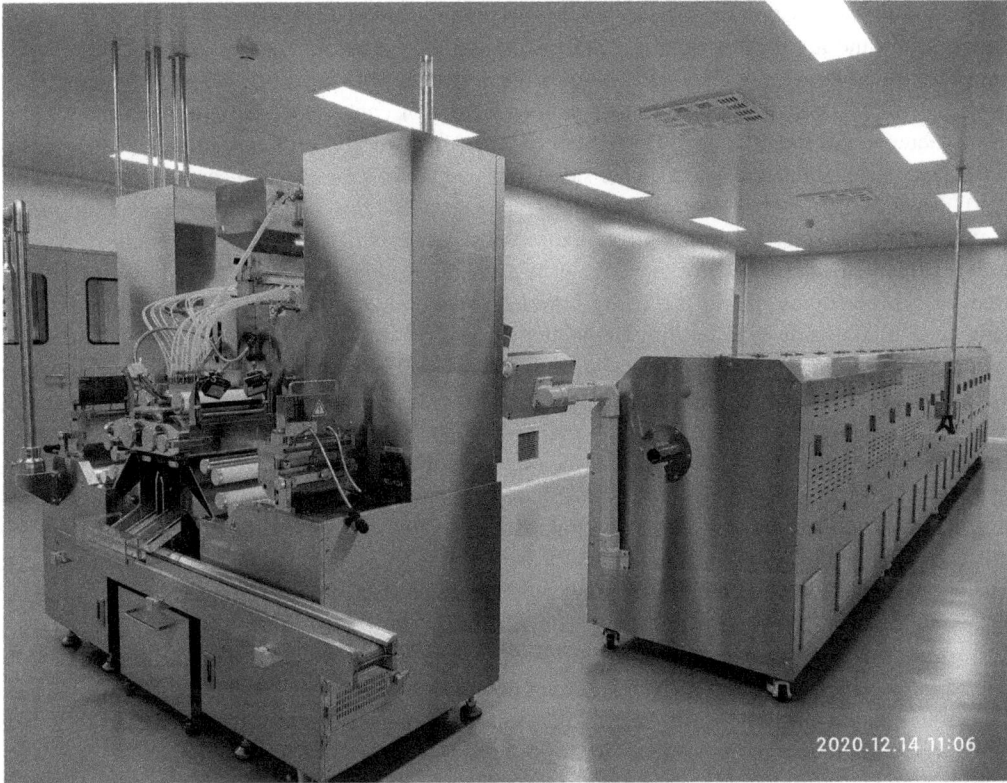

Figure 4.5 Soft-gel production line. By courtesy of GANOHERB GROUP.

general, a good rule of thumb is to use *in vitro* studies as a guide for more applicable clinical studies. Unfortunately, many nutraceutical-based products on the market today claim immune and against free radical protection, even though such claims are based on *in vivo* analyses conducted on mice solely. In some cases, their formulas include mushrooms known not to have such therapeutical properties but probably included to be used as formula enhancers. A common practice by the industry is to use a very small quantity of a particular highly therapeutic medicinal mushroom. And state with big letters on the label its presence in the formula (*fairy dusting*).

Complex combinations including several mushrooms and their body parts are even harder to evaluate due to a synergistic effect although an opposite might occur, like in the case of antioxidants. Usually, mushroom compounds act synergistically as demonstrated by Ribeiro et al. (2015). When added into a cottage cheese extracts of *Suillus luteus* and *Coprinopsis atramentaria* exhibited increased antioxidant activity. An adverse effect is also possible, when compounds are added in a high amount because antioxidants act like pro-oxidants, weather in food matrices or intracellular (Novakovic et al., 2020). This is the case with phenolic compounds that are present in mushroom extracts (Jomová et al., 2019). Rather than species-dependent or brand of preparation-dependent intracellular reactive oxygen species generation is determined by the concentration of the mixture of mushroom polysaccharide and phenolics (Wei & Van Griensven, 2008).

Several of the mushroom polysaccharide compounds have passed through phase I, II, and III clinical trials and are used extensively and successfully as drugs in Asia to treat various cancers and other diseases. A new class of antitumor medicinal mushroom drugs is called *biological response modifiers*, which have been used as a new kind of cancer treatment, along with surgery, chemotherapy, and radiotherapy. These treat cancers, focusing on improving the patient's quality of life since they significantly reduce the side effects and help overcome cancer growth (Lam et al., 2020). Most of them activate natural immune responses of the host and can be used as a supportive treatment for cancer prevention and, in some cases, alone with conventional therapies (Wasser, 2011). Effects are described to be variable in a dose-dependent manner, according to the form used, or powder blend, being positively correlated to the administered time. Literature data on daily dose intake, side effects, drug/herb interactions, or long-term period effects in humans is scarce. The form in which the product is administered and its potency in humans and animals is still in debate and accurate data is missing; however, extracts are known to contain concentrated active ingredients and therefore the daily intake should be somewhat lower compared to powdered fruit body or mycelia unless there are no side effects in the case of an overdose. Caution should be taken on synergistic drug/herb interactions of some of the medicinal fungi but also their antagonistic effects with such herbs. The daily oral intake depends on the condition to be treated and this tells us the range of dosage recommended for best benefits. Data is also missing on the long-term intake of various forms of medicinal fungi or there is no data on safety regarding the intake of dietary supplements in young children (Table 4.1).

Despite the lack of consistent data on human clinical trials for fungi like *G. applanatum, G. tsugae,* or *C. militaris* dietary supplements originating from these medicinal fungi are available on the markets across Europe, North America, or Asia and are gaining rapid popularity in other parts of the world. More and more products are formulated in combination with herbal adaptogens such as ashwagandha (*Withania somnifera*), different ginseng species (*Panax ginseng, Eleutherococcus senticosus,* etc.), schisandra (*Schisandra chinensis*), licorice (*Glycyrrhiza glabra*), or astragalus (*Astragalus propinquus*) and other medicinal herbs. Sometimes medicinal mushrooms are part of formulations of up to 100 herbs or more for one product – with a clear message that more variety is better. This is not necessarily true because of possible unknown antagonistic reactions that such herbs may cause in a dose and time-dependent manner. In medicinal mushrooms, we may observe similar therapeutic effects and very few mild side effects, which resonate with formulations using a blend (complex) of several medicinal mushrooms in one product. Among medicinal herbs there is a contrast among the therapeutic effects, the dosage needed for a particular response and some of them have serious side effects. This is why such medicinal mushroom-herb formulations should be carefully analyzed and evaluated before launching such products on the market. This point of view is also supported by Mehta (2010) and Helal et al. (2019).

4.10 MUSHROOM DIETARY SUPPLEMENTS PRODUCTION AND ETHICS

Like in any other branch of life and business, the nutraceutical industry is not free from ethics violations. Among already mentioned issues like mushroom identification, especially common is *fairy dusting*. The term is not present in the scientific literature, but it is very descriptive and may be used by both, producers and consumers. This term refers to cases where one or several ingredients are added in a formula in such small quantities that it disturbs the overall effectiveness of the final product. A double-blinded, placebo-controlled, cross-over intervention study of Wachtel-Galor et al. (2004) examined biomarkers for health parameters and conditions often used for

Table 4.1 Recommended dosage, possible side effects, and drug/herb interactions in various forms of *Ganoderma applanatum*, *Ganoderma tsugae*, *Cordyceps militaris*, and *Agaricus subrufescens* (as *A. brasiliensis*)

Fungus	Form	Dosage	Side effects	Drug/herb interactions	References
Ganoderma applanatum	mycelia powder	30 g/day in tea or water extract	No data	No data	Hobbs, 2003
Ganoderma applanatum, G. tsugae	tincture	1–3 drops/1–3 × per day	Safe		Pursell, 2015
Ganoderma applanatum, G. tsugae	mycelia powdertincture	1 g/3 × per day3–6 g/day – in chronic diseases9–15 g/day – in acute conditions2–4 tbs/3 × per day	Contraindicated in case of obstructed bile duct. Occasionally, skin, rash, loose stool, dry mouth, sleepiness, bloating, frequent urination, sweating, nausea may occur	Synergistic with cefazolin, interferon-alpha, interferon-gamma, acyclovir. Caution should be exercised with immunosuppressive drugs	Buhner, 2012
Ganoderma tsugae	extractfruit body	2–5 g/day3 g/day	No data	No data	Powell, 2014
Cordyceps militaris	mycelia	2 capsules twice/day	Overdose- Immune suppression, headache, edema, anxiety, nosebleeds	Not to use with immunosuppressive medication	Kuhn & Winston, 2008
Agaricus subrufescens	mycelia powder hot-water extract	2–4 of 400 mg × 2/day 1600-3200mg/day	Non-toxic	Not to use with immunosuppressive agents	Stengler, 2005

commercial purposes in mushroom nutraceuticals (antioxidant status, cardiovascular disease risk, DNA damage, inflammation, immune status, and markers of renal and liver toxicity). In this study, the beneficial effect of supplementation with a commercial product containing 1.44 g of *G. lucidum* extract was expressed after four weeks. In one example, a commercial product presenting a complex of different mushrooms, in which the amount of *G. lucidum* mycelium (not extract!) was 110 mg only. The same product contained raw materials of 15 mushroom species and only one was produced from the fruit body. The amount of this ingredient was 16 mg in a total of 1,050 mg. A clinically relevant dose cannot be obtained by the consumption of this product. But, how is it possible that fairy dusting passes? This is more of a question to be addressed to psychologists and marketing teams of nutraceutical companies. The poor knowledge about mushroom dietary supplements, the desperation and trust among consumers, the lack of proper legislation, and poor ethics leads to low-quality products.

Some regulatory procedures important to ensure quality control standards are still missing in the process of obtaining dietary supplements from medicinal mushrooms. Attempts to establish such standards are currently ongoing and are the major focus of the *International Medicinal Mushroom Conference* scientific community interested to solve existing current issues. However, there is still work to do not only on setting up reliable standards that would work internationally, but also there is an effort to be invested in their implementation by regulatory organizations, in particular countries, as rules to be followed by producers. Until then, the chaos persists across the mushroom nutraceuticals industry and among the products available on the markets. Human philosophy and nature may be blamed for the existence of doubtful products on the market today. Products derived from sources with no respect for ethics and of questionable quality produced with the sole intent to market them are not few. On the other hand, we may notice very few companies in the world today that struggle to get on the market the best possible mushroom-based products, which in terms of potency and safety are at the top – such products are manufactured regardless of cost and the steps involved in labor, show solid production protocols, and highly skilled personnel together with production transparency, consistency in their product quality, and scientific proof supporting their claims and strain identification. The adulteration of medicinal mushroom products is common among producers and it was emphasized by Wasser (2011) – for example different *Ganoderma* species for *G. lucidum*; *Stereum* species for *T. versicolor*; teleomorphs of different *Cordyceps* species, and even different anamorphs for *O. sinensis*. This definitely puts particular mushroom-based producers in a doubtful position easily recognized by competitors due to bad publicity. Several companies look for obtaining raw material from China because of low cost and availability. Unfortunately, when it is about medicinal mushroom raw material, China has a bad reputation among the pharmaceutical industry but also product consumers. Not China has to be blamed for this, but those who purchase raw materials without knowing where they come from, the way they were produced and deposited. This is valid for any location in the world – this is another ethics violation by some companies involved with formulation, extraction and product launch on the market. Not few are the mushroom farmers interested to cut corners in order to get better yields with the risk of using *non-organic* substrates but with mushrooms sold as certified *organic*. Some of the substrates they use state as 100% natural, but not certified organic. Such practices when combined with lack of knowledge and poor experience turn mushroom nutraceuticals, that ironically, should help the population, into a dangerous journey for those who purchase them. Of course, there will always be an ugly truth about these products and producers that want to risk, but the main point here is for the regulatory organizations to establish reliable protocols as part of their good manufacturing practice regulations able to reduce the number of such events.

4.11 PUSH IT TO THE MARKET: ARE WE READY FOR MUSHROOM NUTRACEUTICALS?

Theoretically, we are. Just in the USA the nutraceutical market worth greatly exceeded pharmaceuticals in 2014 (250 billion and 150 billion US$, respectively) (Shinde et al., 2014). Globally, the worth of this market in 2019 was 382 billion US$ (Global Nutraceutical Market Growth Analysis Report, 2020–2027, 2020). Ten percent of the nutraceutical market belongs to mushroom nutraceuticals (Wasser, 2011; Morris et al., 2016). However, these are global numbers and peer into the reports enable an understanding of how the acceptance of macrofungi dietary supplements is distributed around the world. So, the exact answer depends on who are *we*, and where we come from. Asia Pacific is the major market followed by the USA and Canada. The same was confirmed by the chapter's authors through the talk with nutraceutical market participants.

At the beginning of May 2019, Geneva, Europe: with more than 1,300 exhibitors and 24,000 visitors, the *Vitafoods Europe* expo was the biggest and most prestigious one for the nutraceutical and dietary supplement industry in the old continent. The exhibition hall was crowded with people, representatives of distributors, medical doctors, pharmacists, company owners, representatives, and R&D officials – all in search of new exotic ingredients, interesting, and potentially more effective products. The expo was so big and people were in a hurry to see as much as possible. But the sight of oval, shiny, brown-reddish, quite big object, lying on the booth showcase immediately attracted attention. People looked at it with curiosity and finally asked: – What is it? So strange… – It is a mushroom. – A mushroom? You must be kidding? The model you have is very interesting. – It is not a model, but the real fruit body. This mushroom is known as reishi, lingzhi, or *G. lucidum*. The astonishment grew in a second. – Can I touch it? – Yes, of course. It is harmless. – And what is it for? – It is medicinal. – Really? In three days of Vitafoods Europe expo, many similar conversations occurred.

The same year, in October, a different continent, *SupplySide West* expo in Las Vegas, USA: again, more than 1,300 exhibitors displaying ingredients for nutraceuticals and dietary supplements. Two days in crowded Mandalay Bay Convention Center. People were approaching the booth with a paper map: – I see you have medicinal mushrooms. Certified organic? What else do you have, except reishi? Do you have lion's mane? Is it organic as well? Do you perform extraction? What kind? … and many more specific questions showing that they know a lot of details concerning medicinal mushrooms.

The situation is similar in retail stores. While in Europe only two or three mushroom-based supplements can be found on average, the retail racks in the USA offer a significant diversity in species, form of drug delivery, and producers (VanderMolen et al., 2017). In China, special shops, selling only mushroom-based supplements, already exist. There, a person can discuss its condition with a medical doctor in the shop, and with his help choose the right formulation. It appears that the stage of development of the mushroom nutraceutical market depends on which continent you look at. Europe seems to be in the infant stage of this knowledge. On the other hand producers and consumers from the USA and Asia are well informed and much engaged in mushroom exploitation. And although the openness toward mushrooms as a preventive treatment in the West has progressed (personal observation based on commercial exhibitions and talk with visitors) the research dealing with this topic is scarce. As Lai et al. (2004) reported there is an increasing interest in the marketing of lingzhi (*G. lucidum*) dietary supplements in New Zealand. A telephone survey the authors had conducted showed how consumers use lingzhi nutraceuticals, for how long, and if the product met their expectations based on the efficacy. Interestingly, they observed the lack of public acceptance.

Away from industry and research experts, general consumers are not knowledgeable enough about mushroom nutraceuticals. The main problems arise from vast terminology, poor knowledge and understanding of fungi, mycophobia, the lack of public acceptance, non-transparency in labeling, and marketing spinning.

4.12 GENERAL STRATEGIES WHEN FORMULATING MUSHROOM NUTRACEUTICALS

Like in any other industry there is a difference between objective goals, recommended practice, and on-sight doing. The actual formulation depends on many parameters, with a strong influence of the economical aspect. Ideally, novel products should merge the raw material with scientific and clinical study supported background and technology that will minimally affect the beneficial characteristics. Moreover, the purpose of technology is to preserve and even potentiate the health-promoting effect of the starting material (Ribeiro et al., 2015). But the moment you step out from the research facility and enter the highly competitive world of this industry and market things are not that straightforward. Depending on resources as well as ethics, the strategies to consider when formulating a mushroom nutraceutical vary greatly. The cheapest option is to have the cheapest product. This does not mean that the product is not safe, but rather ineffective. Effectiveness, besides safety, is the most important aspect of the nutraceutical product. Getting high-quality raw material is the starting point. Which species to choose, which supplier, which part of the mushroom is the best, how much to add, bioavailability and stability, which drug delivery system to apply, and how to manage a safe product – are the issues to be addressed when developing any nutraceutical. After all, the product's label will contain some health-regarding claims, so it is recommended to be able to support them. Another option is to make a fair product with effective ingredients, meaning to add it in a clinically relevant dose. Finally, nutraceutical may consist of one or more active ingredients, in this case derived from different parts of the organism (mycelium, fruit body, or spores). The diversity of the approach reflects on the number of products present on the market. As reported by Shinde et al. (2014), just in the USA 1,000 new dietary supplements appear on the market every year. The in-market search conducted by VanderMolen et al. (2017) disclosed 70 products with some version of chaga on its label, 315 products mentioning reishi, 223 with shiitake (*L.edodes*), 267 containing maitake (*Grifola frondosa*) ingredient in some form, and 39 with turkey tail (*T. versicolor*).

4.13 OPPORTUNITIES AND PROSPECTS

The nutraceutical market seems mature with well-established sub-categories and market analysis predicts its stable growth for about 8.4% in the next 6 years (Global Nutraceutical Market Growth Analysis Report, 2020–2027). Mushroom products are already a part of the nutraceutical market although not properly visible across the global scene. With competitive players from other categories of raw materials, what are the opportunities for mushroom dietary supplements? One of the focal points should be the mushroom's exotic nature. This aspect, besides new delivery formats and improved bioavailability, represents one of the differentiation offerings. New mushroom species, eco-friendly cultivation techniques, and a unique story to tell are additional benefits for new products. With typical consumers being the aging population and millennials the focus is expanded from treatment to prevention of diseases (Childs, 2000). With mushrooms' immunomodulatory ability substantiated by new clinical trials, it is the objective to expect to fulfill these needs. Advisable is to focus on further segmentation inside the product categories, like unmet consumer needs (e.g. menopause), female energy, and especially wellness and beauty. The latter exploded in the last couple of years since big players, such as Nestle and Unilever, entered the sector. Their new products are based on *beauty from within,* offering treatments for skin or hair through nutrition. Since mushrooms are already recognized as cosmeceuticals it should not be unlogical to expect their increased usage in these formulations (Taofiq et al., 2016). Mushrooms can reduce the severity of inflammatory skin disease and help in hyperpigmentation disorders. The additional prospect of beauty through food is that it is often presented as a powder that should be mixed with food or beverage. This would rid the mushroom nutraceutical players of the need to pack their products inside of gel capsules, thus decreasing the overall cost. Also, mushrooms can be mixed with flavorings of natural origin. This can improve the consumption experience and make it more enjoyable, especially when mushrooms rich in bitter compounds like *G. lucidum* are part of the formulation. Macrofungi already contain umami compounds making them suitable as a supplement or spice to be added to dishes. An interesting palette of mushroom nutraceutical serving the beauty and protection segments can derive from their melanin content. Melanin acts protective against UV light, an asset heavily exploited as a part of sun creams. A study conducted by Oak et al. (2021) demonstrated that consumption of grape for two weeks increased the natural protection of the skin by 74.8%. Prados-Rosales et al. (2015) reported that melanin pigment from *Auricularia auricula-judae* mushroom can protect mice against ionizing radiation. The authors even proposed mushroom-derived melanin in the design of edible electronic devices due to the semiconductor properties of melanin.

Another still-under-developed sector for mushroom nutraceutical use is the field of neuro-protection. Mushrooms' ability to act neuroregenerative and neuroprotective was recently thoroughly reviewed by Remya et al. (2019). The main mechanisms of action are the effect on mitochondria and endoplasmatic reticulum, and the decrease of oxidative stress. Species *Hericium erinaceus*, *G. lucidum*, *Pleurotus cornucopiae*, *Polyozellus multiplex*, *Armillaria mellea*, and *Pleurotus giganteus* were noted as most promising in this regard. With neurological disorders being the second leading cause of death globally in 2016 (around 9 million/year) and an increasing burden on the economy, healthcare system, and society, there is a huge interest and market possibility for mushroom nutraceuticals (Naghavi, 2019).

Another area of opportunity for mushroom-based supplements is their use in sleep deprivation disorders. The interest for nutraceutical solutions to sleeping disorders is quite significant, having in mind that 50% of just the U.S. population will suffer from it during their life span. No wonder that in the year 2020 the global market of sleep supplements reached 67.37 billion US$ worth (Sleep Aids Market Size, Share & Industry Growth, 2020–2025)! Some of the natural compounds already in use, but without a definite conclusion about their effectiveness, are caffeine, chamomile, cherries, L-tryptophan, marijuana, and valerian (Yurcheshen et al., 2015). Hu et al. (2018) described several natural ingredients with prospects of being used for treating different sleep disorders, like magnolia (*Magnolia officianalis*), rosemary (*Salvia rosmarinus*), longan (*Dimocarpus longan*), ginseng (*P. ginseng*), the dried seed of suanzaoren (*Zizyphus jujube* var. *spinose*), sinomenine, decursinol, apigenin, and epigallocatechin-3-O-gallate. Among promising natural products, mushrooms were also evaluated. Hu et al. (2013) suggested that cordycepin (adenosine analogue from *O. sinensis*) may be helpful for sleep-disturbed subjects. In their study cordycepin reduced sleep-wake cycles and increased non-rapid eye movement sleep. Sleep regulation was also improved by an increase in the protein levels of adenosine receptors in the rat hypothalamus. Another species, *Wolfiporia cocos*, is known as a TCM herb for treating sleeping problems like insomnia (Singh & Zhao, 2017). Pachymic acid, a triterpenoid from W. *cocos*, was tested in a mouse model, proving its hypnotic effect as an enhancer of pentobarbital-induced sleeping behaviors via $GABA_A$-ergic mechanisms (Shah et al., 2014). Furthermore, Vigna et al. (2019) found *H. erinaceus* to be effective in improving mood and sleep disorders tested on 77 volunteers who were also suffering from overweight or obesity. The same mushroom even finds its way as a component of a commercial product marketed in Canada (U~Dream[®]). A commercial product, Amyloban[®] 3399, based on *H. erinaceus,* was tested in a pilot study among eight postgraduate female students in Japan. The product has been administered for four weeks and the researchers reported an increase in salivary-MHPG, corresponding to an improvement in anxiety and quality of sleep (Okamura et al., 2015). Although the results from these studies are promising, there is a lack of evidence-based clinical studies and pilot studies involving a larger number of participants.

Scientific findings from the latest period point to the importance of gut microbiota, consisting of more than 10^{14} microorganisms (Thursby & Juge, 2017). Microbes living in the human gastrointestinal tract are affecting immune and metabolic homeostasis and neurobehavioural traits. Specific products of microbial fermentations directly affect human health. Dietary fibers and endogenous intestinal mucus are substrates for microbial activity (Valdes et al., 2018). Besides depending on heritage, microbiota is strongly affected by environmental factors, especially diet (Rothschild et al., 2018). That is the reason for the emergence of many dietary supplements aiming to improve the content of microbiota as well as their number. These products are known as probiotics and prebiotics. Since mushrooms and yeasts contain glucans, which are dietary fibers, mushroom nutraceuticals can offer benefits in this health-related area as well. For example, Chang et al. (2015) demonstrated the effectiveness of *G. lucidum* polysaccharide as a prebiotic agent to reduce obesity in mice by modulating their gut microbiota. The authors also suggested mushroom polysaccharides as a prebiotic agent in the prevention of gut dysbiosis and obesity-related metabolic disorders in human subjects. Another study reveals that chaga's polysaccharide induced change in the gut microbiota toward a healthy bacterial profile (Hu et al., 2017). A positive impact on gastrointestinal microorganism alterations was observed in the case *T. versicolor*, *T. linteus*, *H. erinaceus*, and *L. edodes*, as reviewed by Jayachandran et al. (2017). A study conducted by Sawangwan et al. (2018) investigated the prebiotic properties of several edible mushrooms (*A. auricula-judae*, *L. edodes*, *Pleurotus citrinopileatus*, *P. djamor*, *P. pulmonarius*, and *P. ostretus*). The prebiotic properties were evaluated based on probiotic growth stimulation, pathogenic inhibition, and gastrointestinal tolerance. The highest probiotic growth stimulation was observed for *Lactobacillus acidophilus* cultured with *L. edodes* as well as for *Lactobacillus plantarum* cultured with *P. pulmonarius*. The highest survival in gastrointestinal conditions was noted in the case of *P. djamor* extract cultured with *L. acidophilus*.

People vary in gut microbiota, so the diet may not be universal but rather the individual in the coming era. As already discussed, mushrooms differ in their polysaccharide content and structure, which perfectly combines with the variations of microbiota in individuals.

4.13.1 Mushroom nutraceuticals and COVID-19: Efficient antiviral or another murky field?

With the emergence of the coronavirus pandemic at the beginning of 2019, the dietary supplement market experienced a period of significant increase. Simultaneously, adverse sources of information (governments, nutritional associations, health organizations, supplement producers, and marketers) bombarded the population with nutritional information on how to prevent COVID-19 (de Faria Coelho-Ravagnani et al., 2020). However, the information appears to be deduced from the pre-pandemic knowledge rather than current evidence. The Food and Agriculture Organization of the United Nations stated that no food or dietary supplements can prevent COVID-19 infection while at the same time insisting on a healthy diet to support a strong immune system (Maintaining a Healthy Diet during the COVID-19 Pandemic, 2020). However, lockdown measures affected the nutrition behavior of young adults. As identified by Huber et al. (2020), food consumption increased in the stressful time, and it was triggered by the consumption of bread and confectionery. This study demonstrated that although some nutritional recommendations exist, they are hard to follow concerning other aspects in connection with pandemic (e.g. lockdown and restricted social life). Due to vague information about disease prevention, restricted approach to the burdened public healthcare system, and strong marketing of alternative measures, the population turned to over-the-counter products, like nutraceuticals and dietary supplements.

As stated by market analysis companies, two-thirds of the U.S. population have taken dietary supplements since the pandemic started (Largest Coronavirus-Related Growth in Dietary Supplements Use U.S. 2020, 2020). The same growing interest and use of supplements was confirmed by Hamulka et al. (2020), who performed Google Trends analysis and cross-sectional studies during two waves of the pandemic. The researchers reported increased interest in the use of vitamins and nutraceuticals, which started during the first COVID-19 wave. Consumer's main goal was to strengthen immune system since no proper or effective measures were identified toward the virus. These beliefs were not supported by any scientific evidence so far. It has been stated that such a situation increased the risk of elevated intake of supplements due to irrational behavior as a consequence of the non-existing education of consumers.

SARS-CoV-2 belongs to β-coronaviruses, and infections caused by this group are characterized by cytokine storm and lymphopenia (Li, Liu, et al., 2020). In severe cases, which account for about 20%, additional lung infections occur due to invasive devices, sedation, and impairment of mucociliary clearance (Maes et al., 2021). Although no effective treatment has been developed, patients are treated with non-corticosteroid anti-inflammatory immunosuppressing agents and corticosteroids, whose aim is to lessen the uncontrollable release of pro-inflammatory cytokines (Shahzad et al., 2020). Since the infection is characterized by cytokine storm, different studies examined the significance of pro-inflammatory cytokines as well as the moment of their release. Therapeutic effect has been observed when IL-1 family members and IL-6 cytokines were suppressed (Murphy et al., 2020). At the same time, in patients who did not survive COVID-19 infection, an increased level of IL-6 was measured.

To identify possible treatment, different approaches have been evaluated, among them Chinese medicine including compounds like glycyrrhizin, hesperetin, baicalin, and quercetin (Li, Liu, et al., 2020). Moreover, a clinical study, based on Ayurvedic medicine, was registered at https://clinicaltrials.gov/ct2/show/NCT04530617. In this case, the medicinal plant *Artemisia annua* was tested against COVID-19 in a randomized, double-blind, placebo-controlled, multi-arm, multi-center, phase II trial. Known for its ability to inhibit viral entry this plant's active compound, artemisinin was effective in decreasing a composite outcome of hospitalization and improve viral clearance after 14 days of treatment (Romero et al., 2006).

Based on the evidence of these traditional herbs, it is logical to expect benefits from medicinal mushrooms too. Besides being a part of a long tradition in Asia, novel studies confirmed several crucial activities when it comes to COVID-19. Mushrooms, especially their β-glucans, are immunomodulators and anti-inflammatory agents. This fact brings them under the focus since they can possibly affect a delicate mechanism of virus-induced inflammation: up- and down-regulation of cytokine release.

Based on this logic Murphy et al. (2020) tested the immunomodulatory and cytoprotective properties of lentinan (β-glucan) isolated from *L. edodes* and commercial lentinan. The study was performed using in vitro models of lung injury and macrophage phagocytosis. Both compounds expressed varying immunomodulatory activities, thus, becoming relevant as agents for targeting COVID-19 induced cytokine storm. Furthermore, patients with COVID-19 develop secondary bacterial infections (>10%), often caused by antibiotic-resistant bacterial strains (Mahmoudi, 2020). At the same time, macrofungi act as antibacterial agents and in some cases interact with antibiotics creating a multidrug-resistance inhibiting effect, which is known as single or multi-target action (Yamaç & Bilgili, 2006; Pandey et al., 2020). Masterson et al. (2019) tested β-glucan's (from *L. edodes*) ability to decrease the number of clinically isolated *Klebsiella pneumoniae* using *in vivo* lung infection model. Application of shiitake's glucan reduced bacterial load and improved physiological parameters.

Another biological activity expressed by mushrooms is antiviral. They have been tested against different viruses, like infectious hematopoietic necrosis virus, herpes simplex type 1 virus, hepatitis B, HIV, and influenza. Due to significant antiviral effect, two studies performed screening of available literature evidence and hypothesized about mushrooms' possible antiviral activity against COVID-19 (Hetland et al., 2020; Shahzad et al., 2020). As perspective mushroom candidates *A. subrufescens* (as *A. blazei*), *I. obliquus*, *G. frondosa*, and *H. erinaceus* were selected. From experiments with other viruses, these species exhibited several mechanisms of action, e.g. direct inactivation and inhibition of virus replication, upregulation of antiviral cytokines (IFN-α and IFN-γ), synergy with IFN α-2b, bypassing the gap between innate and adaptive immunity (Hetland et al., 2020).

Although scarce, the existing literature on mushroom prophylactic and therapeutic use in COVID-19 sounds promising. This creates an opportunity for the study-based commercialization of mushroom nutraceuticals. However, it can easily slip into the murky area of other clinical study–unsubstantiated claims related to these products, in which case the chance for fair market expansion can be missed if mushrooms prove ineffective. This sensitive balance has been examined by Louca et al. (2020), who surveyed 1.4 million users of the *COVID Symptom Study App* with a questionnaire concerning the use of supplements. The authors investigated whether the regular use of dietary supplements reduces the risk of testing positive for SARS-CoV-2 infection. Among reported supplements, 6%–14% or risk lowering was in the case of omega-3 fatty acids, multivitamins, or vitamin D, especially in women, but not in men. No effect was observed for vitamin C, zinc, or garlic supplements. On the other hand, Majeed et al. (2021) found that patients with COVID-19 from southern India have a significantly lower level of selenium in serum compared with the control population and suggested selenium supplementation. In another multicenter, placebo-controlled, randomized clinical trial, Ashraf et al. (2020) found that treatment with honey and black cumin alleviated COVID-19 symptoms within 6 days. These examples prove how sensitive the topic of nutraceuticals and disease is. This is also a matter of culture: while the USA increased consumption of vitamins and minerals during the pandemic, consumers from Asia placed more trust in Traditional Chinese Medicine (TCM) products based on a survey posted by the New Zealand Institute for Plant and Food Research (Daily News on Nutraceuticals, Functional Foods, Health Food and Ingredients in Asia and Oceania.). This correlates with the fact that the Chinese government has officially added three TCM preparations as part of standard therapy for COVID-19. All three were tested in a clinical setting (Lian-Hua Qing-Wen Capsule, 2020, http://www. chictr.org.cn/showproj.aspx?proj=48889). As for the mushrooms, all patented antiviral products were applied by Stamets (2006), comprising of mycelium, extracts, and derivatives of mushrooms from several genera, which are listed in the patent as *Fomitopsis*, *Piptoporus*, *Ganoderma*, and a blend of medicinal mushroom species. The combination is stated as effective for preventing and treating influenza, avian influenza, rhinovirus, New World and Old World arenaviruses, hantavirus, *Orthopox* viruses, Venezuelan equine encephalitis, yellow fever, West Nile, Dengue, Rift Valley fever, sandfly fever, SARS, and other viruses. However, the patent has been abandoned in 2016, and a similar one was registered in France. It included the same mushroom species as in the previous patent, but claiming to be effective only against two viruses, pox and HIV (Stamets, 2005). Korean inventors included the fruiting body of *Porodaedalea pini* in a form of pharmaceutical composition claiming it to be effective against Coxsackie, herpes simplex, and influenza viruses (Kim et al., 2011).

With such scarce literature, evidence, and absence of clinical studies to support the effectiveness of mushroom nutraceuticals against COVID-19, the existing formulations are hard to evaluate. Deductive speculations can generate review papers thus pointing to the direction for future studies while the real market prospects still have to wait for the trial data. An increased number of clinical trials is still needed in order to build a solid base for the medicinal mushroom market worldwide.

4.14 CONCLUSION

With the outlook of the existing mushroom nutraceutical commercialization and market, it is evident that this field has a dual nature. While being traditionally used by many cultures, mushrooms have not been that welcome among Western world countries. The situation started to change among the US consumers driven by the expensive health-care system and greater awareness of preventive health approaches. The mushroom nutraceutical sector is about to open even more once definitive clinical studies elucidate their effect as well as the side effects. From an industry perspective, drug delivery systems have to be improved. However, maybe the most important is the transparency in the process of raw material production and formulations, followed by the quality control system development and standardization. Otherwise, this young nutraceutical sub-sector might miss exploiting in total its starting positive assets. The chemical nature of mushrooms provides multiple applications as evident in the case of gut microbiota modulation, sleep and mood disorders, neurodegenerative conditions, as well as in the case of constantly emerging viruses.

ACKNOWLEDGMENTS

This work was supported by the Ministry of Education, Science and Technological Development, Republic of Serbia (Grant No: Contact No. 451-03-68/2022-14/200051). The authors would like to thank GANOHERB GROUP (Fuzhou, China) for providing photos and protocol from their production. We thank M.Sc. Ivanka Milenković (Ekofungi, Serbia) for some information on organic cultivation of mushrooms. We are also grateful to Prof. Dr. Solomon P. Wasser (Institute of Evolution, University of Haifa, Israel) for providing us literature on human clinical studies of medicinal mushrooms and their compounds. Dr. Tuula Niskanen (Jodrell Laboratory, Royal Botanic Gardens, Kew, UK) gave some valuable comments regarding mushroom identification.

REFERENCES

Abdullah, N. R., Sharif, F., Azizan, N. H., et al. (2020). Pellet Diameter of *Ganoderma lucidum* in a Repeated-Batch Fermentation for the Trio Total Production of Biomass-Exopolysaccharide-Endopolysaccharide and Its Anti-Oral Cancer Beta-Glucan Response. *AIMS Microbiology*, 6(4), 379–400. 10.3934/microbiol.2020023

Akavia, E., Beharav, A., Wasser, S. P., & Nevo, E. (2009). Disposal of Agro-Industrial By-Products by Organic Cultivation of the Culinary and Medicinal Mushroom *Hypsizygus Marmoreus*. *Waste Management*, 29(5), 1622–1627. 10.1016/j.wasman.2008.10.024

Anderson, J. B., & Ullrich, R. C. (1979). Biological Species of *Armillaria Mellea* in North America. *Mycologia*, 71(2), 402. 10.2307/3759160

Announcement of the State Administration for Market Regulation on the Issuance of the Third Phase of Certification Risk Warning for 2020. (2020). *State Administration for Market Regulation*. Accessed December 31. http://gkml.samr.gov.cn/nsjg/rzjgs/202101/t20210104_324916.html

Aronson, J. K. (2016). Defining 'Nutraceuticals': Neither Nutritious nor Pharmaceutical. *British Journal of Clinical Pharmacology*, 83(1), 8–19. 10.1111/bcp.12935

Ashraf, S., Ashraf, S., Ashraf, M., et al. (2020). Honey and Nigella Sativa against COVID-19 in Pakistan (HNS-COVID-PK): A Multi-Center Placebo-Controlled Randomized Clinical Trial, November. 10.1101/2020.10.30.20217364

Assadpour, E., & Jafari, S. M. (2019). Advances in Spray-Drying Encapsulation of Food Bioactive Ingredients: From Microcapsules to Nanocapsules. *Annual Review of Food Science and Technology*, 10(1), 103–131. 10.1146/annurev-food-032818-121641

Aygün, A., Özdemir, S., Gülcan, M., Cellat, K., & Şen, F. (2020). Synthesis and Characterization of Reishi Mushroom-Mediated Green Synthesis of Silver Nanoparticles for the Biochemical Applications. *Journal of Pharmaceutical and Biomedical Analysis*, 178, 112970. 10.1016/j.jpba.2019.112970

Badalyan, S. M., Barkhudaryan, A., & Rapior, S. (2019). Recent Progress in Research of the Pharmacological Potential of Mushrooms and Prospects for Their Clinical Application. In: D. C. Agrawal & M. Dhanasekaran (Eds.), *Medicinal Mushrooms*. Springer Nature, pp. 1–70.

Bai, H., & Du, Y. (1997). Method for Breaking Cell Wall of *Ganoderma lucidum* Spore. issued 1997. https://patents.google.com/patent/CN1165032A/en

Bao, D., Kinugasa, S., & Kitamoto, Y. (2004). The Biological Species of Oyster Mushrooms (*Pleurotus* Spp.) from Asia Based on Mating Compatibility Tests. *Journal of Wood Science*, *50*(2), 162–168. 10.1007/s1 0086-003-0540-z

Barbosa, J. R., dos Santos Freitas, M. M., da Silva Martins, L. H., & de Carvalho, R. N. (2020). Polysaccharides of Mushroom *Pleurotus* Spp.: New Extraction Techniques, Biological Activities and Development of New Technologies. *Carbohydrate Polymers*, *229*, 115550. 10.1016/j.carbpol.2019.115550

Barcode of Life Data System V4. (2021). *Boldsystems.org*. Accessed March 19. http://www.boldsystems.org.

Barros, L., Dueñas, M., Ferreira, I. C. F. R., Baptista, P., & Santos-Buelga, C. (2009). Phenolic Acids Determination by HPLC–DAD–ESI/MS in Sixteen Different Portuguese Wild Mushrooms Species. *Food and Chemical Toxicology*, *47*(6), 1076–1079. 10.1016/j.fct.2009.01.039

Bernstein, B. J., & Grasso, T. (2001). Prevalence of Complementary and Alternative Medicine Use in Cancer Patients. *Oncology (Williston Park)*, *15*(10), 1267–1272.

Betz, J. M., Brown, P. N., & Roman, M. C. (2011). Accuracy, Precision, and Reliability of Chemical Measurements in Natural Products Research. *Fitoterapia*, *82*(1), 44–52. 10.1016/j.fitote.2010.09.011

Bhanja, S. K., Samanta, S. K., Mondal, B., et al. (2020). Green Synthesis of Ag@Au Bimetallic Composite Nanoparticles Using a Polysaccharide Extracted from Ramaria Botrytis Mushroom and Performance in Catalytic Reduction of 4-Nitrophenol and Antioxidant, Antibacterial Activity. *Environmental Nanotechnology, Monitoring & Management*, *14*(December), 100341. 10.1016/j.enmm.2020.100341

Bin, L., Jin-Ping, Z., Wei-Guo, H., Sheng, Y., & Smith, D. L. (2008). PCR-Based Sensitive Detection of the Edible Fungus *Boletus edulis* from RDNA ITS Sequences. *Electronic Journal of Biotechnology*, *11*(3), 102–109.

Boidin, J. (1986). Intercompatibility and the Species Concept in the Saprobic Basidiomycotina. *Mycotaxon*, *26*, 319–336.

Brandenburg, W. E., & Ward, K. J. (2018). Mushroom Poisoning Epidemiology in the United States. *Mycologia*, *110*(4), 637–641. 10.1080/00275514.2018.1479561

Bresinsky, A., Hilber, O., & Molitoris, H. P. (1977). The Genus *Pleurotus* as an Aid for Understanding the Concept of Species in Basidiomycetes. In: H. Clémençon (Ed.), *The Species Concept in Hymenomycetes*. J. Cramer, pp. 209–250.

Bruns, T. D., White, T. J., & Taylor, J. W. (1991). Fungal Molecular Systematics. *Annual Review of Ecology and Systematics*, *22*, 525–564.

Buchalo, A., Mykchaylova, O., Lomberg, M., & Wassser, S. P. (2009). *Microstructures of Vegetative Mycelium of Macromycetes in Pure Cultures*, P. Volz & E. Nevo (Eds.), Alterpress.

Buchanan, P. K. (1993). Identification, Names and Nomenclature of Common Edible Mushrooms. In: S. Chang, J. A. Buswell, & S. Chiu (Eds.), *Mushroom Biology and Mushroom Products*. Chinese University Press, pp. 21–32.

Buhner, S. H. (2012). *Herbal Antibiotics: Natural Alternatives for Treating Drug-Resistant Bacteria*. Storey Publishing.

Burden of Foodborne Illness: Overview. (2019). *Estimates of Foodborne Illness in the United States*. Centers for Disease Control and Prevention. https://www.cdc.gov/foodborneburden/estimates-overview.html

Buyck, B., Kauff, F., Eyssartier, G., Couloux, A., & Hofstetter, V. (2014). A Multilocus Phylogeny for Worldwide *Cantharellus* (Cantharellales, Agaricomycetidae). *Fungal Diversity*, *64*(1), 101–121. 10.1 007/s13225-013-0272-3

Calderón-Jiménez, B., Johnson, M. E., Bustos, A. R. M., Murphy, K. E., Winchester, M. R., & Baudrit, J. R. V. (2017). Silver Nanoparticles: Technological Advances, Societal Impacts, and Metrological Challenges. *Frontiers in Chemistry*, *5*(February), 6. 10.3389/fchem.2017.00006

Cardwell, G., Bornman, J., James, A., & Black, L. (2018). A Review of Mushrooms as a Potential Source of Dietary Vitamin D. *Nutrients*, *10*(10), 1498. 10.3390/nu10101498

Castro-Puyana, M., Marina, M. M., & Plaza, M. (2017). Water as Green Extraction Solvent: Principles and Reasons for Its Use. *Current Opinion in Green and Sustainable Chemistry*, *5*, 31–36. 10.1016/j.cogsc. 2017.03.009

Cellini, M., Attipoe, S., Seales, P., et al. (2013). Dietary Supplements: Physician Knowledge and Adverse Event Reporting. *Medicine and Science in Sports and Exercise*, *45*(1), 23–28. 10.1249/MSS.0b013e31 8269904f

Chakravarty, B. (2011). Trends in Mushroom Cultivation and Breeding. *Australian Journal of Agricultural Engineering*, 2(4), 102–109.

Chang, S. T., & Buswell, J. A. (1996). Mushroom Nutriceuticals. *World Journal of Microbiology & Biotechnology*, 12(5), 473–476. 10.1007/bf00419460

Chang, C. J., Lin, C. S., Lu, C. C., et al. (2015). Ganoderma Lucidum Reduces Obesity in Mice by Modulating the Composition of the Gut Microbiota. *Nature Communications*, 6(1), 7489. 10.1038/ncomms8489

Chang, S. T., & Wasser, S. P. (2012). The Role of Culinary-Medicinal Mushrooms on Human Welfare with a Pyramid Model for Human Health. *International Journal of Medicinal Mushrooms*, 14(2), 95–134. 10.1615/intjmedmushr.v14.i2.10

Cheek, M., Lughadha, E. N., Kirk, P., et al. (2020). New Scientific Discoveries: Plants and Fungi. *Plants, People, Planet*, 2(5), 371–388. 10.1002/ppp3.10148

Childs, N. M. (2000). Nutraceutical Industry Trends. *Journal of Nutraceuticals, Functional & Medical Foods*, 2(1), 73–85. 10.1300/j133v02n01_07

Comunian, T. A., Silva, M. P., & Souza, C. J. F. (2021). The Use of Food By-Products as a Novel for Functional Foods: Their Use as Ingredients and for the Encapsulation Process. *Trends in Food Science & Technology*, 108, 269–280. 10.1016/j.tifs.2021.01.003

Cör, D., Botić, T., Gregori, A., Pohleven, F., & Knez, Ž. (2017). The Effects of Different Solvents on Bioactive Metabolites and 'in Vitro' Antioxidant and Anti-Acetylcholinesterase Activity of Ganoderma Lucidum Fruiting Body and Primordia Extracts. *Macedonian Journal of Chemistry and Chemical Engineering*, 36(1). 10.20450/mjcce.2017.1054

Cordyceps: Side Effects, Dosages, Treatment, Interactions, Warnings. (2021). *RxList*. Accessed February 23. https://www.rxlist.com/consumer_cordyceps/drugs-condition.htm.

Dailu, P., Santini, A., & Novellino, E. (2018). From Pharmaceuticals to Nutraceuticals: Bringing Disease Prevention and Management. *Expert Review of Clinical Pharmacology*, 12(1), 1–7.

D'Archivio, M., Filesi, C., Varì, R., Scazzocchio, B., & Masella, R. (2010). Bioavailability of the Polyphenols: Status and Controversies. *International Journal of Molecular Sciences*, 11(4), 1321–1342. 10.3390/ijms11041321

Delcassian, D., & Patel, A. K. (2020). Nanotechnology and Drug Delivery. In: S. Ladame & J. Y. H. Chang (Eds.), *Bioengineering Innovative Solutions for Cancer*. Academic Press, pp. 197–219.

Dheyab, M. A., Owaid, M. N., Rabeea, M. A., Aziz, A. A., & Jameel, M. S. (2020). Mycosynthesis of Gold Nanoparticles by the Portabello Mushroom Extract, Agaricaceae, and Their Efficacy for Decolorization of Azo Dye. *Environmental Nanotechnology, Monitoring & Management*, 14, 100312. 10.1016/j.enmm.2020.100312

Dias, D. R., Botrel, D. A., De Barros Fernandes, R. V., & Borges, S. V. (2017). Encapsulation as a Tool for Bioprocessing of Functional Foods. *Current Opinion in Food Science*, 13, 31–37. 10.1016/j.cofs.2017.02.001

Dodge, T. (2015). Consumers' Perceptions of the Dietary Supplement Health and Education Act: Implications and Recommendations. *Drug Testing and Analysis*, 8(3–4), 407–409. 10.1002/dta.1857

Doroški, A., Klaus, A., Kozarski, M., et al. (2020). The Influence of Grape Pomace Substrate on Quality Characterization of Pleurotus Ostreatus—Total Quality Index Approach. *Journal of Food Processing and Preservation*, 45(1), e15096. 10.1111/jfpp.15096

Dowlati, M., Sobhi, H. R., Esrafili, A., FarzadKia, M., & Yeganeh, M. (2021). Heavy Metals Content in Edible Mushrooms: A Systematic Review, Meta-Analysis and Health Risk Assessment. *Trends in Food Science & Technology*, 109(March), 527–535. 10.1016/j.tifs.2021.01.064

Duff, M. C., & Ramsey, M. L. (2008). Accumulation of Radiocesium by Mushrooms in the Environment: A Literature Review. *Journal of Environmental Radioactivity*, 99(6), 912–932. 10.1016/j.jenvrad.2007.11.017

Dwyer, J. T., Coates, P. M., & Smith, M. J. (2018). Dietary Supplements: Regulatory Challenges and Research Resources. *Nutrients*, 10(1), 41. 10.3390/nu10010041

Eger, G. (1965). Untersuchungen über die Bildung und Regeneration von Fruchtkörpern bei Hutpilzen. I. *Pleurotus* Florida. *Archiv für Mikrobiologie*, 50, 343–356.

Eger, G., Li, S. F., & Leal-Lara, H. (1979). Contribution to the Discussion on the Species Concept in the *Pleurotus ostreatus* Complex. *Mycologia*, 71, 577–588.

El Enshasy, H., Elsayed, E. A., Aziz, R., & Wadaan, M. A. (2013). Mushrooms and Truffles: Historical Biofactories for Complementary Medicine in Africa and in the Middle East. *Evidence-Based Complementary and Alternative Medicine, 2013*, 1–10. 10.1155/2013/620451

Ellor, T. (2021). Mushrooms and Organic Mushrooms: A Specialty within a Specialty. Accessed February 23. https://www.usda.gov/sites/default/files/documents/Ellor.pdf

de Faria Coelho-Ravagnani, C., Corgosinho, F. C., Sanches, F. L. F. Z., Prado, C. M. M., Laviano, A., & Mota, J. F. (2020 July). Dietary Recommendations during the COVID-19 Pandemic. *Nutrition Reviews, 79*(4), 382–393. 10.1093/nutrit/nuaa067

Gao, Y., & Zhou, S. (2004). Chemopreventive and Tumoricidal Properties of Ling Zhi Mushroom Ganoderma Lucidum (W.Curt.: Fr.) Lloyd (Aphyllophoromycetideae). Part II. Mechanism Considerations (Review). *International Journal of Medicinal Mushrooms, 6*(3), 219–230. 10.1615/intjmedmushr.v6.i3.20

GenBank Overview. (2013). *nih.gov.* https://www.ncbi.nlm.nih.gov/genbank/

Global Nutraceutical Market Growth Analysis Report, 2020–2027. (2020). Www.grandviewresearch.com. Accessed February 10. https://www.grandviewresearch.com/industry-analysis/nutraceuticals-market# :~:text=The%20global%20nutraceutical%20market%20is%20expected%20to%20grow%20at%20a

Gong, P., Wang, S., Liu, M., et al. (2020). Extraction Methods, Chemical Characterizations and Biological Activities of Mushroom Polysaccharides: A Mini-Review. *Carbohydrate Research, 494*, 108037. 10.1 016/j.carres.2020.108037

Goodfellow, J., Bellamy, M. F., Ramsey, M. W., Jones, C. J. H., & Lewis, M. J. (2000). Dietary Supplementation with Marine Omega-3 Fatty Acids Improve Systemic Large Artery Endothelial Function in Subjects with Hypercholesterolemia. *Journal of the American College of Cardiology, 35*(2), 265–270. 10.1016/s0735-1097(99)00548-3

Hamulka, J., Jeruszka-Bielak, M., Górnicka, M., Drywień, M. E., & Zielinska-Pukos, M. A. (2020). Dietary Supplements during COVID-19 Outbreak. Results of Google Trends Analysis Supported by PLifeCOVID-19 Online Studies. *Nutrients, 13*(1), 54. 10.3390/nu13010054

Han, M. D., Han, Y. S., Hyun, S. H., & Shin, H. W. (2008). Solubilization of Water-Insoluble β-Glucan Isolated from Ganoderma Lucidum. *Journal of Environmental Biology, 29*(2), 237–242.

Hassan, N. A., Supramani, S., Sohedein, M. N. A., et al. (2019). Efficient Biomass-Exopolysaccharide Production from an Identified Wild-Serbian Ganoderma Lucidum Strain BGF4A1 Mycelium in a Controlled Submerged Fermentation. *Biocatalysis and Agricultural Biotechnology, 21*, 101305. 10.101 6/j.bcab.2019.101305

Hawksworth, D., & Lücking, R. (2017). Fungal Diversity Revisited: 2.2 to 3.8 Million Species. *Microbiology Spectrum, 5*(4), 79–95. 10.1128/microbiolspec.funk-0052-2016

Helal, N. A., Eassa, H. A., Amer, A. M., et al. (2019). Nutraceuticals' Novel Formulations: The Good, the Bad, the Unknown and Patents Involved. *Recent Patents on Drug Delivery & Formulation, 13*(2), 105–156. 10.2174/1872211313666190503112040

Hetland, G., Johnson, E., Bernardshaw, S. V., & Grinde, B. (2020 July). Can Medicinal Mushrooms Have Prophylactic or Therapeutic Effect against COVID-19 and Its Pneumonic Superinfection and Complicating Inflammation? *Scandinavian Journal of Immunology, 93*(1), e12937. 10.1111/sji.12937

Hilber, O. (1982). Die Gattung *Pleurotus. Bibliotheca Mycologica, 87*, 1–448.

Hilber, O. (1989). Valid, Invalid and Confusing Taxa of the Genus *Pleurotus. Mushroom Science, 12*, 241–248.

Hilber, O. (1997). *The Genus Pleurotus (Fr.) Kummer (2)*. O. Hilber.

Hilton, J. (2017). Growth Patterns and Emerging Opportunities in Nutraceutical and Functional Food Categories: Market Overview. In: D. Bagchi & S. Nair (Eds.), *Developing New Functional Food and Nutraceutical Products*. Academic Press, pp. 1–28.

Hobbs, C. (2003). *Medicinal Mushrooms: An Exploration of Tradition, Healing, and Culture*. Botanica Press.

Hu, Z., Lee, C. I., Shah, V. K., et al. (2013). Cordycepin Increases Nonrapid Eye Movement Sleep via Adenosine Receptors in Rats. *Evidence-Based Complementary and Alternative Medicine, 2013*, 1–8. 10.1155/2013/840134

Hu, Z., Oh, S., Ha, T. W., Hong, J. T., & Oh, K. W. (2018). Sleep-Aids Derived from Natural Products. *Biomolecules & Therapeutics, 26*(4), 343–349. 10.4062/biomolther.2018.099

Hu, Y., Teng, C., Yu, S., et al. (2017). *Inonotus Obliquus* Polysaccharide Regulates Gut Microbiota of Chronic Pancreatitis in Mice. *AMB Express, 7*(1). 10.1186/s13568-017-0341-1

Huber, B. C., Steffen, J., Schlichtiger, J., & Brunner, S. (2020 December). Altered Nutrition Behavior during COVID-19 Pandemic Lockdown in Young Adults. *European Journal of Nutrition, 60*(5), 2593–2602. 10.1007/s00394-020-02435-6

Index Fungorum Home Page. (2021). www.indexfungorum.org. Accessed March 6. http://www. indexfungorum.org/.

Instituto Nacional de Ciencias Medicas y Nutricion Salvador Zubiran. (2021). Randomized, Double-Blind, Placebo-Controlled, Multicenter, Multi-Arm, Phase II Trial of Novel Agents for the Treatment of Mild to Moderate COVID-19 Positive Outpatients. *Clinicaltrials.gov.* Accessed February 10. https://clinicaltrials.gov/ct2/show/NCT04530617

Islam, T., Yu, X., & Xu, B. (2016). Phenolic Profiles, Antioxidant Capacities and Metal Chelating Ability of Edible Mushrooms Commonly Consumed in China. *LWT – Food Science and Technology, 72,* 423–431. 10.1016/j.lwt.2016.05.005

Jaloot, A. S., Owaid, M. N., Naeem, G. A., & Muslim, R. F. (2020). Mycosynthesizing and Characterizing Silver Nanoparticles from the Mushroom Inonotus Hispidus (Hymenochaetaceae), and Their Antibacterial and Antifungal Activities. *Environmental Nanotechnology, Monitoring & Management, 14,* 100313. 10.1016/j.enmm.2020.100313

Janiak, M. C., Chaney, M. E., & Tosi, A. J. (2017). Evolution of Acidic Mammalian Chitinase Genes (CHIA) Is Related to Body Mass and Insectivory in Primates. *Molecular Biology and Evolution, 35*(3), 607–622. 10.1093/molbev/msx312

Jarić, S., Mitrović, M., Djurdjević, L., et al. (2011). Phytotherapy in Medieval Serbian Medicine according to the Pharmacological Manuscripts of the Chilandar Medical Codex (15–16th Centuries). *Journal of Ethnopharmacology, 137*(1), 601–619. 10.1016/j.jep.2011.06.016

Jayachandran, M., Xiao, J., & Xu, B. (2017). A Critical Review on Health Promoting Benefits of Edible Mushrooms through Gut Microbiota. *International Journal of Molecular Sciences, 18*(9), 1934. 10.3390/ijms18091934

Jess, S., Kirbas, J. M., Gordon, A. W., & Murchie, A. K. (2017). Potential for Use of Garlic Oil to Control *Lycoriella Ingenua* (Diptera: Sciaridae) and *Megaselia Halterata* (Diptera: Phoridae) in Commercial Mushroom Production. *Crop Protection, 102*(December), 1–9. 10.1016/j.cropro.2017.08.003

Jomová, K., Hudecova, L., Lauro, P., et al. (2019). A Switch between Antioxidant and Prooxidant Properties of the Phenolic Compounds Myricetin, Morin, 3′,4′-Dihydroxyflavone, Taxifolin and 4-Hydroxy-Coumarin in the Presence of Copper(II) Ions: A Spectroscopic, Absorption Titration and DNA Damage Study. *Molecules, 24*(23), 4335. 10.3390/molecules24234335

Karbownik, M. S., Paul, E., Nowicka, M., et al. (2019). Knowledge about Dietary Supplements and Trust in Advertising Them: Development and Validation of the Questionnaires and Preliminary Results of the Association between the Constructs. *PLOS ONE, 14*(6), e0218398. 10.1371/journal.pone.0218398

Kelsall, A., FitzGerald, A. J., Howard, C. V., et al. (2002). Dietary Lectins Can Stimulate Pancreatic Growth in the Rat. *International Journal of Experimental Pathology, 83*(4), 203–208. 10.1046/j.1365-2613.2002.00230.x

Kerrigan, R. W. (2005). *Agaricus subrufescens*, a Cultivated Edible and Medicinal Mushroom, and Its Synonyms. *Mycologia, 97*(1), 12–24. 10.3852/mycologia.97.1.12

Kim, S., Kim, W., Park, Y., Lee, S., Lee, Y., & Jang, W. J. (2011). Pharmaceutical Composition with Antiviral Activity Extracted from Deciduous Fungi. South Korea, issued September 8, 2011. https://patents.google.com/patent/KR101063825B1/en?oq=KR101063825B1

Klaus, A., Kozarski, M., Vunduk, J., et al. (2015). Biological Potential of Extracts of the Wild Edible Basidiomycete Mushroom *Grifola Frondosa*. *Food Research International, 67,* 272–283. 10.1016/j.foodres.2014.11.035

Klaus, A., Petrovic, P., Vunduk, J., Pavlovic, V., & van Griensven, L. J. L. D. (2020). Antimicrobial Properties of Silver-Nanoparticles of *Agaricus Bisporus, Agaricus Brasiliensis* and *Phellinus Linteus*. *International Journal of Medicinal Mushrooms, 22*(9), 869–883. 10.1615/intjmedmushrooms.2020035988

Klaus, A., Wan-Mohtar, W. A. A. Q. I., Nikolić, B., Cvetković, S., & Vunduk, J. (2021). Pink Oyster Mushroom *Pleurotus Flabellatus* Mycelium Produced by an Airlift Bioreactor—The Evidence of Potent in Vitro Biological Activities. *World Journal of Microbiology and Biotechnology, 37*(1), 17. 10.1007/s11274-020-02980-6

Kõljalg, U., Larsson, K. H., Abarenkov, K., et al. (2005). UNITE: A Database Providing Web-Based Methods for the Molecular Identification of Ectomycorrhizal Fungi. *New Phytologist, 166*(3), 1063–1068. 10.1111/j.1469-8137.2005.01376.x

Kõljalg, U., Nilsson, R. H., Abarenkov, K., et al. (2013). Towards a Unified Paradigm for Sequence-Based Identification of Fungi. *Molecular Ecology, 22*(21), 5271–5277. 10.1111/mec.12481

Kozarski, M., Klaus, A., Jakovljevic, D., et al. (2015). Antioxidants of Edible Mushrooms. *Molecules, 20*(10), 19489–19525. 10.3390/molecules201019489

Kuang, H., Xia, Y., Liang, J., Yang, B., Wang, Q., & Sun, Y. (2011). Fast Classification and Compositional Analysis of Polysaccharides from TCMs by Ultra-Performance Liquid Chromatography Coupled with Multivariate Analysis. *Carbohydrate Polymers, 84*(4), 1258–1266. 10.1016/j.carbpol.2011.01.014

Kuhn, M. A., & Winston, D. (2008). *Winston & Kuhn's Herbal Therapy & Supplements: A Scientific & Traditional Approach*. Wolters Kluwer/Lippincott Williams & Wilkins.

Lai, T., Gao, Y., & Zhou, S. (2004). Global Marketing of Medicinal Ling Zhi Mushroom *Ganoderma Lucidum* (W.Curt.:Fr.) Lloyd (Aphyllophoromycetideae) Products and Safety Concerns. *International Journal of Medicinal Mushrooms, 6*(2), 189–194. 10.1615/intjmedmushr.v6.i2.100

Lakhanpal, T. N., & Monika, R. (2005). Medicinal and Nutraceutical Genetic Resources of Mushrooms. *Plant Genetic Resources, 3*(2), 288–303. 10.1079/PGR200581

Lam, C. S., Cheng, L. P., Zhou, L. M., Cheung, Y. T., & Zuo, Z. (2020). Herb–Drug Interactions between the Medicinal Mushrooms Lingzhi and Yunzhi and Cytotoxic Anticancer Drugs: A Systematic Review. *Chinese Medicine, 15*(1), 75. 10.1186/s13020-020-00356-4

Largest Coronavirus-Related Growth in Dietary Supplements Use U.S. 2020. (2020). *Statista*. Accessed February 27. https://www.statista.com/statistics/1180337/top-dietary-supplement-usage-growth-due-to-covid-us-adults/.

Li, X. P., Li, J., Li, T., Liu, H., & Wang, Y. (2020). Species Discrimination and Total Polyphenol Prediction of Porcini Mushrooms by Fourier Transform Mid-Infrared (FT-MIR) Spectrometry Combined with Multivariate Statistical Analysis. *Food Science & Nutrition, 8*(2), 754–766. 10.1002/fsn3.1313

Li, H., Liu, S. M., Yu, X. H., Tang, S. L., & Tang, C. K. (2020). Coronavirus Disease 2019 (COVID-19): Current Status and Future Perspective. *International Journal of Antimicrobial Agents, 55*(5), 105951. 10.1016/j.ijantimicag.2020.105951

Li, H., Tian, Y., Menolli, N., et al. (2021). Reviewing the World's Edible Mushroom Species: A New Evidence-Based Classification System. *Comprehensive Reviews in Food Science and Food Safety, 20*(2), 1982–2014. 10.1111/1541-4337.12708

Li, S., Xiong, Q., Lai, X., et al. (2015). Molecular Modification of Polysaccharides and Resulting Bioactivities. *Comprehensive Reviews in Food Science and Food Safety, 15*(2), 237–250. 10.1111/1541-4337.12161

Li, S. P., Yang, F. Q., & Tsim, K. W. K. (2006). Quality Control of *Cordyceps Sinensis*, a Valued Traditional Chinese Medicine. *Journal of Pharmaceutical and Biomedical Analysis, 41*(5), 1571–1584. 10.1016/j.jpba.2006.01.046

Lian-Hua Qing-Wen Capsule. (2020).中国临床试验注册中心 - 世界卫生组织国际临床试验注册平台一级注册机构. www.chictr.org.cn. Accessed February 27. http://www.chictr.org.cn/showproj.aspx?proj=48889

Lin, Z. B. (2009). *Lingzhi from Mistery to Science*. Peking University Medical Press.

Liu, H., Hu, H., Chu, C., Li, Q., & Li, P. (2011). Morphological and Microscopic Identification Studies of *Cordyceps* and Its Counterfeits. *Acta Pharmaceutica Sinica B, 1*(3), 189–195. 10.1016/j.apsb.2011.06.013

Louca, P., Murray, B., Klaser, K., et al. (2020 November). Dietary Supplements during the COVID-19 Pandemic: Insights from 1.4M Users of the COVID Symptom Study App – A Longitudinal App-Based Community Survey. 10.1101/2020.11.27.20239087. https://search.bvsalud.org/global-literature-on-novel-coronavirus-2019-ncov/resource/en/ppmedrxiv-20239087

Ma, L., Chen, H., Zhu, W., & Wang, Z. (2013). Effect of Different Drying Methods on Physicochemical Properties and Antioxidant Activities of Polysaccharides Extracted from Mushroom *Inonotus Obliquus*. *Food Research International, 50*(2), 633–640. 10.1016/j.foodres.2011.05.005

Ma, T., Fan, W., Pan, X., & Luo, Y. (2021). Estimating the Risks from Phthalate Esters and Metalloids in Cultivated Edible Fungi from Jingmen, Central China. *Food Chemistry, 348*, 129065. 10.1016/j.foodchem.2021.129065

Ma, G., Yang, W., Zhao, L., Pei, F., Fang, D., & Hu, Q. (2018). A Critical Review on the Health Promoting Effects of Mushrooms Nutraceuticals. *Food Science and Human Wellness*, 7(2), 125–133. 10.1016/j.fshw.2018.05.002

Maes, M., Higginson, E., Pereira-Dias, J., et al. (2021). Ventilator-Associated Pneumonia in Critically Ill Patients with COVID-19. *Critical Care*, 25, 25. 10.1186/s13054-021-03460-5

Mahmoudi, H. (2020). Bacterial Co-Infections and Antibiotic Resistance in Patients with COVID-19. *GMS Hygiene and Infection Control*, 15(Doc. 35). 10.3205/dgkh000370

Maintaining a Healthy Diet during the COVID-19 Pandemic. (2020). www.fao.org. Accessed February 27. http://www.fao.org/publications/card/en/c/CA8380EN/

Maity, P., Sen, I. K., Chakraborty, I., et al. (2021). Biologically Active Polysaccharide from Edible Mushrooms: A Review. *International Journal of Biological Macromolecules*, 172, 408–417. 10.1016/j.ijbiomac.2021.01.081

Majeed, M., Nagabhushanam, K., Gowda, S., & Mundkur, L. (2021). An Exploratory Study of Selenium Status in Normal Subjects and COVID-19 Patients in South Indian Population: Case for Adequate Selenium Status. *Nutrition, November*, 111053. 10.1016/j.nut.2020.111053

Masterson, C. H., Murphy, E., Major, I., et al. (2019). Purified Beta-Glucan from the Lentinus Edodes Mushroom Attenuates Antibiotic Resistant Klebsiella Pneumoniae-Induced Pulmonary Sepsis. *American Journal of Respiratory and Critical Care Medicine*, 199, A1222. 10.1164/ajrccm-conference.2019.199.1_meetingabstracts.a1222

Mazzutti, S., Ferreira, S. R. S., Riehl, C. A. S., Smania, A., Smania, F. A., & Martínez, J. (2012). Supercritical Fluid Extraction of Agaricus Brasiliensis: Antioxidant and Antimicrobial Activities. *The Journal of Supercritical Fluids*, 70, 48–56. 10.1016/j.supflu.2012.06.010

McRorie, J. W. (2015). Evidence-Based Approach to Fiber Supplements and Clinically Meaningful Health Benefits, Part 2. *Nutrition Today*, 50(2), 90–97. 10.1097/nt.0000000000000089

Mehta, J. J. (2010). Practical Challenges of Stability Testing of Nutraceutical Formulations. In: K. Huynh-Ba (Ed.), *Pharmaceutical Stability Testing to Support Global Markets: Pharma*. American Association of Pharmaceutical Scientists, pp. 85–91.

Melgar, M. J., Alonso, J., & García, M. A. (2014). Total Contents of Arsenic and Associated Health Risks in Edible Mushrooms, Mushroom Supplements and Growth Substrates from Galicia (NW Spain). *Food and Chemical Toxicology*, 73, 44–50. 10.1016/j.fct.2014.08.003

Meng, X., Liang, H., & Luo, L. (2016). Antitumor Polysaccharides from Mushrooms: A Review on the Structural Characteristics, Antitumor Mechanisms and Immunomodulating Activities. *Carbohydrate Research*, 424, 30–41. 10.1016/j.carres.2016.02.008

Mengesha, Z. T., & Chaithanya, K. (2020). Historical Perspectives of Mushroom Cultivation in the World. *Drug Invention Today*, 13(4), 593–595.

Ministry of Agriculture and Rural Affairs of the People's Republic of China. (n.d.). 对十三届全国人大二次会议第4493号建议的答复. www.moa.gov.cn. http://www.moa.gov.cn/gk/tzgg_1/tz/201912/t20191210_6333030.htm

Mooralitharan, S., Hanafiah, Z. M., Manan, T. S. B. A., et al. (2021 February). Optimization of Mycoremediation Treatment for the Chemical Oxygen Demand (COD) and Ammonia Nitrogen (AN) Removal from Domestic Effluent Using Wild-Serbian *Ganoderma Lucidum* (WSGL). *Environmental Science and Pollution Research*. 10.1007/s11356-021-12686-3

Moradali, M. F., Mostafavi, M., Ghods, S., & Hedjaroude, G. A. (2007). Immunomodulating and Anticancer Agents in the Realm of Macromycetes Fungi (Macrofungi). *International Immunopharmacology*, 7(6), 701–724. 10.1016/j.intimp.2007.01.008

Morris, H. J., Llauradó, G., Beltrán, Y., et al. (2016). The Use of Mushrooms in the Development of Functional Foods, Drugs, and Nutraceuticals. In: I. C. F. R. Ferreira, P. Morales, & L. Barros (Eds.), *Wild Plants, Mushrooms and Nuts: Functional Food Properties and Applications*. John Wiley & Sons, pp. 123–157.

Mukai, H., Watanabe, T., Ando, M., & Katsumata, N. (2006). An Alternative Medicine, *Agaricus Blazei*, May Have Induced Severe Hepatic Dysfunction in Cancer Patients. *Japanese Journal of Clinical Oncology*, 36(12), 808–810. 10.1093/jjco/hyl108

Murphy, E. J., Masterson, C., Rezoagli, E., et al. (2020). β-Glucan Extracts from the Same Edible Shiitake Mushroom *Lentinus Edodes* Produce Differential In-Vitro Immunomodulatory and Pulmonary Cytoprotective Effects—Implications for Coronavirus Disease (COVID-19) Immunotherapies. *Science of the Total Environment*, *732*, 139330. 10.1016/j.scitotenv.2020.139330

Mushrooms%20biological%20activity – Search Results – PubMed. (2021). *PubMed*. Accessed February 9. https://pubmed.ncbi.nlm.nih.gov/?term=Mushrooms%20biological%20activity&timeline=expanded.

Mycobank Database. (2021). www.mycobank.org. Accessed March 1. https://www.mycobank.org/

Nabipour, H., & Hu, Y. (2020). Sustainable Drug Delivery Systems through Green Nanotechnology. In: M. Mozafari (Ed.), *Nanoengineered Biomaterials for Advanced Drug Delivery*. Elsevier, pp. 61–98.

Nagajyothi, P. C., Sreekanth, T. V. M., Lee, J., & Lee, K. D. (2014). Mycosynthesis: Antibacterial, Antioxidant and Antiproliferative Activities of Silver Nanoparticles Synthesized from *Inonotus Obliquus* (Chaga Mushroom) Extract. *Journal of Photochemistry and Photobiology B: Biology*, *130*, 299–304. 10.1016/j.jphotobiol.2013.11.022

Naghavi, M. (2019 February). Global, Regional, and National Burden of Suicide Mortality 1990 to 2016: Systematic Analysis for the Global Burden of Disease Study 2016. *BMJ*, *364*, 194. 10.1136/bmj.l94. https://www.bmj.com/content/364/bmj.l94

Narayanan, K. B., Park, H. H., & Han, S. S. (2015). Synthesis and Characterization of Biomatrixed-Gold Nanoparticles by the Mushroom *Flammulina Velutipes* and Its Heterogeneous Catalytic Potential. *Chemosphere*, *141*, 169–175. 10.1016/j.chemosphere.2015.06.101

National Certification and Accreditation Information Public Service Platform. (2021). *cx.cnca.cn*. Accessed March 27. http://cx.cnca.cn

Nguyen, A. N. H., Capone, D. L., Johnson, T. E., Jeffery, D. W., Danner, L., & Bastian, S. E. P. (2019). Volatile Composition and Sensory Profiles of a Shiraz Wine Product Made with Pre- and Post-Fermentation Additions of *Ganoderma Lucidum* Extract. *Foods*, *8*(11), 538. 10.3390/foods8110538

Nguyen, A. N. H., Johnson, T. E., Jeffery, D. W., Capone, D. L., Danner, L., & Bastian, S. E. P. (2020). Sensory and Chemical Drivers of Wine Consumers' Preference for a New Shiraz Wine Product Containing *Ganoderma Lucidum* Extract as a Novel Ingredient. *Foods*, *9*(2), 224. 10.3390/foods9020224

Novakovic, S., Djekic, I., Klaus, A., et al. (2020). Application of Porcini Mushroom (*Boletus Edulis*) to Improve the Quality of Frankfurters. *Journal of Food Processing and Preservation*, *44*(8), e14556. 10.1111/jfpp.14556

Oak, Al. S.W., Shafi, S., Elsayed, M., et al. (2021 January). Dietary Table Grape Protects against UV Photodamage in Humans: 2. Molecular Biomarker Studies. *Journal of the American Academy of Dermatology*, *85*(4), 1032–1134. 10.1016/j.jaad.2021.01.036

Obodai, M., Mensah, D. L. N., Fernandes, Â., et al. (2017). Chemical Characterization and Antioxidant Potential of Wild *Ganoderma* Species from Ghana. *Molecules*, *22*(2), 196. 10.3390/molecules22020196

Okamura, H., Anno, N., Tsuda, A., Inokuchi, T., Uchimura, N., & Inanaga, K. (2015). The Effects of *Hericium Erinaceus* (Amyloban® 3399) on Sleep Quality and Subjective Well-Being among Female Undergraduate Students: A Pilot Study. *Personalized Medicine Universe*, *4*, 76–78. 10.1016/j.pmu.2015.03.006

Organic Mushroom Production. (2021). Accessed February 27. http://www.fungifun.org/docs/mushrooms/Organic%20mushroom%20production%20in%20the%20EU.pdf.

Oshiman, K., Fujimiya, Y., Ebina, T., Suzuki, I., & Noji, M. (2002). Orally Administered β-1,6-D-Polyglucose Extracted from *Agaricus Blazei* Results in Tumor Regression in Tumor-Bearing Mice. *Planta Medica*, *68*(7), 610–614. 10.1055/s-2002-32904

Owaid, M. N. (2019). Green Synthesis of Silver Nanoparticles by *Pleurotus* (Oyster Mushroom) and Their Bioactivity: Review. *Environmental Nanotechnology, Monitoring & Management*, *12*, 100256. 10.1016/j.enmm.2019.100256

Owaid, M. N., Rabeea, M. A., Aziz, A. A., Jameel, M. S., & Dheyab, M. A. (2019). Mushroom-Assisted Synthesis of Triangle Gold Nanoparticles Using the Aqueous Extract of Fresh *Lentinula Edodes* (Shiitake), Omphalotaceae. *Environmental Nanotechnology, Monitoring & Management*, *12*, 100270. 10.1016/j.enmm.2019.100270

Palacios, I., Lozano, M., Moro, C., et al. (2011). Antioxidant Properties of Phenolic Compounds Occurring in Edible Mushrooms. *Food Chemistry*, *128*(3), 674–678. 10.1016/j.foodchem.2011.03.085

Pandey, A. T., Pandey, I., Hachenberger, Y., et al. (2020). Emerging Paradigm against Global Antimicrobial Resistance via Bioprospecting of Mushroom into Novel Nanotherapeutics Development. *Trends in Food Science & Technology*, *106*, 333–344. 10.1016/j.tifs.2020.10.025

Petersen, R. H., & Hughes, K. W. (1993). Intercontinental Interbreeding Population of *Pleurotus Pulmonarious*, with Notes on *P. Ostreatus* and Other Species. *Sydowia*, *45*, 139–152.

Petrova, R. D., Mahajna, J., Reznick, A. Z., Wasser, S. P., Denchev, C. M., & Nevo, E. (2006). Fungal Substances as Modulators of NF-KB Activation Pathway. *Molecular Biology Reports*, *34*(3), 145–154. 10.1007/s11033-006-9027-5

Petrović, P., Vunduk, J., Klaus, A., et al. (2019). From Mycelium to Spores: A Whole Circle of Biological Potency of Mosaic Puffball. *South African Journal of Botany*, *123*(July), 152–160. 10.1016/j.sajb.2019.03.016

Petrovic, P., Ivanovic, K., Octrue, C., et al. (2019). Immobilization of Chaga Extract in Alginate Beads for Modified Release: Simplicity Meets Efficiency. *Chemical Industry*, *73*(5), 325–335. 10.2298/hemind190819028p

Pettinato, M., Trucillo, P., Campardelli, R., Perego, P., & Reverchon, E. (2020). Bioactives Extraction from Spent Coffee Grounds and Liposome Encapsulation by a Combination of Green Technologies. *Chemical Engineering and Processing –Process Intensification*, *151*, 107911. 10.1016/j.cep.2020.107911

Powell, M. (2014). *Medicinal Mushrooms: A Clinical Guide*. Mycology Press, An Imprint of Bamboo Publishing.

Prados-Rosales, R., Toriola, S., Nakouzi, A., et al. (2015). Structural Characterization of Melanin Pigments from Commercial Preparations of the Edible Mushroom *Auricularia Auricula*. *Journal of Agricultural and Food Chemistry*, *63*(33), 7326–7332. 10.1021/acs.jafc.5b02713

Prasad, S., Rathore, H., Sharma, S., & Yadav, A. S. (2015 October). Medicinal Mushrooms as a Source of Novel Functional Food. *International Journal of Food Science, Nutrition and Dietetics*, *4*(5), 221–225. 10.19070/2326-3350-1500040

Publication Card | FAO | Food and Agriculture Organization of the United Nations. (2021). www.fao.org. Accessed February 27. http://www.fao.org/publications/card/en/c/CA8380EN/.

Pursell, J. J. (2015). *The Herbal Apothecary: 100 Medicinal Herbs and How to Use Them*. Timber Press.

Rafeeq, C. M., Paul, E., Saagar, E. V., & Ali, P. P. M. (2021 January). Mycosynthesis of Zinc Oxide Nanoparticles Using *Pleurotus Floridanus* and Optimization of Process Parameters. *Ceramics International*, *47*, 12375–12380. 10.1016/j.ceramint.2021.01.091

Raja, H. A., Baker, T. R., Little, J. G., & Oberlies, N. H. (2017). DNA Barcoding for Identification of Consumer-Relevant Mushrooms: A Partial Solution for Product Certification? *Food Chemistry*, *214*(January), 383–392. 10.1016/j.foodchem.2016.07.052

Raja, H. A., Miller, A. N., Pearce, C. J., & Oberlies, N. H. (2017). Fungal Identification Using Molecular Tools: A Primer for the Natural Products Research Community. *Journal of Natural Products*, *80*(3), 756–770. 10.1021/acs.jnatprod.6b01085

Ratnasingham, S., & Hebert, P. D. N. (2007). BOLD: The Barcode of Life Data System. *Molecular Ecology Notes*, *7*(3), 355–364. http://www.barcodinglife.org.

Reis, F. S., Martins, A., Vasconcelos, M. H., Morales, P., & Ferreira, I. C. F. R. (2017). Functional Foods Based on Extracts or Compounds Derived from Mushrooms. *Trends in Food Science & Technology*, *66*(August), 48–62. 10.1016/j.tifs.2017.05.010

Remya, V. R., Chandra, G., & Mohanakumar, K. P. (2019). Edible Mushrooms as Neuro-Nutraceuticals: Basis of Therapeutics. In: D. C. Agrawal & M. Dhanasekaran (Eds.), *Medicinal Mushrooms*. Springer Nature Singapore, pp. 71–101.

Ribeiro, A., Ruphuy, G., Lopes, J. C., et al. (2015). Spray-Drying Microencapsulation of Synergistic Antioxidant Mushroom Extracts and Their Use as Functional Food Ingredients. *Food Chemistry*, *188*, 612–618. 10.1016/j.foodchem.2015.05.061

Richter, C., Wittstein, K., Kirk, P. M., & Stadler, M. (2014). An Assessment of the Taxonomy and Chemotaxonomy of *Ganoderma*. *Fungal Diversity*, *71*(1), 1–15. 10.1007/s13225-014-0313-6

Romero, M., Serrano, M., Vallejo, M., Efferth, T., Alvarez, M., & Marin, H. (2006). Antiviral Effect of Artemisinin from *Artemisia Annua* against a Model Member of the Flaviviridae Family, the Bovine Viral Diarrhoea Virus (BVDV). *Planta Medica*, *72*(13), 1169–1174. 10.1055/s-2006-947198

Rop, O., Mlcek, J., & Jurikova, T. (2009). Beta-Glucans in Higher Fungi and Their Health Effects. *Nutrition Reviews*, *67*(11), 624–631. 10.1111/j.1753-4887.2009.00230.x

Rothschild, D., Weissbrod, O., Barkan, E., et al. (2018). Environment Dominates over Host Genetics in Shaping Human Gut Microbiota. *Nature*, *555*(7695), 210–215. 10.1038/nature25973

Rowan, N. J., & J. E. Smith. (2002). *Medicinal Mushrooms: Their Therapeutic Properties and Current Medical Usage with Special Emphasis on Cancer Treatments*. Cancer Research UK, University of Starchylde.

Santana, Á. L., & Macedo, G. A. (2019). Challenges on the Processing of Plant-Based Neuronutraceuticals and Functional Foods with Emerging Technologies: Extraction, Encapsulation and Therapeutic Applications. *Trends in Food Science & Technology*, *91*, 518–529. 10.1016/j.tifs.2019.07.019

Santini, A., Cammarata, S. M., Capone, G., et al. (2018). Nutraceuticals: Opening the Debate for a Regulatory Framework. *British Journal of Clinical Pharmacology*, *84*(4), 659–672. 10.1111/bcp.13496

Sawangwan, T., Wansanit, W., Pattani, L., & Noysang, C. (2018). Study of Prebiotic Properties from Edible Mushroom Extraction. *Agriculture and Natural Resources*, *52*(6), 519–524. 10.1016/j.anres.2018.11.020

Schoch, C. L., Seifert, K. A., Huhndorf, S., et al. (2012). Nuclear Ribosomal Internal Transcribed Spacer (ITS) Region as a Universal DNA Barcode Marker for Fungi. *Proceedings of the National Academy of Sciences*, *109*(16), 6241–6246. 10.1073/pnas.1117018109

Shah, V. K., Choi, J. J., Han, J. Y., Lee, M. K., Hong, J. T., & Oh, K. W. (2014). Pachymic Acid Enhances Pentobarbital-Induced Sleeping Behaviors via GABAA-Ergic Systems in Mice. *Biomolecules & Therapeutics*, *22*(4), 314–320. 10.4062/biomolther.2014.045

Shahzad, F., Anderson, D., & Najafzadeh, M. (2020). The Antiviral, Anti-Inflammatory Effects of Natural Medicinal Herbs and Mushrooms and SARS-CoV-2 Infection. *Nutrients*, *12*(9), 2573. 10.3390/nu12092573

Shao, P., Xuan, S., Wu, W., & Qu, L. (2019). Encapsulation Efficiency and Controlled Release of *Ganoderma Lucidum* Polysaccharide Microcapsules by Spray Drying Using Different Combinations of Wall Materials. *International Journal of Biological Macromolecules*, *125*, 962–969. 10.1016/j.ijbiomac.2018.12.153

Shinde, N., Bangar, B., Deshmukh, S., & Kumbhar, P. (2014). Nutraceuticals: A Review on Current Status. *Research Journal of Pharmacology and Technology*, *7*(1), 110. 113.

Shwet, K., Sharma, V. P., Gupta, M., Barh, A., & Singh, M. (2019). Genetics and Breeding of White Button Mushroom, *Agaricus Bisporus* (Lange.) Imbach. – A Comprehensive Review. *Mushroom Research*, *28*(1). 10.36036/mr.28.1.2019.91938

Siddiqui, R. A., & Moghadasian, M. H. (2020). Nutraceuticals and Nutrition Supplements: Challenges and Opportunities. *Nutrients*, *12*(6), 1593. 10.3390/nu12061593

Singh, M., Kamal, S., & Sharma, V. P. (2017). Status and Trends in World Mushroom Production-I. *Mushroom Research*, *26*(1), 1–20.

Singh, A., & Zhao, K. (2017). Treatment of Insomnia with Traditional Chinese Herbal Medicine. *International Review of Neurobiology*, *135*, 97–115. 10.1016/bs.irn.2017.02.006

Sleep Aids Market Size, Share & Industry Growth. (2020–2025). *Market Data Forecast*. Accessed February 28. https://www.marketdataforecast.com/market-reports/global-sleep-aids-market

Sliva, D., Sedlak, M., Slivova, V., Valachovicova, T., Lloyd, F. P., & Ho, N. W. Y. (2003). Biologic Activity of Spores and Dried Powder from *Ganoderma Lucidum* for the Inhibition of Highly Invasive Human Breast and Prostate Cancer Cells. *The Journal of Alternative and Complementary Medicine*, *9*(4), 491–497. 10.1089/107555303322284776

Stalpers, J. A. (1978). Identification of Wood-Inhabiting Aphyllophorales in Pure Culture. In: *Studies in Mycology*. Centraalbureau voor Schimmelcultures.

Stamets, P. (2005). Antimicrobial Activity from Medicinal Mushrooms. WIPO (PCT), issued July 28, 2005. https://patents.google.com/patent/WO2005067955A1/en?oq=WO2005067955A1

Stamets, P. (2006). Antiviral Activity from Medicinal Mushrooms. United States, issued August 3, 2006. https://patents.google.com/patent/US20060171958A1/en?oq=US20060171958A1

Stengler, M. (2005). *The Health Benefits of Medicinal Mushrooms*. Basic Health Publications.

Sun, L., & Niu, Z. (2020). A Mushroom Diet Reduced the Risk of Pregnancy-Induced Hypertension and Macrosomia: A Randomized Clinical Trial. *Food & Nutrition Research*, *64*. 10.29219/fnr.v64.4451

Sweet, E. S., Standish, L. J., Goff, B. A., & Andersen, M. R. (2013). Adverse Events Associated with Complementary and Alternative Use in Ovarian Cancer Patients. *Obstetrical & Gynecological Survey*, *68*(11), 741–742. 10.1097/01.ogx.0000438239.12283.98

Synytsya, A., & Novák, M. (2013). Structural Diversity of Fungal Glucans. *Carbohydrate Polymers*, *92*(1), 792–809. 10.1016/j.carbpol.2012.09.077

Taofiq, O., González-Paramás, A. M., Martins, A., Barreiro, M. F., & Ferreira, I. C. F. R. (2016). Mushrooms Extracts and Compounds in Cosmetics, Cosmeceuticals and Nutricosmetics—A Review. *Industrial Crops and Products*, *90*, 38–48. 10.1016/j.indcrop.2016.06.012

The Crisis Facing Organic Food Certification. (2015). www.foodmate.net. Accessed October 24. http://www.foodmate.net/zhiliang/youji/165205.html

Thursby, E., & Juge, N. (2017). Introduction to the Human Gut Microbiota. *Biochemical Journal*, *474*(11), 1823–1836. 10.1042/bcj20160510

Tian, F., Qiao, C., Luo, J., et al. (2020). Development and Validation of a Method for the Analysis of Five Diamide Insecticides in Edible Mushrooms Using Modified QuEChERS and HPLC-MS/MS. *Food Chemistry*, *333*, 127468. 10.1016/j.foodchem.2020.127468

Trutmann, P. (2012). *The Forgotten Mushrooms of Ancient Peru*. Guardamunt Center Publications.

Tuli, H. S., Sandhu, S. S., & Sharma, A. K. (2014). Pharmacological and Therapeutic Potential of Cordyceps with Special Reference to Cordycepin. *3 Biotech*, *4*(1), 1–12. 10.1007/s13205-013-0121-9

Valdes, A. M., Walter, J., Segal, E., & Spector, T. D. (2018). Role of the Gut Microbiota in Nutrition and Health. *BMJ*, *361*, k2179. 10.1136/bmj.k2179

VanderMolen, K. M., Little, J. G., Sica, V. P., et al. (2017). Safety Assessment of Mushrooms in Dietary Supplements by Combining Analytical Data with in Silico Toxicology Evaluation. *Food and Chemical Toxicology*, *103*, 133–147. 10.1016/j.fct.2017.03.005

Vigna, L., Morelli, F., Agnelli, G. M., et al. (2019). *Hericium Erinaceus* Improves Mood and Sleep Disorders in Patients Affected by Overweight or Obesity: Could Circulating Pro-BDNF and BDNF Be Potential Biomarkers? *Evidence-Based Complementary and Alternative Medicine*, *2019*, 1–12. 10.1155/2019/7861297

Vilgalys, R. (1991). Speciation and Species Concepts in the *Collybia Dryophila* Complex. *Mycologia*, *83*(6), 758. 10.2307/3760433

Vilgalys, R., Smith, A., Sun, B. L., & Miller, Jr. O. K. (1993). Intersterility Groups in the *Pleurotus Ostreatus* Complex from the Continental United States and Adjacent Canada. *Canadian Journal of Botany*, *71*(1), 113–128. 10.1139/b93-013

Vunduk, J., Anita Klaus, A., Kozarski, M., et al. (2015). Did the Iceman Know Better? Screening of the Medicinal Properties of the Birch Polypore Medicinal Mushroom, *Piptoporus Betulinus* (Higher Basidiomycetes). *International Journal of Medicinal Mushrooms*, *17*(12), 1113–1125. 10.1615/intjmedmushrooms.v17.i12.10

Vunduk, J., Wan-Mohtar, W. A. A. Q. I., Mohamad, S. A., et al. (2019). Polysaccharides of *Pleurotus Flabellatus* Strain Mynuk Produced by Submerged Fermentation as a Promising Novel Tool against Adhesion and Biofilm Formation of Foodborne Pathogens. *LWT*, *112*, 108221. 10.1016/j.lwt.2019.05.119

Wachtel-Galor, S., Tomlinson, B., & Benzie, I. F. F. (2004). *Ganoderma Lucidum* ('Lingzhi'), a Chinese Medicinal Mushroom: Biomarker Responses in a Controlled Human Supplementation Study. *The British Journal of Nutrition*, *91*(2), 263–269. England. 10.1079/BJN20041039

Wang, Q., & Ellis, P. R. (2014). Oat β-Glucan: Physico-Chemical Characteristics in Relation to Its Blood-Glucose and Cholesterol-Lowering Properties. *British Journal of Nutrition*, *112*(S2), S4–S13. 10.1017/s0007114514002256

Wang, Y., Liu, Y., Yu, H., et al. (2017). Structural Characterization and Immuno-Enhancing Activity of a Highly Branched Water-Soluble β-Glucan from the Spores of *Ganoderma Lucidum*. *Carbohydrate Polymers*, *167*(July), 337–344. 10.1016/j.carbpol.2017.03.016

Wang, Q., Sheng, X., Shi, A., et al. (2017). β-Glucans: Relationships between Modification, Conformation and Functional Activities. *Molecules*, *22*(2), 257. 10.3390/molecules22020257

Wasser, S. P. (2011). Current Findings, Future Trends, and Unsolved Problems in Studies of Medicinal Mushrooms. *Applied Microbiology and Biotechnology*, *89*(5), 1323–1332. 10.1007/s00253-010-3067-4

Wasser, S. P. (2017). Medicinal Mushrooms in Human Clinical Studies. Part I. Anticancer, Oncoimmunological, and Immunomodulatory Activities: A Review. *International Journal of Medicinal Mushrooms*, *19*(4), 279–317. 10.1615/intjmedmushrooms.v19.i4.10

Wasser, S. P., & Akavia, E. (2008). Regulatory Issues of Mushrooms as Functional Foods, Dietary Supplements: Safety and Efficacy. In: P. C. K. Cheung (Ed.), *Mushrooms as Functional Foods*. John Wiley & Sons, pp. 199–228.

Wasser, S. P., & Weis, A. L. (1999). Medicinal Properties of Substances Occurring in Higher Basidiomycetes Mushrooms: Current Perspectives (Review). *International Journal of Medicinal Mushrooms*, *1*(1), 31–62. 10.1615/intjmedmushrooms.v1.i1.30

Wei, S., & Van Griensven, L. J. L. D. (2008). Pro- and Antioxidative Properties of Medicinal Mushroom Extracts. *International Journal of Medicinal Mushrooms*, *10*(4), 315–324. 10.1615/intjmedmushr.v1 0.i4.30

What You Need to Know about Dietary Supplements. (2019). *US Food and Drug Administration*. https:// www.fda.gov/food/buy-store-serve-safe-food/what-you-need-know-about-dietary-supplements

Witkowska, A. M. (2014). Selenium-Fortified Mushrooms – Candidates for Nutraceuticals? *Austin Therapeutics*, *1*(2).

Wu, T. N., Yang, K. C., Wang, C. M., et al. (1996). Lead Poisoning Caused by Contaminated Cordyceps, a Chinese Herbal Medicine: Two Case Reports. *Science of the Total Environment*, *182*(1–3), 193–195. 10.1016/0048-9697(96)05054-1

Xiang, L., Song, J., Xin, T., et al. (2013 August). DNA Barcoding the Commercial Chinese Caterpillar Fungus. *FEMS Microbiology Letters*, *347*(2), 156–162. 10.1111/1574-6968.12233

Xu, J. (2016). Fungal DNA Barcoding. S. Adamowicz (Ed.), *Genome*, *59*(11), 913–932. 10.1139/gen-2016-0046

Yamaç, M., & Bilgili, F. (2006). Antimicrobial Activities of Fruit Bodies and/or Mycelial Cultures of Some Mushroom Isolates. *Pharmaceutical Biology*, *44*(9), 660–667. 10.1080/13880200601006897

Yates, J. S., Mustian, K. M., Morrow, G. R., et al. (2005). Prevalence of Complementary and Alternative Medicine Use in Cancer Patients during Treatment. *Supportive Care in Cancer*, *13*(10), 806–811. 10.1 007/s00520-004-0770-7

Yurcheshen, M., Seehuus, M., & Pigeon, W. (2015). Updates on Nutraceutical Sleep Therapeutics and Investigational Research. *Evidence-Based Complementary and Alternative Medicine*, *2015*, 1–9. 10.1155/2015/105256

Zajac, I. T., Barnes, M., Cavuoto, P., Wittert, G., & Noakes, M. (2020). The Effect of Vitamin D-Enriched Mushrooms and Vitamin D3 on Cognitive Performance and Mood in Healthy Elderly Adults: A Randomized, Double-Blinded, Placebo-Controlled Trial. *Nutrients*, *12*, 3847. 10.3390/nu12123847

Zamora, J. C., Svensson, M., Kirschner, R., et al. (2018). Considerations and Consequences of Allowing DNA Sequence Data as Types of Fungal Taxa. *IMA Fungus*, *9*(1), 167–175. 10.5598/imafungus.2018.09.01.10

Zervakis, G. I. (1998). Mating Competence and Biological Species within the Subgenus Coremiopleurotus. *Mycologia*, *90*(6), 1063. 10.2307/3761281

Zhang, M., Cheung, P. C. K., Zhang, L., Chiu, C. M., & Ooi, V. E. C. (2004). Carboxymethylated Beta-Glucans from Mushroom Sclerotium of *Pleurotus Tuber-Regium* as Novel Water-Soluble Anti-Tumor Agent. *Carbohydrate Polymers*, *57*(3), 319–325. 10.1016/j.carbpol.2004.05.008

Zhang, L., Li, C. G., Liang, H., & Reddy, N. (2017). Bioactive Mushroom Polysaccharides: Immunoceuticals to Anticancer Agents. *Journal of Nutraceuticals and Food Science*, *2*(2). https://nutraceuticals.imedpub. com/bioactive-mushroom-polysaccharides-immunoceuticals-to-anticancer-agents.php?aid=20119

Zhang, Z. Q., Liu, L. P., Lei, L., Wang, C. N., Tang, Q. L., & Wu, T. X. (2019). Antioxidative and Immunomodulatory Activities of the Exopolysaccharides from Submerged Culture of Hen of the Woods or Maitake Culinary-Medicinal Mushroom, *Grifola Frondosa* (Agaricomycetes) by Addition of Rhizoma Gastrodiae Extract and Its Main Components. *International Journal of Medicinal Mushrooms*, *21*(8), 825–839. 10.1615/intjmedmushrooms.2019031597

Zhang, F. L., Yang, X. F., Wang, D., et al. (2020). A Simple and Effective Method to Discern the True Commercial Chinese Cordyceps from Counterfeits. *Scientific Reports*, *10*(1), 2974. 10.1038/s41598-02 0-59900-9

Zhang, W., Zhang, X., Li, M., et al. (2017). Identification of Chinese Caterpillar Medicinal Mushroom, *Ophiocordyceps Sinensis* (Ascomycetes) from Counterfeit Species. *International Journal of Medicinal Mushrooms*, *19*(12), 1061–1070. 10.1615/intjmedmushrooms.2017024823

Zhao, S., Gao, Q., Rong, C., et al. (2020). Immunomodulatory Effects of Edible and Medicinal Mushrooms and Their Bioactive Immunoregulatory Products. *Journal of Fungi*, *6*(4), 269. 10.3390/jof6040269

Zheng, S., Hu, Y., Zhao, R., et al. (2020). Quantitative Assessment of Secondary Metabolites and Cancer Cell Inhibiting Activity by High Performance Liquid Chromatography Fingerprinting in Dendrobium Nobile. *Journal of Chromatography B, 1140*, 122017. 10.1016/j.jchromb.2020.122017

Zhou, Y., Wang, M., Zhang, H., Huang, Y., & Ma, J. (2019). Comparative Study of the Composition of Cultivated, Naturally Grown *Cordyceps Sinensis*, and Stiff Worms across Different Sampling Years. B. T. Šiler (Ed.), *PLOS ONE, 14*(12), e0225750. 10.1371/journal.pone.0225750

Zou, H., Zhou, C., Li, Z., et al. (2019). Occurrence, Toxicity, and Speciation Analysis of Arsenic in Edible Mushrooms. *Food Chemistry, 281*, 269–284. 10.1016/j.foodchem.2018.12.103

PART II

Health aspects of wild mushrooms

CHAPTER 5

Mushroom ingestion for mineral supplementation

Somanjana Khatua[1,2] and Krishnendu Acharya[1]
[1]Molecular and Applied Mycology and Plant Pathology Laboratory, Centre of Advanced Study, Department of Botany, University of Calcutta, Kolkata, West Bengal, India
[2]Department of Botany, Krishnagar Government College, West Bengal, India

CONTENTS

5.1 INTRODUCTION

Minerals are fundamentally metals and other inorganic substances, present in all body tissues and fluids, and necessary for the maintenance of certain physicochemical processes essential to life (Gupta & Gupta, 2014). These micronutrients constitute approximately 4% of animal body weight (Radwińska & Żarczyńska, 2014). They are also called essential nutrients because they cannot be synthesized in the body and therefore must be taken through foods, water, or, in rare cases, supplements (Awuchi, 2020). Thus, a poorly balanced diet such as low intake of cereals, vegetables, legumes, and fruit as well as inadequate health care represents the most important contributor for mineral deficiency. Apart from that, certain diseases leave individuals less able to absorb minerals and more vulnerable to further illness (Gómez–Galera et al., 2010; Bhandari & Banjara, 2015). The situation has become an endemic ailment in many countries, particularly in low- and middle-income

DOI: 10.1201/9781003152583-7

nations because animal protein is in short supply due to viral diseases, drought, scarcity, and high cost of feed (Ndimele et al., 2017). Such micronutrient deficit, also known as hidden hunger, afflicts billions of individuals, or one in six people, globally (FAO, 2009). The effect can be distressing, leading to poor health, mental impairment, low productivity, and even death. Besides, vitamin absorption and function would be hampered in the absence of the specific mineral in the appropriate amount (Gupta & Gupta, 2014). It is thus suggested that patients who have a subclinical shortage of minerals may be at risk of impaired immune function and an increased chance of viral infections including the novel coronavirus (Shenkin, 2006; Mahluji et al., 2021). Consequently, mushrooms are considered a "powerhouse of nutrition" as they contain a lot of essential minerals. Wild, edible macrofungi may contain a relatively high concentration of crucial elements, even better than inhabiting soil due to a large surface area to volume ratio that promotes high nutrient uptake. This contrasts with vascular plants proving mushrooms as good sources of many mineral elements (Jedidi et al., 2017; Gałgowska & Pietrzak–Fiećko, 2020). Generally, the ash content of mushrooms that provides a measure of the total amount of minerals ranges between 60 and 120 g/kg dry matter, which is somewhat higher than or comparable to those of many vegetables (Kalač, 2013). A few minerals, like potassium (K), phosphorus (P), iron (Fe), and zinc (Zn) are abundantly present in mushroom fruiting bodies (Wang et al., 2017). The other detected elements include calcium (Ca), magnesium (Mg), sodium (Na), manganese (Mn) copper (Cu), and selenium (Se). Along with these benefits, some toxic minerals like chromium (Cr), lead (Pb), cadmium (Cd), and arsenic (As) may also be detected in wild-grown fruiting bodies. The contents of these detrimental elements in cultivated mushrooms are generally low, particularly due to the use of unpolluted substrates (Falandysz & Borovička, 2013; Egbuna & Tupas, 2020; Rasalanavhoa et al., 2020). Subsequently, it is worth mentioning that fruiting bodies not only accumulate bioelements but can release minerals effectively into artificial digestive juices, ideal for human consumption (Zajac et al., 2015; Kała et al., 2016; Kała et al., 2020; Muszyńska et al., 2020). Currently, knowledge on the mineral profile in wild and cultivatable mushrooms is relatively extensive, although fragmented. In many countries, mainly from the Northern Hemisphere, numerous studies have been conducted on the elemental composition of various edible macrofungi to more accurately estimate their nutritional value demanding a systematic and efficient congregation. In this review, we have tried to compile all the previously published reports to make an overview that might help future scientists to think differently on the exploitation of mineral enriched mushrooms in a better and effective way.

5.2 MINERALS

Minerals are mainly located in the skeleton, enzymes, hormones, and vitamins. They usually function as cofactors of enzymes and their presence at a balanced concentration is needed for numerous physiological processes including muscle contraction, nerve conduction, heart rhythm, and acid-base balance homeostasis. They also have structural functions particularly important for teeth and bones (Varela-López et al., 2016). Minerals can be classified as either major minerals (that are required in the diet each day in amounts of >100 mg) or trace elements (that are required in the diet each day in amounts of <100 mg). The major minerals include sodium, potassium, calcium, magnesium, and phosphorus, which are present in edible mushrooms in sufficient quantity, as summarized in Table 5.1. On the other hand, the trace minerals encompass iron, zinc, iodine, selenium, copper, cobalt, chromium, manganese, and molybdenum (Gupta & Gupta, 2014; Varela-López et al., 2016; Awuchi, 2020). Among them, iron, copper, zinc, manganese, and selenium are essential metals since they play a vital role in biological systems (Sesli et al., 2008) and mushrooms stand out for being an imperative source of these components (Table 5.2).

Table 5.1 Five major mineral compositions (mg/kg dry weight) of edible mushrooms grown naturally in a non-polluted area (W) or cultivated (C) or purchased from the market (P)

Mushroom species	W/C/P	Ca	K	Mg	Na	P	Origin	References
Agaricus arvensis	W	550	33,400	1,210	527	10,700	Turkey	Ayaz et al., 2011
	W	340	19,000	827	43.1	15,800	Poland	Mleczek, Budka, et al., 2021
Agaricus bisporus	W	74.5	35,000	1,200	96.3	NA	Turkey	Demirbaş, 2001
	W	NA	32,569.54	NA	NA	NA	Turkey	Turhan et al., 2010
	W	936	7,624	2,095	NA	NA	Turkey	Keleş & Gençelep, 2020
	C	230.9	46,900.22	1,250	428.05	NA	Tunisia	Jedidi et al., 2017
Agaricus campestris	W	50	59,000	2,300	NA	20,000	Poland	Falandysz et al., 2001
Agaricus langei	W	1,872	86,000	2,140	NA	NA	Turkey	Sarikurkcu et al., 2012
	W	238	5,386	1,273	NA	NA	Turkey	Keleş & Gençelep, 2020
Agaricus silvicola	W	81.6	42,000	980	103	NA	Turkey	Demirbaş, 2001
Agaricus subrufescens	C	958–1,520	28,440–30,711	995–1,167	140.9–188.6	6,792–11,775	Hungary	Györfi et al., 2010
Agrocybe cylindracea	W	1,162	37,600	2,160	NA	NA	Turkey	Sarikurkcu et al., 2012
	C	972–1,178	23,147–23,511	630–851	183–190	8,567–8,685	Poland	Siwulski et al., 2019
Agrocybe dura	W	1,542	12,066	1,507	NA	NA	Turkey	Keleş & Gençelep, 2020
Agrocybe praecox	W	573	7,786	979	NA	NA	Turkey	
Amanita caesaria	W	NA	NA	833.1	NA	NA	Greece	Ouzouni et al., 2009
Amanita rubescens	W	400–700	43,500–55,100	600–1,100	NA	5,900–10,500	Poland	Rudawska & Leski, 2005
Amanita ovoidea	W	4,380	64,800	3,560	NA	NA	Turkey	Sarikurkcu et al., 2012
Armillaria mellea	W	78	53,000	1,500	70	9,000	Poland	Falandysz et al., 2001
	W	NA	NA	1,063.1	NA	NA	Greece	Ouzouni et al., 2009
	W	58.3	24,600	485	62.8	5,300	Poland	Mleczek, Gąsecka, Budka, Siwulski, et al., 2021
Armillaria ostoyae	W	100	8,636	732	NA	NA	Turkey	Keleş & Gençelep, 2020
Armillaria tabesceus	W	NA	NA	1,150.7	NA	NA	Greece	Ouzouni et al., 2009
Amanita vaginata	W	92	48,000	1,300	120.8	NA	Turkey	Demirbaş, 2001
Auricularia polytricha	W	886.2	2,940	835.4	109.1	6,239.6	Cameroon	Teke et al., 2021
Boletus aureus	W	NA	NA	755.1	NA	NA	Greece	Ouzouni et al., 2009
Boletus aereus	W	638–973	18,512–14,947	773–1,269	389–503	NA	China	Liu et al., 2016

(Continued)

Table 5.1 (Continued) Five major mineral compositions (mg/kg dry weight) of edible mushrooms grown naturally in a non-polluted area (W) or cultivated (C) or purchased from the market (P)

Mushroom species	W/C/ P	Ca	K	Mg	Na	P	Origin	References
Boletus badius	W	3,896–4,228	6,722–7,268	271–309	328–385	4,236–4,588	Poland	Mleczek et al., 2016
	P	54.7	36,001	526	568	NA	Poland	Gałgowska & Pietrzak–Fiećko, 2020
Boletus edulis	W	39	3,400	1,100	180	10,000	Poland	Falandysz et al., 2001
	W	NA	33,561.72	NA	NA	NA	Turkey	Turhan et al., 2010
	W	53–150	470–790	16,000–25,000	160–340	NA	China	Zhang et al., 2010
	W	268	21,800	680	501	6,090	Turkey	Ayaz et al., 2011
	W	384–863	23,358–5,44	574–1,083	482–1184	NA	China	Liu et al., 2016
	P	75.3	29,136	566	653	NA	Poland	Gałgowska & Pietrzak–Fiećko, 2020
	W	26.47–168.29	18,180–27,090	540–860	300–1050	NA	South Africa	Rasalanavhoa et al., 2020
Boletus griseus	W	73.7	19,000	380	463	6,120	Poland	Mleczek, Budka, et al., 2021
	W	440	4,600	200	670	NA	China	Liu, Zhang, et al., 2012
Boletus mirabilis	W	14.8–166.17	16,460–35,000	530–820	330–1770	NA	South Africa	Rasalanavhoa et al., 2020
Boletus speciosus	W	38	2,500	110	160	NA	China	Liu, Zhang, et al., 2012
Boletus violaceofuscus	W	593	10,685	619	1275	NA	China	Liu et al., 2016
Cantharellus cibarius	W	340	56,000	1,200	83	5,700	Poland	Falandysz et al., 2001
	W	NA	NA	866.3	NA	NA	Greece	Ouzouni et al., 2009
	W	NA	50,027.51	NA	NA	NA	Turkey	Turhan et al., 2010
	W	722	32,500	815	550	3,850	Turkey	Ayaz et al., 2011
	W	190–400	24,000–64,000	550–1,300	68–160	3,000–4,800	Poland	Falandysz et al., 2012
	W	439	15,747	686	NA	NA	Turkey	Keleş & Gençelep, 2020
	P	211	46,024	842	142	NA	Poland	Gałgowska & Pietrzak–Fiećko, 2020
Cantharellus tubaeformis	W	47.2	39,600	572	181	4,690	Poland	Mleczek, Budka, et al., 2021
Catathelasma ventricosum	W	1,600	33,300	561	669	2,590	Turkey	Ayaz et al., 2011
Cerioporus squamosus	W	1,973	27,230	1,538	349	4,820	China	Liu, Sun, et al., 2012
	W	102	7,384	1,574	NA	NA	Turkey	Keleş & Gençelep, 2020

Species							Country	Reference
Clavulina cinerea	W	184	15,737	412	NA	NA	Turkey	Keleş & Gençcelep, 2020
Clavulina rugosa	W	564	28,900	812	336	4,150	Turkey	Ayaz et al., 2011
Clitocybe maxima	W	962	26,430	520	1692	5,390	China	Liu, Sun, et al., 2012
	C	2,840–3,455	12,243–13,250	1,846–2,289	263–365	9,490–9,721	Poland	Siwulski et al., 2019
Clitocybe subconnexa	W	926.8	20,232.5	1,488.6	7220.4	NA	Portugal	Heleno, Ferreira, et al., 2015
Collybia dryophila	W	2,380	53,400	1,284	NA	NA	Turkey	Sarikurkcu et al., 2012
Coprinus comatus	W	2,778	9,548	1,553	NA	NA	Turkey	Keleş & Gençcelep, 2020
	C	NA	NA	1,334	NA	NA	Serbia	Stilinović et al., 2020
	W	57.5	41,400	973	286	12,800	Poland	Mleczek, Budka, et al., 2021
Craterellus cornucopioides	W	560	50,000	700	53	2,300	Poland	Falandysz et al., 2001
	W	NA	50,977.48	NA	NA	NA	Turkey	Turhan et al., 2010
	W	1,255	36,620	978	1185	7,130	China	Liu, Sun, et al., 2012
Cyclocybe cylindracea	W	160	9,264	544	NA	NA	Turkey	Keleş & Gençcelep, 2020
Fistulina hepatica	W	NA	NA	898.3	NA	NA	Greece	Ouzouni et al., 2009
Flammulina velutipes	C	1,271–1,845	22,488–22,843	1,058–1,143	257–269	9,025–9,308	Poland	Siwulski et al., 2019
	W	24.6	32,000	1,770	190	8,890	Poland	Mleczek, Gąsecka, Budka, Siwulski, et al., 2021
Ganoderma lucidum	W	1092	7,421	891	205	5,025	Pakistan	Sharif et al., 2016
	C	998–1,354	6,876–7,227	427–632	136–164	8,160–8,631	Poland	Siwulski et al., 2019
Helvella leucopus	W	3,620	125,400	1,756	NA	NA	Turkey	Sarikurkcu et al., 2012
Hericium cirrhatum	W	121	32,300	609	283	6,000	Poland	Mleczek, Gąsecka, Budka, Siwulski, et al., 2021
Hericium coralloides	W	837.5	17,784.7	1,340	6780.4	NA	Portugal	Heleno, Barros, et al., 2015
Hericium erinaceum	W	443.5	11,880.5	855.7	5867.8	NA	Portugal	Heleno, Barros, et al., 2015
	P	110.02	29,123	758.1	320	7,708	Pakistan	Sharif et al., 2016
Hydnum repandum	W	68.5	36,000	1,030	92.6	NA	Turkey	Demirbaş, 2001
	W	461	38,300	670	611	5,230	Turkey	Ayaz et al., 2011
	W	83.73	28,052.05	3,236.08	5365.99	NA	Tunisia	Jedidi et al., 2017
	W	103	6,334	481	NA	NA	Turkey	Keleş & Gençcelep, 2020
Hygrophorus russula	W	NA	NA	758.4	NA	NA	Greece	Ouzouni et al., 2009
Laccaria amethistina	W	46	51,000	1,300	360	6,700	Poland	Falandysz et al., 2001
Laccaria amethystea	W	2,004	25,290	1,482	361	5,040	China	Liu, Sun, et al., 2012
Laccaria laccata	W	1,050	30,200	964	601	6,140	Turkey	Ayaz et al., 2011

(Continued)

Table 5.1 (Continued) Five major mineral compositions (mg/kg dry weight) of edible mushrooms grown naturally in a non-polluted area (W) or cultivated (C) or purchased from the market (P)

Mushroom species	W/C/P	Ca	K	Mg	Na	P	Origin	References
Lactarius deliciosus	W	590	33,000	1,200	490	8,300	Poland	Falandysz et al., 2001
	W	1,900–5,300	12,100–29,300	600–1,100	NA	2,300–6,600	Poland	Rudawska & Leski, 2005
	W	262.17	6859.1	1,624.68	3261.19	NA	Tunisia	Jedidi et al., 2017
	W	78.37–165.16	16,290–20,750	830–1,110	140–430	NA	South Africa	Rasalanavhoa et al., 2020
	W	222	7,121	579	NA	NA	Turkey	Keleş & Gençelep, 2020
	W	172	16,000	521	65.4	5,760	Poland	Mleczek, Budka, et al., 2021
Lactarius hygrophoroides	W	46	3,600	140	190	NA	China	Liu, Zhang, et al., 2012
Lactarius rufus	W	400–800	28,600–31,200	400–800	NA	3,300–5,100	Poland	Rudawska & Leski, 2005
Lactarius salmonicolor	W	237	4,170	529	NA	NA	Turkey	Keleş & Gençelep, 2020
Laetiporus sulphureus	W	130.4	4,336.2	138.5	42	5,428. 8	Cameroon	Teke et al., 2021
	W	82.7	20,700	547	99.4	3,780	Poland	Mleczek, Gąsecka, Budka, Siwulski, et al., 2021
Leccinum aurantiacum	W	66.6	35,400	689	78.6	9,610	Poland	Mleczek, Budka, et al., 2021
Leccinum griseum	W	38	40,000	1,200	830	7,300	Poland	Falandysz et al., 2001
Leccinum scabrum	W	71	40,000	1,200	290	7,700	Poland	Falandysz et al., 2001
	W	800–1,000	10,300–31,300	300–400	NA	1,200–1,600	Poland	Rudawska & Leski, 2005
	W	178	8,040	752	NA	NA	Turkey	Keleş & Gençelep, 2020
Leccinum versipelle	W	133	17,400	265	277	5,920	Poland	Mleczek, Budka, et al., 2021
	W	25	27,000	1,300	NA	9,000	Poland	Falandysz et al., 2001
Lentinula edodes	C	1,749	13,020	407	3274	7,699	India	Mallikarjuna et al., 2013
	P	121.06	21,740.8	1,020.1	168.1	8,674	Pakistan	Sharif et al., 2016
	C	1,021–1,620	18,345–19,501	856–942	190–254	8,091–8,327	Poland	Siwulski et al., 2019
Lepista nuda	W	NA	NA	949.8	NA	NA	Greece	Ouzouni et al., 2009
	W	731	27,800	1,200	433	12,400	Turkey	Ayaz et al., 2011
Lepista personata	W	531	33,500	836	96.6	13,700	Poland	Mleczek, Budka, et al., 2021
Lepista saeva	W	97.1	26,100	1,060	172	12,600	Poland	Mleczek, Budka, et al., 2021
	W	55	54,000	2,100	170	25,000	Poland	Falandysz et al., 2001
Leucopaxillus giganteus	W	468	27,900	1,140	339	14,000	Turkey	Ayaz et al., 2011

Species							Country	Reference
Lyophyllum decastes	W	240	1,400	84	350	NA	China	Liu, Zhang, et al., 2012
	W	1,026	81,600	1,518	NA	NA	Turkey	Sarikurcu et al., 2012
Macrocybe gigantea	W	470	1,300	550	580	NA	China	Liu, Zhang, et al., 2012
Macrolepiota procera	W	60	28,000	1,400	69	12,000	Poland	Falandysz et al., 2001
	W	52.6	31,700	816	288	8,210	Poland	Mleczek, Budka, et al., 2021
Marasmius oreades	W	NA	40,308.84	NA	NA	NA	Turkey	Turhan et al., 2010
	W	761	12,124	888	NA	NA	Turkey	Keleş & Genççelep, 2020
Melanoleuca arcuata	W	260	1,300	230	450	NA	China	Liu, Zhang, et al., 2012
Melanoleuca excissa	W	1,144	33,800	1,940	NA	NA	Turkey	Sarikurcu et al., 2012
Meripilus giganteus	W	213	22,100	1,860	190	10,700	Poland	Mleczek, Gąsecka, Budka, Siwulski, et al., 2021
Morchella angusticeps	W	2,500	66,200	1,442	NA	NA	Turkey	Sarikurcu et al., 2012
Morchella deliciosa	W	240	2,100	130	440	NA	China	Liu, Zhang, et al., 2012
Morchella esculanta	W	NA	46,349.87	NA	NA	NA	Turkey	Turhan et al., 2010
	W	1,112	33,400	1,476	NA	NA	Turkey	Sarikurcu et al., 2012
	W	416	14,400	704	74.4	9,410	Poland	Mleczek, Budka, et al., 2021
Morchella eximia	W	4,700	46,000	1,852	NA	NA	Turkey	Sarikurcu et al., 2012
Mycena haematopus	W	120	3,200	270	190	NA	China	Liu, Zhang, et al., 2012
Paxillus involutus	W	600–1,200	39,000–43,800	700–1,100	NA	2,000–7,900	Poland	Rudawska & Leski, 2005
Pleurotus citrinopileatus	C	6,230	17,740	1235	NA	NA	Poland	Krakowska et al., 2020
Pleurotus cystidious	C	697–1,220	14,000–24,300	647–801	167–215	3,014–3,127	Poland	Mleczek, Gąsecka, Budka, Niedzielski, et al., 2021
Pleurotus djamor	W	342	36,340	316	616	7,432	India	Mallikarjuna et al., 2013
	C	4,600	18,790	851	NA	NA	Poland	Krakowska et al., 2020
	C	1,080–1,910	20,000–23,900	523–706	171–196	2,838–2,983	Poland	Mleczek, Gąsecka, Budka, Niedzielski, et al., 2021
Pleurotus eryngii	C	843–988	13,927–14,870	699–1,228	306–337	8,572–8,851	Poland	Siwulski et al., 2019
	W	205	7,839	1,838	NA	NA	Turkey	Keleş & Genççelep, 2020
	C	6,190	17,000	1,110	NA	NA	Poland	Krakowska et al., 2020
Pleurotus florida	C	82.7	24,720	359	305	6,402	India	Mallikarjuna et al., 2013
	C	3,880	21,540	890	NA	NA	Poland	Krakowska et al., 2020
Pleurotus ostreatus	W	106	51,000	1,280	133	NA	Turkey	Demirbaş, 2001
	C	613.3	23,950.4	1,254	3958	8,330	Pakistan	Sharif et al., 2016

(Continued)

Table 5.1 (Continued) Five major mineral compositions (mg/kg dry weight) of edible mushrooms grown naturally in a non-polluted area (W) or cultivated (C) or purchased from the market (P)

Mushroom species	W/C/ P	Ca	K	Mg	Na	P	Origin	References
	W	465	9,975	1,514	NA	NA	Turkey	Keleş & Gençelep, 2020
	C	5,270	25,160	956	NA	NA	Poland	Krakowska et al., 2020
	C	1,040–1,630	24,800–29,400	683–798	200–293	2,791–2,898	Poland	Mleczek, Gąsecka, Budka, Niedzielski, et al., 2021
Pleurotus ostreatus var. florida	C	444–958	17,800–20,600	582–662	183–231	3,000–3,146	Poland	Mleczek, Gąsecka, Budka, Niedzielski, et al., 2021
Pleurotus pulmonarius	C	4,210	18,390	128	NA	NA	Poland	Krakowska et al., 2020
	C	702–980	19,500–22,991	681–861	186–200	3,003–3,267	Poland	Mleczek, Gąsecka, Budka, Niedzielski, et al., 2021
Pleurotus sajor–caju	C	748–1,310	21,200–24,900	716–814	190–206	2,792–3,153	Poland	Mleczek, Gąsecka, Budka, Niedzielski, et al., 2021
Podaxis pistillaris	W	64.85	8,554.86	6,275.66	NA	2,058.03	India	Sai & Basavarju, 2020
Polyporus dictyopus	W	653.1	2,394.5	644.7	79.5	6,842.1	Cameroon	Teke et al., 2021
Polyporus septosporus	W	35.2	26,100	1,480	64.4	10,900	Poland	Mleczek, Gąsecka, Budka, Siwulski, et al., 2021
Polyporus tenuiculus	W	909.5	4,284.1	944.8	97	5,922.5	Cameroon	Teke et al., 2021
Pulveroboletus ravenelii	W	260	2,700	210	250	NA	China	Liu, Zhang, et al., 2012
Ramaria largentii	W	NA	NA	837.5	NA	NA	Greece	Ouzouni et al., 2009
Rhizopogon roseolus	W	1,380	38,800	1,256	NA	NA	Turkey	Sarikurcu et al., 2012
Russula chloroides	W	2,200	27,400	1,502	NA	NA	Turkey	Sarikurcu et al., 2012
Russula cyanoxantha	W	86.3	46,000	1,160	110	NA	Turkey	Demirbaş, 2001
Russula delica	W	72.8	34,000	1,060	82.9	NA	Turkey	Demirbaş, 2001
	W	1,416	402,000	1,438	NA	NA	Turkey	Sarikurcu et al., 2012
	W	119	5,198	295	NA	NA	Turkey	Keleş & Gençelep, 2020
Rusula delica var chloroides	W	NA	NA	688.7	NA	NA	Greece	Ouzouni et al., 2009
Russula griseocarnosa	W	850–3,690	16,800–19,800	380–570	530–1340	1,650–3,420	China	Chen et al., 2010
Russula rosea	W	776	26,800	739	361	3,980	Turkey	Ayaz et al., 2011
Russula vesca	W	31,000	2,200	14,000	120	NA	Nigeria	Adejumo & Awosanya, 2005
Sparassis crispa	W	22.7	18,800	496	73	4,950	Poland	Mleczek, Gąsecka, Budka, Siwulski, et al., 2021
Stropharia coronilla	W	NA	32,518.74	NA	NA	NA	Turkey	Turhan et al., 2010

Species							Country	Reference
Stropharia rugoso-annulata	W	1,371	16,320	1,135	411	7,290	China	Liu, Sun, et al., 2012
	W	34	30,400	555	55.6	6,910	Poland	Mleczek, Budka, et al., 2021
Suillellus luridus	W	731–1,482	21,654–9,167	748–1,583	365–639	NA	China	Liu et al., 2016
Suillus bovinus	W	94	26,000	930	920	6,700	Poland	Falandysz et al., 2001
	W	519–763	12,434–9,863	897–963	402–479	NA	China	Liu et al., 2016
	W	147	15,000	316	66.3	6,280	Poland	Mleczek, Budka, et al., 2021
Suillus granulatus	W	277	5,618	593	NA	NA	Turkey	Keleş & Gençcelep, 2020
Suillus luteus	W	60	39,000	1,200	55	8,700	Poland	Falandysz et al., 2001
	W	500–800	25,300–34,100	600–1,100	NA	4,500–7,600	Poland	Rudawska & Leski, 2005
	W	184	6,449	736	NA	NA	Turkey	Keleş & Gençcelep, 2020
Suillus variegatus	W	98	35,000	1,300	40	1,200	Poland	Falandysz et al., 2001
Termitomyces heimii	W	144.14	9,015.33	6,283.39	W	2,050.97	India	Sai & Basavaraju, 2020
Termitomyces microcarpus	W	374.7	11,127.6	390.3	129.1	8,981.7	Cameroon	Teke et al., 2021
Termitomyces striatus	W	263.9	14,504.4	284.7	123.1	7,390.6	Cameroon	Teke et al., 2021
Tricholoma auratum	W	1,504	49,400	1,174	NA	NA	Turkey	Sarikurkcu et al., 2012
Tricholoma equestre	W	77.46	18,123.11	2,414.55	4239.08	NA	Tunisia	Jedidi et al., 2017
Tricholoma flavovirens	W	150	70,000	1,400	60	8,300	Poland	Falandysz et al., 2001
Tricholoma matsutake	W	270	2,400	370	270	NA	China	Liu, Zhang, et al., 2012
Tricholoma saponaceumvar. saponaceum	W	657	39,800	776	487	4,180	Turkey	Ayaz et al., 2011
Volvopluteus gloiocephalus	W	735.2	46,925.5	2,001.9	7905.6	NA	Portugal	Heleno, Ferreira, et al., 2015
	W	1,368	45,000	2,220	NA	NA	Turkey	Sarikurkcu et al., 2012
Volvariella volvacea	C	328	35,470.1	1,456	421	12,210.1	Pakistan	Sharif et al., 2016
Xerocomus badius Xerocomus	W	30	43,000	1,100	470	8,300	Poland	Falandysz et al., 2001
badius	W	600–900	33,300–36,500	700–1,200	NA	4,700–6,600	Poland	Rudawska & Leski, 2005
Xerecomus chrysenteron	W	18	4,300	1,000	70	9,300	Poland	Falandysz et al., 2001
Xerocomus chrysenteron	W	200–400	32,200–38,900	600–900	NA	6,700–8,800	Poland	Rudawska & Leski, 2005

NA: Not analyzed.

Table 5.2 Minor mineral composition (mg/kg dry weight) of edible mushrooms grown naturally in a non-polluted area (W) or cultivated (C) or purchased from the market (P)

Mushroom species	W/C/P	Cu	Fe	Mn	Zn	Origin	References
Agaricus arvensis	W	70.6	232	52.9	92.8	Turkey	Ayaz et al., 2011
Agaricus bisporus	W	23.8	103	20.4	252	Poland	Mleczek, Budka, et al., 2021
	W	5.22	126	22.3	17.8	Turkey	Demirbaş, 2001
	W	65.8–72.81	NA	NA	62.44–75.83	Spain	Alonso et al., 2003
	W	39.54	691.03	21.9	266.17	Turkey	Turhan et al., 2010
	C	26.11	4.8	NA	16.2	Tunisia	Jedidi et al., 2017
	P	1.54	5.14	NA	2.55	Spain	Rubio et al., 2018
	C	40.8	NA	5.91	63.4	Poland	Mirończuk–Chodakowska et al., 2019
	W	69.19	217.99	33.07	61.06	Turkey	Keleş & Gençcelep, 2020
Agaricus campestris	W	240	ND	16	210	Poland	Falandysz et al., 2001
	W	104.2–126.8	NA	NA	149.3–215	Spain	Alonso et al., 2003
	W	38	280	23	81	Turkey	Sarikurkcu et al., 2012
	W	0.75	7.12	0.48	4.13	Turkey	Gezer et al., 2015
Agaricus cupreobrunneus	W	27.4	159	49.4	81	Greece	Ouzouni et al., 2007
Agaricus langei	W	46.85	104.18	13.98	47.19	Turkey	Keleş & Gençcelep, 2020
Agaricus macrosporus	W	193–242.4	NA	NA	175.7–267	Spain	Alonso et al., 2003
Agaricus silvicola	W	6.24	59.3	3	25.6	Turkey	Demirbaş, 2001
	W	129.6–193.5	NA	NA	130.8–209.4	Spain	Alonso et al., 2003
Agaricus subrufescens	C	73.4–151.2	99–181.1	6.08–8.8.4	144.3–254.7	Hungary	Győrfi et al., 2010
Agrocybe cilindrica	W	32.07–42.24	NA	NA	53.34–79.29	Spain	Alonso et al., 2003
	W	43	72	9	108	Turkey	Sarikurkcu et al., 2012
	C	15.6–19.2	28.9–40.2	9.7–11.5	107–130	Poland	Siwulski et al., 2019
Agrocybe dura	W	13.4	102	18.6	33.9	Turkey	Soylak et al., 2005
	W	17.6	395.66	170.25	40.06	Turkey	Keleş & Gençcelep, 2020
Agrocybe praecox	W	24	133.54	14.69	38.34	Turkey	Keleş & Gençcelep, 2020
Amanita caesaria	W	19.32	356.9	47.99	65.65	Greece	Ouzouni et al., 2009
Amanita franchetii	W	22	257	38.9	96.9	Greece	Ouzouni et al., 2007
Amanita vaginata	W	5.11	58.1	10.5	19.6	Turkey	Demirbaş, 2001
Amanita ovoidea	W	19	710	33	83	Turkey	Sarikurkcu et al., 2012

Species						Country	Reference
Amanita rubescens	W	49.8–63.93	NA	NA	133–195.9	Spain	Alonso et al., 2003
	W	NA	61.2–69.7	29.7–31.2	154–176	Poland	Rudawska & Leski, 2005
Amanita rubescens var. *rubescens*	W	39.2	105	33.6	52.3	India	Lalotra et al., 2018
Armillaria mellea	W	ND	120	19	75	Poland	Falandysz et al., 2001
	W	15.6	510	49.1	43.5	Turkey	Sesli et al., 2008
	W	17.38	499	55.59	54.12	Greece	Ouzouni et al., 2009
	W	62.47	63.5	NA	60.85	Turkey	Kalyoncu et al., 2010
	W	23	NA	24.7	68.3	Poland	Mironiczuk–Chodakowska et al., 2019
	W	4.31	350	19.2	46.6	Poland	Mleczek, Gąsecka, Budka, Siwulski, et al., 2021
Armillaria ostoyae	W	31.11	242.11	32.89	57.13	Turkey	Keleş & Gençelep, 2020
Armillaria tabesceus	W	17.47	60.4	11.18	64.45	Greece	Ouzouni et al., 2009
Auricularia auricula	W	1.78	123	15.6	8.75	Nigeria	Nnorom et al., 2020
Auricularia polytricha	W	1.4	176.4	NA	15.1	Cameroon	Teke et al., 2021
Boletus aereus	W	68.18–80.07	NA	NA	96.54–160	Spain	Alonso et al., 2003
	W	29–44	97–351	24–63	19–117	China	Liu et al., 2016
Boletus appendiculatus	W	18	1040	35	63	Turkey	Turkekul et al., 2004
Boletus aureus	W	41.47	112.8	18.31	89.45	Greece	Ouzouni et al., 2009
Boletus badius	W	1.31–1.55	1313–1569	23–25	16–21	Poland	Mleczek et al., 2016
	P	29.7	38.8	11.9	163	Poland	Gałgowska & Pietrzak–Fiećko, 2020
Boletus edulis	W	62	50	9.6	290	Poland	Falandysz et al., 2001
	W	51.99–85.76	NA	NA	63.7–133.4	Spain	Alonso et al., 2003
	W	16.04	175.03	24.9	69.54	Turkey	Turhan et al., 2010
	W	31.8	74	14.1	125	Turkey	Ayaz et al., 2011
	W	11–42	31–49	10–18	58–120	China	Zhang et al., 2010
	W	38.2	NA	22.5	131	Italy	Giannaccini et al., 2012
	W	19–73	221–524	28–69	76–159	China	Liu et al., 2016
	W	34.4	812	54.4	96.3	India	Lalotra et al., 2018
	W	19	NA	20.1	170	Poland	Mironiczuk–Chodakowska et al., 2019
	P	23.4	48.9	11.3	158	Poland	Gałgowska & Pietrzak–Fiećko, 2020
	W	39.71–101.09	20–130	3.53–22.69	53.63–107.07	South Africa	Rasalanavhoa et al., 2020
	W	22.6	57.1	115	185	Poland	Mleczek, Budka, et al., 2021
Boletus griseus	W	52	47	63	94	China	Liu, Zhang, et al., 2012
	W	29.1–35.3	484–523	16.2–16.6	75.9–121.5	China	Wang et al., 2017

(Continued)

Table 5.2 (Continued) Minor mineral composition (mg/kg dry weight) of edible mushrooms grown naturally in a non-polluted area (W) or cultivated (C) or purchased from the market (P)

Mushroom species	W/C/P	Cu	Fe	Mn	Zn	Origin	References
Boletus impolitus	W	9.8–16.2	672–1831	16.2–41.5	38.7–78.1	China	Wang et al., 2017
Boletus luridus	W	26.6–34.2	288–371	16.2–16.6	118–164.5	China	Wang et al., 2017
Boletus mirabilis	W	37.5–154.9	30–3160	5.42–16.33	47.39–130.44	South Africa	Rasalanavhoa et al., 2020
Boletus pinophilus	W	49.84–85.76	NA	NA	81.44–146.4	Spain	Alonso et al., 2003
Boletus reticulatus	W	52.75–69.56	NA	NA	119.9–195.4	Spain	Alonso et al., 2003
Boletus speciosus	W	16.5–33.8	516–714	29.8–34.7	41.7–108.5	China	Wang et al., 2017
	W	28	78	2	50	China	Liu, Zhang, et al., 2012
	W	32.5–70.4	568–581	12.5–13.3	115.9–187.4	China	Wang et al., 2017
Boletus subtomentosus	W	18.5	NA	16.3	112	Poland	Mirończuk–Chodakowska et al., 2019
Boletus umbriniporus	W	36.3–62.6	591–816	29.2–36.4	71.1–172.9	China	Wang et al., 2017
Boletus violaceofuscus	W	33	88	28	47	China	Liu et al., 2016
Bovista plumbea	W	42	2340	36	42	Turkey	Turkekul et al., 2004
Calvatia excipuliformis	W	29.71	244.66	24.54	49.55	Jordan	Semreen & Aboul–Enein, 2011
	W	25.2	350	75.1	52.7	Turkey	Sesli et al., 2008
Calvatia utriformis	W	219.2–251.9	NA	NA	250.5–281.1	Spain	Alonso et al., 2003
	W	25	924	28	58	Turkey	Turkekul et al., 2004
	W	21.51	211.77	38.08	35.98	Jordan	Semreen & Aboul–Enein, 2011
Cantharellus cibarius	W	ND	140	30	100	Poland	Falandysz et al., 2001
	W	52.7–70.39	NA	NA	71.41–108.2	Spain	Alonso et al., 2003
	W	32.6	119	22.1	54.1	Greece	Ouzouni et al., 2007
	W	15.5	1741	131	72.5	Turkey	Sesli et al., 2008
	W	32.49	118.2	22.09	54.29	Greece	Ouzouni et al., 2009
	W	22.76	2970.56	71.29	ND	Turkey	Turhan et al., 2010
	W	37.3	130	25.2	71.5	Turkey	Ayaz et al., 2011
	W	30–52	43–180	21–32	57–97	Poland	Falandysz et al., 2012
	W	69	NA	32.5	121	Poland	Mirończuk–Chodakowska et al., 2019
	W	46.91	174.42	18.73	82.22	Turkey	Keleş & Gençcelep, 2020
	P	48.4	58.9	23.7	113	Poland	Gałgowska & Pietrzak–Fiećko, 2020
	W	34.8	142	62.2	108	Poland	Mleczek, Budka, et al., 2021

Species						Country	Reference
Cantharellus tubaeformis	W	63.4	205	87.4	106	Turkey	Sesli et al., 2008
Catathelasma ventricosum	W	44.6	166	48.4	57.5	Turkey	Ayaz et al., 2011
Cerioporus squamosus	W	38	673	9	88	China	Liu, Sun, et al., 2012
Clavulina cinerea	W	15.16	50.25	4.7	44.47	Turkey	Keleş & Gençcelep, 2020
	W	135.35	355.2	35.84	49.75	Turkey	Keleş & Gençcelep, 2020
Clavulina rugosa	W	321	829	76.5	77.5	Turkey	Ayaz et al., 2011
Clitocybe gibba	W	52.4	406	36.9	65.7	Turkey	Sesli et al., 2008
Clitocybe maxima	W	52	308	33	127	China	Liu, Sun, et al., 2012
	C	3.1–5.5	436–730	14.3–21.3	30–34	Poland	Siwulski et al., 2019
Clitocybe nebularis	W	72.54–92.35	NA	NA	100.6–158.3	Spain	Alonso et al., 2003
Clitocybe subconnexa	W	52.2	65.3	1	63.7	Portugal	Heleno, Ferreira, et al., 2015
Collybia dryophila	W	32.6	852	28.6	55.9	Turkey	Sesli et al., 2008
	W	26	398	77	98	Turkey	Sarikurkcu et al., 2012
Coprinus atramentarius	W	57.12	1183.6	64.2	288.4	Turkey	Bengu, 2019
Coprinus comatus	W	95.32–147.3	NA	NA	88.18–139.7	Spain	Alonso et al., 2003
	W	47.26	277.66	20.18	48.34	Turkey	Keleş & Gençcelep, 2020
	C	10.17	1471	NA	31.73	Serbia	Stilinović et al., 2020
	W	37.4	96.4	10.8	86.4	Poland	Mleczek, Budka, et al., 2021
Cortinarius caperatus	W	57.1	NA	41	90.2	Poland	Mirończuk–Chodakowska et al., 2019
Craterellus cornucopioides	W	26	260	56	100	Poland	Falandysz et al., 2001
	W	73.8	502	145	167	Turkey	Sesli et al., 2008
	W	24.47	185.01	29.14	128.03	Turkey	Turhan et al., 2010
	W	43	413	27	61	China	Liu, Sun, et al., 2012
Cyclocybe cylindracea	W	3.58	300.87	7.23	9.74	Turkey	Gezer et al., 2015
Fistulina hepatica	W	19.09	259.91	21.31	30.1	Turkey	Keleş & Gençcelep, 2020
	W	32.33–39.51	NA	NA	35.83–50.33	Spain	Alonso et al., 2003
Flammulina velutipes	W	7.38	38.9	7.19	34.43	Greece	Ouzouni et al., 2009
	C	3.8–5.8	53.4–63.7	7.8–8.9	71–114	Poland	Siwulski et al., 2019
	W	42.9	160	17.9	66.1	Poland	Mleczek, Gąsecka, Budka, Siwulski, et al., 2021
Ganoderma lucidum	W	12	121	11	22	Pakistan	Sharif et al., 2016
	C	13–14	23.7–29.7	5.8–8.3	54–79	Poland	Siwulski et al., 2019
Geopora arenicola	W	88.8	267	50.5	36.5	India	Lalotra et al., 2018

(Continued)

Table 5.2 (Continued) Minor mineral composition (mg/kg dry weight) of edible mushrooms grown naturally in a non-polluted area (W) or cultivated (C) or purchased from the market (P)

Mushroom species	W/C/P	Cu	Fe	Mn	Zn	Origin	References
Helvella lacunosa	W	13.22	44.17	2.12	10.16	Turkey	Gezer et al., 2015
Helvella leucopus	W	31	242	11	354	Turkey	Sarikurcu et al., 2012
Hericium cirrhatum	W	10.3	48.4	10.9	51.3	Poland	Mleczek, Gąsecka, Budka, Siwulski, et al., 2021
Hericium coralloides	W	7.2	779.6	3.1	47.6	Portugal	Heleno, Barros, et al., 2015
Hericium erinaceum	W	2.2	67.7	0.9	21.1	Portugal	Heleno, Barros, et al., 2015
	P	9	112	8	34.1	Pakistan	Sharif et al., 2016
Hydnum repandum	W	6.84	33.5	3.12	14.1	Turkey	Demirbaş, 2001
	W	35.38–42.83	NA	NA	30–50.5	Spain	Alonso et al., 2003
	W	24.3	317	26.3	35.9	Greece	Ouzouni et al., 2007
	W	11.2	199	20.8	36.2	Turkey	Ayaz et al., 2011
	W	2.76	2.12	NA	3.82	Tunisia	Jedidi et al., 2017
	W	11.55	1,121.53	16.82	34.68	Turkey	Keleş & Gençcelep, 2020
Hygrophorus chrysodon	W	4.65	180	48.7	76.3	Greece	Ouzouni et al., 2007
Hygrophorus eburneus	W	16	193	100	82.6	Greece	Ouzouni et al., 2007
Hygrophorus russula	W	9.44	300.7	34.14	57.01	Greece	Ouzouni et al., 2009
Imleria badia	W	41.3	NA	27.6	144	Poland	Mirończuk–Chodakowska et al., 2019
Infundibulicybe geotropa	W	74.67	112.33	NA	87.4	Turkey	Kalyoncu et al., 2010
	W	30.37	63.7	NA	61.24	Turkey	Sevindik et al., 2020
Laccaria amethistina	W	75	83	34	110	Poland	Falandysz et al., 2001
Laccaria amethystea	W	32.8	780	125	89.2	Turkey	Sesli et al., 2008
	W	36	211	35	59	China	Liu, Sun, et al., 2012
Laccaria laccata	W	72.9	360	34.9	241	Turkey	Ayaz et al., 2011
Lactarius deliciosus	W	ND	50	10	180	Poland	Falandysz et al., 2001
	W	18.55–32.62	NA	NA	152.2–309.8	Spain	Alonso et al., 2003
	W	NA	24.3–33.2	17.5–19.2	146–213	Poland	Rudawska & Leski, 2005
	W	15.49	216.83	5.98	123.57	Serbia	Kosanić et al., 2016
	W	11.86	2.39	NA	7.4	Tunisia	Jedidi et al., 2017
	P	1.64	10.9	NA	2.32	Spain	Rubio et al., 2018

Species							Country	Reference
	W	10.6	NA	13	129		Poland	Mirończuk–Chodakowska et al., 2019
	W	14.85–22.14	40–80	3.85–11.91	59.37–150.53		South Africa	Rasalanavhoa et al., 2020
Lactarius hygrophoroides	W	8.85	144.01	12.45	57.12		Turkey	Keleş & Gençcelep, 2020
Lactarius rufus	W	10.9	85.5	18.1	124		Poland	Mleczek, Budka, et al., 2021
	W	28	28	3.7	16		China	Liu, Zhang, et al., 2012
Lactarius salmonicolor	W	NA	118–142	28.2–35.1	63.8–83.5		Poland	Rudawska & Leski, 2005
	W	6.15	239	20.8	94.5		Greece	Ouzouni et al., 2007
	W	13.7	73.93	17.45	34.5		Turkey	Keleş & Gençcelep, 2020
Laetiporus sulphureus	W	11.5	86.9	NA	26.6		Cameroon	Teke et al., 2021
	W	5	162.92	19.36	28.36		Turkey	Bengu, 2019
	W	3.35	129	5.56	57.9		Poland	Mleczek, Gąsecka, Budka, Siwulski, et al., 2021
Lactarius triviralis	W	8	1230	120	NA		Nigeria	Adejumo & Awosanya, 2005
Leccinum aurantiacum	W	10.7	83.9	13.7	102		Poland	Mleczek, Budka, et al., 2021
Leccinum griseum	W	ND	30	7	150		Poland	Falandysz et al., 2001
Leccinum rufum	W	50.9	NA	15.1	115		Poland	Mirończuk–Chodakowska et al., 2019
Leccinum rugosiceps	W	15.1–35.7	323–449	19.3–19.7	62.1–91.6		China	Wang et al., 2017
Leccinum scabrum	W	ND	20	12	220		Poland	Falandysz et al., 2001
	W	41.88–49.67	NA	NA	58.53–142.8		Spain	Alonso et al., 2003
	W	NA	7.5–25.2	8.2–11.7	21.5–35.7		Poland	Rudawska & Leski, 2005
	W	16.6	NA	13.5	87.7		Poland	Mirończuk–Chodakowska et al., 2019
	W	34.59	98.14	3.82	48.57		Turkey	Keleş & Gençcelep, 2020
	W	6.2	66.2	16.4	54.2		Poland	Mleczek, Budka, et al., 2021
Leccinum versipelle	W	84	ND	ND	240		Poland	Falandysz et al., 2001
Lentinula edodes	C	14.8	148	10	94.4		India	Mallikarjuna et al., 2013
	P	11.01	69.01	13	67.1		Pakistan	Sharif et al., 2016
	P	1.53	10.5	NA	2.23		Spain	Rubio et al., 2018
	C	7.3	NA	27.1	61.4		Poland	Mirończuk–Chodakowska et al., 2019
	C	2.1–2.5	25.5–33.9	14.2–25.7	84–92		Poland	Siwulski et al., 2019
Lentinus cladopus	W	9.7	353	5.4	15.8		India	Mallikarjuna et al., 2013
Lentinus crinitus	C	11.23	129.26	120.34	59.59		Colombia	Dávila et al., 2020
Lentinus squarrosulus	W	4.36	123	24.29	38.85		Nigeria	Nnorom et al, 2020

(Continued)

Table 5.2 (Continued) Minor mineral composition (mg/kg dry weight) of edible mushrooms grown naturally in a non-polluted area (W) or cultivated (C) or purchased from the market (P)

Mushroom species	W/C/ P	Cu	Fe	Mn	Zn	Origin	References
Lepista nuda	W	117.7–119.2	NA	NA	108.9–182.1	Spain	Alonso et al., 2003
	W	20	568	16	45	Turkey	Turkekul et al., 2004
	W	20.1	258	52.4	89.3	Turkey	Sesli et al., 2008
	W	75.06	74.6	33.65	98.99	Greece	Ouzouni et al., 2009
	W	51.84	317.54	32.87	58.77	Jordan	Semreen & Aboul-Enein, 2011
	W	74.9	135	68.3	86.2	Turkey	Ayaz et al., 2011
	W	86.8	425	49.9	140	Poland	Mleczek, Budka, et al., 2021
Lepista personata	W	135	179	15.4	180	Poland	Mleczek, Budka, et al., 2021
Lepista saeva	W	120	90	88	150	Poland	Falandysz et al., 2001
Leucopaxillus giganteus	W	43.8	257	21.9	84.4	Turkey	Ayaz et al., 2011
	W	50	510	60	85	China	Liu, Zhang, et al., 2012
Lycoperdon perlatum	W	56.2	950	102	205	Turkey	Sesli et al., 2008
Lyophyllum decastes	W	10.6	85	12	46	Turkey	Sarikurkcu et al., 2012
Macrocybe gigantea	W	13	79	5.9	160	China	Liu, Zhang, et al., 2012
Macrolepiota procera	W	130	80	14	90	Poland	Falandysz et al., 2001
	W	199.9–235.8	NA	NA	78.18–106.8	Spain	Alonso et al., 2003
	W	158	NA	20.1	124.7	Italy	Giannaccini et al., 2012
	W	1.31	10.75	0.93	10.77	Turkey	Gezer et al., 2015
	W	109.57	89.53	9.38	53.85	Serbia	Kosanić et al., 2016
	W	123	NA	16.3	84.1	Poland	Mironczuk–Chodakowska et al., 2019
	W	204	77.5	96.8	135	Poland	Mleczek, Budka, et al., 2021
Marasmius oreades	W	107.3–116.1	NA	NA	84.06–152.9	Spain	Alonso et al., 2003
	W	50.6	227	25.1	53.4	Turkey	Soylak et al., 2005
	W	30.5	150	130	135	Turkey	Sesli et al., 2008
	W	29.31	230.01	7.44	119.04	Turkey	Turhan et al., 2010
	W	58.48	262.92	30.78	91.76	Turkey	Keleş & Gençcelep, 2020
Melanoleuca arcuata	W	22	22	1.4	38	China	Liu, Zhang, et al., 2012
Melanoleuca excissa	W	45	139	53	83	Turkey	Sarikurkcu et al., 2012
Meripilus giganteus	W	70.17	84.33	NA	75.63	Turkey	Kalyoncu et al., 2010

Species						Country	Reference
Morchella angusticeps	W	40.1	208	89	181	Poland	Mleczek, Gąsecka, Budka, Siwulski, et al., 2021
Morchella costata	W	16.3	324	22	94	Turkey	Sarikurkcu et al., 2012
Morchella deliciosa	W	9.21	10.4	0.83	10.72	Turkey	Gezer et al., 2015
	W	55	42	70	58	China	Liu, Zhang, et al., 2012
	W	33.4	213	53.3	117	India	Lalotra et al., 2018
Morchella esculanta	W	22.43	360.04	31.24	112.01	Turkey	Turhan et al., 2010
	W	16.4	148	27	84	Turkey	Sarikurkcu et al., 2012
	W	19.2	237	44.4	173	Poland	Mleczek, Budka, et al., 2021
Morchella eximia	W	14.1	460	41	115	Turkey	Sarikurkcu et al., 2012
Mycena aetites	C	45.7	743	85.6	93.7	Turkey	Sesli et al., 2008
Mycena haematopus	W	23	180	24	54	China	Liu, Zhang, et al., 2012
Paxillus involutus	W	NA	44.8–70.5	9.9–24.4	128–139	Poland	Rudawska & Leski, 2005
Pholiota nameko	P	1.73	10.9	NA	1.93	Spain	Rubio et al., 2018
Pleurotus citrinopileatus	C	7.7	43.8	9.2	47.9	Poland	Krakowska et al., 2020
Pleurotus cystidious	C	6.14–10.7	20.8–34.3	5.89–6.04	92.8–117	Poland	Mleczek, Gąsecka, Budka, Niedzielski, et al., 2021
Pleurotus djamor	W	14.5	148	11.2	92.1	India	Mallikarjuna et al., 2013
	C	5.8	69	11.3	94.1	Poland	Krakowska et al., 2020
	C	9.43–14.5	26.2–60	11.7–13.4	98.4–123	Poland	Mleczek, Gąsecka, Budka, Niedzielski, et al., 2021
Pleurotus eryngii	C	0.9–1.4	27.9–36.4	10.4–10.7	55–92	Poland	Siwulski et al., 2019
	C	5.6	50.9	7.5	47.4	Poland	Krakowska et al., 2020
	W	9.39	103.86	8.15	56.69	Turkey	Keleş & Gençcelep, 2020
Pleurotus florida	C	10.6	62.7	6.2	50.6	India	Mallikarjuna et al., 2013
	C	9.2	78.9	10.7	60.1	Poland	Krakowska et al., 2020
Pleurotus ostreatus	W	13.6	86.1	6.27	29.8	Turkey	Demirbaş, 2001
	W	24.16–26.28	NA	NA	68.88–96.56	Spain	Alonso et al., 2003
	C	14.2	102	4	46	Pakistan	Sharif et al., 2016
	p	1.99	11	NA	2.91	Spain	Rubio et al., 2018
	C	14.1	NA	12.9	72.4	Poland	Mirończuk-Chodakowska et al., 2019
	C	26.9	84.1	9.6	88.2	Poland	Krakowska et al., 2020
	W	5.89	242.09	10.57	55.29	Turkey	Keleş & Gençcelep, 2020
	C	6.1–14.61	25.3–52	6.87–7.34	79.9–108	Poland	Mleczek, Gąsecka, Budka, Niedzielski, et al., 2021

(Continued)

Table 5.2 (Continued) Minor mineral composition (mg/kg dry weight) of edible mushrooms grown naturally in a non-polluted area (W) or cultivated (C) or purchased from the market (P)

Mushroom species	W/C/P	Cu	Fe	Mn	Zn	Origin	References
Pleurotus ostreatus var. *florida*	C	4.41–13	25.5–33.7	7.35–8.7	66.4–78.2	Poland	Mleczek, Gąsecka, Budka, Niedzielski, et al., 2021
Pleurotus pulmonarius	C	3.4	31.5	9.9	43.8	Poland	Krakowska et al., 2020
	C	2.65–7.98	34–63.1	10.5–14.1	89.2–110	Poland	Mleczek, Gąsecka, Budka, Niedzielski, et al., 2021
Pleurotus sajor–caju	C	5.04–11	32.8–87.5	11.3–16	96.5–105	Poland	Mleczek, Gąsecka, Budka, Niedzielski, et al., 2021
Pleturotus tuber–reguim	W	4.39	74.08	15.88	28.47	Nigeria	Nnorom et al., 2020
Podaxis pistillaris	W	625.45	128.04	61.08	549.09	India	Sai & Basavarju, 2020
Polyporus dictyopus	W	7.8	117.6	NA	13.1	Cameroon	Teke et al., 2021
Polyporus frondosus	W	41	2003	28	122	Turkey	Turkekul et al., 2004
	W	37.61	269.51	26.41	37.75	Jordan	Semreen & Aboul–Enein, 2011
Polyporus septosporus	W	18.5	73.2	8.71	61.1	Poland	Mleczek, Gąsecka, Budka, Siwulski, et al., 2021
Polyporus tenuiculus	W	8.6	69.2	NA	48.2	Cameroon	Teke et al., 2021
Pulveroboletus ravenelii	W	58	370	58	34	China	Liu, Zhang, et al., 2012
Ramaria largentii	W	17.79	302.1	62.63	46.33	Greece	Ouzouni et al., 2009
Ramaria stricta	W	95.54	451.21	NA	39.19	Ukraine	Krupodorova & Sevindik, 2020
Rhizopogon roseolus	W	9	764	29	47	Turkey	Sarikurkcu et al., 2012
Russula chloroides	W	2.19	39.97	8.49	11.02	Turkey	Gezer et al., 2015
Russula cyanoxantha	W	24	500	48	52	Turkey	Sarikurkcu et al., 2012
	W	18.9	63.2	5.42	21.7	Turkey	Demirbaş, 2001
	W	59.58–85.18	NA	NA	80.46–112.5	Spain	Alonso et al., 2003
	W	8.87–9.37	85.3–340.34	NA	89.46–99.62	Romania	Busuioc et al., 2011
Russula delica	W	13.6	74.8	6.62	32.6	Turkey	Demirbaş, 2001
	W	58.41	288.54	36.55	41.44	Jordan	Semreen & Aboul–Enein, 2011
	W	37	470	66	52	Turkey	Sarikurkcu et al., 2012
	W	26.9	106.39	8.13	22.99	Turkey	Keleş & Genççelep, 2020
Russula delica var chloroides	W	51.71	81.8	16.61	56.58	Greece	Ouzouni et al., 2009
Russula foetens	W	6.37–7.86	59.02–63.06	NA	24.26–62.81	Romania	Busuioc et al., 2011
Russula griseocarnosa	W	34–48	500–954	22–23	72–88	China	Chen et al., 2010

Species						Country	Reference
Russula heterophylla	W	19.9	NA	33.7	115.9	Poland	Mirończuk–Chodakowska et al., 2019
Russula nigrescens	W	5.73–7.15	58.83–65.03	NA	19.7–23.6	Romania	Busuioc et al., 2011
Russula rosea	W	39.4	212	62.2	101	Turkey	Ayaz et al., 2011
Russula vinosa	W	77.2	NA	27.8	68.9	Poland	Mirończuk–Chodakowska et al., 2019
Russula virescens	W	5.73–7.15	58.83–65.03	NA	69.07–87.28	Romania	Busuioc et al., 2011
Sarcodon squamosus	W	1.72	12.95	0.58	29.16	Turkey	Gezer et al., 2015
Sparassis crispa	W	54.30	54	NA	47.4	Turkey	Kalyoncu et al., 2010
	W	2.55	7.7	0.68	14.69	Turkey	Gezer et al., 2015
	W	25.5	555	41	137	India	Lalotra et al., 2018
	W	7.88	80	21.4	62.7	Poland	Mleczek, Gąsecka, Budka, Siwulski, et al., 2021
Stropharia coronilla	W	16.9	1580	65.7	74.5	Turkey	Soylak et al., 2005
Stropharia rugoso-annulata	W	30.86	122.56	13.9	107.64	Turkey	Turhan et al., 2010
	W	29	195	59	102	China	Liu, Sun, et al., 2012
	W	13.6	185	35.9	62.3	Poland	Mleczek, Budka, et al., 2021
Suillellus luridus	W	32–58	94–342	22–69	21–93	China	Liu et al., 2016
Suillus bovinus	W	ND	53	ND	87	Poland	Falandysz et al., 2001
	W	48–51	56–286	29–35	9–46	China	Liu et al., 2016
	W	15.2	NA	19.9	81.6	Poland	Mirończuk–Chodakowska et al., 2019
	W	4.38	61.5	14.5	43.5	Poland	Mleczek, Budka, et al., 2021
Suillus granulatus	W	35.8	658	77.5	59.8	Argentina	Arce et al., 2008
	W	16.72	166.02	5.18	40.31	Turkey	Keleş & Gençcelep, 2020
Suillus grevillei	W	29.3	NA	12.2	93.9	Poland	Mirończuk–Chodakowska et al., 2019
Suillus luteus	W	27	ND	14	160	Poland	Falandysz et al., 2001
	W	NA	33.8–42.8	10.7–16.1	32.3–73.1	Poland	Rudawska & Leski, 2005
	W	17.8	NA	20.3	101	Poland	Mirończuk–Chodakowska et al., 2019
	W	13.36	283.24	22.84	118.84	Turkey	Bengu, 2019
	W	14.84	113.96	7.53	52.3	Turkey	Keleş & Gençcelep, 2020
Suillus variegatus	W	34	3600	14	150	Poland	Falandysz et al., 2001
Termitomyces heimii	W	372.3	501.84	264.72	906.9	India	Sai & Basavaiju, 2020
Termitomyces microcarpus	W	39	208.6	NA	81.3	Cameroon	Teke et al., 2021
Termitomyces striatus	W	24.1	277.7	NA	49	Cameroon	Teke et al., 2021
Thelephora ganbajun	W	NA	75.8	5.25	17.3	China	Chen et al., 2021

(Continued)

Table 5.2 (Continued) Minor mineral composition (mg/kg dry weight) of edible mushrooms grown naturally in a non-polluted area (W) or cultivated (C) or purchased from the market (P)

Mushroom species	W/C/ P	Cu	Fe	Mn	Zn	Origin	References
Tricholoma anatolicum	W	1.48	14.72	0.82	9.27	Turkey	Gezer et al., 2015
Tricholoma argyraceum	W	13.9	216	15.5	89.5	Turkey	Soylak et al., 2005
Tricholoma auratum	W	19	388	49	356	Turkey	Sarikurkcu et al., 2012
Tricholoma columbetta	W	61.11–91.68	NA	NA	166–238	Spain	Alonso et al., 2003
Tricholoma equestre	W	34.24–72.14	NA	NA	106.1–233.5	Spain	Alonso et al., 2003
	W	3.88	4.92	NA	36.23	Tunisia	Jedidi et al., 2017
	W	25.2	NA	15.9	176	Poland	Mirończuk–Chodakowska et al., 2019
Tricholoma flavovirens	W	55	43	36	460	Poland	Falandysz et al., 2001
Tricholoma matsutake	W	20	34	3	62	China	Liu, Zhang, et al., 2012
Tricholoma portentosum	W	48.18–66.76	NA	NA	83.56–164.6	Spain	Alonso et al., 2003
	W	15	NA	14.9	112.6	Poland	Mirończuk–Chodakowska et al., 2019
Tricholoma saponaceum var. saponaceum	W	26.8	229	28.4	99.7	Turkey	Ayaz et al., 2011
Tuber aestivum	W	2.55	4.28	1.22	6.24	Turkey	Gezer et al., 2015
Volvariella volvacea	C	19	177	6	75	Pakistan	Sharif et al., 2016
Volvopluteus gloiocephalus	W	50.1	699.1	1.3	108.9	Portugal	Heleno, Ferreira, et al., 2015
	W	94	232	84	136	Turkey	Sarikurkcu et al., 2012
Xerocomus badius	W	82	33	16	180	Poland	Falandysz et al., 2001
	W	48.39–61.58	NA	NA	162.3–225.7	Spain	Alonso et al., 2003
	W	NA	48.9–122	14.4–14.9	136–157	Poland	Rudawska & Leski, 2005
Xerecomus chrysenteron	W	ND	ND	ND	80	Poland	Falandysz et al., 2001
	W	66.09–77.56	NA	NA	111.9–162.3	Spain	Alonso et al., 2003
	W	NA	47.5–59.7	17.1–26.8	91.6–164	Poland	Rudawska & Leski, 2005
	W	3.8	46.3	11.3	78.5	Greece	Ouzouni et al., 2007
Xerocomellus chrysenteron	W	36.8	NA	21.7	184	Poland	Mirończuk–Chodakowska et al., 2019
Xerocomus spadiceus	W	19.8–28	1,315–6,762	29–103.9	47.5–77.8	China	Wang et al., 2017

NA: Not analyzed, ND: Not detected.

5.2.1 Calcium

Calcium, the most abundant mineral in humans, is associated mainly with the formation and metabolism of bone. More than 99% of total body Ca is found as calcium hydroxyapatite in teeth and bones, providing hard tissues. The element is also involved in muscle contraction, cell differentiation, enzyme activation, immune response, neuronal activity, programmed cell death, vasodilatation, intracellular signaling, and hormonal secretion. However, the human body system cannot produce Ca, but expends up to 100 mg of Ca in a day resulting in the deficiency of the mineral (Ogidi et al., 2020). The scarcity could cause various diseases such as osteoporosis, rickets, hypertension, anemia, and colorectal cancer (Beto, 2015; Pu et al., 2016). Supplementation of Ca into various food products has been an advantageous way to improve mineral dose in humans (Ogidi et al., 2020).

Based on previous publications, the mushroom could be considered as an excellent food concerning contribution with Ca for the human body. However, the amount varies considerably in wild samples ranging from 22.7 mg/kg dry weight (DW) in the case of *Sparassis crispa* (Mleczek, Gąsecka, Budka, Siwulski, et al., 2021) to 4,700 mg/kg DW in *Morchella eximia* (Sarikurkcu et al., 2012). A similar trend has also been observed in cultivated samples where the amount ranged between 53.7 mg/kg DW in *Pleurotus ostreatus* (Sharif et al., 2016) to 6,190 mg/kg DW as found in *Pleurotus eryngii* (Krakowsk et al., 2020).

5.2.2 Potassium

Potassium, the third most abundant mineral in the body, plays an important role in assisting nerve function and maintaining the balance of the physical fluid system. The element is also related to heart activity muscle contraction (Martínez–Ballesta et al., 2010). However, nowadays, intake of K has decreased due to processing of food and reduction in consumption of fruit and vegetables that in turn has greatly increased salt intake elevating the risk of high blood pressure (He & MacGregor, 2008). The deficiency may also result in fatigue, muscle weakness, cramping legs, acne, slow reflexes, mood changes, dry skin, and irregular heartbeat. Moreover, a reduced level of K causes alkalosis that makes the kidney less able to retain the mineral (Martínez–Ballesta et al., 2010). The danger can be diminished by an adequate intake of potassium that may help reduce the risk of cardiovascular disease and stroke (He & MacGregor, 2008).

High potassium content is characteristic of mushrooms where the amount varied from 20–40 g/kg DW, as reported by Kalač (2013). Keleş and Gençcelep (2020) reported mineral composition of 20 wild macrofungi where the amount was quite lower ranging from 5.2 g/kg to 15.75 g/kg DW. In contrast, some other naturally grown mushrooms namely *Agaricus campestris* (Falandysz et al., 2001; Sarikurkcu et al., 2012), *Armillaria mellea* (Falandysz et al., 2001), *Craterellus cornucopiodes* (Turhan et al., 2010; Liu, Sun, et al., 2012), *Lepista saeva* (Falandysz et al., 2001), *Amanita ovoidea*, *Collybia dryophila*, *Lyophyllum decastes*, *M. eximia*, *Morchella angusticeps* (Sarikurkcu et al., 2012), and *P. ostreatus* (Demirbaş, 2001) were reported to contain the mineral higher than 50 g/kg DW.

5.2.3 Magnesium

Magnesium is the second and fourth most abundant cation in the intracellular compartment and the whole body respectively. It acts as a cofactor for over 300 enzymes, regulating several fundamental functions such as muscle contraction, glycemic control, neuromuscular conduction, energy production, nerve transmission, active transmembrane transport of other ions, synthesis of nuclear materials, myocardial contraction, and bone development (Al Alawi et al., 2018). The mineral deficiency is linked with aging and age-related disorders, insulin resistance, intensification of the stress response, calcium cell overload, ischemic heart disease, and endothelial dysfunction (Martínez–Ballesta et al., 2010; Podkowa et al., 2021). Besides, the element shows antioxidant activity and thus the shortage is related to the development of oxidative stress (Podkowa et al., 2021). The recommended daily intake is 200–400 mg; but fruits and vegetables contain, in general, Mg^{2+} in the range of 5.5–191 mg/100 g fresh weight. Moreover, the Western diet encompasses more processed food and refined grains where 80%–90% of Mg^{2+} is lost during food processing. As a result, a significant number of people suffer from Mg^{2+} deficiency, which may comprise around 60% of critically ill patients (de Baaij et al., 2015).

Chen et al. (2010) has suggested that magnesium deficiency could be overcome, or prevented, by regular consumption of mushrooms. Indeed, the element has been found in every studied sample in appreciable quantity. A large amount of magnesium has been reported in *Boletus edulis* (Zhang et al., 2010), *Podaxis pistillaris*, *Termitomyces heimii* (Sai & Basavarju, 2020), and *Russula vesca* (Adejumo & Awosanya, 2005) where the content exceeded 4,000 mg/kg DW. However, the result contradicts Liu, Zhang, et al. (2012) depicting <200 mg/kg DW of Mg in the case of *Boletus speciosus*, *Leucopaxillus giganteus,* and *Morchella deliciosa*.

5.2.4 Sodium

Sodium is the main cation of extracellular liquid and intake of the mineral is undoubtedly indispensable for normal body functions. The essential nutrient is required for maintaining cellular homeostasis, controlling blood pressure, and regulating fluid as well as electrolyte balance (Ciudad-Mulero et al., 2021). Apart from that, the element is equally important for the excitability of muscle and nerve cells, carrying nutrients through plasma membranes and maintaining extracellular fluid volume. However, a clinically relevant food deficit of sodium is extremely unlikely in healthy individuals due to the presence of added salt in commonly used food products (Strazzullo, 2014; Chawla et al., 2019).

In general fruit, vegetables, oils, and cereals are low in Na with their content ranges from traces to ~20 mg/100 g contributing negligible total sodium intake (Kilcast & Angus, 2007). Conversely, mushrooms may be considered a better reservoir of the element where the quantity varied from 43.1 mg/kg DW (Mleczek, Budka, et al., 2021) to 6,780.4 mg/kg DW (Heleno, Barros, et al., 2015). However, as shown in Table 5.1, the study on Na content in macrofungi is still limited.

5.2.5 Phosphorus

Phosphorus is another essential nutrient for the body and routinely consumed through food. In humans, 70% of ingested phosphorus is absorbed and the element makes up about 1% to 1.4% of fat-free mass. It is generally absorbed through the intestine, transported in the bloodstream, and deposited in bones and teeth (Elekes & Busuioc, 2010). The element is an indispensable component of bones, teeth, and nucleic acids. In the form of phospholipids, phosphorus is also an ingredient of cell membrane structure and ATP, the body's key energy source. In addition, the nutrient plays a key role in the regulation of gene transcription, maintenance of normal pH in extracellular fluid, and activation of enzymes (Heaney, 2012). An increase in serum levels of inorganic phosphate diminishes serum levels of ionic calcium that usually leads to hypophosphatemia. In contrast, dietary phosphate deficiency, mostly due to malnutrition, can impair the bone mineralization process and eventually lead to the development of rickets (Chen et al., 2016).

In this context, mushrooms could be considered as good sources of phosphorus, although the content of P varies in a wide range (1.2 g/kg–25 g/kg DW) in the analyzed species. Overall, the nutrient concentration was described to be the lowest in the case of *Leccinum scabrum*

(Rudawska & Leski, 2005), while the highest level was detected in the case of *L. saeva* (Falandysz et al., 2001).

5.2.6 Copper

As an essential metal, copper is required for adequate growth, development, and survival. It is associated with myriad functions including cardiovascular integrity, free radical eradication, lung elasticity, energy production, neovascularization, neuroendocrine function, alterations in cholesterol metabolism, connective tissue formation, and metabolism of oxygen and iron (Failla et al., 2001). Shortage of the mineral has been linked with bone deformity during development contributing osteoporosis in adults. Marginal copper deficiency may participate in increased cardiovascular risk and alterations in cholesterol metabolism (Araya et al., 2007). Copper has also been related to Fe deficiency anemia, altered immune response and augmented rate of infections (Gupta & Gupta 2014; Mahluji et al., 2021). The deficiency also promotes development of oxidative stress-related disorders due to inhibition of superoxide dismutase or ceruloplasmin. Shortage of the mineral has frequently been reported in developed countries addressing supplementation as the right approach (Podkowa et al., 2021).

Previous publications have reported that mushrooms contain superior extent of Cu and thus can be a better option for the micronutrient supplementation as compared to that of products of plant origin (Mirończuk–Chodakowska et al., 2019). To date, minimum 163 macrofungal species have been reported to comprise the element where most of the values mainly varied between 10–80 mg/kg DW. However, a higher amount is also evident as reported in case of *P. pistillaris*, *T. heimii* (Sai & Basavarju, 2020), *M. procera* (Mleczek, Budka, et al., 2021), *C. rugosa* (Keleş & Gençcelep, 2020), and *Calvatia utriformis* (Alonso et al., 2003) where the amount exceeded 200 mg/kg DW. Opposite to that, a few mushrooms, *A. auricula* (Nnorom et al., 2020), *A. polytricha* (Teke et al., 2021), *Pholiota nameko* (Rubio et al., 2018), and *Sarcodon squamosus* (Gezer et al., 2015), have been depicted to contain less amounts of copper.

5.2.7 Iron

Iron is an essential element for almost all living organisms as it participates in a wide range of metabolic processes. As a component of hemoglobin and myoglobin, it functions as a carrier of oxygen in the blood and muscles. The metal also plays an important role during synthesis of DNA, electron transport, inflammation, and immune response to infection (Mahluji et al., 2021). Disorders related to Fe metabolism are among the most common health problems and cover a broad spectrum of diseases including iron deficiency and neurodegenerative disorders (Sousa et al., 2019). Iron-deficiency anemia may be caused due to low dietary intake, excessive blood loss, inadequate intestinal absorption, and/or increased needs. The shortage occurring particularly during pregnancy may represent a high risk for fetal growth retardation, preterm delivery, inferior neonatal health, and low birth weight (Gupta & Gupta, 2014).

According to previous studies, mushrooms are good source of iron. Up to now, more than 157 species have been investigated for iron content where around 140 taxa showed presence of the mineral in the range of 100–900 mg/kg DW. Rubio et al. (2018) reported quite lower value in case of *A. bisporus*, *L. deliciosus*, *L. edodes*, *P. nameko,* and *P. ostreatus.* The report of Gezer et al. (2015) described less extent of Fe in *Morchella costata*, *S. crispa,* and *T. aestivum.* In contrast, iron content greater than 1 g/kg DW has been found in *Boletus appendiculatus* (Turkekul et al., 2004), *C. cibarius* (Sesli et al., 2008; Turhan et al., 2010), *Coprinus atramentarius* (Bengu, 2019), *Lactarius triviralis* (Adejumo & Awosanya, 2005), *S. variegatus* (Falandysz et al., 2001), and *Xerocomus spadiceus* (Wang et al., 2017).

5.2.8 Manganese

Manganese is an essential nutrient in the human body, necessary for a variety of metabolic functions including synthesis and activation of certain enzymes mostly antioxidants, catalysis of hematopoiesis, energy metabolism, regulation of endocrine, neurotransmitter synthesis, improvement in immune function, and acceleration in synthesis of protein, vitamin B, as well as C (Li & Yang, 2018). The element also plays an essential role in the regulation of bone and connective tissue growth and blood clotting (Avila et al., 2013). Manganese deficiency is quite rare in humans; thus, symptoms of the shortage are relatively elusive. However, it has been suggested that lacking the metal might cause poor growth in children, bone demineralization, altered mood, decreased serum cholesterol, change in lipid and carbohydrate metabolism, abnormal glucose tolerance, and increased premenstrual pain in women (Nielsen, 2012).

In this context, many macrofungal species have been reported to contain Mn. Generally, the amount ranged between 5–90 mg/kg DW. In turn, a lower extent of manganese was found in *H. erinaceum* (Heleno, Barros, et al., 2015), *Tricholoma anatolicum*, *T. aestivum* (Gezer et al., 2015), and *Volvopluteus gloiocephalus* (Heleno, Ferreira, et al., 2015). Sesli et al. (2008) analyzed the level of certain metals, including Mn, in several wild mushrooms growing in Turkey, where most of the members were found to contain the element at the level of >100 mg/kg DW. A similar observation has also been reported by some other groups (Adejumo & Awosanya, 2005, Ouzouni et al., 2007, Sai & Basavarju, 2020).

5.2.9 Zinc

Zinc, an essential nutrient for human, is extensively involved in lipid, protein, nucleic acid metabolism, cell division and transcription. It is required for the metabolic activity of over 300 enzymes, including those exhibiting antioxidant effects such as zinc-copper superoxide dismutase and nicotinamide adenine dinucleotide phosphate (NADPH) oxidase. The mineral also plays an extensive role in reproduction, wound repair, immune function, and complement activity (Gupta & Gupta, 2014). Zinc deficiencies, prevalent in developing countries, are estimated to affect approximately 17% of the world's population, encompassing around 82% of pregnant women with inadequate zinc intake (Das et al., 2019). The mineral shortage during early life stages causes growth failure as gastrointestinal, epidermal, immune, skeletal, central nervous, and reproductive systems become the most affected. Therefore, it is important to supplement this metal in the daily diet to diminish the risk of zinc depletion (Roohani et al., 2013).

The level of zinc absorption has been found to be greater from protein-rich food and this is why vegetarians suffer from a deficiency of Zn. Consequently, studies revealed that edible mushrooms contain higher amounts of Zn than that of fruits, herbs, crops, and vegetables. Indeed, zinc is one of the dominant elements in macrofungi and thus intake of 100 g of dried fruiting bodies provides about half of the daily Zn requirement. These bioresources could be an alternative for a vegetarian diet (Zajac et al., 2015; Mirończuk–Chodakowska et al., 2019). Among mushrooms, *C. atramentarius* (Bengu, 2019); *P. pistillaris*, *T. heimii* (Sai & Basavarju, 2020), *H. leucopus*, *T. auratum* (Sarikurkcu et al., 2012), *L. versipelle*, and *T. flavovirens* (Falandysz et al., 2001) have been characterized by a high content of zinc.

5.2.10 Selenium

Another essential trace element for human health is selenium which is an indispensable component for several enzymes, such as glutathione peroxidase, thioredoxin reductase, deiodinases, and iodothyronine 5-deiodinase. Selenium has other vital beneficial health effects, particularly in relation to cancer prevention and immune response (Rayman, 2000). Recent studies have indicated that Se intake level below 30 μg per day can lead to increased aging and risk of cardiovascular and other degenerative disorders. Thus, proper dietary supplementation of Se could reduce the danger of these diseases (Dong et al., 2021).

Studies of Se content in mushrooms is relatively scarce as most of the edible species examined until now are selenium-poor (<1 mg/kg DW) (Falandysz, 2008) amongst the studied taxa, *P. florida, L. edodes, H. erinaceus, A. bisporus,* and *A. auricula* are known to be good selenium accumulators that can transform inorganic selenium into abundant selenoamino acids (SeMet, SeCys, and MeSeCys) (Dong et al., 2021). In this context, an excellent review has been published by Falandysz (2008), presenting Se content in 190 species belonging to 21 families and 56 genera, where most of the members are considered edible.

5.3 FACTORS AFFECTING MINERAL PROFILE

As presented previously, the mineral profile in mushrooms greatly varies in comparison to plants. Various fungal factors that control the composition include development of mycelium, fungal structure, nutritional need, substrate decomposition activity, and morphological portion (Kułdo et al., 2014; Wang et al., 2017; Jedidi et al., 2017). Since each fruiting body can develop after crossbreeding of different hyphae, thus it presents a distinct genotype resulting in a wide variation in mineral content (Gałgowska & Pietrzak–Fiećko, 2020). In the case of mycorrhizal species, mycelium is characterized by the dense interlacement of hyphae, high surface to volume ratio, and grow up to 50 cm deep in the soil covering an area up to 100 m^2 (Falandysz et al., 2001). Sometimes this absorbing device has a life span of more than 100 years and it is suggested that the level of elements in the fruit body significantly augments with the age of mycelium (Kalač, 2013). Besides, most of the elements in mushrooms are not evenly distributed in the whole basidiocarps. For instance, the highest concentration of K has been found in caps, followed by stipes, spore-forming parts, and spores (Zocher et al., 2018). Chen et al. (2010) have reported that *R. griseocarnosa* contains Na, K, P, Mg, Zn, and Cu in abundance in the pileus except Ca and Fe, which are present predominately in the stipe. The observation was in accord with Rudawska and Leski (2005), revealing minerals except Ca are preferably being translocated into the cap in eight studied fungal species. A similar trend has also been reported in case of *B. edulis* where several minerals (K, Mg, Cu, Fe, and Zn) were found to be present in higher quantities in the cap than the stipe (Zhang et al., 2010). Wang et al. (2017) collected eight samples from the wild and performed a comparative study on mineral profiles. Their research depicted the presence of Cu, Fe, and Zn in superior quantity in the cap; while Co, Mn, and Ni were detected in higher amounts in the stipe. In another study, hymenophore in mushrooms under investigation showed better metal levels than the rest of the fruit bodies (Alonso et al., 2003).

Apart from species types, environmental conditions greatly affect the trace element level in fruit bodies as well (Gałgowska & Pietrzak–Fiećko, 2020). Principal environmental factors influencing such accumulation are acidic and organic matter content, biochemical composition, metallic elements, metalloids, soil pH, environmental pollution and texture, and other chemical elements in abundance (Kułdo et al., 2014; Jedidi et al., 2017; Wang et al., 2017). As such, some macromycetes can absorb heavy metals and accumulate them in fruiting bodies at concentrations higher than the substrate on which they grew (Kojta et al., 2012). Special care, hence, should be taken during harvesting and subsequent consumption in places where metal ores have been mined and processed for several years, leading to severe environmental contamination (Pająk et al., 2020). Attention must also be paid when using industrial wastes/sludge for mushroom cultivation, as toxic substances can be translocated from substrates to mushrooms (Falandysz et al., 2012; Yildiz et al., 2019). Indeed, previous reports have shown that a correlation between heavy metal concentrations of wild fungi and sources of metal pollution such as smelters, metal ore mines, or polluted urbanized areas and consumption of these foods could be noxious (Işıloğlu et al., 2001; Falandysz & Borovička, 2013; Mleczek et al., 2015; Mleczek et al., 2016). The content of these elements in the fruiting bodies thus may be a determinant of environmental purity (Mirończuk–Chodakowska et al., 2019).

Eventually, mineral content in the fruiting body depends on the type of substrate used for mushroom cultivation. For development, fungi require macronutrients, including C, P, N, Mg, and K, along with some trace elements such as Se, Fe, Mn, Zn, Mo, and Cu (Mleczek, Gąsecka, Budka, Niedzielski, et al., 2021). Some elements (Zn, Cu, and Mn) play an important role in the elimination of quinine, a substance that inhibits fructification, and thus their inclusion in compost stimulates fruit body production (Calvalcante et al., 2008). Generally, lignocellulosic materials are a weak source of mineral content, and to increase the efficiency of mushroom yield, minerals are added in form of mineral salt and/or fertilizer. A recent study showed that the macro elemental composition of different species of *Pleurotus* increased in the case of K, Mg, Na, and P when fertilizer was added (Mleczek, Gąsecka, Budka, Niedzielski, et al., 2021). Various studies have tested the effect of different types of substrate or their supplementation with mineral salts on the outcome of the cultivation of *Pleurotus* mushrooms. *Pleurotus ostreatus* cultivated on sawdust of *Pycnanthus ongoleubis* showed an increase in mineral content, such as K, Na, Ca, Mg, and P (Oyetayo & Ariyo, 2013). In another study, six species of *Pleurotus* were cultivated on the iron-fortified agro-waste substrate to enrich mineral content. Estimated iron uptake in the investigated taxa ranged from 37.8 µg/g to 96.6 µg/g indicating cultivation of edible macrofungi enriched with iron could improve the socio-economic status (Ogidi et al., 2016). A similar trend has also been reported by Miletić et al. (2020) depicting mineral enrichment in *Coriolus versicolor* during submerged cultivation in a selenium-fortified medium. Machado et al. (2016) depicted enrichment of minerals in *Lentinus citrinus* when cultivated on *Theobroma grandiflorum* exocarp mixed with rice bran in the ratio of 2:1. Exogenous Se application mainly in form of selenite is useful for fruit body production of *L. edodes* (Zhou et al., 2018). Besides, the bioavailability of minerals also depends on the food processing method. Dried materials have been demonstrated to contain higher quantities of elements than fresh or frozen material, which is directly related to low water content. Lyophilization and sun-drying methods have been suggested to be more advantageous than that of drying in a dryer in terms of release of elements. Similarly, fresh material was found as a more valuable source of minerals than frozen material (Kała et al., 2020). Mineral content in mushrooms is linked to the cooking process as well. The highest reduction in mineral content (K, Mg, P, Mn, Na, and Cu) was observed in boiled shiitake mushrooms, followed by blanching and steaming. However, microwaving and roasting treatments resulted in greater retention of all minerals in comparison to other treatments. The observation could be justified as loss of minerals by leaching into cooking water which can be restored by microwaving or roasting in the case of *L. edodes* (Lee et al., 2019). Karun et al. (2018) depicted that cooking adversely affects sodium, potassium, calcium, magnesium, phosphorus, iron, and copper in *A. auricula*. Nevertheless, the Na/K ratio (0.32 uncooked, 0.47 cooked) and Ca/P ratio (2.01 uncooked, 1.88 cooked) in *A. auricula* were found to be favorable for human health. As such, Ca/P ratio >1 helps to prevent loss of calcium in the urine and Na/K ratio <1 helps to control high blood pressure.

5.4 MINERAL FORTIFICATION

Several strategies have been employed to supplement micronutrients to women and children. These include education, dietary modification, food rationing, supplementation, and fortification. Food fortification is the process of adding micronutrients to foods and has been practiced in developed countries for well over a century now (Das et al., 2019). Certain types of fortification are more accurately called enrichment in which micronutrients added to food are those that are lost during processing (Whiting et al., 2016). In recent times, scientists are emphasizing artificially enriching mushrooms with different mineral compositions to make them a valuable product for patients suffering from different diseases and mineral deficiency. In this context, several researchers have prepared mushroom fortified foods and evaluated the effect on the nutritional property (Salehi et al., 2016). Kumar and Barmanray (2007) reported that incorporation of 20% button mushroom powder into wheat flour could increase ash content by around three times. In separate research, quality characteristics of mushroom-wheat-oats enriched flour have been evaluated. Results showed that the use of different levels of *P. ostreatus* for the production of fortified food augmented the ash content of the preparation along with the elevated level of Na, Fe, K, Ca, and Mn (Farzana et al., 2019). A similar observation has also been reported by Parvin et al. (2020), reporting *P. ostreatus* fortified noodles contained a higher level of Fe, Ca, and K than the noodles made from wheat only as well as locally branded noodles. The same genus has been used by another group of scientists (Bello et al., 2017) to prepare biscuits and investigated nutrient composition and sensory characteristics of the fortified food. Results showed that the formula prepared by mixing the mushroom powder (*P. sajur-caju*) with wheat in the ratio of 3:7 contained a superior quantity of K, Na, Ca, Mg, P, Fe, Cu, Zn, and Mn. Alongside Ca/P ratio value was found to be within the range of 0.59–1.03, indicating that the preparation would be a good source of minerals for bone formation. Ibrahium and Hegazy (2014) performed an extensive study to evaluate the effect of partial replacement of wheat flour by different levels (10%, 20%, and 30%) of the mixture of mushroom (*Pleurotus plumonarius*) powder and sweet potato flour at equal rates on chemical and sensory characteristics for prepared biscuits. The outcome revealed that incorporation of the preparation into the formula caused escalation in total ash content, including Fe, Ca, K, and P. Singh et al. (2016) depicted that the addition of *L. edodes* powder for preparation of biscuits improved nutrient content (P and Ca). A novel formulation of sponge cake production with button mushroom has been developed where the mushroom powder was substituted at the 5%–15% elevated ash content (Salehi et al., 2016). In another study, *P ostreatus* and *Calocybe indica* were used at the level ranging from 0%–20% to substitute wheat flour in bread formulation. The attempt increased K, Na, Ca, Mg, Mn, Cu, Zn, and Fe content of the bread significantly with an increase in the mushroom powder (Oyetayo & Oyedeji, 2017). Dong et al. (2021) showed that Se biofortification increased biomass yield of the fruiting body of *F. velutipes* and an elevated content of several minerals (Fe, Ca, and Cu) and thus enhanced the nutritional value. Enrichment of growth medium with Se and Zn was also found to be effective in biofortification in both mycelium and fruiting bodies of *P eryngii* (Zięba et al., 2020). All these findings suggest that the fortified mushroom has the chance for improving nutritional values with consumer acceptance and commercial preference for the manufacturers.

5.5 CONCLUSION AND FUTURE PERSPECTIVES

All in all, edible mushrooms could be considered as a reservoir of a range of minerals, as revealed by extensive studies from all over the world, particularly from the Northern Hemisphere. Several species were found to contain a significant quantity of major minerals presented mainly in the order of K> P> Mg> Ca> Na. Attention has also been paid to procure trace elements profile where Fe, Zn, and Cu remain mainly in focus; while studies on other essential elements such as Se are limited. Unreasonable use of pesticides and fertilizers, as well as extensive industrial and urban pollution, can affect the mineral composition of edible macrofungi, even though health risk has been assessed as negligible. Research should also be conducted to investigate the effect of substrate type on mineral profiling of cultivatable species so that such nutrient-enriched taxa can further be utilized to develop fortified food as supplementary foods to the populations largely dependent on a cereal diet. In this context, the release of minerals from fruit bodies or bioaccessibility must be extensively investigated for effective downstream application.

REFERENCES

Adejumo, T. O., & Awosanya, O. B. (2005). Proximate and mineral composition of four edible mushroom species from south western Nigeria. *African Journal of Biotechnology*, *4*(10), 1084–1088.

Al Alawi, A. M., Majoni, S. W., & Falhammar, H. (2018). Magnesium and human health: Perspectives and research directions. *International Journal of Endocrinology*, *2018*, 9041694.

Alonso, J., García, M. A., Pérez–López, M., & Melgar, M. J. (2003). The concentrations and bioconcentration factors of copper and zinc in edible mushrooms. *Archives of Environmental Contamination and Toxicology*, *44*, 180–188.

Araya, M., Olivares, M., & Pizarro, F. (2007). Copper in human health. *International Journal of Environmental Research and Public Health*, *1*(4), 608–620.

Arce, S., Cerutti, S., Olsina, R., Gomez, M. R., & Martínez, L. D. (2008). Trace element profile of a wild edible mushroom (*Suillus granulatus*). *Journal of AOAC International*, *91*, 853–857.

Avila, D. S., Puntel, R. L., & Aschner, M. (2013). Manganese in health and disease. *Metal Ions in Life Sciences*, *13*, 199–227.

Awuchi, C. G. (2020). Health benefits of micronutrients (vitamins and minerals) and their associated deficiency diseases: A systematic review. *International Journal of Food Sciences*, *3*(1), 1–32.

Ayaz, F. A., Torun, H., Colak, A., Sesli, E., Millson, M., & Glew, R. H. (2011). Macro– and microelement contents of fruiting bodies of wild–edible mushrooms growing in the east black sea region of Turkey. *Food and Nutrition Sciences*, *2*, 53–59.

Bello, M., Oluwamukomi, M. O., & Enujiugha, V. N. (2017). Nutrient composition and sensory properties of biscuit from mushroom–wheat composite flours. *Archives of Current Research International*, *9*, 1–11.

Bengu, A. S. (2019). Some elements and fatty acid profiles of three different edible mushrooms from Tokat province in Turkey. *Progress in Nutrition*, *21*, 189–193.

Beto, J. A. (2015). The role of calcium in human aging. *Clinical Nutrition Research*, *4*, 1–8.

Bhandari, S., & Banjara, M. R. (2015). Micronutrients deficiency, a hidden hunger in Nepal: prevalence, causes, consequences, and solutions. *International Scholarly Research Notices*, *2015*, Article ID 276469.

Busuioc, G., Elekes, C. C., Stihi, C., Lordache, S., & Ciulei, S. C. (2011). The bioaccumulation and translocation of Fe, Zn, and Cu in species of mushrooms from *Russula* genus. *Environmental Science and Pollution Research*, *18*, 890–896.

Calvalcante, J. L. R., Gomes, V. F. F., Filho, J. K., Minhoni, M. T. D. A., & De Andrade, M. C. N. (2008). Cultivation of *Agaricus blazei* in the environmental protection area of the Baturité region under three types of casing soils. *Maringá*, *30*(4), 513–517.

Chawla, P., Kumar, N., Kaushik, R., & Dhull, S .B. (2019). Synthesis, characterization and cellular mineral absorption of gum arabic stabilized nanoemulsion of *Rhododendron arboreum* flower extract. *Journal of Food Science and Technology*, *56*(12), 5194–5203.

Chen, L., Tian, F., & Sun, Z. (2016). Phosphorus nutrition and health: Utilization of phytase–producing Bifidobacteria in food industry. In: V. Rao & G. L. Rao (Eds.), *Probiotics and Prebiotics in Human Nutrition and Health*. London: IntechOpen, pp. 263–267.

Chen, X., Xia, L., Zhou, H., & Qiu, G. (2010). Chemical composition and antioxidant activities of *Russula griseocarnosa* sp. nov. *Journal of Agricultural and Food Chemistry*, *58*, 6966–6971.

Chen, L., Zhu, H., Li, Y., Zhang, Y., Zhang, W., Yang, L., Yin, H., Dong, C., & Wang, Y. (2021). Combining multielement analysis and chemometrics to trace the geographical origin of *Thelephora ganbajun*. *Journal of Food Composition and Analysis*, *2021*, 103699.

Ciudad-Mulero, M., Matallana-González, M. C., Callejo, M. J., Carrillo, J. M., Morales, P., & Fernández-Ruiz, V. (2021). Durum and bread wheat flours. Preliminary mineral characterization and its potential health claims. *Agronomy*, *11*, 108.

Das, J. K., Salam, R. A., Mahmood, S. B., Moin, A., Kumar, R., Mukhtar, K., Lassi, Z. S., & Bhutta, Z. A. (2019). Food fortification with multiple micronutrients: Impact on health outcomes in general population. *Cochrane Database of Systematic Reviews*, *12*(12), CD011400.

Dávila, G. L. R., Murillo, A. W., Zambrano, F. C. J., Suárez, M. H., & Méndez, A. J. J. (2020). Evaluation of nutritional values of wild mushrooms and spent substrate of *Lentinus crinitus* (L.) Fr. *Heliyon*, *6*(3), e03502.

de Baaij, J. H., Hoenderop, J. G., & Bindels, R. J. (2015). Magnesium in man: Implications for health and disease. *Physiological Reviews*, *95*(1), 1–46.

Demirbaş, A. (2001). Concentrations of 21 metals in 18 species of mushrooms growing in the East Black Sea region. *Food Chemistry*, *75*, 453–457.

Dong, Z., Xiao, Y., & Wu, H. (2021). Selenium accumulation, speciation, and its effect on nutritive value of *Flammulina velutipes* (Golden needle mushroom). *Food Chemistry*, *350*, 128667.

Egbuna, C. & Tupas, G. D. (Eds.). (2020). *Functional Foods and Nutraceuticals*. Cham: Springer, pp. 139–147.

Elekes, C. C., & Busuioc, G. (2010). The biomineral concentrations and accumulation in some wild growing edible species of mushrooms. *Annals: Food Science and Technology*, *11*(1), 74–78.

Failla, M. L., Johnson, M. A., & Prohaska, J. R. (2001). Copper. In: B. Marriott, D. Birt, D. Stalling, & A. Yates (Eds.), *Present Knowledge in Nutrition*. Washington DC: Life Sciences Institute Press, pp. 373–383.

Falandysz, J. (2008). Selenium in edible mushrooms. *Journal of Environmental Science and Health Part C*, *26*, 256–299.

Falandysz, J., & Borovička, J. (2013). Macro and trace mineral constituents and radionuclides in mushrooms: Health benefits and risks. *Applied Microbiology and Biotechnology*, *97*, 477–501.

Falandysz, J., Drewnowska, M., Jarzyńska, G., & Dan, Z. (2012). Mineral constituents in common chanterelles and soils collected from a high mountain and lowland sites in Poland. *Journal of Mountain Science*, *9*, 697–705.

Falandysz, J., Szymczyk, K., Ichihashi, H., Bielawksi, L., Gucia, M., Frankowska, A., & Yamasaki, S. (2001). ICP/MS and ICP/AES elemental analysis (38 elements) of edible wild mushrooms growing in Poland. *Food Additives & Contaminants*, *18*, 503–513.

FAO. (2009). *The State of Food Insecurity in the World 2009*. Rome: Food and Agriculture Organization of the United Nations. Available from http://www.fao.org/publications/sofi/en/

Farzana, T. T. N. Orchy, S. Mohajan, N. C. Sarkar, & Kakon, A. J. (2019). Effect of incorporation of mushroom on the quality characteristics of blended wheat and oats flour. *Archives of Nutrition and Public Health*, *1*(1), 1–7.

Gałgowska, M., & Pietrzak–Fiećko, R. (2020). Mineral composition of three popular wild mushrooms from Poland. *Molecules*, *25*, 3588.

Gezer, K., Kaygusuz, O., Eyupoglu, V., Surucu, A., & Doker, S. (2015). Determination by ICP/MS of trace metal content in ten edible wild mushrooms from turkey. *Oxidation Communications*, *38*(1A), 398–407.

Giannaccini, G., Betti, L., Palego, L., Mascia, G., Schmid, L., Lanza, M., Mela, A., Fabbrini, L., Biondi, L., & Lucacchini, A. (2012). The trace element content of top–soil and wild edible mushroom samples collected in Tuscany, Italy. *Environmental Monitoring and Assessment*, *184*(12), 7579–7595.

Gómez–Galera, S., Rojas, E., Sudhakar, D., Zhu, C., Pelacho, A. M., Capell, T., & Christou, P. (2010). Critical evaluation of strategies for mineral fortification of staple food crops. *Transgenic Research*, *19*, 165–180.

Gupta, U. C., & Gupta, S. C. (2014). Sources and deficiency diseases of mineral nutrients in human health and nutrition: A review. *Pedosphere*, *24*(1), 13–38.

Györfi, J., Geösel, A., & Vetter, J. (2010). Mineral composition of different strains of edible medicinal mushroom *Agaricus subrufescens* Peck. *Journal of Medicinal Food*, *13*(6), 1510–1514.

He, F. J., & MacGregor, G. A. (2008). Beneficial effects of potassium on human health. *Physiologia Plantarum*, *133*(4), 725–735.

Heaney, R. P. (2012). Phosphorus. In: J. W. Erdman, I. A. Macdonald, & S. H. Zeisel (Eds.), *Present Knowledge in Nutrition*. 10th ed. Washington, DC: Wiley-Blackwell, pp. 447–458.

Heleno, S. A., Barros, L., Martins, A., Queiroz, M. J. R., Morales, P., Fernández–Ruiz, V., & Ferreira, I. C. F. R. (2015). Chemical composition, antioxidant activity and bioaccessibility studies in phenolic extracts of two *Hericium* wild edible species. *LWT–Food Science and Technology*, *63*, 475–481.

Heleno, S. A., Ferreira, R. C., Antonio, A. L., Queiroz, M. R., Barros, L., & Ferreira, I. C. F. R. (2015). Nutritional value, bioactive compounds and antioxidant properties of three edible mushrooms from Poland. *Food Bioscience*, *11*, 48–55.

Ibrahium, M. I., & Hegazy, A. I. (2014). Effect of replacement of wheat flour with mushroom powder and sweet potato flour on nutritional composition and sensory characteristics of biscuits. *Current Science International*, *3*(1), 26–33.

Işıloğlu, M., Yılmaz, F., & Merdivan, M. (2001). Concentrations of trace elements in wild edible mushrooms. *Food Chemistry*, *73*(2), 169–175.

Jedidi, I. K., Ayoub, I. K., Philippe, T., & Bouzouita, N. (2017). Chemical composition and nutritional value of three Tunisian wild edible mushrooms. *Food Measure*, *11*, 2069–2075.

Kała, K., Krakowska, A., Zięba, P., Opoka, W., & Muszyńska, B. (2020). Effect of conservation methods on the bioaccessibility of bioelements from *in vitro*–digested edible mushrooms. *Journal of the Science of Food and Agriculture*. 10.1002/jsfa.10979 (in press).

Kała, K., Maślanka, A., Sułkowska–Ziaja, K., Rojowski, J., Opoka, W., & Muszyńska, B. (2016). *In vitro* culture of *Boletus badius* as a source of indole compounds and zinc released in artificial digestive juices. *Food Science and Biotechnology*, *25*(3), 829–837.

Kalač, P. (2013). A review of chemical composition and nutritional value of wild-growing and cultivated mushrooms. *Journal of the Science of Food and Agriculture*, *93*, 209–218.

Kalyoncu, F., Ergönül, B., Yildiz, H., Kalmiş, E., & Solak, M. H. (2010). Chemical composition of four wild edible mushroom species collected from southwest Anatolia. *Gazi University Journal of Science*, *23*(4), 375–379.

Karun, N. C., Sridhar, K. R., & Ambarish, C. N. (2018). Nutritional potential of *Auricularia auricula-judae* and *Termitomyces umkowaan* – The wild edible mushrooms of south-western India. In: V. K. Gupta, H. Treichel, V. Shapaval, L. Antonio de Oliveira, & M. G. Tuohy (Eds.), *Microbial Functional Foods and Nutraceuticals*. First ed. Chichester, UK: John Wiley & Sons, pp. 281–301.

Keleş, A., & Gençcelep, H. (2020). Determination of elemental composition of some wild growing edible mushrooms. *The Journal of Fungus*, *11*(2), 129–137.

Kilcast, D. & Angus, F. (Eds.). (2007). *Reducing Salt in Foods. Practical Strategies*. Cambridge, UK: CRC Press, Woodhead Publishing.

Kojta, A. K., Jarzyńska, G., & Falandysz, J. (2012). Mineral composition and heavy metal accumulation capacity of bay bolete (*Xerocomus badius*) fruiting bodies collected near a former gold and copper mining area. *Journal of Geochemical Exploration*, *121*, 76–82.

Kosanić, M., Ranković, B., Rančić, A., & Stanojković, T. (2016). Evaluation of metal concentration and antioxidant, antimicrobial, and anticancer potentials of two edible mushrooms *Lactarius deliciosus* and *Macrolepiota procera*. *Journal of Food and Drug Analysis*, *24*(3), 477–484.

Krakowska, A., Zięba, P., Włodarczyk, A., Kała, K., Sułkowska–Ziaja, K., Bernaś, E., Sękara, A., Ostachowicz, B., & Muszyńska, B. (2020). Selected edible medicinal mushrooms from Pleurotus genus as an answer for human civilization diseases. *Food Chemistry*, *327*, 127084.

Krupodorova, T., & Sevindik, M. (2020). Antioxidant potential and some mineral contents of wild edible mushroom *Ramaria stricta*. *AgroLife Scientific Journal*, *9*(1), 186–191.

Kułdo, E., Jarzyńska, G., Gucia, M., & Falandysz, J. (2014). Mineral constituents of edible parasol mushroom *Macrolepiota procera* (Scop. ex Fr.) Sing and soils beneath its fruiting bodies collected from a rural forest area. *Chemical Papers*, *68*, 484–492.

Kumar, K., & Barmanray, A. (2007). Nutritional evaluation and storage studies of button mushroom powder fortified biscuits. *Mushroom Research*, *16*(1), 31–35.

Lalotra, P., Bala, P., Kumar, S., & Sharma, Y. (2018). Biochemical characterization of some wild edible mushrooms from Jammu and Kashmir. *Proceedings of the National Academy of Sciences, India Section B*, *88*, 539–545.

Lee, K., Lee, H., Choi, Y., Kim, Y., Jeong, H. S., & Lee, J. (2019). Effect of different cooking methods on the true retention of vitamins, minerals, and bioactive compounds in shiitake mushrooms (*Lentinula edodes*). *Food Science and Technology Research*, *25*(1), 115–122.

Li, L., & Yang, X. (2018). The essential element manganese, oxidative stress, and metabolic diseases: Links and interactions. *Oxidative Medicine and Cellular Longevity*, *2018*, 7580707.

Liu, Y., Chen, D., You, Y., Zeng, S., Li, Y., Tang, Q., Han, G., et al. (2016). Nutritional composition of boletus mushrooms from Southwest China and their antihyperglycemic and antioxidant activities. *Food Chemistry*, *211*, 83–91.

Liu, Y., Sun, J., Luo, Z., Rao, S., Su, Y., Xu, R., & Yang, Y. (2012). Chemical composition of five wild edible mushrooms collected from Southwest China and their antihyperglycemic and antioxidant activity. *Food and Chemical Toxicology*, *50*, 1238–1244.

Liu, H., Zhang, J., Li, T., Shi, Y., & Wang, Y. (2012). Mineral element levels in wild edible mushrooms from Yunnan, China. *Biological Trace Element Research*, *147*, 341–345.

Machado, A. R., Teixeira, M. F., L. de Souza Kirsch, Campelo Mda, C., & de Aguiar Oliveira, I. M. (2016). Nutritional value and proteases of *Lentinus citrinus* produced by solid state fermentation of lignocellulosic waste from tropical region. *Saudi Journal of Biological Sciences*, *23*(5), 621–627.

Mahluji, S., Jalili, M., Ostadrahimi, A., Hallajzadeh, J., Ebrahimzadeh–Attari, V., & Saghafi–Asl, M. (2021). Nutritional management of diabetes mellitus during the pandemic of COVID–19: A comprehensive narrative review. *Journal of Diabetes & Metabolic Disorders*, *2021*, 1–10.

Mallikarjuna, S. E., Ranjini, A., Haware, D. J., Vijayalakshmi, M. R., Shashirekha, M. N., & Rajarathnam, S. (2013). Mineral composition of four edible mushrooms. *Journal of Chemistry*, *2013*, Article ID 805284.

Martínez–Ballesta, M. C., Dominguez–Perles, R., Moreno, D. A., Muries, B., Alcaraz–López, C., Bastías, E., García–Viguera, C., & Carvajal, M. (2010). Minerals in plant food: Effect of agricultural practices and role in human health. A review. *Agronomy for Sustainable Development*, *30*, 295–309.

Miletić, D., Pantić, M., Sknepnek, A., Vasiljević, I., Lazović, M., & Nikšić, M. (2020). Influence of selenium yeast on the growth, selenium uptake and mineral composition of *Coriolus versicolor* mushroom. *Journal of Basic Microbiology*, *60*(4), 331–340.

Mirończuk–Chodakowska, I., Socha, K., Zujko, M. E., Terlikowska, K. M., Borawska, M. H., & Witkowska, A. M. (2019). Copper, manganese, selenium and zinc in wild–growing edible mushrooms from the eastern territory of "Green Lungs of Poland": Nutritional and toxicological implications. *International Journal of Environmental Research and Public Health*, *16*(19), 3614.

Mleczek, M., Budka, A., Kalač, P., Siwulski, M., & Niedzielski, P. (2021). Family and species as determinants modulating mineral composition of selected wild–growing mushroom species. *Environmental Science and Pollution Research*, *28*, 389–404.

Mleczek, M., Gąsecka, M., Budka, A., Niedzielski, P., Siwulski, M., Kalač, P., Mleczek, P., & Rzymski, P. (2021). Changes in mineral composition of six strains of *Pleurotus* after substrate modifications with different share of nitrogen forms. *European Food Research and Technology*, *247*, 245–257.

Mleczek, M., Gąsecka, M., Budka, A., Siwulski, M., Mleczek, P., Magdziak, Z., Budzyńska, S., & Niedzielski, P. (2021). Mineral composition of elements in wood–growing mushroom species collected from of two regions of Poland. *Environmental Science and Pollution Research*, *28*, 4430–4442.

Mleczek, M., Magdziak, Z., Gąsecka, M., Niedzielski, P., Kalač, P., Siwulski, M., Rzymski, P., Zalicka, S., & Sobieralski, K. (2016). Content of selected elements and low–molecular–weight organic acids in fruiting bodies of edible mushroom *Boletus badius* (Fr.) Fr. from unpolluted and polluted areas. *Environmental Science and Pollution Research*, *23*, 20609–20618.

Mleczek, M., Siwulski, M., Mikołajczak, P., Gąsecka, M., Sobieralski, K., Szymańczyk, M., & Goliński, P. (2015). Content of selected elements in *Boletus badius* fruiting bodies growing in extremely polluted wastes. *Journal of Environmental Science and Health, Part A*, *50*, 767–775.

Muszyńska, B., Kała, K., Włodarczyk, A., Krakowska, A., Ostachowicz, B., Gdula–Argasińska, J., & Suchocki, P. (2020). *Lentinula edodes* as a source of bioelements released into artificial digestive juices and potential anti–inflammatory material. *Biological Trace Element Research*, *194*(2), 603–613.

Ndimele, C. C., Ndimele, P. E., & Chukwuka, K. S. (2017). Accumulation of heavy metals by wild mushrooms in Ibadan, Nigeria. *Journal of Health & Pollution*, *7*(16), 26–30.

Nielsen, F. H. (2012). Manganese, Molybdenum, Boron, Chromium, and Other Trace Elements. In: John W. Erdman, I. A. MacDonald, & S. H. Zeisel (Eds.), *Present Knowledge in Nutrition*. 10th ed. Wiley-Blackwell, pp. 586–607.

Nnorom, I. C., Eze, S. O., & Ukaogo, P. O. (2020). Mineral contents of three wild–grown edible mushrooms collected from forests of south eastern Nigeria: An evaluation of bioaccumulation potentials and dietary intake risks. *Scientific African*, *8*, e00163.

Ogidi, C. O., Akindulureni, E. D., Agbetola, O. Y., Akinyele, B. J. (2020). Calcium bioaccumulation by *Pleurotus ostreatus* and *Lentinus squarrosulus* cultivated on palm tree wastes supplemented with calcium-rich animal wastes or calcium salts. *Waste Biomass Valor*, *11*, 4235–4244.

Ogidi, O. C., Nunes, M. D., Oyetayo, V. O., Akinyele, B. J., & Kasuya, M. C. M. (2016). Mycelial growth, biomass production and iron uptake by mushrooms of *pleurotus* species cultivated on *Urochloa decumbens* (Stapf) R. D. Webster. *Journal of Food Research*, *5*(3), 13–19.

Ouzouni, P. K., Petridis, D., Koller, W., & Riganakos, K. A. (2009). Nutritional value and metal content of wild edible mushrooms collected from West Macedonia and Epirus, Greece. *Food Chemistry, 115,* 1575–1580.

Ouzouni, P. K., Veltsistas, P. G., Paleologos, E. K., & Riganakos, K. A. (2007). Determination of metal content in wild edible mushroom species from regions of Greece. *Journal of Food Composition and Analysis, 20,* 480–486.

Oyetayo, V. O., & Ariyo, O. O. (2013). Micro and macronutrient properties of *Pleurotus ostreatus* (Jacq:Fries) cultivated on different wood substrates. *Jordan Journal of Biological Sciences, 6,* 223–226.

Oyetayo, V. O., & Oyedeji, R. R. (2017). Proximate and mineral composition of bread fortified with mushroom (*Plerotus ostreatus* and *Calocybe indica*). *Microbiology Research Journal International, 19*(4), 1–9.

Pająk, M., Gąsiorek, M., Jasik, M., Halecki, W., Otremba, K., & Pietrzykowski, M. (2020). Risk assessment of potential food chain threats from edible wild mushrooms collected in forest ecosystems with heavy metal pollution in upper Silesia, Poland. *Forests, 11,* 1240.

Parvin, R., Farzana, T., Mohajan, S., Rahman, H., & Rahman, S. S. (2020). Quality improvement of noodles with mushroom fortified and its comparison with local branded noodles. *NFS Journal, 20,* 37–42.

Podkowa, A., Kryczyk–Poprawa, A., Opoka, W., & Muszyńska, B. (2021). Culinary–medicinal mushrooms: A review of organic compounds and bioelements with antioxidant activity. *European Food Research and Technology, 247,* 513–533.

Pu, F., Chen, N., & Xue, S. (2016). Calcium intake, calcium homeostasis and health. *Food Science and Human Wellness, 5,* 8–16.

Radwińska, J., & Żarczyńska, K. (2014). Effects of mineral deficiency on the health of young Ruminants. *Journal of Elementology, 19*(3), 915–928.

Rasalanavhoa, M., Moodley, R., & Jonnalagadda, S. B. (2020). Elemental bioaccumulation and nutritional value of five species of wild growing mushrooms from South Africa. *Food Chemistry, 319,* 126596.

Rayman, M. P. (2000). The importance of selenium to human health. *Lancet, 356,* 233–241.

Roohani, N., Hurrell, R., Kelishadi, R., & Schulin, R. (2013). Zinc and its importance for human health: An integrative review. *Journal of Research in Medical Sciences, 18*(2), 144–157.

Rubio, C., Martínez, C., Paz, S., Gutiérrez, A. J., González–Weller, D., Revert, C., Burgos, A., & Hardisson, A. (2018). Trace element and toxic metal intake from the consumption of canned mushrooms marketed in Spain. *Environmental Monitoring and Assessment, 190,* 237.

Rudawska, M., & Leski, T. (2005). Macro– and microelement contents in fruiting bodies of wild mushrooms from the Notecka forest in west–Central Poland. *Food Chemistry, 92,* 499–506.

Sai, I., & Basavarju, R. (2020). Nutritional composition and antimicrobial activity of two wild edible mushrooms from Andhra Pradesh. *Research Journal of Agricultural Sciences, 11*(6), 1388–1394.

Salehi, F., Kashaninejad, M., Asadi, F., & Najafi, A. (2016). Improvement of quality attributes of sponge cake using infrared dried button mushroom. *Journal of Food Science and Technology, 53,* 1418–1423.

Sarikurkcu, C., Tepe, B., Solak, M. H., & Cetinkaya, S. (2012). Metal concentrations of wild edible mushrooms from Turkey. *Ecology of Food and Nutrition, 51,* 346–363.

Semreen, M. H., & Aboul–Enein, H. Y. (2011). Determination of heavy metal content in wild–edible mushroom from Jordan. *Analytical Letters, 44,* 932–941.

Sesli, E., Tuzen, M., & Soylak, M. (2008). Evaluation of trace metal contents of some wild edible mushrooms from Black sea region, Turkey. *Journal of Hazardous Materials, 160,* 462–467.

Sevindik, M., Akgul, H., Selamoglu, Z., & Braidy, N. (2020). Antioxidant and antigenotoxic potential of *Infundibulicybe geotropa* mushroom collected from Northwestern Turkey. *Oxidative Medicine and Cellular Longevity, 2020,* Article ID 5620484.

Sharif, S., Mustafa, G., Munir, H., Weaver, C. M., Jamil, Y., & Shahid, M. (2016). Proximate composition and micronutrient mineral profile of wild *Ganoderma lucidum* and four commercial exotic mushrooms by ICP–OES and LIBS. *Journal of Food and Nutrition Research, 4*(11), 703–708.

Shenkin, A. (2006). Micronutrients in health and disease. *Postgraduate Medical Journal 82,* 559–567.

Singh, J., Sindhu, S. C., Sindhu, A., & Yadav, A. (2016). Development and evaluation of value added biscuits from dehydrated shiitake (*Lentinus Edodes*) mushroom. *International Journal of Current Research, 8,* 27155–27159.

Siwulski, M., Rzymski, P., Budka, A., Kalač, P., Budzyńska, S., Dawidowicz, L., Hajduk, E., Kozak, L., Budzulak, J., & Sobieralski, K. (2019). The effect of different substrates on the growth of six cultivated mushroom species and composition of macro and trace elements in their fruiting bodies. *European Food Research and Technology*, *245*, 419–431.

Sousa, C., Moutinho, C., Vinha, A. F., & Matos, C. (2019). Trace minerals in human health: Iron, zinc, copper, manganese and fluorine. *International Journal of Science and Research Methodology*, *13*(3), 57–80.

Soylak, M., Saracoglu, S., Tuzen, M., & Mendil, D. (2005). Determination of trace metals in mushroom samples from Kayseri, Turkey. *Food Chemistry*, *92*, 649–652.

Stilinović, N., Čapo, I., Vukmirović, S., Rašković, A., Tomas, A., Popović, M., & Sabo, A. (2020). Chemical composition, nutritional profile and *in vivo* antioxidant properties of the cultivated mushroom *Coprinus comatus*. *Royal Society Open Science*, *7*, 200900.

Strazzullo, P. (2014). Sodium. *Advances in Nutrition*, *5*(2), 188–190.

Teke, A. N., Bi, M. E., Ndam, L. M., & Kinge, T. R. (2021). Nutrient and mineral components of wild edible mushrooms from the Kilum–Ijim forest, Cameroon. *African Journal of Food Science*, *15*(4),152–161.

Turhan, S., Zararsız, A., & Karabacak, H. (2010). Determination of element levels in selected wild mushroom species in turkey using non–destructive analytical techniques. *International Journal of Food Properties*, *13*, 723–731.

Turkekul, I., Elmastas, M., & Tüzen, M. (2004). Determination of iron, copper, manganese, zinc, lead and cadmium in mushrooms samples from Tokat. *Food Chemistry*, *84*, 389–392.

Varela-López, A., Giampieri, F., Bullón, P., Battino, M., & Quiles, J. L. (2016). A systematic review on the implication of minerals in the onset, severity and treatment of periodontal disease. *Molecules*, *21*(9), 1183.

Wang, X., Liu, H., Zhang, J., Li, T., & Wang, Y. (2017). Evaluation of heavy metal concentrations of edible wild–grown mushrooms from China. *Journal of Environmental Science and Health, Part B*, *52*, 178–183.

Whiting, S. J., Kohrt, W. M., Warren, M. P., Kraenzlin, M. I., & Bonjour, J. (2016). Food fortification for bone health in adulthood: A scoping review. *European Journal of Clinical Nutrition*, *70*, 1099–1105.

Yildiz, S., Gürgen, A., & Çevik, U. (2019). Accumulation of metals in some wild and cultivated mushroom species. *Sigma Journal of Engineering and Natural Sciences*, *37*(4), 1371–1380.

Zajac, M., Muszynska, B., Kala, K., Sikora, A., & Opoka, W. (2015). Popular species of edible mushrooms as a good source of zinc to be released to artificial digestive juices. *Journal of Physiology and Pharmacology*, *66*(5), 763–769.

Zhang, D., Frankowska, A., Jarzyńska, G., Kojta, A. K., Drewnowska, M., Wydmańska, D., Bielawski, L., Wang, J., & Falandysz, J. (2010). Metals of king bolete (*Boletus edulis*) bull.: Fr. collected at the same site over two years. *African Journal of Agricultural Research*, *5*, 3050–3055.

Zhou, F., Yang, W., Wang, M., Miao, Y., Cui, Z., Li, Z., & Liang, D. (2018). Effects of selenium application on Se content and speciation in *Lentinula edodes*. *Food Chemistry*, *265*, 182–188.

Zięba, P., Kała, K., Włodarczyk, A., Szewczyk, A., Kunicki, E., Sękara, A., & Muszyńska, B. (2020). Selenium and zinc biofortification of *Pleurotus eryngii* mycelium and fruiting bodies as a tool for controlling their biological activity. *Molecules*, *25*(4), 889.

Zocher, A. L., Kraemer, D., Merschel, G., & Bau, M. (2018). Distribution of major and trace elements in the bolete mushrooms *Suillus luteus* and the bioavailability of rare earth elements. *Chemical Geology*, *483*, 491–500.

CHAPTER 6

Edible mushrooms: A source of quality protein

Jaspreet Kaur[1], Jyoti Singh[1], Vishesh Bhadariya[2], Simran Gogna[1], Sapna Jarial[3], Prasad Rasane[1], and Kartik Sharma[4]

[1]Department of Food Technology and Nutrition, School of Agriculture, Lovely Professional University, Punjab, India

[2]Department of Chemical and Petroleum Engineering, School of Chemical Engineering and Physical Sciences, Lovely Professional University, Phagwara, Punjab, India

[3]Department of Agricultural Economics & Extension, School of Agriculture, Lovely Professional University, Punjab, India

[4]International Center of Excellence in Seafood Science and Innovation, Faculty of Agro-Industry, Prince of Songkla University, Thailand

CONTENTS

6.1 INTRODUCTION

Edible mushrooms are the delicacies in the different food cuisines because of its medicinal, nutritional, and organoleptic characteristics and the commonly consumed ones in the whole world are *Agaricus bisporus, Pleurotus* spp., and *Lentinus edodes* (Roncero-Ramos & Delgado-Andrade, 2017). The largest producer of mushrooms in the world is China, followed by the United States, the

DOI: 10.1201/9781003152583-8

Netherlands, Poland, Spain, and France (Erjavec et al., 2012). Due to their high dietary fiber, low-fat content, and low-calorie content, edible mushrooms are considered good for human consumption. The protein content in edible mushrooms varies from 19% to 37% of dry weight along with the presence of essential amino acids (Bach et al., 2017). The presence of enzymes and bioactive compounds in edible mushrooms other than nutritional and gastronomic value makes it valuable for pharmaceutical industries. Various researchers have explored the anti-hypertensive, anti-cancer, anti-microbial, anti-viral, anti-oxidant, anti-inflammatory, hypolipidemic, hypoglycemic, and immune-modulatory properties (Carrasco-González et al., 2017 and Rathore et al., 2017). The increase in population worldwide has increased the demand for food and the estimated increase in population by 2050 is more than 9 billion, which demands an increase in agricultural production by 70% as stated by Food and Agriculture Organization. Henceforth, finding alternative sources of protein needs the production of high-quality protein in less space and time. Protein is an important nutrient for the human body and plays an important role in the maintenance of body tissues as enzymes and hormones (Bhutta et al., 2013). Edible mushrooms are considered a source of high-quality protein because of the presence of most of the essential amino acids and due to their fast production and low monetary expenditure on production other than that of animal and plant proteins. These can be grown on various substrates like waste from agricultural industries, paper, and wood, which makes them eco-friendly (Lavelli et al., 2018). Mushrooms possess anti-microbial properties because of the presence of secondary metabolites including steroids, terpenes, benzoic acid derivatives, and quinolones and primary metabolites like oxalic acid proteins and peptides. *Lentinus edodes* is the commonly studied species of mushroom for its anti-microbial action against both gram-positive and gram-negative bacteria (Alves et al., 2012). Various varieties of mushrooms have been utilized for the treatment and prevention of diseases due to their pharmaceutical and immunomodulatory properties since ancient times (Patel & Goyal, 2012). The medicinal functions imparted by the mushrooms include anti-diabetic, anti-allergic, cardiovascular protector, anti-bacterial, anti-parasitic, hepato-protective, anti-cholesterolemic, anti-cancer, and also the protection from tumor development and anti-inflammatory processes (Chang & Wasser., 2012; Finimundy et al., 2013; Zhang et al., 2011).

6.2 VARIOUS SPECIES OF EDIBLE MUSHROOMS ALL OVER THE WORLD

The different varieties of edible mushrooms grown worldwide are shown in Table 6.1. *A. bisporus* is the most cultivated mushroom in the world and is widely studied for its therapeutic and

Table 6.1   Various species of edible mushrooms grown worldwide

Name of the species	Country/region	References
Agaricus bisporus	Worldwide	Ho et al., 2020, Wasser, 2002
Lentinus edodes (shiitake mushroom)	Worldwide	Al-Dbass et al., 2012
Flammulina velutipes (enoki mushroom)	Taiwan	Yeh et al., 2014
Ganoderma sp.Omphalotus oleariusHebeloma mesophaeum	NigeriaChinaJapanKorea	Aremu et al., 2009, Valverde et al., 2015
Tricholoma matsutake, Lactarius hatsudake and *Boletus aereus*	China	Wang et al., 2014
Agaricus blazei	Brazil	Hakime-Silva et al., 2013
A. subrufescens	United States of America	Wisitrassameewong et al., 2012
L. polychrous	Northern and Northeastern Thailand	Thetsrimuang et al., 2011

medicinal properties (Lee et al., 2013). *A. blazei* is grown in Brazil and is also cultivated in Japan. It is also known as "sun mushroom" and nowadays it is also used in tea and food due to its medicinal properties. It also possesses antimutagenic, immunostimulatory, and anticarcinogenic properties (Hakime-Silva et al., 2013). *A. subrufescens* is also known as "almond mushroom" for its almond taste and is grown in the USA. It is rich in bioactive compounds that help in treating diabetes, hyperlipidemia, arteriosclerosis, cancer, and hepatitis (Wisitrassameewong et al., 2012). *L. edodes,* or "shiitake mushroom", has been used for many years for its excellent effect on human health. It has been proven to treat the common cold by Mattila et al. (2001). *L. polychrous* grown in Thailand is used in medicine to treat dyspepsia or envenomation caused by snake or scorpion bites. *Pleurotus,* or "oyster mushroom", has approximately 40 species and is used for its health-promoting properties due to the presence of compounds like lectins which possess immunomodulatory and antitumor properties. It has also shown hematological, antibacterial, hypocholesterolemic, and antioxidant properties (Makropoulou et al., 2012). *Ganoderma,* also popular as the "mushroom of immortality", has been used in Chinese medicine for the treatment of neurasthenia, carcinoma, hypertension, and hepatopathy (Mahajna et al., 2008).

6.2.1 Nutritional profiling of edible mushrooms

Edible mushrooms are high in protein and carbohydrates and are also a rich source of vitamins and minerals (Kayode et al., 2015; Han et al., 2016). They are considered useful for vegetarian people as they provide the essential amino acids and, besides that, they also contain different bioactive compounds with various health benefits (Valvarde et al., 2015). The nutritional value is highly influenced by growth conditions, characteristics, and postharvest conditions, and also the great variations are seen among and within species (Kalač, 2013; Reis et al., 2012).

6.2.2 Energy

The energy content of *Ganoderma* spp. and *Hebeloma mesophaeum* was calculated by Aremu et al. (2009) and the values reported were 1,476.7 kJ/100 g and 1,513.5 kJ/100 g, respectively, which indicated that the mushrooms are the denser source of energy as compared to cereals. Mushrooms are considered a low-calorie food, as they are low in fat content, i.e. 20–30 g/kg of dry matter (Guillamón et al., 2010).

6.2.3 Carbohydrates

Mushrooms are a good source of both digestible carbohydrates (glucose, glycogen, trehalose, and mannitol) and non-digestible carbohydrates (chitin, β-glucan, and mannans), which form the larger portion of total carbohydrates (Samsudin & Abdullah, 2019). The quality of powder processed from an oyster mushroom, a variety of *Pleurotus sajor-caju* (PSC), was studied by Han et al. (2016) and the result stated that PSC powder showed high content of carbohydrates (60.47 g/ 100 g). Polysaccharides including dietary fibers are present in PSC, which provides medicinal and pharmacological properties to it (Elleuch et al., 2011).

6.2.4 Proteins

Different parts of a mushroom, the cap, stalk, and cap with stalk, have a difference in nutritional content, especially protein. The cap of an oyster mushroom contains 34.19/100 g, cap with stalk contains 30.48 g/100 g, and stalk contains 20.96 g/100 g of protein (Oluwafemi et al., 2016). Three different varieties of edible mushroom grown in Nigeria, namely *Ganoderma* sp., *Omphalotus olearius*, and *Hebeloma mesophaeum,* as shown in Table 6.1, were studied by Aremu et al. (2009)

and reported the protein content in the range of 18–5–21.5/100 g. Maize flour was fortified with mushroom flour from *Agaricus bisporus* and *Pleurotus ostreatus* (Ishara et al., 2018) and the result reported the protein content of maize flour was increased from 6.9 g/100 g to 15.87 g/100 g (*Agaricus bisporus*) and up to 19.32 g/100 g (*Pleurotus ostreatus*). They also contain essential amino acids including threonine, histidine, glutamic acid, aspartic, lysine, phenylalanine, isoleucine, leucine, and methionine (Passari et al., 2016; Wang et al., 2014), and out of these, glutamic acids and aspartic acids are responsible for imparting umami taste to the mushrooms (Tsai et al., 2008).

6.2.5 Lipids

The major fatty acid reported in mushrooms is palmitic, oleic, and linoleic acids (Valverde et al., 2015). Polyunsaturated fatty acids are mostly found in edible mushrooms, which further helps in lowering cholesterol (Guillamón et al., 2010). Ergosterol is the major sterol produced by the edible mushrooms and it has been stated that sterol helps in preventing cardiovascular diseases (Kalač, 2013).

6.2.6 Vitamins

Various varieties of edible mushrooms, namely *Pleurotus* sp., *Hygrocybe* sp., *Hygrophorus* sp., *Schizophyllum* commune, and *Polyporus tenuiculus,* were studied for the distribution of vitamins including vitamin A, E, C, B1, and B2 by Chye et al. (2008). The result reported that *Schizophyllum commune* has the highest content of vitamin A (2,711.30 mg/g fresh weight) and vitamin E (85.08 mg/g fresh weight). Keegan et al. (2013) have stated that edible mushrooms can act as a biological precursor to vitamin D2 due to the presence of ergosterol.

6.2.7 Minerals

Alexopoulos et al. (1996) stated that *Marasmius oreades* (fairy mushrooms) are a good source of zinc, folic acid, iron, and copper and *Lentinula edodes* have a low amount of glucose and sodium, which makes it the perfect choice for the diabetic population. Phosphorous, potassium, and magnesium are the minerals that were found abundantly in mushrooms with 497.35 mg/100 g in the cap, 340.59 mg/100 g stalk, and 466.24 mg/100 g caps with a stalk (Oluwafemi et al., 2016). Potassium is the major element found in mushrooms and is unevenly distributed too in different parts of the fruit body. The high concentration is found in the cap followed by the stipe, spore-forming part, and the least in spores. Magnesium is the second highest mineral found abundantly in mushroom species (Kalač et al., 2010).

6.3 COMPARATIVE ANALYSIS OF EDIBLE MUSHROOMS WITH OTHER PROTEIN SOURCES

The amino acid content from different sources should be balanced to meet the requirement of the consumer. The protein content of *Terfezia claveryi* (desert truffle) was found to be the highest (62.1 g/100 g) among the species provided in the data, while *Letinus edodes* (shiitake mushroom) had the lowest protein content (2.2 g/100 g) (Ho et al., 2020). When compared with other sources of protein, it was found that *T. claveryi* and *Arthrospira platensis* (spirulina) had almost the same amount of protein content, that is 62.1 g and 63 g, respectively (Roberto, 2015). Therefore, a complete analysis of protein content present in mushrooms is represented with different sources of proteins in Table 6.2.

Table 6.2 Comparative analysis of edible mushrooms with other protein sources

Mushroom species			
S.no.	Sources	Protein content	References
1.	*Agaricus bisporus*	3.09 g/10 0g	Ho et al. (2020)
2.	*Flammulina velutipes*	2.66 g/100 g	
3.	*Letinus edodes*	2.2 g/100 g	
4.	*Pleurotus ostreatus*	3.3 g/100 g	
5.	*Pleurotus djamor*	22.5% d.m.	González et al. (2020)
6.	*Agaricus brasilensis*	33.3% d.m.	
7.	*Pleurotus eryngii*	16.4% d.m.	
8.	*Pleurotus ostreatus (white oyster)*	22.5% d.m.	
9.	*Pleurotus ostreatus (black oyster)*	36.9% d.m.	
10.	*Terfezia claveryi*	62.1 g/100 g	Dabbour and Takruri (2002)
11.	*Tricholoma terreum*	55.5 g/100 g	
12.	*Agaricus macrosporus*	35.3 g/100 g	
13.	*Hypsizygus tessellatus*	33.8 g/100 g	Chauhan et al. (2017)
14.	*Calvatia gigantea*	34.37% d.m.	Kivrak et al. (2016)
15.	*Auricularia auricula-judae*	12.5 % d.m.	Bandara et al. (2019)
16.	*Sparassis crispa*	13.4 g/100 g	Kimura (2013)
Other protein sources			
S.no.	Sources	Protein content	References
1.	Milk	3.2 g/100 g	Pereira (2014)
2.	Egg	12.5 g/100 g	Réhault-Godbert et al. (2019)
3.	Spirulina	63 g/100 g	Roberto (2015)
4.	Soybean	36.4 g/100 g	Rizzo and Baroni (2018)
5.	Chickpeas	20.4 g/100 g	
6.	Fava beans	26.1 g/100 g	
7.	Navy beans	22.3 g/100 g	
8.	Kidney beans	23.5 g/100 g	
9.	Lentils	24.6 g/100 g	
10.	Peanuts	25.8 g/100 g	
11.	Quinoa	15% d.m.	Filho et al. (2015)
12.	Walnuts	14.3 g/100 g	Cannella and Dernini (2004)
13.	Pumpkin seeds	30.23 g/100 g	Syed et al. (2019)
14.	Sesame seeds	18.35 g/100 g	Elleuch et al. (2007)
15.	Sunflower seeds	20.78 g/100 g	Anjum et al. (2012)

6.4 PROTEIN DIGESTIBILITY OF EDIBLE MUSHROOMS

Protein that contains all essential amino acids is termed a complete protein, while that lacks one or more essential amino acids is termed an incomplete protein. Protein digestibility mentions the amount of protein available for absorption after the digestion process has finished. It has been investigated that lysine, methionine, cysteine, threonine, and tryptophan are capable of limiting the quality of protein for human consumption as some essential amino acids can be synthesized from precursors, such as branched-chain keto acids and homocysteine (Reinert et al., 2020). The Essential Amino Acid Score (EAAS) measures the proportion of each essential amino acid in a test protein to determine if the protein is complete, where a score ≥ of 1.0 indicates that the mushroom

Table 6.3 Protein-efficiency ratio of different species of edible mushrooms

Mushroom species	PER	Reference
Agaricus bisporus (Champignon)	0.074	Bach et al. (2017)
Agaricus brasilensis	0.090	
Flammulina velutipes	0.051	
Letinus edodes	0.051	
Pleurotus djamor	0.061	
Pleurotus ostreatus (black oyster)	0.098	
Pleurotus eryngii	0.059	

has all the essential amino acids. The limiting amino acids for *A. brasilensis* were isoleucine and valine, while for *P. eryngii*, *P. ostreatus* (black oyster), and *F. velutipes* was leucine. The species which contained all essential amino acids were *A. bisporus* (champignon and portobello), *L. edodes*, *P. djamor,* and *P. ostreatus* (white oyster) (Reinert et al., 2020). Some of the EAAS of certain edible mushroom species are provided in Table 6.3.

The apparent digestibility (AD) measures the amount of protein available for absorption after the ingestion of food products. The fecal nitrogen loss in the case of adults is measured to be 12 mg/kg/day while for infants it is as high as 20 mg/kg/day. True protein digestibility (TPD) measures the amount of fecal nitrogen that is derived from endogenous intestinal losses. The most protein assimilation was noted to be of *Agaricus macrosporus* with 80.50% score, while the least protein digestibility was measured in *Pleurotus sajor-caju* with a 43.38% score (Dabbour & Takruri, 2002). The biological value (BV) of protein is determined by measuring the amount of nitrogen consumed and excreted, which is mainly the consumption of amounts of urinary and fecal nitrogen followed by the losses of urinary and fecal nitrogen. The BV of *Lentinus Lepidus* was 63.72, *P. sajor-caju* had 71.94, *P. ostreatus* had 74.82, and *L. edodes* had 77.18 (Cuptapun et al., 2010). The protein digestibility corrected amino acid score (PDCAAS) is a scale that measures the nutritional quality of protein with the content and profile of essential amino acids in the test protein when compared with the reference protein. The PDCAAS value of 0.70 was found in *Tricholoma terreum* (*in vitro*), which had only tryptophan as the limiting amino acids. The lowest PDCAAS value of 0.38 was found in *Lentinus edodes* (in vitro), which also had only tryptophan as the limiting amino acid. When compared with the animal protein then, the meat had the PDCAAS value of 0.94 (Diez & Alvarez, 2001; Cuptapun et al., 2010).

Protein efficiency ratio (PER) refers to the assessment of the quality of proteins examined by the weight gain of the subject concerning protein intake (Gonzalez et al., 2020). The highest PER is shown by the species *P. ostreatus* (black oyster), while the lowest PER is shown by *F. velutipes* and *L. edodes*. PER of different species of edible mushrooms is provided in Table 6.4.

6.5 HEALTH BENEFITS AND PHARMACOLOGICAL PROPERTIES OF MUSHROOMS

Mushrooms offer different benefits, which are contributed by different bioactive compounds present in them such as β-D-glucans, grifolan, proteoglycan, heteroglycan, galactomannan, glucoxylan, agaricoglyceride, lentinan, hericenons, erinacines, hericerins, resorcinols, steroids, mono-terpenes, diterpenes, pyrogallol, hydroxybenzoic, and many more (Yeh et al., 2011; Sanchez et al., 2017). These compounds are responsible for various roles in the body, including antioxidant action, anti-carcinogenic property, immunomodulating property, antibacterial and antiviral mechanisms, hemagglutinating effects, hypolipidaemic properties, anti-diabetic effects, as an anticoagulant, anti-tumor effect, anti-thrombotic activity, and anti-allergic actions (Wong et al., 2007;

Table 6.4 Pharmaceutical and therapeutic roles exhibited by edible mushrooms

S.no.	Species of edible mushroom	Health benefits and pharmaceutical potential	Significant findings	References
1.	*Grifola frondosa*	• Antioxidant role • Anti-inflammatory role	• The bioactive components present in *G. frondosa* include grifolan, proteoglycan, heteroglycan, galactomannan, glucoxylan, mannogalactofucan, fucomannogalactan, and agaricoglyceride. • The free radicals produced in the body can increase the risk of many chronic diseases such as cancer, rheumatoid arthritis, and atherosclerosis, thus due to the presence of high reducing power serves as an indicator of its antioxidant property including chain initiation, binding of transition metal ion catalysts, decomposition of peroxides, reductive capacity, and radical scavenging mechanisms. • The chelating characteristic was higher on ferrous ion, which may be effective in reducing the concentration of transition metals that may act as catalysts to generate free radicals. Therefore, it can be concluded that this species can be used as an accessible source of natural antioxidants. • The anti-inflammatory action is enhanced by the presence of proteoglycan, heteroglycan, and galactomannan, thus reducing the risk of degenerative illnesses.	Yeh et al. (2011)
2.	*Mushroom sclerotium*	• Immuno-modulating • Anti-tumor effect	• Due to the presence of fibrillary and matrix components such as chitin and the polysaccharides including β-D-glucans and mannans, in their cell wall, they exhibit various health benefits. • The immune modulating properties are mediated by the production of cytokines. • The anti-tumor activity is achieved by the inhibition of cancer cells by the mode of action of cell cycle arrest and cytotoxicity.	Cheung (2013)
3.	*Lentinus edodes*	• Anti-carcinogenic properties • Immune system enhancers • Antibacterial and antiviral properties • Hemagglutinating effects • Hypolipidemic properties	• Due to the presence of bioactive compounds such as Lentinan, KS-2-alpha-mannan-peptide, LEM, LAP, and EP3, it exhibits anti-tumor effects against allogenic, synergic, and autochthonous tumors. If combined with 5-Fluorouracil, it may reduce colon cancer. • Mannoglucan, polysaccharide, and lentinan in mushrooms has been proved to show immune modulating effects by increasing the activities of natural killer cells, cytotoxic T lymphocytes, and lymphokine activated killer cells. By inducing the responses of Interleukin-12 (IL-12), T-cells and macrophage-dependent immune system, it helped in showing anti-tumor effects. • Lentinamicin and lentinan possess antibacterial and antiviral properties, respectively, mainly against gram-positive bacteria such as *Bacillus subtilis* and *Staphylococcus aureus*. Lentinan present enhanced the host resistance against viral infections, by reducing the toxicity of azidothymidine AZT (drug for treating HIV/AIDS).	Bisen et al. (2010)

(Continued)

Table 6.4 (Continued)   Pharmaceutical and therapeutic roles exhibited by edible mushrooms

S.no.	Species of edible mushroom	Health benefits and pharmaceutical potential	Significant findings	References
4.	*Pleurotus tuber-regium*	• Anti-cancer property	• Lectin possesses hemagglutinating effects, by inducing the formation of mycelial mat. • Eritadenine, lentinacin, and lentysine exhibit cardioprotective effects, and thus help in lowering serum cholesterol levels (both high density lipoprotein and low density lipoprotein), by accelerating its excretion. • They exhibit anti-cancer properties by inducing the apoptosis of acute promyelocytic leukemic cells (HL-60), thereby reducing its activity. • Stronger anti-proliferative activity was exerted at higher concentrations of 200 and 400 Ig/mL. • The main regulators of apoptosis include Bcl-2 family proteins, such as antiapoptotic Bcl-2 and proapoptotic Bax proteins. The increase in the ratio of Bcl-2/Bax helps to recognize the susceptibility of cells to apoptosis and the therapeutic response to the chemotherapy.	Wong et al. (2007)
5.	*Cantharellus cibarius*	• Anti-oxidant property • Cytotoxic activity • Blood pressure regulatory effects	• It possesses free radical scavenging activity, chelating effect on iron and inhibits the lipid peroxidation, when extracted by methanol. It causes the inhibition of breakdown of lipid hydro peroxides to unwanted volatile products and regenerates primary antioxidants. • It exhibits promising cytotoxic effects of its extract against cancer cells such as human cervix adenocarcinoma HeLa, human myelogenous leukaemia K562, and human breast carcinoma cells. • On the same note, it exhibits angiotensin converting enzyme (ACE) inhibitory function and helps in the maintenance of arterial blood pressure through the renin angiotensin aldosterone system.	Kozarski et al. (2015)
6.	*Craterellus tubaeformis*	• Anti-inflammatory • Antioxidant • Antimicrobial effect	• Polysaccharides present in mushrooms acts as bioactive compounds and possess beneficial effects. • Its antioxidative effect is seen by its effectiveness to scavenge free radicals and chelate iron. • It has shown effective mode of action against bacteria such as *Listeria innocua, Bacillus cereus, Escherichia coli, Pseudomonas aeruginosa, Staphylococcus aureus,* and *Candida albicans* as an antimicrobial effect.	Vamanu and Voica (2017)
7.	*Clitocybe nuda*	• Antioxidant effects • Antidiabetic property • Anti-hyperlipidaemic effects	• The ethanol extracts of this species showed flavonoid content to be 8.21 µg/mg. Flavonoids exhibit antimicrobial, antithrombotic effects, and also scavenge free radicals by terminating the chain reactions, thereby exerting antioxidative effects.	Chen et al. (2014)

8.	*Boletus edulis*	• Anti-inflammatory properties • Antioxidant • Antimicrobial action	• 3-keto-drimenol, 3 beta-hydroxydrimenol, and 3 beta, 11, 12-trihydroxydrimene possess inhibitory function against 11 beta-hydroxysteroid dehydrogenases, which acts as a treatment for metabolic syndrome. • It also helps in stimulating the expression of glucose transporter 4 (GLUT 4) leading to translocation of its contents to plasma membrane. It further has been proven to activate AMP-activated protein kinase (AMPK), which inhibits the synthesis of fatty acids and triglycerides (TGL), thereby improving the response of insulin. • Reductions in the levels of TGL and glucose, while increment in the expression of hepatic PPAR-α was observed. • The methanolic extract of *B. edulis* showed lower levels of TNF-α, and inhibited the expression of mRNA of the other pro-inflammatory cytokines such as Interleukin-1b (IL-1b) and IL-6. • It exhibited antioxidant action and was found effective in reducing the oxidative stress. • Polysaccharides present as a bioactive compound, possess anti-inflammatory effect against various non-communicable diseases. • It was also found effective against certain bacteria such as *E. coli* and *K. pneumoniae*, thus preventing the body's immune system from illnesses and infections.	Moro et al. (2012); Rosa et al. (2020)
9.	*Hericium erinaceus*	• Antibiotic properties • Immunomodulatory activity • Anti-tumor effects • Cardioprotective effects • Hepatoprotective function • Nephroprotective function • Neuroprotective function	• Some of the bioactive components present include hericenons, erinacines, hericerins, resorcinols, steroids, mono-terpenes, diterpenes, heteroglycan peptide, b-1,3 branched-b-1,2-mannan, and lectin. • It has shown antibiotic activity against the bacteria methicillin-resistant *Staphylococcus aureus* (MRSA), which otherwise can cause food-borne illnesses and infections, and against *Salmonella typhimurium* via the mode of action of activation of the innate macrophage immune cells, which could otherwise lead to liver necrosis and cancer. It also works against *Helicobacter pylori* (*H. pylori*), which can cause gastrointestinal ulcers. • It causes the inhibition of cancer cells including human leukemia (K562) and human prostate cancer (LANCAP) cells and therapeutic effects against drug-resistant human liver carcinoma. • It acts by increasing the pro-inflammatory cytokines tumour necrosis factor-α (TNF- α) interleukin-1β, and interleukin-6 and reducing the blood vessels inside the tumor by 50%–60%. It induces the activity of natural killer cells and inhibits angiogenesis. • The bioactive components present cause the secretion of nerve growth factor (NGF), and enhanced the secretion of the NGF protein.	Friedman (2015)
10.	*Tricholoma matsutake*	• Anti-inflammatory properties • Immunostimulatory effects	• The bioactive compound present in it includes the polysaccharide (β-glucan). • The demonstration of its immunological and anti-inflammatory effects is due to the production of nitric oxide (NO) from macrophages which also led to increase in inducible nitric oxide synthase (iNOS) expression.	Kim et al. (2008)

(Continued)

Table 6.4 (Continued) Pharmaceutical and therapeutic roles exhibited by edible mushrooms

S.no.	Species of edible mushroom	Health benefits and pharmaceutical potential	Significant findings	References
11.	*Agaricus bisporus*	• Anti-inflammatory properties • Antibacterial and antifungal properties • Anti-diabetic • Antioxidant • Anti-cancer • Immunomodulating properties	• β-glucan upregulates the innate immune responses of macrophages. • The expression of TNF-α, IL-1b, IL-6, and IL-12 were significantly increased thereby exhibiting immunomodulatory effects. • The bioactive compound responsible for its role includes pyrogallol, hydroxybenzoic acid derivatives, and flavonoids. • It inhibits the activity of acetylcholinesterase and butyrylcholinesterase, thereby exerting promising effects in treating Alzheimer's disease. • The polysaccharides present contribute to immunomodulating effects and helps in the reduction of tumor size. Its ingestion also helps in accelerating the secretion of Immunoglobulin-A (IgA). It suppresses aromatase, which helps in decreasing the risk of breast cancer. • The levels of TGL and low density lipoproteins (LDL) are reduced when the cholesterol absorption is reduced by the phytosterols present in them. It was also stated that *A. bisphorus* also consists of 565.4 mg/kg of lovastatin, which lowers cholesterol and exhibits anti-carcinogenic effects. • Alpha-glucans present in this variety lowers blood glucose levels (by 69%) by producing lipopolysaccharide-induced TNF-α. • The presence of the phenolic compounds such as gallic acid, protocatechuic acid, catechin, caffeic acid, ferulic acid, and myricetin exhibited the anti-oxidative effects, thereby acting as free radical scavengers.	Volman et al. (2010); Chen et al. (2006); Liu et al. (2013); Atila et al. (2021)
12.	*Volvariella volvacea*	• Antioxidant properties • Anticoagulant • Anti-inflammatory role • Anti-hypertensive action • Immunomodulatory effects	• The immune enhancing action is contributed by the presence of a fungal immunomodulatory protein Fip-vvo. • The promising antioxidative effects of *V. volvacea* are demonstrated by the presence of certain bioactive compounds as they possess the free radical scavenging activity, and also prevent lipid peroxidation. • They exhibit phenolic compounds, which are known to prevent various disorders including cardiovascular disorders, hypertension, and cancer, thereby inducing its anti-inflammatory effects.	Dulay et al. (2016)
13.	*Flammulina velutipes*	• Anti-inflammatory • Antibacterial and antiviral • Anti-tumor • Anti-hyperlipidaemic effects • Antioxidant	• Peptidoglycan, polysaccharides, and flammulin are the bioactive components present in *F. velutipes*, known to exhibit various health effects. • Enokipodins, isolated from *F. velutipes*, have been known to possess antibacterial effects and anti-proliferative activity several tumors. Promising effects were also shown against the activity of the malarial parasite *Plasmodium falciparum*. • Flamvelutpenoids and flamulinol A (cucumane-type sesquiterpene) have also been proven to show antibacterial effects against *B. subtilis*, *E. coli*, and MRSA.	Wang et al. (2012); Tabuchi et al. (2020)

	Species	Bioactive roles	Health effects	References
14.	*Ganoderma lucidum*	• Anti-inflammatory role • Antibacterial and antifungal properties • Anti-tumor effect • Hepato-protective action • Antioxidant • Hypoglycaemic effects	• Cytotoxic effects against HeLa cells and HepG2 cells, which are known to cause cancer, were also observed thereby implementing its anti-carcinogenic activity. • The health effects are exhibited by some bioactive compounds such as ganoderic acids, ganoderiol, ganodermanontriol, ganoderan a and b, ganopoly, lucidenic acids, ganoderic acids, and ling zhi-8 (protein). • It possesses powerful immunomodulating property due to the presence of various bioactive compounds, thereby aiding in the treatment of leukemia, carcinoma, hepatitis, and diabetes. • It can inhibit gram-positive as well as gram-negative bacteria, one of them being *Bacillus subtilis*. *Ganodermin* was known to possess anti-fungal properties by inhibiting the growth of *Botrytis cinerea* and *Fusarium oxysporum*. • The polysaccharide (active β-D-glucans) present shows anti-tumor action. It also helps in inhibiting the action of hepatocellular cancer cells and gastric cancer cells. • The promising hypoglycaemic effects correspond to insulin-releasing activity, which is caused by the facilitation of calcium flow towards pancreatic cells. Sterols helps in reducing cholesterol synthesis, increase the high density lipoprotein (HDL), thereby producing hypolipidaemic effects.	Gao et al. (2004); Wang and Ng (2010); Batra et al. (2013)
15.	*Auricularia auricula-judae*	• Anti-hyperglycaemic • Immunomodulatory • Anti-tumor • Anti-inflammatory role • Antimicrobial action	• Glucan is the main bioactive component present in this specie. • It is proven to supress the elevated fasting level and non-fasting blood glucose level and on the same note also reduces (glycosylated haemoglobin) HbA1c levels. It has had promising effects in improving glucose urinary concentrations and glucose tolerance as well. Therefore, it helps in maintaining normal blood glucose levels. • Due to the presence of fatty acid, steric acid, oleic acid, and palmitic acid, it exhibits anti-inflammatory action and also prevents liver damage by alcohol. It also reduces the risk of cardiovascular diseases. The considerable amount of fibre present possesses anti-inflammatory and immunomodulatory roles. • It has proved effective antimicrobial action against both gram-positive and gram-negative bacteria such as *E. coli* and *S. aureus* and prevents pneumonia, urinary tract infection, and diarrhea.	Yuan et al. (1998); Cai et al. (2015); Karun et al. (2018)
16.	*Sparassis crispa*	• Immunomodulatory role • Anti-tumor role • Anti-oxidant effect • Anti-thrombotic activity	• The main bioactive component present in this variety known for its effective action is β-glucan. • As a source of prebiotics, they enhance the function of *Bifidobacterium* and *Lactobacillus* species, which help in boosting the immune system and digestive function. • They exhibit a powerful action in suppressing the tumour due to the presence of the bioactive component (1,3-β-glucan) which helps in stimulating the response of body to various infections and diseases. It is also found effective against angiogenesis and metastasis, and implements its anti-carcinogenic and anti-tumor effect.	Harada et al. (2002); Hyun et al. (2006); Puttaraju et al. (2006); Yamamoto et al. (2009)

(Continued)

Table 6.4 (Continued) Pharmaceutical and therapeutic roles exhibited by edible mushrooms

S.no.	Species of edible mushroom	Health benefits and pharmaceutical potential	Significant findings	References
17.	*Pleurotus eryngii*	• Anti-hypercholesteraemic effect • Anti-viral effect • Anti-carcinogenic role	• Due to the presence of gallic acid, protocatechuic acid, gentisic acid, and coumaric acid, it possesses antioxidant function and thus acts as free radical scavengers and reduces oxidative stress. It has been investigated that it can inhibit platelet aggregation and thus, exhibit anti-thrombotic effect. • Laccase, pleureryn, lovastatin, and eryngeolysin are the main bioactive components present in this variety known for its physiological role. • Reduction in the body weight and levels of cholesterol, TGL, LDL, VLDL, as well as HMG-CoA reductase activity in the liver which was achieved by supressing the synthesis of cholesterol. Therefore, its ingestion has promising effects on lipid profile. • The protein in this species has anti-inflammatory properties, and exhibits promising effects against colon cancer by the mode of action involving apoptosis followed by downregulation of cell cycle–related signalling proteins such as cyclin B and cyclin E. • Lovastatin present helps to decrease cholesterol level in blood and thus maintain normal lipid profiles. While laccase shows inhibitory activity toward HIV-1 reverse transcriptase, and possess anti-viral mode of action.	Stajić et al. (2009); Alam et al. (2011); Yuan, Ma, et al. (2017)
18.	*Morchella esculenta var. rotunda*	• Antioxidative effects • Anti-tumor • Anti-hyperglycaemic effects • Immune system function	• Heteroglycan, galactomannan, and β-1,3-D-glucan are the bioactive compounds present in *Morchella esculenta*. • It exhibits effective role in scavenging free radicals due to the presence of β-carotene, and therefore neutralises the linoleate free radical. It also reduces the oxidative stress in the body and prevents lipid peroxidation by its antioxidative mechanism. • Due to the presence of beta-glucan, which is a polysaccharide, it has potential effect in decreasing blood cholesterol levels and blood glucose levels, thus procuring anti-inflammatory role. It also supports immunomodulatory role to fight against bacterial infections.	Gursoy et al. (2009)
19.	*Hypsizygus tessellatus*	• Antioxidant • Anti-bacterial action • Anti-allergic • Anti-inflammatory effect	• Some of the active compounds known for its potential role include ergosterol, mannitol, trehalose, methionine, and marmorin. • It has shown potential anti-bacterial effects against gram positive bacteria such as *E. coli* and *S. marscenscens* as well as against gram- negative bacteria such as *B. subtilis* and *S. aureus*. • The presence of gallic acid, mannitol, and quercetin executes antioxidant, anti-inflammatory, and antineoplastic properties which possess therapeutic effects in cardiovascular diseases, neuropsychological, and metabolic disorders.	Chien et al. (2016); Shah et al. (2018)

No.	Species	Properties	Description	Reference
20.	*Pleurotus ostreatus*	• Immunomodulatory • Anti-tumor • Antioxidant • Antiviral and antibacterial action • Anti-hyperglycaemic role	• The anti-tumor activity is depicted by the compound marmorin, by inhibiting angiogenesis, as the mode of action. • The effectiveness of its anti-inflammatory action could be seen by its mechanism of reducing the production of nitric oxide (NO), IL-6, TNF-α, as well as IL-1B (proinflammatory cytokines), and increasing the production of IL-10 (anti-inflammatory cytokine). • The bioactive components which serve its function are pleuran, β-glucan, proteoglycan, laccase, protocatechuic acid, gallic acid, rutin, myrictin, chrysin, naringin, α-tocopherol, γ- tocopherol, ascorbic acid, β-carotene, and pleurostrin. • Its antibacterial role is categorized by its ability to inhibit the growth of gram-positive and gram-negative bacteria. It also inhibits the replication of viruses without having any effect on normal cell division, thus manifesting its antiviral characteristics. • It possesses the ability to reduce high blood-glucose levels and improve the lipid profile as well, which is induced by the presence of proteoglycan. • It can inhibit lipid peroxidation, and reduce oxidative stress by ameliorating the activities of enzymatic and non-enzymatic antioxidants. It has helped in decreasing the toxicity in the body and has improved immune resistance due to the presence of polysaccharides.	Deepalakshmi and Sankaran (2014)
21.	*Calvatiagigantea*	• Anti-tumor mechanism • Antioxidant • Anti-hyperglycaemic effect • Anti-cancer	• Calvacin is the major bioactive component present in this variety responsible for its anti-tumor mechanism. • Benzoic acid, cinnamic acid, protocatechuic acid, gallic acid, gentisic acid, vanillic acid, and syringic acid are the antioxidants identified and are responsible for reducing the risks of chronic illnesses such as atherosclerosis, cancer, diabetes, and other degenerative diseases. • They are known to have potential chain breaking function, inhibit lipid peroxidation, and may scavenge free radicals, thus promoting antioxidant activity. • It was also found to inhibit the proliferation of A549 lung cancer cells by the mode of action of cell cycle arrest and cell apoptosis by decreasing the expressions of cyclin D1 and cyclin D2, cyclin-dependent kinases (CDK4), and protein kinase B (Akt).	Eroğlu et al. (2016); Kivrak et al. (2016)

Kim et al., 2008; Bisen et al., 2010; Moro et al., 2012; Cheung, 2013; Kozarski et al., 2015; Kivrak et al., 2016). They exhibit the power to scavenge free radicals, prevent lipid peroxidation, and chelate iron. Moreover, due to the presence of polysaccharides, they have been known to possess an anti-inflammatory mechanism and, therefore, prevent the risk of certain chronic diseases such as cardiovascular diseases, atherosclerosis, cancer, rheumatoid arthritis, hypertension, and neurodegenerative disorders as well (Deepalakshmi & Sankaran, 2014; Chien et al., 2016; Shah et al., 2018). A detailed description of various pharmaceutical and therapeutic roles exhibited by mushrooms is provided in Table 6.4.

6.6 ROLE OF EDIBLE MUSHROOMS AS FUNCTIONAL FOODS

Mushroom cultivation is global. Edible mushrooms are tasty food, a source of non-animal protein which fits well with the social and religious well-being of people (S Li et al., 2017). Various types of nutritious mushroom dishes are consumed around the globe. Mushrooms are excellent functional food. As per the Oxford Dictionary, functional foods are "food that has had substances that are good for your health, especially added to it". Nutrients in natural foods make all food functional, yet in functional food, some processed foods are added with specific ingredients that benefit human health. Therefore, functional foods are foods with positive health benefits that exceed those attributable to the nutritional value of the food, promote optimal health, and reduce the risk of many diseases (Bains & Chawla, 2020). Evidence suggests that edible mushrooms contain many compounds with biological properties such as "antimicrobial, antioxidant, antitumor, antidiabetic, immunomodulatory, hepatoprotective and hypocholesterolemic activities" (Chang & Miles, 2004). Because of the nutritional and medicinally important components, mushrooms are functional foods. Mushrooms enhance the immune system and natural defenses against several diseases. (Raghavendra et al., 2017). They release bioactive compounds, for instance, b-glucans, a polysaccharide constituent of the cell wall of mushrooms, resists gastrointestinal enzymes, aiding in the increase of feces volume and intestinal mobility and resists peptides, chitinous substances, terpenes, sterols, and phenolic compounds as well as absorb less toxic and harmful chemicals; thereby cancer incidence is low (Wani et al., 2010; Cheung, 2013; Deepalakshmi & Sankaran, 2014; Rosli et al., 2015; Ruthes et al., 2015). There is evidence that mushroom and polysaccharide intake improves cardiometabolic fitness, vitamin D status, and immune function (Fritz et al., 2015; Cashman et al., 2016; Dicks & Ellinger, 2020). Mushrooms contain glutamine and glutamate, which are raw material for functional foods (Raghavendra et al., 2017). According to Ahmad and Singh (2016), white bread nutritional quality improves by adding mushroom species A. bisporus with dried dates; also, Arora et al. (2017) found sponge cake nutritional quality and sensory acceptability improved with powdered A. bisporus.

Not only that, mushroom intake could be inversely associated with cancer risk, particularly breast cancer (Shin et al., 2010; Li et al., 2014; Ba et al. 2020). Mushroom consumption can protect against obesity-related hypertension and dyslipidemia by mediating antioxidant and anti-inflammatory pathways (Ganesan & Xu, 2018; Grotto et al., 2019). Mushroom polysaccharides can function as prebiotics in the digestive system because they have antioxidative, anti-inflammatory, and immunomodulating properties (Friedman, 2016). Further, the Mushrooms and Health Summit Proceedings proposed benefits for memory, breast cancer risk reduction, weight loss, and oral health (Feeney et al., 2014). It is well known that consumption of edible mushrooms results in reduced plasma cholesterol, low-density lipoprotein, and increased high-density lipoprotein as the positive improvements in blood lipid levels (Turnbull et al., 1990; Ishikawa, 1994; Nakamura et al., 1994; Homma et al., 1995; Ruxton & McMilan, 2010). The plasma lipidome (the entirety of lipids in cells) is modulated by regular mycoprotein intake over a week, supporting the hypothesis that mycoprotein favorably modulates lipid regulation (Coelho, 2020). Research indicated that glycemia and insulinemia markers improved, insulin levels decreased, and hyperinsulinemia and hyperaminoacidemia were maintained (Turnbull & Ward, 1995; Bottin et al., 2016; Dunlop et al., 2017). Although more research is required to analyze habitual intake of mushrooms as protein sources in relation to specific health markers, as observational studies often neglect this significant dietary component, evidence shows that eating 226 g of roasted *Agaricus bisporus* mushrooms increases average stool weight and results in a greater abundance of *Bacteroidetes* (beneficial gut bacteria) and reduced levels of less favorable firmicutes (Hess et al., 2018). In healthy lean, overweight, and obese adults, mycoprotein consumption was linked to lower insulin levels and appeared to reduce ad libitum energy intake (Cherta-Murillo et al., 2020).

6.7 ROLE OF EDIBLE MUSHROOMS AS A NUTRACEUTICAL SOURCE

Mushrooms possess nutritional and nutraceutical properties because of their natural composition (Atri & Singh, 2017). They are high in vital nutrients such as carbohydrates, proteins, vitamins, minerals, antioxidants, and bioactive components (Nyman 1994; Chang & Miles, 2004; Voet &Voet, 2004; Prabu & Kumuthakalavalli, 2016). For instance, Ashraf et al. (2020) reported that cordycepsis is a rare, naturally occurring entomopathogenic fungus that can be found at high altitudes on the Himalayan plateau and is used in traditional Chinese medicine. Cordyceps contains several bioactive elements, the most important of which is cordycepin, which has tremendous therapeutic and nutraceutical potential. Cordycepin is also a bioactive component because of its structural similarity to adenosine. Nutraceuticals based on cordycepin would attract a lot of attention and money in the global market. Nutraceuticals can dramatically reduce antibiotic use, which is becoming more prevalent as life expectancy increases and lifestyle diseases are also on the rise. Researchers can identify and characterize mushrooms for their bioactive compounds using advanced technologies such as attenuated total reflectance-Fourier transform infrared (ATR-FTIR) spectroscopy. ATR-FTIR is a potential process analytical technology tool in the mushroom industry. Traditional analytical techniques/assays for measuring bioactive compounds in mushrooms can be time-consuming, labor-intensive, and costly. Such a tool (i.e. predicting the content of mushrooms in selected constituents) could be handy in the mushroom industry, both for farmers as a way of promoting a high-value nutraceutical commodity and for companies processing mushrooms to make health-promoting foods or drugs/cosmetics (Bekiaris et al., 2020). Venditti et al. (2016) identified 13 compounds in Suillus bellinii by using classical chromatographic methods as well as spectroscopic techniques (nuclear magnetic resonance and mass spectroscopy). *S. bellinii* has been used for health and nutraceutical purposes, based on the identified components and their nutritional/healthy properties. *Suillus bellinii* mushroom's applications range from medicine (due to suillin's high cytotoxic and antitumor properties) to nutrition (due to the presence of essential nutrients such as amino acids and pre-vitamin D2 (ergosterol), as well as low-calorie sweeteners). *Cantharellus cibarius,* another edible mushroom, contains bioactive properties contain useful phytochemicals such as phenols and flavonoids. The high flavonoid levels discovered could help protect against diseases caused by oxidative stress. *In vitro*, the *Cantharellus cibarius* mushroom has an antihypertensive function, selective cytotoxicity, antioxidant activities against lipid peroxidation, and antibacterial activity, making it a promising source of nutraceuticals (Kozarski et al., 2015). Bioactive mushroom-based ingredients should be integrated into stable food items that are eaten widely by the general population to optimize their possible health benefits (Yuan, Zhao, et al., 2017). In Odisha, in similipal biosphere reserve (SBR) variety of edible mushrooms is growing, for instance, *Russula vesca, Russula delica,* and *Termitomyces eurrhizus* which are rich sources of proteins (22.82–35.17 g/100 g) and carbohydrates (45.68–63.27 g/100 g) and low contents in fats (2.03–4.62 g/100 g). These mushrooms possess moderate antibacterial properties and can be used in the human diet as nutraceuticals/functional foods to preserve and promote health, longevity, and life quality, because they are a source of nutrients and molecules with medicinal potential (Singdevsachan et al., 2014). *Agaricus* spp., *Pleurotus* spp., and *Lentinula edodes* are a few edible mushrooms that are commonly cultivated, and efforts are underway to use them in the development of selenium-enriched food or nutraceuticals (via mycelia). Lithium enrichment is also attempted (Falandysz & Borovička, 2012). Although the white *Agaricus bisporus* is the world's most common and widely consumed edible mushroom species, it is popular not only because of its flavor, but also because of its high nutritional content, which includes dietary fiber (chitin), essential and semi-essential amino acids, unsaturated fatty acids such as linoleic and linolenic acids, easily digestible proteins, sterols, phenolic, and indole compounds, and vitamins, especially provitamin D2 and B1, B2, B6, B7, and C. *A. bisporus* is also a rich source of selenium, zinc, and other elements such as magnesium, copper, iron, potassium, sodium, calcium, phosphorus, sulfur, and manganese. Antioxidant, antibacterial, anti-inflammatory, antitumor, and immunomodulatory activity are all found in the fruiting bodies of *A. bisporus*. The antioxidant ergothioneine (has antimutagenic, chemo-, and radioprotective properties) is present. The existence of these biologically active compounds and elements in *A. bisporus* fruiting bodies confirms their nutraceutical and medicinal properties (Muszyńska et al., 2017).

6.8 PERISHABILITY OF EDIBLE MUSHROOMS

Mushrooms are perishable products that begin to deteriorate soon after they are harvested. Due to the enzymatic activity of polyphenol oxidase, on mushroom caps' surface a brown discoloration develops and quickly becomes soft at high temperatures. Mushrooms have a short shelf life, so dehydration tends to be a promising, cost-effective preservation option for Indian environments. Dehydrated mushrooms are easier to transport than dried, pickled, or frozen mushrooms (Chandra & Samsher, 2006). It is critical to developing adequate storage and post-harvest technologies to increase their marketability and availability to customers in both fresh and processed forms. Appropriate storage and preservation technology must be developed to increase their marketability and availability for consumption in both fresh and processed forms. Due to market gluts and the highly perishable nature of the commodity, its preservation into more value-added, stable goods is critical during production's peak cycle. Mushrooms are a new edible fungus with a soft texture and a limited shelf life due to their high moisture content (91%) and perishability. To increase their marketability and availability for consumption in both fresh and processed forms, adequate storage and preservation technologies must be developed. Dehydration, steeping in a chemical solution, blanching in boiling water and chemical solution, canning, freezing, freeze drying, vacuum cooling, osmotic dehydration, pickling, dielectric heating, irradiation, micro-oven heating, low-temperature storage, and adjusted and regulated atmospheric packaging are some of the techniques for the preservation of fresh mushrooms in various processed ways. It is well established that low-temperature storage of produce is critical for preventing post-harvest losses. The use of refrigerated vans to collect produce directly from farmers and distribute it to local markets and processors will increase mushroom production and make high-quality products more accessible to consumers. Individual fast freezing, freeze drying, and vacuum drying are all essential technologies for the export of processed mushrooms. Many export-oriented units with modern preservation and packaging methods have sprung up in recent years, generating considerable interest among entrepreneurs. Compared to some industrialized countries, proper infrastructure construction for post-harvest handling of these perishable commodities is almost non-existent (Chandra & Samsher, 2006). Interestingly, thermal (drying/freezing), chemical (edible coatings, films, and cleaning solutions), and physical (packing, irradiation, pulsed electric field, and ultrasound) processes are the three types of mushroom preservation methods. The nutritional value and bioactive properties of this product can be altered as a result of these processes (Marcal et al., 2021). Further, extending the shelf life of mushrooms can be done in a variety of ways, including using modified atmosphere packaging (Kim et al., 2006), controlled atmosphere storage (Briones et al., 1992), vacuum cooling technology, coating (Nussinovitch & Kampf, 1993), and refrigeration (Mau et al., 1993). Coatings add sheen and luster to products, making them more desirable and appealing to customers, in addition to extending shelf life and delaying senescence. The use of aloe vera and gum tragacanth edible coatings on mushrooms is effective in increasing shelf life and limiting post-harvest losses (Mohebbi et al., 2011). Therefore, the edible coating is the most prevalent method that can increase the shelf life of mushrooms along with conserving their quality characteristics (Thakur et al., 2021). Pre-cooling is in essence the removal of heat or the reduction in the temperature of the mushrooms as soon as possible after harvest. Packaging technology can improve the shelf life of mushrooms (Singh et al., 2010).

6.9 CONCLUSION

Mushrooms are widely consumed across the world by all age groups. Mushrooms are considered one of the superfoods as they are rich in protein, fiber, vitamins, and minerals. The amount of bioactive components in edible mushrooms is also high, which imparts health-promoting properties like anti-viral, anti-hypertensive, and anti-hypercholesterolemic. The findings mentioned have investigated their ability in applications in food. The addition of mushrooms in food products increases the nutritional value and their physical properties. Future studies on the mechanism of action of the component present in edible mushrooms responsible for treating the health problems will help the researcher for better utilization of the mushrooms in making medicines and value-added products. Further investigations on novel properties of edible mushrooms are still deficient, which can be conducted to enhance their use in the food and pharmaceutical industries.

REFERENCES

Ahmad, K., & Singh, N. (2016). Evaluation of nutritional quality of developed functional bread fortified with mushroom and dates. *Clarion, 5*, 23–28.

Alam, N., Yoon, K. N., Lee, J. S., Cho, H. J., Shim, M. J., & Lee, T. S. (2011). Dietary effect of Pleurotus eryngii on biochemical function and histology in hypercholesterolemic rats. *Saudi Journal of Biological Sciences, 18*(4), 403–409.

Al-Dbass, A. M., Al-Daihan, S. K., & Bhat, R. S. (2012). Agaricus blazei Murill as an efficient hepatoprotective and antioxidant agent against CCl4-induced liver injury in rats. *Saudi Journal of Biological Sciences, 19*(3), 303–309.

Alexopoulos, C. J., Mims, Charles W., & Blackwell, M. (1996). *Introductory Mycology*. 4th ed. New York: John Wiley and Sons.

Alves, M. J., Ferreira, I. C. F. R., Dias, J. F., Teixeira, V., Martins, A., & Pintado, M. (2012). A review on antimicrobial activity of mushroom (Basidiomycetes) extracts and isolated compounds. *Planta Medica, 78*, 1707–1718.

Anjum, F. M., Nadeem, M., Khan, M. I., & Hussain, S. (2012). Nutritional and therapeutic potential of sunflower seeds: A review. *British Food Journal, 114*(4), 544–552.

Aremu, M. O., Basu, S. K., Gyar, S. D., Goyal, A., Bhowmik, P. K., & Datta Banik, S. (2009). Proximate composition and functional properties of mushroom flours from Ganoderma spp., Omphalotus olearius (DC.) Sing. and Hebeloma mesophaeum (Pers.) Quél. sed in Nasarawa State, Nigeria. *Malaysian Journal of Nutrition, 15*(2), 233–241.

Arora, B., Arora, S. K., et al. (2017). Sensory, nutritional and quality attributes of sponge cake supplemented with mushroom (Agaricus bisporus) powder. *Nutrition & Food Science, 47*, 578–590.

Ashraf, S. A., Elkhalifa, A. E. O., Siddiqui, A. J., Patel, M., Awadelkareem, A. M., Snoussi, M., ... Hadi, S. (2020). Cordycepin for health and wellbeing: A potent bioactive metabolite of an entomopathogenic medicinal fungus cordyceps with its nutraceutical and therapeutic potential. *Molecules*, *25*(12), 2735. 10.3390/molecules25122735.

Atila, F., Owaid, M. N., & Shariati, M. A. (2021). The nutritional and medical benefits of Agaricus bisporus: A review. *Journal of Microbiology, Biotechnology and Food Sciences 2021*, *7*(3), 281–286.

Atri, L., & Singh, N. (2017). Amino acid profile of a basidiomycetous edible mushroom-Lentinus sajor-caju. *International Journal of Pharmacy and Pharmaceutical Sciences*, *9*(9). SSN- 0975-1491.

Ba, D., Ssentongo, P., Beelman, R. B., Gao, X., & Richie, J. P. (2020). Mushroom consumption is associated with low risk of cancer: A systematic review and meta-analysis of observation studies. *Current Developments in Nutrition*, *4*, 307.

Bach, F., Helm, C. V., Bellettini, M. B., Maciel, G. M., Haminiuk, C. W. I. (2017). Edible mushrooms: A potential source of essential amino acids, glucans and minerals. *International Journal of Food Science & Technology*, *52*(11), 2382–2392.

Bains, A., & Chawla, P. (2020). In vitro bioactivity, antimicrobial and anti-inflammatory efficacy of modified solvent evaporation assisted Trametes versicolor extract. *3 Biotech*, *10*(9), 1–11.

Bandara, A. R., Rapior, S., Mortimer, P. E., Kakumyan, P., Hyde, K. D., & Xu, J. (2019). A review of the polysaccharide, protein and selected nutrient content of Auricularia, and their potential pharmacological value. *Mycosphere*, *10*(1), 579–607.

Batra, P., Sharma, A. K., & Khajuria, R. (2013). Probing Lingzhi or Reishi medicinal mushroom Ganoderma lucidum (higher Basidiomycetes): A bitter mushroom with amazing health benefits. *International Journal of Medicinal Mushrooms*, *15*(2), 127–143.

Bekiaris, G., Tagkouli, D., Koutrotsios, G., Kalogeropoulos, N., & Zervakis, G. I. (2020). Pleurotus mushrooms content in glucans and ergosterol assessed by ATR-FTIR spectroscopy and multivariate analysis. *Foods*, *9*(4), 535. 10.3390/foods9040535

Bhutta, Z. A., Sadiq, K., & Aga, T. (2013). Protein digestion and bioavailability. *Encyclopedia of Human Nutrition*, *4*, 66–73.

Bisen, P. S., Baghel, R. K., Sanodiya, B. S., Thakur, G. S., & Prasad, G. B. K. S. (2010). Lentinus edodes: A macrofungus with pharmacological activities. *Current Medicinal Chemistry*, *17*(22), 2419–2430.

Bottin, J. H., Swann, J. R., Cropp, E., Chambers, E. S., Ford, H. E., Ghatei, M. A., & Frost, G. S. (2016). Mycoprotein reduces energy intake and postprandial insulin release without altering glucagon-like peptide-1 and peptide tyrosine-tyrosine concentrations in healthy overweight and obese adults: A randomised-controlled trial. *British Journal of Nutrition*, *116*, 360–374.

Briones, G. L., Varoquaux, P., Chambroy, Y., Bouquant, J., Bureau, G., & Pascat, B. (1992). Storage of common mushroom under controlled atmospheres. *International Journal of Food Science & Technology*, *27*(5), 493–505.

Cai, M., Lin, Y., Luo, Y.-L., Liang, H.-H., & Sun, P. (2015). Extraction, antimicrobial, and antioxidant activities of crude polysaccharides from the wood ear medicinal mushroom Auricularia auricula-judae (higher basidiomycetes). *International Journal of Medicinal Mushrooms*, *17*(6), 591–600.

Cannella, C., & Dernini, S. (2004) Walnut: Insights and nutritional value. In: *V International Walnut Symposium, Acta Hortic*, *705*, pp. 547–550.

Carrasco-González, J. A., Serna-Saldívar, S. O., & Gutiérrez-Uribe, J. A. (2017). Nutritional composition and nutraceutical properties of the Pleurotus fruiting bodies: Potential use as food ingredient. *Journal of Food Composition and Analysis*, *58*, 69–81.

Cashman, K. D., Kiely, M., Seamans, K. M., & Urbain, P. (2016). Effect of ultraviolet light-exposed mushrooms on vitamin D status: Liquid chromatography-tandem mass spectrometry reanalysis of biobanked sera from a randomized controlled trial and a systematic review plus meta-analysis. *Journal of Nutrition*, *146*, 565–575.

Chandra, S., & Samsher (2006 July). Nutritional and medicinal aspects of edible mushrooms. *International Journal of Agricultural Sciences*, *2*(2), 647–651.

Chang, S. T., & Miles, G. (2004). The nutritional attributes of edible mushrooms. *Mushroom Cultivation, Nutritional Value, Medicinal Effects and Environmental Impact*. 2nd ed London, New York Washington, DC: Boca Raton, pp. 27–36.

Chang, S.-T., & Wasser, S. P. (2012). The role of culinary-medicinal mushrooms on human welfare with a pyramid model for human health. *International Journal of Medicinal Mushrooms*, *14*(2), 95–134.

Chauhan, G., Prasad, S., Rathore, H., & Sharma, S. (2017). Nutritional profiling and value addition of products from Hypsizygus tessellatus. *Food Biology*, *6*, 1–6.

Chen, M.-H., Lin, C.-H., & Shih, C.-C. (2014). Antidiabetic and antihyperlipidemic effects of Clitocybe nuda on glucose transporter 4 and AMP-activated protein kinase phosphorylation in high-fat-fed mice. *Evidence-Based Complementary and Alternative Medicine*, *2014*, 1–14.

Chen, S., Oh, S.-R., Phung, S., Hur, G., Ye, J. J., Kwok, S. L., Shrode, G. E., Belury, M., Adams, L. S., Williams, D. (2006). Anti-aromatase activity of phytochemicals in white button mushrooms (Agaricus bisporus). *Journal of Cancer Research and Clinical Oncology*, *66*(24), 12026–12034.

Cherta-Murillo, A., Lett, A. M., Frampton, J., Chambers, E. S., Finnigan, T. J. A., & Frost, G. S. (2020). Effects of mycoprotein on glycaemic control and energy intake in humans: A systematic review. *British Journal of Nutrition*, *123*, 1321–1332.

Cheung, P. C. K. (2013). Mini-review on edible mushrooms as source of dietary fiber: Preparation and health benefits. *Food Science and Human Wellness*, *2*(3–4), 162–166.

Chien, R.-C., Yang, Y.-C., Lai, E. I., & Mau, J.-L. (2016). Anti-inflammatory effects of extracts from the medicinal mushrooms Hypsizygus marmoreus and Pleurotus eryngii (Agaricomycetes). *International Journal of Medicinal Mushrooms*, *18*(6), 477–487.

Chye, F. Y., Wong, J. Y., & Lee, J.-S. (2008). Nutritional quality and antioxidant activity of selected edible wild mushrooms. *Food Science and Technology International*, *14*(4), 375–384.

Coelho, M. O. C. (2020). Daily mycoprotein consumption for one week does not affect insulin sensitivity or glycaemic control but strongly modulates the plasma lipoma. *British Journal of Nutrition*, submitted.

Cuptapun, Y., Hengsawadi, D., Mesomya, W., & Yaieiam, S. (2010). Quality and quantity of protein in certain kinds of edible mushroom in Thailand. *Agriculture and Natural Resources*, *44*(4), 664–670.

Dabbour, I. R., & Takruri, H. R. (2002). Protein digestibility using corrected amino acid score method (PDCAAS) of four types of mushrooms grown in Jordan. *Plant Foods for Human Nutrition*, *57*(1), 13–24.

Deepalakshmi, K., & Sankaran, M. (2014). Pleurotus ostreatus: An oyster mushroom with nutritional and medicinal properties. *Journal of Biochemical Technology*, *5*(2), 718–726.

Dicks, L., & Ellinger, S. (2020). Effect of the intake of oyster mushrooms (Pleurotus ostreatus) on cardio-metabolic parameters – A systematic review of clinical trials. *Nutrients*, *12*, 1134.

Dıez, V. A., & Alvarez, A. (2001). Compositional and nutritional studies on two wild edible mushrooms from northwest Spain. *Food Chemistry*, *75*(4), 417–422.

Dulay, R. M. R., Vicente, J. J. A., Cruz, A. G. D., Gagarin, J. M., Fernando, W., Kalaw, S. P., & Reyes, R. G. (2016). Antioxidant activity and total phenolic content of Volvariella volvacea and Schizophyllum commune mycelia cultured in indigenous liquid media. *Mycosphere*, *7*(2), 131–138.

Dunlop, M. V., Kilroe, S. P., Bowtell, J. L., Finnigan, T. J. A., Salmon, D. L., & Wall, B. T. (2017). Mycoprotein represents a bioavailable and insulinotropic non-animal-derived dietary protein source: A dose-response study. *British Journal of Nutrition*, *118*, 673–685.

Elleuch, M., Bedigian, D., Roiseux, O., Besbes, S., Blecker, C., & Attia, H. (2011). Dietary fibre and fibre-rich by-products of food processing: Characterisation, technological functionality and commercial applications: A review. *Food Chemistry*, *124*(2), 411–421.

Elleuch, M., Besbes, S., Roiseux, O., Blecker, C., & Attia, H. (2007). Quality characteristics of sesame seeds and by-products. *Food Chemistry*, *103*(2), 641–650.

Erjavec, J., Kos, J., Ravnikar, M., Dreo, T., & Sabotič, J. (2012). Proteins of higher fungi – From forest to application. *Trends in Biotechnology*, *30*(5), 259–273.

Eroğlu, C., Seçme, M., Atmaca, P., Kaygusuz, O., Gezer, K., Bağcı, G., & Dodurga, Y. (2016). Extract of Calvatia gigantea inhibits proliferation of A549 human lung cancer cells. *Cytotechnology*, *68*(5), 2075–2081.

Falandysz, J., & Borovička, J. (2012). Macro and trace mineral constituents and radionuclides in mushrooms: Health benefits and risks. *Applied Microbiology and Biotechnology*, *97*(2), 477–501. 10.1007/s00253-012-4552-8

Feeney, M. J., Dwyer, J., Hasler-Lewis, C. M., Milner, J. A., Noakes, M., Rowe, S., Wach, M., Beelman, R. B., Caldwell, J., Cantorna, M. T., et al. (2014). Mushrooms and health summit proceedings. *Journal of Nutrition*, *144*, 1128S–1136S.

Filho, A. M. M., Pirozi, M. R., Borges, J. T. D. S., Pinheiro Sant'Ana, H. M., & Chaves, J. B. P., Coimbra, J. S. D. R. (2015). Quinoa: Nutritional, functional, and antinutritional aspects. *Critical Reviews in Food Science and Nutrition*, *57*(8), 1618–1630.

Finimundy, T. C., Gambato, G., Fontana, R., Camassola, M., Salvador, M., Moura, S., Hess, J., Henriques, J. A. P., Dillon, A. J. P., & Roesch-Ely, M. (2013). Aqueous extracts of Lentinula edodes and Pleurotus sajor-caju exhibit high antioxidant capability and promising in vitro antitumor activity. *Nutrition Research*, *33*(1), 76–84.

Friedman, M. (2015). Chemistry, nutrition, and health-promoting properties of Hericium erinaceus (Lion's Mane) mushroom fruiting bodies and mycelia and their bioactive compounds. *Journal of Agricultural and Food Chemistry*, *63*(32), 7108–7123.

Friedman, M. (2016). Mushroom polysaccharides: Chemistry and antiobesity, antidiabetes, anticancer, and antibiotic properties in cells, rodents, and humans. *Foods*, *5*(4), 80.

Fritz, H., Kennedy, D. A., Ishii, M., Fergusson, D., Fernandes, R., Cooley, K., & Seely, D. (2015). Polysaccharide K and Coriolus versicolor extracts for lung cancer: A systematic review. *Integrative Cancer Therapies*, *14*, 201–211.

Ganesan, K., & Xu, B. (2018). Anti-obesity effects of medicinal and edible mushrooms. *Molecules*, *23*, 2880.

Gao, Y., Lan, J., Dai, X., Ye, J., & Zhou, S. H. (2004). A phase I/II study of Ling Zhi mushroom Ganoderma lucidum extract in patients with type II diabetes mellitus. *International Journal of Medicinal Mushrooms*, *6*, 33–39.

Gonzalez, A., Cruz, M., Losoya, C., Nobre, C., Loredo, A., Rodríguez-Jasso, R. M., & Belmares, R. (2020). Edible mushrooms as a novel protein source for functional foods. *Food & Function*, *11*(9), 7400–7414.

Grotto, D., Camargo, I. F., Kodaira, K., Mazzei, L. G., Castro, J., Vieira, R. A. L., Bergamaschi, C. C., & Lopes, L. C. (2019). Effect of mushrooms on obesity in animal models: Study protocol for a systematic review and meta-analysis. *Systematic Reviews*, *8*, 288.

Guillamón, E., García-Lafuente, A., Lozano, M., Rostagno, M. A., Villares, A., & Martínez., J. A. (2010). Edible mushrooms: Role in the prevention of cardiovascular diseases. *Fitoterapia*, *81*(7), 715–723.

Gursoy, N., Sarikurkcu, C., Cengiz, M., & Solak, M. H. (2009). Antioxidant activities, metal contents, total phenolics and flavonoids of seven Morchella species. *Food and Chemical Toxicology*, *47*(9), 2381–2388.

Hakime-Silva, R. A., Vellosa, J. C. R., Khalil, N. M., Khalil, O. A. K., Brunetti, I. L., & Oliveira, O. M. M. F. (2013). Chemical, enzymatic and cellular antioxidant activity studies of Agaricus blazei Murrill. *Anais da Academia Brasileira de Ciencias*, *85*(3), 1073–1082.

Han, N. S., Ahmad, W. A. N. W., & Ishak, W. R. W. (2016). Quality characteristics of Pleurotus sajor-caju powder: Study on nutritional compositions, functional properties and storage stability. *Sains Malaysiana*, *45*(11), 1617–1623.

Harada, T., Miura, N., Adachi, Y., Nakajima, M., Yadomae, T., & Ohno, N. (2002). Effect of SCG, 1,3-beta-D-glucan from Sparassis crispa on the hematopoietic response in cyclophosphamide induced leuko-penic mice. *Biol Pharm Bull*, *25*, 931–939.

Hess, J., Wang, Q., Gould, T., & Slavin, J. (2018). Impact of Agaricus bisporus Mushroom Consumption on Gut Health Markers in Healthy Adults. *Nutrients*, *10*, 1402.

Ho, L.-H., Zulkifli, N. A., & Tan, T.-C. (2020). Edible mushroom: Nutritional properties, potential nu-traceutical values, and its utilisation in food product development. *An Introduction to Mushroom*, 19–38.

Homma, Y., Nakamura, H., Kumagai, Y., Ryuzo, A., Saito, Y., Ishikawa, T., Takada, K., Yamagami, H., Kikuchi, H., & Inadera, H. (1995). Effects of eight week ingestion of mycoprotein on plasma levels of lipids and Apo (Lipo) proteins. *Progress in Medicine*, *15*, 183–195.

Hyun, K. W., Jeong, S. C., Lee, D. H., Park, J. S., & Lee, J. S. (2006). Isolation and characterization of a novel platelet aggregation inhibitory peptide from the medicinal mushroom, Inonotus obliquus. *Peptides*, *27*, 1173–1178.

Ishara, J. R. M., Sila, D. N., Kenji, G. M., & Buzera, A. K. (2018). Nutritional and functional properties of mushroom (Agaricus bisporus & Pleurotus ostreatus) and their blends with maize flour. *American Journal of Food Science and Technology*, *6*(1), 33–41.

Ishikawa, T. (1994). Effect of mycoprotein on serum lipids and apolipoproteins in normolipidemic and hypercholesterolemic subjects. *Atherosclerosis*, *109*, 76.

Kalač, P. (2010). Trace element contents in European species of wild growing edible mushrooms: a review for the period 2000–2009. *Food Chemistry*, *122*(1), 2–15.

Kalač, P. (2013). A review of chemical composition and nutritional value of wild-growing and cultivated mushrooms. *Journal of the Science of Food and Agriculture*, *93*(2), 209–218.

Karun N. C., Sridhar K. R., & Ambarish C. N. (2018). Nutritional potential of Auricularia auricula-judae and Termitomyces umkowaan – The wild edible mushrooms of southwestern India. *Microbial Functional Foods and Nutraceuticals*. Hoboken: Wiley, pp. 281–301.

Kayode, R. M. O., Olakulehin, T. F., Adedeji, B. S., Ahmed, O., Aliyu, T. H., & Badmos, A. H. A. (2015). Evaluation of amino acid and fatty acid profiles of commercially cultivated oyster mushroom (Pleurotus sajor-caju) grown on gmelina wood waste. *Nigerian Food Journal*, *33*(1), 18–21.

Keegan, R.-J. H., Lu, Z., Bogusz, J. M., Williams, J. E., & Holick, M. F. (2013). Photobiology of vitamin D in mushrooms and its bioavailability in humans. *Dermato-Endocrinology*, *5*(1), 165–176.

Kim, J.-Y., Byeon, S.-E., Lee, Y.-G., Lee, J.-Y., Park, J.-S., Hong, E.-K., & Cho, J.-Y. (2008). Immunostimulatory activities of polysaccharides from liquid culture of pine-mushroom Tricholoma matsutake. *Journal of Microbiology and Biotechnology*, *18*(1), 95–103.

Kim, K. M., Ko, J. A., Lee, J. S., Park, H. J., & Hanna, M. A. (2006). Effect of modified atmosphere packaging on the shelf-life of coated, whole and sliced mushrooms. *LWT-Food Science and Technology*, *39*, 364–371.

Kimura, T. (2013). Natural products and biological activity of the pharmacologically active cauliflower mushroom Sparassis crispa. *BioMed Research International*, *2013*, 1–9.

Kivrak, I., Kivrak, S., & Harmandar, M. (2016). Bioactive compounds, chemical composition, and medicinal value of the giant puffball, Calvatia gigantea (higher basidiomycetes), from Turkey. *International Journal of Medicinal Mushrooms*, *18*(2), 97–107.

Kozarski, M., Klaus, A., Vunduk, J., Zizak, Z., Niksic, M., Jakovljevic, D., & Van Griensven, L. J. L. D. (2015). Nutraceutical properties of the methanolic extract of edible mushroom Cantharellus cibarius (Fries): Primary mechanisms. *Food & Function*, *6*(6), 1875–1886. 10.1039/c5fo00312a

Lavelli, V., Cristina Proserpio, F. G., Laureati, M., & Pagliarini, E. (2018). Circular reuse of bio-resources: The Role of Pleurotus spp. in the development of functional foods. *Food & Function*, *9*(3), 1353–1372.

Lee, J., Hong, J.-H., Kim, J.-D., Ahn, B. J., Kim, B. S., Kim, G.-H., & Kim, J.-J. (2013). The antioxidant properties of solid-culture extracts of basidiomycetous fungi. *The Journal of General and Applied Microbiology*, *59*(4), 279–285.

Li, S., Jiang, Z., Xu, W., Xie, Y., Zhao, L., Tang, X., Wang, F., & Xin, F. (2017). FIP-sch2, a new fungal immunomodulatory protein from Stachybotrys chlorohalonata, suppresses proliferation and migration in lung cancer cells. *Applied Microbiology and Biotechnology*, *101*(8), 3227–3235.

Li, J., Zou, L., Chen, W., Zhu, B., Shen, N., Ke, J., Lou, J., Song, R., Zhong, R., & Miao, X. (2014). Dietary mushroom intake may reduce the risk of breast cancer: Evidence from a meta-analysis of observational studies. *PLoS ONE*, *9*, e93437.

Liu, J., Jia, L., Kan, J., & Jin, C. (2013). In vitro and in vivo antioxidant activity of ethanolic extract of white button mushroom (Agaricus bisporus). *Food and Chemical Toxicology, 51,* 310–316.

Mahajna, J., Dotan, N., Zaidman, B.-Z., Petrova, R. D., & Wasser, S. P. (2008). Pharmacological values of medicinal mushrooms for prostate cancer therapy: The case of Ganoderma lucidum. *Nutrition and Cancer*, *61*(1), 16–26.

Makropoulou, M., Aligiannis, N., Gonou-Zagou, Z., Pratsinis, H., Skaltsounis, A.-L., & Fokialakis, N. (2012). Antioxidant and cytotoxic activity of the wild edible mushroom Gomphus clavatus. *Journal of Medicinal Food*, *15*(2), 216–221.

Marçal, S., Sousa, A. S., Taofiq, O., Antunes, F., Morais, A. M. M. B., Freitas, A. C., Barros, L., Ferreira, I. C. F. R., & Pintado, M. (2021). Impact of postharvest preservation methods on nutritional value and bioactive properties of mushrooms. *Trends in Food Science & Technology*, *110*, 418–431. ISSN 0924-2244. 10.1016/j.tifs.2021.02.007

Mattila, P., Könkö, K., Eurola, M., Pihlava, J.-M., Astola, J., Vahteristo, L., Hietaniemi, V., Kumpulainen, J., Valtonen, M., & Piironen, V. (2001). Contents of vitamins, mineral elements, and some phenolic compounds in cultivated mushrooms. *Journal of Agricultural and Food Chemistry*, *49*(5), 2343–2348.

Mau J. L., Miklus M. B., & Beelman R. B. 1993. The shelf life of Agaricus mushrooms. In: C. Charalambous (Ed.), *The Shelf Life of Foods and Beverages*. Amsterdam: Elsevier, pp. 255–288.

Mohebbi, M., Ansarifar, E., Hasanpour, N., & Amiryousefi, M. R. (2011). Suitability of aloe vera and gum tragacanth as edible coatings for extending the shelf life of button mushroom. *Food and Bioprocess Technology*, *5*(8), 3193–3202. 10.1007/s11947-011-0709-1

Moro, C., Palacios, I., Lozano, M., D'Arrigo, M., Guillamón, E., Villares, A., & García-Lafuente, A. (2012). Anti-inflammatory activity of methanolic extracts from edible mushrooms in LPS activated RAW 264.7 macrophages. *Food Chemistry*, *130*(2), 350–355.

Muszyńska, B., Kała, K., Rojowski, J., Grzywacz, A., & Opoka, W. (2017). Composition and biological properties of Agaricus bisporus fruiting bodies – A review. *Polish Journal of Food and Nutrition Sciences*, *67*(3), 173–182. 10.1515/pjfns-2016-0032

Nakamura, H., Ishikawa, T., Akanuma, M., Nishiwaki, M., Yamashita, T., Tomiyasu, K., Yoshida, H., Nishio, E., Hosoai, K., Shiga, H., et al. (1994). Effect of mycoprotein intake on serum lipids of healthy subjects. *Progress in Medicine*, *14*, 1972–1976.

Nussinovitch, A., & Kampf, N. (1993). Shelf life extension and conserved texture of alginate coated mushrooms (Agaricus bisporus). *Journal of Food Technology*, *26*, 469–475.

Nyman, J. (1994). Incorporation of arginine, ornithine and phenylalanine into tropane alkaloids in suspension-cultured cells and aseptic roots of intact plants on Atropa belladonna. *Journal of Experimental Botany*, *45*, 979–986.

Oluwafemi, G. I., Seidu, K. T., & Fagbemi, T. N. (2016). Chemical composition, functional properties and protein fractionation of edible oyster mushroom (Pleurotus ostreatus). *Annals Food Science and Technology*, *17*(1), 218–223.

Passari, A. K., Mishra, V. K., Leo, V. V., Singh, B. P., Meyyappan, G. V., Gupta, V. K., Uthandi, S., & Upadhyay, R. C. (2016). Antimicrobial potential, identification and phylogenetic affiliation of wild mushrooms from two sub-tropical semi-evergreen indian forest ecosystems. *Plos One*, *11*(11), e0166368.

Patel, S., & Goyal, A. (2012). Recent developments in mushrooms as anti-cancer therapeutics: A review. *3 Biotech*, *2*(1), 1–15.

Pereira, P. C. (2014 June). Milk nutritional composition and its role in human health. *Nutrition*, *30*(6), 619–627.

Prabu, M., & Kumuthakalavalli, R. (2016). Antioxidant activity of oyster mushroom (Pleurotus florida [Mont.] Singer) and milky mushroom (Calocybe indica P and C). *International Journal of Current Pharmaceutical Research*, *8*, 48–51.

Puttaraju, N. G., Venkateshaiah, S. U., Dharmesh, S. M., Urs, S. M. N., & Somasundaram, R. (2006). Antioxidant activity of indigenous edible mushrooms. *Journal of Agricultural and Food Chemistry*, *54*, 9764–9772.

Raghavendra, V. B., Venkitasamy, C., Pan, Z., & Nayak, C. (2017). Functional foods from mushroom. In Gupta, V. K., Treichel, H., Shapaval, V. (Olga), de Oliveira, L. A., & Tuohy, M. G., eds, *Microbial Functional Foods and Nutraceuticals*. USA: Wiley, pp. 65–91. 10.1002/9781119048961.ch4

Rathore, H., Prasad, S., & Sharma, S. (2017). Mushroom nutraceuticals for improved nutrition and better human health: A review. *PharmaNutrition*, *5*(2), 35–46.

Réhault-Godbert S., Guyot N., & Nys Y. (2019 March 22). The golden egg: Nutritional value, bioactivities, and emerging benefits for human health. *Nutrients*, *11*(3), 684.

Reinert, J. P., Colunga, K., Etuk, A., Richardson, V., & Dunn, R. L. (2020). Management of overdoses of salvia, kratom, and psilocybin mushrooms: a literature review. Expert Review of Clinical Pharmacology, *13*, 847–856. doi: 10.1080/17512433.2020.1794811

Reis, F. S., Barros, L., Martins, A., & Ferreira, I. C. F. R. (2012). Chemical composition and nutritional value of the most widely appreciated cultivated mushrooms: An inter-species comparative study. *Food and Chemical Toxicology*, *50*(2), 191–197.

Rizzo, G., & Baroni, L. (2018). Soy, soy foods and their role in vegetarian diets. *Nutrients*, *10*(1), 43.

Roberto, P.-S. (2015). Photosynthetic bioenergy utilizing CO2: An approach on flue gases utilization for third generation biofuels. *Journal of Cleaner Production*, *98*, 53–65.

Roncero-Ramos, I., & Delgado-Andrade, C. (2017). The beneficial role of edible mushrooms in human health. *Current Opinion in Food Science*, *14*, 122–128.

Rosa, G. B., Sganzerla, W. G., Ferreira, A. L. A., Xavier, L. O., Veloso, N. C., da Silva, J., Paes de Oliveira, G., Amaral, N. C., de Lima Veeck, A. P., & Ferrareze, J. P. (2020). Investigation of nutritional composition, antioxidant compounds, and antimicrobial activity of wild culinary-medicinal mushrooms boletus edulis and Lactarius deliciosus (Agaricomycetes) from Brazil. *International Journal of Medicinal Mushrooms*, *22*(10), 1243–1259.

Rosli, W. I. W., Maihiza, M. S. N., & Raushan, M. (2015). The ability of oyster mushroom in improving nutritional composition, b-glu- can and textural properties of chicken frankfurter. *International Food Research Journal*, *22*, 311–317.

Ruthes, A. C., Smiderle, F. R., & Iacomini, M. (2015). D-Glucans from edible mushrooms: A review on the extraction, purification and chemical characterisation approaches. *Carbohydrate Polymers*, *117*, 753–761.

Ruxton, C., & McMilan, B. (2010). The impact of mycoprotein on blood cholesterol levels: A pilot study. *British Food Journal*, *112*, 109.

Samsudin, N. I. P., & Abdullah, N. (2019). Edible mushrooms from Malaysia: A literature review on their nutritional and medicinal properties. *International Food Research Journal*, *26*(1), 11–31.

Shah, S. R., Ukaegbu, C. I., Hamid, H. A., & Alara, O. R. (2018). Evaluation of antioxidant and antibacterial activities of the stems of Flammulina velutipes and Hypsizygus tessellatus (white and brown var.) extracted with different solvents. *Journal of Food Measurement and Characterization*, *12*(3), 1947–1961.

Shin, A., Kim, J., Lim, S. Y., Kim, G., Sung, M. K., Lee, E. S., & Ro, J. (2010). Dietary mushroom intake and the risk of breast cancer based on hormone receptor status. *Nutrition and Cancer*, *62*, 476–483.

Singdevsachan, S. K., Patra, J. K., Tayung, K., Sarangi, K., & Thatoi, H. (2014). Evaluation of nutritional and nutraceutical potentials of three wild edible mushrooms from Similipal Biosphere Reserve, Odisha, India. *Journal Für Verbraucherschutz Und Lebensmittelsicherheit*, *9*(2), 111–120. 10.1007/s00003-014-0861-4

Singh, P., Langowski, H.-C., Wani, A. A., & Saengerlaub, S. (2010). Recent advances in extending the shelf life of fresh Agaricus mushrooms: A review. *Journal of the Science of Food and Agriculture*, *90*(9), 1393–1402. 10.1002/jsfa.3971

Stajić, M., Vukojević, J., & Duletić-Laušević, S. (2009). Biology of Pleurotus eryngii and role in biotechnological processes: A review. *Critical Reviews in Biotechnology*, *29*, 55–66.

Syed, Q. A., Akram, M., & Shukat, R. (2019). Nutritional and therapeutic importance of the pumpkin seeds. *Seed*, *21*(2), 15798–15803.

Tabuchi, A., Fukushima-Sakuno, E., Osaki-Oka, K., Futamura, Y., Motoyama, T., Osada, H., et al. (2020). Productivity and bioactivity of enokipodins A–D of Flammulina rossica and Flammulina velutipes. *Bioscience, Biotechnology, and Biochemistry*, *84*, 876–886.

Thakur, R. R., Shahi, N. C., Mangaraj, S., Lohani, U. C., & Chand, K. (2021). Development of an organic coating powder and optimization of process parameters for shelf life enhancement of button mushrooms (Agaricus bisporus). *Journal of Food Processing and Preservation*, *45*(3), e15306.

Thetsrimuang, C., Khammuang, S., Chiablaem, K., Srisomsap, C., & Sarnthima, R. (2011). Antioxidant properties and cytotoxicity of crude polysaccharides from Lentinus polychrous Lév. *Food Chemistry*, *128*(3), 634–639.

Tsai, S.-Y., Tsai, H.-L., & Mau, J.-L. (2008). Non-volatile taste components of Agaricus blazei, Agrocybe cylindracea and Boletus edulis. *Food Chemistry*, *107*(3), 977–983.

Turnbull, W. H., Leeds, A. R., & Edwards, G. D. (1990). Effect of mycoprotein on blood lipids. *The American Journal of Clinical Nutrition*, *52*, 646–650.

Turnbull, W. H., & Ward, T. (1995). Mycoprotein reduces glycemia and insulinemia when taken with an oral-glucose-tolerance test. *The American Journal of Clinical Nutrition*, *61*, 135–140.

Valverde, M. E., Hernández-Pérez, T., & Paredes-López, O. (2015).Edible mushrooms: improving human health and promoting quality life. *International Journal of Microbiology*, *2015*, 1–14.

Vamanu, E., & Voica, A. (2017). Total phenolic analysis, antimicrobial and antioxidant activity of some mushroom tinctures from medicinal and edible species, by in vitro and in vivo tests. *Scientific Bulletin. Series F. Biotechnologies*, *21*, 318–324.

Venditti, A., Frezza, C., Sciubba, F., Serafini, M., & Bianco, A. (2016). Primary and secondary metabolites of an European edible mushroom and its nutraceutical value: Suillus bellinii (Inzenga) Kuntze. *Natural Product Research*, *31*(16), 1910–1919. 10.1080/14786419.2016.1267731

Voet, D., & Voet, J. G. (2004). *Biochemistry*. Hoboken, New Jersey: Wiley.

Volman, J. J., Mensink, R. P., van Griensven, L. J., & Plat, J. (2010). Effects of aglucans from Agaricus bisporus on ex vivo cytokine production by LPS and PHA-stimulated PBMCs; a placebo-controlled study in slightly hypercholesterolemic subjects. *European Journal of Clinical Nutrition*, *64*(7), 720–726.

Wang, Y., Bao, L., Yang, X., Dai, H., Guo, H., Yao, X., et al. (2012). Four new cuparene-type sesquiterpenes from Flammulina velutipes. *Helv Chim Acta*, *95*, 261–267.

Wang, H., & Ng, T. B. (2010). Ganodermin, an antifungal protein from fruiting bodies of the medicinal mushroom Ganoderma lucidum. *Peptides*, *27*, 27–30.

Wang, X.-M., Zhang, J., Wu, L.-H., Zhao, Y.-L., Li, T., Li, J.-Q., Wang, Y.-Z., & Liu, H.-G. (2014). A mini-review of chemical composition and nutritional value of edible wild-grown mushroom from China. *Food Chemistry*, *151*, 279–285.

Wani, B. A., Bodha, R. H., & Wani, A. H. (2010). Nutritional and medicinal importance of mushrooms. *Journal of Medicinal Plants Research*, *4*, 2598–2604.

Wasser, S. P. (2002). Medicinal mushrooms as a source of antitumor and immunomodulating poly-saccharides. *Applied Microbiology and Biotechnology*, *60*(3), 258–274.

Wisitrassameewong, K., Karunarathna, S. C., Thongklang, N., Zhao, R., Callac, P., Moukha, S., Ferandon, C., Chukeatirote, E., & Hyde, K. D. (2012). Agaricus subrufescens: A review. *Saudi Journal of Biological Sciences*, *19*(2), 131–146.

Wong, S., Wong, K., Chiu, L., & Cheung, P. (2007). Non-starch polysaccharides from different develop-mental stages of Pleurotus tuber-regium inhibited the growth of human acute promyelocytic leukemia HL-60 cells by cell-cycle arrest and/or apoptotic induction. *Carbohydrate Polymers*, *68*(2), 206–217.

Yamamoto, K., Kimura, T., Sugitachi, A., & Matsuura, N. (2009). Anti-angiogenic and anti-metastatic effects of beta1,3-D-glucan purified from Hanabiratake, Sparassis crispa. *Biol Pharm Bull*, *32*, 256–263.

Yeh, J.-Y., Hsieh, L.-H., Wu, K.-T., & Tsai, C.-F. (2011). Antioxidant properties and antioxidant compounds of various extracts from the edible basidiomycete Grifola frondosa (Maitake). *Molecules*, *16*(4), 3197–3211.

Yeh, M.-Y., Ko, W.-C., & Lin, L.-Y. (2014). Hypolipidemic and antioxidant activity of enoki mushrooms (Flammulina velutipes). *BioMed Research International*, *2014*, 1–6.

Yuan, Z., He, P., & Takeuchi, H. (1998). Ameliorating effects of water-soluble polysaccharides from woody ear (Auricularia auricula-judge Quel.) in genetically diabetic KK-Ay mice. *Journal of Nutritional Science and Vitaminology*, *44*(6), 829–840.

Yuan, B., Ma, N., Zhao, L., Zhao, E., Gao, Z., Wang, W., Song, M., Zhang, G., Hu, Q., & Xiao, H. (2017). In vitro and in vivo inhibitory effects of a Pleurotus eryngii protein on colon cancer cells. *Food & Function*, *8*(10), 3553–3562.

Yuan, B., Zhao, L., Yang, W., McClements, D. J., & Hu, Q. (2017). Enrichment of bread with nutraceutical-rich mushrooms: Impact of Auricularia auricula (Mushroom) flour upon quality attributes of wheat dough and bread. *Journal of Food Science*, *82*(9), 2041–2050. 10.1111/1750-3841.13812.

Zhang, L., Fan, C., Liu, S., Zang, Z., Jiao, L., & Zhang, L. (2011). Chemical composition and antitumor activity of polysaccharide from Inonotus obliquus. *Journal of Medicinal Plants Research*, *5*(7), 1251–1260.

CHAPTER 7

Health benefits of edible wild mushrooms

Melinda Fogarasi[1], Anca-Corina Fărcas[2], Sonia-Ancuta Socaci[2], Maria-Ioana Socaciu[1], and Cristina Anamaria Semeniuc[1]

[1]Department of Food Engineering, University of Agricultural Sciences and Veterinary Medicine of Cluj-Napoca, Cluj-Napoca, Romania

[2]Department of Food Science, University of Agricultural Sciences and Veterinary Medicine of Cluj-Napoca, Cluj-Napoca, Romania

CONTENTS

7.1 INTRODUCTION

According to the scientific community, there are more and more results that prove and reveal the health benefits of several food categories that influence the development and evolution of the market in the domain of agricultural and biotechnological industries. Moreover, the attention of the consumers has been shifted from the food products enriched with different beneficial bioactive compounds to wild ones grown and produced in uncontaminated, wild areas that can offer a high-quality, natural food (Martins & Ferreira, 2017; Melinda et al., 2020).

There are large numbers of wild, edible food categories including annual and perennial herbs, forbs, ferns, as well as mushrooms, algae, lichens, vines, sedges, rushes, grasses, broad-leaved and needle-like or scale-like leaved shrubs, and trees (Mocan et al., 2018). Among the previously mentioned, wild mushrooms represent an important category considering their medicinal and health-promoting properties. Probably, for this reason, many refer to mushrooms as "flesh of the Gods" or

DOI: 10.1201/9781003152583-9

"the elixir of life". In modern days, a lot of studies highlighted the potential implications of mushrooms in the human diet, considering their nutritional and chemical properties (Valko et al., 2007; Rathore et al., 2017; Reis et al., 2017; Roncero-Ramos & Delgado-Andrade, 2017). It was found that wild edible mushrooms also provide an important amount of fiber and proteins together with other valuable components like essential amino acids, but in contrast to other food products, they have a low-fat content and do not contain cholesterol (Vamanu & Nita, 2013; Bains & Tripathi, 2017; Nagy et al., 2017a; Fogarasi et al., 2018). Due to their flavor and texture, wild edible mushrooms are recognized as a delicacy, especially in mountain areas where they are widely collected considering that many studies pointed out their important nutritional value and the fact that amino acids found in mushrooms are comparable with those of animal origin (Barros et al., 2007).

Based on the literature data, the bioactive compounds identified in wild edible mushrooms, like compounds, terpenoids, unsaturated fatty acids, and carotenoids, make them one of the best natural raw materials for the development of different functional foods. Also, their exceptional chemical characteristics can be valorized in the fabrication of nutraceuticals or pharmaceutical products, exploring the synergies of the large group of bioactive compounds (Barros et al., 2007; Barros et al., 2008b; Reis et al., 2012; Taofiq et al., 2015; Mocan et al., 2018).

The medical interest for the biologically active compounds found in wild edible mushrooms was underlined by many studies. For instance, Ramos R. et al. presents in a few studies the applications of fractions or isolated compounds, extracted from an edible mushroom, in the prevention and treatment of major health issues such as cancer, obesity and hyperlipidemia, hypercholesterolemia, diabetes, hypertension, and neurodegenerative disease (Roncero-Ramos & Delgado-Andrade, 2017). Also, Ramos R. et al. defined the mechanism that involves the stimulating effect of polysaccharides on the immune system inhibiting tumor growth.

It was found that regular consumption of wild edible mushrooms can boost the production of secretory immunoglobulin A, leading to the enhancement of the immune system (Roncero-Ramos & Delgado-Andrade, 2017). The full potential of wild edible mushrooms is not unlocked yet in terms of dietary and medicinal value, considering the numerous biologically active and health-promoting compounds. Figure 7.1 gives a summary of the range of beneficial properties of wild edible mushrooms, such as antioxidative, antibacterial, antiviral, anticancer, and anti-inflammatory properties, strengthening the immune system as well as the ability to improve the functioning of the cardiovascular system (Kalač, 2009; Bains & Tripathi, 2016; Bains & Tripathi, 2017; Muszynska et al., 2018).

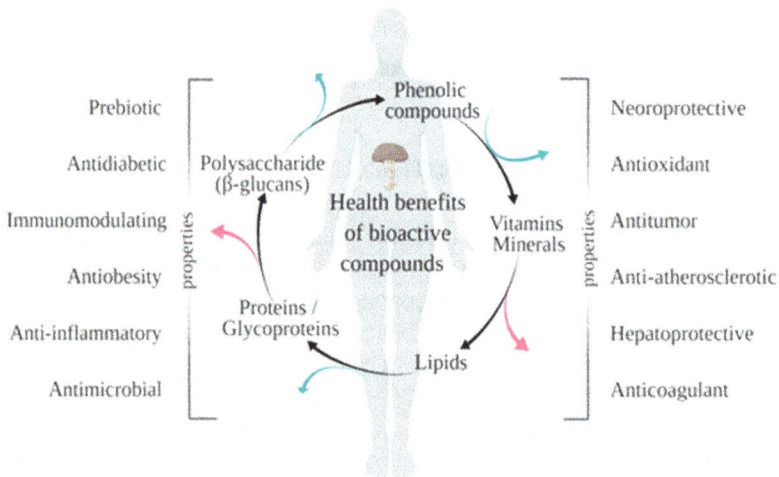

Figure 7.1 Properties of wild edible mushrooms.

This is the reason why wild edible mushrooms are becoming more and more important in the definition of a balanced diet for humans all over the world, achieving exploitation of the health benefits and functioning mechanisms of mushrooms which give good results in the prevention of major diseases, such as cancer, heart, and nervous problems (Ma et al., 2018; Martinez-Medina et al., 2021). The nutraceutical properties assigned to edible mushrooms stand out in Parkinson's or Alzheimer's treatment, inhibition of the tumors' development; reducing inflammatory processes; and through anti-allergic, antiviral, antiparasitic, and hepatoprotective effects (Valverde et al., 2015). Among all these nutritional and biological benefits of edible mushrooms, there is one undesirable aspect of edible mushrooms represented by their ability to accumulate heavy metals and radioactive substances. Regarding the occurrence and toxicity of chemical elements, arsenic, mercury, nickel, and cadmium have been detected in wild edible mushrooms. These elements are frequently accumulated from the natural environment (soil and water) (Zou et al., 2019). Based on the previous discussions, we aim at giving an overview of the nutritional and therapeutic benefits of mushrooms, including the bioactive properties of wild edible mushrooms which could be conducive to better understanding of the correlations between the mushroom consumption and health improvement, and evaluate the advantages of the insertion of mushrooms into the human diet.

7.2 MOLECULES OBTAINED FROM MUSHROOMS WITH BIOACTIVE FUNCTION

Considering the global trends in biotechnology and the features of molecules obtained from mushrooms with bioactive function, many researchers focused on the preparation methods and structure characterizations of these compounds such as polysaccharides, proteins, terpenes, phenolic compounds, and unsaturated fatty acids, to provide the necessary data for the industrial-scale applications in diverse biotechnological processes, even for the development of new intelligent drugs (Wang et al., 2014; Valverde et al., 2015). Bioactive compounds isolated from wild edible mushrooms are gaining interest due to their benefits for human health, considering that they offer protection against chronic-degenerative diseases, cardiovascular and cancer diseases, diabetes mellitus, and neurodegenerative diseases. However, the content of bioactive compounds in wild edible mushrooms depends on a series of factors including the strain, cultivation, developmental stage, age of the fresh mushrooms, storage conditions, and the extraction method (Pérez Montes et al., 2021). In the case of shiitake mushrooms, based on their physicochemical assessment, Lakhanpal et al. found that in addition to proteins, lipids, carbohydrates, fibers, and minerals, they also contain vitamins (B1, B2, C, E, D). Their enzymatic activity helps the infancy leukemia treatment, and minerals as zinc or selenium restore the skin, improve testosterone levels, and have an antioxidant effect (Lakhanpal & Rana, 2007). Results also revealed that the inclusion of *L. edodes* in nutrition can promote blood circulation, preserves health, and prevents cardiovascular diseases. Similarly, another type of mushroom, *A. bisporus,* proved to be a rich source of B-complex vitamins and minerals and act as a hypocholesterolemic agent (Lakhanpal & Rana, 2007). Because of the previous discussions related to the association of different mushrooms with various medical and nutritional applications, it is important to note that there is a keen interest in recent years all over the world, among researchers, to unveil the bioactive compositions and nutritional values of different mushrooms with the ultimate goal to help the elaboration of potential functional foods with the high content of dietary fiber and low level of fat (Ma et al., 2018).

7.2.1 Proteins

It is well known that protein represents an important class of macronutrients for the healthy development and maintenance of the human body, considering its key physiological roles, such as vital performance of hormones and enzyme action. Wild edible mushrooms used as a protein-rich food may be an attractive alternative to conventional protein sources considering their abundance and the fact that there is a growing interest to replace animal proteins with proteins from plant sources as an alternative (Sá et al., 2020; Gonzalez et al., 2021). Recently, proteins of fungal origin have gained the attention of food industry players and the scientific community, due to their high nutritional values associated with the rich level of essential amino acids when compared to vegetables (Bach et al., 2017). An interesting discovery was made by Barros et al. (2008b) when comparing wild edible mushrooms and commercial ones, revealing that wild edible mushrooms possess higher protein and lower fat content than commercial mushrooms (Barros et al., 2008b). According to the literature, edible mushrooms have a protein content on a dry basis between 19% and 39%, being part of a complex fungal cells network (Sánchez, 2017). Among these large numbers of proteins and peptides with remarkable biological performances, wild edible mushrooms contain lectins, fungal immunomodulatory proteins, ribosome-inactivating proteins, antimicrobial proteins, ribonucleases, and laccases (Xu et al., 2011).

Proteins and peptides are significant bioactive nutraceuticals in mushrooms with multiple health benefits, such as the enhancement of the digestion and absorption of exogenous nutritional ingredients, the modulation of the immune function to help the host defending the invasion of pathogens, and the activities' inhibition of some enzymes (Valverde et al., 2015).

7.2.2 Lectins

Lectins found in mushrooms are glycated proteins that can be coupled with high specificity, in a reversible form to carbohydrates (Martinez-Medina et al., 2021). Besides mushrooms, these glycated proteins are present in various species, including plants, algae, other invertebrates, or fungi, but it is important to note that they do not act as an immunological source nor represent catalytic activity (Mishra et al., 2019).

Results indicate that lectins are mainly present in the fruiting bodies of mushrooms and only in very rare cases in mycelia and stem (Singh et al., 2014). Their biological function in mushrooms is associated with the recognition of key structures relating to fungal-host interactions, pathogenic fungal and mycorrhizal processes, and the viral and insect protection system (R"udiger & Gabius, 2001). Mushrooms constitute at least 82% of the lectins which have a molecular weight that varies between 12 and 68 kDa (Varrot et al., 2013). Commonly, lectins are weakly bonded to carbohydrates. Although multiple and identical monomers could be bound to the same carbohydrate to strengthen the bond of lectin-carbohydrate, which could be combined with different types of carbohydrates, such as glucose, lactose, raffinose, turanose, N-acetyl glucosamine or inline (Singh et al., 2014).

7.2.3 Carbohydrates/polysaccharides

Mushroom carbohydrates are polyhydroxylated aldehydes or ketones and their derivatives which, depending on their degrees of polymerization, can be divided into mono-, di-, oligo-, or poly-saccharides. They are constituted as one of the main components in higher fungi and typically in the range from 35% to 80%. Most of the carbohydrates in mushrooms are including monosaccharides such as glucose, fructose, maltose, rhamnose, arabinose, sucrose, mannitol, threalose, and cell wall polysaccharides such as chitin, β-glucans, and mannans (Cheung, 2013; Wang et al., 2014). However, one important fraction in mushrooms is glucans; these are lineal or lineal polysaccharides integrated by glucose with linkage β(1–3), β(1–6), or α(1–3) present as part of fruiting bodies (Martinez-Medina et al., 2021). Nevertheless, the most prominent effectors are the β-glucans, with β (1–3) linkage in its main structure and β (1–6) branching linkage (Wasser, 2002). Carbohydrates mostly occur in the form of polysaccharides in nature. A variety of polysaccharides from different natural sources, such as mushrooms plants, and bacterial extracellular polymeric substances are gaining recognition as supplements to increase the health benefits (Maity et al., 2021).

Lakhanpal et. al. associated certain polysaccharides with immune-modulating and anti-cancer properties, triterpene compounds with the decrease of LDL, and blood pressure; certain peptide fractions to antitumor and anti-angiogenic properties (Lakhanpal & Rana, 2007).

Specific polysaccharides from *Agrocybe aegerita*, *Armillriella mellea*, edible *Auricularia* spp., *Boletus edulis*, *Calvatia gigantea*, *Dendropolyporus umblettus*, *Flammulina veluptipes*, *Hericium erinaceous*, and *Phellinus linteus* are considered active constituents with medicinal value for a wide range of ailments (Lakhanpal & Rana, 2007).

Mushroom polysaccharides have been proved to have an extensive range of antitumor activities. Biological active polysaccharides from diverse edible and medicinal mushrooms such as *Lentinus edodes*, *Grifola frondosa* (Grifolan), *Schizophyllum commune* (Schizo-phyllan), and *Pleurotus ostreus* (Pleuran) have been extensively studied, considering their immune-regulatory activity and anticancer potential (Meng et al., 2016).

Yuan et al. (2020) report the immune stumolator effect in polysaccharides extracted from *Pleurotus fertus* an edible and medicinal mushroom, in their wild and cultivated forms, both polysaccharides possess different compositions and the wild form exhibit a higher immune stimulant effect, promoting the maturation and activation of dendritic cells and macrophagues (Yuan et al., 2020). In addition, Yuan et al. (2020) demonstrated the potential use of mushroom polysaccharides in immune-suppressed people to enhance their response against estranger antigens. Li et al. (2020) performed an in vivo and in vitro study with polysaccharides derived from *Ganoderma lucidum* spores that shows that their consumption could reduce the adverse effects in traditional cancer therapies (Li et al., 2020).

Polysaccharides perform their action against tumor cells mostly through triggering the immune response of the host organism for which they are considered as biological response modifiers (BRMs) (Heleno et al., 2015; Hsin et al., 2016). BRMs has been defined as those agents that modify the host's biological response by stimulation of the immune system, which may result in various therapeutic effects.

Other functionality attributed to polysaccharides consists of their antioxidant potential, which also could indirectly contribute to their anticancer potential (Mingyi et al., 2019). Zhang et al. (2020) report that the selenium quantities present in *Ganoderma* sp. polysaccharide fraction contribute to their scavenging potential (Zhang et al., 2020). Nevertheless, the antioxidant properties also could be related to the fact that selenium acts as a co-factor in enzymes related to control oxidative stress (Li et al., 2019). Finally, their antioxidant properties also could be present in proteoglycan fractions and attributed to the presence of certain amino acids (Mingyi et al., 2019). Crude fiber is a group of indigestible carbohydrates that can improve the function of the alimentary tract and also lower blood glucose and cholesterol levels (Wang et al., 2014).

7.2.4 Lipids/polyunsaturated fatty acids

Fatty acids (FAs) are considered to play a major role in the function of the immune system and the balancing of hormonal levels. The lipid profile of mushrooms reveals highly variable fatty acid profiles, palmitic (16:0), oleic (9-cis 18:1), and linoleic (9-cis,12-cis 18:2) acids are the main FAs found in members of the *Basidiomycetes* (Borthakur & Joshi, 2019). Nutritionally, linoleic and α-linolenic (9-cis,12-cis,15-cis 18:3) acids are essential for basal metabolism in humans, while long-chain polyunsaturated FAs have many beneficial effects on human health, especially contributing to the reduction of serum cholesterol (Karine et al., 2006). Mushrooms have lipid-based compounds at a percentage between 1.18% and 8.39% on a dry weight basis (Martinez-Medina et al., 2021). Mushrooms from the genus *Boletus* are an excellent source of both fatty acids and neutral and polar lipids. The concentrations of oleic acid are higher in *Boletus edulis*, *Boletus piperatus*, *Boletus subglabripes*, *Boletus erythropus*, *Boletus subtomentosus*, and *Boletus variipes* than in other mushrooms, as reported by Hanus et al. (2008). The major sterol produced by mushrooms was detected as ergosterol, which displayed significant antioxidant properties. Moreover, researches have demonstrated the critical role of a sterols-rich diet in the prevention of cardiovascular diseases (Guillamon et al., 2010; Kalac, 2012). Another polyunsaturated fatty acid was detected as tocopherols, which were considered as novel and effective natural antioxidants because their free radical scavenging peroxyl components produced from different reactions. These antioxidants have high biological activity for protection against degenerative malfunctions, microbial, and cardiovascular diseases (Heleno et al., 2015). From the human nutritional point, long-chain polyunsaturated linoleic acid and α-linolenic acid are important for basal metabolism in humans, and mushrooms are known to contain all these nutritional values (Barros et al., 2008b).

7.2.5 Other compounds (vitamins/minerals)

Besides the other molecules previously mentioned, mushrooms have been reported as a source of other compounds with nutritional properties as vitamins and minerals. Mushrooms contain several primary vitamins such as thiamine, riboflavin, pyridoxine, pantothenic acid, nicotinic acid, folic acid, cobalamin, ergosterol, biotin, phytochinone, and tocopherols. Nevertheless, inside their composition, a set of molecules with biological potential are distributed in their structures. Among the bioactive mushrooms' constituents are sterols, with the predominance of ergosterol, the precursor of vitamin D. In mushrooms, ergosterol is converted to vitamin D2 (ergocalciferol) when exposed to UV radiation. Vitamin D2 from fungi and mushrooms serves as the only available dietary source of vitamin D for those who eat no animal products. The crucial role of vitamin D for bone health is well established, while during the last decade its role in immune system modulation and cancer prevention has been recognized (Kalogeropoulos et al., 2013). Another important class of compounds is polyphenols, which normally are developed by mushrooms due to their protective effect against UV and consumers like animals or insects (Martinez-Medina et al., 2021). The phenolic compounds present in macrofungi are classified as phenolic acids, flavonoids, hydroxybenzoic acids, hydroxycinnamic acids, lignan, tannin, stilbene, and oxidized polyphenols (Nowacka-Jechalke et al., 2018). These compounds intervene in mushroom organoleptic properties like flavor or color and also possess the capacity to interact with some macromolecules conferring interesting biological properties (Valdez-Morales et al., 2016).

7.3 FUNCTIONAL AND BIOLOGICAL PROPERTIES OF MUSHROOMS

Mushrooms contain several chemical compounds of nutraceutical importance, such as terpenes, bioactive proteins, and antioxidants, which make them a therapeutically stronger foodstuff in the battle against various degenerative diseases (Rathore et al., 2017).

Mushrooms have a wide variety of compounds operating in their natural environment, but they can be used to ensure or promote human health in the form of nutraceuticals, additives, functional foods, and others. Thus, the creation of a research-oriented field of study for the scientific and novel use of edible or medicinal mushrooms, the exploitation, and promotion of their full use is necessary. Mushrooms have been shown to have antioxidant, antibacterial, antifungal anti-inflammatory, anti-tumor, and anti-viral properties. Some of the major properties of the mushroom have been described in the following (Table 7.1).

7.3.1 Prebiotic properties

The human intestine represents one of the most complex microbiologic environments in the digestive system, where to co-exist a large number of microorganisms, which according to many studies, could include more than 1,000 different species (Roberfroid et al., 2010). Considering its

Table 7.1 Health benefits of different wild edible mushrooms

	Species	Health benefits	References
1	*Agaricus bisporus*	Hepatoprotective, immuno-stimulatory, and antitumor activities;anti-aging activity, protect hepatic and nephric by improving serum enzyme activities, biochemical levels, lipid contents, and antioxidant status	(Huang et al., 2016; Li et al., 2018; Ma et al., 2018)
2	*Agaricus campestris*	Enhance the secretion of insulin, treatment of ulcers	(Popa et al., 2014)
3	*Cantharellus cibarius*	Excellent source of polysaccharides like chitin and chitosan, reduce inflammation, and lower the risk of developing certain cancers	(Barros et al., 2008a)
4	*Hypsizygus marmoreus*	Antifungal and anti-proliferative activities, inhibit the growth of several fungal	(Lam & Ng, 2001)
5	*Lactarius piperatus*	Reduce atherosclerosis, diabetes, cancer, and cirrhosis	(Barros et al., 2007)
6	*Lepista nuda*	Antioxidant and antimicrobial properties, immunologic effects	(Lin et al., 2011)
7	*Leucopaxillus giganteus*	Provide an inhibitory effect in angiotensin I-converting enzyme (ACE), providing hypotension of blood pressure in spontaneously hypertensive rats (SHR)	(Vieira et al., 2016)
8	*Lycoperdon molle*	Antioxidant, antimicrobial, anticancer, antiproliferative, DNA-protective, antiallergic, analgesic, antitumor, immunosuppressive, antiatherogenic, hypoglycemic, anti-inflammatory, hepatoprotective activities	(Bal et al., 2019)
9	*Lyophyllum shimeiji*	Antitumor activity suppresses the proliferation of hepatoma cells and breast cancer	(Zhang et al., 2010)
10	*Macrolepiota procera*	Antioxidant properties, antitumor activity	(Popa et al., 2014; Žurga et al., 2017)
11	*Ramaria botrytis*	Antioxidant, antitumor, and antimicrobial properties	(Kumar Sharma & Gautam, 2017)
12	*Sparassis latifolia*	Antibacterial and antifungal activities against *Escherichia coli,* resistant strains of *Staphylococcus aureus, Pseudomonas aeruginosa*, and *Candida* and *Fusarium* species	(Ma et al., 2018)
13	*Tricholoma acerbum*	Antioxidant and immunomodulating effects, reduce the risk of coronary heart disease	(Zhou & Hu, 2010)
14	*Tuber indicum* (truffle)	Antitumor activity, inhibits the proliferation of hepatoma and human breast cancer cell lines	(Ma et al., 2018)
15	*Lentinula edodes* (shiitake)	Immunoregulatory activity and anticancer potential; lung protection activity, regulate the antioxidant and inflammation status, antitumor activity	(Thangthaeng et al., 2015)
16	*Volvariella volvacea*	Reducing free radicals, strengthening bones, prevent anemia	(Lakhanpal & Rana, 2007)
17	*Pleurotus ostreatus*	Cardiovascular, hypertensive, hypercholesterolemia antioxidant and antimicrobial activities, antidiabetic activity	(Wasser, 2002; Zhang et al., 2010; Vamanu & Nita, 2013)
18	*Pleurotus eryngii*	Hypolipidaemic and hypoglycaemic activities, increase the level of high-density lipoprotein cholesterol and liver glycogen	(Chen et al., 2016)
19	*Pleurotus cornucopiae*	Antiviral and antitumor activities	(Wu et al., 2014)

(Continued)

Table 7.1 (Continued)   Health benefits of different wild edible mushrooms

	Species	Health benefits	References
20	*Aspergillus panamensis*	Immuno-modulatory activity, against trinitrobenzene sulphonic acid-induced ulcerative colitis	(Singh et al., 2017)
21	*Auricularia auricular-judae*	Anticoagulant effects, anti-cholesterol, and cardioprotective effects	(Muszynska et al., 2018)
22	*Coprinus comatus*	Antiviral and antitumor activities, suppress proliferation of tumor cell lines, and inhibit human immunodeficiency virus reverse transcriptase	(Zhao et al., 2014)
23	*Boletus edulis*	Antiviral, antiinflammatory, antimicrobial, antioxidant	(Popa et al., 2014)
24	*Calocybe indica*	Antioxidant and anti-aging activities, neuroprotective activity	(Yuan et al., 2019)
25	*Flammulina velutipes*	Regulation of the immune system, cancer immunotherapy, antioxidantimmuno-modulating activity, stimulate the proliferation of mouse spleen lymphocytes and B lymphocytes	(Popa et al., 2014; Feng et al., 2016)
26	*Morchella fertus*	Antioxidative and anti-inflammatory bioactivities, in addition to immunostimulatory and antitumor properties	(Muszynska et al., 2018)
27	*Morchella elata*	Antioxidants, essential minerals, and vitamins-required for blood cell production (hematopoiesis), and neurotransmission, work as co-factors for enzymes during cellular substrate metabolism inside the human body.	(Beluhan & Ranogajec, 2011)
28	*Lyophyllum decastes*	Antitumor activity, sugar-lowering effects	(Grangeia et al., 2011)
29	*Armillariella mellea*	Meniere's syndrome, vertigo, epilepsy, neurasthenia and hypertension, antioxidant, antimicrobial properties	(Popa et al., 2014)
30	*Schizophyllum commune*	Immune-regulator activity and anticancer potential	(Meng et al., 2016)
31	*Bovista plumbea*	Head affections, diabetes, ovarian cysts, acne	(Popa et al., 2014)
32	*Fistulina hepatica*	Antibacterial, antioxidant potential	(Ribeiro et al., 2006)
33	*Hericium coralloides*	Antibacterial and nematicidal activities, antiinflammatory properties, anti- tumours, nerve regenerator in muscular dystrophy, Parkinson's disease, Alzheimer's, and dementia	(Popa et al., 2014)
34	*Lepista nebularis*	Stops leukemia T cells from proliferating, antimicrobial activity	(Kim et al., 2008)
35	*Laetiporus sulphureus*	Hemolytic and hemagglutination activities, antimicrobial and antioxidant activities	(Popa et al., 2014)
36	*Stropharia* sp.	Edible, but undesirable due to mildly spicy taste, medicinal, antitumor, neuromodulatory effects	(Popa et al., 2014)
37	*Ganoderma lucidum*	Antidiabetic activity, improve insulin sensitivity by regulating inflammatory cytokines and gut microbiota composition, anti-inflammatory activity	(Nagai et al., 2017; Xu et al., 2017)
38	*Ganoderma applanatum*	Antitumor activity, cytotoxic and pro-apoptotic activities against HT-29 colon adenocarcinoma cells	(Kumar Sharma & Gautam, 2017)
39	*Ganoderma atrum*	Antitumor activity induces growth inhibition and cell death in breast cancer cells	(Li et al., 2017)
40	*Inonotus baumii*	Antitumor activity	(Sun et al., 2014)

complex microbiologic features, the human intestine has an important effect on human health that can be influenced by a series of factors and compounds like prebiotics (Yamashiro, 2017). It is well known that prebiotics are substances of varied nature that have the potential to selectively promote the development and function of a select group of bacterial species that exist in the human intestine and which beneficially influence the metabolic process (Gibson et al., 2017). The prebiotic compounds can stimulate some important philological processes such as immune response in the intestine, leading to interaction with certain receptors, or with microorganisms that generate molecules like butyrate or propionate that could influence the expression of cytokines related to diverse proinflammatory procedures. Furthermore, they proved to enhance the absorption of some nutritional components while their long-term consumption could be involved in lipid and carbohydrates metabolism (Khangwal & Shukla, 2019). Recently, many studies revealed that prebiotics obtained from different mushroom species such as *Ganoderma lucidum* and *Poria cocos* and also from co-products derived from *Lentinula edodes*, *P. eryngii,* and *Flammulina velutipes* could positively influence the processing of polysaccharides (Chou et al., 2013; Khan et al., 2018).

7.3.2 Antioxidant properties

Antioxidants are compounds or systems that can safely interact with free radicals and terminate the chain reaction before vital molecules are damaged. Antioxidants (e.g. flavonoids, phenolic acids, tannins, vitamin C, vitamin E) have diverse biological properties, such as anti-inflammatory, anti-carcinogenic, and anti-atherosclerotic effects that reduce the incidence of coronary diseases and contribute to the maintenance of gut health by the modulation of the gut microbial balance (Oroian & Escriche, 2015). Free radicals induce oxidative stress, which is balanced by the body's endogenous antioxidant system with input from cofactors and by the ingestion of exogenous antioxidants. When the generation of free radicals exceeds the protective effect of antioxidants and some cofactors, it can cause oxidative damage, which can result in aging and other diseases such as cardiovascular, cancer, and neurodegenerative disorders (Valko et al., 2007). Recently, many review papers regarding antioxidants from mushrooms and different extraction and quantification procedures have been published (Barros et al., 2009; Keleş et al., 2011; Boonsong et al., 2016; Islam et al., 2016; Kaewnarin et al., 2016).

Their radical scavenging activity has been extensively studied and documented; species like *Pleurotus* spp., *Agaricus* spp., *G. lucidum, B. edulis,* and *L. edodes* are known for their profound antioxidant activities (Da Silva & Jorge, 2011; Gonzalez-Palma et al., 2016; Ramos et al., 2019). The range of antioxidant activity of these edible mushrooms with their DPPH radical-scavenging and chelating activities ranged from 13.63% to 69.67% and 60.25% to 82.7%, respectively (Martinez-Medina et al., 2021). Variations in concentrations of these substances have been influenced by the type of strains, materials, method of cultivation, stage of growth and development, age and freshness of the mushrooms, storage conditions, and method of extraction; in particular, the type of solvent used for extraction (Fogarasi et al., 2018). Generally, mushrooms contain ergothioneine, a naturally occurring and powerful antioxidant that protects the body's cells from generated free radicals as well as boosts immunity (Ming et al., 2015; Kalaras et al., 2017). Ergosterol obtained from fruiting bodies of edible mushrooms, such as *A. bisporus* and *Imleria badia* have been reported to have anti-cancer and anti-inflammatory properties (Baur et al., 2019).

Overall, the mushroom phenolic compounds, mainly including the phenolic acids, flavonoids, hydroxybenzoic acids, hydroxycinnamic acids, lignans, tannins, stilbenes, and oxidized polyphenols, were considered as aromatic hydroxylated compounds with one or more aromatic rings and one or more hydroxyl groups (D'Archivio et al., 2010). Elhadi M. Yahia et al. analyzed the phenolic compounds of seventeen species of wild mushrooms and detected their antioxidant activities by FRAP and DPPH assays, suggesting an effective nutritional and health value of different mushroom species (Yahia et al., 2017). Qiuhui Hu et al. investigated the neuro-protection of six

⟩

components from *F. velutipes*, which were mainly phenolic compounds, on H_2O_2-induced oxidative damage in PC12cells, and demonstrated that most of the components displayed neuroprotective effects, along with their antioxidant activities (Hu et al., 2017). Darija Cor et al. evaluated the anti-acetylcholinesterase activity of *Ganoderma lucidum* extracts, which were mainly composed of phenolic compounds (Cör et al., 2017).

7.3.3 Antimicrobial properties

The term *antimicrobial* comprises a wide variety of pharmaceutical agents that include antibacterial, antifungal, antiviral, and anti-parasitic medicines (Reis et al., 2017). The use of antibiotics is the single most crucial factor leading to increased resistance of pathogenic microorganisms around the world (Borges et al., 2013; Román et al., 2020). Another major factor in the growth of antibiotic resistance is the spread of the resistant strains of bacteria from person to person, or from the non-human sources in the environment, including food (Ashbolt et al., 2013). Natural resources have been taken advantage of over the years, and among them, wild edible mushrooms have vast diversity of active compounds with nutritional and antimicrobials properties (Ren et al., 2014; Smolskaitė et al., 2015; Liu et al., 2017). Mushrooms have long been playing an essential role in several aspects of human activity, like food and medicinal properties (Alves et al., 2012; Bach et al., 2017). Current researches have been focused on searching for new antimicrobial therapeutically potential compounds of edible mushrooms, recognizing that some of these molecules have health beneficial effects, including antimicrobial properties (Borges et al., 2013). Recently, Kosanić et al. state that acetone extract of *Craterellus cornucopioides* has a strong minimum inhibitory concentration (MIC) against gram-positive (*Staphylococcus aureus*, *Bacillus cereus*, and *Bacillus subtilis*) and gram-negative (*Escherichia coli* and *Proteus mirabilis*) bacteria with a range of 0.1–0.2 mg/mL. Interestingly, the effect of feeding C57BL/6 mice *Agaricus bisporous* (white button mushroom) in mice to evaluate the bacterial microflora, urinary metabolome, and resistance to a gastrointestinal (GI) pathogens along with control untreated mushrooms (Kosanić et al., 2019). Chaiharn et al. reported that different types of extracts such as ethyl-acetate, methanol, and ethanol and aqueous solvent of *Flammulina velutipes*, *Ganoderma lucidum*, *Pleurotus ostreatus*, and *Pleurotus pulmonarius* showed significant antibacterial activity against gram-positive and gram-negative bacteria pathogens (Chaiharn et al., 2018).

7.4 APPLICATION OF MUSHROOMS IN FOOD AND HUMAN HEALTH

When we think of a balanced diet and the healthiest means by which to achieve it, plants and plant products immediately come to mind as well as mushrooms. Although their use has been reported for thousands of years, it has only been in recent years that the consumption of mushrooms has increased, mainly due to the increasing awareness that a stable and balanced diet exerts a key role in normal body functioning and sustaining health (Reis et al., 2017). Based on the recent socioeconomic trends, the substitution of edible mushrooms as an essential source of functional ingredients in food products could become a natural adjuvant for the prevention and alleviation of several lifestyle-related diseases. This information could be beneficial for the development of food products with health functionalities, which are of great interest to the medical nutrition industry, which is an industry that emerged from the convergence between the food and pharma industries (Ho et al., 2020). Mushrooms and their extracts have long been used in folk medicine and food due to their low calorific value and pleasant taste and are also reported to have beneficial biological activities; thus, applications in nutraceutical and pharmaceutical products (Mingyi et al., 2019).

Mushroom-based functional foods have gained much attention, particularly in the last few decades; hence, it gives a clear scientific impression towards the increasing demand for mushroom

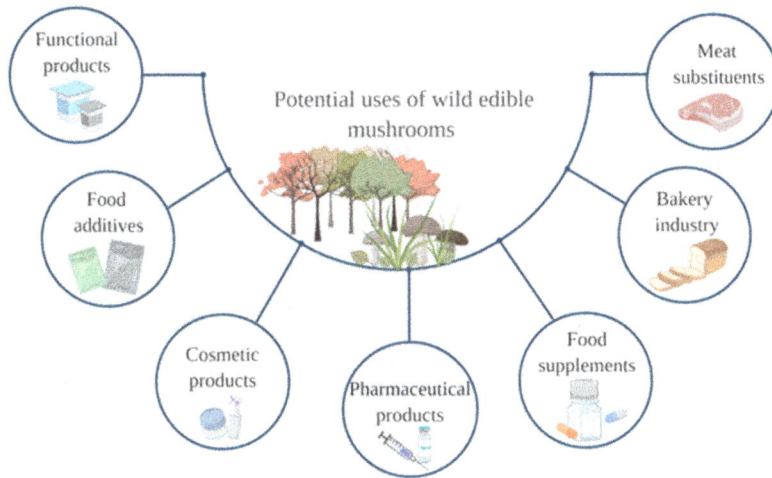

Figure 7.2 Potential uses of wild edible mushrooms.

fruit bodies and their mycelia (Rathore et al., 2019). In recent years, mushroom ingredients were incorporated in various food products to obtain fortified functional foods (Figure 7.2).

Despite the ancient knowledge of mushroom benefits, their use as an ingredient in the elaboration of processed foods is quite recent, mainly as meat, fat, phosphates, salt, and nitrite replacers (Table 7.2). Since countless research was focusing on the investigation of novel and efficient products based on the bioactivities of mushrooms, deeper research is still needed for the demonstration of the correlation between the functional activities and mechanisms, as well as their safety evaluation and safe range of intake.

Table 7.2 Application of mushrooms in functional foods

	Mushroom species	Product	References
1	Shiitake powder	Frankfurter	(Pil-Nam et al., 2015)
2	Mix of *Suillus luteus* and *Coprinopsis atramentaria*	Cottage cheese	(Ribeiro et al., 2015)
3	Mix of *Lentinula edodes, Pleurotus eryngii,* and *Flammulina velutipes*	Yogurt	(Chou et al., 2013)
4	*Suillus luteus*	Cottage cheese	(Ribeiro et al., 2015)
5	*Schizophyllum commune*	Cheese	(Okamura-Matsui et al., 2001)
4	*Pleurotus sajor-caju* stems	Chicken nuggets	(Wan Rosli & Mohsin, 2011)
5	*Tremella fuciformis*	Pork patties	(Cha et al., 2014)
6	*Pleurotus eryngii*	Pork sausages	(Wang et al., 2019a)
7	*Lentinula edodes*	Emulsion-type sausage	(Wang et al., 2019b)
		Beef burgers	(Qing et al., 2021)
		Pork patties	(Chun et al., 2020)
8	*Boletus edulis*	Frankfurters	(Pérez Montes et al., 2021)
	Cantharellus cibarius	Frankfurters	(Pérez Montes et al., 2021)
9	*Agaricus bisporus*	Meat emulsion	(Kurt & Gençcelep, 2018)
		Beef patties	(Cerón-Guevara et al., 2019)
		Smoke sausages	(Nagy et al., 2017b)

(Continued)

Table 7.2 (Continued)   Application of mushrooms in functional foods

	Mushroom species	Product	References
10	*Flamulina velutipes*	Emulsion-type sausage	(Choe et al., 2018)
		Ham	(Jo et al., 2020)
		Chicken sausage	(Jo et al., 2018)
11	*Volvariella volvacea*	Cantonese sausages	(Wang et al., 2018)
		Beef patties	(Qing et al., 2021)
12	*Pleurotus ostreatus*	Frankfurters	(Cerón-Guevara et al., 2019)
13	*Pleurotus sapidus*	Chicken patties	(Wan-Mohtar et al., 2020)
14	*Agaricus bisporus*	Snacks	(Keerthana et al., 2020)
		White bread	(Ma et al., 2018)
		Sponge cake	(Salehi et al., 2016)
15	*Lentinula edodes* (shiitake)	Corn extruded snacks	(Harada-Padermo et al., 2021)
16	*Pleurotus tubberegium*	Cookies	(Kolawole et al., 2020)
17	*Pleurotus sajor-caju*	Biscuit	(Ng et al., 2017)
18	*Pleurotus ostreatus*	Noodles	(Parvin et al., 2020)
19	*Boletus edulis*	Bread	(Alina Vlaic et al., 2019)
20	*Cordyceps militaris*	Extruded product	(Zhong et al., 2017)

7.5 CONCLUSION

Due to the diverse habitats and varied ecological zones across the world, varieties of mushrooms emerge as the next generation's nutraceutical food. Mushrooms are considered to be a complete health food and suitable for all age groups due to their rich protein, mineral, dietary fiber, and vitamin content and because they are a source of biologically active compounds of medicinal importance. Such bioactive molecules are polysaccharides, terpenoids, low molecular weight proteins, glycoproteins, and antioxidants, etc., which have a great role to play in boosting immune strength, lowering risks of cancers, inhibition of tumoral growth, blood sugar maintenance, and much more.

Funding: This work was supported by three grants from the Ministry of Research and Innovation, CNCS-UEFISCDI, project number PN-III-P1-1.1- PD-2019-0475, project number PN-III-P 2-2.1-PED-2019-3622, and 1 project number PN-III-P4-ID-PCE 2020-1847 within PNCDI III.

REFERENCES

Alina Vlaic, R., et al. (2019). Boletus edulis mushroom flour-based wheat bread as innovative fortified bakery product. *Bulletin UASVM Food Science and Technology*, *76*(1), 52–62.

Alves, M. J., et al. (2012). A review on antimicrobial activity of mushroom (Basidiomycetes) extracts and isolated compounds. *Planta Med*, *78*(16), 1707–1718.

Ashbolt, N. J., et al. (2013). Human health risk assessment (HHRA) for environmental development and transfer of antibiotic resistance. *Environ Health Perspect*, *121*(9), 993–1001.

Bach, F., et al. (2017). Edible mushrooms: A potential source of essential amino acids, glucans and minerals. *International Journal of Food Science & Technology*, *52*(11), 2382–2392.

Bains, A., & Tripathi, A. (2016). Antimicrobial and antioxidant activity of aqueous extract of six mushrooms collected from Himachal Pradesh. *IJBPAS*, *5*(7), 1717–1728.

Bains, A., & Tripathi, A. (2017). Evaluation of antioxidant and anti-inflammatory properties of aqueous extract of wild mushrooms collected from Himachal Pradesh. *Asian Journal of Pharmaceutical and Clinical Research*, *10*(3), 467.

Bal, C., AkgÜL, H., & SevİNdİK, M. (2019). The Antioxidant potential of ethanolic extract of edible mushroom Lycoperdon molle Pers. (Agaricomycetes). *Eurasian Journal of Forest Science*, *7*(3), 277–283.

Barros, L., et al. (2007). Fatty acid and sugar compositions, and nutritional value of five wild edible mushrooms from Northeast Portugal. *Food Chemistry*, *105*(1), 140–145.

Barros, L., et al. (2008a). Chemical composition and biological properties of Portuguese wild mushrooms: A comprehensive study. *J. Agric. Food Chem*, *56*(10), 3856–3862.

Barros, L., et al. (2008b). Wild and commercial mushrooms as source of nutrients and nutraceuticals. *Food Chem Toxicol*, *46*(8), 2742–2747.

Barros, L., et al. (2009). Phenolic acids determination by HPLC-DAD-ESI/MS in sixteen different Portuguese wild mushrooms species. *Food Chem Toxicol*, *47*(6), 1076–1079.

Barros, L., Baptista, P., & Ferreira, I. C. (2007). Effect of Lactarius piperatus fruiting body maturity stage on antioxidant activity measured by several biochemical assays. *Food Chem Toxicol*, *45*(9), 1731–1737.

Baur, A. C., et al. (2019). Intake of ergosterol increases the vitamin D concentrations in serum and liver of mice. *J Steroid Biochem Mol Biol*, *194*, 105435.

Beluhan, S., & Ranogajec, A. (2011). Chemical composition and non-volatile components of Croatian wild edible mushrooms. *Food Chemistry*, *124*(3), 1076–1082.

Boonsong, S., Klaypradit, W., & Wilaipun, P. (2016). Antioxidant activities of extracts from five edible mushrooms using different extractants. *Agriculture and Natural Resources*, *50*(2), 89–97.

Borges, A., et al. (2013). Antibacterial activity and mode of action of ferulic and gallic acids against pathogenic bacteria. *Microb Drug Resist*, *19*(4), 256–265.

Borthakur, M., & Joshi, S. R. (2019). Chapter 1 – Wild mushrooms as functional foods: The significance of inherent perilous metabolites. In V. K. Gupta & A. Pandey (Eds.), *New and Future Developments in Microbial Biotechnology and Bioengineering* (pp. 1–12). Amsterdam: Elsevier.

Cerón-Guevara, M. I., et al. (2019). Effect of the addition of edible mushroom flours (Agaricus bisporus and Pleurotus ostreatus) on physicochemical and sensory properties of cold-stored beef patties. *Journal of Food Processing and Preservation*, *44*(3), e14351.

Cha, M.-H., et al. (2014). Quality and sensory characterization of white jelly mushroom (T remella fuciformis) as a meat substitute in pork patty formulation. *Journal of Food Processing and Preservation*, *38*(4), 2014–2019.

Chaiharn, M., et al. (2018). Antibacterial, antioxidant properties and bioactive compounds of Thai Cultivated mushroom extracts against food-borne bacterial strains. *Chiang Mai Journal of Science*, *45*(4), 1713–1727.

Chen, L., et al. (2016). Hypolipidaemic and hypoglycaemic activities of polysaccharide from Pleurotus eryngii in Kunming mice. *Int J Biol Macromol*, *93*(Pt A), 1206–1209.

Cheung, P. C. K. (2013). Mini-review on edible mushrooms as source of dietary fiber: Preparation and health benefits. *Food Science and Human Wellness*, *2*(3–4), 162–166.

Choe, J., et al. (2018). Application of winter mushroom powder as an alternative to phosphates in emulsion-type sausages. *Meat Sci*, *143*, 114–118.

Chou, W. T., Sheih, I. C., & Fang, T. J. (2013). The applications of polysaccharides from various mushroom wastes as prebiotics in different systems. *J Food Sci*, *78*(7), M1041–M1048.

Chun, S., Chambers, E. T., & Chambers, D. H. (2020). Effects of shiitake (Lentinus edodes P.) mushroom powder and sodium tripolyphosphate on texture and flavor of pork patties. *Foods*, *9*(5), 611.

Cör, D., et al. (2017). The effects of different solvents on bioactive metabolites and "in vitro" antioxidant and anti-acetylcholinesterase activity of Ganoderma lucidum fruiting body and primordia extracts. *Macedonian Journal of Chemistry and Chemical Engineering*, *36*(1), 129–141.

D'Archivio, M., et al. (2010). Bioavailability of the polyphenols: Status and controversies. *Int J Mol Sci*, *11*(4), 1321–1342.

Da Silva, A. C., & Jorge, N. (2011). Antioxidant properties of lentinus edodes and Agaricus blazei extracts. *Journal of Food Quality*, *34*(6), 386–394.

Feng, T., et al. (2016). Structural characterization and immunological activities of a novel water-soluble polysaccharide from the fruiting bodies of culinary-medicinal winter mushroom, Flammulina velutipes (Agaricomycetes). *Int J Med Mushrooms*, *18*(9), 807–819.

Fogarasi, M., et al. (2018). Bioactive compounds and volatile profiles of five Transylvanian wild edible mushrooms. *Molecules*, *23*(12).

Gibson, G. R., et al. (2017). Expert consensus document: The International Scientific Association for Probiotics and Prebiotics (ISAPP) consensus statement on the definition and scope of prebiotics. *Nat Rev Gastroenterol Hepatol, 14*(8), 491–502.

Gonzalez, A., et al. (2021). Evaluation of functional and nutritional potential of a protein concentrate from Pleurotus ostreatus mushroom. *Food Chem, 346*, 128884.

Gonzalez-Palma, I., et al. (2016). Evaluation of the antioxidant activity of aqueous and methanol extracts of Pleurotus ostreatus in different growth stages. *Front Microbiol, 7*, 1099.

Grangeia, C., et al. (2011). Effects of trophism on nutritional and nutraceutical potential of wild edible mushrooms. *Food Research International, 44*(4), 1029–1035.

Guillamon, E., et al. (2010). Edible mushrooms: Role in the prevention of cardiovascular diseases. *Fitoterapia, 81*(7), 715–723.

Hanuš, L. O., Shkrob, I., & Dembitsky, V. (2008). Lipids and fatty acids of wild edible mushrooms of the genus boletus. *J. Food Lipids, 15*, 370–383.

Harada-Padermo, S.d.S., et al. (2021). Umami Ingredient, a newly developed flavor enhancer from shiitake byproducts, in low-sodium products: A study case of application in corn extruded snacks. *LWT - Food Science and Technology, 138*, 110806.

Heleno, S. A., et al. (2015). Nutritional value, bioactive compounds, antimicrobial activity and bioaccessibility studies with wild edible mushrooms. *LWT - Food Science and Technology, 63*(2), 799–806.

Ho, Lee-Hoon, Zulkifli, N. A., & Tan, T.-C. (2020). *Edible Mushroom: Nutritional Properties, Potential Nutraceutical Values, and Its Utilisation in Food Product Development.* IntechOpen.

Hsin, I. L., et al. (2016). Immunomodulatory proteins FIP-gts and chloroquine induce caspase-independent cell death via autophagy for resensitizing cisplatin-resistant urothelial cancer cells. *Phytomedicine, 23*(13), 1566–1573.

Hu, Q., et al. (2017). Neuroprotective effects of six components from Flammulina velutipes on H2O2-induced oxidative damage in PC12 cells. *Journal of Functional Foods, 37*, 586–593.

Huang, J., et al. (2016). Hepatoprotective effects of polysaccharide isolated from Agaricus bisporus industrial wastewater against CCl(4)-induced hepatic injury in mice. *Int J Biol Macromol, 82*, 678–686.

Islam, T., Yu, X., & Xu, B. (2016). Phenolic profiles, antioxidant capacities and metal chelating ability of edible mushrooms commonly consumed in China. *LWT - Food Science and Technology, 72*, 423–431.

Jo, K., et al. (2020). Utility of winter mushroom treated by atmospheric non-thermal plasma as an alternative for synthetic nitrite and phosphate in ground ham. *Meat Sci, 166*, 108151.

Jo, K., Lee, J., & Jung, S. (2018). Quality characteristics of low-salt chicken sausage supplemented with a winter mushroom powder. *Korean J Food Sci Anim Resour, 38*(4), 768–779.

Kaewnarin, K., et al. (2016). Phenolic profile of various wild edible mushroom extracts from Thailand and their antioxidant properties, anti-tyrosinase and hyperglycaemic inhibitory activities. *Journal of Functional Foods, 27*, 352–364.

Kalač, P. (2009). Chemical composition and nutritional value of European species of wild growing mushrooms: A review. *Food Chemistry, 113*(1), 9–16.

Kalac, P. (2012). *Chemical composition and nutritional value of european species of wild growing mushrooms.* Nova Science Publishers.

Kalaras, M. D., et al. (2017). Mushrooms: A rich source of the antioxidants ergothioneine and glutathione. *Food Chemistry, 233*, 429–433.

Kalogeropoulos, N., et al. (2013). Bioactive microconstituents and antioxidant properties of wild edible mushrooms from the island of Lesvos, Greece. *Food Chem Toxicol, 55*, 378–385.

Karine, P., et al. (2006). Fatty acid composition of lipids from mushrooms belonging to the family Boletaceae. *Mycol Res, 110*(Pt 10), 1179–1183.

Keerthana, K., et al. (2020). Development of fiber-enriched 3D printed snacks from alternative foods: A study on button mushroom. *Journal of Food Engineering, 287*, 110116.

Keleş, A., Koca, İ., & Gençcelep, H. (2011). Antioxidant properties of wild edible mushrooms. *J Food Process Technol, 2*(6), 1–6.

Khan, I., et al. (2018). Mushroom polysaccharides from Ganoderma lucidum and Poria cocos reveal prebiotic functions. *Journal of Functional Foods, 41*, 191–201.

Khangwal, I., & Shukla, P. (2019). Potential prebiotics and their transmission mechanisms: Recent approaches. *J Food Drug Anal, 27*(3), 649–656.

Kim, M. Y., et al. (2008). Phenolic compound concentration and antioxidant activities of edible and medicinal mushrooms from Korea. *J. Agric. Food Chem, 56*, 7265–7270.

Kolawole, F. L., Akinwande, B. A., & Ade-Omowaye, B. I. O. (2020). Physicochemical properties of novel cookies produced from orange-fleshed sweet potato cookies enriched with sclerotium of edible mushroom (Pleurotus tuberregium). *Journal of the Saudi Society of Agricultural Sciences, 19*(2), 174–178.

Kosanić, M., et al. (2019). Craterellus cornucopioides Edible Mushroom as Source of Biologically Active Compounds. *Natural Product Communications, 14*(5), 1934578X1984361.

Kumar Sharma, S., & Gautam, N. (2017). Chemical and bioactive profiling, and biological activities of coral fungi from Northwestern Himalayas. *Sci Rep, 7*, 46570.

Kurt, A., & Gençcelep, H. (2018). Enrichment of meat emulsion with mushroom (Agaricus bisporus) powder: Impact on rheological and structural characteristics. *Journal of Food Engineering, 237*, 128–136.

Lakhanpal, T. N., & Rana, M. (2007). Medicinal and nutraceutical genetic resources of mushrooms. *Plant Genetic Resources, 3*(2), 288–303.

Lam, S. K., & Ng, T. B. (2001). Hypsin, a novel thermostable ribosome-inactivating protein with antifungal and antiproliferative activities from fruiting bodies of the edible mushroom Hypsizigus marmoreus. *Biochem Biophys Res Commun, 285*(4), 1071–1075.

Li, S., et al. (2017). FIP-sch2, a new fungal immunomodulatory protein from Stachybotrys chlorohalonata, suppresses proliferation and migration in lung cancer cells. *Appl Microbiol Biotechnol, 101*(8), 3227–3235.

Li, S., et al. (2018). Antioxidant and anti-aging effects of acidic-extractable polysaccharides by Agaricus bisporus. *Int J Biol Macromol, 106*, 1297–1306.

Li, J., et al. (2019). A combination of selenium and polysaccharides: Promising therapeutic potential. *Carbohydr Polym, 206*, 163–173.

Li, D., et al. (2020). Polysaccharide from spore of Ganoderma lucidum ameliorates paclitaxel-induced intestinal barrier injury: Apoptosis inhibition by reversing microtubule polymerization. *Biomed Pharmacother, 130*, 110539.

Lin, C. C., et al. (2011). A novel adjuvant Ling Zhi-8 enhances the efficacy of DNA cancer vaccine by activating dendritic cells. *Cancer Immunol Immunother, 60*(7), 1019–1027.

Liu, K., et al. (2017). Polyphenolic composition and antioxidant, antiproliferative, and antimicrobial activities of mushroom Inonotus sanghuang. *LWT - Food Science and Technology, 82*, 154–161.

Ma, G., et al. (2018). A critical review on the health promoting effects of mushrooms nutraceuticals. *Food Science and Human Wellness, 7*(2), 125–133.

Maity, P., et al. (2021). Biologically active polysaccharide from edible mushrooms: A review. *Int J Biol Macromol, 172*, 408–417.

Martinez-Medina, G. A., et al. (2021). Bio-funcional components in mushrooms, a health opportunity: Ergothioneine and huitlacohe as recent trends. *Journal of Functional Foods, 77*, 104326.

Martins, N., & Ferreira, I. C. F. R. (2017). Mountain food products: A broad spectrum of market potential to be exploited. *Trends in Food Science & Technology, 67*, 12–18.

Melinda F., et al. (2020). Elemental Composition, Antioxidant and Antibacterial Properties of SomeWild Edible Mushrooms from Romania. *Agronomy, 10*(1972), 1972.

Meng, X., Liang, H., & Luo, L. (2016). Antitumor polysaccharides from mushrooms: A review on the structural characteristics, antitumor mechanisms and immunomodulating activities. *Carbohydr Res, 424*, 30–41.

Ming Cai, et al. (2015). Extraction, antimicrobial, and antioxidant activities of crude polysaccharides from the wood ear medicinale mushroom Auricularia auricula-judae (HigherBasidiomycetes). *International Journal of Medicinal Mushrooms, 17*(6), 591–600.

Mingyi, Y., et al. (2019). Trends of utilizing mushroom polysaccharides (MPs) as potent nutraceutical components in food and medicine: A comprehensive review. *Trends in Food Science & Technology, 92*, 94–110.

Mishra, A., et al. (2019). Structure-function and application of plant lectins in disease biology and immunity. *Food Chem Toxicol, 134*, 110827.

Mocan, A., et al. (2018). Chemical composition and bioactive properties of the wild mushroom Polyporus squamosus (Huds.) Fr: A study with samples from Romania. *Food Funct, 9*(1), 160–170.

Muszynska, B., et al. (2018). Anti-inflammatory properties of edible mushrooms: A review. *Food Chem, 243*, 373–381.

Nagai, K., et al. (2017). Polysaccharides derived from Ganoderma lucidum fungus mycelia ameliorate indomethacin-induced small intestinal injury via induction of GM-CSF from macrophages. *Cell Immunol*, *320*, 20–28.

Nagy, M., et al. (2017a). Chemical Composition and Bioactive Compounds of Some Wild Edible Mushrooms. *Bulletin of University of Agricultural Sciences and Veterinary Medicine Cluj-Napoca. Food Science and Technology*, *74*(1), 1.

Nagy, M., et al. (2017b). Utilization of brewer's spent grain and mushrooms in fortification of smoked sausages. *Food Science and Technology*, *37*(2), 315–320.

Ng, S. H., et al. (2017). Incorporation of dietary fibre-rich oyster mushroom (Pleurotus sajor-caju) powder improves postprandial glycaemic response by interfering with starch granule structure and starch digestibility of biscuit. *Food Chem*, *227*, 358–368.

Nowacka-Jechalke, N., Olech, M., & Nowak, R. (2018). Chapter 11 – Mushroom polyphenols as chemopreventive agents. In R. R. Watson, V. R. Preedy, & S. Zibadi (Eds.), *Polyphenols: Prevention and Treatment of Human Disease* (Second Edition), pp. 137–150. Academic Press.

Okamura-Matsui, T., Takemura, K., Sera, M., Takeno, T., Noda, H., Fukuda, S., & Ohsugi, M. (2001). Characteristics of a cheese-like food produced by fermentation of the mushroom Schizophyllum commune. *Journal of Bioscience and Bioengineering*, *92*, 30–31.

Oroian, M., & Escriche, I. (2015). *Antioxidants: Characterization, Natural Sources, Extraction and Analysis.* Food Research International.

Parvin, R., et al. (2020). Quality improvement of noodles with mushroom fortified and its comparison with local branded noodles. *NFS Journal*, *20*, 37–42.

Pérez Montes, A., et al. (2021). Edible mushrooms as a novel trend in the development of healthier meat products. *Current Opinion in Food Science. 37*, 118–124.

Pil-Nam, S., et al. (2015). The impact of addition of shiitake on quality characteristics of frankfurter during refrigerated storage. *LWT - Food Science and Technology*, *62*(1), 62–68.

Popa, G., et al. (2014). Studies concerning the in vitro cultivation of some indigenous macromycete species. *Scientific Bulletin. Series F. Biotechnologies*, *XVIII*, 54–59.

Qing, Z., et al. (2021). The effects of four edible mushrooms (Volvariella volvacea, Hypsizygus marmoreus, Pleurotus ostreatus and Agaricus bisporus) on physicochemical properties of beef paste. *LWT - Food Science and Technology*, *135*, 110063.

Ramos, M., et al. (2019). Agaricus bisporus and its by-products as a source of valuable extracts and bioactive compounds. *Food Chem*, *292*, 176–187.

Rathore, H., et al. (2019). Medicinal importance of mushroom mycelium: Mechanisms and applications. *Journal of Functional Foods*, *56*, 182–193.

Rathore, H., Prasad, S., & Sharma, S. (2017). Mushroom nutraceuticals for improved nutrition and better human health: A review. *PharmaNutrition*, *5*(2), 35–46.

Reis, F. S., et al. (2012). Chemical composition and nutritional value of the most widely appreciated cultivated mushrooms: An inter-species comparative study. *Food Chem Toxicol*, *50*(2), 191–197.

Reis, F. S., et al. (2017). Functional foods based on extracts or compounds derived from mushrooms. *Trends in Food Science & Technology*, *66*, 48–62.

Ren, L., et al. (2014). Antibacterial and antioxidant activities of aqueous extracts of eight edible mushrooms. *Bioactive Carbohydrates and Dietary Fibre*, *3*(2), 41–51.

Ribeiro, B., et al. (2006). Contents of carboxylic acids and two phenolics and antioxidant activity of dried Portuguese wild edible mushrooms. *J. Agric. Food Chem*, *54*, 8530–8537.

Ribeiro, A., et al. (2015). Spray-drying microencapsulation of synergistic antioxidant mushroom extracts and their use as functional food ingredients. *Food Chem*, *188*, 612–618.

Roberfroid, M., et al. (2010). Prebiotic effects: Metabolic and health benefits. *British Journal of Nutrition*, *104*, S1–S63.

Román, M. P. G., Mantilla, N. B., Flórez, S. A. C., De Mandal, S., Passari, A. K., Ruiz-Villáfan, B., Rodríguez-Sanoja, R., & Sánchez, S. (2020). *Antimicrobial and Antioxidant Potential of Wild Edible Mushrooms.* IntechOpen, pp.1–18.

Roncero-Ramos, I., & Delgado-Andrade, C. (2017). The beneficial role of edible mushrooms in human health. *Current Opinion in Food Science*, *14*, 122–128.

R¨udiger, H., & Gabius, H.-J. (2001). Plant lectins: Occurrence, biochemistry, functions and applications. *Glycoconjugate Journal, 18*, 589–613.

Sá, A. G. A., Moreno, Y. M. F., & Carciofi, B. A. M. (2020). Plant proteins as high-quality nutritional source for human diet. *Trends in Food Science & Technology, 97*, 170–184.

Salehi, F., et al. (2016). Improvement of quality attributes of sponge cake using infrared dried button mushroom. *J Food Sci Technol, 53*(3), 1418–1423.

Sánchez, C. (2017). Bioactives from mushroom and their application. In M. Puri , Ed., *Food Bioactives: Extraction and Biotechnology Applications* (pp. 23–57). Cham: Springer International Publishing.

Singh, S. S., et al. (2014). Lectins from edible mushrooms. *Molecules, 20*(1), 446–469.

Singh, R. S., et al. (2017). Immunomodulatory and therapeutic potential of a mucin-specific mycelial lectin from Aspergillus panamensis. *Int J Biol Macromol, 96*, 241–248.

Smolskaitė, L., Venskutonis, P. R., & Talou, T. (2015). Comprehensive evaluation of antioxidant and antimicrobial properties of different mushroom species. *LWT – Food Science and Technology, 60*(1), 462–471.

Sun, J., et al. (2014). An extracellular laccase with antiproliferative activity from the sanghuang mushroom Inonotus baumii. *Journal of Molecular Catalysis B: Enzymatic, 99*, 20–25.

Taofiq, O., et al. (2015). The contribution of phenolic acids to the anti-inflammatory activity of mushrooms: Screening in phenolic extracts, individual parent molecules and synthesized glucuronated and methylated derivatives. *Food Res Int, 76*(Pt 3), 821–827.

Thangthaeng, N., et al. (2015). Daily supplementation with mushroom (Agaricus bisporus) improves balance and working memory in aged rats. *Nutr Res, 35*(12), 1079–1084.

Valdez-Morales, M., et al. (2016). Phenolic Compounds, Antioxidant Activity and Lipid Profile of Huitlacoche Mushroom (Ustilago maydis) Produced in Several Maize Genotypes at Different Stages of Development. *Plant Foods Hum Nutr, 71*(4), 436–443.

Valko, M., et al. (2007). Free radicals and antioxidants in normal physiological functions and human disease. *Int J Biochem Cell Biol, 39*(1), 44–84.

Valverde, M. E., Hernandez-Perez, T., & Paredes-Lopez, O. (2015). Edible mushrooms: Improving human health and promoting quality life. *Int J Microbiol, 2015*, 376387.

Vamanu, E., & Nita, S. (2013). Antioxidant capacity and the correlation with major phenolic compounds, anthocyanin, and tocopherol content in various extracts from the wild edible Boletus edulis mushroom. *Biomed Res Int, 2013*, 313905.

Varrot, A., Basheer, S. M., & Imberty, A. (2013). Fungal lectins: Structure, function and potential applications. *Current Opinion in Structural Biology, 23*(5), 678–685.

Vieira, V., et al., (2016). Nutritional and biochemical profiling of Leucopaxillus candidus (Bres.) singer wild mushroom. *Molecules, 21*(1), 99.

Wan Rosli, S. W. I., M. A., & Mohsin, S. S. J. (2011). On the ability of oyster mushroom (Pleurotus sajor-caju) confering changes in proximate composition and sensory evaluation of chicken patty. *International Food Research Journal, 18*(4), 1463–1469.

Wang, X. M., et al. (2014). A mini-review of chemical composition and nutritional value of edible wild-grown mushroom from China. *Food Chem, 151*, 279–285.

Wang, X., et al. (2018). Use of straw mushrooms (Volvariella volvacea) for the enhancement of physico-chemical, nutritional and sensory profiles of Cantonese sausages. *Meat Sci, 146*, 18–25.

Wang, L., et al. (2019a). Production of Pork Sausages Using Pleaurotus eryngii with Different Treatments as Replacements for Pork Back Fat. *J Food Sci, 84*(11), 3091–3098.

Wang, L., et al. (2019b). Roles of Lentinula edodes as the pork lean meat replacer in production of the sausage. *Meat Sci, 156*, 44–51.

Wan-Mohtar, W., et al. (2020). Fruiting-body-base flour from an Oyster mushroom waste in the development of antioxidative chicken patty. *J Food Sci, 85*(10), 3124–3133.

Wasser, S. P. (2002). Medicinal mushrooms as a source of antitumor and immunomodulating polysaccharides. *Appl Microbiol Biotechnol, 60*(3), 258–274.

Wu, X., et al. (2014). A novel laccase with inhibitory activity towards HIV-I reverse transcriptase and antiproliferative effects on tumor cells from the fermentation broth of mushroom Pleurotus cornucopiae. *Biomed Chromatogr, 28*(4), 548–553.

Xu, X., et al. (2011). Bioactive proteins from mushrooms. *Biotechnol Adv, 29*(6), 667–674.

Xu, S., et al. (2017). Ganoderma lucidum polysaccharides improve insulin sensitivity by regulating in-flammatory cytokines and gut microbiota composition in mice. *Journal of Functional Foods*, *38*, 545–552.

Yahia, E. M., Gutierrez-Orozco, F., & Moreno-Perez, M. A. (2017). Identification of phenolic compounds by liquid chromatography-mass spectrometry in seventeen species of wild mushrooms in Central Mexico and determination of their antioxidant activity and bioactive compounds. *Food Chem*, *226*, 14–22.

Yamashiro, Y. (2017). Gut microbiota in health and disease. *Ann Nutr Metab*, *71*(3–4), 242–246.

Yuan, F., et al. (2019). Characterization, antioxidant, anti-aging and organ protective effects of sulfated polysaccharides from Flammulina velutipes. *Molecules*, *24*(19), 104050.

Yuan, P., et al. (2020). Comparison of the structural characteristics and immunostimulatory activities of polysaccharides from wild and cultivated Pleurotus feruleus. *Journal of Functional Foods*, *72*, 104050.

Zhang, R. Y., et al. (2010). A novel ribonuclease with antiproliferative activity from fresh fruiting bodies of the edible mushroom Lyophyllum shimeiji. *Biochem Genet*, *48*(7–8), 658–668.

Zhang, H., et al. (2020). Multivariate relationships among sensory attributes and volatile components in commercial dry porcini mushrooms (Boletus edulis). *Food Res Int*, *133*, 109112.

Zhao, S., et al. (2014). A novel laccase with potent antiproliferative and HIV-1 reverse transcriptase inhibitory activities from mycelia of mushroom Coprinus comatus. *Biomed Res Int*, *2014*, 417461.

Zhong, L., et al. (2017). Evaluation of anti-fatigue property of the extruded product of cereal grains mixed with Cordyceps militaris on mice. *J Int Soc Sports Nutr*, *14*, 15.

Zhou, B., & Hu, X. (2010). Compositional analysis and nutritional studies of Tricholoma matsutake collected from Southwest China. *Journal of Medicinal Plant Research*, *4*(12), 1222–1227.

Zou, H., et al. (2019). Occurrence, toxicity, and speciation analysis of arsenic in edible mushrooms. *Food Chem*, *281*, 269–284.

Žurga, S., et al. (2017). Fungal lectin MpL enables entry of protein drugs into cancer cells and their sub-cellular targeting. *Oncotarget*, *8*, 26896–26910.

CHAPTER **8**

Nature and chemistry of bioactive components of wild edible mushrooms

Predrag Petrović[1] and Jovana Vunduk[2,3]
[1]Innovation Center of the Faculty of Technology and Metallurgy, University of Belgrade, Belgrade, Serbia
[2]Institute of General and Physical Chemistry, Belgrade, Serbia
[3]Ekofungi Ltd., PadinskaSkela bb, Belgrade, Serbia

CONTENTS

8.1 INTRODUCTION

Molds of the genus *Penicillium* did become the most important fungi in medicine and pharmacy during the 20th-century antibiotic revolution, but mushrooms have been used in traditional medicine, particularly in the Far East, for thousands of years. Recent studies confirmed that many mushroom species possess biological potential. Mushrooms are reported to exhibit 126 different medicinal properties (Wasser, 2010), including antitumor, immunomodulating, antimicrobial and antiviral, anti-oxidative, anti-inflammatory, hepatoprotective, anti-neurodegenerative, antihyperglycemic, and antihyperlipidemic, as well as wound healing activity (Lindequist et al., 2005; Vunduk et al., 2015; Petrović et al., 2019a, Petrović, Vunduk et al., 2019). Basidiomycetes, a major group of mushroom-forming fungi, produce a vast number of secondary metabolites (Schüffler & Anke, 2009; Schüffler, 2018), although only a relatively small number of these metabolites have been studied thoroughly. In this chapter, the most important and well-defined pharmacological properties of mushrooms will be

DOI: 10.1201/9781003152583-10

reviewed, as well as compounds produced by mushrooms that have found use in modern medicine or are promising candidates for future treatment of various disorders.

8.2 ANTITUMOR PROPERTIES OF MUSHROOMS

Cancer remains one of the leading causes of death globally, claiming nearly 10 million victims only in 2020 (Sung et al., 2021). The outcome of this disease depends on many factors and the immune response of the organism may be crucial (Maehara et al., 2011). Mushrooms are a source of immunomodulating polysaccharides – β-glucans and products based on β-glucans have been widely studied as adjuvants in cancer treatment, some being clinically proven to be effective in combination with conventional chemotherapeutics. Although β-glucans seem to exhibit various effects on the gut microbiome, lipid and sugar metabolism, etc. (Murphy et al., 2020), their immunomodulatory activity, which is associated with their antitumor activity is the most studied and regarded as the most important. Mushrooms are also a source of secondary metabolites with cytotoxic effects, some of which are promising new compounds for the next generation of antineoplastic drugs.

8.2.1 Immunomodulating activity of mushrooms – β-glucans

Glucans – glucose polymers – are the most abundant polysaccharides in nature and the most abundant organic compounds as well. According to the type of glycosidic bond, they are divided into two large groups, α-D- and β-D-glucans (Ruiz-Herrera & Ortiz-Castellanos, 2019), although mixed α and β-D-glucans are also known (Synytsya & Novák, 2013). They can be further categorized based on the position of the carbon atoms involved in the glycosidic bond formation (Synytsya & Novák, 2013; Ruiz-Herrera & Ortiz-Castellanos, 2019). They can be simple, linear, or branched, and may form complexes with proteins or other polysaccharides like chitin. α-D-glucans, which include starch and glycogen, are mostly energy-reserve polysaccharides, while β-D-glucans, such as cellulose, are structural polysaccharides (Ruiz-Herrera & Ortiz-Castellanos, 2019).

Fungal β-D-glucans, together with chitin and mannans, are the main structural polysaccharides of fungal cell walls (Gow et al., 2017). Although made entirely of glucose units, β-glucans show a great structural variety. The presence of branching, branching degree, the position of glycosidic bonds in the backbone, and the position of branching, size, quaternary structure, and solubility in water are some of the glucan characteristics.

They can be linear but branched β-glucans are the most abundant type, the majority of described being $(1\rightarrow3)(1\rightarrow6)$-β-D-glucans, meaning they have a backbone consisting of glucose units connected via β-$(1\rightarrow3)$ glycosidic bonds, with branches attached via β-$(1\rightarrow6)$ linkage; branches may consist of a single glucose unit or may represent small glucose chains (Synytsya & Novák, 2013). Lentinan, isolated from shiitake (*Lentinula edodes*) is probably the best known $(1\rightarrow3)(1\rightarrow6)$-β-D-glucan; it consists of a β-$(1\rightarrow3)$ backbone with a single glucose unit attached via β-$(1\rightarrow6)$ linkage. Other well-known and named $(1\rightarrow3)(1\rightarrow6)$-β-D-glucans include schyzophyllan from *Schyzophyllum commune*, grifolan from *Grifola frondosa*, scleroglucan from *Athelia rolfsii* (syn. *Sclerotium rolfsii*) (Viñarta et al., 2007), pleuran from *Pleurotus ostreatus* (Selvamani et al., 2018), but many more were characterized from numerous mushroom species such are *Ganoderma lucidum*, *Hericium erinaceus* (Friedman, 2016), *Flammulina velutipes* (Smiderle et al., 2006), *Pleurotus sajor-caju* (Carbonero et al., 2012), *Sparassis crispa* (Tada et al., 2007), *Amanita muscaria* (Kiho et al., 1992), *Lactarius rufus* (Ruthes et al., 2013), etc.

$(1\rightarrow6)(1\rightarrow3)$-β-D-glucans are on the other hand somewhat rarely reported; such β-glucan is known from *Agaricus subrufescens* (syn. *A. brasiliensis*, *A. blazei*) and its structure is proposed to consist of a β-$(1\rightarrow6)$ backbone with a chain of two β-$(1\rightarrow3)$ linked glucose units attached at O-3 of every third backbone unit (Dong et al., 2002). Apart from grifolan, another β-glucan fraction isolated from *G. frondosa* ("MT-2") was shown to be a (1→6)(1→3)-β-D-glucan (Adachi et al., 1987; Hishida et al., 1988). Some rather unusual structures were also reported; although β-(1→4) linkage is a characteristic of cellulose (Klemm et al., 2005), Dong et al. (2012) described a (1→6)(1→4)-β-D-glucan isolated from *G. lucidum*, for which they proposed a structure of a β-(1→6) backbone with two β-(1→4) linked glucose units attached to every second backbone unit. Ma et al. (2008) reported (1→4)(1→6)-β-D-glucan isolated from *Auricularia auricula-judae*, with a β-(1→4) backbone; a significant part of glucose units was found to be oxidized to glucuronic acid (19%). Liu and Wang (2007) further reported a β-glucan from *Phellinus ribis* with a mixed (1→4), (1→6)-linked backbone and β-(1→3) linked branches, and Dai et al. (2012) described a similar structure from *Polyporus umbellatus*. A branched, mixed α, β-glucan was described by Mandal et al. (2012) from *Calocybe indica*. Linear β-glucans can have either β-(1→3) or β-(1→6) structure. Linear glucans are mostly water-insoluble and are usually obtained by alkaline extraction. Pachyman, isolated from *Wolfiporia cocos* sclerotium (syn. *Poria cocos*) is a linear (1→3)-β-D-glucan (Jin et al., 2003). Alquini et al. (2004) characterized another (1→3)-β-D-glucan from *Laetiporus sulfureus* and Chakraborty et al. (2006) from *Termitomyces eurhizus*. Linear (1→6)-β-D-glucans were reported from both *A. subrufescens* (Kawagishi et al., 1989; Kawagishi et al., 1990; Gonzaga et al., 2005) and *A. bitorquis* (Nandan et al., 2008), the latter one being isolated from the water-soluble mushroom extract fraction.

Camelini et al. (2005) found that during maturation of *A. subrufescens* fruiting bodies, β-glucan fraction changes significantly; glucans of young fruiting bodies contain both β-(1→6) and β-(1→3) linkages, but the amount of β-(1→3) linked glucose increases with age. They suggested that, according to the previous research, *A. subrufescens* contains (1→6)(1→3)-β-D-glucans and that during maturation β-(1→3) branching increases. The most common glucan structures found in mushrooms are given in Figure 8.1.

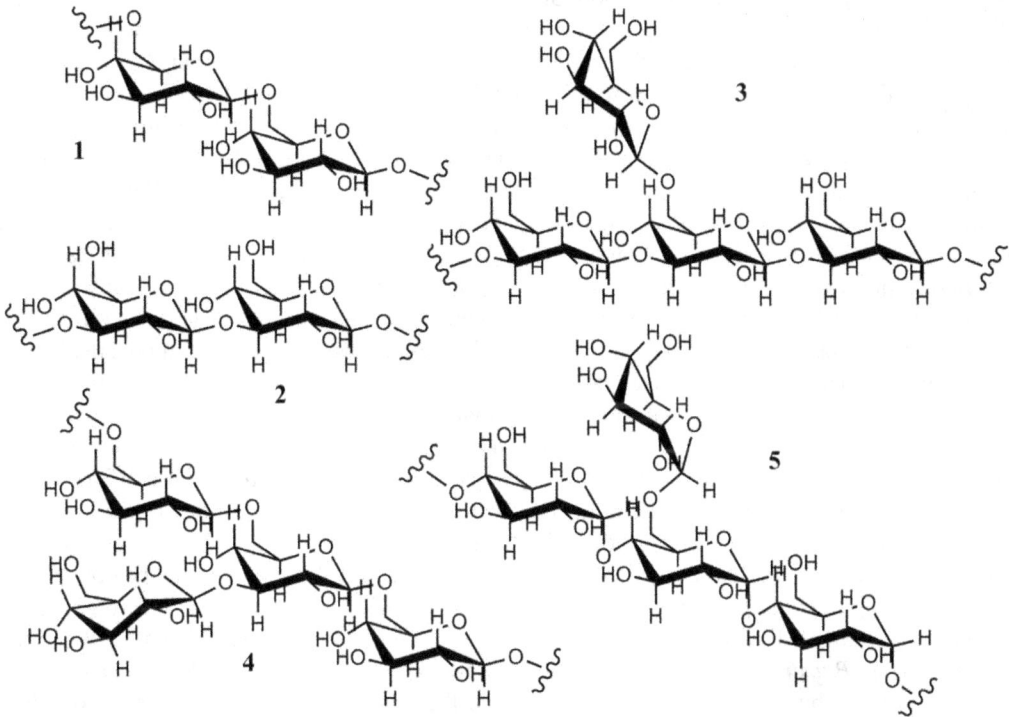

Figure 8.1 Mushroom glucan structures: (1→6)-β-D-glucan (1), (1→3)-β-D-glucan (2), (1→3) (1→6)-β-D-glucan (3), (1→6) (1→3)-β-D-glucan (4), (1→4),(1→6)-α-D-glucan (5).

β-glucans can form complexes with proteins and some of the most studied mushroom-derived immunostimulating products are β-glucan-protein complexes. Linear (1→6)-β-D-glucan described from *A. subrufescens* is a part of such complex, which was found to be very stable; as it is water-insoluble, proteolytic enzymes are ineffective in degrading it (Kawagishi et al., 1989; Kawagishi et al., 1990). "Fraction D" or "D-fraction", isolated from *G. frondosa* ("maitake") is a high molecular-weight-β-glucan-protein complex (1,000 kDa), the glucan part ("MT-2") being a (1→6)(1→3)-β-D-glucan (Adachi et al., 1987; Hishida et al., 1988; Deng et al., 2009). "Protein-bound polysaccharide", "Krestin", or simply PSK and poly-saccharopeptide (PSP), isolated from mycelium of *Trametes versicolor* (syn. *Coriolus versi-color*) are known to represent complexes between polysaccharides and proteins/polypeptides (Ng, 1998), but polysaccharide fractions of these complexes are not fully characterized yet. PSK and PSP both have a molecular weight of about 100 kDa, they contain (15) 25%–38% (Maehara et al., 2011; Friedman, 2016) and ~31% of proteins (Man-Fan Wan, 2013), re-spectively, which are rich in acidic amino acids, glutamic, and aspartic acid (Ng, 1998). The structure of the PSK polysaccharide fraction has been variously interpreted. The earliest structural studies found that protein fraction could not be eliminated by the Sevag method, precipitation with trifluorotrichloroethane, digestion with pronase, or column chromatography. No sugars other than glucose were detected upon hydrolysis of PSK and the polysaccharide part of the complex was suggested to be of β-glucan structure. Presence of β-(1→3), β-(1→4), and β-(1→6) linkages were detected and it was proposed that glucan part consisted of β-(1→4) backbone with both β-(1→3) and β-(1→6) side chains (Hirase et al., 1976a; Hirase et al., 1976b; Tsukagoshi et al., 1984), though experimental data was limited from today's point of view since the configuration of anomeric carbons was not determined using NMR spectro-scopy. Smaller amounts of other sugars (mannose, fucose, xylose, and galactose) were later also found to be part of the PSK carbohydrate fraction (Tsukagoshi et al., 1984). This minor sugar fraction is sometimes said to distinguish PSK and PSP; PSP does not contain fucose, but rhamnose and arabinose instead (Ng, 1998; Man-Fan Wan, 2013). Some sources state that α-(1→4) and β-(1→3) linkages are present in both PSK and PSP (Ng, 1998), while others report the presence of β-(1→3), β-(1→6), α-(1→4), and α-(1→2) glucosidic linkages in both of these complexes (Man-Fan Wan, 2013). PSP was however reported to contain (1→4), (1→2) and (1→3) linked glucose units (anomeric carbon not specified), together with smaller amounts of galactose, mannose, and arabinose (Ng, 1998). PSK was also cited/misinterpreted in literature as a (1→6)(1→3)-β-D-glucan (Yang et al., 2019), which however refers to an exopoly-saccharide produced by *T. versicolor* during submerged cultivation (Rau et al., 2009).

β-glucans can be extracted in several ways; a pretreatment of fungal material with organic solvents such as methanol or ethanol is usually required to remove liposoluble compounds. Water-soluble glucans can be then extracted with water, at room or high temperature. The process of extraction may be repeated several times until exhaustion, to maximize the yield of extraction. High pressure, microwave, and ultrasound-assisted extraction are newer techniques that are shown to be more efficient. Alkaline extraction using aqueous solutions of strong bases (2% NaOH, KOH), at high temperatures is often used. Purification of the isolated β-glucan fraction may be needed and the process may include dialysis, to remove salts and small molecules, as well as deproteinization, although these steps often lead to losses. Removal of proteins can be obtained by using the Sevag method, precipitation with trichloroacetic acid, or by enzymatic deproteinization (Ruthes et al., 2015). Using HCl solution as a solvent was reported to be efficient in β-glucans extraction from *P. ostreatus* with simultaneous deproteinization (Szwengiel & Stachowiak, 2016). The best way to obtain pure β-glucan fraction from the crude extract is by size exclusion chro-matography (Ruthes et al., 2015).

Photograph 8.1. *Trametes versicolor*, one of the best known mushroom species used in cancer treatment. Bor district, Serbia, 2014. Photo: P. Petrović

The mechanism of β-glucans' immunomodulatory activity is not completely understood. Studies on glucans primarily involved glucans with a β-(1→3) backbone. β-glucans are active when administered both systemically and orally (Hino et al., 2020). It is known for certain that they act stimulatory on the host's immune system. β-glucans primarily activate innate immunity but affect adaptive immunity as well (Vannucci et al., 2013; Hino et al., 2020). They produce responses in various types of immune system cells, primarily dendritic cells, macrophages/monocytes, NK ("natural killer") cells, neutrophils, as well as T lymphocytes (Chan et al., 2009) by binding to specific receptors expressed by these cells. Many receptor types are reported to recognize β-glucans: TLR ("*Toll-like receptor*"), CR3 ("*complement receptor 3*"), NLR ("*NOD-like receptor*"), RLR ("*RIG-I-like receptor*"), lactosylceramides (LacCer), scavenger receptors (Sc), and especially Dectin-1 receptor, considered to be the main β-glucan receptor. These receptors initiate an immune response during infection of fungal origin (Vannucci et al., 2013; Dalonso et al., 2015) which is not a coincidence, since β-glucans are the universal building blocks of fungal cell walls, including pathogenic fungi, and are recognized by the immune system as "pathogen-associated molecular patterns" (PAMPs) (Hino et al., 2020). Among these receptors, most studied and best understood are Dectin-1, TLR and CR3. Dectin-1 is found on various immune cells but plays a major role in β-glucan recognition by macrophages and dendritic cells. Once β-glucan binds to Dectin-1, it gets internalized by phagocytosis and induces several signaling pathways (such as NF-κB), leading to activation of reactive oxygen species (ROS) generation. Production and release of proinflammatory cytokines, such as TNF-α and IL-12 is mediated by a collaboration between Dectin-1 receptors and TLRs (Kim et al., 2011; Camilli et al., 2018; Legentil et al., 2015; Hino et al., 2020), although it was shown that in certain immune cell populations Dectin-1 activation alone can induce cytokine production (Goodridge et al., 2009). Cytokines are small signal proteins that represent intercellular messengers of the immune system; they integrate the function of various cell types from different parts of the organism in a coherent response (Gulati et al., 2016; Chawla et al., 2019).

CR3 receptors seem to be crucial for the antitumor activity of β-glucans; these receptors are expressed by NK cells, neutrophils, and lymphocytes. The binding of β-glucans to CR3 receptors enhances the cytotoxic activity of immune cells towards iC3b-opsonized target cells, including tumors that are otherwise resistant to CR3-dependent cytotoxicity (Chan et al., 2009; Camilli et al., 2018; Hino et al., 2020).

Glucans derived from different sources may show great variability in affinity for certain receptors and thus exhibit a weaker or stronger effect on immune cells (Vannucci et al., 2013); for example, it was shown that Dectin-1 receptor can bind glucans with an affinity that ranges from very low (3×10^{-3} M) to very high (2×10^{-12} M), depending on their size and structure (Goodridge et al., 2009, Adams et al., 2008). Zymosan, a $(1\rightarrow3)(1\rightarrow6)$-β-D-glucan, exhibits a higher affinity to Dectin-1 than curdlan, a $(1\rightarrow3)$-β-D-glucan, and pustulan, a $(1\rightarrow6)$-β-D-glucan (Hino et al., 2020). Not only the presence of branches but branching degree as well can affect β-glucans' activity. A lower degree of branching of 0.2–0.33 (branching at every third to fifth backbone glucose unit) is associated with better activity than a high degree of 0.67–0.75. Solubility of β-glucans is also important for their activity; some studies showed that both water-soluble and water-insoluble β-glucans can bind to Dectin-1 receptors, but only insoluble forms stimulate phagocytosis. β-glucans are known to have a quaternary structure under certain conditions and lentinan, a $(1\rightarrow3)(1\rightarrow6)$-β-D-glucan possesses antitumor activity in triple-helical conformation, while unfolded, in the single-chained confirmation it doesn't have this effect (Ruthes et al., 2015).

Many products of medicinal mushrooms have shown antitumor activity in animal models, but only a few were tested in human clinical trials. Complexes of β-glucans with proteins have shown superior activity to pure β-glucans (Wasser, 2017). The most important mushroom-derived β-glucan containing products include those obtained from *L. edodes*, *T. versiclor*, *G. lucidum*, *G. frondosa*, and others (Zhou et al., 2014). Lentinan is probably the most used pure β-glucan derived from mushrooms. It was first isolated back in 1969 (Murphy et al., 2020) and has been approved in both Japan and China as adjuvant agent in cancer treatment since the 1980s. It can be taken both orally, or by intravenous injection (Zhang et al., 2019); a superfine dispersed lentinan is a newer lentinan formulation that was found to have better bioavailability in humans (Vannucci et al., 2017). Clinical trials that evaluated lentinan's efficacy and safety mostly showed a positive effect of lentinan inclusion in the cancer therapy, especially for some lung and gastrointestinal cancers; positive effects include improved response rate, longer survival, reduction of side effects of chemotherapy/radiation therapy, and improvement of life quality (Ina et al., 2013; Vannucci et al., 2017; Antonelli et al., 2020). In clinical trials in China, which included near 10,000 patients, lentinan was used mostly in the treatment of lung cancer, followed by gastric and colorectal cancers. Lentinan was found to promote the efficacy of chemotherapy and radiation therapy, with few side effects (Zhang et al., 2019). However, more studies are needed that would include a broader population, as well as standardization of lentinan production, quality assessment, therapeutic protocols, etc. (Vannucci et al., 2017; Antonelli et al., 2020). Several clinical trials that involved another $(1\rightarrow3)(1\rightarrow6)$-β-D-glucan with a higher branching degree, schizophyllan, were conducted in the 1980s and 1990s; schizophyllan was also used in patients with recurrent and inoperable gastric cancer, together with chemotherapy, leading to increased median survival. It extended survival time in patients with head and neck cancers and had a limited effect on cervical cancer (Zhang et al., 2013).

T. versicolor-derived PSK and PSP have been extensively studied. PSK has been registered as a drug in Japan and is used as an immunotherapeutic adjuvant in cancer treatment. There have been multiple clinical trials in which PSK was used together with chemotherapy in the postoperative treatment of gastric, colorectal, and lung cancer. It was found that patients who received both chemotherapy and PSK lived longer, with a lower incidence of cancer recurrence and/or prolonged remission periods, comparing to those who received chemotherapy alone. PSK was reported to improve the quality of life in patients and is considered to be safe, with few, mild side effects. PSK also showed positive effects as an adjuvant in the treatment of other types of cancers: nasopharyngeal, esophageal, bladder, and breast cancer (Maehara et al., 2011). Numerous studies reported direct cytotoxicity of *T. versicolor* polysaccharide-containing preparations towards cancer cell lines *in vitro*, via apoptosis; these cell lines include breast, cervix, gastric, colon, hepatic, lung cancer cells, etc. (Maehara et al., 2011; Man-Fan Van, 2013; Habtemariam, 2020). D-fraction, another protein-bound polysaccharide isolated from *G. frondosa*, containing $(1\rightarrow6)(1\rightarrow3)$-β-D-glucan ("MT-2"), was evaluated in many animal studies and was reported to repress cancer progression and prevent metastasis, as well as to decrease the effective dose of chemotherapeutics in tumor-bearing mice. Several studies suggested that D-fraction is more effective in combination with vitamin C (He et al., 2019). Inoue et al., (2002) reported that D-fraction can potentiate the activity of helper T cells and affect the balance between Th-1 and Th-2 subtypes. However, clinical studies in humans are still scarce and limited by a low number of participants. A phase I/II clinical trial in 34 breast cancer patients showed that orally administered D-fraction was associated with changes of certain immunological parameters and that D-fraction acted stimulatory on some and suppressive on others in a dose-dependent way. The "optimal dose" for most studied functional parameters was found to be 5–7 mg/kg/day. The authors of the study, however, commented that the clinical significance of the observed changes in immunological parameters was unknown. They noted the increase of both anti-inflammatory IL-10 and immunostimulatory IFN-γ and IL-2; the clinical effect of the balance of pro- and anti-inflammatory cytokines remains to be uncovered. Although *G. lucidum* is one of the most widely used "medicinal mushroom" species, known for its immunostimulating effects, the potential positive effects of *G. lucidum* products on lung cancer, liver cancer, melanoma, leukemia, and colon cancer are known mostly from preclinical studies (Cao et al., 2018) and clinical trials in humans are very limited. A meta-analysis performed by Jin et al. (2016) found that the use of *G. lucidum* as the only cancer treatment option cannot be supported as its antitumor effects were negligible (which is why mushroom products are used in combination with chemoterapeutics).There is evidence that *G. lucidum* may be helpful in the treatment of lung cancer, as part of the therapeutic protocol, improving response to conventional therapy. However, there is a lack of evidence that improvement of tumor response is associated with prolonged survival of the patients. Like in other β-glucan-containing products, the use of *G. lucidum* enhanced the immune system of cancer patients, suggesting that it could be very useful in fighting the common immunosuppressive side effect of chemotherapeutics. *G. lucidum*–based products that have been used in research do not represent standardized, purified glucans, or glucan-protein complexes, but extracts that contain other compounds as well, making it difficult to compare different studies.

Besides β-glucans, α-glucans have been recently reported to exhibit immunostimulatory effects as well. Branched α-glucans are referred to as "glycogens" and may differ in branching degree, depending on the source, and may also form complexes with proteins (Synytsya & Novák, 2013). Masuda et al. (2017) isolated a highly branched $(1\rightarrow4)(1\rightarrow6)$-α-D-glucan from *G. frondosa* ("YM-2A"), which showed immunostimulation-associated antitumor effects in mice with colon and melanoma cancer when administered orally. The authors suggested that YM-2A was resistant to digestive enzymes because of the high branching degree. Glycogens of other plant and animal origins were also tested in the study but did not show the same effect. *In vitro* studies confirmed that YM-2A stimulated proinflammatory cytokine production in macrophages, although mechanisms of its action are yet to be understood. Stimulation of production of TNF-α in the macrophage cell culture was also reported for $(1\rightarrow4),(1\rightarrow6)$-α-D-glucan fraction obtained from *L. edodes* (Kojima et al., 2009), while *in vivo* studies in mice showed that similar α-glucan, isolated from *A. subrufescens* (as *A. blazei*) stimulated the increase of CD4+ and CD8+ lymphocytes (Mizuno et al., 1998). Mannose receptors may also have a similar function as studies in Dectin-1 deficient macrophages showed they were capable of recognizing fungi (Goodridge et al., 2009).

Photograph 8.2. *Ganoderma lucidum*; despite being one of the most popular traditional medicinal mushrooms, it is still insufficiently characterized. Crni vrh, Serbia, 2014. Photo: P. Petrović

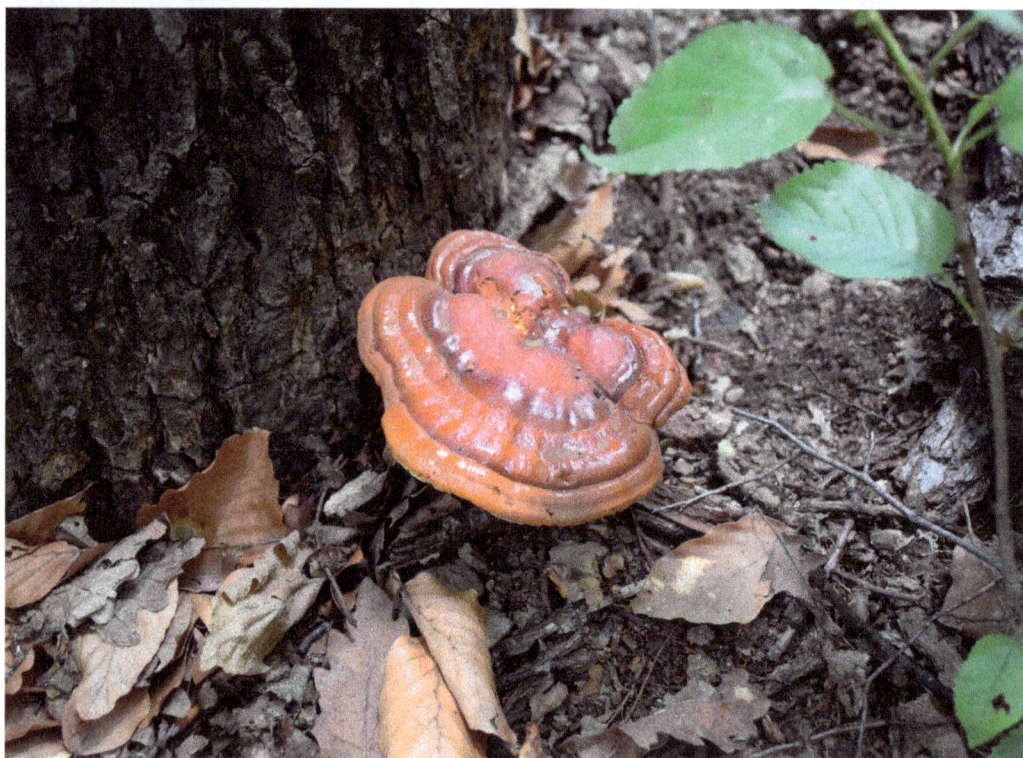

8.2.2 Cytotoxic compounds of mushrooms

Mushrooms contain compounds that exert direct cytotoxicity. Numerous species are reported to possess cytotoxic activity (Bézivin et al., 2002; Tomasi et al., 2004; Petrović et al., 2016), and a relatively great number of cytotoxic/antitumor compounds derived from mushrooms have been isolated. Most of them are however still poorly characterized in respect to their activity, which is, in most cases, known from scarce and limited *in vitro* studies on cancer cell lines or studies in animal models and xenografts. These compounds belong to different classes and include terpenoids (sesqui-, di-, and triterpenoids), polyketides, styrylpyrones, fatty acid derivatives, as well as other polyphenolic and amino compounds of uncertain biogenetic origin (Schüffler & Anke, 2009; Lee & Yun, 2011; Schüffler, 2018). Most of the antitumor mushroom-derived compounds act through apoptosis induction, although other mechanisms of action are seen as well. Some of the best-known cytotoxic compounds isolated from mushrooms, as well as some of the compounds with interesting/unusual mechanisms of antitumor activity, are reviewed below. Mushroom-derived terpenoids have been studied extensively and the great majority of reported cytotoxic compounds from mushrooms belong to this class of molecules.

Illudins are sesquiterpenes isolated from the poisonous mushrooms *Omphalotus lludens* and *O. japonicus* (Figure 8.2) and are among the most powerful natural cytotoxic compounds, active against multi-drug resistant experimental human cancers (Kelner et al., 1997). Illudin M and S were first isolated and named in 1950 as the main responsible antibacterial agents of *O. illudens* (as *Clitocybe illudens*) (Anchel et al., 1950), and their structures were elucidated independently by McMorris and Anchel (1965) and Nakanishi et al. (1965), who coined the name "lampterol" to a

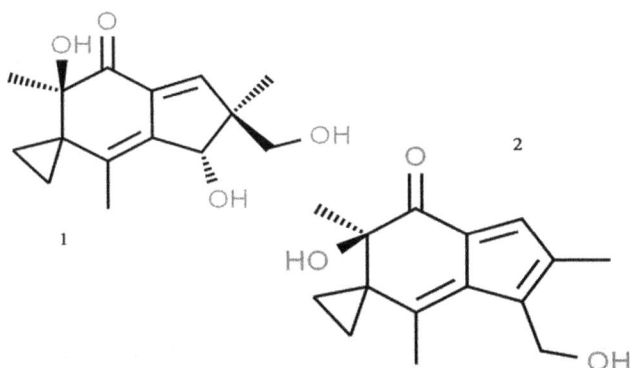

Figure 8.2   Chemical structures of illudin S (1) and irofulven (2).

Source: http://www.chemspider.com.

previously isolated cytotoxic compound from *O. japonicus*, identical to illudin S (Nakanishi et al., 1963). More related compounds were later isolated from *Omphalotus* spp., named illudin A, B, D, F, G, H. Interestingly, species of the distantly related genus *Coprinopsis* (previously included in *Coprinus*), such are *C. episcopalis* and *C. gonophylla,* are also able to produce this class of compounds and several illudins were isolated from them: illudin C, C_2, C_3, I, I_2, J_2 and illudinic acid (Gonzalez Del Val et al., 2003). Most studied, in respect to their antitumor abilities, are however illudin M, and particularly illudin S.

Illudins have a characteristic structure with a cyclopropane ring and a highly reactive α, β-unsaturated ketone group (Tanaka et al., 1994). They act via unstable intracellular metabolites which bind covalently to DNA (Kelner et al., 1994), causing a still poorly characterized lesion in the DNA strain. These lesions are highly specific in the way that they are ignored by all known cell repair systems and can only be repaired when trapped in stalled replication or transcription complexes (Jaspers et al., 2002). The first screening of illudins' antitumor activity was in animal models, where they were shown to significantly prolong the life of rats with leukemia, but had a narrow therapeutical index in the treatment of solid tumors (Kelner et al., 1990). *In vitro* studies showed that both Illudin S and M have very pronounced cytotoxic activity towards various human leukemia cell lines, being active in the range of concentrations between 6 and 100 nM. Myeloid and T-lymphocyte leukemia cells were found to be particularly sensitive to illudin S, which showed inhibitory activity towards certain types of these cells at the same concentrations as ricin, one of the most toxic compounds of plant origin (Kelner et al., 1987). Both healthy bone marrow progenitors and some solid tumors were relatively resistant. Selectivity towards certain leukemias, as well as breast, colon, lung, and ovarian carcinomas is seen in shorter exposure times of cells to illudins (≤2 h); this selectivity is based on an active transport mechanism present in sensitive cells, but absent in cells resistant to illudin S (Kelner et al., 1990; Kelner et al., 1994). Illudins inhibit DNA synthesis, causing a block at interphase (G_1-S phase) of the cell cycle. Ketone moiety was determined to be crucial for the activity, as reduction to the alcohol group leads to 10,000 times lower activity towards HL60 cells. The main problem for the use of illudins *in vivo* is that they have a narrow therapeutic index, meaning there is a little difference between therapeutic and toxic dose (Kelner et al., 1987), and it was soon realized that potential antitumor drugs should be looked for among the semi-synthetic analogues of illudins (Kelner et al., 1994). This led to discovery of a related compound family named acylfulvens (Puyo et al., 2014). Irofulven was ultimately developed and entered several clinical trials to evaluate its efficacy for certain cancer types. Although marked as a promising new agent, it showed no effects in patients with advanced or metastatic renal cell cancer (Berg et al., 2001; Amato et al., 2002), relapsed, or refractory non-small-cell lung cancer (Sherman et al., 2004). It did not demonstrate significant activity for the continuation of research in patients with advanced melanoma (Pierson et al., 2002) and was found to be minimally active in patients with recurrent or persistent endometrial carcinoma while exhibiting significant toxicity (Schilder et al., 2004). No complete or partial responses were observed in pediatric patients with solid tumors treated with irofulven (Bomgaars et al., 2006). Irofulven showed only modest activity in patients with recurrent or persistent intermediately platinum-sensitive ovarian or primary peritoneal cancer (Schilder et al., 2010). The effect of irofulven in patients with acute leukemia was mixed in phase I trial, and at the time, the authors of the study proposed a second phase in patients with better prognosis (Giles et al., 2001). In the phase II trial, irofulven showed activity in patients with metastatic hormone-refractory prostate cancer and had an acceptable safety profile (Senzer et al., 2005), suggesting that it may have at least limited use in the future. The possibility of the use of irofulven with other cytostatics was also explored. Irofulven was shown to be active in the treatment of advanced solid tumors in combinations with cisplatin (Hilgers et al., 2006) and capecitabine (Alexandre et al., 2007) in phase I clinical trials, leading to *in vitro* and *in vivo* studies of the interaction between irofulven and other antitumor drugs against lung carcinoma; it was shown that anti-metabolites gemcitabine, cyclocytidine, cytarabine, fludarabine phosphate, cladribine, and 5-fluorouracil have strong synergistic activity with irofulven when combined (Kelner et al., 2008).

There are numerous examples of mushroom triterpenoids that exhibit antitumor activity. Lignicolous mushrooms seem to be particularly potent producers of a vast variety of triterpenoids and those isolated from cultivated lignicolous mushrooms have been extensively studied. Ganoderic acids are probably the most studied mushroom-derived triterpenoids (Figure 8.3), regarding their antitumor activity. They represent highly oxygenated lanostan-type triterpenoids and are abundant in the genus *Ganoderma* (Min et al., 2000). More than 130 ganoderic acid derivatives and other triterpenoids have been isolated from various *Ganoderma* spp., both from fruiting bodies and mycelium cultures (Keypour et al., 2010), as well as spores (Liang et al., 2019).

Photograph 8.3. *Omphalotus olearius*, a poisonous mushroom species containing powerful cytotoxins, illudins. Bor district, Serbia, 2017. Photo: P. Petrović

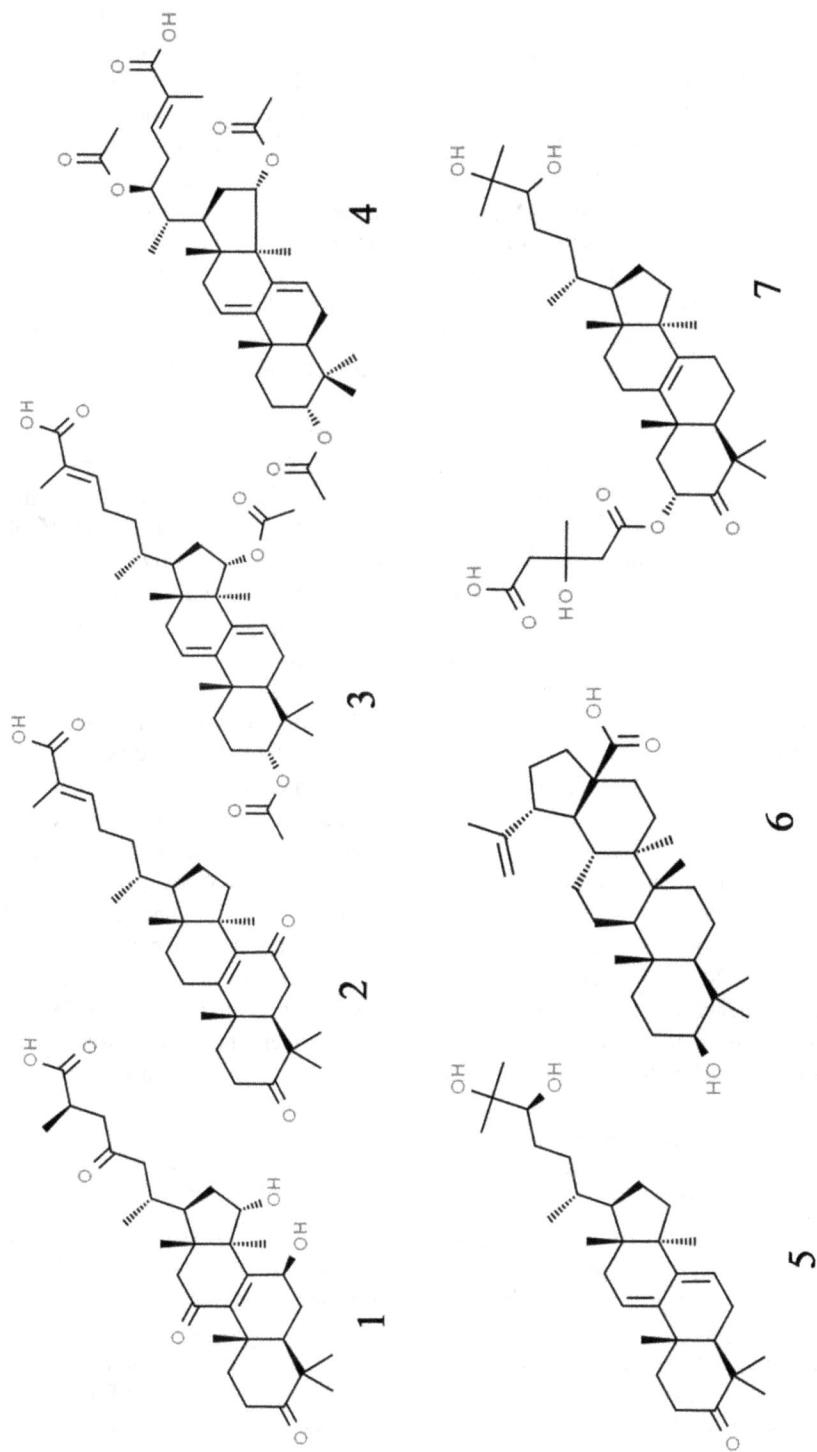

Figure 8.3 Examples of mushroom triterpenoids with cytotoxic/antitumor activity: ganoderic acid A (1), ganoderic acid DM (2), ganoderic acid Me (3), ganoderic acid T (4), ganodermanondiol (5), betulinic acid (6), clavaric acid (7).

Source: http://www.chemspider.com.

Ganoderic acids and related compounds seem to induce mitochondria-mediated apoptosis of cancer cells, but other antitumor mechanisms have been implicated in their *in vivo* activity. Ganoderic acid X (*G. amboniense*) was found to induce apoptosis in human hepatoma HuH-7 cells. The authors noted mitochondrial membrane permeability dysfunction and consequent release of cytochrome-*c* into cytosol and activation of caspase-3. Levels of the anti-apoptotic protein Bcl-xL were found to be decreased (Li et al., 2005). Ganoderic acid T was found to induce mitochondria-mediated apoptosis in lung cancer cells via the same mechanism; the authors further concluded that the apoptotic mechanism does not include caspase-8 activation. Ganoderic acid T caused the upregulation of p53 protein and Bax (while not affecting Bcl-2, another protein involved in apoptotic mechanisms), which consequently lead to changes in the potential of mitochondrial membrane and its dysfunction, causing the release of cytochrome-*c* into cytosol and activation of caspase-3. The same study confirmed the growth inhibition of implanted solid tumors by ganoderic acid T in mice (Tang et al., 2006). Chen et al. (2010) further found that it can effectively inhibit cancer cell invasion *in vitro* and metastasis *in vivo*. Chen and Zhong (2009) also suggested p53 protein to be the target of ganoderic acid Me in an *in vitro* study, while Wang et al. (2007) studied the mechanism of *in vivo* antitumor action of systemically administered ganoderic acid Me in mice bearing Lewis lung carcinoma and suggested that it may also exert an antitumor effect through immunostimulation. They noticed enhanced activity on NK cells in mice treated with the compound and increased expression of IL-2 and INF-γ, as well as upregulation of NF-κB, involved in the production of cytokines. Ganoderic acid Mf and S, two positional isomers, both decreased growth of various human cancer cell lines, but Mf isomers were more selective towards tumor cells when compared to their activity to normal cells. These isomers were further investigated on human cervical carcinoma cells (HeLa) to see whether they act through the same mode of action. They both caused mitochondria-mediated apoptosis, which was characterized by the activation of both caspase-3 and caspase-9, with increased Bax/Bcl2 ratio, but they differed concerning their specificity to cause cell cycle arrest at a certain phase. It was found that ganoderic acid S caused the arrest of cells in the S phase, while ganoderic acid Mf caused the arrest in the G1 phase. Whether this has a clinical significance is yet to be elucidated (Liu & Zhong, 2011). Ganoderic acid A was reported to suppress the proliferation of human hepatocellular carcinoma, by inducing cell cycle arrest at the G0/G1 phase and apoptosis (Wang et al., 2017). Apart from exhibiting cytotoxic activity via apoptosis towards human breast cancer cells (Wu et al., 2012), ganoderic acid DM was found to have anti-androgenic activity and thus to be potentially effective in prostate cancer treatment (Liu et al., 2009). It acts as a 5α-reductase inhibitor, an enzyme that catalyzes the transformation of testosterone to the more active dihydrotestosterone, involved in the development and progression of hormone-dependent prostate cancer. It further binds to androgen receptors, suppressing associated signaling pathways. Apart from being active against both androgen-dependent and independent prostate cancer, ganoderic acid DM was also shown to inhibit osteoclast differentiation, which might be useful in preventing bone metastases that may accompany prostate cancer and which are responsible for a significant percentage of prostate cancer–related deaths (Johnson et al., 2010).

Liu et al. (2012) tested *in vitro* cytotoxicity of several semi-synthetic derivatives of ganoderic acid T towards HeLa cells; they didn't find a difference in the activity of ganoderic acid T and its methyl, ethyl, and propyl esters, but amide analog was shown to be significantly more active, suggesting that carboxylic group is not crucial for the activity of ganoderic acids. This is further backed up by the fact that ganoderic alcohols are also reported to have cytotoxic activity *in vitro*, towards certain cell lines; lucidumols, ganoderiol F, and ganodermanondiol were found to be even more effective against Meth A sarcoma and Lewis lung carcinoma (LLC) cells than ganoderic acids in the study performed by Min et al. (2000).

Betulinic acid is a lupan-type pentacyclic triterpene that has been drawing increasing attention due to its various biological activities, including cytotoxic and antitumor effects (Dehelean et al., 2021; Lou et al., 2021). It is found in the outer bark of birches (*Betula* spp.), but also in various other plant species, although in relatively low amounts (Hordyjewska et al., 2019). On the other hand, betulin, an alcohol-analog and a precursor of betulinic acid, a potent cytotoxic compound as well, can account for up to 30% of bark dry weight of some birch species (Green et al., 2007; Hordyjewska et al., 2019). Both betulin and betulinic acid can however be found in a very specific fungus – *Inonotus obliquus* or Chaga, which grows as a parasite mostly on birch trees (Lee et al., 2008); these compounds have been thought to originate from the host. Chaga causes rupturing of the tree bark and exposes a dark-colored mycelium mass – sclerotium. This sterile "conk" has been used in the folk medicine of northern nations, particularly Russians to treat stomach diseases, liver and heart ailments, as well as different kinds of cancer (Lee et al., 2008; Kim et al., 2011; Petrović et al., 2019b). Chaga extracts were reported to exhibit cytotoxic effects on cervical, colon, and liver cancer cell lines, as well as tumor growth reduction in a melanoma xenograft model, by induction of apoptosis (Blagodatski et al., 2018). Some of Chaga's activity may be associated with the content of betulin and betulinic acid, which however seems to vary with its geographical origin, production/accumulation capacity of the host tree species, and probably other factors as well; thus one study found that French specimens of Chaga contained a significant amount of these compounds, unlike those collected in Ukraine and Canada (Géry et al., 2018). Betulinic acid induces apoptosis in tumor cells by mitochondrial-mediated pathways (Zeng et al., 2019; Dehelean et al., 2021) characterized by disturbance of mitochondrial membrane potential, cytochrome-*c* release, caspase activation, and DNA fragmentation (Fulda et al., 1997; Fulda & Debatin, 2000; Thurnher et al., 2003). Although the first reports on betulinic acid's activity marked it as a melanoma-selective agent (Pisha et al., 1995), consequent research has shown that it is active *in vitro* against several tumor cell lines, including head and neck squamous cell carcinoma (Thurnher et al., 2003), Ewing's sarcoma, neuroblastoma, medulloblastoma, and glioblastoma (Fulda et al., 1997; Fulda et al., 1999), and *in vivo*, in mice ovarian and breast tumor models (Hordyjewska et al., 2019). Betulin was also found to exhibit *in vitro* toxicity towards cell lines of various tumors: neuroblastoma, thyroid, breast, cervical, and skin epidermoid carcinoma, colorectal adenocarcinoma, T lymphoblast leukemia, and others; it was proposed to share the same mode of action as betulinic acid (Król et al., 2015). Betulin's poor solubility in water limits its potential use in cancer treatment, so betulinic acid is regarded as a more suitable candidate for further research (Hordyjewska et al., 2019). Although assumed that betulin and betulinic acid found in Chaga conk are absorbed from the host, several studies on the submerged-cultivated mycelium of *I. obliquus* have demonstrated that these compounds can be produced by the fungus (Bai et al., 2012; Xu et al., 2016); in these studies, however, the identification of the compounds was done using HPLC with UV detector, only by comparing the retention times to those of the standards. Lou et al. (2021) provided more solid evidence of the presence of betulinic acid in the mycelial culture of *I. obliquus* by using HPLC-MS and showed that the production of betulinic acid may be greatly enhanced by altering culturing media and fermentation conditions. If *I. obliquus* is truly confirmed to be a potent producer of betulinic acid, it would make it a much reliable source of this compound, as mycelium is relatively easily grown in laboratory conditions. The fact that *I. obliquus* is capable of producing betulin and betulinic acid may not come as a surprise as it had been already established that some fungal plant pathogens, as well as endophytic fungi, produce the same compounds that are found in the associated plants, probably as a result of genetic exchange between these organisms; the most notable example of such relationship are pacific yew (*Taxus baccata*) and *Taxomyces andreanae*, an endophytic fungus that lives in the tree's bark and can synthesize paclitaxel, the same powerful cytostatic produced by pacific yew (Stierle et al., 1995). Chaga is a source of other biologically active triterpenoids that were found to exhibit *in vitro* antitumor effects: inotodiol, inonotodiol, inonotsuoxides, etc, although these compounds are yet to be thoroughly studied (Blagodatski et al., 2018).

Clavaric acid is a triterpenoid isolated from *Clavariadelphus truncatus* and it inhibits farnesyl-protein transferase (IC_{50}=1.3 µM), which catalyzes the specific transfer of farnesyl group to Ras-peptides, an action that is essential for oncogene-mediated tumors. Inhibition of farnesyl-protein transferase was confirmed to reduce tumor development in mice, thus marking this enzyme as a potential target of antitumor agents (Jayasuriya et al., 1998). Other triterpenoids that were shown to possess cytotoxic activity *in vitro* were isolated from *Lactarius volemus, Antrodia camphorata, Ophiocordyceps sinensis* (Jin-Ming, 2006), *Irpex* sp., *Perenniporia fraxinea, Leucopaxillus gentianeus* (Schüffler & Anke, 2009), *Hypholoma fasciculare* (Schüffler, 2018), *Tricholoma pardinum* (Zhang et al., 2018), and many more.

Styrilpyrones (Figure 8.4) are an interesting class of yellow polyphenol pigments that are known primarily from lignicolous mushrooms, most of them being characterized from *Phellinus* and *Inonotus* spp. (*P. igniarius, P. linteus, P. ribis, I. hispidus, I. obliquus, I. xeranticus,* etc). They are thought to have similar functions as plant phenylpropanoids and flavonoids. They are powerful antioxidants, but exhibit other activities as well, such as anti-inflammatory, anti-platelet aggregation, and anti-diabetic. Some styrilpyrones, hispidin, hispolon, and phelligridins are also shown to possess cytotoxic activity. Hispidin and hispolon have shown significant inhibitory activity towards certain tumor cell lines, while the activity of phelligridins was only moderate. Hispidin was shown to be more cytotoxic towards tumor pancreatic duct cells and keratinocytes than normal cells. Hispolon's cytotoxic ability was further investigated and it was found that it acts through mitochondria-mediated apoptosis induction (Lee & Yun, 2011).

Various amino compounds from mushrooms were reported to possess antitumor activity, with a great variety of mechanisms of action (Figure 8.5).

Clitocine was first isolated from the North American population of *Paralepista flaccida* (as *Clitocybe inversa*) as the main insecticidal principle of the mushroom extract, constituting about

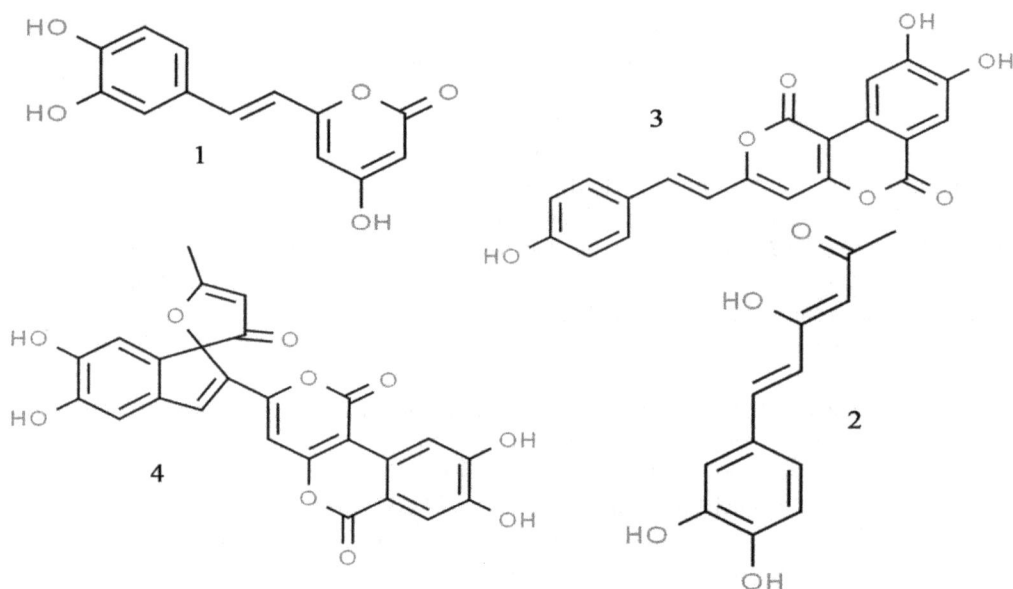

Figure 8.4 Chemical structures of selected styrilpyrones: hispidin (1), hispolon (2), phelligridin C (3), phelligridin E (4).

Source: http://www.chemspider.com.

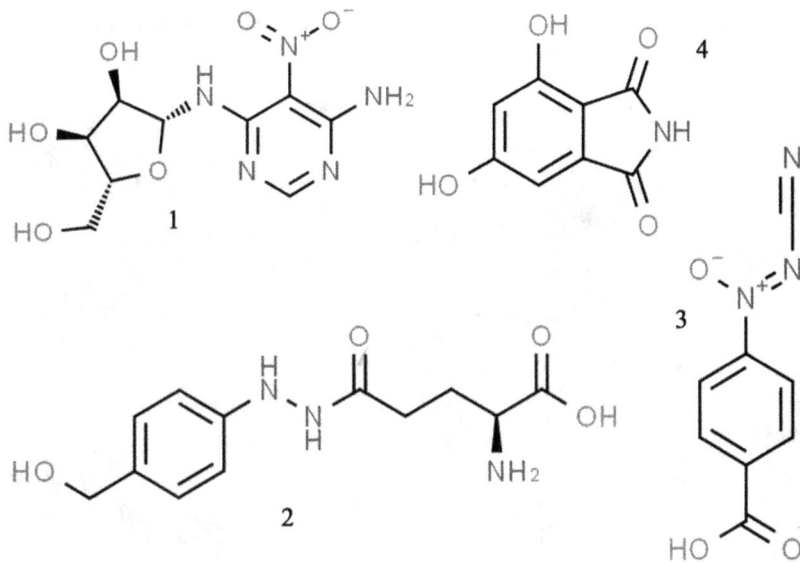

Figure 8.5 Examples of mushroom amino-compounds with citotoxic/antitumor activity: clitocine (1), agaritine (2), calvatic acid (3), 4,6-dihydroxy-1H-isoindole-1,3(2H)-dione (4).

Source: http://www.chemspider.com.

0.001% of the mushroom fresh weight. It represents an amino exocyclic nucleoside, which has structural similarities to adenosine (Kubo et al., 1986). Clitocine was later reported from *Leucopaxillus giganteus* as well (Ren et al., 2008). First studies of clitocine's potential cytotoxic activity towards tumor cell lines were carried by Moss et al. (1988); they tested the compound against murine lymphocytic leukemia cell line (L1210), as well as human B lympho-blastic leukemia (WI-L2) and T lymphoblastic leukemia (CCRF-CEM) and determined relatively low IC_{50} value against all lines of 30 nM. Almost 20 years later, the interest in clitocine rose again as Fortin et al. (2006) isolated the same compound from *Paralepista flaccida* as the compound responsible for the mushroom's pronounced cytotoxic activity. Clitocine exhibited very strong activity towards all tested cancer cell lines, murine (3LL and L1210) and human (DU145, K-562, MCF7, and U251), with IC_{50} values in the nanomolar/sub-millimolar range; the strongest activity was obtained against murine 3LL and L1210 cell lines, with IC_{50} values being 33.1 nM and 22.6 nM, respectively, similar to the values obtained by Moss et al. (1988), establishing clitocine as one of the most potent cytotoxic compounds isolated from a basidiomycete (Fortin et al., 2006). Clitocine has two anomers (α and β) and there was no significant difference in the activity of the mixtures with different anomer ratios. The anomer mixture with α:β ratio 1:4 was further tested for toxicity in healthy mice, as well as 3LL and L1210 tumor-bearing mice. As a daily dosage of ≤6.25 mg/kg was found to be safe during two-week treatment, both tumor-bearing mice lines were treated with clitocine with daily doses of 0.5, 3, or 5 mg/kg. While no effects were noted during the treatment of the 3LL tumor-bearing mice, there was evident prolongation in the survival of L1210 tumor-bearing mice, which was the most pronounced for dosage regime of 3 mg/kg; the *in vivo* experiments confirmed the *in vitro* obtained better activity towards L1210 cell line. The authors proposed that clitocine acts via apoptosis induction; Ren et al. (2008), who investigated clitocine's activity towards HeLa cells, confirmed that it induces apoptosis by activating multiple pathways including death receptor and mitochondrial pathways.

Photograph 8.4. *Paralepista flaccida*, one of the mushroom species clitocine was isolated from. Crni vrh, Serbia, 2013. Photo: P. Petrović

Clitocine was recently found to be a highly potent readthrough agent, allowing the normal translation of a functional protein despite the presence of a nonsense mutation in the DNA which consequently results in a premature stop codon in the mRNA. Nonsense mutations in the p53 tumor suppressor protein are one of the causes of its inactivation, which is one the most prevalent underlying mechanism of human cancer development; mutations in p53 protein may also lower the efficacy of chemotherapeutics. Reactivation of p53 protein would thus have a positive effect on tumor treatment. Clitocine was shown to be able to induce the production of fully functional p53 protein in cells with p53 protein nonsense mutations and consequently prevented the growth of nonsense-containing human ovarian cancer tumors in xenograft tumor models. Incorporation of clitocine in RNA, as adenosine replacement, seems to be the key point of its mechanism of action; when incorporated at the third position of a nonsense codon, it allows the read-through of RNA. The discovery of readthrough agents which allow translation of functional proteins in a presence of a nonsense mutation led to a novel approach in the treatment of both genetic disorders and cancer; ataluren, was the first such drug to be conditionally approved by the European Medicines Agency and clitocine may one day soon find the use for the same indications (Friesen et al., 2017).

Calvatic acid is a phenylazoxycyanide isolated from puffballs *Calvatia lilacina* (Gasco et al., 1974), *C. craniiformis* (Umezawa et al., 1975), and some other related species. Calvatic acid bears a unique azoxycyano-moiety (Blair & Sperry, 2013) and has been a subject of extensive studies concerning its potent antimicrobial activity (see section 8.3). However, it was also found that it possesses antitumor properties as well. It inhibited the growth of Yoshida sarcoma cells, with an

IC_{50} value of 1.56 μg/mL, and prolonged survival of mice with leukemia 1210 when given systemically (Umezawa et al., 1975). Calvatic acid was suspected to inhibit the polymerization of tubulin, the main protein of microtubules, parts of the cytoskeleton, involved in the cell shape maintenance, cell division, and other vital cell processes. While calvatic acid was shown to possess only weak tubulin-polymerization inhibitory activity *in vitro*, it led to the synthesis of compounds with much more pronounced activity, namely a derivate with a *p*-carboxylic group replaced with Cl – *p*-chlorophenylazoxycyanide. Azoxycyano-moiety was shown to be crucial for the activity and mechanism of inhibition might involve reaction with sulfhydryl groups of tubulin (Gadoni et al., 1995). Agaritine, another diazo compound found in members of the genus *Agaricus*, has been suspected to have genotoxic properties by some researchers, although these claims have been heavily criticized by others (Roupas et al., 2010). Endo et al. (2010) then found that agaritine was the responsible agent for the direct antitumor activity of *A. subrufescens* (as *A. blazei*) extracts. It inhibited the proliferation of several leukemia cell lines with IC_{50} values ranging between 2.7 and 16.0 μg/mL, while normal lymphatic cells were not significantly affected when subjected to agaritine concentrations up to 40 μg/mL. Agaritine was shown to induce apoptosis of tumor cells via caspase activation through cytochrome *c* release from mitochondria (Akiyama et al., 2011).

Lü et al. (2013) isolated a simple isoindolone, 4,6-dihydroxy-1H-isoindole-1,3(2H)-dione from *Lasiosphaera fenzlii*; this compound was tested *in vitro* for its potential cytotoxic activity and it did not have any effect on several cancer cell lines. Instead, it inhibited the secretion of vascular endothelial growth factor (VEGF) from human alveolar basal epithelial adenocarcinoma cells (A549). VEGF is of great importance in tumor development as it promotes the growth of new blood vessels and enables the vascularization of tumor tissue (Lugano et al., 2019). It showed better inhibitory activity than thalidomide, which shares a similar structure (Lü et al., 2013). The same compound was isolated from *Calvatia nipponica* and its antiangiogenic activity was also confirmed by Lee et al. (2017). The authors also identified *N*-(4-hydroxyphenyl)-acetamide and nicotinamide as the active principles of the *C. nipponica* extract, although they exhibited less pronounced antiangiogenic activity.

8.3 ANTIBIOTIC COMPOUNDS FROM MUSHROOMS

The search for antimicrobials from *Basidiomycetes* started more than half a century ago when in a period of 10 years (1940–1950) about 2,000 species were screened (Schüffler 2018). During that period, the antibiotic compound pleuromutilin has been discovered. However, it needed to wait another half a century to regain interest (de Mattos-Shipley et al., 2017). Another group of microorganisms, streptomycetes, attracted research focus, as being more productive and easier to manipulate. Recently, interest in macrofungi as a source of antibiotics has been revived mainly because of omnipresent antibiotic resistance. The ability of bacteria to adapt to the presence of antimicrobial compounds was ever-present as an evolutional asset; however, cheap and easily accessible antibiotics, as well as improper use, enabled speeding up the acquired resistance. The pharmaceutical industry was of little help since it was not willing to indulge in the laborious and expensive *bioprospection*. The industry's answer was a chemical modification of already present antibiotics (Zaman et al., 2017). Still, 2019 started *the age of fungiceuticals*, which are drug candidates originating from edible fungi (Pandey et al., 2020). In the meantime, techniques for identification, genetic engineering, metabolomics, and cultivation of mushrooms were developed and improved (Anke, 2020; Klaus et al., 2021). One of the problems with natural antibiotics is that in the material of their origin (mycelium or mushroom fruiting body) they are mostly present in relatively low concentrations (Beattie et al., 2010). So, genetic engineering has to be applied to increase the yield.

The first steps in antibiotic discovery are extraction and screening and the majority of research papers dealing with basidiomycetes examine antimicrobial activity of different types of extracts (Klaus et al., 2015; Klaus et al., 2016; Vunduk et al., 2019; Pandey et al., 2020); those which are promising undergo compound isolation and identification of its active chemical constituents. Suay et al. (2000) published the results of antimicrobial screening of 204 mushroom species from Spain and 109 of them expressed mainly antibacterial activity (against G + species). The strongest effect was observed among the genus *Ganoderma*, *Collybia*, *Polyporus*, and *Psathyrella*. Isolates from these species were active in a range from 75%–100%, with isolates from family Ganodermataceae being effective in most of the cases (73% of isolates). Antifungal activity was observed in isolates mostly from order *Agaricales*.

One of the first promising mushroom-derived antimicrobial candidates was calvatic acid. As it was already stated in section 8.2, it is a phenylazoxycyanide derivate, discovered at the same time by two separate research groups, and from two different mushroom sources, *Calvatia lilacina* and *C. craniiformis* (Gasco et al., 1974; Umezawa et al., 1975). Further research went toward calvatic acid analogs, and phenylazoxycyanide and its synthetic derivatives expressed wider antimicrobial activity (Umezawa et al., 1975; Calvino et al., 1986). At the beginning of the 21st century, Sorba et al. (2001) examined the potential of calvatic acid and some of its analogs as a possible solution for the unresponsive treatment of *Helicobacter pylori* infection. Calvatic acid was found to be very active against metronidazole-resistant *H. pilori* strains, but besides antibiotics, the existing therapy for *H. pilori* infection includes H_2-receptor antagonists or proton pump inhibitors. The authors thus designed a series of compounds by combining calvatic acid with a lamtidine pharmacophoric group, with varying lengths of aliphatic chain spacer between the two moieties. The compounds were tested for their antimicrobial activity and for the ability to irreversibly inactivate H_2-binding sites *in vitro* (using histamine H_2-receptors of guinea pig right atria). The inhibitory activity of the compounds on *H. pilori* growth was significantly lowered in comparison to calvatic acid, but was still similar or better than that of metronidazole; longer spacers were associated with better antimicrobial activity. The compound with the shortest spacer was on the other hand the most efficient as an irreversible H_2-receptor antagonist; the increase of spacer length was accompanied by the decrease of irreversible action. The irreversible activity, which is not seen in lamtidine itself, was explained by the presence of a cyanoazoxy functional group of the calvatic acid moiety which is under physiological condition capable of reacting with the SH group of the receptor, thus covalently binding to it. Moreover, the shorter the spacer, the better is the placement of the reactive cyanoazoxy function for the SH addition.

A widespread and widely studied group of compounds with antibiotic activity isolated from mushroom-producing fungi includes β-methoxyacrylic acid derivatives, strobilurins, and oudemansins (Clough, 1993), first isolated from *Strobilurus tenacellus* and *Oudemansiella mucida*, respectively. These compounds were found to be very active at inhibiting the growth of both yeasts and filamentous fungi by interfering with their respiration process. Both strobilurins and oudemansins are produced by several mushroom species and many different structures have been characterized to date (Anke & Steglich, 1981; Clough, 1993: Sauter et al., 1999). Strobilurin A was found to be the same compound as mucidin, which had been previously isolated from *O. mucida* and patented in 1970, although its structure had not been proposed at the time. "Mucidin" found its use in Czechoslovakia as an antifungal ointment (Mucidermin "Spofa"), but strobilurins have become more important as agricultural fungicides (Sauter et al., 1999).

Photograph 8.5. *Strobilurus tenacellus*; strobilurins, β-methoxyacrylic acid derivatives with antifungal activity were first isolated from this species. Brestovačka banja, Serbia, 2014. Photo: P. Petrović

Other compounds with antibiotic activity isolated from basidiomycetes include scorodonin, a member of the rare group of C_7-acetylenes, produced by *Marasmius scorodonius* (Anke et al., 1980); pterulones and pterulinic acid found in *Pterula* sp., which represent halogenated derivatives of 2,3-dihydro-1-benzoxepine (Lemaire et al., 2003); merulidial, a sesquiterpene dialdehyde from *Merulius tremellosus;* and many others (Anke & Steglich, 1981). Chemical structures of selected mushroom compounds with antibiotic activity are given in Figure 8.6.

Antibiotic compounds are in most cases isolated from culture broth or auxenic cultures (Gasco et al., 1974; Umezawa et al., 1975; Engler et al., 1998). This seems logical having in mind that mycelium develops in soil or substrate where it has to compete with other microorganisms. In the experiment by Engler et al. (1998), it has been proved that mediums richer in glucose and amino acids, like yeast malt glucose medium, provide a wider spectrum of compounds with antibiotic activity. Transferred to natural habitat this means that more competitors are likely to occur when the substrate is abundant in nutrients.

Basidiomycetes can be good sources of compounds effective against so-called *superbugs*, highly resistant bacteria like methicillin-resistant *Staphylococcus aureus* (MRSA). As discovered by Jin and Zjawiony (2006), eight benzendiols isolated from *Merulius incarnatus* mushroom were highly efficient in inhibiting the growth of MRSA. On average IC_{50} values were less than 10 μg/mL.

Compounds isolated from mushrooms can be active against *Mycobacterium tuberculosis*, which is one of the leading causes of infectious disease-related deaths. Multi-drug resistance and a long recovery

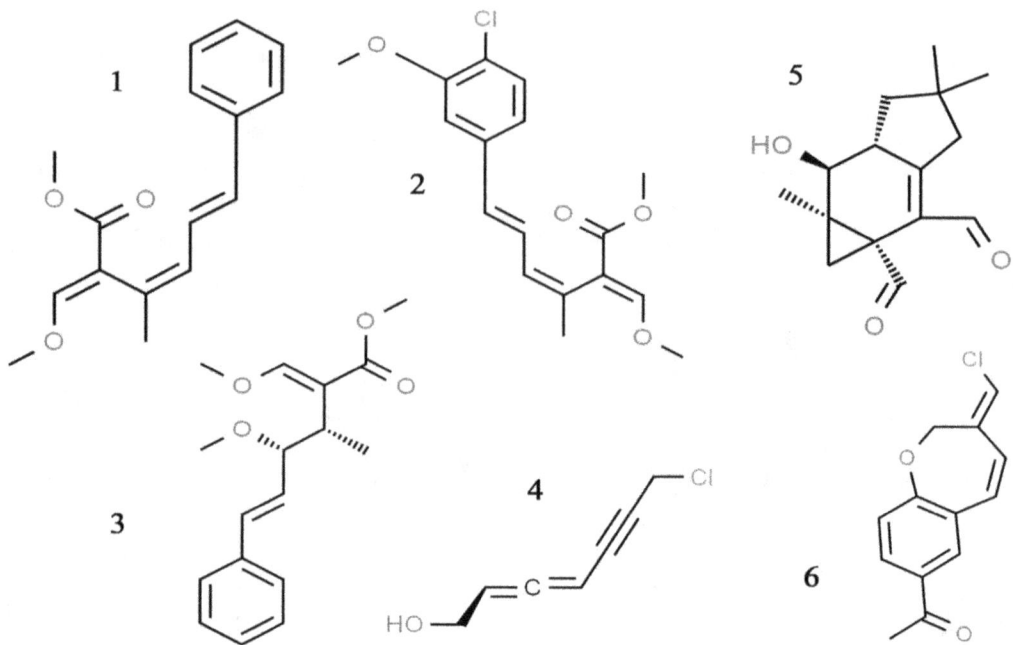

Figure 8.6 Selected mushroom compounds with antibiotic activity: strobilurin A (1), strobilurin B (2), oudemansin (3), scorodonin (4), merulidilal (5), pterulone (6).

Source: http://www.chemspider.com.

are the main problems in the treatment of this pathogen. Two chlorinated coumarins, 6-chloro-4-phenyl-2H-chromen-2-one and ethyl 6-chloro-2-oxo-4-phenyl-2H-chromen-3-carboxylate were isolated from *Fomitopsis officinalis*. These compounds were tested against several pathogenic microbes but showed no activity against *S. aureus*, *Escherichia coli*, *Enterococcus faecalis*, *Pseudomonas aeruginosa*, or *Candida albicans* at the concentration of 100 μg/mL; however, the latter compound was shown to be active against *M. tuberculosis*, as well as synthetic 7-chloro analogs of both of these compounds. The authors noted that these coumarins may have the potential for shortening the duration of therapy due to their efficiency against non-replicating persistor cells (Hwang et al., 2013).

As reviewed by Schüffler (2018) and Pandey et al. (2020), most of the mushroom-derived antibiotics are phenolic compounds, terpenoids – sesquiterpenoids, diterpenoids, sesterpenes, and triterpenoids. Although some of them are very potent in inhibiting microbial growth, they also express cytotoxic activity that prevents their use as commercial antibiotics.

So far, the only antibiotic from a basidiomycete that led to the synthesis of novel drugs approved for use in humans is pleuromutilin (Figure 8.7). It was first isolated in 1951 from *Omphalina mutila* (as *Pleurotus mutilis*, hence the name) and *Clitopilus passeckerianus* (as *Pleurotus passeckerianus*), but it was not until the end of the 20th century that it regained interest as an alternative antibiotic when semi-synthetic derivatives were developed for commercial use (De Mattos-Shipley et al., 2017). Pleuromutilin and its derivatives act as ribosomal protein synthesis inhibitors, by binding to the 50S ribosomal subunit of bacteria (Poulsen et al., 2008) and are mainly active against gram-positive bacteria (Kavanagh et al., 1951). Before the possibility of use of pleuromutilin derivatives in the treatment of human bacterial diseases was explored, semi-synthetic pleuromutilin-derived compounds such as tiamulin and valnemulin became important antibacterial agents in the veterinary practice (Poulsen et al., 2008).

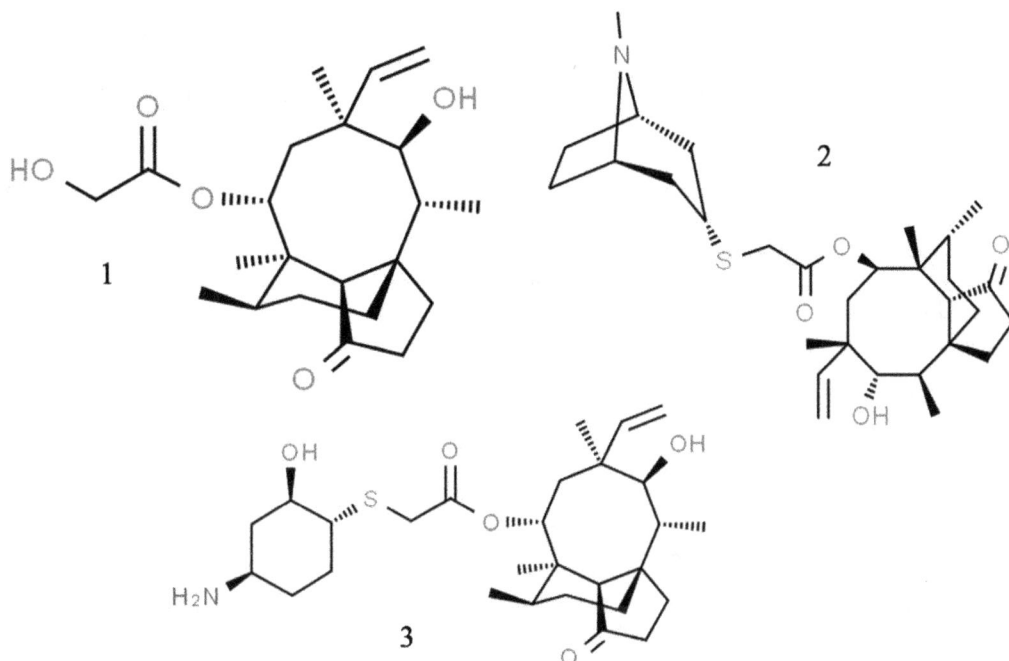

Figure 8.7 Chemical structures of pleuromutilin (1), retapamulin (2), and lefamulin (3).

Source: http://www.chemspider.com.

There are currently two pleuromutilin-class antibiotics with permission for human use. The first one, retapamulin was approved in 2007, by both FDA and EMA, for topical treatment of impetigo caused by *Staphylococcus aureus* and *S. pyogenes*, as 1% ointment (Butler, 2008). Just recently, in the first part of 2020, the Committee for Medicinal Products for Human Use of the European Medicines Agency gave a positive opinion and approved lefamulin (under commercial name Xenleta) (https://www.ema.europa. eu/en/medicines/human/EPAR/xenleta). The approval states that Xenleta is allowed for adult use only to treat community-acquired pneumonia. Lefamulin thus became the first pleuromutilin-type antibiotic approved for systemic administration in humans. The results of a clinical trial reported by Alexander et al. (2019) showed that the early response rate of patients with community-acquired pneumonia was the same for lefamulin and moxifloxacin (90.8%). The administration period of five days with lefamulin was not inferior to seven days of moxifloxacin. In a small number of patients who received lefamulin adverse effects as diarrhea and nausea occurred. Another clinical trial confirmed the efficacy of lefamulin as well as its non-inferiority to moxifloxacin (File et al., 2019). An important difference between these two studies was the primary endpoint for the clinical response (five to seven days and 96 ± 24 hours, set by EMA and FDA). Moreover, in the latter study, a lower percentage of drug discontinuation due to adverse events was observed for lefamulin in comparison with moxifloxacin (2.9% and 4.4%, respectively). When Tang, Lai, and Chao (2020) extracted clinical data for elderly patients from two clinical studies and performed a meta-analysis of the data, the researchers showed that lefamulin has a high potential in being used for treating non-hospital-acquired bacterial pneumonia. This class of antibiotics seems promising for multi-drug resistant and extensively drug-resistant tuberculosis as explained by De Mattos-Shipley, Foster, and Bailey, (2017), since they do not show cross-resistance with other antibiotics.

The synthesis process of products with antibiotic activity when the starting material is mushroom extract is called *mycogenesis* or *mycosynthesis* (Pandey et al., 2020). Mushroom products can be used to synthesize silver nanoparticles, which are widely used due to their strong antimicrobial activity, acting both as stabilizing agents as well as active principles (Petrović et al., 2018; Klaus et al., 2020). Silver nanoparticles prepared with crude polysaccharide extracts of *A. bisporus, A. brasiliensis*, or *Phellinus linteus* expressed a significantly more pronounced anti-bacterial effect than that of amoxicillin. The obtained minimal inhibitory concentrations were in some cases up to 100 times lower than those of antibiotics (Klaus et al., 2020). A recent review by Bhardwaj et al. (2020) summarized progress in this area with various mushroom species including the antibacterial activity of mycosynthesized nanoparticles.

Natural compounds with antimicrobial activity can even contribute to the decrease of the danger of antibiotic resistance acquired through the food chain as one of the main sources of bacterial resistance to antibiotics. This was demonstrated by Muchaamba, Stephen, and Tasara (2020) with *Listeria monocytogenes* introduced on bologna-type sausage and smoked salmon. The same was claimed by Pandey et al. (2020).

8.4 MUSHROOMS IN THE TREATMENT OF METABOLIC DISORDERS

Hypertension, obesity, insulin resistance, and atherogenic dyslipidemia have become major health problems of modern society that are often manifesting together and are together referred to as metabolic syndrome. The development of diabetes and cardiovascular disease, which are among the leading causes of death worldwide (Grover & Joshi, 2014) is strongly as-sociated with metabolic syndrome (Kaur, 2014; Rochlani et al., 2017). Functional foods, such are mushrooms, are a source of biologically active components like vitamins, ω-3 fatty acids, antioxidants, and dietetic fibers, which are recognized as important factors in both prevention and the treatment of chronic diseases (Padmavathi, 2013; Raghavendra et al., 2018). Mushrooms also produce specific compounds that can interfere with metabolic pathways associated with the pathogenesis of these disorders, some of which are reviewed below.

8.4.1 Anti-hypercholesterolemic activity of mushrooms

In addition to antibiotics, fungal metabolites that have found a place in official medicine are also statins – a group of drugs used in the treatment of hypercholesterolemia. High levels of total and LDL cholesterol are one of the main risk factors for the development of diseases of the cardiovascular system and consequent mortality. Statins act as competitive inhibitors of 3-hydroxy-3-methyl-glutaryl CoA reductase (HMGCR), which is a rate-limiting enzyme in cholesterol synthesis, i.e. it dictates the speed of its synthesis. Lovastatin (Figure 8.8), a natural statin, belongs to the polyketide family (Atlı et al., 2015), and is mainly known to be produced by many soil fungi, although *Aspergillus terreus* is the most potent producer. It is also isolated from the mold *Monascus purpureus* (as monacolin K) (Gunde-Cimerman et al., 1993), which is used in the fermentation of red yeast rice, a traditional Asian food (Seenivasan et al., 2018). However, lovastatin has also been found to be a secondary metabolite of some macromycetes as well. Gunde-Cimerman, Plemenitaš, & Cimerman (1993) were first to identify the lovastatin in the fruiting bodies of oyster mushroom (*Pleurotus ostreatus*), using several detection methods (thin layer chromatography, HPLC, and mass spectro-scopy). Alarcón et al. (2003) further reported (by HPLC-UV) that lovastatin content in wild-growing *P. ostreatus* fruiting bodies from Chile can reach 2.8% of the dry weight. Some other studies however failed to detect lovastatin in *P. ostreatus* (Gil-Ramírez et al., 2013), while Tsiantas et al. (2021) reported only 1.11 mg/kg dry weight. Gunde-Cimerman et al. (1993) had previously reported great variability in the production of lovastatin by different *P. ostreatus* strains. They cultured the

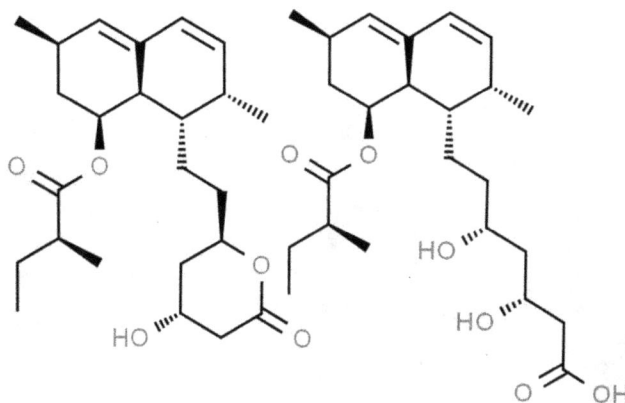

Figure 8.8 Chemical structures of lovastatin (1), and its active form, lovastatin acid (2).
Source: http://www.chemspider.com.

mycelium of 15 *P. ostreatus* strains and found that production of lovastatin ranged between 0.4 and 27.3 mg/L. In the same study, they confirmed that other species of the genus *Pleurotus* also produce lovastatin, *P. saca,* and particularly *P. sapidus* (41 mg/L). Traces of lovastatin (<1 mg/L) were found in the fermentation broths after submerged cultivation of *Agrocybe aegerita, Trametes versicolor, Agaricus bisporus*, and *Volvariella volvacea*. Subsequent studies reported various content of lovastatin from several mushroom species (mycelium and/or fruiting bodies), including *Agaricus blazei, Boletus edulis, Hypsizigus marmoreus, Ophiocordyceps sinensis* (as *Cordyceps sinensis*), *Omphalotus olearius*, and *Bovistella utriformis* (as *Handkea utriformis*), although the identification of the compound was based on HPLC analysis using a UV detector (Chen et al., 2012; Atlı et al., 2015, Petrović, Vunduk et al., 2019). A standardized method for isolation and determination of lovastatin from mushrooms is needed to obtain reliable information in the future.

Lovastatin acts as a prodrug, meaning that it needs to be metabolized to its active, β-hydroxy acid form ("lovastatin acid") (Valentovic, 2007). However, extracts of many mushroom species were shown to inhibit 3-hydroxy-3-methyl-glutaryl CoA reductase (HMGCR) *in vitro*, in a purified enzymatic system. This action cannot be associated with lovastatin, suggesting that there are other active principles with HMGCR inhibitory activity present in mushrooms. Gil-Ramírez et al. (2013) tested extracts obtained from 26 different mushroom species and found that most of them, mainly water extracts, exhibited some HMGCR inhibitory activity. There was no correlation between the phylogenetic relationship of the species and their activity, thus *P. ostreatus* was one of the most active species, while *P. eryngii* showed only slight inhibitory activity. Other species, besides *P. ostreatus*, that exerted significant inhibition of the enzyme included *Lentinula edodes, Amanita ponderosa,* and *Craterellus cornucopioides*. Petrović, Vunduk et al. (2019) tested methanol extracts of mycelium, and both immature and mature fruiting bodies of *B. utriformis* (as *H. utriformis*) and found that the HMGCR inhibitory activity decreased dramatically with the maturation of the fruiting bodies. The activity of all three samples correlated with lovastatin content (determined by HPLC-UV), so the authors suggested that it might be present in the β-hydroxy acid form as well, as seen in *Aspergillus terreus* (Jahromi et al., 2012), and that the correlation between lovastatin and the *in vitro* activity resembled correlation between lovastatin and lovastatin acid content in the extracts.

Photograph 8.6. *Pleurotus ostreatus*, an unexpected natural source of lovastatin. Crni vrh, Serbia, 2015. Photo: P. Petrović

Recently, triterpenoids isolated from *Ganoderma leucocontextum*, named ganoleucoins (Figure 8.9), were shown to exhibit HMGCR inhibitory activity, and some of these compounds were reported to be more active than atorvastatin. In the same study, some other previously known triterpenoids were identified in *G. leucocontextum* and tested for HMGCR inhibitory activity as well and some of them showed the same activity. These include ganoderiol J, ganoderic acid DM, SZ, Y, and ganodermanontriol (Wang et al., 2015; Zhang, Ma et al., 2018). Several meroterpenoids, isolated from *G. leucocontextum* and related species, were also found to be potent inhibitors of this enzyme, particularly the compound named ganomycin I (Wang, Bao et al., 2017).

Since statins inhibit the synthesis of mevalonate, which is a precursor to all isoprenoids, they can cause some side effects, and finding other potential inhibition sites in the cholesterol synthesis pathway has been the focus of research. Apart from their cytotoxic/antitumor activity, oxygenated lanosterol derivatives from *Ganoderma lucidum* were also investigated for their potential hypocholesterolemic ability. Some of the ganoderic alcohols, aldehydes, and acids (ganoderols A and B, ganoderal A, and ganoderic acid Y) were found to be very potent inhibitors of lanosterol 14α-demethylase, which converts 24,25-dihydro-lanosterol to cholesterol, thus being recognized as a promising new group of cholesterol-lowering drugs (Hajjaj et al., 2005).

Other mushroom metabolites were also found to have a cholesterol-lowering effect, which does not include inhibition of cholesterol synthesis. Eritadenine is an acyclic sugar adenosine analog isolated from *L. edodes* and was shown to have potent hypocholesterolemic activity in animal models. It inhibits S-adenosylhomocysteine hydrolase, an enzyme that indirectly affects phospholipid and fatty acid metabolism and consequently lipoprotein metabolism as well. It was shown that eritadenine causes the decrease of concentrations of, both plasma LDL and HDL, with no

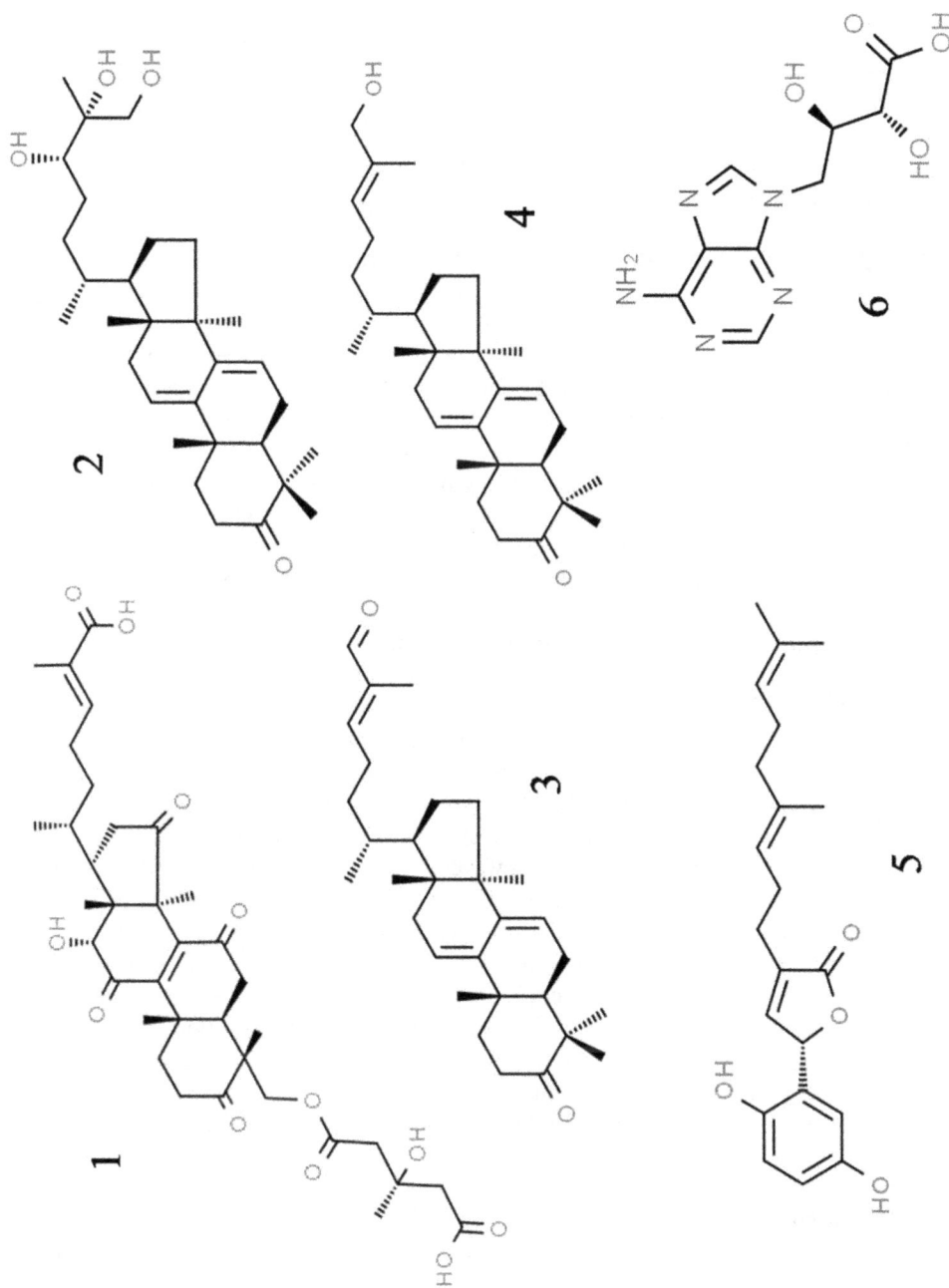

Figure 8.9 Selected mushroom compounds with anti-hypercholesterolemic activity: ganoleucoin K (1), ganodermanontriol (2), ganoderal A (3), ganoderol A (4), ganomycin I (5), eritadenine (6).

Source: http://www.chemspider.com.

effect on VLDL, suggesting that it stimulates tissues to take up cholesterol and phospholipids from higher-density lipoproteins. Many analogues of eritadenine have been synthesized and tested *in vivo*, leading to the discovery of compounds with even higher activity, primarily esters with short-chained hydroxy alcohols. Replacing the carboxyl/ester group with amide moiety leads to an inactive derivative. The analogs where N3 of the adenine moiety was replaced by CH to prolong the half-life of eritadenine were found to have lower S-adenosylhomocysteine hydrolase inhibitory activity, but the same *in vivo* effects, probably due to prolonged effect (Yamada et al., 2007).

Mushrooms are rich in fibers, such as chitin and β-glucans, and thus may also lower the amount of absorption of exogenous cholesterol, leading to a decrease in LDL. Many mushroom species have shown hypocholesterolemic activity *in vivo* studies (Gil-Ramírez et al., 2013), which probably involves several mechanisms of action.

8.4.2 Anti-hyperglicemic properties of mushroom products

Although essential for human organisms carbohydrates can exhibit detrimental effects if consumed in large quantities combined with an unhealthy lifestyle. Processed food makes a dominant part of the diet in many developed countries and even areas characterized by high consumption of local and healthy ingredients (like in Asia and Africa) are exposed to the explosion in consumption of food rich in carbohydrates (Ramachandran et al., 2012; Gill, 2014; Khan et al., 2020). In the long-term run, these life choices disrupt glucose homeostasis. There are two possible scenarios; a) pancreatic cells being unable to produce insulin which is a characteristic of type I diabetes mellitus also known as insulin-dependent type and b) impairment of insulin secretion or ineffective use of insulin are typical for type II diabetes. The latter occurs in 90%–95% of diabetes cases and is characterized by insulin resistance usually appearing after having a meal. As a consequence glucose concentration in blood overgoes its normal level which is known as hyperglycemia. Since type II diabetes usually does not require the administration of insulin, oral medicines are used. There are several types of drugs in use with different mechanisms of action like the increase of insulin sensitivity of target organ, gene expression modulation, the decrease of glucose absorption in the small intestine, attenuation of insulin signaling pathways, β-cell regeneration, and proliferation (Marín-Peñalver et al., 2016). Based on their study with polysaccharide fraction from *Coprinus comatus* on mice induced by alloxan Zhou et al. (2015) proposed that hypoglycemic activity is a consequence of immune stimulation. However, all existing treatments have side effects ranging from mild inconveniences to weight gain to heart failure (Kalsi et al., 2014). To provide effective and safe treatments natural sources are being constantly screened for their anti-hyperglycemic activity and mushrooms present a promising option. Most research papers are still dealing with this activity *in vitro* and by using crude extracts. As a result, there is still an urgent need for mushroom-based studies coupled with active compound isolation, identification, and characterization. So far, only one review paper gathered studies aiming at specific mushroom-derived compounds with anti-hyperglycemic activity dividing them into two major groups: polysaccharides and terpenoids (Aramabašić Jovanović et al., 2021). The same authors discussed mechanisms of mushroom-derived compounds known so far. Polysaccharides are known to act through the enhancement of pancreatic β-cell mass, an increase of insulin signaling, and inhibition of glucose absorption. When it comes to terpenoids anti-hyperglycemic effect is obtained via α-glucosidase inhibition and as insulin sensitizers. Inhibition of enzymes employed in carbohydrate metabolism is one of the most efficient strategies in managing blood level glucose (Su et al., 2013; Stojkovic et al., 2019). Thus, this ability of mushroom extracts became an interesting area of research in connection with diabetes. Mainly, there are two directions, one is to exploit mushrooms as a whole and integrate them into a healthy diet especially in a population with chronic health issues like diabetes type II.

Another one is the development of medications capable of treating the condition, alone or as a combination therapy.

While examining changes of selected functional characteristics of commercial *A. bisporus* (white and brown strain) stored in modified atmosphere packaging, Vunduk et al. (2021) discovered that methanol extracts of this widely popular species exhibit inhibition of two enzymes connected with the digestion of carbohydrates, α-amylase, and α-glucosidase. Contrary to other studies dealing with this topic, commercial button mushroom was better in inhibiting α-amylase (the maximal measured percentage of inhibition was 92.97). However, inhibition activity was dependable on the type of modified atmosphere as well as the storage period, and strain. For clear diet recommendations which will include the strain, moment of purchase, conditions of storage, and preparation method, more studies are necessary. Kumar et al. (2018) demonstrated the same ability of *A. bisporus* to inhibit α-amylase, but in this study the compounds of interest were polysaccharides. On the other hand, Su et al. (2013) reported that different types of extracts from *Grifola frondosa* medicinal mushrooms were better at inhibiting α-glucosidase. The most effective was n-hexane extract and connected the anti-hyperglycemic effect with oleic and linoleic acid. The authors proposed that the activity of these compounds occurred through retardation of the absorption of glucose from food. Other authors who examined the same species pointed to its glucans as being responsible for the anti-diabetic effect. Hong, Xun, and Wutong (2007) demonstrated the anti-diabetic ability of glucan from *G. frondosa* in a mice model with type II diabetes and proposed that it might be related to the effect on insulin receptors/insulin sensitivity. A thorough review of mushroom polysaccharide's antidiabetic activity was published by Ganesan and Xu (2019) with a discussion of the mechanisms of their activity. Mushrooms whose polysaccharides proved to be effective in glucose blood regulation are *L. edodes, Ganoderma atrum, G. frondosa, Agaricus subrufescens, Ganoderma lucidum, Pleurotus tuber-regium,* and *Trametes gibbosa.* Similarly, Fatmawati, Shimizu, and Kondo (2011) reported strong inhibition activity of *G. lucidum* extract toward α-glucosidase. The authors identified ganoderol B (Figure 8.10), a triterpenoid compound from $CHCl_3$ extract as responsible for the anti-hyperglycemic effect. Some novel approaches

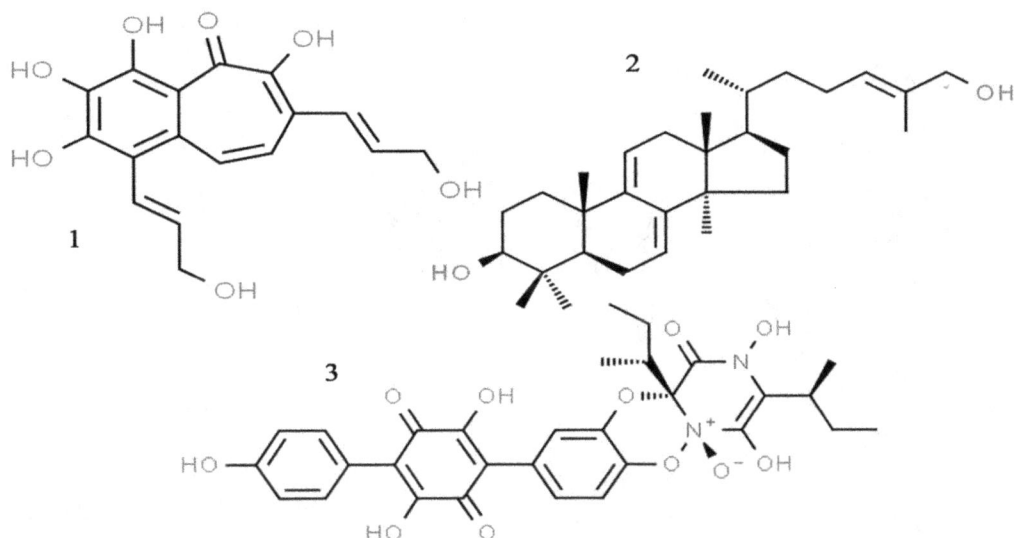

Figure 8.10 Selected mushroom compounds with anti-hyperglicemic activity: fomentariol (1), ganoderol B (2), sarcoviolinβ (3).

Source: http://www.chemspider.com.

concerning diabetes treatment include polyphenols so mushrooms fit into this category as well (Marín-Peñalver et al., 2016). The same anti-α-glucosidase activity was also reported for *Sarcodon leucopus* from Tibet (Ma et al., 2014). This study revealed the connection between the active compound (sarcoviolin *β*) and its chemical structure. Components with more expressed activity in α-glucosidase inhibition had *cis* configuration on N-1*β* and C-2*β*. In addition, the number of phenolic groups in the molecule's structure of p-terphenyl derivates was the most important for α-glucosidase inhibition. The most recent study exploring the ability of mushrooms to inhibit α-amylase and α-glucosidase activity was published by Stojkovic et al. (2019). In this study, methanol extracts of six mushroom species were examined (*A. blazei, C. comatus, Cordyceps militaris, Inonotus obliquus, Morchella conica,* and *Phellinus linteus)*. Activity toward both diabetes-connected enzymes was reported with *C. comatus* being the strongest inhibitor of α-amylase and *I. obliquus* of α-glucosidase. When it comes to the active components present in these extracts the authors proposed phenolic acids, organic acids, and fatty acids. Two more novel compounds with the ability to inhibit α-glucosidase were identified. Fomentariol, a benzotropolone compound isolated from tinder fungus *Fomes fomentarius,* was reported by Maljurić et al. (2018). Another one, phallac acid A and B, isolated from *Phallus luteus* became a new sesquiterpene from mushrooms capable to inhibit α-glucosidase (Lee et al., 2020).

In an extensive review, Gulati, Singh, and Gulati (2019) gave a chart of mushroom species whose antidiabetic properties were examined and, for the majority of species, polysaccharides were the active component.

Photograph 8.7. *Panaeolus cinctulus*, one of the psilocybin containing mushrooms, Zlot, Serbia, 2016. Photo: P. Petrović

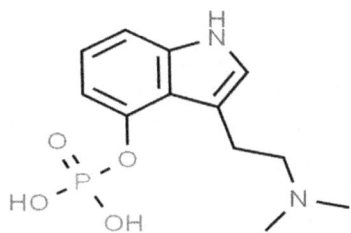

Figure 8.11 Chemical structure of psilocybin.
Source: http://www.chemspider.com.

Concerning the market, commercial products based on mushrooms and claiming to be effective in blood-glucose-lowering are rare. It is no wonder having in mind that clinical reports are scarce and cover several genera including *Agaricus*, *Coprinus*, *Ganoderma*, *Grifola*, and *Pleurotus* (Lindequist & Haertel 2020). One of the products with a relatively long commercial life is based on *C. comatus*. Its active substance was investigated in 2010 by Ding et al. and named comatin. The authors used several instrumental techniques like mass infra-red spectroscopy, nuclear magnetic resonance, und UV analysis, to elucidate the compounds' chemical nature, and identified it as 4,5-dihydroxy-2-methoxy-benzaldehyde. Comatin was tested in an animal model and was effective in maintaining low blood glucose levels as well as improving glucose tolerance.

8.5 MUSHROOM PRODUCTS IN THE TREATMENT OF MAJOR DEPRESSION – PSILOCYBIN

Since its modern-age rediscovery during the 1950s, psilocybin-containing mushrooms faced a scientific and public roller coaster. The reason was its powerful active compound, alkaloid psilocybin, and its analogs. An encounter with Western society attracted scientific pioneers who suspected that this alkaloid might find its place among contemporary pharmacological arsenals in combating psychiatric disorders like depression. Indeed, the research into this field restarted in the early 1990s after several successful experiments with human subjects conducted at the beginning of the second half of the 20th century (Reiche et al., 2018).

Psilocybin (Figure 8.11) is a naturally occurring substance from the tryptamine/indolamine hallucinogens group. Unlike other tryptamines, psilocybin is active when orally ingested since it is a prodrug of psilocin that is formed when psilocybin is hydrolyzed (Nichols, 1986). Psilocybin is a hydrophilic molecule but, once ingested, it undergoes dephosphorylation and, as a result, an increase in lipophilicity enables it to overcome the blood–brain barrier (Nichols, 1986; Tylš, Páleníček, & Horáček, 2014; Patra, 2016). There are several genera of mushrooms that contain this tryptamine: *Agrocybe*, *Conocybe*, *Weraroa*, *Galerina*, *Gymnopilus*, *Hypholoma*, *Inocybe*, *Pluteus*, *Panaeolus*, *Pholiotina*, and particularly *Psilocybe* (Wieczorek et al., 2015). From a biological perspective, psilocybin is a serotonin agonist, which binds to serotonin as well as dopamine receptors with different degrees of affinity. Its activity in humans is mainly expressed through the increase of dopamine release via the $5\text{-}HT_{2A}$ receptor (Tylš, Páleníček, & Horáček, 2014; Reiche et al., 2018). Those findings founded the direction in which to go with further studies. There are several directions in which psilocybin use might be found beneficial, chronic long-term-persistent depression, addiction to alcohol/tobacco, and possessive compulsive disorder (Carhart-Harris et al., 2017). The majority of research currently goes into the direction of treating depression.

According to the WHO, 264 million people are suffering from depression on a global level (https://www.who.int/news-room/fact-sheets/detail/depression). Treatments are currently based mainly on the use of antidepressants, which are often without effect or not efficient enough or express a wide range of side effects and psychotherapy. Under this setting, a new glimpse of hope came from the world of mushrooms. As Carhart-Harris et al. (2017) emphasized, if taken in a clinical setting, under strict professional guidance, and affirmative environment psilocybin treatments, show fast and long after-effects on mental health, even in healthy individuals. When 12 participants with mild to severe depression and a pre-history of failed medicaments-based therapy received psilocybin orally throughout a two-session with guidance and seven days pause between the sessions, a significant decrease in scores for the assessment of depression symptoms was observed (Carhart-Harris et al., 2016). In a later experiment, Carhart-Harris et al. (2017) went further in proving the psilocybin-based therapy of depression. The researchers used neuroimaging to confirm the changes present in the brain cortex. Brain functions were recorded with the help of an MRI scanner before and after psilocybin administration. As reported, all 19 patients had a positive reaction to the therapy. Finally, *the reset mechanism* of psilocybin treatment has been proposed in the meaning that the neuronal network goes through disintegration and desegregation. So, the old unhealthy mental patterns are broken and new healthy ones are established during these guided sessions. More importantly, a long-term effect was confirmed. In addition, treatment efficiency correlated with the increased right amygdala responsiveness to fearful faces one day after the oral ingestion of psilocybin (Roseman et al., 2018). In an extension of this study, Mertens et al. (2019) explained that changes in the functional connectivity between the amygdala and ventromedial prefrontal cortex are the proof of the revival of emotional responsiveness after psychedelic therapy. In a support of this finding, another group of researchers reported that psilocybin therapy affected the change in patients' mindfulness and openness based on the individual change in 5-HT2AR (serotonin receptor) (Madsen et al., 2020). According to the authors, this might represent the possible mechanism of psilocybin's long-term positive effects. Another study, by Barrett et al. (2020), provided the first empirical evidence that psilocybin alters the functional connectivity of a subcortical nucleus (claustrum) with brain networks supporting perception, memory, and attention. This study provided an additional mechanism of psilocybin action.

Most recently, Romeo et al. (2021) summarized the findings of psychedelic clinical studies involving psilocybin and hypothesized several mechanisms of psilocybin action: modulation of the serotoninergic system by 5-HT2A receptors agonism, modulation of the default mode network with an acute modular disintegration of the same network and its subsequent re-integration ending with normal functioning, and anti-inflammatory activity. Benville et al. (2021) conducted a randomized, double-blind, placebo-controlled, crossover-design trial of psilocybin use in cancer patients with a risk factor for suicidal ideation, and reported that even a single dose of psilocybin may cause persisting anti-suicidal effects.

Based on the controlled psilocybin studies, Germann (2020) proposed a bold medical hypothesis that connects psilocybin and genetic aging. Namely, beneficial effects on mental health can be connected with longer telomeres, further implying longer life expectancy. Since brain alterations with psilocybin are proven and on the mental level, it causes life-changing transformative experiences the author postulated that we can expect measurable changes on a genetic level. Furthermore, he proposes the inclusion of quantitative telomere analysis in future psilocybin studies.

8.6 MUSHROOM COMPOUNDS WITH VITAMIN ACTIVITY

The statement *mushrooms are rich, good, or abundant (even excellent) source of vitamins* is often to be found throughout popular literature, marketing purposes, and scientific literature (Mattila et al., 2002; Fulgoni & Agarwal, 2021). These claims are connected with mainly two vitamin groups: vitamins B and D. The question is whether or not these statements are scientifically supported. The answer is not straightforward and depends on the preciseness of the definition of vitamin and its metabolic active form. Moreover, it is not the same if we speak about an average amount of (pro) vitamin in mushroom's fruit body, transformation rut when ingested, and its bioavailability.

Vitamin D is important for many functions in the human body and is directly connected with calcium homeostasis and bone formation and health (Jasinghe et al., 2005). Its deficit occurs in populations all over the world, being more often in areas with little sun and with air pollution (Sławińska et al., 2016; Ložnjak & Jakobsen, 2018). The incidence of the deficiency of vitamin D is so severe that it can be classified as a pandemic, and is certainly connected with the fast industrialization of underdeveloped countries, sunscreen usage, and worldwide air pollution. Classical dietetic supplements are one type of solution but more and more natural sources and fortified food present a good alternative (Salemi et al., 2021). It is well known that food of animal origin serves as a source of vitamin D (Jasinghe & Perera, 2005). However, there are different diet groups, like vegetarian and vegan, as well as people with health conditions like food allergies or lactose intolerance, where this type of food cannot be used (Jasinghe & Perera, 2005). For many people, dietetic supplements (which are mainly based on vitamin D_3 originating from animals) are of no use due to their cost, hypercalcemic effects, and personal principles; thus, other approaches have to be developed (Wu & Ahn, 2014). Mushrooms can be a viable solution to all mentioned issues since they are the only non-animal natural source of vitamin D. The kingdom of fungi synthesizes ergosterol (via mevalonate pathway), which is a pro-vitamin of vitamin D_2 (Hashim et al., 2016). Similar to humans, whose skin synthesizes vitamin D_3 (cholecalciferol) from cholesterol when exposed to the sun's UV light, mushroom's ergosterol transforms to ergocalciferol (vitamin D_2) as a result of UV exposure and thermal reaction (Salemi et al., 2021). Although some mushroom species are rich in ergosterol this does not necessarily mean that they will contain vitamin D_2. Species growing in nature and exposed to the sun will contain this vitamin, but the amount will depend on the initial ergosterol level, geographical location, species and strain, duration of sun exposure, maturity level, and temperature (Barreira, Oliveira, & Ferreira, 2013; Villares et al., 2014; Jiang, Zhang, & Mujumdar, 2020). Interestingly, Barreira, Oliveira, and Ferreira (2013) reported that cultivated mushrooms tend to synthesize more ergosterol, especially *Agaricus bisporus*. Among cultivated species, *A. bisporus*, *Lentinula edodes,* and *Pleurotus* spp. are considered as having the highest content of ergosterol (Jiang, Zhang, & Mujumdar, 2020). To use this potential of mushrooms a UV-irradiation technique has been widely explored. Different authors reported the use of UV light within several wavelength ranges with UV-B as the most effective in transforming ergosterol to ergocalciferol (Salemi et al., 2021). Other important factors in the photothermal transformation of ergosterol are the time of exposure, distance from the irradiation source, mushroom specific surface, humidity, fruit body part and orientation, as well as pretreatment (Jasinghe & Perera, 2005; Jiang, Zhang, & Mujumdar, 2020). Moreover, it has been proven that irradiating sliced and dried mushrooms are even more effective since conditions present during drying are naturally beneficial for ergosterol transformation (Sławińska et al., 2016). Exposure of ergosterol to UV light stimulates a photochemical reaction in which previtamin D_2 is created. Finally, thermal reaction provides isomerization to vitamin D_2. However, the process does not necessarily stop here. If the conditions are inadequate degradation will occur. Irradiating mushrooms during the first phase of drying (while the humidity level is still high) prevents unwanted enzyme-induced (monooxygenase) vitamin transformation (Sławińska et al., 2016).

As being highly efficient and safe, mushroom irradiation even became recognized and accepted, so it is a part of EU regulative for novel food, Regulation (EC) No. 258/97. The regulation applies for commercially grown *A. bisporus* specifically. Defined by this regulation mushrooms are irradiated after being harvested, with UV light within the range of 200–800 nm. The regulation also defines the maximum vitamin D_2 level and labeling requirements.

Another still not thoroughly explored issue is mushroom-originating vitamin D_2 bioavailability . There are not many studies dealing with this topic and the results are still not conclusive. The majority of them indicate that D_2 from mushrooms is bioavailable in most of the examined cases. Vitamin D_2 bioavailability indicator is 25-hydroxyvitamin D. This is its main form in blood serum and has a half-life of seven days (Jasinghe, Perera, & Barlow, 2005; Keegan et al., 2013). A study on rats showed that a diet based on irradiated mushrooms stimulated an increase in 25-hydroxyvitamin D serum level (Babu et al., 2013). Studies with humans reported both positive outcomes as well as the lack of indicator serum level increase. In the first case, mushrooms were prepared as soup and administered to 26 healthy young subjects over a five weeks course. Urbain et al. (2011) reported that vitamin D_2 from irradiated mushrooms was bioavailable. Moreover, it was not different from the vitamin D_2 supplement. The same has been confirmed by Keegan et al. (2013), but in this study, *L. edodes* was used instead of a white button mushroom. The authors also proved that mushrooms can produce vitamin D_4, which is obtained from the UV irradiation of provitamin D_4 or 22,23-dihydroergosterol. Studies conducted by Stephensen et al. (2012) and Bogusz, Pagonis, and Holick (2013) also confirmed an increase in indicator serum level after consumption of irradiated mushrooms. In the case of a review of randomized controlled trials conducted by Cashman et al. (2016) increase in 25-hydroxyvitamin D serum level appeared in cases with previously low vitamin D status. For those individuals with a high level of vitamin D, supplementation with irradiated mushrooms did not show any significant increase. Another study dealing with vitamin D_2 bioavailability examined 25-hydroxyvitamin D serum levels in prediabetic adult subjects over 16 weeks (Mehrotra et al., 2014). Mushrooms were given in a form of a meal with no effect on indicator serum increase. However, as the authors stated by themselves, the study had many limitations like the meal preparation and administration procedure which included meal freezing and reheating in a microwave. In addition, the study had a small sample size, an unknown amount of vitamin D_2 level in the dish after being cooked, and one group of subjects received a supplement with a significantly higher dose of vitamin D_3.

It should be mentioned that orally taken ergosterol does not possess vitamin D activity, since it is metabolized to brassicasterol and not transported to skin where it could be transformed to its active form by UV light (Tsugawa et al., 1992).

Tocopherols (vitamin E) are also present in mushrooms, but the information about their content is limited. Barros et al. (2008) and Heleno et al. (2010) examined tocopherol profile and concentration in various wild-growing mushrooms and found that total tocopherol content was only 0.016–8 mg/kg of mushrooms fresh weight. β-tocopherol was found to be the main tocopherol in most mushroom samples, while μ-and α-tocopherol were found in lesser amounts. Petrović, Vunduk et al. (2019). There are several vitamins from this group that examined α-tocopherol content in *Bovistella utriformis* (as *Handkea utriformis*), as the biologically most active form, and found that during maturation of this puffball species the extractable α-tocopherol content increases. The authors suggested that this is due to the specific maturation process of puffballs as they go through autolysis, which may lead to the liberation of tocopherols and other lipid membrane constituents and easier extraction.

Although there is a general opinion that mushrooms are good sources of vitamins of the B complex, particularly B_1 and B_2, there are very few studies devoted specifically to these physiologically important molecules. There are several vitamins from this group that were identified in mushrooms: thiamine (B_1), riboflavin (B_2), pyridoxine (B_6), and cyanocobalamin (B_{12}). *Pleurotus ostreatus* is considered as having the highest level of vitamin B_1, 0.9–4.25 mg/100 g of dry matter (Bernaś, Jaworska, & Lisiewska, 2006). In another study, thiamine level was assessed in six cultivated *Pleurotus* species (Krakowska et al., 2020). Contrary to the previous report significantly lower amounts of vitamin B_1 were measured, ranging from 0.15 to 1.08 mg/100 g of dry weight. Moreover, the same study showed that mycelium of the same species had a bit higher thiamine content, 0.17–2.01 mg/100 g of dry weight. According to Chye et al. (2008), wild mushrooms are not so abundant in vitamin B_1. On the contrary, fresh Suillus luteus contained a very high amount of this vitamin (11.2 mg/100 g of dry weight) (Jaworska et al., 2014). Even blanched, these samples still had more thiamine than reported for other species in other studies. However, other processes like canning, storage, and freezing, significantly lowered the content of vitamins B_1, B_2, B_3, and B_6 in cultivated mushrooms (Furlani & Godoy, 2008). In general, mushrooms are modest sources of thiamine, comparable to most vegetables. Similarly, mushrooms might contribute as sources of riboflavin, better than most vegetables. B_2 content varies from species to species, and the type of substrate also affects it (Chye et al., 2008; Furlani & Godoy, 2008; Jaworska et al., 2014). This vitamin undergoes degradation when

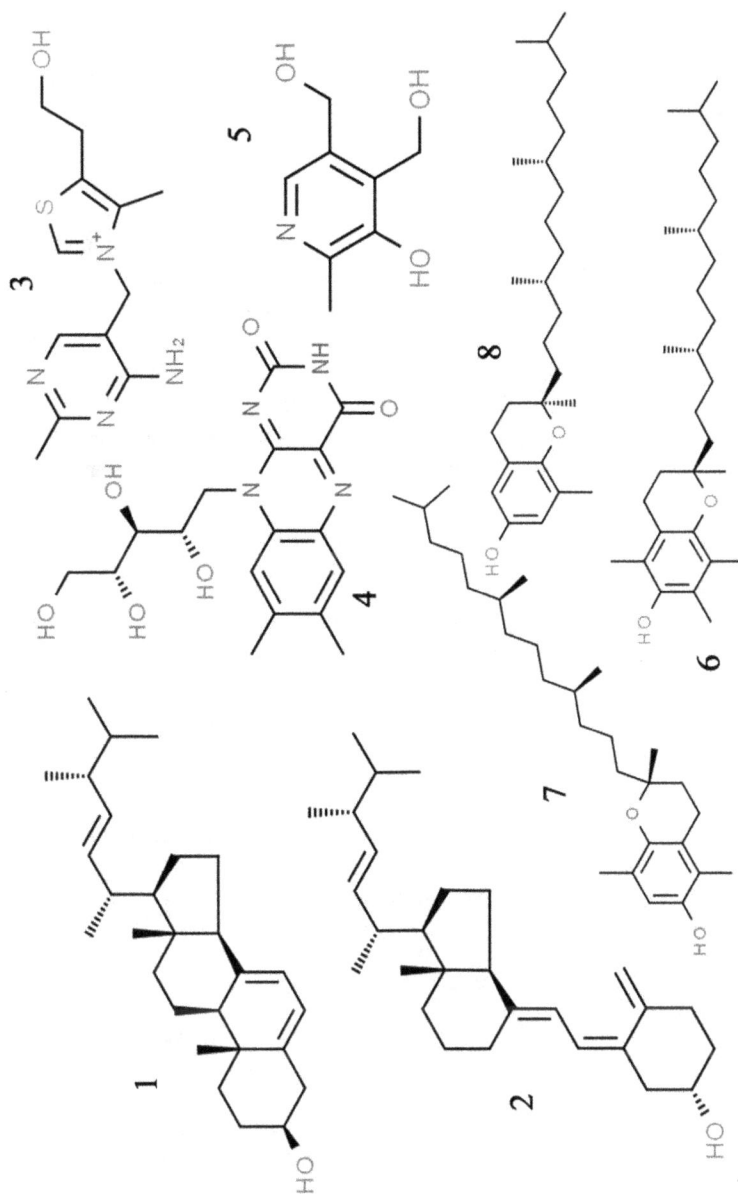

Figure 8.12 (Pro)vitamins produced by mushrooms: ergosterol (1), ergocalciferol, thiamine (3), riboflavin (4), pyridoxine (5), α-tocopherol (6), β-tocopherol (7), γ-tocopherol.

Source: http://www.chemspider.com.

mushrooms are exposed to heat (like during cooking) or just by being stored. Freezing did not show a negative effect on the B_2 level in *S. luteus* (Jaworska et al., 2014).

Vitamin B_{12} (cobalamin)vegetables and fermented food is produced by bacteria exclusively and appears mainly in food of animal origin and insignificant amounts in plant-based food. It acts as a co-enzyme. Deficiency might occur in the elder population andvegetarians and among solutions are consumption of fortified cereals or vegetables and fermented food (Watanabe et al., 2014). In addition, it might be taken via mushroom-based meals. Research showed that species like *L. edodes, Craterellus cornucopioides*, or *Cantharellus cibarius* might contain this vitamin, probably originating from their surrounding area, substrate, or soil (Bito et al., 2014). When food containing B_{12} is chewed it mixes with saliva containing protein R, which binds to vitamin B_{12}. Next, parietal cells of the stomach secrete intrinsic factor which binds to vitamin B_{12} until it is reabsorbed by the distal ileum (Zik, 2019). There are also inactive forms like $B_{12}[c\text{-lactone}]$ which has low specific protein binding affinity.

The bioavailability of B_{12} from animal sources varies from 40% to 89% (Banjari & Hjartåker, 2018). There are several biologically active forms of vitamin B_{12}, denosylcobalamin, cyanocobalamin, hydroxocobalamin, and methylcobalamin. Mushrooms mainly contain active forms but inactive $B_{12}[c\text{-lactone}]$ was also identified in *Hericium erinaceus* (Watanabe et al., 2014). A study conducted by) showed that vitamin B_{12} maintained high bioavailability when subjects consumed bread prepared from fortified flour. Recently, it was also pointed that B_{12} bioavailability can be improved by using delivery systems like poly(acrylic acid)-cysteine (Sarti et al., 2013). However, no studies deal with *mushroom origin* B_{12} bioavailability. Selected (pro)vitamins found in mushrooms are given in Figure 8.12.

ACKNOWLEDGMENT

This work was supported by the Ministry of Education, Science and Technological Development of the Republic of Serbia (Contract No. 451-03-68/2022-14/200135 & Contract No. 451-03-68/2022-14/200051)

REFERENCES

Adachi, K., Nanba, H., & Kuroda, H. (1987). Potentiation of host-mediated antitumor activity in mice by β-glucan obtained from Grifola frondosa (maitake). *Chemical and Pharmaceutical Bulletin*, *35*(1), 262–270.

Adams, E. L., Rice, P. J., Graves, B., Ensley, H. E., Yu, H., Brown, G. D., Gordon, S., Monteiro, M. A., Papp-Szabo, E., Lowman, D. W., Power, T. D., Wempe, M. F., & Williams, D. L. (2008). Differential high-affinity interaction of dectin-1 with natural or synthetic Glucans is dependent upon primary structure and is influenced by polymer chain length and side-chain branching. *Journal of Pharmacology and Experimental Therapeutics*, *325*(1), 115–123.

Akiyama, H., Endo, M., Matsui, T., Katsuda, I., Emi, N., Kawamoto, Y., Koike, T., & Beppu, H. (2011). Agaritine from Agaricus blazei Murrill induces apoptosis in the leukemic cell line U937. *Biochimica et Biophysica Acta (BBA) – General Subjects*, *1810*(5), 519–525.

Alarcón, J., Águila, S., Arancibia-Avila, P., Fuentes, O., Zamorano-Ponce, E., & Hernández, M. (2003). Production and purification of statins from Pleurotus ostreatus (Basidiomycetes) strains. *Zeitschrift für Naturforschung C*, *58*(1–2), 62–64.

Alexander, E., Goldberg, L., Das, A. F., Moran, G. J., Sandrock, C., Gasink, L. B., Spera, P., Sweeney, C., Paukner, S., Wicha, W. W., Gelone, S. P., & Schranz, J. (2019). Oral Lefamulin vs Moxifloxacin for early clinical response among adults with community-acquired bacterial pneumonia. *JAMA*, *322*(17), 1661.

Alexandre, J., Kahatt, C., Bertheault-Cvitkovic, F., Faivre, S., Shibata, S., Hilgers, W., Goldwasser, F., Lokiec, F., Raymond, E., Weems, G., Shah, A., MacDonald, J. R., & Cvitkovic, E. (2007). A phase I and pharmacokinetic study of irofulven and capecitabine administered every 2 weeks in patients with advanced solid tumors. *Investigational New Drugs*, *25*(5), 453–462.

Alquini, G., Carbonero, E. R., Rosado, F. R., Cosentino, C., & Iacomini, M. (2004). Polysaccharides from the fruit bodies of the basidiomycete Laetiporus sulphureus (Bull.: Fr.) Murr. *FEMS Microbiology Letters*, *230*(1), 47–52.

Amato, R. J., Perez, C., & Pagliaro, L. (2002). Irofulven, a novel inhibitor of DNA synthesis, in metastatic renal cell cancer. *Investigational New Drugs*, *20*, 413–417.

Anchel, M., Hervey, A., & Robbins, W. J. (1950). Antibiotic substances from basidiomycetes: VII. Clitocybe Illudens. *Proceedings of the National Academy of Sciences*, *36*(5), 300–305.

Anke, T. (2020). Secondary metabolites from mushrooms. *The Journal of Antibiotics*, *73*(10), 655–656.

Anke, T., Kupka, J., Schramm, G., & Steglich, W. (1980). Antibiotics from basidiomycetes. X. Scorodonin, a new antibacterial and antifungal metabolite from marasmius scorodonius (Fr.) Fr. *The Journal of Antibiotics*, *33*(5), 463–467.

Anke, T., & Steglich, W. (1981). Screening of basidiomycetes for the production of new antibiotics. *Scientific and Engineering Principles* (pp. 35–40). Ontario: Pergamon Press.

Antonelli, M., Donelli, D., & Firenzuoli, F. (2020). Lentinan for integrative cancer treatment: An umbrella review. *The 1st International Electronic Conference on Biomolecules: Natural and Bio-Inspired Therapeutics for Human Diseases*. At: Online Conference. 10.3390/iecbm2020-08733

Aramabašić Jovanović, J., Mihailović, M., Uskoković, A., Grdović, N., Dinić, S., & Vidaković, M. (2021). The effects of major mushroom Bioactive compounds on mechanisms that control blood glucose level. *Journal of Fungi*, *7*(1), 58.

Atlı, B., Yamaç, M., Yıldız, Z., & Isikhuemnen, O. S. (2015). Enhanced production of lovastatin by Omphalotus olearius (DC.) singer in solid state fermentation. *Revista Iberoamericana de Micología*, *32*(4), 247–251.

Babu, U. S., Balan, K. V., Garthoff, L. H., & Calvo, M. S. (2013). Vitamin D2 from UVB light exposed mushrooms modulates immune response to LPS in rats. *Molecular Nutrition & Food Research*, *58*(2), 318–328.

Bai, Y., Feng, Y., Mao, D., & Xu, C. (2012). Optimization for betulin production from mycelial culture of Inonotus obliquus by orthogonal design and evaluation of its antioxidant activity. *Journal of the Taiwan Institute of Chemical Engineers*, *43*(5), 663–669.

Banjari, I., & Hjartåker, A. (2018). Dietary sources of iron and vitamin B12: Is this the missing link in colorectal carcinogenesis? *Medical Hypotheses*, *116*, 105–110.

Barreira, J. C., Oliveira, M. B., & Ferreira, I. C. (2013). Development of a novel methodology for the analysis of ergosterol in mushrooms. *Food Analytical Methods*, *7*(1), 217–223.

Barrett, F. S., Krimmel, S. R., Griffiths, R. R., Seminowicz, D. A., & Mathur, B. N. (2020). Psilocybin acutely alters the functional connectivity of the claustrum with brain networks that support perception, memory, and attention. *NeuroImage*, *218*, 116980.

Barros, L., Venturini, B. A., Baptista, P., Estevinho, L. M., & Ferreira, I. C. (2008). Chemical composition and biological properties of Portuguese wild mushrooms: A comprehensive study. *Journal of Agricultural and Food Chemistry*, *56*(10), 3856–3862.

Beattie, A. J., Hay, M., Magnusson, B., De Nys, R., Smeathers, J., & Vincent, J. F. (2010). Ecology and bioprospecting. *Austral Ecology*, *36*(3), 341–356.

Benville, J., Agin-Liebes, G., Roberts, D. E., Lo, S., Ghazal, L., Franco-Corso, S. J., & Ross, S. (2021). Effects of psilocybin on suicidal ideation in patients with life-threatening cancer. *Biological Psychiatry*, *89*(9), S235–S236.

Berg, W. J., Schwartz, L., Yu, R., Mazumdar, M., & Motzer, R. J. (2001). Phase II trial of irofulven (6-hydroxymethylacylfulvene) for patients with advanced renal cell carcinoma. *Investigational New Drugs*, *19*(4), 317–320.

Bernaś, E., Jaworska, G., & Lisiewska, Z. (2006). Edible mushrooms as a source of valuable nutritive constituents. *Acta Scientiarum PolonorumTechnologia Alimentaria*, *5*(1), 5–20.

Bhardwaj, K., Sharma, A., Tejwan, N., Bhardwaj, S., Bhardwaj, P., Nepovimova, E., Shami, A., Kalia, A., Kumar, A., Abd-Elsalam, K. A., & Kuča, K. (2020). Pleurotus macrofungi-assisted Nanoparticle synthesis and its potential applications: A review. *Journal of Fungi*, *6*(4), 351.

Bito, T., Teng, F., Ohishi, N., Takenaka, S., Miyamoto, E., Sakuno, E., Terashima, K., Yabuta, Y., & Watanabe, F. (2014). Characterization of vitamin B12 compounds in the fruiting bodies of shiitake mushroom (Lentinula edodes) and bed logs after fruiting of the mushroom. *Mycoscience*, *55*(6), 462–468.

Blagodatski, A., Yatsunskaya, M., Mikhailova, V., Tiasto, V., Kagansky, A., & Katanaev, V. L. (2018). Medicinal mushrooms as an attractive new source of natural compounds for future cancer therapy. *Oncotarget, 9*(49), 29259–29274.

Blair, L. M., & Sperry, J. (2013). Natural products containing a nitrogen–nitrogen bond. *Journal of Natural Products, 76*(4), 794–812.

Bogusz, J., Pagonis, G., & Holick, M. F. (2013). Evaluation of the bioavailability of vitamin D2 in mushrooms in healthy adults. *Experimental Biology 2013 Meeting Abstracts, 27*(S1), 794.4.

Bomgaars, L. R., Megason, G. C., Pullen, J., Langevin, A., Dale Weitman, S., Hershon, L., Kuhn, J. G., Bernstein, M., & Blaney, S. M. (2006). Phase I trial of irofulven (MGI 114) in pediatric patients with solid tumors. *Pediatric Blood & Cancer, 47*(2), 163–168.

Butler, M. S. (2008). Natural products to drugs: Natural product-derived compounds in clinical trials. *Natural Product Reports, 25*(3), 475.

Bézivin, C., Lohézic, F., Sauleau, P., Amoros, M., & Boustie, J. (2002). Cytotoxic activity of Tricholomatales determined with murine and human cancer cell lines. *Pharmaceutical Biology, 40*(3), 196–199.

Calvino, R., Fruttero, R., Gasco, A., Miglietta, A., & Gabriel, L. (1986). Chemical and biological studies on calvatic acid and its analogs. *The Journal of Antibiotics, 39*(6), 864–868.

Camelini, C. M., Maraschin, M., De Mendonça, M. M., Zucco, C., Ferreira, A. G., & Tavares, L. A. (2005). Structural characterization of β-glucans of Agaricus brasiliensis in different stages of fruiting body maturity and their use in nutraceutical products. *Biotechnology Letters, 27*(17), 1295–1299.

Camilli, G., Tabouret, G., & Quintin, J. (2018). The complexity of fungal β-glucan in health and disease: Effects on the mononuclear phagocyte system. *Frontiers in Immunology, 9*. 10.3389/fimmu.2018.00673

Cao, Y., Xu, X., Liu, S., Huang, L., & Gu, J. (2018). Ganoderma: A cancer immunotherapy review. *Frontiers in Pharmacology, 9*. 10.3389/fphar.2018.01217

Carbonero, E. R., Ruthes, A. C., Freitas, C. S., Utrilla, P., Gálvez, J., Silva, E. V., Sassaki, G. L., Gorin, P. A., & Iacomini, M. (2012). Chemical and biological properties of a highly branched β-glucan from edible mushroom Pleurotus sajor-caju. *Carbohydrate Polymers, 90*(2), 814–819.

Carhart-Harris, R. L., Bolstridge, M., Rucker, J., Day, C. M., Erritzoe, D., Kaelen, M., Bloomfield, M., Rickard, J. A., Forbes, B., Feilding, A., Taylor, D., Pilling, S., Curran, V. H., & Nutt, D. J. (2016). Psilocybin with psychological support for treatment-resistant depression: An open-label feasibility study. *The Lancet Psychiatry, 3*(7), 619–627.

Carhart-Harris, R. L., Roseman, L., Bolstridge, M., Demetriou, L., Pannekoek, J. N., Wall, M. B., Tanner, M., Kaelen, M., McGonigle, J., Murphy, K., Leech, R., Curran, H. V., & Nutt, D. J. (2017). Psilocybin for treatment-resistant depression: FMRI-measured brain mechanisms. *Scientific Reports, 7*(1). 10.1038/s41598-017-13282-7

Cashman, K. D., Kiely, M., Seamans, K. M., & Urbain, P. (2016). Effect of ultraviolet light–exposed mushrooms on vitamin D status: Liquid chromatography–tandem mass spectrometry reanalysis of Biobanked sera from a randomized controlled trial and a systematic review plus meta-analysis. *The Journal of Nutrition, 146*(3), 565–575.

Chakraborty, I., Mondal, S., Rout, D., & Islam, S. S. (2006). A water-insoluble (1→3)-β-d-glucan from the alkaline extract of an edible mushroom Termitomyces eurhizus. *Carbohydrate Research, 341*(18), 2990–2993.

Chan, G. C., Chan, W. K., & Sze, D. M. (2009). The effects of β-glucan on human immune and cancer cells. *Journal of Hematology & Oncology, 2*(1). 10.1186/1756-8722-2-25

Chawla, P., Kumar, N., Kaushik, R., & Dhull, S. B. (2019). Synthesis, characterization and cellular mineral absorption of gum arabic stabilized nanoemulsion of Rhododendron arboreum flower extract. *Journal of Food Science and Technology, 56*(12), 5194–5203.

Chen, N., Liu, J., & Zhong, J. (2010). Ganoderic acid T inhibits tumor invasion in vitro and in vivo through inhibition of MMP expression. *Pharmacological Reports, 62*(1), 150–163.

Chen, N., & Zhong, J. (2009). Ganoderic acid me induces G1 arrest in wild-type p53 human tumor cells while G1/S transition arrest in p53-null cells. *Process Biochemistry, 44*(8), 928–933.

Chen, S., Ho, K., Hsieh, Y., Wang, L., & Mau, J. (2012). Contents of lovastatin, γ-aminobutyric acid and ergothioneine in mushroom fruiting bodies and mycelia. *LWT, 47*(2), 274–278.

Chye, F. Y., Wong, J. Y., and Lee, J. S. 2008. Nutritional Quality and Antioxidant Activity of Selected Edible Wild Mushrooms. *Food Science and Technology International, 14*(4): 375–384.

Clough, J. M. (1993). The strobilurins, oudemansins, and myxothiazols, fungicidal derivatives of β-methoxyacrylic acid. *Nat. Prod. Rep*, *10*(6), 565–574.

Czarska-Thorley, D. 2020. Xenleta – European Medicines Agency.European Medicines Agency. May 26, 2020. https://www.ema.europa.eu/en/medicines/human/EPAR/xenleta

Dai, H., Han, X., Gong, F., Dong, H., Tu, P., & Gao, X. (2012). Structure elucidation and immunological function analysis of a novel β--glucan from the fruit bodies of polyporus umbellatus (Pers.) fries. *Glycobiology*, *22*(12), 1673–1683.

Dalonso, N., Goldman, G. H., & Gern, R. M. (2015). β-(1→3),(1→6)-Glucans: Medicinal activities, characterization, biosynthesis and new horizons. *Applied Microbiology and Biotechnology*, *99*(19), 7893–7906.

De Mattos-Shipley, K. M., Foster, G. D., & Bailey, A. M. (2017). Insights into the classical genetics of Clitopilus passeckerianus – the Pleuromutilin producing mushroom. *Frontiers in Microbiology*, *8*. 10.3389/fmicb.2017.01056

Dehelean, C. A., Marcovici, I., Soica, C., Mioc, M., Coricovac, D., Iurciuc, S., Cretu, O. M., & Pinzaru, I. (2021). Plant-derived Anticancer compounds as new perspectives in drug discovery and alternative therapy. *Molecules*, *26*(4), 1109.

Deng, G., Lin, H., Seidman, A., Fornier, M., D'Andrea, G., Wesa, K., Yeung, S., Cunningham-Rundles, S., Vickers, A. J., & Cassileth, B. (2009). A phase I/II trial of a polysaccharide extract from Grifola frondosa (Maitake mushroom) in breast cancer patients: Immunological effects. *Journal of Cancer Research and Clinical Oncology*, *135*(9), 1215–1221.

Depression. (2020, January 30). WHO | World Health Organization. https://www.who.int/news-room/fact-sheets/detail/depression

Dong, Q., Wang, Y., Shi, L., Yao, J., Li, J., Ma, F., & Ding, K. (2012). A novel water-soluble β-D-glucan isolated from the spores of Ganoderma lucidum. *Carbohydrate Research*, *353*, 100–105.

Dong, Q., Yao, J., Yang, X., & Fang, J. (2002). Structural characterization of a water-soluble β-D-glucan from fruiting bodies of Agaricus blazei Murr. *Carbohydrate Research*, *337*(15), 1417–1421.

Endo, M., Beppu, H., Akiyama, H., Wakamatsu, K., Ito, S., Kawamoto, Y., Shimpo, K., Sumiya, T., Koike, T., & Matsui, T. (2010). Agaritine purified from Agaricus blazei Murrill exerts anti-tumor activity against leukemic cells. *Biochimica et Biophysica Acta (BBA) - General Subjects*, *1800*(7), 669–673.

Engler, M., Anke, T., & Sterner, O. (1998). Production of antibiotics by Collybia nivalis, Omphalotus olearius, a Favolaschia and a Pterula species on natural substrates. *Zeitschrift für Naturforschung C*, *53*(5–6), 318–324.

Fatmawati, S., Shimizu, K., & Kondo, R. (2011). Ganoderol B: A potent α-glucosidase inhibitor isolated from the fruiting body of Ganoderma lucidum. *Phytomedicine*, *18*(12), 1053–1055.

File, T. M., Goldberg, L., Das, A., Sweeney, C., Saviski, J., Gelone, S. P., Seltzer, E., Paukner, S., Wicha, W. W., Talbot, G. H., & Gasink, L. B. (2019). Efficacy and safety of intravenous-to-oral Lefamulin, a Pleuromutilin antibiotic, for the treatment of community-acquired bacterial pneumonia: The phase III Lefamulin evaluation against pneumonia (LEAP 1) trial. *Clinical Infectious Diseases*, *69*(11), 1856–1867.

Fortin, H., Tomasi, S., Delcros, J., Bansard, J., & Boustie, J. (2006). In vivo antitumor activity of Clitocine (I), an Exocyclic amino nucleoside isolated from Lepista inversa. *ChemMedChem*, *1*(2), 189–196.

Friedman, M. (2016). Mushroom polysaccharides: Chemistry and antiobesity, antidiabetes, anticancer, and antibiotic properties in cells, rodents, and humans. *Foods*, *5*(4), 80.

Friesen, W. J., Trotta, C. R., Tomizawa, Y., Zhuo, J., Johnson, B., Sierra, J., Roy, B., Weetall, M., Hedrick, J., Sheedy, J., Takasugi, J., Moon, Y., Babu, S., Baiazitov, R., Leszyk, J. D., Davis, T. W., Colacino, J. M., Peltz, S. W., & Welch, E. M. (2017). The nucleoside analog clitocine is a potent and efficacious readthrough agent. *RNA*, *23*(4), 567–577.

Fulda, S., & Debatin, K. (2000). Betulinic acid induces apoptosis through a direct effect on mitochondria in neuroectodermal tumors. *Medical and Pediatric Oncology*, *35*(6), 616–618.

Fulda, S., Friesen, C., Los, M., Scaffidi, C., Mier, W., Benedict, M., Nuñez, G., Krammer, P. H., Peter, M. E., & Debatin, K. (1997). Betulinic acid triggers CD95 (APO-1/Fas)- and p53-independent apoptosis via activation of caspases in neuroectodermal tumors. *Cancer Research*, *57*, 4956–4964.

Fulda, S., Jeremias, I., Steiner, H. H., Pietsch, T., & Debatin, K. (1999). Betulinic acid: A new cytotoxic agent against malignant brain-tumor cells. *International Journal of Cancer*, *82*(3), 435–441.

Fulgoni, V. L., & Agarwal, S. (2021). Nutritional impact of adding a serving of mushrooms on usual intakes and nutrient adequacy using national health and nutrition examination survey 2011–2016 data. *Food Science & Nutrition, 9*(3), 1504–1511.

Furlani, R. P., & Godoy, H. T. (2008). Vitamins B1 and B2 contents in cultivated mushrooms. *Food Chemistry, 106*(2), 816–819.

Gadoni, E., Gabriel, L., Olivero, A., Bocca, C., & Miglietta, A. (1995). Antimicrotubular effect of calvatic acid and of some related compounds. *Cell Biochemistry and Function, 13*(4), 231–238.

Ganesan, K., & Xu, B. (2019). Anti-diabetic effects and mechanisms of dietary polysaccharides. *Molecules, 24*(14), 2556.

Garrod, M. G., Buchholz, B. A., Miller, J. W., Haack, K. W., Green, R., & Allen, L. H. (2019). Vitamin B12 added as a fortificant to flour retains high bioavailability when baked in bread. *Nuclear Instruments and Methods in Physics Research Section B: Beam Interactions with Materials and Atoms, 438*, 136–140.

Gasco, A., Serafino, A., Mortarini, V., Menziani, E., Bianco, M., & Ceruti Scurti, J. (1974). An antibacterial and antifungal compound from Calvatia lilacina. *Tetrahedron Letters, 15*(38), 3431–3432.

Germann, C. B. (2020). The psilocybin-telomere hypothesis: An empirically falsifiable prediction concerning the beneficial neuropsychopharmacological effects of psilocybin on genetic aging. *Medical Hypotheses, 134*, 109406. 10.1016/j.mehy.2019.109406

Giles, F., Cortes, J., Garcia-Manero, G., Kornblau, S., Estey, E., Kwari, M., Murgo, A., & Kantarjian, H. (2001). Phase I study of irofulven (MGI 114), an acylfulvene illudin analog, in patients with acute leukemia, *19*(1), 13–20. 10.1023/a:1006432012394

Gill, G. (2014). Diabetes in Africa – Puzzles and challenges. *Indian Journal of Endocrinology and Metabolism, 18*(3), 249.

Gil-Ramírez, A., Clavijo, C., Palanisamy, M., Ruiz-Rodríguez, A., Navarro-Rubio, M., Pérez, M., Marín, F. R., Reglero, G., & Soler-Rivas, C. (2013). Screening of edible mushrooms and extraction by pressurized water (PWE) of 3-hydroxy-3-methyl-glutaryl CoA reductase inhibitors. *Journal of Functional Foods, 5*(1), 244–250.

Gonzaga, M. L., Ricardo, N. M., Heatley, F., & Soares, S. D. (2005). Isolation and characterization of polysaccharides from Agaricus blazei Murill. *Carbohydrate Polymers, 60*(1), 43–49.

Gonzalez Del Val, A., Platas, G., Arenal, F., Orihuela, J. C., Garcia, M., Hernandez, P., Royo, I., De Pedro, N., Silver, L. L., Young, K., Vicente, M. F., & Pelaez, F. (2003). Novel illudins from Coprinopsis episcopalis (syn. Coprinus episcopalis), and the distribution of illudin-like compounds among filamentous fungi. *Mycological Research, 107*(10), 1201–1209.

Goodridge, H. S., Wolf, A. J., & Underhill, D. M. (2009). β-glucan recognition by the innate immune system. *Immunological Reviews, 230*(1), 38–50.

Gow, N. A., Latge, J., & Munro, C. A. (2017). The fungal cell wall: Structure, biosynthesis, and function. *Microbiology Spectrum, 5*(3). 10.1128/microbiolspec.FUNK-0035-2016

Green, B., Bentley, M. D., Chung, B. Y., Lynch, N. G., & Jensen, B. L. (2007). Isolation of Betulin and rearrangement to Allobetulin. A Biomimetic natural product synthesis. *Journal of Chemical Education, 84*(12), 1985.

Grover, A., & Joshi, A. (2014). An overview of chronic disease models: A systematic literature review. *Global Journal of Health Science, 7*(2), 210–227.

Gulati, J. K., Guhathakurta, S., Joshi, J., Rai, N., & Ray, A. (2016). Cytokines and their role in health and disease: A brief overview. *MOJ Immunology, 4*(2). 10.15406/moji.2016.04.00121

Gulati, V., Singh, M. D., & Gulati, P. (2019). Role of mushrooms in gestational diabetes mellitus. *AIMS Medical Science, 6*(1), 49–66.

Gunde-Cimerman, N., Friedrich, J., Cimerman, A., & Benički, N. (1993). Screening fungi for the production of an inhibitor of HMG CoA reductase: Production of mevinolin by the fungi of the genus Pleurotus. *FEMS Microbiology Letters, 111*(2–3), 203–206.

Gunde-Cimerman, N., Plemenitaš, A., & Cimerman, A. (1993). Pleurotusfungi produce mevinolin, an inhibitor of HMG CoA reductase. *FEMS Microbiology Letters, 113*(3), 333–337.

Géry, A., Dubreule, C., André, V., Rioult, J., Bouchart, V., Heutte, N., Eldin de Pécoulas, P., Krivomaz, T., & Garon, D. (2018). Chaga (Inonotus obliquus), a future potential medicinal fungus in oncology? A chemical study and a comparison of the cytotoxicity against human lung adenocarcinoma cells (A549) and human bronchial epithelial cells (BEAS-2B). *Integrative Cancer Therapies, 17*(3), 832–843.

Habtemariam, S. (2020). Trametes versicolor (Synn. Coriolus versicolor) polysaccharides in cancer therapy: Targets and efficacy. *Biomedicines*, *8*(5), 135.

Hajjaj, H., Mace, C., Roberts, M., Niederberger, P., & Fay, L. B. (2005). Effect of 26-Oxygenosterols from Ganoderma lucidum and their activity as cholesterol synthesis inhibitors. *Applied and Environmental Microbiology*, *71*(7), 3653–3658.

Hashim, S. N., Schwarz, L. J., Danylec, B., Mitri, K., Yang, Y., Boysen, R. I., & Hearn, M. T. (2016). Recovery of ergosterol from the medicinal mushroom, Ganoderma tsugae Var. Janniae, with a molecularly imprinted polymer derived from a cleavable monomer-template composite. *Journal of Chromatography A*, *1468*, 1–9.

He, Y., Zhang, L., & Wang, H. (2019). The biological activities of the antitumor drug Grifola frondosa polysaccharide. *In Progress in Molecular Biology and Translational Science*, *163*, 221–261. 10.1016/bs.pmbts.2019.02.010

Heleno, S. A., Barros, L., Sousa, M. J., Martins, A., & Ferreira, I. C. (2010). Tocopherols composition of Portuguese wild mushrooms with antioxidant capacity. *Food Chemistry*, *119*(4), 1443–1450.

Hilgers, W., Faivre, S., Chieze, S., Alexandre, J., Lokiec, F., Goldwasser, F., Raymond, E., Kahatt, C., Taamma, A., Weems, G., MacDonald, J. R., Misset, J., & Cvitkovic, E. (2006). A phase I and pharmacokinetic study of irofulven and cisplatin administered in a 30-min infusion every two weeks to patients with advanced solid tumors. *Investigational New Drugs*, *24*(4), 311–319.

Hino, S., Nishimura, N., Matsuda, T., & Morita, T. (2020). Intestinal absorption of β-glucans and their effect on the immune system. Preprints 2020. 10.20944/preprints202012.0250.v1

Hirase, S., Nakai, S., Akatsu, T., Kobayashi, A., Oohara, M., Matsunaga, K., Fujii, M., Kodaira, S., Fuji, T., Furusho, T., Ohmura, Y., Wada, T., Yoshikumi, C., Ueno, S., & Ohtsuka, S. (1976a). Structural studies on the anti-tumor active polysaccharides from Coriolus versicolor (Basidiomycetes). I. Fractionation with barium hydroxide. *Yakugaku Zasshi*, *96*(4), 413–418.

Hirase, S., Nakai, S., Akatsu, T., Kobayashi, A., Oohara, M., Matsunaga, K., Fujii, M., Kodaira, S., Fujii, T., Furusho, T., Ohmura, Y., Wada, T., Yoshikumi, C., Ueno, S., & Ohtsuka, S. (1976b). Structural studies on the anti-tumor active polysaccharides from Coriolus versicolor (Basidiomycetes). II. Structures of β-D-Glucan moieties of fractionated polysaccharides. *Yakugaku Zasshi*, *96*(4), 419–424.

Hishida, I., Nanba, H., & Kuroda, H. (1988). Antitumor activity exhibited by orally administered extract from fruit body of Grifola frondosa (Maitake). *Chemical and Pharmaceutical Bulletin*, *36*(5), 1819–1827.

Hong, L., Xun, M., & Wutong, W. (2007). Anti-diabetic effect of an α-glucan from fruit body of maitake (Grifola frondosa) on KK-ay mice. *Journal of Pharmacy and Pharmacology*, *59*(4), 575–582.

Hordyjewska, A., Ostapiuk, A., Horecka, A., & Kurzepa, J. (2019). Betulin and betulinic acid: Triterpenoids derivatives with a powerful biological potential. *Phytochemistry Reviews*, *18*(3), 929–951.

Hwang, C. H., Jaki, B. U., Klein, L. L., Lankin, D. C., McAlpine, J. B., Napolitano, J. G., Fryling, N. A., Franzblau, S. G., Cho, S. H., Stamets, P. E., Wang, Y., & Pauli, G. F. (2013). Chlorinated coumarins from the Polypore MushroomFomitopsis officinalisand their activity againstMycobacterium tuberculosis. *Journal of Natural Products*, *76*(10), 1916–1922.

Ina, K., Kataoka, T., & Ando, T. (2013). The use of Lentinan for treating gastric cancer. *Anti-Cancer Agents in Medicinal Chemistry*, *13*(5), 681–688.

Inoue, A., Kodama, N., & Nanba, H. (2002). Effect of Maitake (Grifola frondosa) D-fraction on the control of the T lymph node Th-1/Th-2 proportion. *Biological and Pharmaceutical Bulletin*, *25*(4), 536–540.

Jahromi, M. F., Liang, J. B., Ho, Y. W., Mohamad, R., Goh, Y. M., & Shokryazdan, P. (2012). Lovastatin production by Aspergillus terreus using agro-biomass as substrate in solid state fermentation. *Journal of Biomedicine and Biotechnology*, *2012*, 196264. 10.1155/2012/196264

Jasinghe, V. J., & Perera, C. O. (2005). Distribution of ergosterol in different tissues of mushrooms and its effect on the conversion of ergosterol to vitamin D2 by UV irradiation. *Food Chemistry*, *92*(3), 541–546.

Jasinghe, V. J., Perera, C. O., & Barlow, P. J. (2005). Bioavailability of vitamin D2 from irradiated mushrooms: An in vivo study. *British Journal of Nutrition*, *93*(6), 951–955.

Jaspers, N. G., Raams, A., Kelner, M. J., Ng, J. M., Yamashita, Y. M., Takeda, S., McMorris, T. C., & Hoeijmakers, J. H. (2002). Anti-tumour compounds illudin S and Irofulven induce DNA lesions ignored by global repair and exclusively processed by transcription- and replication-coupled repair pathways. *DNA Repair*, *1*(12), 1027–1038.

Jaworska, G., Pogoń, K., Bernaś, E., Skrzypczak, A., & Kapusta, I. (2014). Vitamins, phenolics and antioxidant activity of culinary prepared Suillus luteus (L.) Roussel mushroom. *LWT - Food Science and Technology, 59*(2), 701–706.

Jayasuriya, H., Silverman, K. C., Zink, D. L., Jenkins, R. G., Sanchez, M., Pelaez, F., Vilella, D., Lingham, R. B., & Singh, S. B. (1998). Clavaric acid: A triterpenoid inhibitor of farnesyl-protein transferase from Clavariadelphus truncatus. *Journal of Natural Products, 61*(12), 1568–1570.

Jiang, Q., Zhang, M., & Mujumdar, A. S. (2020). UV induced conversion during drying of ergosterol to vitamin D in various mushrooms: Effect of different drying conditions. *Trends in Food Science & Technology, 105*, 200–210.

Jin, W., & Zjawiony, J. K. (2006). 5-Alkylresorcinols from merulius incarnatus. *Journal of Natural Products, 69*(4), 704–706.

Jin, X., Ruiz Beguerie, J., Sze, D. M., & Chan, G. C. (2016). Ganoderma lucidum (Reishi mushroom) for cancer treatment. *Cochrane Database of Systematic Reviews*, (6), CD007731. 10.1002/14651858. CD007731.pub2

Jin, Y., Zhang, L., Chen, L., Chen, Y., Keung Cheung, P. C., & Chen, L. (2003). Effect of culture media on the chemical and physical characteristics of polysaccharides isolated from Poria cocos mycelia. *Carbohydrate Research, 338*(14), 1507–1515.

Jin-Ming, G. (2006). New biologically active metabolites from Chinese higher fungi. *Current Organic Chemistry, 10*(8), 849–871.

Johnson, B. M., Doonan, B. P., Radwan, F. F., & Haque, A. (2010). Ganoderic acid DM: An alternative agent for the treatment of advanced prostate cancer. *The Open Prostate Cancer Journal, 3*(1), 78–85.

Kalsi, A., Singh, S., Taneja, N., Kukal, S., & Mani, S. (2014). Current treatments for type 2 diabetes, their side effects and possible complementary treatments. *International Journal of Pharmacy and Pharmaceutical Sciences, 7*(3), 13–18.

Kaur, J. (2014). A comprehensive review on metabolic syndrome. *Cardiology Research and Practice, 2014*.

Kavanagh, F., Hervey, A., & Robbins, W. J. (1951). Antibiotic substances from basidiomycetes: VIII. Pleurotus Multilus (Fr.) Sacc. and Pleurotus Passeckerianus Pilat. *Proceedings of the National Academy of Sciences, 37*(9), 570–574.

Kawagishi, H., Inagaki, R., Kanao, T., Mizuno, T., Shimura, K., Ito, H., Hagiwara, T., & Nakamura, T. (1989). Fractionation and antitumor activity of the water-in-soluble residue of Agaricus blazei fruiting bodies. *Carbohydrate Research, 186*(2), 267–273.

Kawagishi, H., Kanao, T., Inagaki, R., Mizuno, T., Shimura, K., Ito, H., Hagiwara, T., & Nakamura, T. (1990). Formolysis of a potent antitumor (1 → 6)-β-d-glucan-protein complex from Agaricus blazei fruiting bodies and antitumor activity of the resulting products. *Carbohydrate Polymers, 12*(4), 393–403.

Keegan, R. H., Lu, Z., Bogusz, J. M., Williams, J. E., & Holick, M. F. (2013). Photobiology of vitamin D in mushrooms and its bioavailability in humans. *Dermato-Endocrinology, 5*(1), 165–176.

Kelner, M. J., McMorris, T. C., Beck, W. T., Zamora, J. M., & Taetle, R. (1987). Preclinical Evaluation of Illudins as Anticancer Agents. *Cancer Research, 47*(12), 3186–3189.

Kelner, M. J., McMorris, T. C., & Taetle, R. (1990). Preclinical evaluation of Illudins as anticancer agents: Basis for selective cytotoxicity. *Journal of the National Cancer Institute, 82*(19), 1562–1565.

Kelner, M. J., McMorris, T. C., Estes, L., Rutherford, M., Montoya, M., Goldstein, J., Samson, K., Starr, R., & Taetle, R. (1994). Characterization of illudin S sensitivity in DNA repair-deficient Chinese hamster cells. *Biochemical Pharmacology, 48*(2), 403–409.

Kelner, M. J., McMorris, T. C., Montoya, M. A., Estes, L., Rutherford, M., Samson, K. M., & Taetle, R. (1997). Characterization of cellular accumulation and toxicity of illudin S in sensitive and nonsensitive tumor cells. *Cancer Chemotherapy and Pharmacology, 40*(1), 65–71.

Kelner, M. J., McMorris, T. C., Rojas, R. J., Estes, L. A., & Suthipinijtham, P. (2008). Synergy of Irofulven in combination with various anti-metabolites, enzyme inhibitors, and miscellaneous agents in MV522 lung carcinoma cells: Marked interaction with gemcitabine and 5-fluorouracil. *Investigational New Drugs, 26*(5), 407–415.

Keypour, S., Rafati, H., Riahi, H., Mirzajani, F., & Moradali, M. F. (2010). Qualitative analysis of ganoderic acids in Ganoderma lucidum from Iran and China by RP-HPLC and electrospray ionisation-mass spectrometry (ESI-MS). *Food Chemistry, 119*(4), 1704–1708.

Khan, M. A., Hashim, M. J., King, J. K., Govender, R. D., Mustafa, H., & Al Kaabi, J. (2020). Epidemiology of type 2 diabetes – Global burden of disease and forecasted trends. *Journal of Epidemiology and Global Health, 10*(1), 107–111.

Kiho, T., katsurawaga, M., Nagai, K., Ukai, S., & Haga, M. (1992). Structure and antitumor activity of a branched (1→3)-β-d-glucan from the alkaline extract of amanita muscaria. *Carbohydrate Research, 224*, 237–243.

Kim, H. S., Hong, J. T., Kim, Y., & Han, S. (2011). Stimulatory effect of β-glucans on immune cells. *Immune Network, 11*(4), 191.

Klaus, A., Kozarski, M., Vunduk, J., Todorovic, N., Jakovljevic, D., Zizak, Z., Pavlovic, V., Levic, S., Niksic, M., & Van Griensven, L. J. (2015). Biological potential of extracts of the wild edible Basidiomycete mushroom Grifola frondosa. *Food Research International, 67*, 272–283.

Klaus, A., Kozarski, M., Vunduk, J., Petrovic, P., & Niksic, M. (2016). Antibacterial and antifungal potential of wild basidiomycete mushroom Ganoderma applanatum. *Lekovite sirovine, (36)*, 37–46.

Klaus, A., Petrovic, P., Vunduk, J., Pavlovic, V., & Van Griensven, L. (2020). Antimicrobial properties of silver-nanoparticles of Agaricus bisporus, Agaricus brasiliensis and Phellinus linteus. *International Journal of Medicinal Mushrooms, 22*(9), 869–883. 10.1615/IntJMedMushrooms.2020035988

Klaus, A., Wan-Mohtar, W. A., Nikolić, B., Cvetković, S., & Vunduk, J. (2021). Pink oyster mushroom Pleurotus flabellatus mycelium produced by an airlift bioreactor—the evidence of potent in vitro biological activities. *World Journal of Microbiology and Biotechnology, 37*(1).

Klemm, D., Heublein, B., Fink, H., & Bohn, A. (2005). Cellulose: Fascinating Biopolymer and sustainable raw material. *Angewandte Chemie International Edition, 44*(22), 3358–3393.

Kojima, H., Akaki, J., Nakajima, S., Kamei, K., & Tamesada, M. (2009). Structural analysis of glycogen-like polysaccharides having macrophage-activating activity in extracts of Lentinula edodes mycelia. *Journal of Natural Medicines, 64*(1), 16–23.

Krakowska, A., Zięba, P., Włodarczyk, A., Kała, K., Sułkowska-Ziaja, K., Bernaś, E., Sękara, A., Ostachowicz, B., & Muszyńska, B. (2020). Selected edible medicinal mushrooms from Pleurotus genus as an answer for human civilization diseases. *Food Chemistry, 327*, 127084.

Król, S. K., Kiełbus, M., Rivero-Müller, A., & Stepulak, A. (2015). Comprehensive review on Betulin as a potent Anticancer agent. *BioMed Research International, 2015*, 1–11.

Kubo, I., Kim, M., Hood, W. F., & Naoki, H. (1986). Clitocine, a new insecticidal nucleoside from the mushroom clitocybe inversa. *Tetrahedron Letters, 27*(36), 4277–4280.

Kumar, P. M., Kumar, M. S., Manivel, A., & Mohan, S. C. (2018). Structural characterization and anti-diabetic activity of polysaccharides from Agaricus bisporus mushroom. *Research Journal of Phytochemistry, 12*(1), 14–20.

Lee, I., & Yun, B. (2011). Styrylpyrone-class compounds from medicinal fungi Phellinus and Inonotus spp., and their medicinal importance. *The Journal of Antibiotics, 64*(5), 349–359.

Lee, M., Hur, H., Chang, K., Lee, T., Ka, K., & Jankovsky, L. (2008). Introduction to distribution and ecology of sterile conks ofInonotus obliquus. *Mycobiology, 36*(4), 199.

Lee, S., Park, J. Y., Lee, D., Seok, S., Kwon, Y. J., Jang, T. S., Kang, K. S., & Kim, K. H. (2017). Chemical constituents from the rare mushroom Calvatia nipponica inhibit the promotion of angiogenesis in HUVECs. *Bioorganic & Medicinal Chemistry Letters, 27*(17), 4122–4127.

Lee, S. R., Lee, D., Lee, B. S., Ryoo, R., Pang, C., Kang, K. S., & Kim, K. H. (2020). Phallac acids a and B, new sesquiterpenes from the fruiting bodies of phallus luteus. *The Journal of Antibiotics, 73*(10), 729–732.

Legentil, L., Paris, F., Ballet, C., Trouvelot, S., Daire, X., Vetvicka, V., & Ferrières, V. (2015). Molecular interactions of β-(1→3)-Glucans with their receptors. *Molecules, 20*(6), 9745–9766.

Lemaire, P., Balme, G., Desbordes, P., & Vors, J. (2003). Efficient syntheses of pterulone, pterulone B and related analogues. *Org. Biomol. Chem, 1*(23), 4209–4219.

Li, C., Chen, P., Chang, U., Kan, L., Fang, W., Tsai, K., & Lin, S. (2005). Ganoderic acid X, a lanostanoid triterpene, inhibits topoisomerases and induces apoptosis of cancer cells. *Life Sciences, 77*(3), 252–265.

Liang, C., Tian, D., Liu, Y., Li, H., Zhu, J., Li, M., Xin, M., & Xia, J. (2019). Review of the molecular mechanisms of Ganoderma lucidum triterpenoids: Ganoderic acids a, c2, D, F, DM, X and Y. *European Journal of Medicinal Chemistry, 174*, 130–141.

Lindequist, U., & Haertel, B. (2020). Medicinal mushrooms for treatment of type 2 diabetes: An update on clinical trials. *International Journal of Medicinal Mushrooms, 22*(9), 845–854.

Lindequist, U., Niedermeyer, T. H., & Jülich, W. (2005). The pharmacological potential of mushrooms. *Evidence-Based Complementary and Alternative Medicine, 2*(3), 285–299.

Liu, J., Shiono, J., Shimizu, K., Kukita, A., Kukita, T., & Kondo, R. (2009). Ganoderic acid DM: Anti-androgenic osteoclastogenesis inhibitor. *Bioorganic & Medicinal Chemistry Letters, 19*(8), 2154–2157.

Liu, R., Li, Y., & Zhong, J. (2012). Cytotoxic and pro-apoptotic effects of novel ganoderic acid derivatives on human cervical cancer cells in vitro. *European Journal of Pharmacology, 681*(1–3), 23–33.

Liu, R., & Zhong, J. (2011). Ganoderic acid Mf and S induce mitochondria mediated apoptosis in human cervical carcinoma Hela cells. *Phytomedicine, 18*(5), 349–355.

Liu, Y., & Wang, F. (2007). Structural characterization of an active polysaccharide from Phellinus ribis. *Carbohydrate Polymers, 70*(4), 386–392.

Lou, H., Li, H., Wei, T., & Chen, Q. (2021). Stimulatory effects of Oleci acid and fungal elicitor on Betulinic acid production by submerged cultivation of medicinal mushroom Inonotus obliquus. *Journal of Fungi, 7*(4), 266.

Ložnjak, P., & Jakobsen, J. (2018). Stability of vitamin D3 and vitamin D2 in oil, fish and mushrooms after household cooking. *Food Chemistry, 254*, 144–149.

Lugano, R., Ramachandran, M., & Dimberg, A. (2019). Tumor angiogenesis: Causes, consequences, challenges and opportunities. *Cellular and Molecular Life Sciences, 77*(9), 1745–1770.

Lü, W., Gao, Y., Su, M., Luo, Z., Zhang, W., Shi, G., & Zhao, Q. (2013). Isoindolones fromLasiosphaera fenzlii Reich. and their Bioactivities. *Helvetica Chimica Acta, 96*(1), 109–113.

Ma, K., Han, J., Bao, L., Wei, T., & Liu, H. (2014). Two Sarcoviolins with Antioxidative and α-glucosidase inhibitory activity from the edible Mushroom Sarcodon leucopus Collected in Tibet. *Journal of Natural Products, 77*(4), 942–947.

Ma, Z., Wang, J., & Zhang, L. (2008). Structure and chain conformation of β-glucan isolated from Auricularia auricula-judae. *Biopolymers, 89*(7), 614–622.

Madsen, M. K., Fisher, P. M., Stenbæk, D. S., Kristiansen, S., Burmester, D., Lehel, S., Páleníček, T., Kuchař, M., Svarer, C., Ozenne, B., & Knudsen, G. M. (2020). A single psilocybin dose is associated with long-term increased mindfulness, preceded by a proportional change in neocortical 5-HT2A receptor binding. *European Neuropsychopharmacology, 33*, 71–80.

Maehara, Y., Tsujitani, S., Saeki, H., Oki, E., Yoshinaga, K., Emi, Y., Morita, M., Kohnoe, S., Kakeji, Y., Yano, T., & Baba, H. (2011). Biological mechanism and clinical effect of protein-bound polysaccharide K (KRESTIN®): Review of development and future perspectives. *Surgery Today, 42*(1), 8–28.

Maljurić, N., Golubović, J., Ravnikar, M., Žigon, D., Štrukelj, B., & Otašević, B. (2018). Isolation and determination of Fomentariol: Novel potential Antidiabetic drug from fungal material. *Journal of Analytical Methods in Chemistry, 2018*, 1–9.

Mandal, E. K., Maity, K., Maity, S., Gantait, S. K., Behera, B., Maiti, T. K., Sikdar, S. R., & Islam, S. S. (2012). Chemical analysis of an immunostimulating (1→4)-, (1→6)-branched glucan from an edible mushroom, Calocybe indica. *Carbohydrate Research, 347*(1), 172–177.

Man-Fan Wan, J. (2013). Polysaccharide Krestin (PSK) and Polysaccharopeptide PSP. *Handbook of Biologically Active Peptides* (Second Edition) (pp. 180–184). Cambridge: Academic Press.

Marín-Peñalver, J. J., Martín-Timón, I., Sevillano-Collantes, C., & del Cañizo-Gómez, F. J. (2016). Update on the treatment of type 2 diabetes mellitus. *World Journal of Diabetes, 7*(17), 354–395.

Masuda, Y., Nakayama, Y., Tanaka, A., Naito, K., & Konishi, M. (2017). Antitumor activity of orally administered maitake α-glucan by stimulating antitumor immune response in murine tumor. *PLoS One, 12*(3), e0173621.

Mattila, P., Lampi, A., Ronkainen, R., Toivo, J., & Piironen, V. (2002). Sterol and vitamin D2 contents in some wild and cultivated mushrooms. *Food Chemistry, 76*(3), 293–298.

McMorris, T. C., & Anchel, M. (1965). Fungal metabolites. The structures of the novel Sesquiterpenoids Illudin-S and -M. *Journal of the American Chemical Society, 87*(7), 1594–1600.

Mehrotra, A., Calvo, M. S., Beelman, R. B., Levy, E., Siuty, J., Kalaras, M. D., & Uribarri, J. (2014). Bioavailability of vitamin D2 from enriched mushrooms in prediabetic adults: a randomized controlled trial. *European Journal of Clinical Nutrition, 68*(10), 1154–1160.

Mertens, L., Wall, M., Roseman, L., Demetriou, L., Nutt, D., & Carhart-Harris, R. (2019). Therapeutic mechanisms of psychedelic drugs: Changes in amygdala and prefrontal functional connectivity during emotional processing after psilocybin for treatment-resistant depression. *Journal of Psychopharmacology, 34*(2), 167–180. 10.1177/0269881119895520

Min, B., Gao, J., Nakamura, N., & Hattori, M. (2000). Triterpenes from the spores of Ganoderma lucidum and their cytotoxicity against Meth-A and LLC tumor cells. *Chemical and Pharmaceutical Bulletin*, *48*(7), 1026–1033.

Mizuno, M., Morimoto, M., Minato, K., & Tsuchida, H. (1998). Polysaccharides from Agaricus blazei stimulate lymphocyte T-cell subsets in mice. *Bioscience, Biotechnology, and Biochemistry*, *62*(3), 434–437.

Moss, R. J., Petrie, C. R., Meyer, R. B., Nord, L. D., Willis, R. C., Smith, R. A., Larson, S. B., Kini, G. D., & Robins, R. K. (1988). Synthesis and intramolecular hydrogen bonding and biochemical studies of clitocine, a naturally occurring exocyclic amino nucleoside. *Journal of Medicinal Chemistry*, *31*(4), 786–790.

Muchaamba, F., Stephan, R., & Tasara, T. (2020). β-phenylethylamine as a natural food additive shows antimicrobial activity against listeria monocytogenes on ready-to-Eat foods. *Foods*, *9*(10), 1363.

Murphy, E. J., Rezoagli, E., Major, I., Rowan, N. J., & Laffey, J. G. (2020). β-glucan metabolic and immunomodulatory properties and potential for clinical application. *Journal of Fungi*, *6*(4), 356.

Nakanishi, K., Ohashi, M., Suzuki, N., Tada, M., Yamada, Y., & Inagaki, S. (1963). Isolation of Lampterol from Lampteromyces japonicus (Kawam.) Sing. *Yakugaku Zasshi*, *83*(4), 377–380.

Nakanishi, K., Ohashi, M., Tada, M., & Yamada, Y. (1965). Illudin S (lampterol). *Tetrahedron*, *21*(5), 1231–1246.

Nandan, C. K., Patra, P., Bhanja, S. K., Adhikari, B., Sarkar, R., Mandal, S., & Islam, S. S. (2008). Structural characterization of a water-soluble β-(1→6)-linked D-glucan isolated from the hot water extract of an edible mushroom, Agaricus bitorquis. *Carbohydrate Research*, *343*(18), 3120–3122.

Ng, T. (1998). A review of research on the protein-bound polysaccharide (polysaccharopeptide, PSP) from the mushroom Coriolus versicolor (basidiomycetes: Polyporaceae). *General Pharmacology: The Vascular System*, *30*(1), 1–4.

Nichols, D. E. (1986). Studies of the relationship between molecular structure and hallucinogenic activity. *Pharmacology Biochemistry and Behavior*, *24*(2), 335–340.

Padmavathi, M. (2013). Chronic Disease Management with Nutraceuticals. *Journal of Pharmaceutical Science Invention*, *2*(4), 1–11.

Pandey, A. T., Pandey, I., Hachenberger, Y., Krause, B., Haidar, R., Laux, P., Luch, A., Singh, M. P., & Singh, A. V. (2020). Emerging paradigm against global antimicrobial resistance via bioprospecting of mushroom into novel nanotherapeutics development. *Trends in Food Science & Technology*, *106*, 333–344.

Patra, S. (2016). Return of the psychedelics: Psilocybin for treatment resistant depression. *Asian Journal of Psychiatry*, *24*, 51–52.

Petrović, P., Ivanovic, K., Jovanovic, A., Simovic, M., Milutinovic, V., Kozarski, M., Petkovic, M., Cvetkovic, A., Klaus, A., & Bugarski, B. (2019a). The impact of puffball autolysis on selected chemical and biological properties: Puffball extracts as potential ingredients of skin-care products. *Archives of Biological Sciences*, *71*(4), 721–733.

Petrović, P., Ivanovic, K., Octrue, C., Tumara, M., Jovanovic, A., Vunduk, J., Niksic, M., Pjanovic, R., Bugarski, B., & Klaus, A. (2019b). Immobilization of Chaga extract in alginate beads for modified release: Simplicity meets efficiency. *Chemical Industry*, *73*(5), 325–335.

Petrović, P., Vunduk, J., Klaus, A., Carević, M., Petković, M., Vuković, N., Cvetković, A., Žižak, Ž., & Bugarski, B. (2019). From mycelium to spores: A whole circle of biological potency of mosaic puffball. *South African Journal of Botany*, *123*, 152–160.

Petrović, P., Kostić, D., Klaus, A., Vunduk, J., Nikšić, M., Veljović, Đ., & Griensven, L. V. (2018). Characterisation and antimicrobial activity of silver nanoparticles derived from Vascellum pratense polysaccharide extract and sodium citrate. *Journal of Engineering & Processing Management*, *10*(1). 10.7251/JEPM1810001P

Petrović, P., Vunduk, J., Klaus, A., Kozarski, M., Nikšić, M., Žižak, Ž., Vuković, N., Šekularac, G., Drmanić, S., & Bugarski, B. (2016). Biological potential of puffballs: A comparative analysis. *Journal of Functional Foods*, *21*, 36–49.

Pierson, A. S., Gibbs, P., Richards, J., Russ, P., Eckhardt, S. G., & Gonzalez, R. (2002). A Phase II Study of Irofulven (MGI 114) in Patients with Stage IV Melanoma. *Investigational New Drugs*, *20*(3), 357–362.

Pisha, E., Chai, H., Lee, I., Chagwedera, T. E., Farnsworth, N. R., Cordell, G. A., Beecher, C. W., Fong, H. H., Kinghorn, A. D., Brown, D. M., Wani, M. C., Wall, M. E., Hieken, T. J., Das Gupta, T. K., & Pezzuto, J. M. (1995). Discovery of betulinic acid as a selective inhibitor of human melanoma that functions by induction of apoptosis. *Nature Medicine*, *1*(10), 1046–1051.

Poulsen, S. M., Karlsson, M., Johansson, L. B., & Vester, B. (2008). The pleuromutilin drugs tiamulin and valnemulin bind to the RNA at the peptidyl transferase centre on the ribosome. *Molecular Microbiology*, *41*(5), 1091–1099.

Puyo, S., Montaudon, D., & Pourquier, P. (2014). From old alkylating agents to new minor groove binders. *Critical Reviews in Oncology/Hematology*, *89*(1), 43–61.

Raghavendra, V. B., Venkitasamy, C., Pan, Z., & Nayak, C. (2018). Functional Foods from Mushroom. In *Microbial Functional Foods and Nutraceuticals* (pp. 65–91). Hoboken: John Wiley & Sons.

Ramachandran, A., Snehalatha, C., Shetty, A. S., & Nanditha, A. (2012). Trends in prevalence of diabetes in Asian countries. *World Journal of Diabetes*, *3*(6), 110–117.

Rau, U., Kuenz, A., Wray, V., Nimtz, M., Wrenger, J., & Cicek, H. (2009). Production and structural analysis of the polysaccharide secreted by Trametes (Coriolus) versicolor ATCC 200801. *Applied Microbiology and Biotechnology*, *81*(5), 827–837.

Reiche, S., Hermle, L., Gutwinski, S., Jungaberle, H., Gasser, P., & Majić, T. (2018). Serotonergic hallucinogens in the treatment of anxiety and depression in patients suffering from a life-threatening disease: A systematic review. *Progress in Neuro-Psychopharmacology and Biological Psychiatry*, *81*, 1–10.

Ren, G., Zhao, Y., Yang, L., & Fu, C. (2008). Anti-proliferative effect of clitocine from the mushroom Leucopaxillus giganteus on human cervical cancer Hela cells by inducing apoptosis. *Cancer Letters*, *262*(2), 190–200.

Rochlani, Y., Pothineni, N. V., Kovelamudi, S., & Mehta, J. L. (2017). Metabolic syndrome: Pathophysiology, management, and modulation by natural compounds. *Therapeutic Advances in Cardiovascular Disease*, *11*(8), 215–225.

Romeo, B., Hermand, M., Pétillion, A., Karila, L., & Benyamina, A. (2021). Clinical and biological predictors of psychedelic response in the treatment of psychiatric and addictive disorders: A systematic review. *Journal of Psychiatric Research*, *137*, 273–282.

Roseman, L., Demetriou, L., Wall, M. B., Nutt, D. J., & Carhart-Harris, R. L. (2018). Increased amygdala responses to emotional faces after psilocybin for treatment-resistant depression. *Neuropharmacology*, *142*, 263–269.

Roupas, P., Keogh, J., Noakes, M., Margetts, C., & Taylor, P. (2010). Mushrooms and agaritine: A mini-review. *Journal of Functional Foods*, *2*(2), 91–98.

Ruiz-Herrera, J., & Ortiz-Castellanos, L. (2019). Cell wall glucans of fungi. A review. *The Cell Surface*, *5*, 100022. 10.1016/j.tcsw.2019.100022

Ruthes, A. C., Carbonero, E. R., Córdova, M. M., Baggio, C. H., Santos, A. R., Sassaki, G. L., Cipriani, T. R., Gorin, P. A., & Iacomini, M. (2013). Lactarius Rufus (1→3),(1→6)-β-d-glucans: Structure, antinociceptive and anti-inflammatory effects. *Carbohydrate Polymers*, *94*(1), 129–136.

Ruthes, A. C., Smiderle, F. R., & Iacomini, M. (2015). D-glucans from edible mushrooms: A review on the extraction, purification and chemical characterization approaches. *Carbohydrate Polymers*, *117*, 753–761.

Salemi, S., Saedisomeolia, A., Azimi, F., Zolfigol, S., Mohajerani, E., Mohammadi, M., & Yaseri, M. (2021). Optimizing the production of vitamin D in white Button mushrooms (Agaricus bisporus) using ultraviolet radiation and measurement of its stability. *LWT*, *137*, 110401. 10.1016/j.lwt.2020.110401

Sarti, F., Müller, C., Iqbal, J., Perera, G., Laffleur, F., & Bernkop-Schnürch, A. (2013). Development and in vivo evaluation of an oral vitamin B12 delivery system. *European Journal of Pharmaceutics and Biopharmaceutics*, *84*(1), 132–137.

Sauter, H., Steglich, W., & Anke, T. (1999). Strobilurins: Evolution of a new class of active substances. *Angewandte Chemie International Edition*, *38*(10), 1328–1349.

Schilder, R. J., Blessing, J. A., Pearl, M. L., & Rose, P. G. (2004). Evaluation of irofulven (MGI-114) in the treatment of recurrent or persistent endometrial carcinoma: A phase II study of the Gynecologic oncology group. *Investigational New Drugs*, *22*(3), 343–349.

Schilder, R. J., Blessing, J. A., Shahin, M. S., Miller, D. S., Tewari, K. S., Muller, C. Y., Warshal, D. P., McMeekin, S., & Rotmensch, J. (2010). A phase 2 evaluation of Irofulven as second-line treatment of recurrent or persistent intermediately platinum-sensitive ovarian or primary peritoneal cancer. *International Journal of Gynecological Cancer*, *20*(7), 1137–1141.

Schüffler, A. (2018). Secondary metabolites of basidiomycetes. *The Mycota: Physiology and Genetics* (pp. 231–275). Cham: Springer.

Schüffler, A., & Anke, T. (2009). Secondary metabolites of basidiomycetes. *The Mycota: Physiology and Genetics* (pp. 209–231). Berlin, Heidelberg: Springer-Verlag.

Seenivasan, A., Venkatesan, S., & Panda, T. (2018). Cellular localization and production of Lovastatin from Monascus purpureus. *Indian Journal of Pharmaceutical Sciences, 80*(1), 85–98.

Selvamani, S., El-Enshasy, H. A., Dailin, D. J., Malek, R. A., Hanapi, S. Z., Ambehabati, K. K., Sukmawati, D., Leng, O. M., & Moloi, N. (2018). Antioxidant compounds of the edible mushroom Pleurotus ostreatus. *International Journal of Biotechnology for Wellness Industries, 7*, 1–14.

Senzer, N., Arsenau, J., Richards, D., Berman, B., MacDonald, J. R., & Smith, S. (2005). Irofulven demonstrates clinical activity against metastatic hormone-refractory prostate cancer in a phase 2 single-agent trial. *American Journal of Clinical Oncology, 28*(1), 36–42.

Sherman, C. A., Herndon, J. E., Watson, D. M., & Green, M. R. (2004). A phase II trial of 6-hydroxymethylacylfulvene (MGI-114, irofulven) in patients with relapsed or refractory non-small cell lung cancer. *Lung Cancer, 45*(3), 387–392.

Smiderle, F. R., Carbonero, E. R., Mellinger, C. G., Sassaki, G. L., Gorin, P. A., & Iacomini, M. (2006). Structural characterization of a polysaccharide and a β-glucan isolated from the edible mushroom Flammulina velutipes. *Phytochemistry, 67*(19), 2189–2196.

Sorba, G., Bertinaria, M., Di Stilo, A., Gasco, A., Scaltrito, M. M., Brenciaglia, M. I., & Dubini, F. (2001). Anti-helicobacter pylori agents endowed with H2-antagonist properties. *Bioorganic & Medicinal Chemistry Letters, 11*(3), 403–406.

Stephensen, C. B., Zerofsky, M., Burnett, D. J., Lin, Y., Hammock, B. D., Hall, L. M., & McHugh, T. (2012). Vitamin D2 intake increases 25-hydroxy vitamin D2 but decreases 25-hydroxy vitamin D3 concentration in the serum of healthy adults. *Experimental Biology 2012 Meeting Abstracts 26* (S1), 642.1.

Stierle, A., Strobel, G., Stierle, D., Grothaus, P., & Bignami, G. (1995). The search for a taxol-producing microorganism among the endophytic fungi of the Pacific yew, taxus brevifolia. *Journal of Natural Products, 58*(9), 1315–1324.

Stojkovic, D., Smiljkovic, M., Ciric, A., Glamoclija, J., Van Griensven, L., Ferreira, I., & Sokovic, M. (2019). An insight into antidiabetic properties of six medicinal and edible mushrooms: Inhibition of α-amylase and α-glucosidase linked to type-2 diabetes. *South African Journal of Botany, 120*, 100–103.

Su, C., Lu, T., Lai, M., & Ng, L. (2013). Inhibitory potential of Grifola frondosabioactive fractions on α-amylase and α-glucosidase for management of hyperglycemia. *Biotechnology and Applied Biochemistry, 60*(4), 446–452.

Suay, I., Arenal, F., Asensio, F. J., Basilio, A., Cabello, M. A., Díez, M. T., García, J. B., González del Val, A., Gorrochategui, J., Hernández, P., Peláez, F., & Vicente, M. F. (2000). Screening of Basidiomycetes for Antimicrobial Activities. *Antonie van Leeuwenhoek, 78*, 129–139.

Sung, H., Ferlay, J., Siegel, R. L., Laversanne, M., Soerjomataram, I., Jemal, A., & Bray, F. (2021). Global cancer statistics 2020: Globocan estimates of incidence and mortality worldwide for 36 cancers in 185 countries. *CA: A Cancer Journal for Clinicians, 71*(3), 209–249.

Synytsya, A., & Novák, M. (2013). Structural diversity of fungal glucans. *Carbohydrate Polymers, 92*(1), 792–809.

Szwengiel, A., & Stachowiak, B. (2016). Deproteinization of water-soluble S-glucan during acid extraction from fruiting bodies of Pleurotus ostreatus mushrooms. *Carbohydrate Polymers, 146*, 310–319.

Sławińska, A., Fornal, E., Radzki, W., Skrzypczak, K., Zalewska-Korona, M., Michalak-Majewska, M., Parfieniuk, E., & Stachniuk, A. (2016). Study on vitamin D$_2$ stability in dried mushrooms during drying and storage. *Food Chemistry, 199*, 203–209.

Tada, R., Harada, T., Nagi-Miura, N., Adachi, Y., Nakajima, M., Yadomae, T., & Ohno, N. (2007). NMR characterization of the structure of a β-(1→3)-d-glucan isolate from cultured fruit bodies of Sparassis crispa. *Carbohydrate Research, 342*(17), 2611–2618.

Tanaka, K., Inoue, T., Kanai, M., & Kikuchi, T. (1994). Metabolism of illudin S, a toxic substance ofLampteromyces japonicus.IV. Urinary excretion of an illudin S metabolite in rat. *Xenobiotica, 24*(12), 1237–1243.

Tang, H., Lai, C., & Chao, C. (2020). The clinical efficacy of lefamulin in the treatment of elderly patients with community-acquired bacterial pneumonia. *Journal of Thoracic Disease, 12*(8), 4588–4590.

Tang, W., Liu, J., Zhao, W., Wei, D., & Zhong, J. (2006). Ganoderic acid T from Ganoderma lucidum mycelia induces mitochondria mediated apoptosis in lung cancer cells. *Life Sciences, 80*(3), 205–211.

Thurnher, D., Turhani, D., Pelzmann, M., Wannemacher, B., Knerer, B., Formanek, M., Wacheck, V., & Selzer, E. (2003). Betulinic acid: A new cytotoxic compound against malignant head and neck cancer cells. *Head & Neck*, *25*(9), 732–740.

Tomasi, S., Lohézic-Le Dévéhat, F., Sauleau, P., Bézivin, C., & Boustie, J. (2004). Cytotoxic activity of methanol extracts from Basidiomycete mushrooms on murine cancer cell lines. *Pharmazie*, *59*(4), 290–293.

Tsiantas, K., Tsiaka, T., Koutrotsios, G., Siapi, E., Zervakis, G. I., Kalogeropoulos, N., & Zoumpoulakis, P. (2021). On the identification and quantification of Ergothioneine and Lovastatin in various mushroom species: Assets and challenges of different analytical approaches. *Molecules*, *26*(7), 1832.

Tsukagoshi, S., Hashimoto, Y., Fujii, G., Kobayashi, H., Nomoto, K., & Orita, K. (1984). Krestin (PSK). *Cancer Treatment Reviews*, *11*(2), 131–155.

Tsugawa, N., Okano, T., Takeuchi, A., Kayama, M., & Kobayashi, T. (1992). Metabolism of orally Adiministered ergosterol and 7-Dehydrocholesterol in rats and lack of evidence for their vitamin D biological activity. Journal of Nutritional Science and Vitaminology, 38, 15–25.

Tylš, F., Páleníček, T., & Horáček, J. (2014). Psilocybin – Summary of knowledge and new perspectives. *European Neuropsychopharmacology*, *24*(3), 342–356.

Umezawa, H., Takeuchi, T., Iinuma, H., Ito, M., Ishizuka, M., Kurakata, Y., Umeda, Y., Nakanishi, Y., Nakamura, T., Obayashi, A., & Tanabe, O. (1975). A new antibiotic, calvatic acid. *The Journal of Antibiotics*, *28*(1), 87–90.

Urbain, P., Singler, F., Ihorst, G., Biesalski, H., & Bertz, H. (2011). Bioavailability of vitamin D2 from UV-B-irradiated Button mushrooms in healthy adults deficient in serum 25-hydroxyvitamin D: A randomized controlled trial. *European Journal of Clinical Nutrition*, *65*(8), 965–971.

Valentovic, M. (2007). Lovastatin. In *xPharm: The Comprehensive Pharmacology Reference* (pp. 1–5). Amsterdam: Elsevier.

Vannucci, L., Krizan, J., Sima, P., Stakheev, D., Caja, F., Rajsiglova, L., Horak, V., & Saieh, M. (2013). Immunostimulatory properties and antitumor activities of glucans. *International Journal of Oncology*, *43*(2), 357–364.

Vannucci, L., Sima, P., Vetvicka, V., & Krizan, J. (2017). Lentinan properties in Anticancer therapy: A review on the last 12-Year literature. *American Journal of Immunology*, *13*(1), 50–61.

Villares, A., Mateo-Vivaracho, L., García-Lafuente, A., & Guillamón, E. (2014). Storage temperature and UV-irradiation influence on the ergosterol content in edible mushrooms. *Food Chemistry*, *147*, 252–256.

Viñarta, S. C., François, N. J., Daraio, M. E., Figueroa, L. I., & Fariña, J. I. (2007). Sclerotium rolfsii scleroglucan: The promising behavior of a natural polysaccharide as a drug delivery vehicle, suspension stabilizer and emulsifier. *International Journal of Biological Macromolecules*, *41*(3), 314–323.

Vunduk, J., Klaus, A., Kozarski, M., Petrovic, P., Zizak, Z., Niksic, M., & Van Griensven, L. J. (2015). Did the iceman know better? Screening of the medicinal properties of the birch Polypore medicinal mushroom, Piptoporus betulinus (Higher basidiomycetes). *International Journal of Medicinal Mushrooms*, *17*(12), 1113–1125.

Vunduk, J., Kozarski, M., Djekic, I., Tomašević, I., & Klaus, A. (2021). Effect of modified atmosphere packaging on selected functional characteristics of Agaricus bisporus. *European Food Research and Technology*, *247*(4), 829–838.

Vunduk, J., Wan-Mohtar, W. A., Mohamad, S. A., Abd Halim, N. H., Mohd Dzomir, A. Z., Žižak, Ž., & Klaus, A. (2019). Polysaccharides of Pleurotus flabellatus strain Mynuk produced by submerged fermentation as a promising novel tool against adhesion and biofilm formation of foodborne pathogens. *LWT*, *112*, 108221.

Wang, G., Zhao, J., Liu, J., Huang, Y., Zhong, J., & Tang, W. (2007). Enhancement of IL-2 and IFN-γ expression and NK cells activity involved in the anti-tumor effect of ganoderic acid me in vivo. *International Immunopharmacology*, *7*(6), 864–870.

Wang, F., & Zhang, K. (2010). Hypoglycaemic effect of comatin, an antidiabetic substance separated from Coprinus comatus broth, on alloxan-induced-diabetic rats. *Food Chemistry*, *121*(1), 39–43.

Wang, K., Bao, L., Xiong, W., Ma, K., Han, J., Wang, W., Yin, W., & Liu, H. (2015). Lanostane Triterpenes from the Tibetan medicinal mushroom Ganoderma leucocontextum and their inhibitory effects on HMG-CoA reductase and α-glucosidase. *Journal of Natural Products*, *78*(8), 1977–1989.

Wang, K., Bao, L., Ma, K., Zhang, J., Chen, B., Han, J., Ren, J., Luo, H., & Liu, H. (2017). A novel class of α-glucosidase and HMG-CoA reductase inhibitors from Ganoderma leucocontextum and the anti-diabetic properties of ganomycin I in KK-A Y mice. *European Journal of Medicinal Chemistry*, *127*, 1035–1046.

Wang, X., Sun, D., Tai, J., & Wang, L. (2017). Ganoderic acid a inhibits proliferation and invasion, and promotes apoptosis in human hepatocellular carcinoma cells. *Molecular Medicine Reports*, *16*(4), 3894–3900.

Wasser, S. P. (2010). Current findings, future trends, and unsolved problems in studies of medicinal mushrooms. *Applied Microbiology and Biotechnology*, *89*(5), 1323–1332.

Wasser, S. P. (2017). Medicinal Properties and Clinical Effects of Medicinal Mushrooms. In *Edible and Medicinal Mushrooms: Technology and Applications* (pp. 503–540). Chichester: John Wiley & Sons.

Watanabe, F., Yabuta, Y., Bito, T., & Teng, F. (2014). Vitamin B12-containing plant food sources for vegetarians. *Nutrients*, *6*(5), 1861–1873.

Wieczorek, P. P., Witkowska, D., Jasicka-Misiak, I., Poliwoda, A., Oterman, M., & Zielińska, K. (2015). Bioactive alkaloids of hallucinogenic mushrooms. *Studies in Natural Products Chemistry* (Vol. 46, pp. 133–168). Amsterdam: Elsevier.

Wu, G., Lu, J., Guo, J., Li, Y., Tan, W., Dang, Y., Zhong, Z., Xu, Z., Chen, X., & Wang, Y. (2012). Ganoderic acid DM, a natural triterpenoid, induces DNA damage, G1 cell cycle arrest and apoptosis in human breast cancer cells. *Fitoterapia*, *83*(2), 408–414.

Wu, W., & Ahn, B. (2014). Statistical optimization of ultraviolet irradiate conditions for vitamin D2 synthesis in oyster mushrooms (Pleurotus ostreatus) using response surface methodology. *PLoS One*, *9*(4).

Xu, X., Zhang, X., & Chen, C. (2016). Stimulated production of triterpenoids of Inonotus obliquus using methyl jasmonate and fatty acids. *Industrial Crops and Products*, *85*, 49–57.

Yamada, T., Komoto, J., Lou, K., Ueki, A., Hua, D. H., Sugiyama, K., Takata, Y., Ogawa, H., & Takusagawa, F. (2007). Structure and function of eritadenine and its 3-deaza analogues: Potent inhibitors of S-adenosylhomocysteine hydrolase and hypocholesterolemic agents. *Biochemical Pharmacology*, *73*(7), 981–989.

Yang, D., Zhou, Z., & Zhang, L. (2019). An overview of fungal glycan-based therapeutics. *Progress in Molecular Biology and Translational Science* (Vol. 163, pp. 135–163). Cambridge: Academic Press.

Zaman, S. B., Hussain, M. A., Nye, R., Mehta, V., Mamun, K. T., & Hossain, N. (2017). A review on antibiotic resistance: Alarm bells are ringing. *Cureus*, *9*(6), e1403. 10.7759/cureus.1403

Zeng, A., Hua, H., Liu, L., & Zhao, J. (2019). Betulinic acid induces apoptosis and inhibits metastasis of human colorectal cancer cells in vitro and in vivo. *Bioorganic & Medicinal Chemistry*, *27*(12), 2546–2552.

Zhang, J., Ma, K., Han, J., Wang, K., Chen, H., Bao, L., Liu, L., Xiong, W., Zhang, Y., Huang, Y., & Liu, H. (2018). Eight new triterpenoids with inhibitory activity against HMG-CoA reductase from the medical mushroom Ganoderma leucocontextum collected in Tibetan Plateau. *Fitoterapia*, *130*, 79–88.

Zhang, M., Zhang, Y., Zhang, L., & Tian, Q. (2019). Lentinan as an immunotherapeutic for treating lung cancer: a review of 12 years clinical studies in China. *Progress in Molecular Biology and Translational Science* (Vol. 163, pp. 297–328). Cambridge: Academic Press.

Zhang, S., Li, Z., Stadler, M., Chen, H., Huang, Y., Gan, X., Feng, T., & Liu, J. (2018). Lanostane triterpenoids from Tricholoma pardinum with NO production inhibitory and cytotoxic activities. *Phytochemistry*, *152*, 105–112.

Zhang, Y., Kong, H., Fang, Y., Nishinari, K., & Phillips, G. O. (2013). Schizophyllan: A review on its structure, properties, bioactivities and recent developments. *Bioactive Carbohydrates and Dietary Fibre*, *1*(1), 53–71.

Zhou, S., Liu, Y., Yang, Y., Tang, Q., & Zhang, J. (2015). Hypoglycemic activity of polysaccharide from fruiting bodies of the shaggy ink cap medicinal mushroom, Coprinus comatus (Higher basidiomycetes), on mice induced by Alloxan and its potential mechanism. *International Journal of Medicinal Mushrooms*, *17*(10), 957–964.

Zhou, Z., Han, Z., Zeng, Y., Zhang, M., Cui, Y., Xu, L., & Zhang, L. (2014). Chinese FDA approved fungal glycan-based drugs: An overview of structures, mechanisms and clinical related studies. *Translational Medicine*, *4*(4). 10.4172/2161-1025.1000141

Zik, C. (2019). Late life vitamin B12 deficiency. *Clinics in Geriatric Medicine*, *35*(3), 319–325.

CHAPTER 9

In vitro and in vivo bioactivity of edible wild mushrooms

Anamaria Pop, Adriana Păucean, Simona Maria Man, Maria Simona Chis, Mihaela Mihai, and Sevastița Muste

Department of Food Engineering, Faculty of Food Science and Technology, University of Agricultural Science and Veterinary Medicine, Cluj-Napoca, Romania

CONTENTS

9.1 INTRODUCTION

International trade in edible wild mushrooms has grown spectacularly in recent years. An increasing number of scientific study results show that the driver behind these changes seems to be the shift in world demand for these products, with changes in consumption habits that entail a major diversification in demand (de Frutos, 2020). Deficiencies of antioxidants, vitamins, anti-inflammatory elements (zinc, selenium), as well as physiological processes, such as the age of a person, may cause inefficient resolution of inflammation. Chronic inflammation characterizes autoimmune diseases and it also accompanies diseases of the cardiovascular system, metabolic and neurodegenerative diseases, and cancers (Okin & Medzhitov, 2012). Mushrooms have been used for their nutritional and medicinal properties for centuries. Edible species constitute a good source of carbohydrates, mainly chitin that fulfills the role of dietary fiber. They are valuable sources of proteins containing essential amino acids, and thus they may be considered an alternative to animal

DOI: 10.1201/9781003152583-11

products. Mushrooms are also rich in anti-inflammatory components, such as polysaccharides, phenolic and indolic compounds, mycosteroids, fatty acids, carotenoids, vitamins, and biometals (Muszyńska, Grzywacz-Kisielewska, et al., 2018). Wild edible mushrooms are also regarded as a cheap food source, while also rich in bioactive compounds, especially in phenolic compounds (Vamanu, 2018). Moreover, mushrooms are low in calories, which is due to their low-fat content, yet they are rich in poly-unsaturated fatty acids (PUFAs) that are beneficial for health. The dietary and medicinal value of edible mushrooms is further supported by the fact that they are sources of numerous biologically active and health-promoting compounds (Muszyńska, Grzywacz-Kisielewska, et al., 2018). Nowadays, consumers are interested in those food bioactive compounds that provide beneficial effects to humans in terms of health promotion and disease risk reduction (Salanta et al., 2020). Recently, several studies have shown there has been a growing interest in understanding food's digestion to strengthen the possible effects of food on human health. The quantification of the amount of bioavailable bioactive compounds is more important than determining the quantity of these compounds existent in the foods (Santos et al., 2019). It has been reported that the bioavailability of bioactive compounds may change due to their combination with macronutrients, namely the fiber content of low-processed foods and beverages and the polysaccharides and proteins of processed foods (Dupas et al., 2006). The bioavailability of each polyphenol is distinct and there is no relationship between the amount of polyphenol in the food and its bioavailability in the human body since bioavailability depends on digestive stability, food matrix release (bioaccessibility), and efficiency of the transepithelial passage. In bioavailability tests, it must be confirmed that the compound is efficiently digested, assimilated, and after absorption, that it promotes a positive influence on human health. Bioaccessibility may be characterized as the amount or portion of a compound that is released from the food matrix in the gastrointestinal tract and becomes accessible for absorption (Santos et al., 2019). Moreover, it has been reported by the same authors that methods for measuring bioavailability and/or bioaccessibility of nutrients imply research in humans, mice, pigs, and other animals (*in vivo*) or simulation in laboratory assays (*in vitro*). *In vitro* models are preferred instead of *in vivo* experiments when studying the effects of digestion on different bioactive compounds, the bioaccessibility of these compounds, and their degree of absorption (Socaci et al., 2019). The fate of food in the GIT (gastrointestinal tract) can be studied using several methods or models including static and dynamic *in vitro* models, various cell and *ex vivo* cultures, animals, and humans. Recently, such models have also been suggested as valuable tools for investigating digestion in various populations (Bohn et al., 2018). There are static digestion models, where the quantities and concentrations of materials are preestablished, and dynamic, where continuous digestion is simulated with the implied changes of conditions. The second type manages to identify much better with the *in vivo* models, even though it is more expensive and laborious (Alminger et al., 2014; Martinez-Medina et al., 2021). Evidence shows that despite the simplicity of *in vitro* models, they are often very useful in predicting outcomes of the *in vivo* digestion. Overall, static *in vitro* digestion methods are particularly popular because they are easy to use, cheap and do not require specific equipment. However, a huge number of protocols differ in terms of experimental conditions (pH and duration of the different steps, amount of digestive enzymes and bile, etc.). However, this relies on the complexity of *in vitro* models and their tuning towards answering specific questions related to human digestion physiology, which leaves vast room for future studies and improvements (Bohn et al., 2018).

9.2 *IN VITRO* AND *IN VIVO* ANTIOXIDANT POTENTIAL OF EDIBLE MUSHROOMS – BIOACTIVE COMPOUNDS ASSOCIATED

9.2.1 Polyphenols, phenolic acids, and flavonoids

Mushrooms are a rich source of polyphenols and phenolic acids, compounds that are known as secondary metabolites with powerful antioxidant potential. These compounds have been extensively identified and studied in plant species but represented the focus of researchers' attention for the valuable bioactivity of wild mushrooms, as well. Phenolic acids belong to the polyphenols family, while from a chemical point of view, they contain at least one aromatic ring in which one or more hydrogen is substituted by a hydroxyl group (Heleno et al., 2015). Hydroxybenzoic acids and hydroxycinnamic acids are the two major groups of phenolic acids. Nowadays, it is well known that due to the phenolic hydroxyl groups which are attached to ring structures, they possess important properties like reducing agents, hydrogen donators, singlet oxygen quenchers, superoxide radical scavengers, and metal chelators over hydroxyl and peroxyl radicals, as well as superoxide anions and peroxynitrites (Heleno et al., 2015).

Boletus is a genus of mushrooms, comprising over 100 species and is one of the most harvested wild mushrooms in Europe, North America, and Asia (Heleno et al., 2011). The popularity of *Boletus* ssp. mushrooms are due to their sensorial characteristics, as their aroma, taste, and texture are widely appreciated (Jaworska & Bernaś, 2009). However, alongside the consumers' sensory preference, it is their content in primary and secondary metabolites that must also be known to the general public, especially due to *Boletus* beneficial health effects. In agreement with this, Heleno et al. (2011) identified p-hydroxybenzoic acid, protocatechuic acid, p-coumaric acid, and cinnamic acid in six edible and non-edible *Boletus* species. As for edible mushrooms, *B. aereus* contained the highest concentration of total phenolic acids and cinnamic acid. Even if other authors did not find phenolic compounds in *B. edulis*, probably due to the different methods of extraction, Heleno et al. (2011) reported the presence of all the analyzed phenolic acids, but in much smaller amounts. In the same study, *in vitro* antioxidant activity was investigated by several assays such as scavenging activity on DPPH radicals, reducing power, and inhibition of lipid peroxidation. A strong correlation was found between the content of phenolic acids and the antioxidant activity, emphasizing that *B. aerus*, *B. edulis*, *B. reticulatus,* and the non-edible species revealed significant antioxidant properties. Using a different extraction method, Özyürek et al. (2014) studied the antioxidant/antiradical properties of *B. edulis* mushroom. This approach led to methanolic extracts being obtained by using microwave-assisted extraction (MAE), a technique considered advantageous over others because of its capability to enhance extraction efficiency. In *B. edulis* methanolic extract, the authors identified valuable concentrations of phenolic compounds (gallic acid, protocatechuic acid, vanillic acid, rutin, hesperidin, quercetin, naringenin, apigenin, etc.) that were found in strong positive correlation to the antioxidant/antiradical properties, which were tested by several assays such as: free radical scavenging activity, hydroxyl radical scavenging activity, hydrogen peroxide scavenging activity, superoxide anion radical scavenging activity, EC_{50} values in antioxidant activity, and total antioxidant capacity. *B. edulis,* along with 16 other species of wild mushrooms, were analyzed by Yahia, Gutiérrez-Orozco, and Moreno-Pérez (2017) in order to determine the total content of phenolics, flavonoids, anthocyanins, and antioxidant capacity. As such, liquid chromatography coupled to mass spectrometry (HPLC-MS) was used for the identification and quantification of phenolic compounds. As other studies reported, *B. edulis* was found to possess good antioxidant capacity *in vitro*, determined by DPPH and FRAP assays, due to its content in total soluble phenols (55.6 mg of GAE/100 g fw) and total flavonoid (23.66 mg catechin/100 g fw) compounds. In the case of *B. edulis*, HPLC-ESI-MS analyses of the extract revealed the following phenolic compounds: caffeic acid hexoside, sinapic acid hexoside, cinnamic acid derivatives, the dimer of p-hydroxybenzoic acid, and traces of protocatechuic acid hexoside. P-Hydroxybenzoic acid was found in the highest content among phenolic compounds of *B. edulis*, while organic acids such as malic, citric, and quinic were not detected, as in the case of other analyzed species of wild mushrooms.

Moreover, wild fruiting bodies of *Lactariuspiperatus* (L.) at different stages of maturity were analyzed by Barros, Baptista, and Ferreira (2007) to monitor the dynamics of the phenolic, flavonoid, ascorbic acid, β-carotene, and lycopene contents, as well as their antioxidant activity during these stages. Total phenols, followed by flavonoids, were found in the highest concentrations comparing to ascorbic acid and β-carotene and lycopene, in all stages of fruiting body maturity (Chawla et al., 2019). An interesting result was that the stage of immature spores contained the highest amount of phenols and flavonoids and no the mature ones. The authors concluded that the aging process might induce the reduction of the antioxidant compounds that are reactive oxygen species. According to these results, the highest antioxidant activity was found in the mature stage with immature spores. Different biochemical assays: reducing power, scavenging activity on DPPH radicals, inhibition of the erythrocyte hemolysis mediated by peroxyl free radicals, and lipid peroxidation inhibition by the b-carotene-linoleate system were carried out to evaluate the antioxidant activity of *L. piperatus* extracts. All the correlations reported by the authors support that the antioxidant compounds and the antioxidant activity of the wild mushrooms (i.e. *L. piperatus*) are dependent on the stage of fruiting body maturity. This result underlines the necessity to obtain qualitative functional food or nutraceuticals from mushrooms with high content in phenols and flavonoids (Barros, Baptista, & Ferreira, 2007).

Kosanić and Ranković (2016) studied the *in vitro* antioxidant activity of the acetone and methanol extracts of the *Amanita rubescens* and *Russulacyanoxantha* mushrooms. The antioxidant activity was evaluated by performing several assays: free radical scavenging, reducing power, and determination of total phenolic compounds. *Amanita rubescens* was found with the highest reducing power and strong antioxidant activity. The IC_{50} values were 114.21 µg/mL, for the acetone extract and 185.70 µg/mL in the case of methanol extract. The phenolic contents (µg of pyrocatechol equivalent) were 4.86 and 5.22, respectively.

As for *Ganoderma* species, they are very popular wild mushrooms in Taiwan due to their demonstrated therapeutic effects. The study conducted by Mau et al. (2005) aimed to assess the antioxidant properties of methanolic extracts from *Ganoderma tsugae* by comparing mature and young fruit bodies, mycelia, and filtrate from the submerged culture. Ascorbic acid, β-carotene, tocopherols, and total phenols were quantified to obtain an image of the antioxidant compounds. However, total phenols were found as the major fraction of the analyzed group of compounds with values exceeding 24 mg/g for each tested sample. The rest of the antioxidant compounds were identified in much smaller amounts, with mycelia and filtrate having the smallest or no detected values. Results indicated good antioxidant properties for the tested samples of *G. tsugae*. According to these results, the authors also concluded that *G. tsugae*, in different forms (mature and young fruit bodies, mycelia, and filtrate) could be included in the human diet as a food or food ingredient and could fight against oxidative damage.

Rahman et al. (2018) evaluated *in vitro* bioactive potential of *Lentinula edodes* (shiitake mushroom) using its solvent–solvent partitioned fractions that consisted of methanol:dichloromethane (M:DCM), hexane (HEX), dichloromethane (DCM), ethyl acetate (EA), and aqueous residue (AQ). The results of this study demonstrated the hexane fraction of *L. edodes* extract possesses bioactive food components (α-tocopherol-vitamin E, oleic acid, linoleic acid, ergosterol, and butyric acid) capable of conferring anti-oxidative defense and curtailing LDL oxidation, as well as being potent in inhibiting the activity of the rate-limiting enzyme in cholesterol biosynthesis, thus supporting its use as an anti-atherosclerotic agent.

Additionally, Sasidharan et al. (2010) completed an *in vitro* investigation on the antioxidant activity and the effects of the methanol extracts of *Lentinula edodes* on liver function markers in the serum, as well as on hepatic histopathology of mice liver showing paracetamol-induced hepatotoxicity. The authors reported the amount of total phenolics was estimated to be 70.83 mg gallic acid equivalent (GAE) per gram of dry extract, and the antioxidant activity of the *L. edodes* extract was 39.0% at a concentration of 1 mg/mL. Moreover, the effects of the *Lentinula edodes* extract on serum transaminases (SGOT, SGPT), alkaline phosphatase (ALP), and bilirubin were measured in the paracetamol-induced hepatotoxic mice, and was noticed that it produced significant ($p < 0.05$) hepatoprotective effects by decreasing the activity of serum enzymes and bilirubin.

The study conducted by Vamanu and Nita (2014), aimed to assess *in vitro* antioxidant and anti-inflammatory activities of fluidized bed ethanol extracts of three species *of Pleurotusostreatus* (PQMZ91109, PBS281009, M2191 – obtained from a laboratory greenhouse), comparatively to a commercial species (*Agaricus bisporus*) and three wild edible mushrooms (*Marasmiusoreades*, *Craterelluscornucopioides*, and *Tuber melanosporum*). Following the results, the presence of more tocopherols than polyphenol carboxylic acids and flavones was demonstrated, while tocopherol levels are positively correlated with anti-inflammatory activities and inhibition of lipid peroxidation. Also, HPLC showed that *Tuber melanosporum* extracts had high correlation values between levels of different active compounds and antioxidant assay results. Similarly, aqueous and ethanolic extracts of *Pleurotus giganteus* fruiting bodies were investigated *in vitro* by Phan et al. (2012) for their effects in neurite outgrowth of rats pheochromocytoma (PC12) cells. It was found that aqueous and ethanolic *Pleurotus giganteus* extracts have a high content of phenolic compounds and triterpenoids. The study shows that these extracts have induced neurite outgrowth of PC12 cells in a dose and time-dependant manner with no detectable cytotoxic effect (Phan et al., 2012).

Biological properties of Portuguese wild mushrooms (*Cantharellus cibarius*, *Hypholomafasciculare*, *Lepista nuda*, *Lycoperdonmolle*, *Lycoperdonperlatum*, *Ramaria botrytis*, and *Tricholomaacerbum*) were evaluated *in vitro* by Barros et al. (2008) as they contain very useful phytochemicals such as phenolics, tocopherols, ascorbic acid, and carotenoids. Phenols were the major antioxidant components found in the extracts (1.75–20.32 mg/g), followed by flavonoids (0.47–16.56 mg/g). Ascorbic acid was found in small amounts (0.09–0.40 mg/g), and β-carotene and lycopene were found in only vestigial amounts (<0.08 mg/g). The antioxidant activity measured by four different methods (Table 9.1), has been proven for all species while being more significant for *Romaria botrytis*.

*Armillaria mellea*is one of the well-known wild edible mushrooms, with a rich chemical composition in bioactive compounds such as polysaccharides, vitamins, proteins, aminoacids, and minerals (S. Chen et al., 2020). In a study conducted by R. Chen et al. (2020), two fractions were extracted and purified from the *Armillaria mellea*, AMPs-1-1, and AMPa-2-1, respectively, through anion exchange and gel chromatography. The antioxidant activity of these fractions was determined using FRAP assay (μM/L), DPPH radical scavenging activity (%), ORAC (μM TE/g), and ABTS$^+$ (%). The quantitative differences in antioxidant activity between the aforementioned fractions could be due to their different molecular weight, structure, and chemical composition, highlighting that AMPs-2-1 fraction showed the highest antioxidant activity compared with AMPs-1-1. The study reported that AMPs-2-1 could be successfully used and explored as a novel potential antioxidant insuch industries aspharmaceutical one or even in functional food manufacturing. *Marasmiusoreades* antioxidant activity was analyzed through different methods such as DPPH (2,2-diphenyl-1-picrylhydrazyl) radical-scavenging activity, lipid peroxidation by using TBARS (thiobarbituric acid reactive substances) assay, reducing power and ß -carotene bleaching inhibition (Table 9.1) combined with different proportions of *Boletus edulis* extracts (Vieira et al., 2012). The results showed that the proportion between the combined mushroom samples could influence the antioxidant activity. For instance, the radical scavenging activity reached the highest value (6 mg/mL) when mushroom extracts were combined in a 1:1 ratio (50% *Marasmiusoreades* extract and 50% *Boletus edulis* extract), while the reducing power (1.6 mg/mL) and TBARS inhibition (3 mg/mL) showed the highest value when 100% *Boletus edulis* extract was used. The study concluded that combined methanol extract from *Marasmiusoreades* and *Boletus edulis* could be successfully used as an important source of bioactive molecules. The antioxidant activity of *Marasmiusoreades* extracts was also proved by Shomali et al. (2019), through radical scavenging activity assayed by using the DPPH method. In short, *Marasmiusoreades* ethanolextracts showed a strong radical scavenging activity of 80%, which could be related to its higher polyphenolic content (especially gallic, ferulic, and catechin, vanillic acids). Similarly, to highlight the antioxidant activities of *Marasmiusoreades*, antioxidant enzyme assays likeglutatatione-S-transferase (GST), glutathione peroxidase (GPx), catalase (CAT), and superoxide dismutase (SOD) were used. The results showed that 10 mg/mL

Table 9.1 Bioactivity-based analysis of selected edible wild mushrooms species

Edible wild mushrooms species	Bioactive compounds associated	Methodologies used in research	Bioactivity potential	Digestion models	Others observation	References
Pleurotus ostreatus	a. Total phenolic content b. Selenium-enriched polysaccharide (Se-POP)	a. Freeze-dried (FD) and vacuum-dried (VD) mushrooms/milled and defatted. The extraction solvent was chosen as ethanol: water (70:30, v-v) mixture, due to their different polarity, and acceptability for human consumption/the extract was evaluated as a food supplement b. Hot water extraction from fresh fruiting bodies	a. Antioxidant activities b. Antioxidant activities	a. *in vitro* b. *in vitro*	a. When bioaccessibility index was evaluated, in FD mushrooms, only gallic acid (14.8%) and protocatechuic acid (14%) was bioaccessible, in VD mushrooms except for chlorogenic acid, other identified phenolics were bioaccessible between 17.8% and 41.9%. Regarding the use of dried mushroom powders in the preparation of functional foods, or the recipe of other food products, vacuum drying seems a better alternative for both extractability of phenolics and their fate during *in vitro* gastrointestinal digestion. b. *In vitro* study of DPPH, hydroxyl, and ABTS free radicals, Se-POP had a stronger antioxidant capacity than POP. Compared to POP, Se-POP had a superior ability to reduce hydrogen peroxide (H2O2)-induced oxidative stress and apoptosis in murine skeletal muscle (C2C12) cells.	a. (Ucar & Karadag, 2019) b. (Ma et al., 2018)
Agaricus bisporus	a. Gallic acid, protocatechuic acid, catechin, caffeic acid, ferulic acid, and myricetin b. Phenolic compounds c. Spermidine and amino acids	a. Ethanolic extract b. Metanolic extract c. Fresh *Agaricus bisporus* mushroom was submitted to cooking and canning/UHPLC method was used for the simultaneous determination of 18 free amino acids, 10 biogenic amines, and ammonia	a. Antioxidant activity b. Anti-inflammatory properties c. Bioaccessibility	a. *in vitro* and *in vivo* b. *in vitro* c. *in vitro*	a. Using LPS-activated RAW 264.7 macrophages. Experiments demonstrated the influence of extracts on the expression of inflammation markers, such as IL-1β and IL-6, and the production of nitric oxide (NO). b. Spermidine in A. bisporus was fully bioaccessible, a novelty for amines and mushroom; - emphasizing the higher abundance of glutamic acid, leucine, lysine, arginine and tyrosine after *in vitro* gastric-intestinal digestion.	a. (Cruz et al., 2016) b. (Muszyńska, Kała, et al., 2018) c. (Reis et al., 2020)

Species	Compounds	Extract/Method	Bioactivity	Study type	Notes	Reference
Imleria badia	Phenolic compounds; proto-catechuic, p-hydroxybenzoic, cinnamic, and p-coumaric acids	Methanolic extract	Antioxidant activity	*in vitro*	to 99.2% in the tests of linoleic acid oxidation	(Reis et al., 2020)
Agaricus blazei	Total phenolic, compounds (TPCs), flavonoids	*Agaricus blazei* residue (MAR) from the hydroalcoholic extraction of bioactive compounds/powder mixed with methanol	Antioxidant activity	*in vitro*	*Agaricus blazei* mushroom residue, which is usually discarded, can be used as a potential antioxidant ingredient, capable of enriching foods at a low cost.	(Vital et al., 2017)
Coriolus versicolor	Polysaccharopeptide (PSP)	PSP – is recovered from ethanol precipitation (dark brawn powder)	Antidiabetic activity	*in vitro*	PSP can competitively inhibit tolbutamide 4-hydroxylation in both pooled human liver microsomes and specific human CYP2C9 • No studies were found using biomass from *Coriolus versicolor* mushroom to demonstrate bioactive effects. Two polysaccharides extracted from *C. versicolor* (PSP and PSK) have shown important biological properties.	(Cruz et al., 2016)
Agrocybe aegerita	Ceramide; methyl-b-D-glucopyranoside a-D-glucopyranoside, linoleic acid, and its methyl ester.	Methanolic extract from fruiting bodies (MeOH (1 l) (8 hours × 3) at 25°C). The structure elucidation was accomplished by NMR and mass spectral methods.	Anti-cancer potential	*in vitro*	• Is the first report of the isolation of ceramide from *A. aegerita* and its COX -1 and -2 enzymes and tumor cell proliferation inhibitory activities.	(Diyabalanage et al., 2008)
Morchella esculenta	Eight main compounds (1–8), including three fatty acids and five sterols: (1eO-octadecanoyl-sn-glycerol (1), (Z,Z)-9,12-octadecadienoic acid (2) (3β,5α,8α,22E,24S)-5,8-epidioxyergosta-6,9(11),22-trien-3-ol (3) (3β,5α,8α,22E,24S)-5,8-epidioxyergosta-6,22-dien-3-ol (4) (3β,5α,22E)-Ergosta-7,22,24(28)-trien-3-ol (5), 1-octadecanoic acid (6), (22E,24S)-Ergosta-4,22-dien-3-one (7), and (3β,5α,8α,22E)-5,8-epidioxyergosta-6,22-dien-3-yl β-D-glucopyranoside (8)	Methanolic extract from fruiting bodies (the extract was fractionated into hexane-, CH2Cl2-, EtOAc-, and n-BuOH-soluble fractions/resolved by chromatography on a silica gel (230–400 mesh) column with hexane–EtOAc (30:1–1:1, gradient system) to yield five fractions (A–E)/repeated column chromatography and HPLC purification	Antitumor activity	*in vivo and in vitro*	Tested on four human lung adenocarcinoma cell lines. Bioactivity-guided fractionation of the MeOH extract followed by chemical investigation of its most cytotoxic hexane-soluble fraction led to the isolation of eight compounds (1–8), including three fatty acids and five sterols. Among the eight isolated compounds tested, compounds 1, 3, and 5 exhibited the most potent cytotoxicity to human lung cancer cell lines, with IC50 values ranging from 156.9 to 278 M; this activity was mediated by induction of apoptosis.	(S. R. Lee et al., 2018)
Boletus edulis	a. Phenolics and phenolic acids p-Hydroxybenzoic acid Protocatechuic acid p-Coumaric acid Cinnamic acid b. Phenolics Polysaccharides- c1. Arabinose Xylose Mannose; c2. Glucose Galactose Rhamnose c3. Heteroglycan c4. Glucan polysaccharides	a. HPLC-DAD, colorimetric Folin Ciocalteu assay, scavenging activity on DPPH radicals, reducing power, and inhibition of lipid peroxidation. b. Methanolic extract using microwave-assisted extraction (MAE) c1, c2. Water-soluble polysaccharides	a. Antioxidant activity b. Antioxidant activity c1, c2, c3, c4. Antioxidant/antitumor/immunomodulatory activities d. Anti-tumor activity e. Effect on microbial fingerprint (*Lactobacillus* and	a. *in vitro* b. *in vitro* c1., c2. *in vitro* c3. *in vivo* c4. *in vivo* d. *in vitro/in vivo* e. *in vitro simulated*	a. Good antioxidant properties but lower EC50 values were found comparing to samples from other countries b. Good antioxidant/antiradical properties were found by testing free radical scavenging activity, hydroxyl radical scavenging activity, hydrogen peroxide scavenging activity,	a. (Heleno et al., 2011) b. (Özyürek et al., 2014) c.1. (L. Zhang et al., 2018) c.2. (D. Wang et al., 2014) c.3. (Lemieszek et al., 2013)

(Continued)

Table 9.1 (Continued) Bioactivity-based analysis of selected edible wild mushrooms species

Edible wild mushrooms species	Bioactive compounds associated	Methodologies used in research	Bioactivity potential	Digestion models	Others observation	References
	Proteins d. Lectins (BEL) Anti-A549 active protein (BEAP) e. Effect on microbial fingerprint (Lactobacillus and Bifidobacterium) in the descending colon	c3. Hot water extract Glc:Gal:Rha:Arab = 2.9:3.2:1.3:1.6 c4. Hot water soluble d. Purification on a column of human erythrocytic stroma incorporated into a polyacrylamide gel followed by X-ray investigations of the structure e. Whole mushroom dried in an oven 50°C	Bifidobacterium) in the descending colon	digestion in the stomach and small intestine	superoxide anion radical scavenging activity, EC_{50} values in antioxidant activity, total antioxidant capacity c1, c2,c3,c4. concentrations of polysaccharides varied with climatic and geographic conditions' d. p16/cyclin D1/CDK4-6/pRb pathway decreased Cell cycle arrest in G0/G1 phase increased e. The ratio of microorganisms was generally stable during mushroom consumption -positive effect on the population of lactobacilli, bifidobacteria	c4. (P. Maity et al., 2021) d. (Bovi et al., 2013) e. (Vamanu & Pelinescu, 2017)
Boletus aereus	a. Phenolics and phenolic acids p-Hydroxybenzoic acid p-Coumaric acid Cinnamic acid b. Polysaccharides- Arabinose Xylose Mannose Glucose Galactose Rhamnose c. Exopolisaccharides (EPS) Fr-I, Fr-II, Fr-III	a. Scavenging activity on DPPH radicals, reducing power, and inhibition of lipid peroxidation. b. Water-soluble polysaccharides c. EPS were obtained from the 3-liter fermentation broth from submerged culture of *B. aereus* by the method of ethanol precipitation.	a. Antioxidant activity b. Antioxidant activity of polysaccharides c. Antioxidant activity DPPH and ABTS radical scavenging assays	a. *in vitro* b. *in vitro* c. *in vitro*	a. DPPH radical scavenging activity showed significant correlations with arabinose and galactose content, and the galactose contents also showed a significant correlation with ferrous ion reducing power. b. Xylose content had a significant influence on metal chelating activity. Polysaccharides with a beta configuration in the pyranose form have higher antioxidant activity. c. The antioxidant activity of polysaccharide varied with structural characteristics, molecular weight, monosaccharide constituent, configuration of chain • Antioxidant activity may originate from the hydrogen atom-donating ability to a radical • Fr-I (glucose/manose) showed a remarkably superior antioxidant among the three EPS fraction	a. (Heleno et al., 2011) b. (L. Zhang et al., 2018) c. (J. Q. Zheng et al., 2014)

Mushroom	Bioactive compound	Extraction	Bioactivity	Study type	Results	References
Boletus pinophilus		a. Whole mushroom dried in an oven 50°C	Effect on microbial fingerprint (Lactobacillus and Bifidobacterium) in the descending colon	a. in vitro simulated digestion in the stomach and small intestine	a. The ratio of microorganisms was generally stable during mushroom consumption • Positive effect on the population of lactobacilli, bifidobacteria	a. (Vamanu & Pelinescu, 2017)
Echinodontium tinctorium	a. Polysaccharide B-glucan, heteroglycan b. Polysaccharide β-glucan polysaccharide rich in glucuronic acid	a. 80% ethanol, 50% methanol, water, and 5% NaOH b. 80% ethanol, 50% methanol, water	a. Immunostimulatory and anti-inflammatory activities b. Immunostimulatory activity	a. in vitro/in vivo b. in vitro	a. Crude alkali extract of 5% NaOH showed significant bioactivities b. Significantly enhance of cytokines productions (IL-6, MIP-2, G-CSF, GM-CSF, LIF, MCP-1, MIP-1α, MIP-1β, and RANTES in macrophage cells)	a. (Javed et al., 2019) b. (Zeb et al., 2021)
Auricularia auricula-judae	a. Fibres b1. Polysaccharide AAG ((1→4)-linked D-glucopyranosyl main chain with (1→6)-linked D-glucopyranosyl branch at O-6) b2. AF1 ((1→3)-β-D-glucan main chain with two (1→6)-β-D-glucosyl, residues for every three glucose residues)	• Total dietary fiber (TDF) content using the Association of Official Analytical Chemists (AOAC) method and for dietary fibre content and composition using the Uppsala method. b1. Semi-stiff chains in 0.1M NaCl aqueous solution 65.5%, degree of branching, 34–288 k Da b2. Aqueous solution	a. Prevention of cardiovascular diseases Lower LDL cholesterol and serum total cholesterol Lower cholesterol levels, Lower Phospholipids of plasma Modification of Terpenes hepatic phospholipids metabolism Hyperhomocysteinemic effect b1. b2. Antitumor effect	a. in vivo (rats) b. in vitro/in vivo	a. Mechanism: inhibition of the synthesis of hepatic triglycerides by increasing the short chain fatty acids production (acetate, propionate and butyrate) during the dietary fiber fermentation by colonic microflora b.1.b2. Induced apoptosis of cancer cell	a1. (Cheung, 2010) a2. (Guillamón et al., 2010) b1. (Meng, Liang, & Luo, 2016) b2. (X. Xu, Yan, & Zhang, 2012)
Lactarum piperatus	Effect on microbial fingerprint (Lactobacillus and Bifidobacterium) in the descending colon b. Phenolic compounds Phenols, flavonoids, ascorbic acid, b-carotene, and lycopene	a. Whole mushroom dried in an oven 50°C b. Methanol extraction	a. Effect on microbial fingerprint (Lactobacillus and Bifidobacterium) in the descending colon b. Antioxidant activity DPPH radical-scavenging activity Inhibition of erythrocyte hemolysis mediated by peroxyl free radicals Inhibition of lipid peroxidation using the b-carotene linoleate model system	a. in vitro simulated digestion in the stomach and small intestine b. in vitro	a. The ratio of microorganisms was generally stable during mushroom consumption • Positive effect on the population of lactobacilli, bifidobacteria b. The antioxidative components production by wild mushrooms and their antioxidant properties depends on the stage of fruiting body maturity. • The highest antioxidant contents and the lowest EC50 values for antioxidant activity were obtained in the mature stage	a. (Vamanu & Pelinescu, 2017) b. (Barros, Baptista, & Ferreir,a 2007)
Amanita rubescens (edible after 20 min cooking)	a. Phenolics b. Lipid extracts	a. Acetone and methanol extracts b. Hexane extracts were evaporated and dissolved in dimethyl sulfoxide (DMSO)	a. Antioxidant activity Antimicrobial activity b. Anticancer activities	a. in vitro b. in vitro	a. Extracts had effective reducing power and strong antimicrobial activity against the tested microorganisms.	a. (Kosanić & Ranković, 2016) b. (Şahin et al., 2014)

(Continued)

Table 9.1 (Continued) Bioactivity-based analysis of selected edible wild mushrooms species

Edible wild mushrooms species	Bioactive compounds associated	Methodologies used in research	Bioactivity potential	Digestion models	Others observation	References
Pleurotus djamor	a. Polysaccharides (1 → 3)-β-D Glucan b. Mycelia zinc polysaccharides (MZPS)-rhamnose, mannose, and glucose c. Polysaccharides (EnPPs)-rhamnose, xylose, mannose, galactose, glucose	a. Alkaline extract b. Liquid fermentation technology was used to produce *P. djamor* zinc mycelia Polysaccharide isolation: filtration, con-centration, sterilization and lyophilization, the powered of mycelia were extracted twice with proper water for 4 hours (60°C). The super-natant liquid was obtained by centrifugation (3,600 × g, 15 minutes), and precipitated with ethanol (1:3, v/v) at 4°C overnight c. Enzyme assisted extraction procedure; *P. djamor* mycelium was dried to constant weight in vacuum oven at 50°C. EnPPs extraction was performed in a water bath; at the end of the extraction process, the supernatant liquids were centrifuged, filtered, concentrated, and precipitated by ethanol precipitation (1:3, v/v) at 4°C overnight. The polysaccharide precipitate was collected by centrifugation (3,600 r/minutes, 10 minutes), deproteinated by employing and lyophilized	a. Antitumor activity b. Antioxidant activity c. Antioxidant activity, antihyperglycemic	a. *in vitro* b. *in vitro/in vivo* c. *in vivo*	b. Results showed that hexane extracts of this three species were found inactive against HT29 cell line a. Cytotoxic effect against PA1, ovarian carcinoma cells b. Free radicals being released in the liver were effectively scavenged or the oxidative chain reaction was success-fully blocked by MZPSs, suggesting that the MZPSs could effectively protect against the hepatic injury from hepatotoxicity by mediating antioxidant and free radical scavenging activities c. Severe cerebral, hepatic, renal and pancreatic lesions induced by STZ were considerably and dose-dependently prevented by the administration of EnPPs at three dosage (800, 400, and 200 mg/kg) • EnPPs could protect these tissues from acute STZ-intoxication • EnPPs-could be used to prevent oxidative stress damages	a1. (P. Maity et al., 2021) a2. (G. N. Maity et al., 2019) b. (J. Zhang et al., 2016) c. (Jiao et al., 2017)
Ganoderma tsugae	a. *Ganoderma tsugae* extracts b. *Ganoderma tsugae* (GT)- extract c. Polysaccharides d. Polysaccharide mannitol/myo-inositol/trehalose/xylose e. Triterpenoids and alkamide (tsugaric acid and palmitamide isolated and lanostanoid derivatives synthetized	a. Methanol extraction, dissolution in methanol for *in vitro* tests and in ethanol for *in vivo* tests b. Ethanol-extracted c. Water and alkali extracted from *Ganoderma tsugae* Murrill d. Hot-water extracted and hot-alkali extracted polysaccharides from *Ganoderma tsugae* Murrill e. The air-dried fruit bodies were extracted with CHCl3 and then	a. Antitumor effects b. Antitumor activity c. Antioxidant activity d. Anti-proliferation of cancer cells (antitumor) e. Anticancer activity f1, f2. Anti-inflammatory properties g. Antioxidant activity	a. *in vivo and in vitro* b. *in vitro* c. *in vitro* d. *in vitro* e. *in vitro/in vivo* f. *in vitro/f.2. in vivo* g. *in vitro*	a. The GT extracts inhibit colorectal cancer cell proliferation caused by accumulating cells in G2/M phase b. GT can activate cytoprotective autophagy against apoptosis in human leukemia K562 cancer cells c. Scavenging ability on DPPH radicals, scavenging ability on	a. (Hsu et al., 2008) b. (Hseu et al., 2019) c. (Tseng, Yang, & Mau, 2008) d. (Chien et al., 2015) e. (K. W. Lin et al., 2016) f.1, f2. (Ko et al., 2008)

Mushroom	Compound	Methodology	Activity		Findings	References
	f. Triterpenoids and steroids g. containing ascorbic acid, tocopherols, and total phenols	chromatographed on silica gel column. Elution with n-hexane-acetone at different ratios and acetone yield fractions resulted. For obtaining tsugaric acid, n-hexane-CHCl3 elution was used, while CHCl3-MeOH elution was used for palmitamide; lanostanoid derivatives were synthesized f. CHCl3 extracts of *G. tsugae* were subjected to silica gel chromatography; elution of CHCl3 extract with CHCl3-EtOAc ,CHCl3-MeOH and cyclo-hexane-CHCl3-MeOH yielded to tsugaric acid , 3-oxo-5a-lanosta-8,24-dien-21-oic acid g. Air-dried or freeze-dried powder form of mushrooms samples were extracted with methanol and rotary-evaporated at 40°C			- OH radicals, chelating ability on ferrous ions d. The hot-alkali extracted polysaccharides were more effective in anti-proliferation of IMR32 cells than the hot-water extracted polysaccharides, e. Significant inhibitory on xanthine oxidase activity • Potent inhibitory effect on superoxide anion generation in rat • Cancer chemopreventive agents f1. Significant inhibitory effects on fMLP/CB-induced f2. Protective effect against photodamage induced by ultraviolet B (UVB) light g. Scavenging abilities on DPPH and hydroxyl radicals and chelating ability on ferrous ions	g.. (Mau et al., 2005)
Hericium erinaceus	a. β-mannan b. Heteroglycan Man:Glc: Gal = 6.42:67.87:1.00 c. Polysaccharide glucose:galactose:xyl se:fucose d. Polysaccharide (HIPS 1,2) from mycelia of *H. erinaceus* SG-02	a. Hot water extraction and ethanol precipitation of fruiting body and fractionation by DEAE-cellulose and sepharose CL-6B column chromatography b. Water extraction of mushroom mycelium, twice extraction of residues, concentration and precipitation with ethanol to final concentration at 80%;, fractionation by ultrafiltration and the permeate fraction was applied to a DEAE-Sephadex column and eluted stepwise with distilled water, 0.2 M sodium chloride, and 2.0 M sodium chloride c. Methodology: tot water extraction d. Distilled water at 80°C for 3 hours/anhydrous ethanol	a,b. Antitumor activity c. Antioxidant d. Antihyperglycemic and protective	a,b. *in vitro* c. *in vivo-* d. *in vivo-*	a. HEB-AP Fr I was found to act as an immunostimulant through the activation of macrophages b. Anti-chronic atrophic gastritis activity c. Increased the antioxidant enzymes activities in skin, and enhanced the collagen protein levels d. Protective effects on the pancreas, liver and kidney in STZ-induced diabetic	a. (J. S. Lee & Hong, 2010) b1. (M. Wang et al., 2015) b2. (P. Maity et al., 2021) c. (H. Xu et al., 2010) d. (C. Zhang, Li, et al., 2017)
Lentinus edodes	a. Polysaccharides b. Total phenolic content c. Lentinan d. β glucan heteroglycan Glc:Man: Gal = 19.26:1.20:1.00 e. Lentinula edodes derived polysaccharide L2	a. Acetone/water/acetic acid (70:29.5:0.5) extraction of mushroom samples b. Water and methanolic extracts c. ((1→3)-β-glycan main chain with two (1→6)-β-Dglucopyranoside branches for every five (1→3)-β-Dglucopyranoside linear linkages)	a. Inhibition of oxidative damage b. Antioxidant activity c. Immune response recovered the activation of immune cells; increased the expression of	a,b. *in vitro* c. *in vivo* d. *in vivo* e. *in vivo* f. *in vivo* g. *in vitro* RAW264.7 cells	a,b. Mechanism: decreased the expression level of VCAM-1mRNA of thoracic aorta endothelial cell in rats, with an increased antioxidant enzyme activity and significantly reduced serum total cholesterol, triglyceride, lipoprotein cholesterol and	a. (Guillamón et al., 2010) b1. (C. Xu et al., 2008) b2. (L. M. Cheung and Cheung 2005) c. (Ooi & Liu, 2012)

(Continued)

Table 9.1 (Continued) Bioactivity-based analysis of selected edible wild mushrooms species

Edible wild mushrooms species	Bioactive compounds associated	Methodologies used in research	Bioactivity potential	Digestion models	Others observation	References
	f. Polysaccharides-RPS-hydrolyzed residue polysaccharides g. Heteropolisaccharide L2–calcium complex (glucose:galactose:arabinose)	d. Hot water extract e. Boiling water extraction f. Enzymatic and acidic hydrolyze g. Extracted with boiling water	cytokines, etc. d. antitumor activity e. Immunity and gut health f. Antitumor, anti-inflammatory, antioxidant g. Antitumor activities		c. inhibited the oxidative injury caused by oxidation stress c. Inhibition of hematopoietic necrosis virus d. Antitumor effects without toxic influence on mice through immunoregulatory and mitochondria apoptotic pathways were found e. Exhibits significant immuno-stimulating activities involving TLR2; its immuno-stimulating properties remained stable at pH ranging from 4.0 to 10.0, and at processing temperatures below 121°C f. Significantly decreases of pulmonary SOD, CAT, GSH-Px and total antioxidant activities good antiinflamatory potential g. L2–calcium complex increased the secretions of cytokines while complex with calcium ion decreased their secretion	d. (Yu Zhang et al., 2015) (X. Xu & Zhang, 2015) d1. (P. Maity et al., 2021) e. (X. Xu, Yan, & Zhang, 2012) f. (Ren et al., 2018) g. (Cui, Yan, & Zhang, 2015)
Russula virescens	a. Polysaccharide sulphate derivatives of (1-3)-β-D-glucan, named as RVS3-II (water insoluble polysaccharide) b. Water soluble polysaccharide (RVP) and sulfated polysaccharide SRVP	a. Methodology: defatted, boiled in water, sulfating of dried powder b. lyophilized and powdered water-soluble polysaccharide from fruiting bodies was sulphated using sulphur trioxide pyridine complex method under different reaction conditions (1:5, 1:10, 1:15, 1:20, 1:25 g/g)	a. Antitumor activities b. Anticoagulant, antioxidant, antibacterial, and antitumor activities	a. *in vitro and in vivo* b1. *in vitro* b2. *in vitro*	a. Sulfate derivatives of (1-3)-b-D-glucan could improve its antitumor activity. b1. Sulfate derivative could improve the antioxidant b2. Anticoagulant, antibacterial, and antitumor activities of RVP b3. SRVP1–25 exhibited the strongest radical scavenging capability and anticoagulant activity, while SRVP1–20 showed the best antibacterial activity and antitumor activity against Caco-2 cells	a. (Z. Sun et al., 2009) b1, b2. (Huaping Li et al., 2021)
Cordyceps sinensis	a. Polysaccharide (CSP) b. Exopolysaccharide (EPS)-glucose,mannose, galactose and ribose in complex structure with protein	a. Water extraction and ethanol fractionation precipitation b. The EPS was isolated from the Cs-HK1mycelial fermentation broth by ethanol precipitation	a. Colon antitumor effects b. Anti-inflammatory c. Anti-tumor effect through immunomodulatory effects	a. *in vitro* b. *in vitro/in vivo* c. *in vivo/in vitro*	a. CSP inhibited the proliferation of colon cancer cells by inducing apoptosis and autophagy b. EPS significantly inhibited lipopolysaccharide induced	a. (Qi et al., 2020) b. (L.Q. Li et al., 2020) c. (J. Li et al., 2020) d. (Bai et al., 2020)

Compound/Species	Extraction	Bioactivity	Model	Reference
c. Cordyceps sinensis (WECS) with 20.63% (w/w) content of polysaccharides d. Cordyceps sinensis extract (CSE) e. Cordycepin, (3′-deoxyadenosine) f. Polysaccharide (NCSP-50)-glucan (1/4)-linked-a-DGlcp with a single a-D-Glcp branch substituted at C-6 g. Polysaccharides (CSP)	and purified by deproteinization and dialysis c. Water extracts d. Water dissolution, ultrasonically extracted for 90 minutes at 4°C. e. Crude mushroom was extracted with 800 mL distilled water, centrifuged. The supernatant was applied into G150 gel filtration column with 50 mM acetonitrile buffer at pH 6.0 f. Water extraction and ethanol fractionation precipitation g. Water extraction and alcohol-precipitation	d. Relieve cerebral ischemia injury e. Stimulation of reproduction function in male (steroidogenesis) f. Immunostimulatory effect g. Effect of CSP on the intestinal mucosal immunity and gut microbiota	d. in vivo/in vitro e. in vitro and in vivo f. in vitro g. in vivo	inflammatory responses of the cells In the murine model of LPS-induced acute intestinal injury, the oral administration of EPS to the animals effectively suppressed the expression of major inflammatory cytokinesTNF-α, IL-1β, IL-10, and iNOS and alleviated the intestinal injury. c. WECS can inhibit the growth of breast cancer in vivo and in vitro through promoting polarization of macrophage toward M1 phenotype and producing inflammatory cytokines d. Protective effect on cerebral ischemia caused by middle cerebral artery occlusion (MCAO) in rats by improving the antioxidant effect • CSE can relieve cerebral ischemia injury and exhibit protective effects via modulating the mitochondrial respiratory chain and inhibiting the mitochondrial apoptotic pathway. e. Increase the secretion of adrenal hormones f. Significantly stimulate the proliferation of macrophages, promote nitric oxide production and enhance cytokine secretion • Could be effective as immune-potentiating agent g. Regulating Th cells differentiation via stimulating cytokines secretion and transcription factors production, which maybe associate with TLRs and NF-κB pathway • Potential of CSP as a prebiotics to reduce side effects of Cy on intestinal mucosal immunity and gut microbiota	e. (Y. C. Chen et al., 2017) f. (J. Wang et al., 2017) g. (Ying et al., 2020)
Romaria botrytis a. Phenolic compounds b. Total phenolic, tocopherol, ascorbic acid, β-carotene c. Phenolic compounds, steroids, oxalic acid, sesquiterpenoids, and epipolythiopiperazine-2,5-diones	a. Methanol/water (80:20; 30 mL) extract • Microdilution method was used to determine the minimum inhibitory concentration (MIC) and the	a. Antimicrobial activity b. Antioxidant properties c. Antimicrobial activity d. Antioxidant activity e. Antitumour, hemagglutination,	a. in vitro b. in vitro c. in vitro d. in vitro e. in vitro	a. *Ramaria botrytis* extract showed inhibitory activity against *Enterococcus faecalis* and *Listeria monocytogenes*, being bactericide for *Pasteurella multocida*,	a. Alves et al., 2012 b. Barros et al., 2008 c. Barros et al., 2008 d. (R. Zhou et al., 2017)

(Continued)

Table 9.1 (Continued) Bioactivity-based analysis of selected edible wild mushrooms species

Edible wild mushrooms species	Bioactive compounds associated	Methodologies used in research	Bioactivity potential	Digestion models	Others observation	References
	d. Polysaccharides (RBP) e. Ubiquitin–protein (RBUP)	minimum bactericidal concentration (MBC). • Also, the rapid INT colorimetric were performed by MIC b. Measured by four different methods: 1. DPPH Radical-Scavenging ActiVity (RSA). 2. Reducing Power 3. Inhibition of β-carotene Bleaching 4. Inhibition of lipid peroxidation using Thiobarbituric Acid ReactiVe Substances (TBARS). c. A screening of antibacterial activities against the gram-negative and gram-positive bacteria and fungi was performed • The minimal inhibitory concentration (MIC) was determined by an adaptation of the agar streak dilution method based on radial diffusion. • The MIC was considered to be the lowest concentration of the tested sample able to inhibit the growth of bacteria or fungi, after 24 hours. d. Water extract • The total content of polysaccharide in *R. botrytis* (RBP) was analyzed by phenol–sulfuric acid method using glucose as standard • The crude polysaccharide was purified sequentially by DEAE-52 cellulose and Sephadex G-100 filtration chromatography • The monosaccharide composition of RBP-1, RBP-	deoxyribonuclease (DNase) activities		*Streptococcus agalactiae*, and *Streptococcus pyogenes*. b. Of all the species studied, *Romaria botrytis* proved to have an antioxidant activity more significant (lower EC50 values) c. *Romaria botrytis* revealed antimicrobial activity against gram-positive bacteria (*B. cereus, B. subtilis, S. aureus*) with very low the minimum inhibitory concentration (MIC) d. The antioxidant activity tests showed that RBP-4 had strong assay of reducing power and high scavenging activity on DPPH radical, while RBP-3 exhibited the strongest ability of hydroxyl radical scavenging activity. • All the results implied that RBP could be a promising new natural antioxidant in food industry or drug therapies. e. It is a protein exhibiting strong anticancer activity towards A549 cells. • It was estimated that RBUP exerted its antitumor effect by inducing apoptosis. • RBUP displayed hemagglutinating and deoxyribonuclease activities; A temperature of 40°C and pH of 7.0 were required for optimal DNase activity. Confirming the antitumor activity of RBUP *in vivo* and determining the detailed mechanism should be required.	

Mushroom	Compounds	Methods	Activity	In vitro/in vivo	Results	References
		2, RBP-3 and RBP-4 were analyzed by high performance anion exchange chromatography (Dionex ICS-3000, Sunnyvale, CA, USA) in combination with a carbopac PA-1 ion exchange column (4 x 250 mm). • Antioxidant activities of RBPs was measured by: 1. DPPH radical-scavenging activity 2. Hydroxyl radical-scavenging activity 3. Reducing power e. The protein was isolated with a purification protocol involving ion exchange chromatography on DEAE-Sepharose fast flow and gel filtration on Sephadex G-75. SDS-PAGE, Native-PAGE -ultracentrifugation analysis • Disclosed that RBUP was a monomeric protein				e. Barros et al., 2008
Lycoperdon molle and *Lycoperdon perlatum*	Phenolic compounds, steroids, oxalic acid, sesquiterpenoids, and Epipolythiopiperazine-2,5-diones	• A screening of antibacterial activities against the gram-negative and gram-positive bacteria and fungi was performed • The minimal inhibitory concentration (MIC) was determined by an adaptation of the agar streak dilution method based on radial diffusion. • The MIC was considered to be the lowest concentration of the tested sample able to inhibit the growth of bacteria or fungi, after 24 hours.	Antimicrobial activity	in vivo	• *Lycoperdon molle* were resistant to all of the tested microorganisms: *B. cereus*, *B. subtilis*, *S. aureus* (gram-positive), *E. coli*, *P. aeruginosa*, *K. peumoniae* (gram-negative) bacteria, *C. albicans*, and *C. neoformans* (fungi).	
Pleurotus eryngii *Pleurotus ostreatus* and *Cyclocybe cylindracea*	a. β-glucan- (acetate, propionate and butyrate, aromatic amino acids (Phe, Tyr), trimethylamine (TMA) and gamma-aminobutyric acid). b. Polysaccharide extracted from the stalk residue c. Polysaccharide	a. T The β-glucan content was calculated as the difference between the total and α-glucans using the mushroom and yeast Beta-Glucan assay kit (Megazyme Int., Bray, Ireland) • Mushrooms, were fermented in vitro using fecal inocula from healthy volunteers	a. Genoprotective effects b. Antioxidant and antitumor effect c. Hypolipidaemic and hypoglycemic activities d. Antioxidant activity	a. *in vitro* b. *in vitro* c. *in vivo* d. *in vitro*	*In vitro* fermentation supernatants exhibited a clear protective effect against BOOH-induced DNA damage. • The production of several fermentation-specific metabolites was considerably enhanced: short chain fatty acids (acetate, propionate), butyrate, aromatic amino acids,	a. Boulaka et al., 2020 b. Zheng et al., 2020 c. Chen et al., 2016 d. He et al., 2016

(Continued)

Table 9.1 (Continued) Bioactivity-based analysis of selected edible wild mushrooms species

Edible wild mushrooms species	Bioactive compounds associated	Methodologies used in research	Bioactivity potential	Digestion models	Others observation	References
	d. Polysaccharide from spent substrate	• The cytotoxic and anti-genotoxic properties of the fermentation supernatants were investigated in Caco-2 human colon adenocarcinoma cells b. Ethanolic extracts • Crude polysaccharide was purified by DEAE Sepharose CL-6B ion exchange chromatography and Sepharose CL-6B size-exclusion chromatography. • Structural features were investigated by gas chromatography (GC), gel permeation chromatography (GPC), methylation analysis, and Fourier Transform Infrared Spectrum (FT-IR) c. Aqueous extracts with hot water d. Extracted by hot alkali liquor • A polysaccharide-protein complex (RPS) was purified from crude polysaccharide (CPS)			trimethylamine, gamma-aminobutyric acid. • The genotoxicity studies, which revealed that mushrooms fermentation supernatants have the ability to protect Caco-2 cells against tert-butyl hydroperoxide (t-BOOH), a known genotoxic agent. b. *In vitro* antioxidant assay showed a high scavenging effects on hydroxyl radical and *in vitro* antitumor assay showed it had a dose-dependent antiproliferative effect against human gastric MGC-803 cancer cells and human epithelial Hela cancer cells. • Polysaccharide extracted from the stalk residue has the similar structure and bioactivities as that from fruit-body of the mushroom. • It could be potentially used as a natural source for the development of health-care food.	
Pleurotus giganteus	Phenolic compounds and triterpenoids	• Aqueous and ethanolic extracts • A rat pheochromocytoma cell line was assessed by using 3-[4,5-dimethylthiazol-2-yl]-2,5-diphenyltetrazolium bromide (MTT) assay. • Neurite outgrowth stimulation assay was carried out with nerve growth factor (NGF) as control.	Neurite outgrowth activity		d. Pharmacological experiments *in vitro* su • Both aqueous and ethanolic extracts induced neurite outgrowth of Pheochromocytoma (PC12) cells in a dose- and time-dependent manner with no detectable cytotoxic effect; • This mushroom may be developed as a nutraceutical for the mitigation of neurodegenerative diseases.	Phan et al., 2012
Lactariu deliciosus	a. Heteropolysaccharide	a. Aqueous and ethanolic extracts b. Ethanol and aqueous extracts	a. Antitumor activity	a. *in vivo* b. *in vitro* c. *in vivo*	a. The study showed that *L. deliciosus* Gray polysaccharide (of high purity) consisted of two	a. (Ding, Hou, & Hou, 2012)

Species	Compound	Method	Activity	in vitro/in vivo	Findings	Reference
	b. Phenolic compounds, α-amylase, and β-glucosidase inhibitory activities c. Polysaccharide	• Tests by through DPPH and ABTS radical scavenging and ferric ion reducing (FRAP) • α-amylase and β-glucosidase inhibitory activities c. Ethanolic extracts	b. Antioxidant and antihyperglycemic activity c. Immunostimulant activity		monosaccharides, namely L-Man and D-Xyl and their ratio was 3:1 by GC–MS. • *Lactarius deliciosus* Gray polysaccharide (LDG-A) exhibited significant antitumor activates *in vivo*, on male mice. b. Potential as natural antioxidant and anti-hyperglycemic agents in food and pharmaceutical industry. c. immune activities of a novel polysaccharide (LDG-A) isolated from Lactarius deliciosus Gray were investigated at 20, 40, and 80 mg/kg dose levels for mice. • the purified polysaccharide of *L. deliciosus* Gray is a potential source of natural immune-stimulating substances.	b. (Z. Xu et al., 2019) c. (Hou et al., 2013)
Grifola frondosa	a. Polysaccharide peptide-chemically modified maitake (MPSP) b. Polysaccharide c. Phenolic (GFP) d. Exopolysaccharide (EPS), purified chemically modified to obtain its carboxymethylated (CM-EPS) and selenising derivatives (Se-EPS). e. Protein hydrolysates, (peptide-Fe (II) complex) f. Polysaccharide g. Intracellular zinc polysaccharides (IZPS) h. Polysaccharides i. Protein hydrolysates-GFP (GFHT-1, GFHT-2, GFHT-3. GFHT-4 with trypsin hydrolysate) j. Sulfated polysaccharide (S-GFB) k. Polysaccharides (GFP)	a. Water extract • Crude MPSP was purified from the Maitake D-fraction by applying the water-soluble polysaccharide fraction to the diethylaminoethyl-Sepharose fast-flow anionic resin column and Sephadex G-100 using an AKTA purifier • The fractions with the highest readings, through an UV spectrophotometer at 230 nm from both assays, were chosen for further concentration using the Centricon® Plus-80. • The crude PS [composed of 89% polysaccharides and 11% peptides was freeze-dried to complete dryness and underwent different chemical modifications. • Various forms of MPSP were characterized by SDS-PAGE before the *in vitro* and *in vivo* experiments. • Marker sc-2360 and bovine serum albumin were used as reference and control, respectively.	a. Anticancer activities, adjuvant effect b. Anticancer activities c. Antioxidant activity d. Antioxidant and antitumour activities e. Immunomodulatory activity f. Antioxidant and antitumor activity g. Antioxidant, antibacterial, and anti-aging activities h. Antitumor and antioxidant activities i. Antioxidant activities j. Anticoagulant activity k. Antitumor activity	a. *in vitro/in vivo* b. *in vitro/in vivo* c. *in vitro* d. *in vitro* e. *in vitro* f. *in vitro* g. *in vitro/in vivo* h. *in vitro* i. *in vitro* j. *in vitro* k. *in vitro*	a. MPSP could significantly improve anticancer activities (*in vivo* adjuvant effect and *in vitro* growth inhibition effect) in comparison with the original crude MPSP. • The same rank order of anticancer activities enhancement was observed in both the *in vivo* and *in vitro* assays. b. GFW and GFW-GF inhibited phosphatidylinositol 3-kinase (PI3K) and stimulated c-Jun N-terminal kinase (JNK) pathways, thereby inducing autophagy. • GFW and GFW-GF inhibited proliferation, induced cell cycle arrest, and apoptosis in Hep3B hepatoma cells; decreased the expression levels of the anti-apoptotic proteins protein kinase B and extracellular signal-regulated kinase. • GFW significantly inhibited tumor growth in nude mice implanted with Hep3B cells. c. Three main fractions, GFP-1, GFP-2, and GFP-3, were obtained through the isolation and purification steps.	a. (Chan et al., 2011) b. (C. H. Lin et al., 2016) c. (Chen et al., 2012) d. (W. Zhang et al., 2016) e. (Yuan et al., 2019) f. (Chen et al., 2019) g. (Zhang, Gao, et al., 2017) h. (Ji et al., 2019) i. (Dong et al., 2015) j. (Cao et al., 2010) k. (Li & Liu, 2020)

(Continued)

Table 9.1 (Continued) Bioactivity-based analysis of selected edible wild mushrooms species

Edible wild mushrooms species	Bioactive compounds associated	Methodologies used in research	Bioactivity potential	Digestion models	Others observation	References

		b. Cold-water extract of Grifola frondosa (GFW) and its purified active fraction (GFW-GF)
		  • The polysaccharide contents of GFW (100 mg/mL) and GFW-GF (100 mg/mL) were estimated using sugar analysis and compared with the glucose standard solutions
		c. Ethanolic extracts
		  • Purified by DEAE cellulose-52 chromatography and Sephadex G-100 size-exclusion chromatography
		d. Aqueous extracts
		  • Chemical characteristics of the polysaccharides were determined based on high-performance liquid chromatography (HPLC) and infrared analysis (IR) extract
		e. Deionized water extract at pH 9.0 adjusted using 5 M HCl and 5 M NaOH
		  • Protein was hydrolyzed using Alcalase (two main fractions, named GFP-1 and GFP-2 were obtained. CGFP-Fe and GFP-Fe were synthesized using ferrous chelation; CGFP-Fe and GFP-Fe were synthesized using ferrous chelation).
		f. Hot water extract
		  • The optimum conditions were as follows: extraction time 3.5 hours, extraction temperature 75°C, ratio of raw material to water 1:15 and ultrasound assisted.
		g. Aqueous extracts
		  • IZPS or IPS polysaccharide was determined by phenol-sulfuric acid method, using Glu as standard
					• GFP-1, GFP-2, and GFP-3 possessed significant inhibitory effects on 1,1-diphenyl-2-picrylhydrazyl (DPPH) radical, hydroxyl radical and superoxide radical.
					• The reducing power, the ferrous ions chelating effect, and the ability to inhibit the rat liver lipid oxidation were strong.
					• The G. frondosa polysaccharides (GFP) can be used as an easily accessible source of natural antioxidants, as a food supplement, or in the pharmaceutical and medical industries.
					d The antioxidant capacity of CM-EPS and Se-EPS were much better than that of EPS.
					• The SeEPS could significantly inhibit the proliferation of Hela cells compared with EPS.
					• Carboxymethylation and selenylation played a great influence on enhancing the antioxidant and antitumor activities of exopolysaccharides from G. frondosa
					e. The GFP-2 and GFP-Fe had good immune-enhancing activity.
					• GFP-Fe showed an ability to promote the proliferation of splenocytes and peritoneal macrophages.
					• The results suggested that GFP-Fe might be beneficial as a new iron supplement and as an immunological enhancement.	

(Continued)

- The bioactivities including antioxidant and antibacterial activities *in vitro* and antiaging properties *in vivo* of IZPS were investigated comparing with the IPS

h. Distilled water ethanol extract
i. Distilled water extract
- Then GFP was hydrolyzed by six different proteases separately (papain, neutrase, alcalase, trypsin, pepsin, and Protamex)
- DPPH, ferrous ion chelating effect, reducing power, and ability to inhibit the autoxidation of linoleic acid

j. Ethanol extract polysaccharide of Grifola frondosa (GFB)
- (S-GFB) was obtained by sulfation with hlorosulfonic acid (20 mL) and dry pyridine (40 mL)

k. Distilled water extract
- Antitumor activity - using 3-(4,5-dimethyl-2-thiazolyl)-2,5-diphenyl-2-H-tetrazolium bromide (MTT)

f. Polysaccharide from *Ggrifola frondosa* had great antioxidant effect, especially scavenging activity for reducing power and DPPH radical and has antitumor activity, which can inhibit the proliferation of HepG2 cells.
- Polysaccharides from Grifola frondosa can be explored as a novel natural antioxidant and antitumor drug.

g. ZPS had superior antioxidant and anti-aging activities by scavenging the hydroxyl and DPPH radicals, increasing enzyme activities, decreasing the malondialdehyde contents and ameliorating the anile condition of mice.
- IZPS also showed potential antibacterial activities.
- IZPS might be a potential source of natural antioxidant, antibacterial agent and anti-aging agent.

h. Results of 3-(4,5-dimethyl-2-thiazolyl)-2,5-diphenyl-2-H-tetrazolium bromide (MTT) assay showed that GFAP could significantly inhibit the proliferation of Hepatocellular Carcinoma cell line (HepG2) in time-and dose-dependent manners.
- Possessed excellent scavenging effects on ABTS (+) (61.2%), DPPH- (90.3%) and HO (49.2%) at the concentration of 2 mg/mL
- Polysaccharides from Grifola frondosa have strong antitumor and antioxidant activities.

i. All fractions are effective antioxidants and comparably GFHT-4 has the highest antioxidant activity.

j. Activated partial thromboplastin time (APTT) and thrombin time (TT) were effectively prolonged, but no clotting inhibition was observed in prothrombin time (PT) assay at different concentrations.

278 WILD MUSHROOMS

Table 9.1 (Continued) Bioactivity-based analysis of selected edible wild mushrooms species

Edible wild mushrooms species	Bioactive compounds associated	Methodologies used in research	Bioactivity potential	Digestion models	Others observation	References
					• Sulfated polysaccharide of *Grifola frondosa* (S-GFB) had an important effect on decompounding fibrin polymerization through the fibrinolytic. k Results showed that GFP-4 exhibited the highest inhibitory effects owing to the presence of (1→3) α-D-Galp and the higher content of (1→3,4) α-D-Galp (molecular weight: 1.05 × 106 Da) • After heat treatment, different levels of degeneration and degradation were reduced to different levels, resulting in lower antitumor effects.	a. (Chen, 2010) b. (Q. Wu et al., 2007) d. (Zhang et al., 2014)
Tremella fuciformis	a. Polysaccharides b. Natural *Tremella fuciformis* polysaccharide (TP) • Low molecular weight *Tremella fuciformis* polysaccharide (LTP) • Sulfated low molecular-weight *Tremella fuciformis* polysaccharides (SLTP) c. Polysaccharide	a. Boiling water extract, optimum extraction for 4.5 hours at 100°C. b. Microwave-assisted extraction for TP • Acid hydrolysis for LTP (HCl is 0.7 M, 80°C of water bath for 2 hours) • SLTP was synthesized by the chlorosulfonic acid-pyridine method c. Distilled water extract • The combination of Fe^{2+}, ascorbic acid and H_2O_2 was used as degradation regents in order to obtain the lower molecular weight polysaccharide.	a. Antioxidant and antitumor activities b. Antioxidant activities c. Antioxidant activities	a. *in vitro* b. *in vitro* c. *in vitro*	a. Antitumor activities of *Tremella fuciformis* polysaccharides increased from 73.4% to 92.1% with increasing concentration of polysaccharides. • Pharmacology experiment indicated that *Tremella fuciformis* polysaccharides was useful to the therapy of free radical injury and cancer diseases. b. Compared to natural *Tremella fuciformis* polysaccharide (TP) and low molecular weight *Tremella fuciformis* polysaccharide (LTP), sulfated LTP (SLTP) exhibited stronger scavenging activity towards superoxide anion, DPPH and hydroxyl radicals. • In all the cases the effect was found to be dose dependent. • LTP and SLTP can be taken as potential natural resource of antioxidant.	

Species	Bioactive compounds	Methods	Activities	Study type	Findings	References
Lentinula edodes (shiitake)	a. Polysaccharide (with different molecular weights) b. Purified polysaccharides (Isolation of α- and β-D-glucans) -β-(1→6), β-(1→3),(1→6) and α-(1→3) c. a-tocopherol (vitamin E), oleic acid, linoleic acid, ergosterol, and butyric acid d. Phenolic compounds	a. Water extracts b. Water extract (98°C, 1 hour) • GC-MS, FT-IR, NMR, SEC and colorimetric/fluorimetric determinations. c. Methanol, dichloromethane, ethyl acetate, hexane, and water extract • GC-MS analyses of the hexane fraction d. Wister albino mice of either sex were used to study the hepatoprotective activity of the *L edodes* extract	a. Immunomodulatory effects b. Hypocholesterolemic, antitumoral, anti-inflammatory, and antioxidant activities c. Anti-oxidative and anti-atherosclerotic potential d. Antioxidant activity and hepatoprotective effects against paracetamol-induced hepatotoxicity	a. *in vivo* b. *in vitro* c. *in vitro* d. *in vitro*	c. The result ascertained oxidative-reduce degradation did not change the main structure of polysaccharides in the test conditions. • The degraded sample with lower molecular weight possessed the higher antioxidant activities. • Free-radical degradation is an effective way for enhancing antioxidant activity to decrease molecular weight of polysaccharides. a. The smaller molecular weight may positively affect their immunomodulatory activity *in vivo* (for immunosuppressed mice). • Polysaccharide with the lowest molecular weight was the most effective in regulating immunity in the mice. b. Different chemical structures observed for each glucan, such as α/β-configuration and branching degree may highly influence their solubility, tridimensional conformation and also their interaction with cells and consequently provide different biological outcomes. • The most significant differences on the biological effects exerted by the three glucans were observed on the scavenging activity and the inhibition of tumor cell growth. c. *L. edodes* (shiitake mushroom) has possessing bioactive food components capable of conferring anti-oxidative defence and curtailing low density lipoproteins oxidation (ox-LDL), as well as being potent in inhibiting the activity of the rate limiting enzyme in cholesterol biosynthesis (anti-atherosclerotic agent). • Bio-components present in *L. edodes* enabled it to act as an anti-atherosclerotic agent.	a. (S. Chen et al., 2020) b. (Morales et al., 2020) c. (Rahman, Abdullah, & Aminudin, 2018) d. (Sasidharan et al., 2010)

(Continued)

Table 9.1 (Continued) Bioactivity-based analysis of selected edible wild mushrooms species

Edible wild mushrooms species	Bioactive compounds associated	Methodologies used in research	Bioactivity potential	Digestion models	Others observation	References
					d. *L. edodes* extract could perhaps protect liver cells from paracetamol-induced liver damage by its antioxidative effect on hepatocytes, hence diminishing or eliminating the harmful effects of toxic metabolites of paracetamol.	
Fistulina hepatica	a. Phenolic compounds b. Phenolic compounds and organic acids	a. Methanol/water extract • Microdilution method was used to determine the minimum inhibitory concentration (MIC) and the minimum bactericidal concentration (MBC) b. Aqueous extract • The phenolic compounds and the organic acids composition as determined by HPLC/DAD and HPLC/UV, respectively • DPPH, use for antioxidant activity	a. Antimicrobial activity b. Antioxidant potential	a. *in vitro* b. *in vitro*	a. Extracts inhibited the growth of gram-negative (*Escherichia coli, Morganella morganni,* and *Pasteurella multocida*) and gram-positive (*Staphylococcus aureus, Enterococcus faecalis, Listeria monocytogenes, Streptococcus agalactiae,* and *Streptococcus pyogenes*) bacteria. b. *Fistulina hepatica* extract may constitute a source of natural antioxidants, namely phenolic compounds and organic acids. Can be used as a possible food supplement or as an antioxidative agent in the food industry.	a. Alves et al., 2012 b. (Ribeiro et al., 2007)
Tuber aestivum	Phenolics compounds and flavonoids	Water and methanol extract immature and mature summer truffles (*Tuber aestivum*); phenolics and flavonoids were detected by liquid chromatography–mass spectroscopy (LC–MS) analysis.	Antioxidant potential	*in vitro*	• Mature truffles demonstrated higher total phenolics compounds, *in vitro* antioxidant activity, and lower tannins than immature truffle extracts. • Easy identification of maturity of truffles at the appropriate stage for desired use, either culinary or medicinal.	Shah et al., 2020
Tuber melanosporum	Tocopherols	Fluidized bed ethanol extracts;free radical scavenging activities, metal chelating effects, inhibition of lipid peroxidation (inhibition of peroxyl radicals), xanthine oxidase, and lipoxygenase, and identification of antioxidant compounds	Antioxidant and anti-inflammatory activities	*in vitro*	• *T. melanosporum* extracts had high correlation values between levels of different active compounds and antioxidant assay results. • Tocopherol levels were highly correlated with anti-inflammatory activities and inhibition of lipid peroxidation.	Vamanu & Nita, 2014

	Compounds	Methods	Study type	Activities	Results	References
Armillaria mellea	a. Polysaccharides (AMPs-1-1 and AMPs-2-1 purified fractions) b. Water soluble polysaccharide (AMP) c. Ergothioneine and adenosine d. Aromatic esters: protoilludane sesquiterpenoid from *Armillaria mellea* (PSAM) e. Xylosyl 1,3- galactofucan (AMPS-III) f. Armillarikin	a. Extracted through UDE (ultrasonic disruption extraction), purified by anion-exchange and gel chromatography. Total antioxidant activity by FRAP, ORAC assays, scavenging activity on DPPH, ABTS$^+$ Immunomodulatory activity by MTT (3-(4,5- dimethylthiazol-2-yl)-2,5diphenyltetrazolium bromide) assay. b. Extraction in water and ethanol and isolated through DEAE anion exchange cellulose and gel-permeation chromatography c. Water extract of *Armillaria mellea* (AP), ergothioneine and adenosine were analyzed through HPLC, serotonin by reverse-phrase HPLC d. UV-Vis spectrophotometry method e. Hot water extract (80°C) f. The extraction was made in EtOH; the purification of armillarikin was made on a silica gel column and Sephadex LH-20 columns.	a. *in vivo and in vitro* b. *b.in vitro* c. *in vivo* d. *in vivo* e. *in vivo* f. *in vivo*	a. Antioxidant, immunomodulatory activities b. Tumor inhibition effect on A549 cells, the human non-small-cell lung cancer (NSCLC) c. Antidepressant actions, anti-inflammatory effects d. Antidepressant effect e. Anti-inflammatory effect f. Anticancer effects	• Fluidized bed ethanol extracts from dried mushroom fruit bodies can be a source of products rich in tocopherols, and can constitute an important ingredient in food supplements. a. AMPs-2-1 showed higher antioxidant and immunomodulatory activities compared with AMPs-1-1. b. 200 μg/mL of AMP could highly inhibit the growth of cancer cells ending with apoptosis and cell cycle arrest of A549 cells. c. swimming rats test model and unpredictable chronic mild stress (UCMS) rodent models were used in order to prove the AP depression-like behaviors and its possible anti-inflammatory effects d. SPT, OFT, EPM, TST, FST tests were performed in order to evaluated the lack of pleasure, the anxiety behaviors in rats, the effects of depression models and antidepressants, the mice depression and the degree depression, respectively. PSAM could be have antidepressant effects and might be used as a material for the manufacturing of functional food. e. AMPS-III highlighted anti-inflammatory activity against tumor necrosis factor-α and MCP-1 from the murine macrophage cell line RAW264.7 f. Armillarikin could induce apostosis in human leukemia K562, U937, and HL-60 cells.	a. (R. Chen et al., 2020) b. (J. Wu et al., 2012) c. (Y. E. Lin et al., 2021b) d. (X. Sun et al., 2020) e. (Chang et al., 2018) f. (Y. J. Chen et al., 2014)
Marasmius oreades	a. Total phenolic amount b. Individual phenolic compounds: gallic acid, vanillic acid, ferulic, catechin c. Marasmius oreades agglutinin (lecitin)-MOA d. Protein kappa B (IkB α)	a. Antioxidant activity evaluated through (DPPH) radical-scavenging activity, total phenols were analyzed through Folin-Ciocalteu method, reducing power through a 48-well micro plate, inhibition of β	a. *in vivo* b. *In vivo* c. *in vivo* d. *in vivo*	a. Antioxidant activity b. Antioxidant and anticancer activity c. extending binding site properties	a. *Marasmius oreades* was combined with Boletus edulis aiming to enhance its antioxidant effects. Combined mushrooms are highly recommended to be used in	a. (Vieira et al., 2012) b. (Shomali et al., 2019). c. (Grahn et al., 2009); (Winter,

(Continued)

Table 9.1 (Continued) Bioactivity-based analysis of selected edible wild mushrooms species

Edible wild mushrooms species	Bioactive compounds associated	Methodologies used in research	Bioactivity potential	Digestion models	Others observation	References
		carotene bleaching, inhibition of lipid peroxidation by using TBARS (thiobarbituric acid reactive substances) b. Folin Ciocalteu, UPLC, DPPH radical scavenging activity, colorimetric MTT assay c. Isolation, purification and crystallization of MOA was made with SDS PAGE/EDTA and 0.1 M Hepes buffer, respectively d. Liquid chromatography, trypan blue exclusion method (cytotoxicity assay), SDS-PAGE and western blot analysis	d. Positive influence on the breast cancer tumorigenesis process		diets as a o source of antioxidants. b. *Marasmius oreades* showed moderate effect on the growth of human MCF-7 and MDA-MB-231 (breast cancer) and HT-29 (colorectal carcinoma) cell lines c. Lecitin extracted from the fruiting bodies of the mushroom had the only structure (MOA) that could recognized blood group B determinants. d. SiF2 and SiF3 fractions could cause the inhibition of IKK activity, leading to the apostosis of MCF7 cells.	Mostafapour, & Goldstein, 2002) d. (Petrova et al., 2009)
Lyophyllum shimeji	a. Protein extract b. Fibrinolytic enzyme c. Lyophyllin – a protein	a. DNA fragmentation, Western blotting LIVE-DEAD cell staining b. Purification, molecular weight, biological activity of the purified enzyme were analyzed through anion exchange chromatography, SDS-PAGE and by using fibrin plates c. The protein purification was made on a ion exchange chromatography on CM–cellulose d. Bioactive compounds Methodology: antiviral assay against H1N1 virus were investigated in MDCK cells	a. Anti-cancer extract effects b. Positive effect on the prevention of thrombotic disease c. Antifungal effect d. Antiviral activity	a. *in vivo* b. *in vivo* c. *in vivo* d. *in vivo*	a. KB human oral squamous cell carcinoma growth was inhibited by extract from *L. shimeji* leading to cells apostosis. b. The purified enzyme could be successfully used in thrombolytic therapy and in the abnormal blood clots formation. c. *P. piricola* and *C. comatus* growth was inhibited by the ribosome inactivating protein. d. *Lyophyllum shimeji* could have antiviral activity against H1N1 virus.	a. (정종윈 et al., 2014) – Jung b. (Moon et al., 2014) c. (Lam & Ng, 2001) d. (Krupodorova, Rybalko, & Barshteyn, 2014)
Volvariella volvacea	a. rFIP-SJ75 a novel immunomodulatory protein gene, a recombined DNA sequence b. Tannins, flavonoids, triterpenoids, glycosides, and alkaloids c. Glucan (degree of branching, 1 : 5) d. NADPH oxidases - vvnoxa, vvnoxb, vvnoxr, vvbema, vvrac1 and vvcdc24 (genes)	a. Purification was made on a nickel-nitrilotriacetic acid agarose resin column and the purity was analyzed through SDS-PAGE; protein digestion was carried out on a thrombin kit; the immunomodulatory effect was analyzed by RT-qPCR. b. TLC (Thin-Layer Chromatography) of methanol extracts; antimicrobial activity was carried out by using agar well diffusion method; MIC	a. Immunomodulator activity, enhanced the proliferation of RAW264.7 cells; agglutination effect on mouse erythrocytes b. Antimicrobial activities c. Antitumor activity d. Positive effects on ROS; aging and injury stresses e. Radical scavenging activity and antimicrobial effects	a. *in vitro* b. *in vitro* c. *in vivo* d. *in vivo* e. *in vivo*	a. FIP – a fungal immunomodulatory protein; rFIP-SJ75 proved through *in vitro* bioactivity assay that is able to enhanced the phagocytosis of RAW264.7 cells. b. The methanol extracts antimicrobial activity could be due to the presence of secondary metabolites; inhibited the growth of *Staphylococcus aureus*, *Pseudomonas aeruginosa*,	a. (Shao et al., 2019) b. (Appiah, Boakye, & Agyare, 2017) c. (Misaki et al., 1986) d. (Yan et al., 2020) e. (da Silva et al., 2010)

Mushroom	Compounds	Methods		Activity	Results	References
	e. Active subfractions AFC3-10, AFC11-14 and AFC15-19 from the ATCC62 890 identified	(Minimum Inhibitory Concentration) analyzed through the broth microdilution method c. Cold alkali-extracted d. RT-PCR assay was used to detected the gene expression and histochemical detection of ROS (Reactive Oxygen Species) was carried out by using NBT e. Methanol extraction and analyze through TLC; radical scavenging activity was analyzed through DPPH; antimicrobial activity by using paper disc diffusion method			*Klebsiella pneumoniae, Streptococcus pyogenes.* c. The antitumor activity was measured against Sarnoma 180 solid tumor implanted in rats. d. ROS might have a positive influence on the regulation of NADPH oxidase subunit genes which could be linked to oxidative burst caused by injury stress. e. The fractions extracted in hexane and chloroform showed higher inhibition against *Staphylococcus aureus*, compared to those extracted in methanol. The ATCC62890 strain showed higher antioxidant activity, meanwhile, R83 strain exhibited higher antibacterial activity.	
Leccinum scabrum	a. Phenolic acids (Protocatechuic acid, 4-OH-benzoic acid, p-Coumaric acid) b. Phenolics (2,5-Dihydroxybenzoic, 4-Hydroxybenzoic, Protocatechuic, t-Cynnami) ergosterol, flavonoids (catechin, kaempferol, rutin) c. Total phenolics, FRAP, DPPH	a. Extraction with ethanol; total phenolic content was analyzed through Folin Ciocalteu method, individual phenolic acids through LC-ESI-MS/MS; radical scavenging activity by using DPPH assay; antibacterial effect through micro-broth dilution method b. Methanolic extraction; radical scavenging activity was analyzed through DPPH; phenolics, ergosterol, flavonoids were quantified through chromatographic analysis c. Methanolic extraction; antioxidant activity through DPPH, FRAP, and Folin Ciocalteu method for total phenols	a. *in vitro* b. *in vitro* c. *in vitro*	a. Antioxidant, antimicrobial, and antibacterial effects b. Antioxidant activity c. Antioxidant activity	a. The phenolic acids from the extracts showed strong antioxidant activity ($IC_{50} = 23.89$ mg/mg DPPH); the extract from *Leccinum scabrum* exhibited antibacterial effect on gram-positive and gram-negative bacteria. b. Bioactive compounds of seventeen mushrooms species from Poland were analyzed. The highest value of radical scavenging activity, total phenolic and flavonoids contents were reached by *Leccinum scabrum* extract (87%, 9.24 and 0.77 mg/g dry weight, respectively). c. Methanolic extract could enhanced highly antioxidant activity against numerous *in vitro* antioxidant systems, mainly due to their phenolic content.	a. (Nowacka et al., 2014) b. (Gąsecka, Siwulski, & Mleczek, 2018) c. (Koca, Keleş, & Gençcelep, 2011)
Tricholoma giganteum	a. Novel tripeptide (ACE inhibitor) b. Antifungal protein:trichogin c. PSPC (polysaccharide-protein complex) glycoprotein d. Fa, Fb and Fc polysaccharide fractions	a. Isolated from the water extract at 30°C for 3 hours b. Purification was made in distilled water at 4°C and antifungal activity was carried	a. *in vitro* b. *in vitro* c. *in vitro* d. *in vitro* e. *in vivo*	a. Antihypertensive effect b. Antifungal activity c. Antitumoral activity d. Antioxidant activity	a. At a dose of 1 mg/kg of the rats body weight, ACE inhibitor showed highly antihypertensive effect and positive influence on systolic blood pressure, better than a commercial	a. (J. Zhou et al., 2020); (D. H. Lee et al., 2004) b. (Guo, Wang, & Ng, 2005)

(Continued)

Table 9.1 (Continued) Bioactivity-based analysis of selected edible wild mushrooms species

Edible wild mushrooms species	Bioactive compounds associated	Methodologies used in research	Bioactivity potential	Digestion models	Others observation	References
	e. Fa polysaccharide fraction	out in Petri plates with potato dextrose agar c. Flow cytometric study was used to showed that PSPC could induce the HL-60 cell apostosis *in vivo* d. Extraction was made in ethanol; the antioxidant activity was evaluated against DPPH radical, superoxide radical, hydroxyl radical, ferrous ion chelating ability and reducing power e. The study was made on EAC cells isolated from tumor mice peritoneal cavity; detection of apostosos was analyzed through flow cytometry and gene expression analysis through TR-PCR	e. Anticancer effect – chemopreventive cancer agent active		antihypertensive drug (captopril) b. The antifungal protein identified in water extract showed antifungal activity against *F. oxysporum*, *Physalospora piricola*, and *M. arachidicola*, respectively; cowpea, chestnut, chickpea exhibited lower antifungal activity compared with trichogen. c. PSC was able to increase the macrophages function on the tumor-bearing rats and to release niitic oxide and tumor necrosis factor alfa, acting as mediators. d. The higher antioxidant activity was highlighted by Fa polysaccharide fraction. e. Fa fraction could induce apostosis of EAC (Ehrlich's ascites carcinoma)	c. (Moradali et al., 2007); (Ooi & Liu, 2000); (Chatterjee et al., 2011) d. (Chatterjee et al., 2013)
Morchella elata	Antioxidants, total phenols, flavonoids	Extraction of the bioactive compounds was made in methanol; total antioxidant activity was measured through β-carotene–linoleic acid method; radical scavenging activity by using DPPH; total phenols were analyzed by using Folin Ciocalteu method; flavonoids were determined with aluminum trichloride.	Antioxidant effect	*in vitro*	Morchella elata methanol extract was able to scavenge 48.07% of the DPPH radical at concentration of 4.5 mg/mL. the antioxidant effects of methanolic extract are generally related especially to phenolic fractions.	(Gursoy et al., 2009)
Russula delica	a. Phenolic compounds b. Phenolic compounds c. Oleanoic acid and ursolic d. Total phenols, lycopene, ascorbic acid, and β-carotene e. Antiproliferative ribonuclease f. Lectin g. Total phenols, flavonoids, ascorbic acid, lycopene, and β carotene h. Flavonoids	a. Methanolic extraction; antioxidant activity by DPPH and FRAP; total phenolics through Folin Ciocalteu method b. Methanol/water extract • Microdilution method was used to determine the minimum inhibitory concentration (MIC) and the minimum bactericidal concentration (MBC)	a. Antioxidant activity b. Antimicrobial activity c. Anti-HIV protease and anti-inflammatory activities (ergosterol); COVID-19 treatment d. Antioxidant, antimicrobial effects e. Anti-cancer activity f. Antiproliferative effect on cancer cells and	a. *in vitro* b. *in vitro* c. *c.in vitro* d. *in vitro* e. *in vitro* f. *in vitro* g. *in vitro* h. *in vitro* i. *in vitro*	a. Due to the ability to donate hydrogen, due to metal-chelating capacity and the ability to be good scavengers of free radicals and superoxide, gave various antioxidants mushrooms extract mechanisms. b. *Russula delica* extract showed inhibitory activity against *Enterococcus faecalis* and	a. (Koca, Keleş, & Gençcelep, 2011) b. (Alves et al., 2012) c. (Rangsinth et al., 2021) d. (Yaltirak et al., 2009).

i. Ergosterol, phenolic acids (p-OH-Benzoic acid, p-OH-Phenylacetic acid, syringic acid, vanillic acid), Chlorogenic acid, feluric acid, o-Coumaric acid, tyrosol, vanillin, flavonoids

c. In silico ADME analysis

d. Bioactive compounds were extracted in ethanol; antimicrobial activity was evaluated through agar well diffusion method; HPLC was used to identified individual phenolic composition, and the "stable" DPPH for evaluating the radical scavenging activity

e. The isolation was carried out in distilled water and Tris-HCl buffer; amino acid sequence analysis was analyzed using Hewlett-Packard HP G1000A Edman degradation unit

f. The purification of lecithin was made in distilled water and chromatographed on DEAE-cellulose column and for molecular weight determination SDS-PAGE was used; the assay for HIV was tested using an ELISA kit and MCF-7 and HepG2 cells were used for antiproliferative cells assay

g. The extraction of bioactive compounds was made in ethanolic extract; radical scavenging activity was analyzed through DPPH assay; total polyphenol content was determined by using Folin-Ciocalteu reagent; flavonoids content was estimated through aluminum nitrate reagent and β-carotene and lycopene through spectrophotometric measurements

h. The extraction was made in methanol; total antioxidant activity was measured through β-carotene/linoleic acid method; radical scavenging activity was measured through DPPH radical; total flavonoids assay was measured by using aluminum trichloride

HIV-1 reverse transcriptase inhibitory activities.

g. Antioxidant activity

h. Antioxidant activity

i. Antioxidant effect; ferrous ion chelating capacity

Listeria monocytogenes, being bactericide for *Pasteurella multocida*, *Streptococcus agalactiae*, and *Streptococcus pyogenes*.

c. Olosso- lactone VIII, colossolactone E colossolactone G, ergosterol, helian- triol F and velutinacid have been listed as anti-HIV bioactive compounds and could be successfully used to fight against SARS-CoV-2 infection.

d. The mainly phenol identified in ethanolic extracts was catechin, followed by rutin and gallic acid, respectively. With respect to antimicrobial activity, the highest inhibitory activity was against *Shigella sonnei* RSKK 8177, followed by *Yersinia enterocolitica* ATCC 1501, *Bacillus cereus* RSKK 867, *Listeria monocytogenes* ATCC 7644, *Escherichia coli* O157:H7.

e. The ribonuclease was able to inhibit proliferation of cancer cells such as HepG2 and MCF-7 but it was not able to have inhibitory effect on HIV-1.

f. The antiproliferative activity was studied on MCF-7 (breast adenocarcinoma) and HepG2 (hepatoma) cells, respectively. Lecithin was able to exhibited proliferation of MCF-7 and HepG2 cancer cells and it had highly inhibitory activity toward HIV-1 reverse transcriptase.

g. Phenols and flavonoids are the mainly bioactive compounds involved in antioxidant activity of ethanolic extracts.

h. *Russula delica* methanolic extract analyzed through β-carotene/linoleic acid method, showed highly antioxidant activity (80% inhibition).

e. (Zhao, Zhao, Li, Zhang, et al., 2010)

f. (Zhao, Zhao, Li, Zhao, et al., 2010)

g. (Khatua et al., 2013)

h. (Gursoy et al., 2010)

i. (Kalogeropoulos et al., 2013)

(Continued)

Table 9.1 (Continued) Bioactivity-based analysis of selected edible wild mushrooms species

Edible wild mushrooms species	Bioactive compounds associated	Methodologies used in research	Bioactivity potential	Digestion models	Others observation	References
					meanwhile, radical scavenging activity measured through DPPH inhibition (5) was moderate (27.43%). i. Ergosterol was the mainly sterol identified in methanol extract (12.51 mg/100 g fresh weight), followed by Ergosta-7-enol 91.9 mg/100 g fresh weight) and Ergosta-7,22-dienol (1.33 mg/100 g fresh weight), respectively. It was found a strong correlation between phenolic acid content, flavonoids, and antioxidant activity.	
Suillus bovinus	a. Polyphenols, licopene and β-carotene b. Polysaccharides (arabinose, xylose, mannose, glucose, galactose)	a. Extraction was made in water and methanol, respectively; total phenols were analyzed by using Folin ciocalteu assay; total flavonoids content was measured by aluminum chloride colorimetric assay; β-carotene and lycopene were estimated through spectrophotometrically method b. The water soluble polysaccharides were soaked in methanol and extracted with hot water. Antioxidant activity was measured through DPPH radical; ferrous ion reducing power and methal chelating activity have been also analyzed	a. Antioxidant activity b. Antioxidant activity	a. *in vitro* b. *in vitro*	a. Suillus bovinus cap showed higher total phenolics amount in aqueous extract (5.48 µg of gallic acid/mg dry weight), compared to methanolic one (4.00 µg of gallic acid/mg dry weight). On the other side, the flavonoids amount in aqueous extract was 3.33 µg quercetin/mg dry weight, meanwhile in methanolic extract registered a value of 3.02 µg quercetin/mg dry weight. b. The antioxidant activity from stipes showed a positive correlation with monosaccharide composition.	a. (Robaszkiewicz et al., 2010) b. (L. Zhang et al., 2018)
Suillus granulatus	Phenolic compounds and organic acids	The extraction of the bioactive compounds was made in methanol. Individual phenolic compounds and organic acids were identified through HPLC. Radical scavenging activity was measured by using DPPH assay. HPLC-DAD was used to identified quercetin.	Antioxidant activity	*in vitro*	It was concluded that even if methanol extract was rich in phenolic compounds, no positive relation between antioxidant activity and phenolic content could be identified.	(Ribeiro et al., 2006)

| *Tricholoma equestre* | a. Phenols, fatty acids
b. Polysaccharides
c. Antioxidant compounds
d. Flavomannin-6,6'-dimethylether
e. Antimicrobial compounds | a. Aqueous extraction; radical scavenging activity was measured through DPPH assay; total phenolic content was measured spectrophotometrically; fatty acid composition by GC-MS and antimicrobial activity through broth microdilution method
b. The female rats brain slices (hippocampus) were used in the study; the polysaccharides extract were added to ACSF solution
c. The extraction was made in 95% methanol; radical scavenging activity was evaluated by using DPPH assay and with nitric oxide procedure
d. MTT assay was used to analyzed the effect of flavomannin-6,6'-dimethylether on human adenocarcinoma colorectal cells Caco-2
e. Extraction was made in methanol, hexane, aqueous and ethyl acetate and analyzed by using disk diffusion assay | a. Antioxidant activity; pro-inflammatory effect; antimicrobial activity
b. Antioxidant properties
c. Antioxidant effect
d. Antitumoral properties
e. Antimicrobial activity | a. *in vitro*
b. *in vivo*
c. *in vivo*
d. *in vitro*
e. *in vitro* | a. *Tricholoma equestre* extract could have pro-inflammatory effect on the epithelial cells A549 (human lung carcinoma). It was concluded that the consume of *Tricholoma equestre* should be limited to small quantities. On the other side, the content in sterols, antioxidant activity and fatty acids showed that this mushroom it a nutritional source of food. *Tricholoma equestre* extract was active against *Staphylococcus aureus* strain and *Klebsiella pneumoniae* (Rzymski, Klimaszyk, & Benjamin, 2019) stated that the study made by Muszyńska, Kala, et al. (2018) it isn't reliable enough to prove the pro-inflammatory effect of *Tricholoma equestre* and there is still a necessity to exploit it in further studies.
b. The study showed that polysaccharides could be involved in the inhibition of neuronal ROS and lowering their amount.
c. The radical scavenging activity of the methanol extract increased with increasing concentration of the extract. For instance, at 20 µL of extract the RSA value was 13% and 79% when 50 µL were added. With respect to nitric oxide scavenging activity.
d. This study concluded that flavomannin-6,6'-dimethylether caused the growth inhibition on human adenocarcinoma colorectal Caco-2 cells, without genotoxic effect. | a. (Muszyńska, Kala, et al., 2018); (Rzymski, Klimaszyk, & Benjamin, 2019)
b. (Pessoa et al., 2020)
c. (Ragupathi et al., 2018)
d. (Pachón-Peña et al., 2009).
e. (Venturini et al., 2008) |

(Continued)

Table 9.1 (Continued) Bioactivity-based analysis of selected edible wild mushrooms species

Edible wild mushrooms species	Bioactive compounds associated	Methodologies used in research	Bioactivity potential	Digestion models	Others observation	References
Tricholoma terreum	a. Meroterpenoids, terreumols A–D (1–4): terreumol A (1), terreum B (2), terreum C (3), terreum D (4) b. Linoleic acid and S-coriolic acid c. Triterpenoids, saponaceolides Q–S (1–3)	a. Extraction was made in CHCl3-MeOH; cytotoxicity assay was analyzed on human cancer cell lines b. The extraction of the fruit bodies was carried out in distilled ethyl acetate; the antimicrobial activity was analyzed by using agar diffusion assay c. Extraction was made with chloroform; reverse-phrased CC; Sephadex LH-20 (MeOH)	a. Anticancer activity b. Nematicidal effect c. Anticancer effects	a. *in vitro* b. *in vitro* c. *in vitro*	e. In methanol and aqueous extracts *Tricholoma equestre* could have antimicrobial activity for *Listeria monocytogenes*. a. Compounds 1, 3, and 4 exhibited cytotoxicities against five human cancer cell lines b. Due to the capacity of Tricholoma terreum extract to produce metabolites such as linoleic acid and S- coriolic acid the nematicidal activity increased being accompanied by antibiotic effects. c. In order to study the cytotoxicity assay, lung cancer A-549 cells, human myeloid leukemia HL-60, MCF-7 breast cancer, hepatocellular carcinoma SMMC-7721, and colon cancer SW480 cell lines, were used. Compound 1 was able to showed certain cytotoxicities against four human tumor cell lines.	a. (Yin et al., 2013) b. (Feng et al., 2015)

Marasmiusoreades ethanol extract proved inhibitory effects on CAT, GPx, and GST activities with values such as 19%, 20%, and 7%, respectively. Furthermore, the extract's inhibitory activity on cell lines was measured as the IC_{50} and suggested that *Marasmiusoreades* could have anticancer effects on the aforementioned cells. *Volvariellavolvacea* represents one of the six mushroom species cultivated for food, all over the world. ATCC62890 and R83 *Volvariellavolvacea* strains were extracted in methanol and the scavenging activity was evaluated through DPPH assay-TLC analysis. The inhibition percentage of the mushroom strains was expressed by their capacity to scavenge DPPH. The TLC chromatograms identified five to six subfractions of the *V. volvacea* ATCC62890 and R83 strains with DPPH. ATCC62890 *Volvariellavolvacea* strain exhibited the highest radical scavenging activity (higher than 80%), meanwhile, R83*Volvariellavolvacea* strain reached a value of 65% (da Silva et al., 2010). Furthermore, Nowacka et al. (2014) studied the chemical composition of 19 Poland mushrooms species, highlighting their phenolic composition. The fruiting bodies of the edible mushrooms were extracted in pure ethanol, analyzedusing the Folin Ciocalteu method for the total phenolic amount and LC-ESIMS/MS (Reversed-Phase High-Performance Liquid Chromatography and Electrospray Ionization Mass Spectrometry) for individual phenols, respectively. The radical scavenging activity was assessed through a DPPH assay. *Leccinumscabrum* extract had a total phenols content of 3.80 mg gallic acid/g dry weight meanwhile protocatechuic, caffeic, ferulic, and salicylic acids were individual phenolic compounds identified in the *Leccinumscabrum* extract. IC_{50} of *Leccinumscabrum* extract registered a value of 23.89 mg dry extract/mg DPPH and the antiradical efficiency was 0.04. In conclusion, the 19 Poland mushrooms species could be successfully used as natural antioxidants in the human diet. Seventeen mushrooms species from Poland were analyzed regarding their chemical composition in bioactive compounds such as minerals, ergosterol (D_2), phenolic acids, and ascorbic acid (Gąsecka, Siwulski, & Mleczek, 2018). *Leccinumscabrum* was one of the analyzed Poland mushrooms through methods like Folin Ciocalteu for total phenol content, spectrophotometric aluminum chloride method for total flavonoid content, chromatographic analysis for individual phenolic acids, individual flavonoids, and ergosterol, respectively. The radical scavenging activity was measured through a DPPH assay. From the seventeen mushrooms, *Leccinumscabrum* registered the highest total phenol content (9.24 mg/g extract), the highest radical scavenging activity (87%), and total content of 0.77 mg/g extract of total flavonoids. The highest radical scavenging activity of *Leccinumscabrum* is mainly due to its rich chemical composition in bioactive compounds 2,5-Dihydroxybenzoic, 4-Hydroxybenzoic, Protocatechuic, t-Cynnami phenolic acids, ergosterol, catechin, kaempferol, and rutin, respectively. One of the most easily recognizable and studied species of mushrooms in the world is *Morchella esculenta* (L.) Pers. (*Morchellaceae*), commonly known as morel mushroom or sponge morel. It is also one of the most highly prized wild edible mushrooms. Its widest distribution covers the fields and mountains of Korea, China, Japan, and Europe, as well. It is used in gourmet cuisine due to its special favor and nutrition (S. R. Lee et al., 2018). The latter group of researchers demonstrated that compounds 1, 3, and 5, which are derivatives of ergosterol and octadecanoic acid that were first identified in *M. esculenta* fruiting bodies, are highly cytotoxic to human lung cancer cells (Table 9.1). The activity of compounds 1, 3, and 5 was mediated by the induction of apoptosis. Their cytotoxicity is further supported by previous studies reporting that ergosterol and octadecanoic acid derivatives from natural resources, including mushrooms, are cytotoxic to various human cancer cell lines *in vitro* (Lee et al., 2015; Lee et al., 2018).

The antioxidant activity of the *Leccinumscabrum* mushroom was also confirmed *in vitro* by (Koca, Keleş, & Gençcelep, 2011) who reported that the extract of *Leccinumscabrum* showed the biggest radical scavenging activity (96.96%) out of 24 wild edible mushrooms analyzed. (Koca, Keleş, & Gençcelep, 2011) explained this high value due to the high phenolic bioactive compounds of *Leccinumscabrum*. Similarly, three different polysaccharide fractions were isolated from *Tricholomagiganteum* (Fa, Fb, and Fc) and their antioxidant activity has been studied by Chatterjee et al. (2011), following the methodology described in Table 9.1. Concerning the scavenging ability of hydroxyl radicals, IC_{50} highest value was registered by Fa polysaccharide fraction, followed by Fc and Fb, respectively. On the other hand, regarding the superoxide radical scavenging activity, the Fb fraction showed the biggest inhibitory concentration 50% of superoxide radical scavenging activity, followed by Fc and Fa, respectively. From the three fractions, Fa showed the highest radical scavenging activity value and could be used in medicine, while also possessing a therapeutic value. *Morchellaelata* antioxidant activity was evaluated by Gursoy et al. (2009). Seven species (*Morchella rotunda, Morchella crassipes, Morchella esculentavar. umbrina, Morchella deliciosa, Morchellaelata, Morchellaconica Pers, Morchellaangusticeps*) were considered in the study and analyzed by using the methodology briefly described in Table 9.1. The highest scavenging effect was registered by *Morchellaconica* Pers., with a value of 85.36%, followed by *Morchella crassipes* and *Morchella esculentavar. umbrina* with values of 65.56% and 62.57%, respectively. The antioxidant activity is correlated with the chemical composition of mushrooms, such as phenolic compounds and flavonoids. Alternatively, the antioxidant potential of *Rusulladelica* mushroom has been intensively studied. For instance Koca, Keleş, and Gençcelep (2011) showed that the radical scavenging activity of *Rusulladelica* methanolic extract was 37.10%, the total phenolic content reached a value of 2,020 mg/kg and FRAP (Ferric Antioxidant Reducing Power) registered a value of 1,160 µmol/g. The high antioxidant potential of *Rusulladelica* extract could be linked to its higher phenolic content (Koca, Keleş, & Gençcelep, 2011). In this respect, Yaltirak et al. (2009) showed that the ethanolic extract of *Rusulladelica* reached an antioxidant activity of 44 mg/mL DPPH IC_{50} and the metal chelating activity IC_{50} was 4 mg/mL. These values are linked to the higher value of total phenols content of the extract (6.23 mg/g) and probably to ascorbic acid content (2.93 mg/g). These results are also supported by Khatua et al. (2013) who reported values of superoxide radical scavenging, reducing power, and chelating ability of ferrous ion assays of 0.465 mg/mL, 0.56 mg/mL, and 0.59 mg/mL, respectively. An EC_{50} of 1.2 mg/mL for the DPPH radical scavenging method was reported and a strong correlation ($R^2 = 0.987$) between the total phenols content and antioxidant activity was established.

The antioxidant activity of *Rusulladelica* methanolic extract was also highlighted by Gursoy et al. (2010), who tested the total antioxidant activity by β-carotene/linoleic acid method and its radical scavenging effect on DPPH. They determined a linoleic acid inhibition capacity of 80.97% and a moderate radical scavenging activity (27.43%), emphasizing once again the strong relationship between antioxidant activity and total phenol content. In this line, Kalogeropoulos et al. (2013) showed that *Rusulladelica* is a rich source in phenolic acids (p-OH-Benzoic acid, p-OH-Phenylacetic acid, syringic acid, vanillic acid, chlorogenic acid, ferulic acid, o-Coumaric acid, tyrosol, vanillin), flavonoids, ergosterol, and an *in vitro* antioxidant properties of 26.37 µmol TE/100 g fresh weight.

Suillusbovinus antioxidant activity was emphasized by Robaszkiewicz et al. (2010) and L. Zhang et al. (2018). The polyphenols, lycopene, flavonoids, and β-carotene were the mainly bioactive secondary metabolites identified in *Suillusbovinus* extracts, and a strong correlation between phenolic compounds and antioxidant potential was established (Robaszkiewicz et al., 2010). Furthermore, L. Zhang et al. (2018) showed that the content in polysaccharides and especially monosaccharides could exhibit the higher antioxidant activity of the *Suillusbovinus* extract. In addition, the *Suillusgranulatus* antioxidant potential was studied by Ribeiro et al. (2006) who identified such phenolic acids as fumaric, malic, quinic, citric, and oxalic in the *Suillusgranulatus* extract. These authors explained the antioxidant potential of the mushroom extract through its high amount in the aforementioned phenolic acids and also due to its quercetin content. *Tricholomaequestre* is another mushroom where chemical composition underwent significant study by Muszyńska, Kala, et al. (2018). As such, the authors showed that the *Tricholomaequestre* extract showed strong antioxidant potential, while gallic acid exhibited the highest antioxidant activity. Moreover, the total phenolic content of the methanol extract (14.07 µg gallic acid/mg of extract) could also play a key role in improving its antioxidant potential. The antioxidant properties of *Tricholomaequestre* fruiting bodies were also studied by Pessoa et al. (2020). Pessoa et al. (2020) concluded that the antioxidant potential of *Tricholomaequestre* extract could be due to their polysaccharide content. The authors stated that polysaccharides do not have any effect on the reactive oxygen species signal, but, were able to decrease the signal amplitude and the concentration. On the other hand, Ragupathi et al. (2018) studied the antioxidant potential of *Tricholomaequestre* methanolic extract through nitric oxide scavenging activity. The methanol extract of *Tricholomaequestre* recorded a percentage of 92% nitric oxide scavenging activity at a concentration of 50 µg/mL. *Armillaria mellea* (Vahl) P. Kumm. (AM) is a fungus that belongs to Basidiomycota, also named honey mushroom, while used for centuries in the treatment of headaches, insomnia, neurological disorders, anxiety, and depression (Y. E. Lin et al., 2021). The study of Y. E. Lin et al. (2021) aimed to prove the possible antidepressant effect of water extract from *Armillaria mellea* (Vahl) P. Kumm. Forced swimming tests and unpredictable chronic mild stress were protocols used on rats that were orally administrated with AM water extract. To monitor the effect of the AM water extract, the cerebral serotonin and metabolites from rats' frontal cortex weremeasured. Lin et al. (2021) validated the antidepressant action of AM water extract and explained that the mechanism could occur via anti-inflammatory activities of the bioactive compounds ergothioneine and adenosine.

9.2.2 Polysaccharides

Polysaccharides from wild mushrooms drew increased attention in many research reports, especially due to their properties and bioactivities. Polysaccharides are the most abundant primary metabolites in wild mushrooms and a large body of scientific literature reports about their structure and antioxidant potential (D. Wang et al., 2014; L. Zhang et al., 2018). Nowadays, mushroom polysaccharides are considered low-cost compounds with a great potential for the food industry as functional ingredients, as well as in the prevention or treatment of the diseases induced by oxidative stress (Jiao et al., 2017). The antioxidant properties of polysaccharides appear to be related to their structural characteristics, especially to the ratio between the different constituent monosaccharides (L. Zhang et al., 2018). Thirteen species belonging to *Boletus* family were subjected to extraction to obtain and characterize their water-soluble polysaccharides. For a complex and complete characterization, several analytical methods were used, such as gas chromatography (GC), high performance, liquid chromatography (HPLC), and Fourier-transform infrared spectroscopy (FT-IR). The higher contents in the analyzed monosaccharide samples were arabinose, xylose, mannose, glucose, and galactose. In the case of *B. edulis*, the polysaccharides had mannose as the major constituent (27%–37%), while for *B. aereusarabinose* was found in the higher proportion (10%–12%). It was reported that the monosaccharide content varied with the geographic and climatic conditions but also between the cap and stripes of mushrooms (L. Zhang et al., 2018). The analysis of the FT-IR spectra revealed that *B. edulis* polysaccharides did not contain a pyranose ring, as it was found for the rest of the tested mushrooms. The molecular weight (Mws) of polysaccharides is another important parameter in relation to their bioactivity. High-Mw polysaccharides have less antioxidant activity than low-Mw polysaccharides. It seems that this last feature is caused by poor penetration capability on cell membranes in case of the high-Mws polysaccharides but also, due to the fact that lower Mw have more reductive hydroxyl groups able to donate hydrogen to reduce the free radical species and therefore have a higher antioxidant capacity (Sheng & Sun, 2014; D. T. Wu et al., 2016). The tested polysaccharides were able to scavenge DPPH radicals, but a lower scavenging ability in the case of one sample of *B. aereus* was noticed. It was concluded that DPPH radical scavenging activity showed significant correlations with the arabinose and galactose content. Regarding the ferrous ion reducing power, the results correlated with the radical scavenging activity. For the metal chelating activity, most of the polysaccharides from *Boletus* mushrooms had relatively poor results but it was considered that a hot water extract of *B. edulis* was effective in ferrous ions chelating (L. Zhang et al., 2018).

Due to their valuable biologically active potential, polysaccharides from mushrooms have gained research interest in studying the capability of exopolysaccharides (EPS) to produce bioactive metabolites in submerged cultures. J. Q. Zheng et al. (2014), proposed a method for the isolation and purification of the EPS from the culture broth of *B. aereus* by gel filtration chromatography. The GC-MS analysis of the three fractions (Fr-I, Fr-II, and Fr-III) of EPS revealed that glucose was the major monosaccharide in Fr-I, while a combination of glucose and mannose was found in the other fractions (Fr-II,III). Regarding the molecular parameters of EPS, results pointed out that such fractions do not form large aggregates in an aqueous solution. For Fr-I, a random coil form, while for Fr-II, Fr-III rigid rod forms were attributed. As previously mentioned, these characteristics are relevant due to their influence on polysaccharides bioactivity. Concerning the *in vitro* antioxidant activity assay, it was found that the polysaccharide fraction (Fr-I) containing mainly glucose, with the highest molecular weight showed a superior antioxidant potential among EPS fractions.

Recent studies have reported that polysaccharide's functionality could be increased significantly by replacing the hydroxyl group from the branched-chain with a free phosphate group. The reaction is called phosphorylation and it was proved that the modified polysaccharides are not non-toxic (Huaping Li et al., 2021). A lyophilized powder of phosphorylated mycelia polysaccharide (PMPS) from *Pleurotusdjamor* with a content of 15.22% phosphate was obtained to test its *invitro* and *in vivo* antioxidant ability. The analysis performed to characterize the compositional structure of PMPS showed α-pyranose structure and the presence of glucose (85.82%) and galacturonic acid (13.01%) as the majority of constituents. The reducing power, scavenging DPPH, and superoxide anion radical rates were the assays performed for *in vitro* analysis of antioxidant activity and the results were compared to a control sample (BHT antioxidant). PMPS showed good antioxidant properties *in vitro* but a dose-dependent effect. For the *in vivo* antioxidant activity, the superoxide dismutase (SOD), glutathione peroxidase (GSH-Px), and catalase (CAT), as well as the malondialdehyde (MDA) content of the kidneys were monitored on Kunming mice. These assays were also performed on un-phosphorylated polysaccharides (MPS). According to the authors, the PMPS group showed a better antioxidant effect *in vivo* than MPS. Zinc-enriched polysaccharides were studied by J. Zhang et al. (2016) as modified polysaccharides, extracted from *P. djamor*, with higher antioxidant activities than the regular polysaccharides. By submerged fermentation with zinc acetate, a complex compound was obtained, which multiplies the antioxidant effect of the polysaccharides with the numerous health benefits of zinc. Structural investigations revealed that mycelia zinc polysaccharides (MZPS) and their fractions were composed of high rhamnose and xylose contents (in the case of one fraction). Both monosaccharides were found to be effective antioxidants due to their ability to transfer electrons and reduce the reactive oxygen species. Furthermore, FT-IR analysis showed that bond types of C-H, S=O, and C=O had an impact onbioactivity. Assays performed to explore the *in vitro* antioxidant effect of MZPS indicated good antioxidant capacity but one that is dose-dependent. For *in vivo* studies, it was also revealed that MZPS could be effective against toxicity and could block the oxidative chain reaction. Sulfation of (1–3)-β-D-glucan extracted from fruiting bodies of *Russulavirescens* was proposed by Hui Li et al. (2020) as a way to enhance the bioactivity of the water-soluble polysaccharide due to the presence of sulfated groups which may increase the ability to provide hydrogen atoms and exhibit antioxidant potential. The sulfated derivatives demonstrated improved results in the case of ABTS radical scavenging, superoxide radical scavenging, and hydroxyl radical scavenging activities. The compound with more sulfate groups was more effective in scavenging activity against hydroxyl radicals. In the case of the superoxide radical, the derivative with a moderate degree of sulfation showed a higher capacity.

Due to the limitation of the traditional extraction method of mushroom polysaccharides, enzyme-assisted extraction (EAE) was proposed, which was found to be more advantageous in terms of extraction yield, costs, and energy requirements. By this approach, Jiao et al. (2017) obtained an enzymatic-extractable polysaccharide (EnPPs) from the mycelium of *Pleurotusdjamor* mushrooms, composed of rhamnose, xylose, mannose, galactose, and glucose (with a molar ratio of 7.96:1.00:2.05:5.90:2). *In vivo* inhibiting, activities against oxidative stress were measured for EnPPs. For this purpose, the enzymatic activities of SOD, GSH-Px, and CAT, and the non-enzymatic activities of total antioxidant activity in the brain, liver, kidney, and pancreas homo-genate were determined. Results indicated that the EnPPs could protect the tissues against the induced lesions preventing oxidative stress.

A comparison of the antioxidant activity for hot water extracted and hot alkali extracted polysaccharides from *G. tsugae* was run by Tseng, Yang, and Mau (2008). The polysaccharides from *Ganoderma tsugae* Murrill were obtained from four forms: mature, young Ling Chih, my-celia, and fermentation filtrate. The hot alkaline treatment led to a degradation of the cell wall and water-insoluble components and an increment of the extracted polysaccharides molecular weight was found. Both extracts of polysaccharides showed good antioxidant properties and the authors concluded that these extracts could be used as functional food ingredients/supplements due to their capability to reduce oxidative damage in the human body. The efficiency of the hot-water extracted polysaccharide from *Hericiumerinaceum* was investigated by H. Xu et al. (2010) to determine its antioxidant properties *in vivo* on rats' skin. The outer layer of the skin accumulates organic per-oxide during the aging process and antioxidant supplementations could be a way to restore skin properties. The HPLC analysis revealed that the polysaccharide was formed from glucose and galactose as major monosaccharides, while xylose and fucose were present in small contents. *In vivo* tests showed that *H. erinaceum* polysaccharide increased the enzyme's antioxidant activity significantly, as well as collagen production and, thus, it was concluded that this polysaccharide may delay skin aging. Polysaccharides isolated from the residue of *Lentinula edodes* by enzymatic (ERPS) and acidic (ARPS) hydrolysis were investigated by Ren et al. (2018) to establish their antioxidant potential *in vivo* on lipopolysaccharide-induced lung injured mice. Previously, the authors showed that polysaccharides modified by enzymatic or acidic approach possess superior antioxidant activity compared to the unmodified form. Structural differences were found between the two modified polysaccharides. As such, ERPS was composed of rhamnose, arabinose, man-nose, galactose, and glucose, while ARPS contained rhamnose, arabinose, galactose, and glucose and the mass percentages were significantly different. For both polysaccharides, rhamnose was the main monosaccharide, present in a content of 81.97% for ERPS and 32.47% for ARPS. The results showed significant decreases in pulmonary SOD, CAT, GSH-Px, and total antioxidant activities, as well as increases in pulmonary lipid peroxidation in surveyed mice. As it was reported by other studies (X. Xu & Zhang, 2015), rhamnose is associated with important antioxidant properties and could contribute to the antioxidant activity of the polysaccharides. Chemically modified maitake polysaccharide-peptides (MPSP) and their adjuvant effect (*in vivo*, on rats) and anticancer activity (*in vitro* growth inhibitory effect) compared to crude MPSP from *Grifolafrondosa* were in-vestigated by Chan et al. (2011). The authors showed that chemical phosphorylation could markedly enhance both adjuvant effects and growth inhibitory effects.Furtheremore, Lin et al. (2016), studied the effects of the cold-water extract of *Grifolafrondosa* (GFW) and its active fraction (GFW-GF) on autophagy and apoptosis, and the underlying mechanisms *in vitro* and *in vivo*. *Grifolafrondosa* is known for its antitumor activity, which has been targeted by scientific and clinical research. The present study showed that GFW and GFW-GF inhibited cell proliferation. Moreover, *in vivo* results showed that GFW inhibited cancer growth, activated the mitochondria-dependent apoptotic pathway, and promoted autophagy. The results demonstrate that GFW and GFW-GF may represent remarkable cytotoxic agents at low concentrations, both *in vitro* and *in vivo*. Also, GFW and GFW-GF can serve as promising anticancer agents that target both autophagy and apoptosis. In addition, Li (2017), studied the purification of the fractions of water-soluble *Ramaria botrytis* polysaccharides (RBP) considered an efficient natural antioxidant. Crude *Ramaria botrytis* polysaccharide was eluted and purified by two-column chromatography of DEAE-52 and Sephadex G-100 successively, four purified fractions resulted. RBP-1, RBP-2, RBP-4 were mainly composed of glucose, while RBP-3 contained 41.36% mannose and 28.96% glucose. These four fractions were tested for antioxidant activities *in vitro*, where RBP-4 exhibited a strong assay of reducing power and high scavenging activity on DPPH radical, while RBP-3 showed the stronger ability of hydroxyl radical scavenging activity (Hua Li, 2017). Another study by He et al. (2016), shows the antioxidant activity of the spent substrate of *Pleurotuseryngii*. A crude polysaccharide (CPS) was obtained by extraction with hot alkali liquor. This polysaccharide purified itself and resulted in the polysaccharide-protein complex (RPS). Monosaccharide composition analysis by GC–MS confirmed that RPS is a novel acid heteropolysaccharide and it is mainly composed of xylose, glucose, and arabinose. Moreover, FT-IR spectral analysis of RPS revealed prominent characteristic polysaccharide and protein groups. By using different *in vitro* models (ABTS, superoxide, hydroxyl, and DPPH), this study shows that both CPS and RPS exhibited strong antioxidant activities in a dose-dependent manner. Also, the antioxidant activities for three isolated and purified fractions of polysaccharides of *Grifolafrondosa* (GFP-1, GFP-2,and GFP-3) were investigated *in vitro* by Chen et al. (2012). In this study, the results from different *in vitro* assay systems, including the scavenging effects on DPPH radical, hydroxyl radical, and superoxide radical, the reducing power, the ferrous ions chelating effect, and the ability to inhibit the rat liver lipid oxidation, demonstrated that the purified fractions of *Grifolafrondosa* polysaccharides (GFP), especially GFP-2, have effective antioxidant activities. The antioxidant activity of hydrolysates of *Grifolafrondosa* protein (GFP) was obtained with six different proteases (papain, neutrase, alcalase, trypsin, pepsin, and Protamex) and of fractionated hydrolysates (GFHT-1, GFHT-2, GFHT-3, GFHT-4) prepared using ultrafiltration (UF) membranes was investigated by Dong et al. (2015). Among the six proteases, the trypsin hydrolysate possessed the strongest antioxidant potential. The *in vitro* antioxidant activities of the four fractions were evaluated, in this study, including the scavenging effect on DPPH, ferrous ion chelating effect, reducing power, and ability to inhibit the autoxidation of linoleic acid. The results demonstrated that all fractions are effective antioxidants and comparably GFHT-4 has the highest antioxidant activity (Dong et al., 2015).

Yuan et al. (2019), studied the immunomodulatory activities of iron-chelating *Grifolafrondosa* peptides on immunocytes proliferation and cytokine (IL-6 and TNF-α) secretion were determined using an *in vitro* digestion model. The *Grifolafrondosa* protein hydrolysates were purified using ultrafiltration and Sephadex gel chromatography. Two main fractions (GFP-1 and GFP-2) resulted. CGFP-Fe and GFP-Fe were synthesized using ferrous chelation. This study reported that the iron-chelating peptide of *Grifolafrondosa* still maintained its immune-enhancing activity after *in vitro* gastrointestinal digestion and suggested that GFP-Fe might be beneficial as a new iron supplement and asan immunological enhancement.

Sulfated low molecular-weight *Tremella fuciformis* polysaccharides (SLTP) with different sulfate contents were synthesized by Wu et al. (2007) to monitor their antioxidant activities *in vitro*, including superoxide anion radical, 1,1-diphenyl-2-picryl-hydrazyl (DPPH), radical and hydroxyl radical scavenging activities. The authors claim that the degradation of *Tremella fuciformis* polysaccharides (TP) could improve its water solubility and radical scavenging activity. Sulfation of low molecular-weight *Tremella fuciformis* polysaccharides (LTP) results in more increased antioxidant activity than the natural LTP and TP. The results of this study support the fact that modification of LTP by sulfation will augment the antioxidant activity and LTP and SLTP can be considered as potential natural resources of antioxidants.In the study conducted by Zhang et al. (2014), the free-radical degradation and antioxidant activity of polysaccharides from *Tremella fuciformis* were investigated. The combination of Fe^{2+}, ascorbic acid, and H_2O_2 was used as a degradation agent to obtain the lower molecular weight product. The authors selected five degraded polysaccharides to evaluate their antioxidant activities *in vitro* and found that the degraded sample with lower molecular weight possessed the higher antioxidant activities. Zhang, Chen, et al. (2017) used *Grifolafrondosa* as a vector of zinc biotransformation to produce the intracellular zinc polysaccharides (IZPS). IZPS antioxidant and antibacterial activities *in vitro*, as well as their anti-aging properties *in vivo*, were investigated compared to intracellular polysaccharides (IPS). IZPS had superior antioxidant and anti-aging activities., al. by scavenging the hydroxyl and DPPH radicals, increasing enzyme activities, decreasing the malondialdehyde (MDA) contents, and ameliorating the anile condition of mice. Moreover, IZPS showed potential antibacterial activities. Also, the results of this study showed that zinc-enrichment enhanced the antibacterial activities of IPS, and the superior performance of IZPS might be attributed to its monosaccharide compositions.One of the most common phenomena in the world is senescence, often named aging, as well. It seems that aging is a result of cellular stress and ROS (Reactive Oxidative Species) generated by NADPH (Nicotinamide Adenine Dinucleotide Phosphate) oxidase, which could play an important role in positively influencing the senescence process (Yan et al., 2020). The NADPH encoding genes vvnoxa, vvrac1, vvbema, and vvcdc24 extracted from *Volvariellavolvacea* complete the genomic sequence and might be positively involved in the aging response. Gene identification of *Volvariellavolvacea* NADPH oxidative-encoding genes was made by DNA sequences and bioinformatics analysis was used to predict the transmembrane helix regions. RT-PCR assay was used to identify the gene expression patterns. DPI (diphenyleneiodonium chloride) and GST (glutathione) were used to highlight the relationship between NADPH, ROS signaling molecules, and mechanical injury stress. DPI and GSH were able to eliminate ROS during mechanical injury and lead to repressing the oxidative subunit genes NADPH. Therefore, ROS could positively influence the NADPH oxidative subunit genes. Similarly, Xylosyl 1,3-galactofucan (AMPS-III) polysaccharide was isolated and purified from *Armillaria mellea*. The sugar composition (PS) of *Armillaria mellea* polysaccharides was analyzed through acid hydrolysis (Chang et al., 2018) and an important galactose and fucose amount was highlighted. To determine the effects of fractioned PS, RAW264.7 macrophages were used in the study. The results demonstrated that AMPS-III was able to significantly reduce the release of tumor necrosis factor-α (TNF-α) and cytokine monocyte chemotactic protein in the aforementioned macrophages. TNF-α plays an essential role in anti-inflammatory activities and it might have been inhibited by the AMPS-III which consisted of a novel 1,3-linked α-D-galactosyl-interlaced α-L-fucan with partial 4-O-xylosyl substituents. Two polysaccharides extracted from *C. versicolor* (PSP and PSK) have shown important biological properties and can be a useful adjunct to cancer therapy. In certain cases, however, further research is necessary to prove some of the effects that have been observed *in vitro* and experimental animal studies (Table 9.1). The antidiabetic, antimicrobial, and antioxidant activity demonstrated by studies carried out in *C. versicolor* indicate that this is a potential source of bioactive compounds that can bring important health benefits (Cruz et al., 2016).

9.2.3 Proteins

Lectins are proteins that possess the property to selectively bind carbohydrates without any enzymatic modification. Lectins from mushrooms were found to have important bioactive properties due to their capacity to bind carbohydrates and form glycoproteins. In this regard, an antineoplastic lectin (BEL) was isolated from *B.edulis*. After purification on a column of the human erythrocytic stroma, it was incorporated into a polyacrylamide gel and its structure was investigated by X-ray (Bovi et al., 2013). This investigation revealed that there is a β-trefoil fold configuration present in the form of two protein monomers. It was also found that BEL β-trefoil lectin has a distinct structure compared to other lectins isolated from *B. edulis*, for example, the saline soluble one. Using the X-ray diffraction of single crystals, the interaction of the lectin with lactose, galactose, N-acetylgalactosaminewas studied, alongside the interaction with the T-antigen disaccharide and the T-antigen, the disaccharide linked to serine, which is considered the most probable mediator of the anti-proliferative effect in the case of this compound. The interactions between these saccharides and lectin forming crystals were found similar, revealing the ligand properties of the BEL β-trefoil lectin. The antitumor proliferation activity of the lectin isolated from *B. edulis* was tested on nine different human cancer cell lines. Potent antineoplastic activity of BEL β-trefoil lectin was found, which was attributed to its capacity to penetrate the cell. Furthermore, A 16.7-kDa protein (BEAP) was isolated by Yang Zhang et al. (2021) from fruit bodies of *B. edulis* mushroom. In terms of molecular weight, it is considered to be a smaller protein than lectin isolated by *B. edulis*. BEAP exhibited anti-proliferation activity against A549 non-small cell lung cancer *in vitro* and stopped the growth of A549 solid tumors *in vivo*. As such, a *Marasmiusoreades* extract was highly studied by Winter, Mostafapour, and Goldstein (2002) and Grahn et al. (2009) due to its lectin content, a novel heterodimeric protein of 50 kD molecular weight, with novel carbohydrate-binding activity. They concluded that lectin could have a high general affinity for α -galactosides and could also react with murine laminin from the Engelbreth-Holm-Swarm sarcoma and bovine thyroglobulin, respectively. Furthermore, lectin was involved in the human B erythrocytes agglutination. In this regard, Shao et al. (2019) proved that FIP-SJ75, a novel fungal immunomodulatory protein from *G. lucidum, F. velutipes,* and *V. volvacea,* could act as a possible activator of mouse macrophages. Modern methods such as liquid chromatography coupled with quadrupole time-of-flight mass spectrometry (LC/Q-TOF MS) were used for recombinant protein identification (rFIP-SJ75). The immunomodulatory effect of rFIP-SJ75 on macrophage RAW264.7 cells was studied through RT-qPCR and showed that rFIP-SJ75 might be a potential immunomodulatory agent for some inflammatory diseases.

9.2.4 Other bioactive compounds

A lanostanoid, called tsugaric acid, and a palmitamide were obtained from the CHCl3 extract of *Ganodermatsugae*fruit bodies. Also, to enhance the bioactive properties, an investigation was conducted into 3-oxo-5α-lanosta-8,24-dien-21-oic acid which was used to synthesize lanostoid derivatives tested for anticancer activity against human prostate cancer cells. The lanostoid derivatives showed weak cytotoxic activities and the author concluded that they may be used as chemopreventive agents (C. H. Lin et al., 2016).

9.3 ANTITUMOR, ANTI-CANCER, ADJUVANT EFFECT, AND IMMUNOSTIMULATORY ACTIVITY DUE TO POLYSACCHARIDES FROM EDIBLE MUSHROOMS

Among other bioactive properties, polysaccharides were found and used as anticancer agents because of their effect to activate various immune responses in the host. Therefore, a lot of scientific studies were conducted on the immunostimulating properties of polysaccharides from wild mushrooms. Generally, the polysaccharides with antitumor properties are homo/heteroglycans, which are transformed into glycoproteins or their derivative forms when interacting with proteins (Meng, Liang, & Luo, 2016). In this regard, Wang et al. (2014) have isolated a water-soluble polysaccharide from the fruiting bodies of *B. edulis*. The structural characterization of the polysaccharide was performed by gas chromatography and revealed that the component monosaccharides were glucose, galactose, rhamnose, and arabinose. Glucose was the principal constituent. The anticancer effect of the polysaccharide was tested *in vivo* by using a renal cancer cell line (Renca) which was implanted subcutaneously into the rump of BALB/c mice. It was found that the oral administration of 100 and 400 mg/kg polysaccharides extracted from *B. edulis* could significantly inhibit the growth of tumor cells in mice. Also, up to a concentration of 2000 mg/kg body weight, no toxicity was reported. The hot-water extract of *B. edulis* was also investigated by Lemieszek et al. (2013) aiming to examine its anti-proliferation action on human colon cancer cells. It was found that the extract (WSB) contained a polysaccharide fraction. Its main constituent is monosaccharide glucose, followed by galactose, fucose, and mannose. Furthermore, WSB waspurified by anion exchange chromatography, thus resulting in five fractions (BE1, BE2, BE3, BE4, BE5). BE1 was not retained by the column and was mainly composed of polysaccharides with a lower amount of mannose than WSB. BE2 was a glycoprotein. BE3 and BE4 contained lower amounts of sugars but with relatively similar compositions. The fractions isolated from *B. edulis* showed a significant antiproliferative effect in colon cancer cells but the best results were obtained in the case of the BE3 fraction of the polysaccharide. A recent study by Zheng et al. (2020) suggested that the polysaccharide extracted from the stalk residue of *Pleurotuseryngii* has a similar structure and bioactivities as that from the fruitbody of the mushroom. The results showed that it was a heteropolysaccharide mainly composed of glucose (82.4%). Furthermore, *in vitro* antioxidant assay showed effects on the hydroxyl radical, while the *in vitro* antitumor assay showed it had a dose-dependent antiproliferative effect against human gastric MGC-803 cancer cells and human epithelial Hela cancer cells. The hypolipidemic and hypoglycaemic activities of polysaccharides extracted from *Pleurotuseryngii* (PEP) were investigated *in vivo* by Chen et al. (2016). The studies were performed on KKAy mice, which were divided into control and PEP groups. The latter were fed with high fat and PEP + high fat, respectively, for six weeks. Oral administration of PEP decreased body weight gain, the levels of plasma insulin, serum triglyceride, cholesterol low-density lipoprotein cholesterol, and blasting blood glucose in mice. In addition, the PEP diet increased the level of high-density lipoprotein cholesterol and liver glycogen. The authors claim that *Pleurotuseryngii* polysaccharide extract could be explored as a possible therapeutic agent for hyperlipidemia and hyperglycemia. In the study by W. Zhang et al. (2016), the exopolysaccharides extracted from liquid-culture *Grifolafrondosa* (EPS) were first successfully chemically modified to obtain its carboxymethylated (CM-EPS) and selenylizing derivatives (Se-EPS). Also, the antioxidant and antitumor activities of EPS and two derivatives were evaluated. The results indicated that carboxymethylated and selenylation modification of EPS could significantly enhance the antioxidant activity *in vitro*. Moreover, the authors argued that Se-EPS exhibited stronger inhibitory activity and can be considered a promising antioxidant or antitumor agent with potential value for food and pharmaceutical industries. Moreover, Chen et al. (2019), optimized the technology of extraction of polysaccharides from *Grifolafrondosa*, through an ultrasonic-assisted and microwave-assisted process, followed by an evaluation of the antioxidant and antitumor activity. Upon analyzing the antioxidant activity of DPPH radical scavenging, as well as the reducing power and hydroxyl radical scavenging, the authors reported that polysaccharides from *GgrifolaFrondosa* had powerful antioxidant activities. Also, in this study, *in vitro* antitumor activity of polysaccharides was evaluated proving that polysaccharide extract can inhibit the proliferation of HepG2 cells. Furthermore, *in vitro* antitumor and antioxidant activities of alcohol-soluble polysaccharides from the *Grifolafrondosa* (GFAP) fungus have been studied by Ji et al. (2019). The polysaccharides extraction process has been optimized by the orthogonal test. The results of this study assay showed that GFAP could significantly inhibit the proliferation of Hepatocellular Carcinoma cell line (HepG2) in time- and dose-dependent manners. Furthermore, *in vitro* antioxidant results showed that GFAP possessed excellent scavenging effects on ABTS+, DPPH, and HO at 2 mg/mL. concentration. As well, Chen (2010), investigated the *in vitro* antioxidant and antitumor activities of *Tremella fuciformis* polysaccharides. The authors used response surface methodology (RSM) to determine the optimum extraction conditions (extraction temperature, time, and ratio of solvent to raw material) for maximum polysaccharides yield. The authors reported that *Tremella fuciformis* polysaccharides could scavenge superoxide anion and hydroxyl radicals. Moreover, it was found that antitumor activities of *Tremella fuciformis* polysaccharides increased from 73.4% to 92.1% with an increasing concentration of polysaccharides.

Moreover, the authors claim the pharmacology experiment indicated that *Tremella fuciformis* polysaccharides were useful to the therapy of free radical injury and cancer diseases. A novel polysaccharide isolated from *Lactarius deliciosus* Gray was studied *in vivo* by Ding, Hou, and Hou (2012), as a possible valuable source that contributes to exhibit unique antitumor properties. According to the results of this study, it was concluded that the novel polysaccharide obtained from *Lactarius deliciosus* Gray was a heteropolysaccharide, namely LDG-A, while the purified polysaccharide (LDG-A) was confirmed to be of high purity. Moreover, the study showed that LDG-A consisted of two monosaccharides, namely l-mannose (l-Man) and d-xylose (d-Xyl) and their ratio was 3:1 by GC–MS. To detect the antitumor activity of LDG-A *in vivo*, mice transplanted S180 were used. It was concluded that LDG-A exhibited significant antitumor activates *in vivo* (Ding, Hou, & Hou, 2012).

The immune activities of a novel polysaccharide (LDG-A) isolated from *Lactarius deliciosus* Gray were investigated *in vivo* by Hou et al. (2013) on mice. This study claims that the antitumor activity of the LDG-A was believed to be a consequence of the stimulation of the cell-mediated immune response because it can significantly promote the lymphocyte. S180 tumor cells (3 × 106) were implanted subcutaneously into the right hind groin of the mice and one day following inoculation, LDG-A was dissolved into distilled water and administered intraperitoneally to the mice at doses of 20, 40, and 80 mg/kg, respectively. The authors state that the inhibitory rate in mice treated with 80 mg/kg LDG-A can reach 68.422%, purified polysaccharide of *Lactarius deliciosus* Gray, being a potential source of natural immune-stimulating substances. *In vitro* antitumor activity of the *Grifolafrondosa* (GFP) polysaccharides was recently investigated by Li & Liu (2020). Theyevaluated their inhibitory effects on SPC-A-1 human lung cancer cells. Additionally, the relationship between heat treatment on structural properties and antitumor activity of polysaccharidesunderwent evaluation. For this aim cold-water-soluble polysaccharides from *Grifolafrondosa* were extracted at 4°C (GFP-4) and purified. GFP-4-30, GFP-4-60, and GFP-4-90 were obtained from GFP-4 after treatment at 30°C, 60°C, or 90°C, respectively, for 6 hours. The results of this study showed that after heat treatment, different levels of degeneration and degradation occurred in GFP-4-30, GFP-4-60, and GFP-4-90, and their degrees of branching were reduced to different levels, resulting in lower antitumor effects. Furthermore, Zhou et al. (2017), conducted an *in vitro* investigation of a novel ubiquitin-like antitumor protein (RBUP), isolated from fruiting bodies of the *Ramaria botrytis* edible mushroom by a purification protocol involving ion-exchange chromatography on DEAE-Sepharose fast flow and gel filtration on Sephadex G-75. SDS-PAGE, Native-PAGE, and ultracentrifugation analysis disclosed that RBUP is a monomeric protein, while ESI–MS/MS demonstrated that it shared 69% amino acid sequence similarity with *Coprinellus congregates* ubiquitin. The authors reported that this protein exhibited strong anticancer activity towards A549 cells. Also, RBUP displayed hemagglutinating and deoxyribonuclease activities (DNase) at a temperature of 40°C and pH of 7.0, where DNase activity is optimal. In an *in vitro* assay, J. Wu et al. (2012) demonstrated that *Armillaria mellea*, a famous traditional Chinese edible mushroom could exhibit an inhibitory action against A549 cells (human nonsmall-cell lung cancer), through a water-soluble polysaccharide (AMP). Briefly, the AMP was extracted, purified, and analyzed using methods like DEAE anion exchange cellulose and gel-permeation chromatography. The chemical composition of AMP has mainly formed of 94.8% carbohydrate (^{D}-glucose), 2.3% uronic acid, and 0.5% protein. According to J. Wu et al. (2012), AMP was able to show an inhibition effect on A549 cells through apoptosis and cell cycle arrest of A549. A symbiotic relationship between *Armillaria mellea* mushroom and *Gastrodia elata* Blume, an orchid was intensively studied in Chinese medicine, as *Tianma*, literally translated "heavenly hemp". Also, Chen et al. (2014) identified armillarikin, a bioactive compound from *Armillaria mellea* as possessing antiproliferative and apoptotic activities on human leukemia K562, U937, and HL-60 cells. Briefly, *Armillaria mellea* fermented mycelia were purified in ethanol extracts and the armillarikin fraction was identified through silica gel and Sephadex LH-20 columns. The armillarikin growth inhibition on leukemia cells was measured by alamar Blue assay. Furthermore, ROS (reactive oxygen species) production was identified inside the armillarikin-treated cells, playing a mediating key role in the apoptosis process. Another fungus species, *Marasmiusoreades*, a delicious edible mushroom, showed a moderate effect on HT-29 cells (colorectal carcinoma), and MCF-7, MDA-MB-231 cells (breast cancer) respectively, through MTT colorimetric assay (Shao et al., 2019).

Cancer is one of the main threats to human health, due to cancer cells' ability of mutation and resistance to available drugs. Nuclear transcription factor kappa B (NF-kB) has been identified as one of the responsible factors for the cells' chemoresistance ability. In this line, Petrova et al. (2009) proved that *Marasmiusoreades* could have a positive influence on the MCF7 human breast cancer cell line thanks to SiF2 and SiF3 low molecular fractions through the inhibition of IKK (major cellular signaling pathway) activity. SiF2 and SiF3 fractions showed the highest negative influence on the growth of MCF7 cells. The SiF3 fraction is more efficient in inhibiting the IKK activity and enhancing by 60% the apoptotic cells. In addition, Jung et al. (2014) studied the influence of *Lyophyllum shimeji* protein extract on the KB human oral squamous cell carcinoma and hinted that the extract could exhibit cell cytotoxicity, mainly through apoptosis. To achieve this goal, methods such as Western blotting, DNA fragmentation, and LIVE-DEAD cell staining were used. Cleaved caspases-7 and -9 and cleaved poly-ADP-ribose polymerase were activated through *Lyophyllum shimeji* extract and the LIVE-DEAD cell staining method suggested that dead cell staining was positively influenced by the extract. *Volvariellavolvacea* is another mushroom specieswherecold alkali-extracted glucan was identified as exhibiting antitumor activity against Sarcoma 180, a solid tumor implanted in rats (Misaki et al., 1986).Furthermore,a polysaccharide-protein complex (PSPC) was successfully isolated from *Tricholomagiganteum* culture filtrates, followed by *in vitro* study on mice investigate to study its influence on the development of different tumors. Briefly, Ooi and Liu (2000) and Moradali et al. (2007) showed that the growth of Sarcoma 180 in mice, HL-60, H3B, PU5-1.8 melanoma could be inhibited by *Tricholomagiganteum* PSPC, without any toxicity effect. It was also able to release nitric oxide (NO) and tumor necrosis factor alfa (α), which could act as mediators by activating macrophages and exhibiting indirect cytotoxicity against P815 and L929, respectively. Moreover, the *Tricholomagiganteum* apoptogenic effects were studied by Chatterjee et al. (2013) who showed that Fa fraction could induce an apoptogenic signal in Ehrlich's ascites carcinoma, acting as a chemopreventive agent in cancer therapy. Fa also positively influenced the growth inhibition, cell cycle deregulation, leading to EAC cells apoptosis.

The *Rusulladelica* fruiting body contains an antiproliferative ribonuclease with a unique N-terminal sequence. It was proved by Zhao, Zhao, Li, Zhang, et al. (2010) that the newly identified ribonuclease could have positive effects on cell lines MCF7 (breast adenocarcinoma) and hepatoma (HepG2), inhibiting their proliferation.

In another study by Zhao, Zhao, Li, Zhao, et al. (2010), a novel lectin with 60 kDa molecular weight has been isolated from the *Rusulladelica*fruiting body. Lectin is a non-immunogenic protein, also called glycoprotein which could have antitumor, antiproliferative, and im-munomodulatory effects. The proliferation of MCF-7 and HepG2 cancer cells was extensively inhibited by 2 µM lectin with a percentage of 90.4% and 71.1%, respectively. Similarly, *Tricholomaequestre* exhibited antiproliferative activity against human adenocarcinoma colorectal Caco-2 cells due to the presence of flavomannin-6,6'-dimethylether (Pachón-Peña et al., 2009), without a genotoxic effect. *Tricholomaterreum*, a wild mushroom, has been studied by Yin et al. (2013) for its positive effect on five human cancer cell lines, as follows: lung cancer A-549 cells, MCF-7 breast cancer, hepatocellular carcinoma SMMC-7721, colon cancer SW480, and human myeloid leukemia HL-60, respectively. The authors identified four new meroterpenoids, such as terreumols A–D (1–4): terreumol A (1), terreumol B (2), terreum C (3), terreum D (4), where-compounds 1, 3, and 4exhibitedinhibitory effects on the aforementioned cancer cell lines. The anticoagulant activity of a chemically sulfated polysaccharide of *Grifolafrondosa* (S-GFB), which was derived from water-insoluble polysaccharide of *Grifolafrondosa* (GFB) but soluble in ethanol, was studied by Cao et al. (2010). In this study, the anticoagulant activity was investigated for the first time, by measuring activated partial thromboplastin time (APTT), thrombin time (TT), and prothrombin time (PT), and fibrinolytic activities of the sulfated polysaccharide preparations from *Grifolafrondosa*. Results have proven that S-GFB had an important effect on decompounding fibrin polymerization through a fibrinolytic. Another polysaccharide isolated from another type of edible mushroom, *Lentinula edodes* is *Lentinan,* which was used as an immune adjuvant for stomach cancer treatment along with chemotherapy in Japan since 1985. To the relation between its structure and antitumor activity, it was reported that lentinan is a β-glucan type molecule that impairs the immunostimulating activity through the helix structure and especially the distribution of the branch units along the chain (Ooi & Liu, 2012). The lentinan degree of branching (DB) was found to be 40%. This aspect is important in view of recent findings which are reported for 32%DB a maximal antitumor activity. Regarding lentinan's immunomodulating and antitumor properties, results showed its capacity to stimulate the production of the biologically active compounds responsible for the body's resistance to malignant transformation (i), to activate the immune cells at a normal level, stimulating the production of cytokines (ii) and to inhibit Treg cells activity (iii) (Meng, Liang, & Luo, 2016).

Furthermore, Xu et al. (2012) reported the extraction of a novel heteropolysaccharide (L2) with immunostimulating activity, which was isolated from the fruit body of *L. edodes*. L2 (26 kDa) was composed of glucose, galactose, and arabinose with glucose as the major monosaccharide (87.5%). Although L2 does not have a triple-helical conformation as it was found in the case of lentinan and other polysaccharides extracted from *L. edodes*, L2 showed significant immune activity. Another study conducted by Cui et al. (2015), started from L2 polysaccharide and prepared the L2-calcium complex. The immunostimulating activity was assessed by determining the two cytokines production, TNF-α, and IL-6 in RAW264.7 cells. Results showed significant stimulation of these cytokines compared to L2 (control sample). Furthermore, polysaccharides isolated from the residue of *L. edodes* by enzymatic (ERPS) and acidic (ARPS) hydrolysis were investigated by Ren et al. (2018). Alongside its antioxidant properties, good anti-inflammatory and lung-protective effects were found on the lipopolysaccharide-injection-induced injured lung. A recent *in vivo* study on *Lentinula edodes*, conducted by Chen et al. (2020) aimed at the effects of polysaccharide *Lentinula edodes* on cellular immunity, humoral immunity, and cytotoxicity in NK cells from immunosuppressed mice. Three polysaccharide fractions (F1, F2, and F3) were isolated from *Lentinula edodes* water extracts. The authors concluded that F1 had significant effects only in enhancing cellular immunity, while F2 and F3 improved cellular immunity, humoral immunity, and innate immunity. Also, the immunomodulatory effects became more significant in the order F1 < F2 < F3, inferring that molecular weight was an important factor for the bioactivity of the *Lentinula edodes* polysaccharide fractions (Chen et al., 2020). *Hericiumerinaceus* mushroom was another species used to isolate powerful antitumoral polysaccharides. The investigation of their bioactivities revealed the immunostimulant effect in the case of HEB-AP Fr I, a polysaccharide fraction with β-mannan conformation (J. S. Lee & Hong, 2010) and a good antitumor property (Meng, Liang, & Luo, 2016). As in the case of other properties (i.e. antioxidant capacity) the molecular weight of the polysaccharide is considered a critical parameter for higher bioactivity. In this regard, the glucan fraction (>100 kDa) isolated from *H. erinaceus* demonstrated high antitumor and immunostimulating activity in the case of an artificial pulmonary tumor. A water-insoluble polysaccharide was also isolated from *Pleurotusdjamor* and its anti-proliferation effect against ovarian carcinoma PA1 cells was tested. Regarding the chemical structure, it was concluded that the isolated polysaccharide is a β-glucan with a molecular weight of 9.16×10^4 Da and with a linear (1→3)-β-D-glucan configuration. The cytotoxic effect against ovarian carcinoma cell line from Caucasian ethnicity was tested by the MTT [3-(4,5-dimethylthiazol-2-yl)-2,5-diphenyltetrazoliumbromide] method and compared with the soluble β-glucan effect in similar conditions. Results for the insoluble β-glucan showed the lack of toxicity at a concentration of 250 μg/mL, with cell survival at a 102.9% level. Furthermore, the insoluble polysaccharide revealed a significant anticancer effect on PA1 cells. This assay was also performed in comparison to the soluble β-glucan (2×10^5 Da) and the significant difference which was found was attributed to the high molecular weight of the insoluble polysaccharide (Maity et al., 2019). A large body of research indicates that *Ganoderma* mushrooms possess antitumor effects due to their various bioactive compounds such as polysaccharides, triterpenes, immunomodulatory proteins (Hsu et al., 2008; Ko et al., 2008; Chien et al., 2015; K. W. Lin et al., 2016; Hseu et al., 2019). A methanolic extract from fruiting bodies of *Ganoderma tsugae* showed antitumor effects *in vitro* and *in vivo* on colorectal adenocarcinoma cells by inducing G2/M cell cycle arrest and no toxicities were noticed on the animal model (Hsu et al., 2008). Recently, Hseu et al. (2019) used the ethanolic extract of *G. tsugae* to investigate its antitumor activities on human chronic myeloid leukemia cells. Results showed that the extract induced apoptosis and autophagy in human leukemic cells. Another approach in testing the anticancer activity of *G. tsugae* was to isolate polysaccharides by hot-water and hot-alkali extractions (Chien et al., 2015). As in the case of other studies, mature and young Ling Chih, mycelium and filtrate were used for extraction and eight samples were obtained to be studied. Mannitol, myoinositol, and trehalose were found in higher amounts in all samples, with mannitol as the major component. Due to the different cultivation conditions, between the hot-water and hot-alkali samples, differences in molecular weight and polysaccharides composition were noticed. The hot-alkali extracted polysaccharides showed a higher efficiency against the anti-proliferation of human neuroblastoma IMR 32 cell line. But the hot-water extracts demonstrated effectiveness against anti-proliferation of human hepatoma SK-Hep-1 cell line. The results also showed that samples from fruiting bodies were more powerful.

Z. Sun et al. (2009) proposed the modification by sulfation of (1-3)-β-D-glucan extracted as a water-insoluble polysaccharide from fruiting bodies of *Russulavirescens*. This approach relied on other studies reporting the enhancement of the biological activity of polysaccharides by sulfation. By varying reactionconditions, six sulfate derivatives with different molecular weights and degrees of sulfation were obtained. For five of these compounds *in vitro* and *in vivo* antitumor activities against Sarcoma 180 tumor cells were investigated. The tested hypothesis proved to be true and the sulfate derivatives showed improved antitumor activities. Hui Li et al. (2020) conducted a similar study on the bioactivity of the water-soluble polysaccharide (RVP) extracted from *R. virescens* and its sulfated derivatives (SRVP). RVP was identified as a heteropolysaccharide with a molecular weight of 4×10^5 Da and composed of xylose, mannose, glucose, galactose, arabinose, and fucose. Glucose was the major component of the identified monosaccharides. The sulfated derivatives (SRVP) showed decreases in the molecular weight and contents of glucose and galactose, probably due to the chain modifications. *In vitro* assays revealed that SRVP was more effective as antitumor agents again CaCo-2 cancer cells than RVP. These findings suggest that sulfate modification of polysaccharides extracted from wild mushrooms may serve biomedical use.

Water extract of *Cordyceps sinensis* (WECS) with a content of 20.63% polysaccharide was tested *in vivo* and *in vitro* against breast cancer. Results indicated that WECS could inhibit tumor growth by promoting macrophage polarization and producing inflammatory cytokines (J. Li et al., 2020).

Cordyceps sinensis was also used by J. Wang et al. (2017) to isolate a water-soluble polysaccharide (NCSP-50) to test its immunostimulatory activity. The molecular weight of NCSP-50 was found to be 9.76×10^5 Da, while from a structural point of view, a homogenous glucan containing only glucosewas considered. The immunostimulatory activity of NCSP-50 was analyzed *in vitro* on murine macrophage cell line RAW 264.7. It was reported that NCSP-50 significantly stimulates the proliferation of macrophages, promotes nitric oxide production, and enhances cytokine secretion. Recently, another group of researchers isolated a polysaccharide from mycelial strains of *C. sinensis* (by the same extraction method proposed by J. Wang et al. (2017) and its antitumor effect was tested against human colon cancer cells (Qi et al., 2020). The capacity of CSP to inhibit colon cancer cells proliferation by inducing apoptosis and autophagy was reported.

In a recent study led by Morales et al. (2020), a polysaccharide-enriched extract obtained from *Lentinula edodes* (LPS) was submitted to several purification steps to separate three different D-glucans with β-(1→6), β-(1→3), (1→6) and α-(1→3) linkages, being characterized through GC-MS, FT-IR, NMR, SEC, and colorimetric/fluorometric determinations. These extracts were tested for hypocholesterolemic, antitumoral, anti-inflammatory, and antioxidant activities, *in vitro*. The authors concluded that isolated glucans showedHMGCR (3-hydroxy-3-methylglutaryl coenzyme A reductase) inhibitory activity, but only β-(1→6) and β-(1→3), (1→6) fractions showed DPPH scavenging capacity. Also, glucans were able to lower IL-1β and IL-6 secretion by LPS activated THP-1/M cells and showed a cytotoxic effect on a breast cancer cell line that was not observed on normal breast cells (Morales et al., 2020).

The possible anti-cancer effect on *Tricholomaequestre* was further supported by Feng et al. (2015) who isolated saponaceolides Q–S (1–3), three rare triterpenoids, where compound 1 showed moderate activity on MCF-7 (breast cancer), SW480 (colon cancer), SMMCC-7721 (carcinoma) and HL-60 (human myeloid leukemia) cells.

Agrocybeaegerita is a popular delicacy edible mushroom with unique flavor and taste, with reported antitumor properties. A bioactivity-guided investigation by Diyabalanage et al. (2008) gave isolation of ceramide from *A. aegerita* and its COX and tumor cell proliferation inhibitory activities. Their structure elucidation was accomplished by NMR and mass spectral methods. It was reported that ceramide (1) inhibited cyclooxygenase enzymes, COX-1 and -2, by 43% and 92.3%, respectively at 25 lg/mL (34.4 lM). The 50% inhibition concentration (IC50) of compound 1 against COX-2 was 5.3 lg/mL (7.3 lM). Similarly, its anti-cancer potential was investigated against five human cancer cell lines *in vitro* and it was found to inhibit the proliferation of stomach, breast, and CNS cancer cell lines at 26.9%, 23.2%, and 39.1%, respectively, at 100 lg/mL (139 lM) concentration. This research group suggested that the consumption of *A. aegerita* would assist in alleviating inflammatory conditions, as well as reducing the development of the above-mentioned cancers.

9.4 ANTI-INFLAMMATORY, THROMBOTIC, AND GENOPROTECTIVE EFFECT OF EDIBLE MUSHROOMS

Inflammation is a natural response of the immune system to damaging factors, e.g. physical, chemical, and pathogenic. Deficiencies of antioxidants, vitamins, and microelements, as well as physiological processes, such as aging, can affect the body's ability to resolve inflammation (Muszyńska, Grzywacz-Kisielewska, et al., 2018). Three types of sequential extractswere obtained from *Echinodontiumtinctorium* wild mushroom, using 80% ethanol, 50% methanol, and 5% NaOH to investigate its immuno-stimulatory and anti-inflammatory activities. Firstly, the anti-inflammatory effect was assessed *in vitro* on a lipopolysaccharide-induced RAW264.7 mouse macrophage cell line. The best results were obtained for 5% NaOH, from which a polysaccharide was extracted as the compound with an anti-inflammatory effect. It was characterized as containing glucosein its majority (88.6%) and smaller amounts of galactose, mannose, xylose, and fucose. Therefore, it exhibited the main conformation of β-glucan. *In vivo* anti-inflammatory effect of purified polysaccharide confirmed its potential on the animal model (Javed et al., 2019).

Lyophyllum shimeji is one of the famous Japanese edible mushrooms with a novel fibrinolytic enzyme able to be successfully used in thrombolytic therapy and prevention of thrombotic disease, respectively (Moon et al., 2014). Several materials and methods were used for the purification, molecular weight and fractionation of the extracts such as SDS-PAGE and Fast Protein Liquid Chromatography (FPLC), respectively. The study hinted that the purified enzyme was able to act as a fibrinolytic agent, having a key role in the prevention and formation of abnormal blood clots in veins and arteries. Genoprotective effects of edible mushrooms produced by *Pleurotuseryngii*, *Pleurotusostreatus,*and *Cyclocybecylindracea* (Basidiomycota) were investigated for the first time by Boulaka et al. (2020). The authors fermented the mushrooms *in vitro* from selected species using fecalinocula from eight elderly (>65 years), healthy volunteers. The production of several fermentation-specific metabolites (acetate, propionate, and butyrate, aromatic amino acids (Phe, Tyr), trimethylamine (TMA), and gamma-aminobutyric acid) was considerably enhanced, presenting a protective effect against BOOH-induced DNA damage. Moreover, this study provides substantial evidence that edible mushrooms may contain ingredients protecting genome integrity, which is fundamental, especially for healthy aging.

9.5 ANTIMICROBIAL AND ANTIFUNGAL ACTIVITY OF EDIBLE MUSHROOMS

The antibacterial effectsof native and sulfated polysaccharides extracted from*R.virescens*against gram-negative (*E. coli*) and gram-positive (*S. aureus*) bacteria were tested by Huaping Li et al. (2021). As in the case of other bioactivities, the sulfate derivatives showed better activity. While the native polysaccharide was ineffective against the abovementioned bacteria, the introduction of the sulfate groups into the polysaccharide structure improved the antibacterial activity. It was alsonoticed, that sulfate derivatives could have a higher antibacterial effect against gram-positive bacteria. Even if the exact mechanism of the antibacterial effect is not well known, the authors considered that sulfated polysaccharides may lead to cell death by disruption of the cell walls and cytoplasmic membranes, leading to the dissolution of the essential molecules.

Kosanić (2016) tested acetone and methanol extracts of the mushrooms *Amanita rubescens* and *Russulacyanoxantha* for their antimicrobial activity. The results showed that these extracts had strong antimicrobial activity against the tested microorganisms (*Staphylococcus aureus* (ATCC 25923), *Escherichia coli* (ATCC 25922), *Klebsiella pneumoniae* (ATCC70063), *Pseudomonas aeruginosa* (ATCC 27853), *Enterococcus faecalis (ATCC 29212) and Aspergillus flavus (ATCC 9170), Aspergillus fumigatus (DBFS 310), Candida albicans (IPH 1316), Paecilomycesvariotii (ATCC 22319),* and *Penicillium purpurescens (DBFS 418).* However, the results indicated stronger antibacterial rather than an antifungal activity for the tested mushroom extracts.

Alves et al. (2012), claim that phenolic compounds, from mushroom extracts, could be used as antimicrobials against pathogenic micro-organisms resistant to conventional treatments. *Fistulina hepatica, Romaria botrytis,* and *Russuladelica* are the most promising species used to this end. The authors studied *in vitro* antimicrobial activities of aqueous methanolic extracts of 13 mushroom species, collected in Bragança, against several clinical isolates obtained in Hospital Centre of Trás-os-Montes and Alto Douro, Portugal. For minimum inhibitory concentration, the authors showed that *Russuladelica* and *Fistulina hepatica* extracts inhibited the growth of gram-negative (*Escherichia coli, Morganellamorganni,* and *Pasteurella multocida*) and gram-positive (*Staphylococcus aureus*, methicillin-resistant (MRSA), *Enterococcus faecalis, Listeria monocytogenes, Streptococcus agalactiae,* and *Streptococcus pyogenes*) bacteria. Also, a bactericide effect of both extracts was observed in *Pasteurella multocida, Streptococcus agalactiae,* and *Streptococcus pyogenes*. Regarding *Ramaria botrytis* extract, this study showed that it presents inhibitory activity against *Enterococcus faecalis* and *Listeria monocytogenes,* as well as abactericide for *Pasteurella multocida, Streptococcus agalactiae,* and *Streptococcus pyogenes*.

Barros et al. (2008), conducted a screening of mushroom extracts against gram-positive bacteria (*Bacillus cereus* CECT148, *Bacillus subtilis* CECT 498, *Staphylococcus aureus* ESA 7 isolated from pus), bacteria gram-negative (*Escherichia coli* CECT 101, *Pseudomonas aeruginosa* CECT 108, *Klebsiella peumoniae*ESA 8 isolated from urine), and fungi (*Candida albicans* CECT 1394 and *Cryptococcus neoformans* ESA 3 isolated from vaginal fluid). Extracts of *Lycoperdon* *mole* and *Lycoperdonperlatum* were resistant to all the tested microorganisms, and the other samples revealed antimicrobial activity selectively against gram-positive bacteria, with very low minimal inhibitory concentration. The authors of this study claim that there is no relationship between phenol content and antimicrobial properties but there are other compounds such as steroids, oxalic acid, sesquiterpenoids, and epipolythiopiperazine-2,5- diones isolated from mushrooms, that proved to have antimicrobial activity (Lindequist, Niedermeyer, & Jülich, 2005).

Lyophyllum shimeji is a well-known Asian mushroom that contains a new ribosome-inactivating antifungal protein with 14 kDa molecular weight (Lam & Ng, 2001). The purification of the new antifungal protein (lyophyllin) from the *Lyophyllum shimeji* fruiting body was made through the following steps: mixing the mushroom fruiting body with ammonium acetate, centrifuging the mixture, applying the supernatant on a CM-cellulose column, and chromatographing it on a Affi-gel Blue Gel column. The antifungal activity was tested against several fungal species, such as *Coprimuscomatus, Colletotrichum gossypii, Physalosporapiricola, Mycosphaerellaarachidicola,* and *Rhizoctonia solani*. From the quantitative point of view, the antifungal activity was tested by using IC_{50}, with three different doses of antifungal protein. To antifungal activity, *Lyophyllum shimeji* was efficient against *Physalosporapiricola* (IC_{50} = 70 nM) and *Coprimuscomatus* but not toward *Mycosphaerellaarachidicola* or *Rhizoctonia solani,* respectively.

Similarly, *Volvariellavolvacea* R83 and ATCC62890 strains were studied regarding their antimicrobial activity by using the paper disc diffusion method (da Silva et al., 2010). Different reagents such as methanol, chloroform, and hexane fractions, as well as ethyl acetate fraction, were used to extract *Volvariellavolvacea* mushrooms. Crude methanol extracts exhibited low to moderate activity against *S. aureus*, while chloroform and hexane extracts showed moderate to high inhibition against *S. aureus* growth. *Volvariellavolvacea* R83 strain showed higher antimicrobial activity against *S. aureus*, compared to *Volvariellavolvacea* ATCC62890 strain, probably due to its active fractions with antimicrobial effects (RFH4-6, RFH7-10, RFH14-19, RFH46-58, RFC2-11).

Leccinumscabrum antibacterial assay was analyzed by Nowacka et al. (2014), through the micro-broth dilution method. The *Leccinumscabrum* extract exhibited antimicrobial activity against gram-positive bacteria such as *S. epidermis* and *M. luteus* but also against gram-negative bacteria such *as E. coli, K. pneumoniae,* and *P. aeruginosa.*

From *Tricholomagiganteum* var. *golden blessings*, an antifungal protein (trichogen), was isolated for the first time, using the methods mentioned in Table 9.1. The trichogen antifungal activity was tested against *Fusarium oxysporum, Physalosporapiricola,* and *Mycosphaerellaarachidicola,* respectively by using Petri plates with potato dextrose agar (Guo, Wang, & Ng, 2005). The results showed positive effects against the mentioned fungus species. Furthermore, it seems that *Tricholomagiganteum* might inhibit HIV-1 reverse transcriptase, highlighting an IC_{50} = 83 nM potency (Guo, Wang, & Ng, 2005).

Rusulladelica mushroom ethanolic extract could also exhibit antimicrobial activity (Yaltirak et al., 2009) against *Shigella sonnei, Yernisia enterocolitica, Bacillus cereus, Listeria monocytogenes, Escherichia coli, Staphylococcus aureus,* and *Proteus vulgaris* with an inhibition zone diameter (mm) ranging from 17 to 7 mm. The antimicrobial activity of *Rusulladelica* extract could be related to its phenolic compounds amount, where catechin, rutin, and caffeic acid showed antimicrobial activity (Yaltirak et al., 2009).

Tricholomaequestre was studied by Venturini et al. (2008) for its antimicrobial activity and it was stated that the extract could exhibit growth of *Listeria monocytogenes* and *Staphylococcus aureus,* having an inhibition zone diameter of 17.9 and 16.8 mm, respectively.

Volvariellavolvacea mushroom belongs to the phylum Basidiomycota and it is currently found in Ghana together with *Trametes gibbose, Schizophyllum commune* Fr., and *Trametes elegans,* respectively (Appiah, Boakye, & Agyare, 2017). The fruit body of *Volvariellavolvacea* was extracted with methanol and the antimicrobial activity was analyzed through the agar well diffusion method. The study showed that 30 mg/mL *Volvariellavolvacea* extract could have a positive effect on the development of gram-positive and gram-negative bacteria, highlighting the mean zone of grown inhibition in a range of 19.50 to 16.50 and 22.50 to 21.17, respectively. However, there was no influence against *Candida albicans*. Herein, *Volvariellavolvacea* extract negatively influenced the growth of *S. aureus, P. aeruginosa, S. pyogenes,* and *K. pneumoniae.*

Tricholomaequestre fruiting bodies extract could also have moderate antimicrobial effects (Muszyńska, Kala, et al., 2018) against *Staphylococcus aureus* and *Klebsiella pneumoniae* strains.

Conversely, Stadler et al. showed that *Tricholomaterreum* could impart nematocidal activity and it was able to produce important amounts of linoleic acid and S-coriolic acid. The identified fatty acids are accompanied by antibiotic effects and seem to be responsible for nematocidal effects.

9.6 ANTI-HIV AND POSSIBLE INHIBITORS OF SARS-COV-2 MAIN PROTEASE

In a recent study by Rangsinth et al. (2021), it was reported that in the race for developing a treatment for coronavirus disease 2019 (COVID-19) anti-HIV protease drugs have tablebeen extensively studied. To find natural sources that could replace conventional drugs, the researchers' attention was focused on mushrooms with anti-HIV protease activity. Bioactive compounds identified in *Rusulladelica* such as oleanolic and ursolic acids could be potential candidates for anti-SAR-CoV-2 agents, but mainly colossolactone VIII, colossolactone E colossolactone G, ergosterol, heliantriol F, and velutin were highlighted to be the best bioactive compounds to fight COVID-19.

9.7 CONCLUSIONS

As reported in this chapter, the statements are similar to statements in other specialized studies, which demonstrates that mushrooms are a rich source of biologically valuable components that offer great therapeutic potential for the prevention and control of several diseases (Gupta et al., 2019). The potential therapeutic implication of mushrooms is tremendous, demonstrated by *in vitro* digestive models, but more research and clinical trials on *in vivo* digestion models need to be carried out to validate wild edible mushrooms as a source of bioactive molecules with medicinal applications. Edible mushrooms produce a vast diversity of bioactive compounds and may be classified as peptides and proteins, phenolic compounds, polysaccharides protein complexes, terpenes, terpenoids, steroids, and lectins, etc. These compounds have a wide range of therapeutic effects, where one can include immunomodulatory, anticarcinogenic, antiviral, antioxidant, and anti-inflammatory agents. Some specific bioactive compounds in edible mushrooms are responsible for improving human health in several ways. Bioactive compounds can be found in mushrooms, as well as their cell wall components as polysaccharides (B-glucan) and proteins or as secondary metabolites such as phenolic compounds, terpenes, and steroids. Phenolic compounds are aromatic hydroxylated compounds. It was been reported that a positive correlation was found between the total phenolic content in the mushroom extract and their antioxidative properties for some edible wild mushroom species: *Pleurotusostreatus, Agaricus bisporus, Imleriabadia, Agaricus blazei, Boletus edulis, Boletus Aereus, Lactarumpiperatus, Amanita rubescens, Lentinus edodes, Romania botrytis, Lactariudeliciosus, Lentinula edodes, Fistulina hepatica, Tuber aestivum, Marasmiusoreades, Leccinumscabrum, Morchellaelata, Russuladelica, Suillusgranulatus,* and *Tricholomaequestr*e, which confirms that edible mushrooms have a potential of a natural antioxidant demonstrated *in vitro* digestion models. β-glucan is one of the key components of several mushrooms cell walls. It was reported that β-glucan is able to enhance the immune system and prevent and treat several common diseases and overall promote the health effect, which was observed in some edible mushrooms such as: *Pleurotusdjamor, Russulavirescens, Lentinula edodes, Echinodontiumtinctorium, Auricularia auricula-judae, Pleurotuseryngii, Pleurotusostreatus, Cyclocybecylindracea. Lentinus edodes,* etc. As such, this chapter represents a comprehensive bioactivity-based analysis of selected edible wild mushroom species from around the world, as they represent a growing segment for their medicinal and therapeutic application potential.

REFERENCES

Alminger, M., Aura, A. M., Bohn, T., Dufour, C., El, S. N., Gomes, A., Karakaya, S., et al. (2014). In Vitro Models for Studying Secondary Plant Metabolite Digestion and Bioaccessibility. *Comprehensive Reviews in Food Science and Food Safety*, *13*(4), 413–436. 10.1111/1541-4337.12081

Alves, M. J., Ferreira, I. C. F. R., Martins, A., & Pintado, M. (2012). Antimicrobial Activity of Wild Mushroom Extracts against Clinical Isolates Resistant to Different Antibiotics. *Journal of Applied Microbiology*, *113*(2), 466–475. 10.1111/j.1365-2672.2012.05347.x

Appiah, T., Boakye, Y. D., & Agyare, C. (2017). Antimicrobial Activities and Time-Kill Kinetics of Extracts of Selected Ghanaian Mushrooms. *Evidence-Based Complementary and Alternative Medicine*, *2017*. Article ID 4534350, 1–15 pages. 10.1155/2017/4534350

Bai, X., Tan, T. Y., Li, Y. X., Li, Y., Chen, Y. F., Ma, R., Wang, S. Y., Li, Q., & Liu, Z. Q. (2020). The Protective Effect of Cordyceps Sinensis Extract on Cerebral Ischemic Injury via Modulating the Mitochondrial Respiratory Chain and Inhibiting the Mitochondrial Apoptotic Pathway. *Biomedicine and Pharmacotherapy*, *124*(11), 1–12. 10.1016/j.biopha.2020.109834

Barros, L., Baptista, P., & Ferreira, I. C. F. R. (2007). Effect of Lactarius Piperatus Fruiting Body Maturity Stage on Antioxidant Activity Measured by Several Biochemical Assays. *Food and Chemical Toxicology*, *45*(9), 1731–1737. 10.1016/j.fct.2007.03.006

Barros, L., Venturini, B. A., Baptista, P., Estevinho, L. M., & Ferreira, I. C. F. R. (2008). Chemical Composition and Biological Properties of Portuguese Wild Mushrooms: A Comprehensive Study. *Journal of Agricultural and Food Chemistry*, *56*(10), 3856–3862. 10.1021/jf8003114

Bohn, T., Carriere, F., Day, L., Deglaire, A., Egger, L., Freitas, D., Golding, M., et al. (2018). Correlation between in Vitro and in Vivo Data on Food Digestion. What Can We Predict with Static in Vitro Digestion Models?. *Critical Reviews in Food Science and Nutrition*, *58*(13), 2239–2261. 10.1080/104 08398.2017.1315362

Boulaka, A., Christodoulou, P., Vlassopoulou, M., Koutrotsios, G., Bekiaris, G., Zervakis, G. I., Mitsou, E. K., et al. (2020). Genoprotective Properties and Metabolites of β-Glucan-Rich Edible Mushrooms Following Their in Vitro Fermentation by Human Faecal Microbiota. *Molecules*, *25*(15), 1–22. 10.33 90/molecules25153554

Bovi, M., Cenci, L., Perduca, M., Capaldi, S., Carrizo, M. E., Civiero, L., Chiarelli, L. R., Galliano, M., & Monaco, H. L. (2013). BEL -Trefoil: A Novel Lectin with Antineoplastic Properties in King Bolete (Boletus Edulis) Mushrooms. *Glycobiology*, *23*(5), 578–592. 10.1093/glycob/cws164

Cao, X. H., Yang, Q. W., Lu, M. F., Hou, L. H., Jin, Y. Y., Yuan, J., & Wang, C. L. (2010). Preparation And Anticoagulation Activity Of A Chemically Sulfated Polysaccharide (S-Gfb) Obtained From Grifola Frondosa. *Journal of Food Biochemistry*, *34*(5), 1049–1060. 10.1111/j.1745-4514.2010.00348.x

Chan, J. Y. Y., Chan, E., Chan, S. W., Sze, S. Y., Chan, M. F., Tsui, S. H., Leung, K. Y., Yat Kan Chan, R., & Ying Ming Chung, I. (2011). Enhancement of in Vitro and in Vivo Anticancer Activities of Polysaccharide Peptide from Grifola Frondosa by Chemical Modifications. *Pharmaceutical Biology*, *49*, (11), 1114–1120. 10.3109/13880209.2011.569557

Chang, C. C., Cheng, J. J., Lee, I. J., & Lu, M. K. (2018). Purification, Structural Elucidation, and Anti-Inflammatory Activity of Xylosyl Galactofucan from Armillaria Mellea. *International Journal of Biological Macromolecules*, *114*, 584–591. 10.1016/j.ijbiomac.2018.02.033

Chatterjee, S., Saha, G. K., & Acharya, K. (2011). Antioxidant Activities of Extracts Obtained by Different Fractionation from *Tricholoma giganteum* Basidiocarps. *Pharmacologyonline*, *3*, 88–97.

Chatterjee, S., Biswas, G., Chandra, S., Saha, G. K., & Acharya, K. (2013). Apoptogenic Effects of Tricholoma Giganteum on Ehrlich's Ascites Carcinoma Cell. *Bioprocess and Biosystems Engineering*, *36*(1), 101–107. 10.1007/s00449-012-0765-6

Chawla, P., Kumar, N., Kaushik, R., & Dhull, S. (2019). Synthesis, Characterization and Cellular Mineral Absorption of Gum Arabic Stabilized Nanoemulsion of *Rhododendron arboreum* Flower Extract. *Journal of Food Science and Technology*, *56*(12), 5194–5203.

Chen, B. (2010). Optimization of Extraction of Tremella Fuciformis Polysaccharides and Its Antioxidant and Antitumour Activities in Vitro. *Carbohydrate Polymers*, *81*(2), 420–424. 10.1016/j.carbpol.2010.02.039

Chen, Y. C., Chen, Y. H., Pan, B. S., Chang, M. M., & Huang, B. M. (2017). Functional Study of Cordyceps Sinensis and Cordycepin in Male Reproduction: A Review. *Journal of Food and Drug Analysis*, *25*(1), 197–205. 10.1016/j.jfda.2016.10.020

Chen, X., Ji, H., Xu, X., & Liu, A. (2019). Optimization of Polysaccharide Extraction Process from Grifola Frondosa and Its Antioxidant and Anti-Tumor Research. *Journal of Food Measurement and Characterization*, *13*(1), 144–153. 10.1007/s11694-018-9927-9

Chen, S., Liu, C., Huang, X., Hu, L., Huang, Y., Chen, H., Fang, Q., et al. (2020). Comparison of Immunomodulatory Effects of Three Polysaccharide Fractions from Lentinula Edodes Water Extracts. *Journal of Functional Foods*, *66*(January), 103791. 10.1016/j.jff.2020.103791

Chen, G. T., Ma, X. M., Liu, S. T., Liao, Y. L., & Zhao, G. Q. (2012). Isolation, Purification and Antioxidant Activities of Polysaccharides from Grifola Frondosa. *Carbohydrate Polymers*, *89*(1), 61–66. 10.1016/j.carbpol.2012.02.045

Chen, R., Ren, X., Yin, W., Lu, J., Tian, L., Zhao, L., Yang, R., & Luo, S. (2020). Ultrasonic Disruption Extraction, Characterization and Bioactivities of Polysaccharides from Wild Armillaria Mellea. *International Journal of Biological Macromolecules*, *156*, 1491–1502. 10.1016/j.ijbiomac.2019.11.196

Chen, Y. J., Wu, S. Y., Chen, C. C., Tsao, Y. L., Hsu, N. C., Chou, Y. C., & Huang, H. L. (2014). Armillaria Mellea Component Armillarikin Induces Apoptosis in Human Leukemia Cells. *Journal of Functional Foods*, *6*(1), 196–204. 10.1016/j.jff.2013.10.007

Chen, L., Zhang, Y., Sha, O., Xu, W., & Wang, S. (2016). Hypolipidaemic and Hypoglycaemic Activities of Polysaccharide from Pleurotus Eryngii in Kunming Mice. *International Journal of Biological Macromolecules*, *93*, 1206–1209. 10.1016/j.ijbiomac.2016.09.094

Cheung, P. C. K. (2010). The Nutritional and Health Benefits of Mushrooms. *Nutrition Bulletin*, *35*(4), 292–299. 10.1111/j.1467-3010.2010.01859.x

Chien, R. C., Yen, M. T., Tseng, Y. H., & Mau, J. L. (2015). Chemical Characteristics and Anti-Proliferation Activities of Ganoderma Tsugae Polysaccharides. *Carbohydrate Polymers*, *128*, 90–98. 10.1016/j.carbpol.2015.03.088

Cruz, A., Pimentel, L., Rodríguez-Alcalá, L. M., Fernandes, T., & Pintado, M. (2016). Health Benefits of Edible Mushrooms Focused on Coriolus Versicolor: A Review. *Journal of Food and Nutrition Research*, *4*(12), 773–781. 10.12691/jfnr-4-12-2

Cui, Y., Yan, H., & Zhang, X. (2015). Preparation of Lentinula Edodes Polysaccharide-Calcium Complex and Its Immunoactivity. *Bioscience, Biotechnology and Biochemistry*, *79*(10), 1619–1623. 10.1080/09168451.2015.1044930

Ding, X., Hou, Y., & Hou, W. (2012). Structure Feature and Antitumor Activity of a Novel Polysaccharide Isolated from Lactarius Deliciosus Gray. *Carbohydrate Polymers*, *89*(2), 397–402. 10.1016/j.carbpol.2012.03.020

Diyabalanage, T., Mulabagal, V., Mills, G., DeWitt, D. L., & Nair, M. G. (2008). "Health-Beneficial Qualities of the Edible Mushroom, Agrocybe Aegerita. *Food Chemistry*, *108*(1), 97–102. 10.1016/j.foodchem.2007.10.049

Dong, Y., Qi, G., Yang, Z., Wang, H., Wang, S., & Chen, G. (2015). Preparation, Separation and Antioxidant Properties of Hydrolysates Derived from Grifola Frondosa Protein. *Czech Journal of Food Sciences*, *33*(6), 500–506. 10.17221/197/2015-CJFS

Dupas, C., Baglieri, A. M., Ordonaud, C., Tomè, D., & Noëlle Maillard, M. (2006). Chlorogenic Acid Is Poorly Absorbed, Independently of the Food Matrix: A Caco-2 Cells and Rat Chronic Absorption Study. *Molecular Nutrition and Food Research*, *50*(11), 1053–1060. 10.1002/mnfr.200600034

Feng, T., He, J., Ai, H. L., Huang, R., Li, Z. H., & Liu, J. K. (2015). Three New Triterpenoids from European Mushroom Tricholoma Terreum. *Natural Products and Bioprospecting*, *5*(4), 205–208. 10.1007/s13659-015-0071-5

Frutos, P. D. (2020). Changes in World Patterns of Wild Edible Mushrooms Use Measured through International Trade Flows. *Forest Policy and Economics*, *112*, (November 2019), 1–7. 10.1016/j.forpol.2020.102093

Gąsecka, M., Siwulski, M., & Mleczek, M. (2018). Evaluation of Bioactive Compounds Content and Antioxidant Properties of Soil-Growing and Wood-Growing Edible Mushrooms. *Journal of Food Processing and Preservation*, *42*(1), 1–10. 10.1111/jfpp.13386

Grahn, E. M., Winter, H. C., Tateno, H., Goldstein, I. J., & Krengel, U. (2009). Structural Characterization of a Lectin from the Mushroom Marasmius Oreades in Complex with the Blood Group B Trisaccharide and Calcium. *Journal of Molecular Biology*, *390*(3), 457–466. 10.1016/j.jmb.2009.04.074

Guillamón, E., García-Lafuente, A., Lozano, M., D̓arrigo, M., Rostagno, M. A., Villares, A., & Martínez, J. A. (2010). Edible Mushrooms: Role in the Prevention of Cardiovascular Diseases. *Fitoterapia*, *81*(7), 715–723. 10.1016/j.fitote.2010.06.005

Guo, Y., Wang, H., & Ng, T. B. (2005). Isolation of Trichogin, an Antifungal Protein from Fresh Fruiting Bodies of the Edible Mushroom Tricholoma Giganteum. *Peptides*, *26*(4), 575–580. 10.1016/j.peptides.2004.11.009

Gupta, S., Summuna, B., Gupta, M., & Annepu, S. K. (2019). Edible Mushrooms: Cultivation, Bioactive Molecules, and Health Benefits. In Mérillon, J.-M., & Ramawat, K. G. (eds.), *Bioactive Molecules in Food, Reference Series in Phytochemistry*. Springer Nature Switzerland AG, pp. 1816–1847. 10.1007/978-3-319-78030-6_86

Gursoy, N., Sarikurkcu, C., Cengiz, M., & Solak, M. H. (2009). Antioxidant Activities, Metal Contents, Total Phenolics and Flavonoids of Seven Morchella Species. *Food and Chemical Toxicology*, *47*(9), 2381–2388. 10.1016/j.fct.2009.06.032

Gursoy, N., Sarikurkcu, C., Tepe, B., & Solak, M. H. (2010). Evaluation of Antioxidant Activities of 3 Edible Mushrooms: Ramaria Flava (Schaef.: Fr.) Quél., Rhizopogon Roseolus (Corda) T.M. Fries., and Russula Delica Fr. *Food Science and Biotechnology*, *19*(3), 691–696. 10.1007/s10068-010-0097-8

He, P., Li, F., Huang, L., Xue, D., Liu, W., & Xu, C. (2016). Chemical Characterization and Antioxidant Activity of Polysaccharide Extract from Spent Mushroom Substrate of Pleurotus Eryngii. *Journal of the Taiwan Institute of Chemical Engineers*, *69*, 48–53. 10.1016/j.jtice.2016.10.017

Heleno, S. A., Barros, L., João Sousa, M., Martins, A., Santos-Buelga, C., & Ferreira, I. C. F. R. (2011). Targeted Metabolites Analysis in Wild Boletus Species. *LWT – Food Science and Technology*, *44*(6), 1343–1348. 10.1016/j.lwt.2011.01.017

Heleno, S. A., Martins, A., Queiroz, M. J. R. P., & Ferreira, I. C. F. R. (2015). Bioactivity of Phenolic Acids: Metabolites versus Parent Compounds: A Review. *Food Chemistry*, *173*, 501–513. 10.1016/j.foodchem.2014.10.057

Hou, Y., Ding, X., Hou, W., Song, B., Wang, T., Wang, F., & Zhong, J. (2013). Immunostimulant Activity of a Novel Polysaccharide Isolated from Lactarius Deliciosus (l. Ex Fr.) Gray. *Indian Journal of Pharmaceutical Sciences*, *75*(4), 393–399. 10.4103/0250-474X.119809

Hseu, Y. C., Shen, Y. C., Kao, M. C., Mathew, D. C., Karuppaiya, P., Li, M. L., & Yang, H. L. (2019). Ganoderma Tsugae Induced ROS-Independent Apoptosis and Cytoprotective Autophagy in Human Chronic Myeloid Leukemia Cells. *Food and Chemical Toxicology*, *124*(October 2018), 30–44. 10.101 6/j.fct.2018.11.043

Hsu, S. C., Ou, C. C., Li, J. W., Chuang, T. C., Kuo, H. P., Liu, J. Y., Chen, C. S., Lin, S. C., Su, C. H., & Kao, M. C. (2008). Ganoderma Tsugae Extracts Inhibit Colorectal Cancer Cell Growth via G2/M Cell Cycle Arrest. *Journal of Ethnopharmacology*, *120*(3), 394–401. 10.1016/j.jep.2008.09.025

Javed, S., Li, W. M., Zeb, M., Yaqoob, A., Tackaberry, L. E., Massicotte, H. B., Egger, K. N., Cheung, P. C. K., Payne, G. W., & Lee, C. H. (2019). Anti-Inflammatory Activity of the Wild Mushroom, Echinodontium Tinctorium, in RAW264.7 Macrophage Cells and Mouse Microcirculation. *Molecules*, *24*(19), 1–15. 10.3390/molecules24193509

Jaworska, G., & Bernaś, E. (2009). The Effect of Preliminary Processing and Period of Storage on the Quality of Frozen Boletus Edulis (Bull: Fr.) Mushrooms. *Food Chemistry*, *113*(4), 936–943. 10.1016/j.foodchem.2008.08.023

Ji, H. Y., Yu, J., Chen, X. Y., & Liu, A. J. (2019). Extraction, Optimization and Bioactivities of Alcohol-Soluble Polysaccharide from Grifola Frondosa. *Journal of Food Measurement and Characterization*, *13*(3), 1645–1651. 10.1007/s11694-019-00081-z

Jiao, F., Wang, X., Song, X., Jing, H., Li, S., Ren, Z., Gao, Z., Zhang, J., & Jia, L. (2017). Processing Optimization and Anti-Oxidative Activity of Enzymatic Extractable Polysaccharides from Pleurotus Djamor. *International Journal of Biological Macromolecules*, *98*, 469–478. 10.1016/j.ijbiomac.2017.01.126

정종원, 김진수 (Jung), Park Bo Ram, S. K., 김춘성, & 문성민 (2014). Anti-Cancer Activity of the Protein Extract from Lyophyllum Shimeji in KB Human Oral Squamous Cell Carcinoma. *Oral Biology Research*, *38*(2), 99–105. 10.21851/obr.38.2.201410.99

Kalogeropoulos, N., Yanni, A. E., Koutrotsios, G., & Aloupi, M. (2013). Bioactive Microconstituents and Antioxidant Properties of Wild Edible Mushrooms from the Island of Lesvos, Greece. *Food and Chemical Toxicology*, *55*, 378–385. 10.1016/j.fct.2013.01.010

Khatua, S., Paul, S., Chatterjee, A., Ray, D., Roy, A., & Acharya, K. (2013). Evaluation of Antioxidative Activity of Ethanolic Extract from Russula Delica: An in Vitro Study. *Journal of Chemical and Pharmaceutical Research*, *5*(9), 100–107.

Koca, İ., Keleş, A., & Gençcelep, H. (2011). Antioxidant Properties of Wild Edible Mushrooms. *Journal of Food Processing & Technology*, *2*(06), 1–6. 10.4172/2157-7110.1000130

Ko, H. H., Hung, C. F., Wang, J. P., & Lin, C. N. (2008). Antiinflammatory Triterpenoids and Steroids from Ganoderma Lucidum and G. Tsugae. *Phytochemistry*, *69*(1), 234–239. 10.1016/j.phytochem.2007.06.008

Kosanić, M., & Ranković, B. (2016). Bioactivity of Edible Mushrooms. *Zbornik Radova 2. / XXI Savetovanje o Biotehnologiji Sa Međunarodnim Učešćem, March*, 645–650. https://www.afc.kg.ac.rs/files/data/sb/zbornik/Zbornik_radova_SB2016_-_2.pdf

Krupodorova, T., Rybalko, S., & Barshteyn, V. (2014). Antiviral Activity of Basidiomycete Mycelia against Influenza Type A (Serotype H1N1) and Herpes Simplex Virus Type 2 in Cell Culture. *Virologica Sinica*, *29*(5), 284–290. 10.1007/s12250-014-3486-y

Lam, S. K., & Ng, T. B. (2001). First Simultaneous Isolation of a Ribosome Inactivating Protein and an Antifungal Protein from a Mushroom (Lyophyilum Shimeji) Together with Evidence for Synergism of Their Antifungal Effects. *Archives of Biochemistry and Biophysics*, *393*(2), 271–280. 10.1006/abbi.2001.2506

Lee, J. S., & Hong, E. K. (2010). Hericium Erinaceus Enhances Doxorubicin-Induced Apoptosis in Human Hepatocellular Carcinoma Cells. *Cancer Letters*, *297*(2), 144–154. 10.1016/j.canlet.2010.05.006

Lee, S. R., Jung, K., Noh, H. J., Park, Y. J., Lee, H. L., Lee, K. R., Kang, K. S., & Kim, K. H. (2015). A New Cerebroside from the Fruiting Bodies of Hericium Erinaceus and Its Applicability to Cancer Treatment. *Bioorganic and Medicinal Chemistry Letters*, *25*(24), 5712–5715. 10.1016/j.bmcl.2015.10.092

Lee, D. H., Kim, J. H., Park, J. S., Choi, Y. J., & Lee, J. S. (2004). Isolation and Characterization of a Novel Angiotensin I-Converting Enzyme Inhibitory Peptide Derived from the Edible Mushroom Tricholoma Giganteum. *Peptides*, *25*(4), 621–627. 10.1016/j.peptides.2004.01.015

Lee, S. R., Roh, H. S., Lee, S., Park, H. B., Jang, T. S., Ko, Y. J., Baek, K. H., Kim, K. H. (2018). Bioactivity-Guided Isolation and Chemical Characterization of Antiproliferative Constituents from Morel Mushroom (Morchella Esculenta) in Human Lung Adenocarcinoma Cells. *Journal of Functional Foods*, *40*(September 2017), 249–260. 10.1016/j.jff.2017.11.012

Lemieszek, M. K., Cardoso, C., Milheiro Nunes, F. H. F., Amorim de Barros, A. I. R. N., Marques, G., Pożarowski, P., & Rzeski, W. (2013). Boletus Edulis Biologically Active Biopolymers Induce Cell Cycle Arrest in Human Colon Adenocarcinoma Cells. *Food and Function*, *4*(4), 575–585. 10.1039/c2fo30324h

Li, H. (2017). Extraction, Purification, Characterization and Antioxidant Activities of Polysaccharides from Ramaria Botrytis (Pers.) Ricken. *Chemistry Central Journal*, *11*(1), 1–9. 10.1186/s13065-017-0252-x

Li, J., Cai, H., Sun, H., Qu, J., Zhao, B., Hu, X., Li, W., et al. (2020). Extracts of Cordyceps Sinensis Inhibit Breast Cancer Growth through Promoting M1 Macrophage Polarization via NF-KB Pathway Activation. *Journal of Ethnopharmacology*, *260*(February), 112969. 10.1016/j.jep.2020.112969

Li, H., Feng, Y., Sun, W., Kong, Y., & Jia, L. (2021). Antioxidation, Anti-Inflammation and Anti-Fibrosis Effect of Phosphorylated Polysaccharides from Pleurotus Djamor Mycelia on Adenine-Induced Chronic Renal Failure Mice. *International Journal of Biological Macromolecules*, *170*, 652–663. 10.1016/j.ijbiomac.2020.12.159

Li, X. Q., & Liu, A. J. (2020). Relationship between Heat Treatment on Structural Properties and Antitumor Activity of the Cold-Water Soluble Polysaccharides from Grifola Frondosa. *Glycoconjugate Journal*, *37*(1), 107–117. 10.1007/s10719-019-09894-y

Li, L. Q., Song, A. X., Yin, J. Y., Siu, K. C., Wong, W. T., & Wu, J. Y. (2020). Anti-Inflammation Activity of Exopolysaccharides Produced by a Medicinal Fungus Cordyceps Sinensis Cs-HK1 in Cell and Animal Models. *International Journal of Biological Macromolecules*, *149*, 1042–1050. 10.1016/j.ijbiomac.2020.02.022

Li, H., Wang, X., Xiong, Q., Yu, Y., & Peng, L. (2020). Sulfated Modification, Characterization, and Potential Bioactivities of Polysaccharide from the Fruiting Bodies of Russula Virescens. *International Journal of Biological Macromolecules*, *154*, 1438–1447. 10.1016/j.ijbiomac.2019.11.025

Lin, C. H., Chang, C. Y., Lee, K. R., Lin, H. J., Lin, W. C., Chen, T. H., & Wan, L. (2016). Cold-Water Extracts of Grifola Frondosa and Its Purified Active Fraction Inhibit Hepatocellular Carcinoma in Vitro and in Vivo. *Experimental Biology and Medicine*, *241*(13), 1374–1385. 10.1177/1535370216640149

Lin, K. W., Maitraie, D., Huang, A. M., Wang, J. P., & Lin, C. N. (2016). Triterpenoids and an Alkamide from Ganoderma Tsugae. *Fitoterapia*, *108*, 73–80. 10.1016/j.fitote.2015.11.003

Lin, Y. E., Wang, H. L., Lu, K. H., Huang, Y. J., Panyod, S., Liu, W. T., Yang, S. H., Chen, M. H., Lu, Y. S., & Sheen, L. Y. (2021). Water Extract of Armillaria Mellea (Vahl) P. Kumm. Alleviates the Depression-like Behaviors in Acute- and Chronic Mild Stress-Induced Rodent Models via Anti-Inflammatory Action. *Journal of Ethnopharmacology*, *265*, (September 2020). 10.1016/j.jep.2020.113395

Lindequist, U., Niedermeyer, T. H. J., & Jülich, W. D. (2005). The Pharmacological Potential of Mushrooms. *Evidence-Based Complementary and Alternative Medicine*, *2*(3), 285–299. 10.1093/ecam/neh107

Ma, L., Zhao, Y., Yu, J., Ji, H., & Liu, A. (2018). Characterization of Se-Enriched Pleurotus Ostreatus Polysaccharides and Their Antioxidant Effects in Vitro. *International Journal of Biological Macromolecules*, *111*, 421–429. 10.1016/j.ijbiomac.2017.12.152

Maity, G. N., Maity, P., Choudhuri, I., Bhattacharyya, N., Acharya, K., Dalai, S., & Mondal, S. (2019). Structural Studies of a Water Insoluble β-Glucan from Pleurotus Djamor and Its Cytotoxic Effect against PA1, Ovarian Carcinoma Cells. *Carbohydrate Polymers*, *222*(April), 114990. 10.1016/j.carbpol.2019.114990

Maity, P., Sen, I. K., Chakraborty, I., Mondal, S., Bar, H., Bhanja, S. K., Mandal, S., & Maity, G. N. (2021). Biologically Active Polysaccharide from Edible Mushrooms: A Review. *International Journal of Biological Macromolecules*, *172*, 408–417. 10.1016/j.ijbiomac.2021.01.081

Martinez-Medina, G. A., Chávez-González, M. L., Verma, D. K., Prado-Barragán, L. A., Martínez-Hernández, J. L., Flores-Gallegos, A. C., Thakur, M., Srivastav, P. P., & Aguilar, C. N. (2021). Bio-Funcional Components in Mushrooms, a Health Opportunity: Ergothionine and Huitlacohe as Recent Trends. *Journal of Functional Foods*, *77*, 1–17. 10.1016/j.jff.2020.104326

Mau, J. L., Tsai, S. Y., Tseng, Y. H., & Huang, S. J. (2005). Antioxidant Properties of Methanolic Extracts from Ganoderma Tsugae. *Food Chemistry*, *93*(4), 641–649. 10.1016/j.foodchem.2004.10.043

Meng, X., Liang, H., & Luo, L. (2016). Antitumor Polysaccharides from Mushrooms: A Review on the Structural Characteristics, Antitumor Mechanisms and Immunomodulating Activities. *Carbohydrate Research*, *424*, 30–41. 10.1016/j.carres.2016.02.008

Misaki, A., Nasu, M., Sone, Y., Kishida, E., & Kinoshita, C. (1986). Comparison of Structure and Antitumor Activity of Polysaccharides Isolated from Fukurotake, the Fruiting Body of Volvariella Volvacea. *Agricultural and Biological Chemistry*, *50*(9), 2171–2183. 10.1271/bbb1961.50.2171

Moon, S. M., Kim, J. S., Kim, H. J., Choi, M. S., Park, B. R., Kim, S. G., Ahn, H., et al. (2014). Purification and Characterization of a Novel Fibrinolytic α Chymotrypsin like Serine Metalloprotease from the Edible Mushroom, Lyophyllum Shimeji. *Journal of Bioscience and Bioengineering*, *117*(5), 544–550. 10.1016/j.jbiosc.2013.10.019

Moradali, M. F., Mostafavi, H., Ghods, S., & Hedjaroude, G. A. (2007). Immunomodulating and Anticancer Agents in the Realm of Macromycetes Fungi (Macrofungi). *International Immunopharmacology*, *7*(6), 701–724. 10.1016/j.intimp.2007.01.008

Morales, D., Rutckeviski, R., Villalva, M., Abreu, H., Soler-Rivas, C., Santoyo, S., Iacomini, M., & Smiderle, F. R. (2020). Isolation and Comparison of α- and β-D-Glucans from Shiitake Mushrooms (Lentinula Edodes) with Different Biological Activities. *Carbohydrate Polymers*, *229*, 115521. 10.1016/j.carbpol.2019.115521

Muszyńska, B., Grzywacz-Kisielewska, A., Kała, K., & Gdula-Argasińska, J. (2018). Anti-Inflammatory Properties of Edible Mushrooms: A Review. *Food Chemistry*, *243*(July 2017), 373–381. 10.1016/j.foodchem.2017.09.149

Muszyńska, B., Kała, K., Radović, J., Sułkowska-Ziaja, K., Krakowska, A., Gdula-Argasińska, J., Opoka, W., & Kundaković, T. (2018). Study of Biological Activity of Tricholoma Equestre Fruiting Bodies and Their Safety for Human. *European Food Research and Technology*, *244*(12), 2255–2264. 10.1007/s00217-018-3134-0

Nowacka, N., Nowak, R., Drozd, M., Olech, M., Los, R., & Malm, A. (2014). Analysis of Phenolic Constituents, Antiradical and Antimicrobial Activity of Edible Mushrooms Growing Wild in Poland. *LWT - Food Science and Technology*, *59*(2P1), 689–694. 10.1016/j.lwt.2014.05.041

Ragupathi, V., Stephen, A., Arivoli, D., & Kumaresan, S. (2018). Antioxidant Activity of Some Wild Mushrooms from Southern Western Ghats, India. *International Journal of Pharmaceutics & Drug Analysis*, *6*(2), 72–79. http://ijpda.com; ISSN: 2348-8948

Okin, D., & Medzhitov, R. (2012). Evolution of Inflammatory Diseases. *Current Biology*, *22*(17), R733–R740. 10.1016/j.cub.2012.07.029

Ooi, V., & Liu, F. (2000). Immunomodulation and Anti-Cancer Activity of Polysaccharide-Protein Complexes. *Current Medicinal Chemistry*, *7*(7), 715–729. 10.2174/0929867003374705

Ooi, V., & Liu, F. (2012). Immunomodulation and Anti-Cancer Activity of Polysaccharide-Protein Complexes. *Current Medicinal Chemistry*, *7*(7), 715–729. 10.2174/0929867003374705

Özyürek, M., Bener, M., Güçlü, K., & Apak, R. (2014). Antioxidant/Antiradical Properties of Microwave-Assisted Extracts of Three Wild Edible Mushrooms. *Food Chemistry*, *157*, 323–331. 10.1016/j.foodchem.2014.02.053

Pachón-Peña, G., Reyes-Zurita, F. J., Deffieux, G., Azqueta, A., López de Cerain, A., Centelles, J. J., Creppy, E. E., & Cascante, M. (2009). Antiproliferative Effect of Flavomannin-6,6'-Dimethylether from Tricholoma Equestre on Caco-2 Cells. *Toxicology*, *264*(3), 192–197. 10.1016/j.tox.2009.08.009

Pessoa, A., Miranda, C. F., Batista, M., Bosio, M., Marques, G., Nunes, F., Quinta-Ferreira, R. M., & Quinta-Ferreira, M. E. (2020). Action of Bioactive Compounds in Cellular Oxidative Response. *Energy Reports*, *6*, 891–896. 10.1016/j.egyr.2019.11.035

Petrova, R. D., Mahajna, J., Wasser, S. P., Ruimi, N., Denchev, C. M., Sussan, S., Nevo, E., & Reznick, A. Z. (2009). Marasmius Oreades Substances Block NF-KB Activity through Interference with IKK Activation Pathway. *Molecular Biology Reports*, *36*(4), 737–744. 10.1007/s11033-008-9237-0

Phan, C. W., Wong, W. L., David, P., Naidu, M., & Sabaratnam, V. (2012). Pleurotus Giganteus (Berk.) Karunarathna & K.D. Hyde: Nutritional Value and in Vitro Neurite Outgrowth Activity in Rat Pheochromocytoma Cells. *BMC Complementary and Alternative Medicine*, *12*(1). 10.1186/1472-6882-12-102

Qi, W., Zhou, X., Wang, J., Zhang, K., Zhou, Y., Chen, S., Nie, S., & Xie, M. (2020). Cordyceps Sinensis Polysaccharide Inhibits Colon Cancer Cells Growth by Inducing Apoptosis and Autophagy Flux Blockage via MTOR Signaling. *Carbohydrate Polymers*, *237*(March), 116113. 10.1016/j.carbpol.2020.116113

Rahman, M. A., Abdullah, N., & Aminudin, N. (2018). Lentinula Edodes (Shiitake Mushroom): An Assessment of in Vitro Anti-Atherosclerotic Bio-Functionality. *Saudi Journal of Biological Sciences*, *25*(8), 1515–1523. 10.1016/j.sjbs.2016.01.021

Rangsinth, P., Sillapachaiyaporn, C., Nilkhet, S., Tencomnao, T., Ung, A. T., & Chuchawankul, S. (2021). Mushroom-Derived Bioactive Compounds Potentially Serve as the Inhibitors of SARS-CoV-2 Main Protease: An in Silico Approach. *Journal of Traditional and Complementary Medicine*, *11*(2), 158–172. 10.1016/j.jtcme.2020.12.002

Reis, G. C. L., Dala-Paula, B. M., Tavano, O. L., Guidi, L. R., Godoy, H. T., & Gloria, M. B. A. (2020). In Vitro Digestion of Spermidine and Amino Acids in Fresh and Processed Agaricus Bisporus Mushroom. *Food Research International*, *137*(August), 109616. 10.1016/j.foodres.2020.109616

Ren, Z., Li, J., Song, X., Zhang, J., Wang, W., Wang, X., Gao, Z., Jing, H., Li, S., & Jia, L. (2018). The Regulation of Inflammation and Oxidative Status against Lung Injury of Residue Polysaccharides by Lentinula Edodes. *International Journal of Biological Macromolecules*, *106*, 185–192. 10.1016/j.ijbiomac.2017.08.008

Ribeiro, B., Rangel, J., Valentão, P., Baptista, P., Seabra, R. M., & Andrade, P. B. (2006). Contents of Carboxylic Acids and Two Phenolics and Antioxidant Activity of Dried Portuguese Wild Edible Mushrooms. *Journal of Agricultural and Food Chemistry*, *54*(22), 8530–8537. 10.1021/jf061890q

Ribeiro, B., Valentão, P., Baptista, P., Seabra, R. M., & Andrade, P. B. (2007). Phenolic Compounds, Organic Acids Profiles and Antioxidative Properties of Beefsteak Fungus (Fistulina Hepatica). *Food and Chemical Toxicology*, *45*(10), 1805–1813. 10.1016/j.fct.2007.03.015

Robaszkiewicz, A., Bartosz, G., Ławrynowicz, M., & Soszyński, M. (2010). The Role of Polyphenols, β -Carotene, and Lycopene in the Antioxidative Action of the Extracts of Dried, Edible Mushrooms. *Journal of Nutrition and Metabolism*, *2010*, Article ID 173274, 9 pages. 10.1155/2010/173274

Rzymski, P., Klimaszyk, P., & Benjamin, D. (2019). Comment on 'Study of Biological Activity of Tricholoma Equestre Fruiting Bodies and Their Safety for Human'. *European Food Research and Technology*, *245*(4), 963–965. 10.1007/s00217-019-03236-w

Şahin, Ö. Y., Dumanlı, M., Türkekul, I., & Tekin, Ş. (2014). Amanita Rubescens, Pleurotus Ostreatus and Verpa Bohemica: Anticancer Activities against HT29 Cell Line. *Journal of Biotechnology*, *185*, S87. 10.1016/j.jbiotec.2014.07.297

Salanță, L. C., Uifălean, A., Iuga, C., Tofană, M., Cropotova, J., Pop, O. L., Pop, C. R., Rotar, M. A., Bautista-Ávila, M., & González, C. V. (2020). Valuable Food Molecules with Potential Benefits for Human Health. *The Health Benefits of Foods – Current Knowledge and Further Development*. IntechOpen. 10.5772/intechopen.91218

Santos, D. I., Alexandre Saraiva, J. M., Vicente, A. A., & Moldão-Martins, M. (2019). *Methods for Determining Bioavailability and Bioaccessibility of Bioactive Compounds and Nutrients. Innovative Thermal and Non-Thermal Processing, Bioaccessibility and Bioavailability of Nutrients and Bioactive Compounds*. Elsevier Inc. 10.1016/B978-0-12-814174-8.00002-0

Sasidharan, S., Aravindran, S., Latha, L. Y., Vijenthi, R., Saravanan, D., & Amutha, S. (2010). In Vitro Antioxidant Activity and Hepatoprotective Effects of Lentinula Edodes against Paracetamol-Induced Hepatotoxicity. *Molecules*, *15*(6), 4478–4489. 10.3390/molecules15064478

Shao, K. D., Mao, P. W., Li, Q. Z., Li, L. D. J., Wang, Y. L., & Zhou, X. W. (2019). Characterization of a Novel Fungal Immunomodulatory Protein, FIP-SJ75 Shuffled from Ganoderma Lucidum, Flammulina Velutipes and Volvariella Volvacea. *Food and Agricultural Immunology*, *30*(1), 1253–1270. 10.1080/09540105.2019.1686467

Sheng, J., & Sun, Y. (2014). Antioxidant Properties of Different Molecular Weight Polysaccharides from Athyrium Multidentatum (Doll.) Ching. *Carbohydrate Polymers*, *108*(1), 41–45. 10.1016/j.carbpol.2014.03.011

Shomali, N., Onar, O., Cihan, A. C., Akata, I., & Yildirim, O. (2019). Antioxidant, Anticancer, Antimicrobial, and Antibiofilm Properties of the Culinary-Medicinal Fairy Ring Mushroom, Marasmius Oreades (Agaricomycetes). *International Journal of Medicinal Mushrooms*, *21*(6), 571–582. 10.1615/IntJMedMushrooms.2019030874

Silva, R.F.d., Barros, A.C.d.A., Pletsch, M., Argolo, A. C. C. M., & Araujo, B.S.d. (2010). Study on the Scavenging and Anti-Staphylococcus Aureus Activities of the Extracts, Fractions and Subfractions of Two Volvariella Volvacea Strains. *World Journal of Microbiology and Biotechnology*, *26*(10), 1761–1767. 10.1007/s11274-010-0355-1

Socaci, S. A., Fărcaş, A. C. , Vodnar, D. C., & Tofană, M. (2017). Food Wastes as Valuable Sources of Bioactive Molecules. In Shiomi, N., & Waisundara, V. (eds.), *Superfood and Functional Food – The Development of Superfoods and Their Roles as Medicine*. IntechOpen. 10.5772/66115

Sun, Z., He, Y., Liang, Z., Zhou, W., & Niu, T. (2009). Sulfation of (1→3)-β-d-Glucan from the Fruiting Bodies of Russula Virescens and Antitumor Activities of the Modifiers. *Carbohydrate Polymers*, *77*(3), 628–633. 10.1016/j.carbpol.2009.02.001

Sun, X., Zhang, T., Zhao, Y., Zhu, H., & Cai, E. (2020). Protoilludane Sesquiterpenoid Aromatic Esters from Armillaria Mellea Improve Depressive-like Behavior Induced by Chronic Unpredictable Mild Stress in Mice. *Journal of Functional Foods*, *66*(December 2019), 103799. 10.1016/j.jff.2020.103799

Tseng, Y. H., Yang, J. H., & Mau, J. L. (2008). Antioxidant Properties of Polysaccharides from Ganoderma Tsugae. *Food Chemistry*, *107*(2), 732–738. 10.1016/j.foodchem.2007.08.073

Ucar, T. M., & Karadag, A. (2019). The Effects of Vacuum and Freeze-Drying on the Physicochemical Properties and in Vitro Digestibility of Phenolics in Oyster Mushroom (Pleurotus Ostreatus). *Journal of Food Measurement and Characterization*, *13*(3), 2298–2309. 10.1007/s11694-019-00149-w

Vamanu, E. (2018). Bioactive Capacity of Some Romanian Wild Edible Mushrooms Consumed Mainly by Local Communities. *Natural Product Research*, *32*(4), 440–443. 10.1080/14786419.2017.1308365

Vamanu, E., & Nita, S. (2014). Biological Activity of Fluidized Bed Ethanol Extracts from Several Edible Mushrooms. *Food Science and Biotechnology*, 23(5), 1483–1490. 10.1007/s10068-014-0203-4

Vamanu, E.,& Pelinescu, D. (2017). Effects of Mushroom Consumption on the Microbiota of Different Target Groups – Impact of Polyphenolic Composition and Mitigation on the Microbiome Fingerprint. *LWT – Food Science and Technology*, 85, 262–268. 10.1016/j.lwt.2017.07.039

Venturini, M. E., Rivera, C. S., Gonzalez, C., & Blanco, D. (2008). Antimicrobial Activity of Extracts of Edible Wild and Cultivated Mushrooms against Foodborne Bacterial Strains. *Journal of Food Protection*, 71(8), 1701–1706. 10.4315/0362-028X-71.8.1701

Vieira, V., Marques, A., Barros, L., Barreira, J. C. M., & Ferreira, I. C. F. R. (2012). Insights in the Antioxidant Synergistic Effects of Combined Edible Mushrooms: Phenolic and Polysaccharidic Extracts of Boletus Edulis and Marasmius Oreades. *Journal of Food and Nutrition Research*, 51(2), 109–116.

Vital, A., Carolina, P., Croge, C., Gomes-da-Costa, S. M., & Matumoto-Pintro, P. T. (2017). Effect of Addition of Agaricus Blazei Mushroom Residue to Milk Enriched with Omega-3 on the Prevention of Lipid Oxidation and Bioavailability of Bioactive Compounds after in Vitro Gastrointestinal Digestion. *International Journal of Food Science and Technology*, 52(6), 1483–1490. 10.1111/ijfs.13413

Wang, M., Gao, Y., Xu, D., & Gao, Q. (2015). A Polysaccharide from Cultured Mycelium of Hericium Erinaceus and Its Anti-Chronic Atrophic Gastritis Activity. *International Journal of Biological Macromolecules*, 81, 656–661. 10.1016/j.ijbiomac.2015.08.043

Wang, J., Nie, S., Cui, S. W., Wang, Z., Phillips, A. O., Phillips, G. O., Li, Y., & Xie, M. (2017). Structural Characterization and Immunostimulatory Activity of a Glucan from Natural Cordyceps Sinensis. *Food Hydrocolloids*, 67, 139–147. 10.1016/j.foodhyd.2017.01.010

Wang, D., Sun, S. Q., Wu, W. Z., Yang, S. L., & Tan, J. M. (2014). Characterization of a Water-Soluble Polysaccharide from Boletus Edulis and Its Antitumor and Immunomodulatory Activities on Renal Cancer in Mice. *Carbohydrate Polymers*, 105(1), 127–134. 10.1016/j.carbpol.2013.12.085

Winter, H. C., Mostafapour, K., & Goldstein, I. J. (2002). The Mushroom Marasmius Oreades Lectin Is a Blood Group Type B Agglutinin That Recognizes the Galα1,3Gal and Galα1,3Gal,B1,4GlcNAc Porcine Xenotransplantation Epitopes with High Affinity. *Journal of Biological Chemistry*, 277(17), 14996–15001. 10.1074/jbc.M200161200

Wu, D. T., Lam, S. C., Cheong, K. L., Wei, F., Lin, P. C., Long, Z. R., Lv, X. J., Zhao, J., Ma, S. C., & Li, S. P. (2016). Simultaneous Determination of Molecular Weights and Contents of Water-Soluble Polysaccharides and Their Fractions from Lycium Barbarum Collected in China. *Journal of Pharmaceutical and Biomedical Analysis*, 129, 210–218. 10.1016/j.jpba.2016.07.005

Wu, Q., Zheng, C., Ning, Z. X., & Yang, B. (2007). Modification of Low Molecular Weight Polysaccharides from Tremella Fuciformis and Their Antioxidant Activity in Vitro. *International Journal of Molecular Sciences*, 8(7), 670–679. 10.3390/i8070670

Wu, J., Zhou, J., Lang, Y., Yao, L., Xu, H., Shi, H., & Xu, S. (2012). A Polysaccharide from Armillaria Mellea Exhibits Strong in Vitro Anticancer Activity via Apoptosis-Involved Mechanisms. *International Journal of Biological Macromolecules*, 51(4), 663–667. 10.1016/j.ijbiomac.2012.06.040

Xu, Z., Fu, L., Feng, S., Yuan, M., Huang, Y., Liao, J., Zhou, L., Yang, H., & Ding, C. (2019). Chemical Composition, Antioxidant and Antihyperglycemic Activities of the Wild Lactarius Deliciosus from China. *Molecules*, 24(7), 1357. 10.3390/molecules24071357

Xu, C., HaiYan, Z., JianHong, Z., & Jing, G. (2008). The Pharmacological Effect of Polysaccharides from Lentinus Edodes on the Oxidative Status and Expression of VCAM-1mRNA of Thoracic Aorta Endothelial Cell in High-Fat-Diet Rats. *Carbohydrate Polymers*, 74(3), 445–450. 10.1016/j.carbpol.2008.03.018

Xu, H., Wu, P. r., Shen, Z. y., & Chen, X. d. (2010). Chemical Analysis of Hericium Erinaceum Polysaccharides and Effect of the Polysaccharides on Derma Antioxidant Enzymes, MMP-1 and TIMP-1 Activities. *International Journal of Biological Macromolecules*, 47(1), 33–36. 10.1016/j.ijbiomac.2010.03.024

Xu, X., Yan, H., & Zhang, X. (2012). Structure and Immuno-Stimulating Activities of a New Heteropolysaccharide from Lentinula Edodes. *Journal of Agricultural and Food Chemistry*, 60(46), 11560–11566. 10.1021/jf304364c

Xu, X., & Zhang, X. (2015). Lentinula Edodes-Derived Polysaccharide Alters the Spatial Structure of Gut Microbiota in Mice. *PLoS ONE*, 10(1), 1–15. 10.1371/journal.pone.0115037

Yahia, E. M., Gutiérrez-Orozco, F., & Moreno-Pérez, M. A. (2017). Identification of Phenolic Compounds by Liquid Chromatography-Mass Spectrometry in Seventeen Species of Wild Mushrooms in Central Mexico and Determination of Their Antioxidant Activity and Bioactive Compounds. *Food Chemistry*, *226*, 14–22. 10.1016/j.foodchem.2017.01.044

Yaltirak, T., Aslim, B., Ozturk, S., & Alli, H. (2009). Antimicrobial and Antioxidant Activities of Russula Delica Fr. *Food and Chemical Toxicology*, *47*(8), 2052–2056. 10.1016/j.fct.2009.05.029

Yan, J. J., Tong, Z. J., Liu, Y. Y., Lin, Z. Y., Long, Y., Han, X., Xu, W. N., Huang, Q. H., Tao, Y. X., & Xie, B. G. (2020). The NADPH Oxidase in Volvariella Volvacea and Its Differential Expression in Response to Mycelial Ageing and Mechanical Injury. *Brazilian Journal of Microbiology*, *51*(1), 87–94. 10.1007/s42770-019-00165-4

Yin, X., Feng, T., Li, Z. H., Dong, Z. J., Li, Y., & Liu, J. K. (2013). Highly Oxygenated Meroterpenoids from Fruiting Bodies of the Mushroom Tricholoma Terreum. *Journal of Natural Products*, *76*(7), 1365–1368. 10.1021/np400359y

Ying, M., Yu, Q., Zheng, B., Wang, H., Wang, J., Chen, S., Nie, S., & Xie, M. (2020). Cultured Cordyceps Sinensis Polysaccharides Modulate Intestinal Mucosal Immunity and Gut Microbiota in Cyclophosphamide-Treated Mice. *Carbohydrate Polymers*, *235*(February), 115957. 10.1016/j.carbpol.2020.115957

Yuan, B., Zhao, C., Cheng, C., Huang, D.c., Cheng, S.j., Cao, C.j., & Chen, G.t. (2019). A Peptide-Fe(II) Complex from Grifola Frondosa Protein Hydrolysates and Its Immunomodulatory Activity. *Food Bioscience*, *32*(September), 100459. 10.1016/j.fbio.2019.100459

Zeb, M., Tackaberry, L. E., Massicotte, H. B., Egger, K. N., Reimer, K., Lu, G., Heiss, C., Azadi, P. & Lee, C. H. (2021). Structural Elucidation and Immuno-Stimulatory Activity of a Novel Polysaccharide Containing Glucuronic Acid from the Fungus Echinodontium Tinctorium. *Carbohydrate Polymers*, *258*(May 2020), 117700. 10.1016/j.carbpol.2021.117700

Zhang, C., Gao, Z., Hu, C., Zhang, J., Sun, X., Rong, C., & Jia, L. (2017). Antioxidant, Antibacterial and Anti-Aging Activities of Intracellular Zinc Polysaccharides from Grifola Frondosa SH-05. *International Journal of Biological Macromolecules*, *95*, 778–787. 10.1016/j.ijbiomac.2016.12.003

Zhang, L., Hu, Y., Duan, X., Tang, T., Shen, Y., Hu, B., Liu, A., Chen, H., Li, C., & Liu, Y. (2018). Characterization and Antioxidant Activities of Polysaccharides from Thirteen Boletus Mushrooms. *International Journal of Biological Macromolecules*, *113*, 1–7. 10.1016/j.ijbiomac.2018.02.084

Zhang, C., Li, J., Hu, C., Wang, J., Zhang, J., Ren, Z., Song, X., & Jia, L. (2017). Antihyperglycaemic and Organic Protective Effects on Pancreas, Liver and Kidney by Polysaccharides from Hericium Erinaceus SG-02 in Streptozotocin-Induced Diabetic Mice. *Scientific Reports*, *7*(1), 1–13. 10.1038/s41598-017-11457-w

Zhang, Y., Li, Q., Shu, Y., Wang, H., Zheng, Z., Wang, J., & Wang, K. (2015). Induction of Apoptosis in S180 Tumour Bearing Mice by Polysaccharide from Lentinus Edodes via Mitochondria Apoptotic Pathway. *Journal of Functional Foods*, *15*, 151–159. 10.1016/j.jff.2015.03.025

Zhang, J., Liu, M., Yang, Y., Lin, L., Xu, N., & Zhao, H., Jia, L. (2016). Purification, Characterization and Hepatoprotective Activities of Mycelia Zinc Polysaccharides by Pleurotus Djamor. *Carbohydrate Polymers*, *136*, 588–597. 10.1016/j.carbpol.2015.09.075

Zhang, W., Lu, Y., Zhang, Y., Ding, Q., Hussain, S., Wu, Q., Pan, W., Chen, Y. (2016). Antioxidant and Antitumour Activities of Exopolysaccharide from Liquid-Cultured Grifola Frondosa by Chemical Modification. *International Journal of Food Science and Technology*, *51*(4), 1055–1061. 10.1111/ijfs.13059

Zhang, Z., Wang, X., Zhao, M., & Qi, H. (2014). Free-Radical Degradation by Fe2+/Vc/H2O2 and Antioxidant Activity of Polysaccharide from Tremella Fuciformis. *Carbohydrate Polymers*, *112*, 578–582. 10.1016/j.carbpol.2014.06.030

Zhao, S., Zhao, Y. C., Li, S. H., Zhang, G. Q., Wang, H. X., & Ng, T. B. (2010). An Antiproliferative Ribonuclease from Fruiting Bodies of the Wild Mushroom Russula Delica. *Journal of Microbiology and Biotechnology*, *20*(4), 693–699. 10.4014/jmb.0911.11022

Zhao, S., Zhao, Y., Li, S., Zhao, J., Zhang, G., Wang, H., & Bun Ng, T. (2010). A Novel Lectin with Highly Potent Antiproliferative and HIV-1 Reverse Transcriptase Inhibitory Activities from the Edible Wild Mushroom Russula Delica. *Glycoconjugate Journal*, *27*(2), 259–265. 10.1007/s10719-009-9274-5

Zhang, Y., Zhou, R., Liu, F., & Ng, T. B. (2021). Purification and Characterization of a Novel Protein with Activity against Non-Small-Cell Lung Cancer in Vitro and in Vivo from the Edible Mushroom Boletus Edulis. *International Journal of Biological Macromolecules*, *174*, 77–88. 10.1016/j.ijbiomac.2021.01.149

Zheng, H. G., Chen, J. C., Weng, M. J., Ahmad, I., & Zhou, C. Q. (2020). Structural Characterization and Bioactivities of a Polysaccharide from the Stalk Residue of Pleurotus Eryngii. *Food Science and Technology*, *40*(June), 235–241. 10.1590/fst.08619

Zheng, J. Q., Wang, J. Z., Shi, C. W., Mao, D. B., He, P. X., & Xu, C. P. (2014). Characterization and Antioxidant Activity for Exopolysaccharide from Submerged Culture of Boletus Aereus. *Process Biochemistry*, *49*(6), 1047–1053. 10.1016/j.procbio.2014.03.009

Zhou, J., Chen, M., Wu, S., Liao, X., Wang, J., Wu, Q., Zhuang, M., & Ding, Y. (2020). A Review on Mushroom-Derived Bioactive Peptides: Preparation and Biological Activities. *Food Research International*, *134*(March), 109230. 10.1016/j.foodres.2020.109230

Zhou, R., Han, Y. J., Zhang, M. H., Zhang, K. R., Ng, T. B., & Liu, F. (2017). Purification and Characterization of a Novel Ubiquitin-like Antitumour Protein with Hemagglutinating and Deoxyribonuclease Activities from the Edible Mushroom Ramaria Botrytis. *AMB Express*, *7*(1), 1–11. 10.1186/s13568-017-0346-9

PART III

Analysis of mushroom

Oxidative stress prevention by edible mushrooms and their role in cellular longevity

Maja Kozarski[1] and Leo J. L. D. van Griensven[2]

University of Belgrade, Faculty of Agriculture, Republic of Serbia

Mushrooms 4 Health B.V., Horst, The Netherlands

CONTENTS

10.1 INTRODUCTION

The holistic concept of oxidative stress was developed in the 80s of the last century and is defined as any imbalance between prooxidants and antioxidants (Sies, 1985). It is connected with increasedproduction of oxidizing species or a significant decrease in the effectiveness of normal antioxidant defense mechanisms, leading to a disruption of redox signaling and control, andin-molecular damages (Sies, 1985). Following this concept, the pathogenesis and progression of many diseases, including cardiovascular and neurodegenerative diseases, immune disorders, and cancer as well as of the aging process, have been linked, directly or indirectly, to increased oxidative damage (Dalle-Donne et al., 2006; Liu & Zhang, 2018; Dhama et al., 2019). The impact of oxidative stress depends upon the magnitude of thereactive oxygen and nitrogen species (RONS)

DOI: 10.1201/9781003152583-13

status changes. Oxidative misbalance is a normal phenomenon in the body and cells are being able to overcome small disorders by antioxidant defense systems and regain their original state (Kozarski et al., 2015a; Dhama et al., 2019). If antioxidant defenses are deficient then damage mayoccur in a variety of tissues. However, moderate oxidation can induce ferroptosis and apoptosis, intense may lead to necrosis (He et al., 2017; Liu & Zhang, 2018; Dhama et al., 2019).

RONS are products of cellular aerobic metabolism and are primarily formed during oxidative phosphorylation by electron leakage of mitochondrial electron carriers and enzymes (Giustarini et al., 2009). Under pathophysiological conditions, RONS can be endogenous by produced at increased rates by numerous different processes, as shown in Figure 10.1. Likewise, oxidative species may be generated in the body in response to exposure to environmental stressors as electromagnetic radiation, oxidizing pollutants such as ozone and nitrogendioxide, and toxic chemicals (Figure 10.1). Foreign microbes invading the body and ingested foods with low nutrient value can lead to the production of cell-damaging oxidants by disturbing immune responses (Herman et al., 2018).

Oxidative stress levels can be monitored by the quantitative andqualitative measurement of biomarkers. RONS are reactive, having a short half-life and they cannot be measured directly in cells and body fluids (Giustarini et al., 2009). On the other hand, molecular products formed from

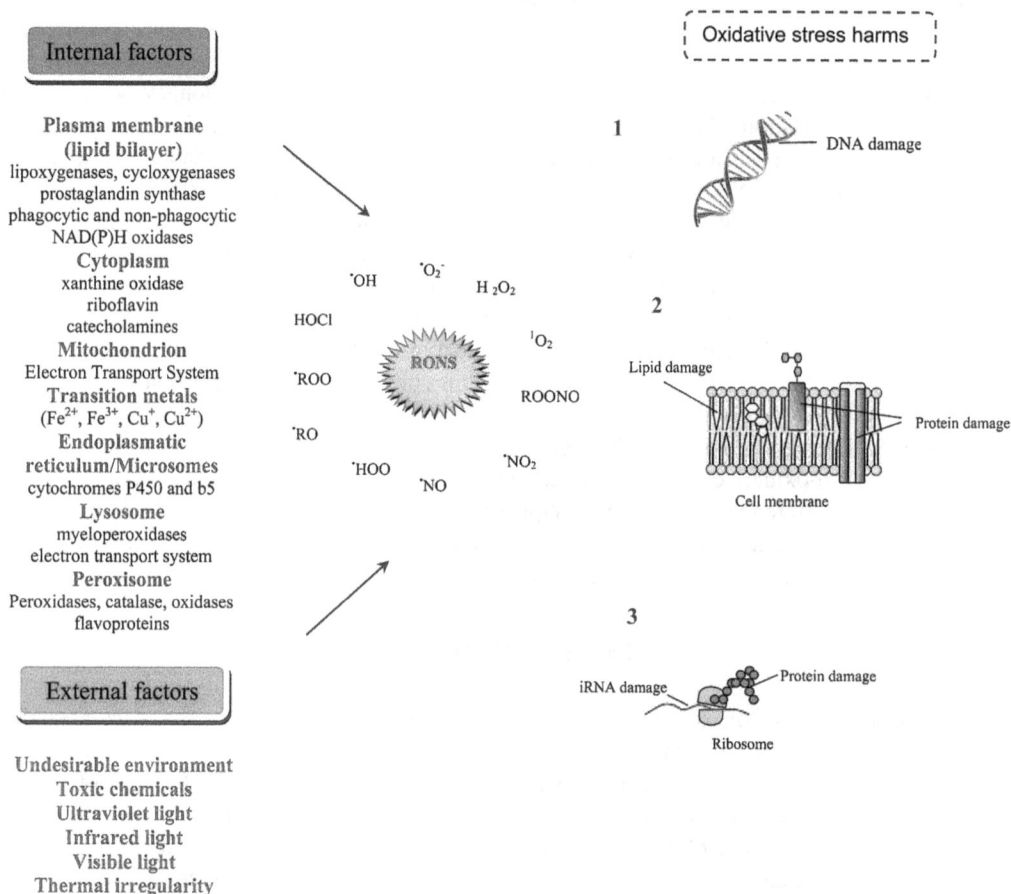

Figure 10.1 Various endogenous and exogenous factors act as stressors and lead to the generation of RONS and oxidative harm to (1,3) DNA/RNA, (2,3) proteins, and (3) lipids.

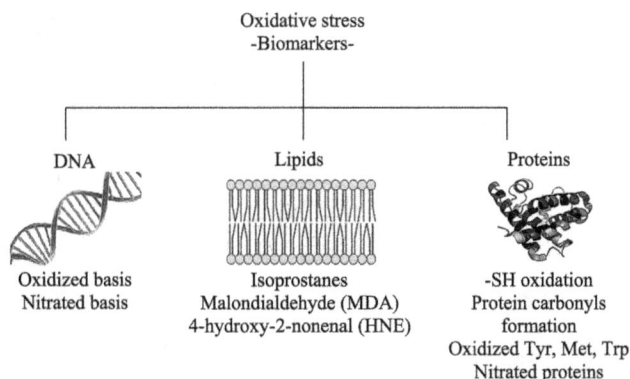

Figure 10.2 Biomarkers of oxidative stress commonly used in the study of the role of oxidative stress in diseases.

the reaction of RONS with various biomolecules are generally stable, and oxidative species have been traced by measuring their oxidation products/biomarkers (Figure 10.2), and/or active antioxidants levels, e.g. thiols, antioxidant enzymes. RONS can stimulate the generation of numerous inflammatory mediators that might stimulate immunological body defense systems and also serve as stress biomarkers (Dhama et al., 2019; Yatoo et al., 2019).

10.2 MUSHROOM STRATEGIES TO PREVENT OXIDATIVE STRESS AND PROLONG CELLULAR LONGEVITY

Mushrooms have been proposed as a novel therapy that may prevent oxidative stress damages, improve cellular longevity, and life span (Stamets & Zwickey, 2014; Kozarski et al., 2015a, Hyde et al., 2019; Podkowa et al., 2021). They represent a unique branch of medicine. Mushrooms demonstrate a magnitude and diversity of pharmacological effects that appear far greater than the responses to plants (Stamets & Zwickey, 2014). Phylogenetically, the animal kingdom is more closely related to mushrooms than it is to plants, and it can be speculated that this relationship is responsible for the enhanced medicinal benefits observed (Crespo et al., 2014).

Mushrooms metabolites: phenolics, flavonoids, glycosides, polysaccharides, sterols, terpenoids, tocopherols, ergothioneine, carotenoids, ascorbic acid, and various others exert numerous beneficial bioactive actions for humans and animals in the prevention and treatment of oxidative stress harms (Chang & Miles, 2004; Van Griensven, 2009; Ferreira et al., 2009; Khatua et al., 2013; Klaus et al., 2015; Kozarski et al., 2015b; Vunduk et al., 2015; Gargano et al., 2017; Sanchez, 2017). They can demonstrate their protective properties at different stages of the oxidation process and by different mechanisms:

1. As primary antioxidants (chain breaking), they can directly scavenge reactive RONS (Hunyadi, 2019);
2. As secondary antioxidants, they are involved in deactivation of metals, inhibition or breakdown of lipid hydroperoxides, regeneration of primary antioxidants, singlet oxygen ($^1O_2$) quenching, etc. (Hunyadi, 2019);
3. Having the potential to interact with various redox signaling pathways by modulation of the activity of redox enzymes and the generation of bioactive secondary metabolites (Hunyadi, 2019);
4. Having the potential of immune regulatory activity. The disruption of oxidative balance may be linked to the immune system since oxidative stress and inflammatory damage are multistage processes. Mushroom antioxidants may produce significant immunomodulatory mechanisms to confer

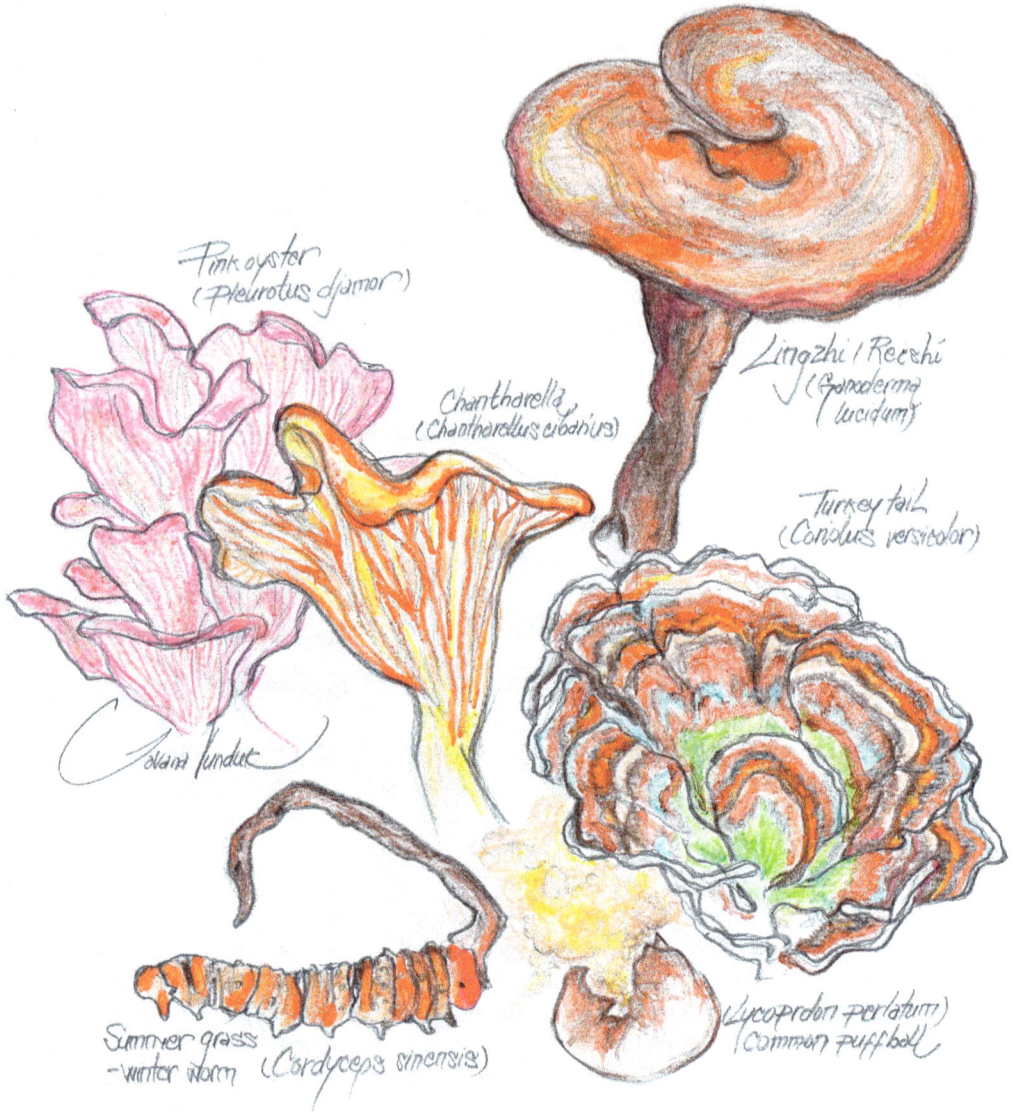

Figure 10.3 Diversity of mushrooms, painting author Dr. Jovana Vunduk.

a better oxidant/antioxidant profile. Strategic modulation of Th1/Th2 immune responses by anti-oxidant molecules may be used in therapeutic as well as prophylactic management of diseases (Ajith et al., 2017).

Various methods are used to measure the antioxidative properties of mushroom compounds or extracts that are appropriate for various levels of antioxidative activity, such as methods based on the transfer of electrons and hydrogen atoms, measurement of thermodynamic parameters: bond dissociation enthalpy, ionization potential, proton dissociation enthalpy, proton affinity, and electron-transfer enthalpy, the ability to chelate transition metal ions, the electronspin resonance (ESR) method, erythrocyte hemolysis, monitoring of enzymes activity e.g. superoxide dismutase (SOD), catalase (CAT) and glutathione peroxidase (GPx), and the quantitative and qualitative measurement of biomarkers (Markovic,2016; Dhama et al., 2019; Yatoo et al., 2019).

Table 10.1 Some studies of antioxidative properties of wild and commercially cultivated mushrooms

Mushroom species	Active constituents	References
Agaricus bisporus	polyphenols, flavonoids, ascorbic acid, nano-coated mushrooms	Zhai et al., 2021; Sami et al., 2021; Vunduk et al., 2021
Agaricus brasiliensis	aqueous extract polyphenols, glucans, polysaccharides	Kozarski et al., 2014a; Navegantes-Lima et al., 2020; Zhai et al., 2021
Cantharellus cibarius	polyphenols, flavonoids, vitamins, proteins, β-carotene, lycopene, sterols, tannins, terpenoids, anthraquinones	Kozarski et al., 2015b; Fogarasi et al., 2020; Ozturk et al., 2021
Cordycepsspp. (among most famous: *Cordyceps sinensis* grows on insect pupae at high altitude in Tibet, *Cordyceps militaris* grows on different substrates *in vitro*)	polysaccharides, proteins, plyphenols, flavonoids, water and ethanol extracts, nucleosides, sterols, cyclic peptides; bioxanthracenes; polyketides, alkaloids	Yamaguchi et al., 2000; Li et al., 2001; Fogarasi et al., 2020; Das et al., 2021
Ganoderma lucidum	polyphenols, polysaccharides, triterpenoids, sterols, α-tocopherol	Kozarski et al., 2019; Yang et al., 2019; Liu et al., 2020
Grifola frondosa	polyphenols, polysaccharides, ergothioneine	Klaus et al., 2015; Yu et al., 2020; Meng et al., 2021
Hericium erinaceus	erinacine A, polyphenols, flavonoids, polysaccharides	Chang et al., 2016; Lew et al., 2020; Valu et al., 2020
Lentinula edodes	polysaccharides (e.g. lentinan) polyphenols, flavonoids, triterpenoids, tocopherols, sterols, vitamin D2, ergothioneine	Kozarski et al., 2011; Chien et al., 2017; Spim et al., 2017; Song et al., 2020
Pleurotus ostreatus	polyphenols, ergosterol, polysaccharides, proteins	Barbosa et al., 2020; Doroski et al., 2021; Sabino Ferrari et al., 2021

10.2.1 Regulation of redox imbalance

Edible mushrooms, wild or cultivated, might be used in direct interactions with RONS as enhancers of cell antioxidant defenses and to reestablish redox balance through the primary and secondary mechanisms. The importance of mushroom antioxidant compounds is related to both their concentration and reactivity (Kozarski et al., 2015a; Bains & Chawla, 2020). The polarity of the environment and the pH value play an important role in these reactions (Markovic, 2016). In this regard, different RONS can react via different primary and secondary mechanisms in different physiological media.

10.2.1.1 Primary mechanism of action of mushroom antioxidants

The anti-RONS properties of mushrooms primary antioxidants are based on several mechanisms of the scavenging actions of free radicals ($^{\bullet}$R): hydrogen atom transfer (HAT), proton-coupled electron transfer (PCET), radical adduct formation (RAF), sequential electron-proton transfer (SEPT), sequential proton loss electron transfer (SPLET), and sequential proton loss hydrogen atom transfer (SPLHAT) (Ferreira et al., 2009; Khatua et al., 2013; Kozarski et al., 2014b; Kozarski et al. 2015a). In these reactions, a newly formed $^{\bullet}$R, which is more stable and less reactive than the previous one, is generated. In a primary mode, mushrooms act as $^{\bullet}$R deactivating antioxidants, that inhibit the initiation stage and interrupt the propagation stage by capturing $^{\bullet}$R before they reach target cells (Butnariuand Grozea, 2012).

10.2.1.1.1 Hydrogen atom transfer (HAT) and proton coupled electron transfer (PCET) mechanisms

In the HAT and PCET mechanisms, a hydrogen atom (H) is transferred to a $^\bullet R$, and the reaction can be schematically represented as:

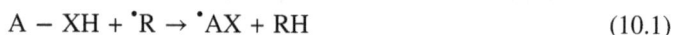

$$A - XH + {}^\bullet R \rightarrow {}^\bullet AX + RH \tag{10.1}$$

where A-XH represents an antioxidant,-XH a bond or group that supplies the H atom, $^\bullet R$ represents free radical, $^\bullet AX$ and RH represent the obtained products, respectively. $^\bullet AX$ is[a] newly formed radical, which is more stable and less reactive than the previous one. The reaction mechanism of HAT and PCET is characterized by the homolytic bond dissociation enthalpy (BDH) of the selected -XH group. A higher BDE value is attributed to a lower ability for donating an H atom (Markovic, 2016).

The HAT and PCET is the most likely mechanism for the hydroxyl radical ($^\bullet OH$) scavenging with mushroom polysaccharides. $^\bullet OH$ is the most reactive $^\bullet R$ and can be formed from superoxide radical ($^\bullet O_2^-$) and hydrogen peroxide (H_2O_2) in the presence of metal ions such as copper (Cu) or iron (Fe) (Santos-Sanchez et al., 2019). Electron paramagnetic resonance (EPR) spin trapping demonstrated high antioxidative activity against $^\bullet OH$ and $^\bullet O_2^-$ of *Agaricus brasiliensis* linear $(1\rightarrow6)$-β-D-glucan (Kozarski et al., 2014a). The results indicated that antioxidant activity against $^\bullet OH$ radicals in the Fenton system was achieved through direct scavenging. The direct RONS scavenging with carbohydrates may occur *in vivo* in higher aerobic organisms, e.g. at gastro-intestinal (GI) tract cells (Van den Ende et al., 2011). RONS are produced within the GI tract. Despite the protective barrier provided by the mucosa, ingested materials, and microbial pathogens can induce oxidative injury and GI inflammatory responses involving the epithelium and immune/inflammatory cells. High antioxidative capacities of polysaccharides from edible mushrooms can prevent lipid peroxidation (LPO) (Kozarski et al., 2011) and the pathogenesis of various GI diseases including peptic ulcers, GI cancers, and inflammatory bowel disease which is in part due to oxidative stress (Cipriani et al., 2006). Furthermore, mushroom polysaccharides via direct RONS scavenging could strengthen the skin's barrier function. It is known that despite their considerable molecular weight, they enter the *stratum corneum* and epidermis, penetrating deep into the dermis (Pillai et al., 2005). Polysaccharides do not directly enter the cell but penetrate the skin via the intercellular space. It has been suggested that they form a thin film above the *stratum corneum* and epidermis and within the dermis, they can protect from harmful effects of RONS (Pillai et al., 2005).

The ability of the polysaccharide molecules to scavenge $^\bullet R$ via HAT mechanism is conditioned by the presence of hydrogen from specific, certain monosaccharide units, and the type of their binding in sidebranches of the main chain (Tsiapali et al., 2001). The enhanced antioxidant activity of the polymers over the monomeric form may be due to the greater ease of abstraction of the anomeric hydrogen from one of the internal monosaccharide units rather than from the reducing end, shown in Figure 10.4. The HAT reaction is occurring in the neutral polysaccharides, while the PCET mechanism usually occurs in the acidic polysaccharides (Kishk & Al-Sayed, 2007). The major antioxidant effects of mushrooms are attributed to β-glycans. Except for reported antioxidant properties, α-glycans are eukaryotic nutrient components and are easily degraded by mammalian enzymes (Kozarski et al., 2015a).

Mushroom phenolic antioxidants might also be scavenging $^\bullet R$ through HAT and PCET mechanism (Urbaniak, Molski, & Szelag 2012). HAT is often assumed to be the predominant mechanism. O-H bond of hydroxyl group directly connected to a benzene ring is a preferred place of $^\bullet R$ attack, shown in Figure 10.5 (Urbaniak, Molski, & Szelag 2012; Badhani & Kakkar, 2018). The additional investigation suggested that the $^\bullet O_2^-$ scavenging reaction of the natural poly-phenolic compounds proceeds efficiently with the one-step concerted PCET or sequential PCET

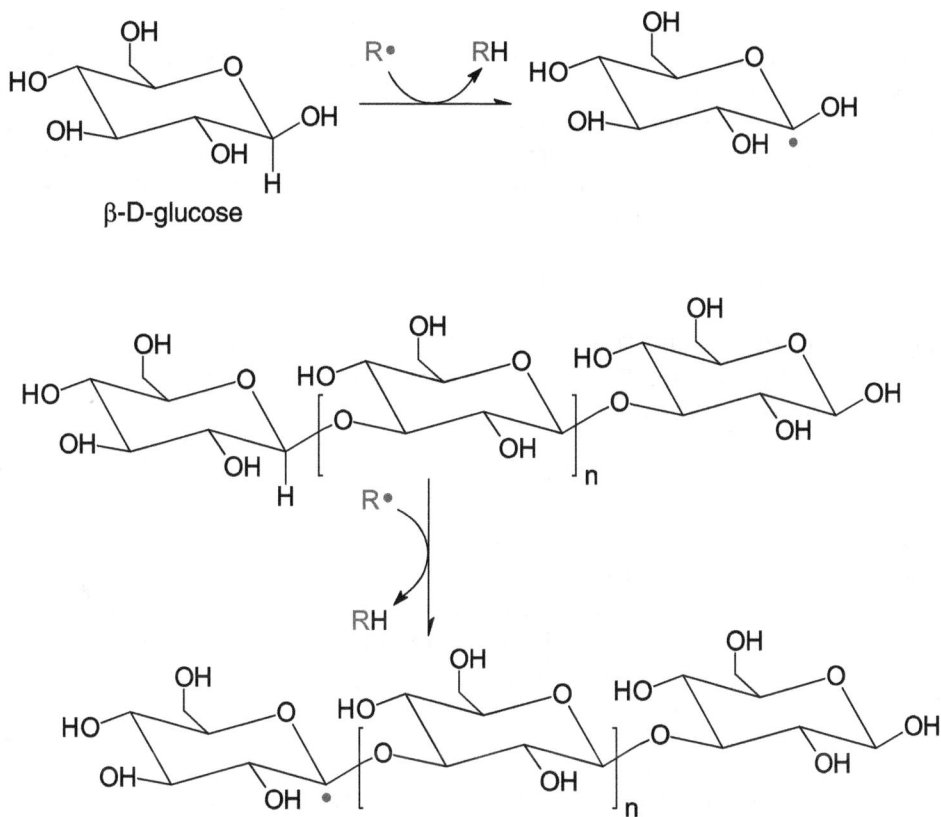

Figure 10.4 Proposed model for the increased antioxidant ability of β-1,3 glucan vs. β-D-glucose.

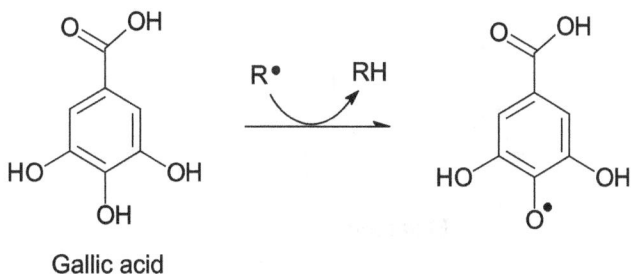

Figure 10.5 HAT route for free radical scavenging activity of gallic acid.

mechanism (Nakayama & Uno, 2015). It has been demonstrated that the catechol moiety in the polyphenols is essential to scavenge $^{\bullet}O_2^-$ via PCET (Nakayama & Uno, 2015).

The main antioxidant phenolic compounds found in mushrooms are phenolic acids (Ferreira et al., 2009). Mushroom phenolic acids can be divided into two major groups, hydroxybenzoic and hydroxycinnamic acids derivates. Hydroxybenzoic acid derivatives commonly occur in the bound form and are typically a component of a complex structure like lignins and hydrolyzable tannins. They can also be found linked to sugars or organic acids. Hydroxycinnamic acid derivatives are mainly present in the bound form, linked to cell-wall structural components, such as cellulose, lignin, and proteins, as well as associated with organic acids, such as tartaric or quinic acids (i.e.

chlorogenic acids), through ester bonds (Ferreira et al., 2009, Kozarski et al., 2015a). The most common benzoic acid derivatives with antioxidant potential found in mushrooms are reported to be *p*-hydroxybenzoic, protocatechuic, gallic, gentisic, homogentisic, vanillic, 5-sulphosalicylic, syringic, veratric, vanillin (Ferreira et al., 2009). The majority of identified cinnamic acid derivatives with antioxidant potential in mushrooms are *p*-coumaric, *o*-coumaric, caffeic, ferulic, sinapic, 3-*o*-caffeoylquinic, 4-*o*-caffeoylquinic, and 5-*o*-caffeoylquinic (Ferreira et al., 2009). Besides, the presence of ellagic and tannic acids is observed (Ferreira et al., 2009).

10.2.1.1.2 Radical adduct formation (RAF)

Opposed to the HAT and PCET mechanisms, the antioxidant does not provide its H atom but forms the radical adduct with the •R. The RAF mechanism of the reaction between antioxidant and the •R is schematically represented as:

$$A - XH + {}^{\bullet}R \rightarrow {}^{\bullet}[A - XH - R] \tag{10.2}$$

where •[A-XH-R] represents the obtained radical adduct, respectively.

This mechanism depends on the structure of the mushroom antioxidant and •R. If the mushroom antioxidant has multiple bonds e.g. phenolic acids and flavonoids (Ferreira et al., 2009), then the RAF is a possible reaction path. In addition, the properties of the •R play an important role; the electrophilic •R has the greatest potential for participation in this type of reaction. The RAF is the most likely mechanism for the scavenging of •OH with gentisic acid (Puttaraju et al., 2006; Joshi et al., 2012) and mushroom melatonin shown in Figure 10.6 (Reiter et al., 2003, Galano, 2011; Meng et al., 2017). Scavenging of a single •OH by melatonin generates the indolyl radical, which then detoxifies a second •OH to produce cyclic 3-hydroxymelatonin, as shown in Figure 10.6. The quantity of this product formed *in vivo* represents the number of •OH scavenged *in vivo*. After its formation *in vivo*, cyclic 3-hydroxymelatonin is excreted in the urine and is used as a biomarker of •OH generation (Reiter et al., 2003). Evidence is suggesting that melatonin can protect cell

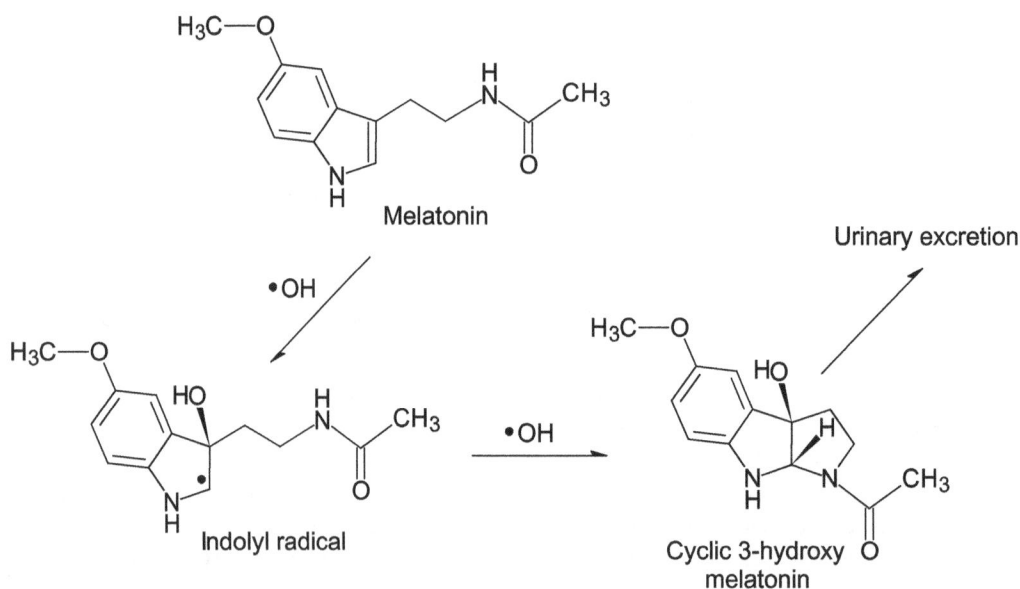

Figure 10.6   RAF pathways for the scavenging of the •OH by melatonin.

membranes against LPO twofold as effective as vitamin E, and it is five times superior to glutathione (GSH) in scavenging ˙OH (Solkoff et al., 1998). It has been shown that melatonin can increase the efficiency of mitochondrial electron transport chain (ETC) and, as a result, reduce electron leakage and ˙R formation (Reiter et al., 2003; Anisimov et al., 2006).

Melatonin may also scavenge H_2O_2 with the production of N1-acetyl-N2-formyl-5-methoxykynuramine (AFMK) which is then enzymatically converted, by CAT, to N1-acetyl-5-methoxykynuramine. These are also potential urinary excretionproducts (Reiter et al., 2003).

10.2.1.1.3 Sequential electron proton transfer (SEPT)

This mechanism is characteristic of mushrooms' phenolic antioxidants and the presence of hydroxyl groups (-OH). In the first step of this mechanism a phenolic compound loses an electron (e^-), and yields the corresponding radical cation, $Ph\text{-}OH^{•+}$:

$$Ph - OH \rightarrow Ph - OH^{•+} + e^- \tag{10.3}$$

The second step of this mechanism is deprotonation of $Ph\text{-}OH^{•+}$:

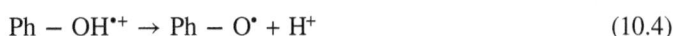

$$Ph - OH^{•+} \rightarrow Ph - O^{•} + H^+ \tag{10.4}$$

The polar solvent is necessary for the first step of this mechanism to stabilize the resulting radical cation; the polar solvent should be also protic due to the nature of thesecond step of this mechanism. Protic solvent has O-H or N-H bonds and can serve as a source of H^+. This mechanism is much less present among mushroom antioxidant compounds in comparison to HAT and PCET mechanisms because the first step is very slow. On the other hand, the radical cation can easily lose H^+ in the second step e.g. quercetin, Figure 10.7 (Justino & Vieira, 2009; Markovic, 2013). SEPT mechanism plays an important and main role in the oxidative damage of biomolecules by highly reactive radicals such as ˙OH (Litwinienkoand Ingold, 2007; Justino & Vieira, 2009).

Electron removal from the neutral quercetin, leading to the formation of the radical cations, is the first step of the SEPT mechanism shown in Figure 10.7. Ionization of quercetin requires an electron acceptor, thus being more likely to occur in the presence of such acceptors, as are proteins, or, in the presence of polar solvents, preferentially those able to establish H bonds with the flavonoid molecules, thus further stabilizing the radical cation (Litwinienko & Ingold, 2007). The radical cations of quercetin undergo favorable deprotonation in an aqueous solution in the second step. The protons involved in deprotonation in quercetin occur from the 4'-OH and the 3'-OH groups (Justino & Vieira, 2009). It is accepted that the capacity of flavonoids to scavenge RONS is governed by the presence and position of the multiple-OH groups in their structure. A double bond and carbonyl function in the heterocycle or polymerization of the nuclear structure, as it occurs in condensed mushroom tannins, increases activity by affording a more stable flavonoid radical through conjugation and electron delocalization (Kozarski et al., 2015a; Gonzalez-Paramas et al., 2018).

Flavonoids as natural mushroom antioxidants are among the essential components of the human diet. A genome-wide survey across the genome sequences of the fungal kingdom shows the presence of all the gene/protein sequences associated with the biosynthesis of flavonoids (Mohanta 2020). They included phenylalanineammonia-lyase, chalcone synthase, chalcone isomerase, flavonol reductase, dihydroflavonal-4-reductase, isoflavone reductase, leucoanthocyanidin reductase, quercetin 2,3-dioxygenase, quercetin 3-O-methyltransferase, dihydrokaempferol 4-reductase, myricetin O-methyl transferase, naringenin, naringenin 3-dioxygenase, naringenin 2,oxoglutarate 3-dioxygenase, gallate precursor, rutin-alpha-L-rhamnosidase, caffeoyl-CoA-O-methyltransferase, etc. (Mohanta, 2020).

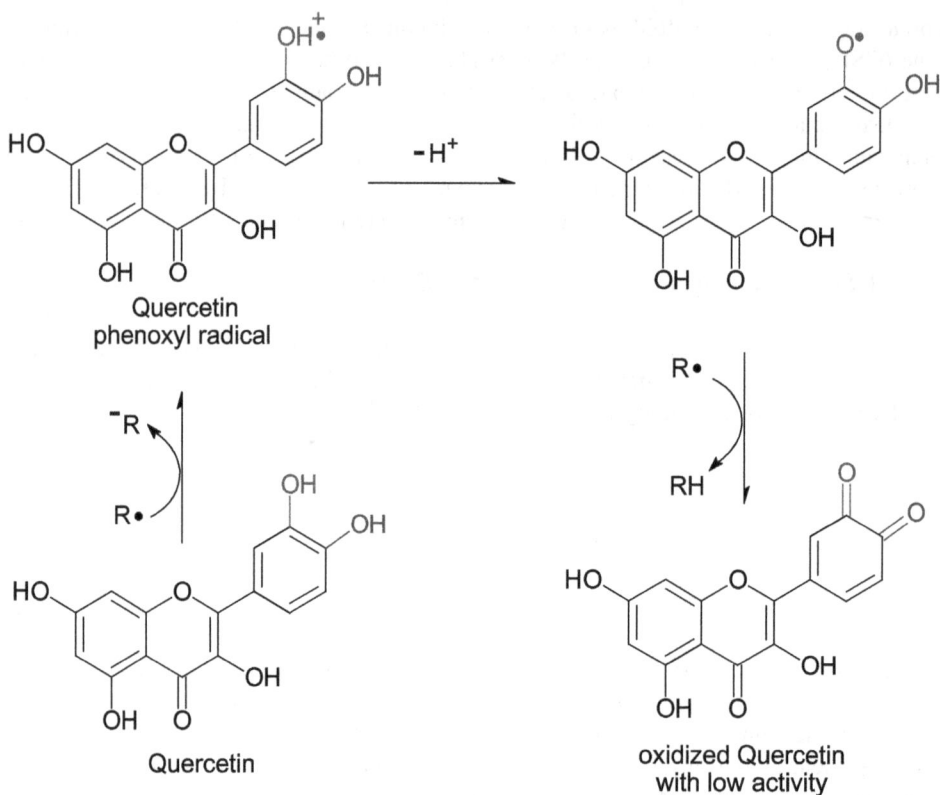

Figure 10.7 Hypothetical SEPT mechanism for ·R scavenging activity of quercetin.

10.2.1.1.4 Sequential proton loss electron transfer (SPLET)

Except for SEPT, the SPLET mechanism is particularly important for the explanation of the antioxidant activity of phenolic compounds (Foti, 2007). The SPLET mechanism is a two-step mechanism (Markovic, 2013; Litwinienko & Ingold, 2004; Litwinienko & Ingold, 2005). The first step is deprotonation of the corresponding antioxidant, followed by the antioxidant anion formation:

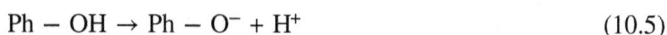

$$Ph - OH \rightarrow Ph - O^- + H^+ \qquad (10.5)$$

The anion on further on loses an electron:

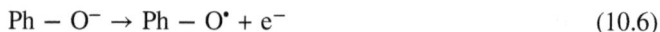

$$Ph - O^- \rightarrow Ph - O^\bullet + e^- \qquad (10.6)$$

There are two important characteristics of mushroom's phenolic antioxidants acting by SPLET: it is acid dissociation constant (K_a) and the electron-donating ability of the deprotonated antioxidant (Markovic et al., 2012). The solvent in SPLET should be polar and protic and be able to provide good solvation of the anion formed. It is expected that the SPLET mechanism is dominant in water but not in the lipid phase characteristic for biological systems (Markovic et al., 2012). SPLET has being identified as a crucial mechanism in the scavenging activity exerted by numerous mushroom phenolic and flavonoid compounds in polar environments e.g. hydroxybenzoic and dihydroxybenzoic acids (Ferreira et al., 2009; Markovic et al., 2014; Gonzalez-Paramas et al., 2018), morin

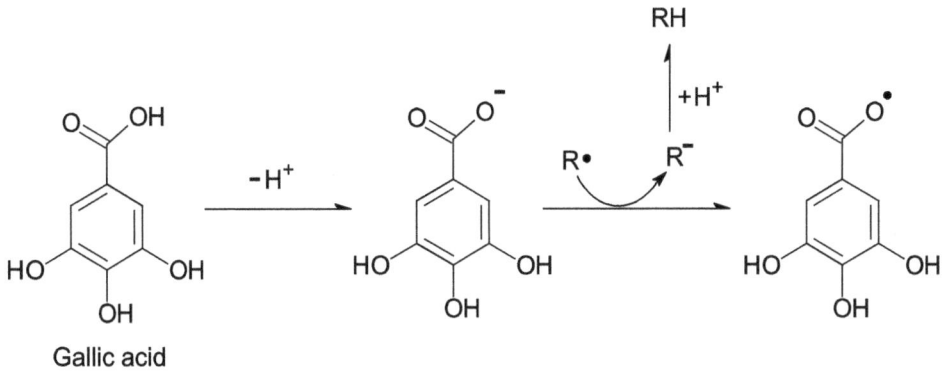

Figure 10.8 Free radical scavenging activity of gallic acid by SPLET mechanism.

(Markovic et al., 2012; Saltarelli et al., 2015), quercetin (Markovic et al., 2010; Nimseand Pal, 2015), kaempferol (Markovic, 2016), gallic acid (Badhaniand Kakkar, 2018; Djorovic et al., 2014), and vitamin E (Markovic, 2016). $^\bullet$R scavenging activity of gallic acid by SPLET mechanism is shown in Figure 10.8.

10.2.1.1.5 Sequential proton loss hydrogen atom transfer (SPLHAT)

The first step of this mechanism is identical to that one of the SPLET mechanisms (Markovic, 2016), deprotonation of the corresponding antioxidant[A-(OH)OH], followed by the antioxidant anion formation:

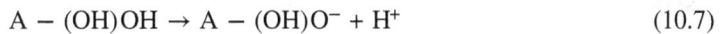

$$A - (OH)OH \rightarrow A - (OH)O^- + H^+ \tag{10.7}$$

while the second one differs. In this step the antioxidant anion further loses an H atom:

$$A - (OH)O^- + {}^\bullet R \rightarrow A - OO^{-\bullet} + RH \tag{10.8}$$

Free radical scavenging activity of gallic acid by the SPLHAT mechanism (Badhani & Kakkar, 2018) is shown in Figure 10.9.

10.2.1.2 Secondary mechanism of action of mushroom antioxidants

The secondary mushroom antioxidant mechanisms may include the deactivation of metals, inhibition of lipid hydroperoxides, regeneration of primary antioxidants, elimination of singlet oxygen ($^1O_2$) by acting as an oxygen capture, and inhibition of some oxidative enzymes such as polyphenol oxidase or lipoxygenase, which prevent oxidation reactions catalyzed by these enzymes (Ferreira et al., 2009; Butnariuand Grozea, 2012; Kozarski et al., 2015a).

10.2.1.2.1 Deactivation of metals

Mushroom antioxidants acting as metal chelators may be of beneficial use in the treatment of neurodegenerative diseases, such as Alzheimer's disease (Collin, 2019). Transition metal ions such as Cu^+ and Fe^{2+} are known to aggravate oxidative stress. These metal ions react with H_2O_2, which is a product formed by the dismutation of the $^\bullet O_2^-$ by enzyme SOD, to produce the highly reactive$^\bullet$OH (Collin, 2019).

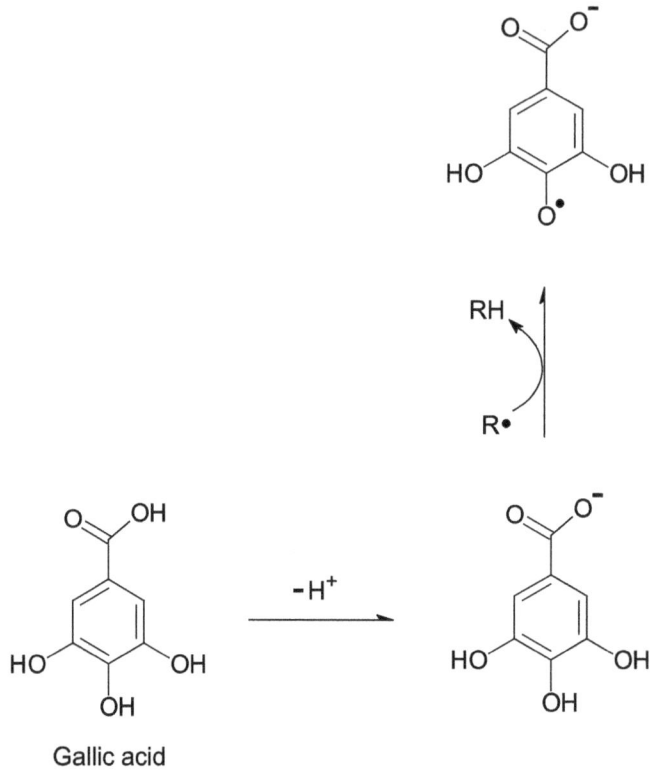

Figure 10.9 R scavenging activity of gallic acid by SPLHAT mechanism.

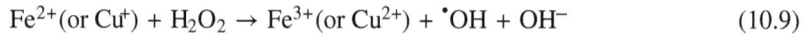

$$Fe^{2+}(or\ Cu^+) + H_2O_2 \rightarrow Fe^{3+}(or\ Cu^{2+}) + {}^{\bullet}OH + OH^- \tag{10.9}$$

Cu^{2+} generates more ${}^1O_2$ than ${}^{\bullet}OH$ upon its reaction with H_2O_2. ${}^1O_2$ is generated via two single-electron oxidation steps from HO_2^-, product of H_2O_2 deprotonation, by Cu^+ where ${}^{\bullet}O_2^-$ is an intermediate (Carrier et al., 2018).

$$Cu^{2+} + HO_2^- \rightarrow Cu^+ + {}^{\bullet}O_2^- + H^+ \tag{10.10}$$

$$Cu^{2+} + {}^{\bullet}O_2^- \rightarrow Cu^+ + {}^1O_2 \tag{10.11}$$

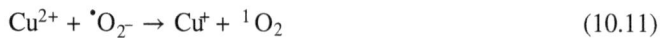

Prooxidative transition metal reactivity may be inhibited by different mushroom chelators e.g. phenolic acids, flavonoids, proteins, ascorbic acid, ergotheneine, and carbohydrates (Kozarski et al., 2015a). Mushroom chelators that exhibit antioxidative properties inhibit metal-catalyzed reactions by one or more of the following mechanisms: the deactivation of metals by the occupation of all-metal coordination sites thus inhibits the transfer of electrons, prevention of metal redox cycling, formation of insoluble metal complexes, stearic hindrance of interactions between metals, and oxidizable substrates (e.g. peroxides). The antioxidative properties of chelators can often be dependent on both metal and chelator concentrations (Kasprzak et al., 2015; Gonzalez-Paramas et al., 2018).

The most potent chelators of the transition metals among edible mushroom compounds are flavonoids and phenolic acids (Ferreira et al., 2009). Due to the presence of carbonyl (>C=) and -OH groups can coordinate metal ions and form complexes (Figure 10.10) (Kasprzak et al., 2015). In the case of flavonoids, chelating complexes with divalent cations may be formed between the 3-OH or 5-OH and 4-oxo group, or between the 6- and 7-OH in A ring or 3'-and 4'-OH in B ring, shown in Figure 10.10 (Gonzalez-Paramas et al., 2018). These complexes do not necessarily render flavonoids inactive, as the complexes retain ROS scavenging activity (Gonzalez-Paramas et al., 2018).

Quercetin, common mushroom flavonoids (Ferreira et al., 2009), possess iron-chelating andiron-stabilizing properties due to their potential binding sites via the 3-hydroxyl and 4-carbonyl group of the C ring, the 4-carbonyl-5-hydroxyl site of the A and C rings, or via the catechol moiety of the B ring (Figure 10.11), suggesting little difference between aglycones and glycosides in the ability to complex metals (Gonzalez-Paramas et al., 2018). The catecholmoiety in the B ring of quercetin is the majorcontributory site for Cu^{2+}-chelate formation. The binding of Cu^{2+} to the quercetin leads to the rapid reduction of Cu^{2+} to Cu^+accompanied by the oxidation of the quercetin tobenzoquinone-type products (Pekal et al., 2011). The preferred binding site depends on the flavonoid, the metal ion, and the pH value (Kasprzak et al., 2015).

Additional to polyphenols, ergothioneine, the betaine of thiolhistidine i.e. a compound which has both + and − charge in its structure, is an effective chelator of divalent metal ionsi ncluding Cu^{2+}, Hg^{2+}, Zn^{2+}, Cd^{2+}, Co^{2+}, Fe^{2+}, and Ni^{2+} (Cheah & Halliwell, 2012). Mushrooms are a primary source of ergothioneine containing from 400 to 2,500 mg/kg DW (Kozarski et al., 2015a). The most stable ergothioneine complex is with Cu^{2+} and has the highest complex formation constant (Cheah & Halliwell, 2012). In contrast to the generation of RONS by GSH in the presence of Cu^{2+} via the formation of a redox-active Cu(I)-[GSH]$_2$ complex, the complex of ergothioneine with Cu^{2+} is relatively stable and hence does not decompose to generate radicals (Zhu et al., 2011). Ergothioneine is concentrated in mitochondria, suggesting a specific role in protecting mitochondrial components, such as DNA, from oxidative damage associated with the mitochondrial generation of $^\bullet O_2^-$ (Kozarski et al., 2015a).

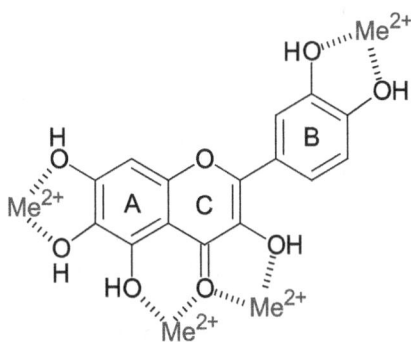

Figure 10.10 Structural requirements associated with antioxidant activity of flavonoids as metal chelators.

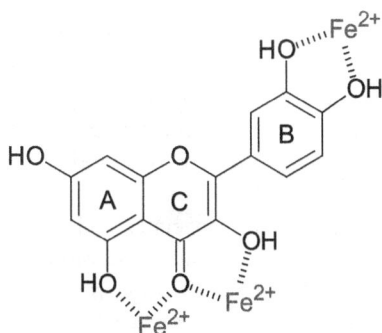

Figure 10.11 Possible chelating sites of quercetin; flavonoids can coordinate metal ions in their neutral (as shown) or anionic form.

10.2.1.2.2 Inhibition of oxidative enzymes

Mushroom polyphenols can also regulate the oxidative status of the cell by inhibiting oxidative enzymes responsible for $^{\bullet}O_2^-$ production, such as cyclooxygenase, lipoxygenase, microsomal succinoxidase, and NADPH oxidase (Gonzalez-Paramas et al., 2018). For example, the inhibition of protein kinase C was suggested to be amechanism of inhibition of NADPH by flavonoid molecules possessing a planar benzopyrone ring system with freehydroxyl substituents at the 3′, 4′ and 7-positions, such as quercetin (Gonzalez-Paramas et al., 2018).

10.2.1.2.3 Inhibition of lipid peroxidation (LPO)

LPO has been implicated in various diseases and aging, including atherosclerosis, cataract, rheumatoid arthritis, and neurodegenerative disorders (Dalle-Donne et al., 2006; Dhama et al., 2019). Consequently, the role of antioxidants has received extensive attention.

Mushrooms are a rich source of antioxidative vitamins E, D, and C, and carotenoids which are effective inhibitors in a process of LPO (Kozarski et al., 2015a). The term "vitamin E" does not refer to a single molecule. It is a family of chemically related compounds, namely tocopherols and tocotrienols, which share a common structure with a chromanol ring and isoprenic side chain. α, β, γ, and δ Tocopherols were identified and quantified in edible mushrooms (Kozarski et al., 2015a). Biologically the most active form of vitamin E in humans is α tocopherol whose main role is to protect cell membranes from LPO (Mustacich et al., 2007). The other forms of vitamin E are poorly recognized by the hepatic α tocopherol transfer protein (TTP), and they are not converted to α tocopherol by humans (Mustacich et al., 2007). α Tocopherol terminates the activity of LPO by scavenging lipid peroxyl radical ($^{\bullet}$LOO), but during this reaction is itself converted into a less reactive radical ($^{\bullet}$α Tocopherol) (Mustacich et al., 2007; Traber & Stevens, 2011):

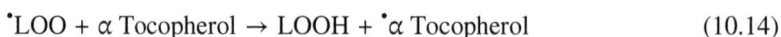

$$LH + \text{Oxidant initiator} \rightarrow {^{\bullet}}L + H_2O \tag{10.12}$$

$$^{\bullet}L + O_2 \rightarrow {^{\bullet}}LOO \tag{10.13}$$

$$^{\bullet}LOO + \alpha\ \text{Tocopherol} \rightarrow LOOH + {^{\bullet}}\alpha\ \text{Tocopherol} \tag{10.14}$$

Ascorbic acid has been shown as an effective inhibitor of LPO. In studies with human plasma lipids, it appeared that ascorbate was far more effective in inhibiting LPO initiated by a $^{\bullet}$LOO initiator than other plasma components, such as protein thiols, urate, bilirubin, and vitamin E (Traber & Stevens, 2011). In the aqueous phase, ascorbic acid can protect biomembranes against peroxidative damageby efficiently trapping $^{\bullet}$LOO before they can initiate LPO (Traber & Stevens, 2011). Ascorbic acid was detected among cultivated and wild mushrooms (Kozarski et al., 2015a).

β-carotene and lutein were found in several mushroom species (Echavarri-Erasunand Johnson, 2002; Ferreira et al., 2009). Carotenoids found in the pink-red *Cantharellus cinnabarinus* and the orange *Cantharellus friesii* contain high amounts of canthaxanthin, a pigment also found in salmon (Pilz et al., 2003). It might explain the use of chanterelles by Chinese herbalists in treating night blindness (Pilz et al., 2003). Canthaxanthin is reported to protect human tissues from oxidative damage and is sold as an antioxidant (Pilz et al., 2003).

Carotenoids act as chain-breaking antioxidants in a lipid environment, especially under low oxygen partial pressure. The extensive systems of double bonds make carotenoids susceptible to attack $^{\bullet}$LOO, resulting in the formation of inactive products (Fiedor & Burda, 2014). Carotenoids reactivity depends on the length of the conjugated double bonds chain and the characteristics of the end groups (Fiedor & Burda, 2014). Carotenoid radicals are stable by virtue of the delocalization of the unpaired electron over the conjugated polyene chain of the molecules. This delocalization also allows additional reactions that occur at many sites on the radical. The carotenoid radicals are very short-lived species (Fiedor & Burda, 2014).

Vitamin D_2 and its active metabolite 1,25-dihydroxycholecalciferol are membrane antioxidants and inhibit Fe-dependent liposomal LPO (Outila et al., 1999). These highly lipophilic compounds may accumulate in membranes and inhibit LPO. Furthermore, decreasing membrane fluidity by the membrane interaction is thought to lead to the observed inhibition of iron-dependent liposomal LPO (Outila et al., 1999). Vitamin D_2 is derived predominantly from mushrooms and yeast (Outila et al., 1999; Phillips et al., 2012). The vitamin D_2 content of mushrooms can be increased dramatically by ultraviolet (UV) irradiation, whereby it is formed from ergosterol that is present in large amounts in its cell wall (Phillips et al., 2012). It was already shown that ergocalciferol was well absorbed in humans from lyophilized and homogenized wild edible mushrooms (Outila et al., 1999). Some recent literature reports have demonstrated a possible relationship between vitamin D2 deficiency and the severe course and lethal consequences in coronavirus disease 2019 (COVID-19) (Laird et al., 2020).

10.2.1.2.4 Singlet oxygen ($^1O_2$) quenching

Carotenoids are very potent natural $^1O_2$ quenchers (Rao & Rao, 2007; Fiedor & Burda, 2014). Due to their energy levels lying close to that of $^1O_2$, they belong to the most efficient physical quenchers of $^1O_2$, both *in vitro* and *in vivo* (Fiedor & Burda, 2014). The process of $^1O_2$ quenching is very efficient, especially for carotenoids having 11 conjugated double bonds as well as lycopene, astaxanthin, β- and γ-carotene common carotenoids in *Basidiomycetes* mushrooms (Echavarri-Erasun & Johnson, 2002). In general, $^1O_2$ deactivation is based on the conversion of an excess of energy to heat via the carotenoid lowest excited triplet state ($^3Crt^*$). The possible damaging effects of excited carotenoids might be ignored mostly because of their low energy and short lifetimes (Fiedor & Burda, 2014).

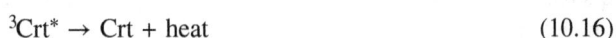

$$^1O_2 + Crt \rightarrow {}^3O_2 + {}^3Crt^* \tag{10.15}$$

$$^3Crt^* \rightarrow Crt + heat \tag{10.16}$$

where Crt represents carotenoids. Carotenoids can also act as chemical quenchers of $^1O_2$, undergoing modifications, such as oxidation or oxygenation (Fiedor & Burda, 2014).

Furthermore, lycopene was shown to cross the blood-brain barrier and be present in the central nervous system in low concentrations as a potential $^1O_2$ quencher (Rao & Rao, 2007).

10.2.1.2.5 Synergism-regeneration of primary antioxidants

Some mushroom antioxidants when acting in synergism are more effective than acting alone. Synergism means that the antioxidant effect of two or several antioxidant close molecules is higher than the arithmetical sum of the effects of all molecules (Butnariu & Grozea, 2012).

α Tocopherols, the main/primary lipid-soluble antioxidant, could be regenerated by ascorbic acid. Ascorbic acid reacts rapidly with the α tocopherol radical by reducing the ascorbate radical, via semidehydroascorbate, to ascorbate by NADH and NADPH dependent semidehydroascorbate reductase (Butnariu and Grozea, 2012; Kozarski et al., 2015a):

$$\text{Ascorbic acid} + {}^\bullet\alpha\text{ Tocopherol} \rightarrow {}^\bullet\text{Ascorbate} + \alpha\text{ Tocopherol} \tag{10.17}$$

$$\text{\textbullet Ascorbate} + NADH + H^+ \rightarrow Ascorbate + NAD^+ \tag{10.18}$$

The mechanism of synergistic action of the two antioxidants involves one that reacts with the $^\bullet R$ and a second that regenerates the first, effectively sparing. Besides synergistic action of tocopherols and ascorbic acid, mushrooms phenolic antioxidants and ascorbic acid appear to work synergistically in this way (Brewer, 2011):

$$^\bullet R + Ph - H \rightarrow R - H + Ph^\bullet \tag{10.19}$$

$$Ph^\bullet + Asc - H \rightarrow Ph - H + Asc^\bullet \tag{10.20}$$

where Ph-H represents a phenolic antioxidant, Asc-H ascorbic acid, $Ph^\bullet$ phenolic radical, and $Asc^\bullet$ stable ascorbic radical, more stable and less reactive than the previous one.

Additionally, mushroom anthocyanins can prevent the oxidation of ascorbic acids by metalions such as Cu^+ (Brewer, 2011). Anthocyanins not only chelate metal ions but also form an ascorbic acid complex that may be the basis for its antioxidative activity. The antioxidative capacity of anthocyanin is dependent on the anthocyanin itself, number and location of -OH groups, the pH of the surrounding environment, and the components of the system e.g. metals, continuous phase (Brewer, 2011).

10.2.1.3 Interaction with various redox signaling pathways

Besides the chemical structure of mushroom antioxidants that makes them able to directly scavenge RONS, the *in vivo* antioxidant action of these molecules can also be linked to their potential to interact with various redox signaling pathways by modulation of the activity of redox enzymes and/or molecules (Li et al., 2013; Lee et al., 2019).

10.2.1.3.1 Activation of nuclear factor (erythroid-derived 2)-like 2

Mushroom bioactive compounds, e.g. glucans, triterpeneganodermanondiol, ferulic acid, and quercetin, have the potential to stimulate the synthesis ofendogenous antioxidant molecules in cells via activating the nuclear factor (erythroid-derived 2)-like 2/antioxidant response element (Nrf2/ARE) pathway (Li et al., 2013; Smith et al., 2016; Meng et al., 2021). Nrf2 serves as the chief transcription factor orchestrating antioxidant response in terms of binding to ARE located in the promoter region, transcribing genes encoding phase II detoxifying antioxidant enzymes and several detoxifying proteins, shown in Figure 10.12. This upregulation includes several antioxidant enzymes such as CAT, GPx (glutathione-S-transferase), GST, paraoxonases (PONs), glutathione reductase (GR), and γ-glutamylcysteine synthetase (γ-GCS) (Lee et al., 2019).

Kelch-like enoyl-CoA hydratase-associated protein 1 (Keap1) is the redox sensor of the Nrf2/ARE system (Smith et al., 2016). The reactive sulfhydryls in the cysteine residues of Keap1 can sense oxidative stress (Figure 10.12). Once it is released from the cytosolic complex with Keap1, Nrf2 becomes phosphorylated, and it can enter the nucleus.

Cells respond to mushrooms' bioactive compounds mainly through direct interactions with receptors or enzymes involved in signal transduction, which may result in modification of the redox status of the cell and may trigger a series ofredox-dependent reactions (Kozarski et al., 2015a). Antioxidants derived from Nrf2 are defined as indirect antioxidants in that their physiological effects last longer than those being exerted by direct antioxidants, suggesting that with a relatively low dosage, they can exert sufficient efficacy. Among mushroom compounds activating the Nrf2/ARE pathway are: ergothioneine (Hseu et al., 2015; Hseu et al., 2020), many polyphenols,

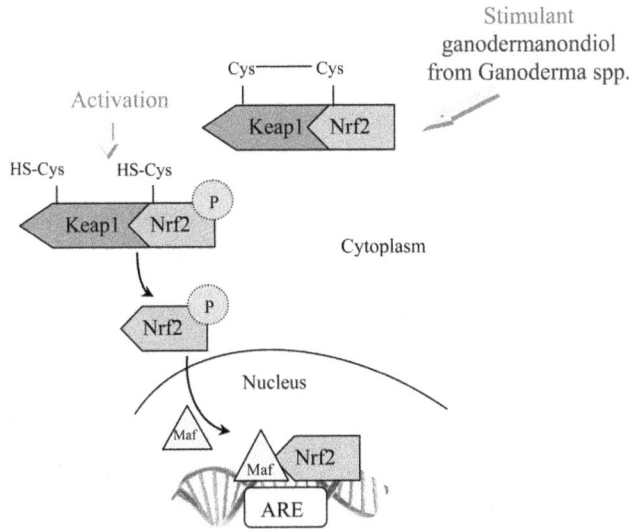

Figure 10.12 Activation of Nrf2, the chief transcription factor orchestrating antioxidant response; ARE-antioxidant response element, Maf (musculoaponeurotic fibrosarcoma) protein-transcription factor.

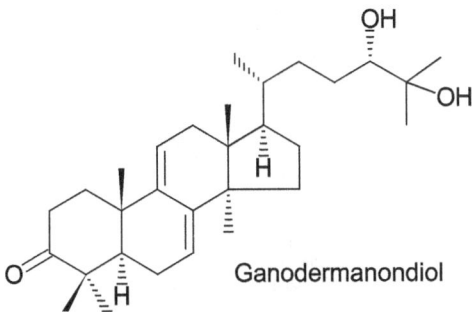

Ganodermanondiol

Figure 10.13 Ganodermanondiol from *Ganoderma lucidum* and *Ganoderma tsugae* activates Nrf2/ARE pathway.

e.g. catechin, epicatechin, quercetin, coumaric acid, ferulic acid (Ferreira et al., 2009), and biologically actively triterpenes e.g. ganodermanondiol isolated from *Ganoderma lucidum* and *Ganoderma tsugae* (Li et al., 2013). Treatment with ganodermanondiol (Figure 10.13) gradually increased GSH levels in human hepatoma cell line HepG2 via antioxidant enzyme-heme oxygenase-1 (HO-1) expression (Li et al., 2013). Ganodermanondiol significantly increased Nrf2 levels and efficiently promoted the translocation of Nrf2 into the nucleus in HepG2 cells, inducing HO-1 expression (Li et al., 2013).

10.2.1.3.2 Activation of sirtuin1 (SIRT1)

Mushrooms polyphenolic and flavonoids e.g. quercetin, and catechin, can modulate the activity of SIRT1which exerts neuroprotective effects (Chung et al., 2010). SIRT1 acts as a "rescue gene", able to repair damages caused by the action of free radicals and preventing premature death of cells. The gene also affects the mitochondria to produce greater amounts of energy what is typical for the metabolism of younger cells (Villalba & Alcain, 2012).

SIRT1 is a crucial component of multiple interconnected regulatory networks that modulate dendritic and axonal growth, as well as survival against stress. This neuronal cell-autonomous activity of SIRT1 is also important for neuronal plasticity, cognitive functions, as well as protection against aging-associated neuronal degeneration and cognitive decline (Ng et al., 2015). As a result, SIRT1 is believed to be a principal regulator of life span.

10.2.1.3.3 Suppression of nuclear factor kappa B

Among the pleiotropic activity of mushroom phenolic compounds is a modulation activity of nuclear factor kappa-light-chain-enhancer of activated B cells (NF-κB). They possess the potentiality to suppress the NF-κB signaling pathway and may suppress oxidative stress effects (Chang et al., 2016).

Inflammatory responses to a wide variety of stimuli mainly attribute to the upregulation of the proinflammatory transcription factor, NF-κB. Since it is a kind of redox-sensitive transcription factor, NF-κB responds to several stimuli including RONS (Lee et al., 2019). During homeostasis, NF-κB is sequestered in the cytoplasm by binding to the inhibitory protein called inhibitor of kappa B (IκB) (Figure 10.14). The IκB kinase (IKK) complex is the signal integration hub for NF-κB activation. It catalyzes the phosphorylation of various IκB, which leads to activation of NF-κB. After being activated by stress, e.g. diet alteration, RONS, inflammatory stimuli, cytokines, and the presence of carcinogens, NF-κB translocates to the nucleus and then induces the expression of different inflammatory cytokines and chemokines, enzymes such as cyclooxygenase 2 (COX2) and nitric oxide synthase (iNOS), and many other genes related to cellular transformation, invasion, metastasis, and inflammation (Lee et al., 2019).

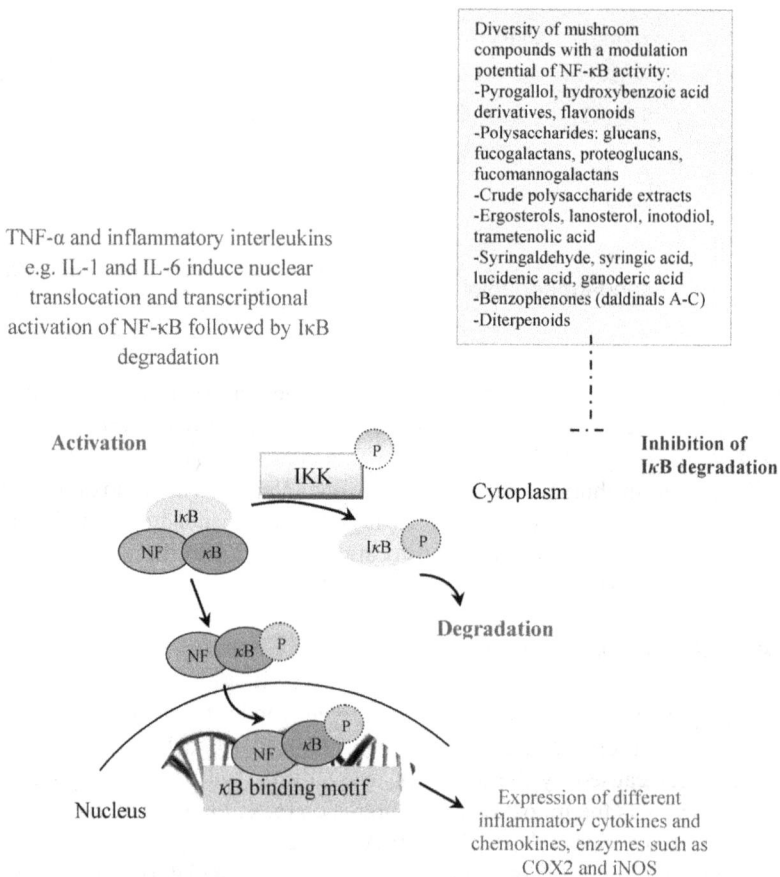

Figure 10.14 Mushroom compounds with a modulation potential of NF-κB transcription.

Mushroom phenolics and flavonoids might regulate tumor necrosis factor (TNF)-α-induced nuclear translocation and transcriptional activation of NF-κB followed by suppression of IκB degradation (Singh et al. 2013). As an example, *Hericium erinaceus* ethanol extract in a dose-depend manner remarkably inhibits the increase of the intracellular ROS production upon TNF-α-stimulation (Chang et al., 2016). *H. erinaceus* treatment suppressed overexpression of matrix metalloproteinase-9 (MMP-9) and intercellular adhesion molecule-1 (ICAM-1). It may contribute as an antioxidant through modulation of MMP-9/NF-κB pathway and shutting down of pro-inflammatory Th1 response (Chang et al., 2016).

TNF-α has been shown to both be secreted by endothelial cells and to induce intracellular RONS formation (Chen et al., 2008). These observations provide a potential mechanism by which TNF-α may activate and injure endothelial cells resulting in endothelial dysfunction (ED). ED has been implicated in atherosclerosis, hypertension, coronary artery disease, vascular complications of diabetes, chronic renal failure, insulin resistance, and hypercholesterolemia (Chen et al., 2008).

Besides phenols, mushroom glucans, fucogalactans, proteoglucans, fucomannogalactans, ergosterol, lanosterol, inotodiol, trametenolic acid, syringaldehyde, syringic acid, lucidenic acid, ganoderic acid, benzophenones (daldinals A-C), and terpenoids have a potential to shutting down NF-κB activation and to inhibit the increase of the intracellular RONS (Figure 10.14) (Elsayed et al., 2014).

10.2.2 Stimulation of the immune system in oxidative stress conditions

The immune system is extremely vulnerable to oxidant and antioxidant balance as uncontrolled ·R production can impair its function and defense mechanism (Aslani & Ghobadi, 2016). In addition to classical, reactive oxygen metabolites generated during protective function against external pathogens, activated phagocytes, neutrophils, and monocytes release the hemoprotein myeloperoxidase (MPO) into the extracellular space, where it catalyzes the oxidation of chloride anion (Cl^-) by H_2O_2 to yield hypochlorous acid (HClO) (Brambilla et al., 2008). HClO is a non-specific oxidizing and chlorinating agent that reacts rapidly with a variety of biological compounds, such as sulphydryls, polyunsaturated fatty acids (PUFAs), DNA, pyridinenucleotides, aliphatic and aromatic amino acids, and nitrogen-containing compounds (Brambilla et al., 2008). Moreover, in the reaction between chlorinated oxidants and plasma proteins catalyzed by a neutrophil enzyme, MPO advanced oxidation protein products (AOPPs) are generated. In humans, AOPPs have been linked to several diseases like chronic renal failure, diabetes mellitus, diabetic nephropathy, coronary artery diseases, and obesity (Brambilla et al., 2008).

During the inflammatory process, activation of phagocytes through the interaction of pro-inflammatory mediators, or bacterial products with specific receptors results in the assembly of the multicomponent enzyme systems NADPH oxidases that catalyze the production of large quantities of the $·O_2^-$ (Brambilla et al., 2008). Immune cells are atypical, as compared to other somatic cells; the plasma membrane of the immune cells contains high concentrations of PUFAs, which make them so susceptible to $·O_2^-$ (Chew & Park, 2004; Brambilla et al., 2008).

So, the antioxidant contents in immune cells have an important role in preserving them in a reduced environment and in protecting them against oxidative damages and immunosuppressio nas well as maintaining their suitable function.

10.2.2.1 *Mushrooms as a source of trace elements in oxidative stress prevention and maintaining efficient immune response*

Mushrooms are generally capable of accumulating trace elements e.g. zinc (Zn), copper (Cu), manganese (Mn), iron (Fe), and selenium (Se), and then become their source in the food chain (Falandysz & Borovicka, 2013). Trace elements are cofactors of antioxidant enzymes: SOD,

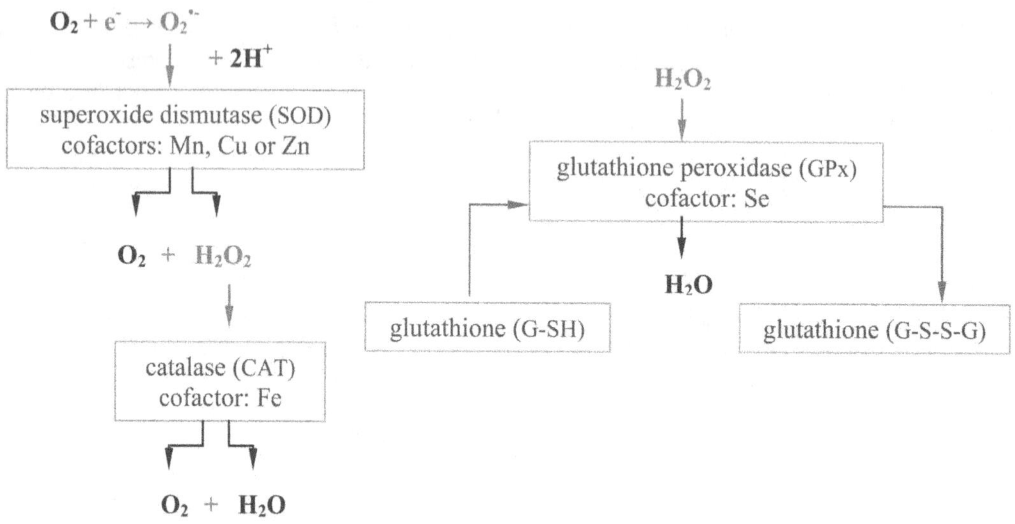

338 WILD MUSHROOMS

Figure 10.15 Mushrooms are capable of accumulating trace element cofactors of antioxidant enzymes: SOD-Zn or Cu or Mn, CAT-Fe, and GPx-Se.

MnSOD, and Cu/ZnSOD involved in scavenging of $^\bullet O_2^-$, and CAT, Fe is a cofactor involved in scavenging H_2O_2. Se, in a form of selenocysteine, is a component of the active site of GPx (Figure 10.15). In mushrooms, several Se compounds have been identified, including selenomethionine, selenocysteine, Se-methylselenocysteine, selenite, and Se-polysaccharides (Falandysz, 2008). They are designated as antioxidant micronutrients.

A particularly rich source of Se could be obtained from Se-enriched mushrooms that are cultivated on a substrate fortified with Se, as inorganic salt or selenized yeast (Falandysz, 2008). Se is essential for optimum immune responses and accumulates in tissues participating in immune responses including lymph nodes, spleen, and liver (Aslani & Ghobadi, 2016). Through the action of GPx that removes excessive reactive species produced during oxidative stress in immune cells, Se has a crucial role in the redox regulation and antioxidant function. Se enhanced the phagocytic and bactericidal activities of neutrophils in humans (Aslani & Ghobadi, 2016). In Se deficiency, neutrophils can swallow microbes but they exert less ability to kill them compared to in Se sufficient cells (Aslani & Ghobadi, 2016). It seems that decreased activity of cytosolic GPx in neutrophils is accounted for in this imperfect mechanism, which leads to the killing of neutrophils themselves by produced $^\bullet R$ during respiratory burst (Arthur et al., 2003). Moreover, Se deficiency influences the humoral system since it can reduce titers of IgM and IgG in humans (Arthur et al., 2003). Also, Se deficiency can impair the metabolism of thyroid hormone. As a result of thyroid hormone production, the high concentrations of generated H_2O_2 can be neutralized by the activity of extracellular GPx present in significant amounts in thyroid tissue (Aslani & Ghobadi, 2016). Hypothyroidism greatly influences immune function. It usually weakens neutrophils' ability to react to a challenge or foreign organisms (Arthur et al., 2003). Zn is another essential element in oxidative stress prevention (Aslani & Ghobadi, 2016). Mushrooms are capable of growing at higher levels of Zn than most other organisms (Chauhan, 2015). Zn is urgent for normalactivity of the immune system which affects both innate and acquired immune functions (Ibs & Rink, 2003). It has a role in cytosolic defense against oxidative stress through enzyme SOD activity and is a necessary cofactor forthymulin, a thymus hormone, which modulates cytokine release and induces proliferation (Aslani & Ghobadi, 2016). *In vivo*, decreased amounts of Zn can impair natural killer cell activity, phagocytosis of macrophages and neutrophils, and generation of the oxidative burst (Ibsand Rink, 2003). Moreover, it has been shown that the number of granulocytes decreases during Zn deficiency. In addition, it has been reported that Zn participates in the development of T cells since thymic atrophy occurs as a result of Zn deficiency (Ibs & Rink, 2003).

10.2.2.2 Activation of antioxidant enzymes by mushroom compounds in the prevention of autoimmunity or transplant rejection

Autoimmunity has been for decades considered the result of a breakdown in self-tolerance. Apart from the genetic defects that may predispose to autoimmune diseases, one must take into account the environmental factors that are implicated in the development of such pathologies. Among them, an important role had xenobiotics such as chemicals, drugs, and metals e.g. Fe, aluminum (Al), and manganese (Mg) (Rahal et al., 2014). Under oxidative stress, cells produce an excess of RONS that react and modify proteins. The oxidative modification of the proteins not only changes the antigenic profile of the latter but also enhances the antigenicity as well. There exist several examples of autoimmune diseases resulting from oxidative modifications of self-proteins, namely, systemic lupus erythematosus (60 kD Ro ribonucleoprotein), diabetes mellitus (high molecular weight complexes of glutamic acid decarboxylase), and diffuse scleroderma (oxidation of beta-2-glycoprotein) (Rahal et al., 2014).

A novel possible approach to modulate the immune system and preventing autoimmunity or transplant rejection is the activation of cytoprotective and antioxidant enzymes such as HO-1. HO-1, the inducible isoform of HO, is a key protein in the cell stress response and its upregulation is a common event during pro-inflammatory conditions (Hahn et al., 2020). Key proteins that regulate transcription factors of HO-1 include extracellular signal-regulated kinase (ERK), c-Jun N-terminal kinase (JNK), p38, and Akt, shown in Figure 10.16. Transcription factors that bind upstream of the initiation site of HO-1 to stimulate its mRNA expression include Nrf2, NF-κB, and activator protein 1 (AP-1) (Hahn et al., 2020). After being expressed by the regulator proteins, HO-1 affects downstream elements such as heme, bilirubin, and carbon monoxide (CO) (Hahn et al., 2020).

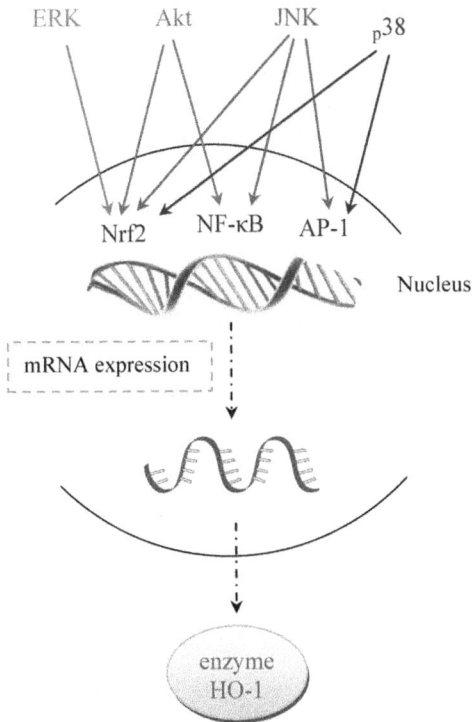

Figure 10.16 Key proteins that regulate transcription factors of HO-1 include extracellular signal-regulated kinase (ERK), c-Jun N-terminal kinase (JNK), p38, and Akt.

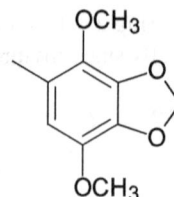

Figure 10.17 A hydrophobic benzenoid (4,7-dimethoxy-5-methyl-1,3-benzodioxole) isolated from *Antrodia camphorata* mushroom.

Under prooxidant conditions regulatory T-cells overexpress HO-1 and release CO. CO may inhibit the proliferation of effector T-cells, thus reducing the immune response and prevent autoimmunity and/or graft reaction (Aslani & Ghobadi, 2016). A hydrophobic benzenoid isolated from *Antrodia camphorata,* a mushroom used for pharmaceutical purposes, 4,7-dimethoxy-5-methyl-1,3-benzodioxole shown in Figure 10.17, exerts anti-inflammatory activity via increasing HO-1 expression without cytotoxic effects (Chen et al., 2007). It attenuates the lipopolysaccharides (LPS) induced proinflammatory factors and inducible enzyme iNOS and Toll-like receptor 4 (TLR4) protein levels expressed on $CD4^+T$ cells which play an important role in autoimmune diseases (Hahn et al., 2020; Raphael et al., 2020). *A. camphorata* hydrophobic benzenoid inhibits the NF-κB pathway, which is dependent on LPS, a prototypical TLR4 agonist (Hahn et al., 2020).

10.3 FACTORS THAT INFLUENCE THE OUTCOME OF MUSHROOM ANTIOXIDANT THERAPY

10.3.1 Physicochemical and biological properties of antioxidants and RONS

Regarding the choice and use of mushroom antioxidants, some important aspects should be carefully considered. Mushroom antioxidants differ in their physicochemical and biological properties (Chang & Miles, 2004; Ferreira et al., 2009; Van Griensven,2009; Khatua et al., 2013; Klaus et al., 2015; Kozarski et al., 2015a; Vunduk et al., 2015; Sanchez, 2017). Besides, different RONS have different affinities for substrates and, consequently, the effectiveness of a given antioxidant may depend on the RONS involved. For example, vitamin E may exert potent antioxidant activity against oxidation induced by metal ions and peroxynitrite (NO_3^-), but it may have little effect on protein modification induced by HClO (Hazell & Stocker, 1997). Likewise, carotenoids are highly efficient antioxidants when the oxidizing species is $^•O_2^-$, but their effectiveness against other RONS is rather questionable (Rao & Rao, 2007).

Moreover, the efficacy of $^•R$ scavenging depends on the distribution and localization of the antioxidant within the body; not all antioxidants can reach the same intra- or extra-cellular sites and cell membranes, because of their different chemical properties (Niki, 2000; Giustarini et al., 2009). For example, vitamin C is a potent scavenger for hydrophilic but not for lipophilic $^•R$. On the contrary, mushroom lipophilic antioxidants structurally based on caffeic, hydrocaffeic, ferulic, and hydroferulic acids mix very well between membrane lipids and lipoproteins; their levels in tissues may be higher than those detected in blood (Roleira et al., 2010).

10.3.2 Pro-oxidative effect of antioxidants

One of the essential aspects of antioxidant therapy is the potential pro-oxidative effect of antioxidants. Besides antioxidant properties, antioxidants may possess pro-oxidative capacity based upon their structure and the cellular redox context that may include increased levels of oxidant scavenging proteins or decreased levels of oxidized proteins and lipids (Wei & Van Griensven, 2008; Kozarski et al., 2015a). For instance, phenolics act as effective antioxidants but

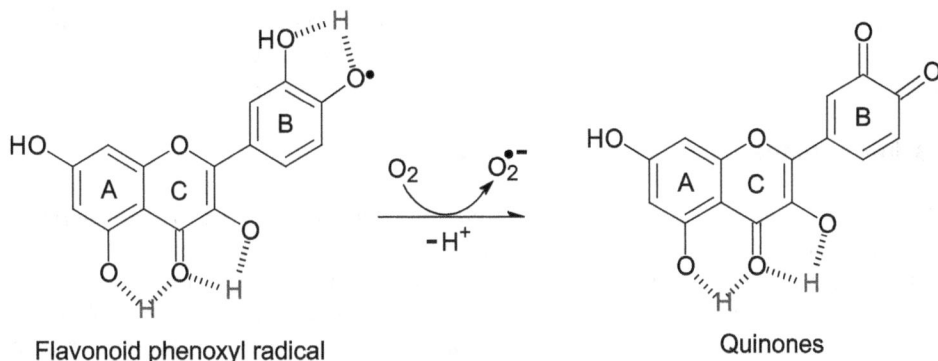

Figure 10.18 Pro-oxidant activity of flavonoids.

can also promote oxidative reactions (Decker, 1997). As pro-oxidants phenols might chelate metals in a manner that maintains or increases their catalytic activity or by their reduction, thus increasing their ability to form $\cdot$R from peroxides. In addition, most of the phenols accelerated the oxidation of glutamine synthetase in the presence of Fe^{3+}, presumably through their ability of Fe^{3+} reduction (Decker, 1997). The pro-oxidative activity of phenols through their ability to promote the oxidization of DNA and act as reductants has also been observed for quercetin and rutin, often present in mushrooms (Decker, 1997).

Polyphenols e.g. quercetin can act as photosensitizers in the generation of $^1O_2$ (Choi et al., 2003; Lagunes & Trigos, 2015). Besides the quenching ability of $^1O_2$, it has also been reported as a molecule capable of absorbing ultraviolet-visible (UV-Vis) light and able to transfer it to O_2 to generate $^1O_2$ by which quercetin would favor a pro-oxidant effect against normal cells (Choi et al., 2003; Lagunes & Trigos, 2015).

At higher doses, flavonoids may act as pro-oxidants based upon a different number of -OH groups in the B ring and the presence of transient metal ions (Amic et al., 2007). The flavonoid phenoxyl radical could interact with O_2, generating quinones and $\cdot O_2^-$, rather than terminating the chain reaction, shown in Figure 10.18. This reaction may take place in the presence of high levels of transient metal ions and may be responsible for the undesired pro-oxidant effect of flavonoids (Amic et al., 2007).

Pro-oxidant activity of flavonoids might also be induced by the formation of a complex with Fe^{3+} (El Amrani et al., 2006). The Fe^{3+}-flavonoid complex might induce DNA cleavage and damage (El Amrani et al., 2006). The damage mechanism is explained by the electron transfer in which Fe^{3+} takes one electron of the carbon atom vicinal to the 3-hydroxy-4-keto moiety, generating the Fe^{2+}-flavonoid complex, as shown in Figure 10.19. Then, the complex binds to DNA, which results in the formation of reactive oxygen species and DNA cleavage (El Amrani et al., 2006).

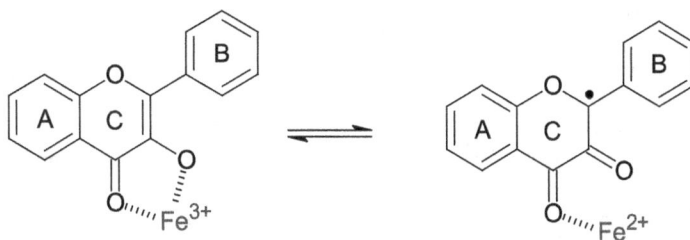

Figure 10.19 Proposed route for the generation of the Fe^{2+}-flavonoid complex.

Prooxidant activity of flavonoids may play an important role in their potency for biological action such as angiogenesis and immune-endothelial cell adhesion, which, respectively, are important processes in the development of cancer and atherosclerosis (Kim et al., 2006).

10.3.3 Interference of antioxidant supplementation with drugs

The potential interference of antioxidants with drug metabolism is one of the important facts that should be taken into account during its supplementation. It has been reported that vitamin E, vitamin C, and β-carotene administration to transplanted, cyclosporine-treated patients decreases the plasma concentration of the drug, thus diminishing its action (Blackhall et al., 2005). Besides beneficial effects in a reduction of oxidative stress, antioxidant treatment as adjuvant therapy might compromise radiation treatment efficacy in cancer patients (Harvie, 2014). For example, β-carotene increases the risk of lung and stomach cancer, vitamin E increases prostate cancer and colorectal adenoma, and Se reduces gastric and lung cancer in populations with low Se levels but increases rates in those with higher levels. Both β-carotene and vitamin E supplementation increase overall mortality in cancer patients (Harvie, 2014).

10.4 CONCLUSION

Mushrooms, distinct from plants and animals, possess unique primary and secondary metabolites that may prevent oxidative stress damages and improve cellular longevity and life span. The efficiency of oxidative stress prevention of mushroom metabolites is linked to their potential to interact with various reactive species and redox signaling pathways by modulation of the activity of redox enzymes and the generation of bioactive secondary metabolites. Mushroom antioxidant compounds have the potential for immune regulatory activity. The unique attributes in oxidative stress prevention are synergistic action when they combine.

As a consequence of an increasing world, population mankind is faced with the problem of diminishing quality of human health. Mushroom bioactive compounds are particularly of interest to the present generation because they have the potential to substantially reduce the expensive, high-tech, disease treatment approaches presently being employed in health care.

Standardization of antioxidant dietary supplements from mushrooms is in rapid progress. There is intensive research to determine which antioxidant components are more effective, less costly, and have a higher safety profile. Mushrooms have been used for pharmacological purposes for centuries by various cultures, but only recently has science begun to recognize what the ancient people knew long ago.

REFERENCES

Ajith, Y., Dimri, U., Dixit, S. K., et al. (2017). Immunomodulatory basis of antioxidant therapy and its future prospects: An appraisal. *Inflammopharmacol, 25,* 487–498.

Amic, D., Davidovic-Amic, D., Beslo, D.,Rastija, V., Lucic, B., & Trinajstic, N. (2007). SAR and QSAR of the antioxidant activity of flavonoids. *Curr Med Chem, 14,* 827–845.

Anisimov, V. N., Popovich, I. G., Zabezhinski, M. A., Anisimov, S. V., Vesnushkin, G. M., & Vinogradova, I. A. (2006). Melatonin as antioxidant, geroprotector and anticarcinogen. *Biochim Biophys Acta, 1757,* 573–589.

Arthur, J. R., McKenzie, R. C., & Beckett, G. J. (2003). Selenium in the immune system. *J Nutr, 133,* 1457S–1459SS.

Aslani, B. A., & Ghobadi, S. (2016). Studies on oxidants and antioxidants with a brief glance at theirrelevance to the immune system. *Life Sci*, *146*, 163–173.

Badhani, B., & Kakkar, R. (2018). Influence of intrinsic and extrinsic factors on the antiradical activity of gallic acid: A theoretical study. *Struct Chem*, *29*, 359–373.

Bains, A., & Chawla, P. (2020). In vitro bioactivity, antimicrobial and anti-inflammatory efficacy of modified solvent evaporation assisted Trametes versicolor extract. *3 Biotech*, *10*(9), 1–11.

Barbosa, J. R., Freitas, M. M. S., Oliveira, L. C., et al. (2020). Obtaining extracts rich in antioxidant poly-saccharides from the edible mushroom *Pleurotusostreatus* using binary system with hot water and supercritical CO_2. *Food Chem*, *330*, 127173.

Blackhall, M. L., Fassett, R. G., Sharman, J. E., Geraghty, D. P., & Coombes, J. S. (2005). Effects of antioxidant supplementation on blood cyclosporin A and glomerular filtration rate in renal transplant recipients. *Nephrol Dial Transplant*, *20*, 1970–1975.

Brambilla, D., Mancuso, C., Scuderi, M. R., et al. (2008). The role of antioxidant supplements in immune system, neoplastic, and neurodegenerative disorders: A point of view for an assessment of the risk/ benefit profile. *Nutr J*, *7*, 29.

Brewer, M.S. (2011). Natural antioxidants: Sources, compounds, mechanisms of action, and potential ap-plications. *Compr Rev Food Sci Food Saf*, *10*, 221–247.

Butnariu, M., & Grozea, I. (2012). Antioxidant (antiradical) compounds. *J BioequivAvailab*, *4*, 6.

Carrier, A., Nganou, C., Oakley, D., et al. (2018). Selective generation of singlet oxygen in chloride ac-celerated copper Fenton chemistry. *ChemRxiv*. Preprint. 10.26434/chemrxiv.7364225.v1

Chang, S-T., & Miles, P. G. (2004). *Mushrooms: Cultivation, Nutritional Value, Medicinal Effect, and Environmental Impact*. Boca Raton: CRC Press/Taylor & Francis.

Chang, H. C., Yang, H. L., Pan, J. H., et al. (2016). *Hericiumerinaceus*inhibits TNF-α-induced angiogenesis and ROS generation through suppression of MMP-9/NF-κB signaling and activation of Nrf2-mediated antioxidantgenes in human EA.hy926 endothelial cells. *Oxid Med Cell Longev*, *2016*, 8257238.

Chauhan, M. (2015). Estimation of accumulation of zinc content in mushroom and soil by atomic absorption spectroscopy. *IJISET*, *2*, 424–428.

Cheah, I. K., & Halliwell, B. (2012). Ergothioneine; antioxidant potential, physiological function and role in disease. *BiochimBiophys Acta1*, *822*, 784–793.

Chen, J. J., Lin, W. J., Liao, C. H., & Shieh, P. C. (2007). Anti-inflammatory benzenoids from *Antrodiacamphorata*. *J Nat Prod*, *70*, 989–992.

Chen, X., Andresen, B. T., Hill, M., Zhang, J., Booth, F., & Zhang, C. (2008). Role of reactive oxygen species in tumor necrosis factor-alpha induced endothelial dysfunction. *CurrHypertens Rev*, *4*, 245–255.

Chew, B. P., & Park, J. S. (2004). Carotenoid action on the immune response. *J Nutr*, *134*, 257S–261SS.

Chien, R. C., Yang, S. C., Lin, L. M., & Mau, J. L. (2017). Anti-inflammatory and antioxidant properties of pulsed light irradiated *Lentinula edodes*. *JFood ProcessPreserv*, *41*, e13045.

Choi, E. J., Chee, K. M., & Lee, B. H. (2003). Anti- and prooxidant effects of chronic quercetin adminis-tration in rats. *Eur J Pharmacol*, *482*, 281–285.

Chung, S., Yao, H., Caito, S., Hwang, J. W., Arunachalam, G., & Rahman, I. (2010). Regulation of SIRT1 in cellular functions: Role of polyphenols. *Arch BiochemBiophys*, *501*, 79–90.

Cipriani, T. R., Mellinger, C. G., de Souza, L. M., et al. (2006). A polysaccharide from a tea (infusion) of *Maytenusilicifolia*leaves with anti-ulcer protective effects. *J Nat Prod*, *69*, 1018–1021.

Collin, F. (2019). Chemical basis of reactive oxygen species reactivity and involvement in neurodegenerative diseases. *Int J Mol Sci*, *2019*, *20*, 2407.

Crespo, A., Divakar, P. K., & Lumbsch, T. (2014). Fungi: Hyperdiversity closer to animals than to plants. In: *The Tree of Life*, P. Vargas & R. Zardoya, ed. Sunderland: Sinauer Associates, pp. 170–181.

Dalle-Donne, I., Rossi, R., Ceciliani, F., Giustarini, D., Colombo, R., & Milzani, A (2006). Proteins as sensitive biomarkers of human conditions associated with oxidative stress. In: *Redox Proteomics: From Protein Modifications to Cellular Dysfunction and Diseases*, I. Dalle-Donne, A. Scaloni, & D. A. Butterfield, ed. Hoboken: John Wiley & Sons, pp. 487–525.

Das, G., Shin, H. S., & Leyva-Gomez, G. (2021). *Cordyceps* spp.: A review on its immune-stimulatory and other biological potentials. *Front Pharmacol*, *11*, 602364.

Decker, E. A. (1997). Phenolics: Prooxidants or antioxidants? *Nutr Rev*, *55*, 396–398.

Dhama, K., Latheef, S. K., Dadar, M., et al. (2019). Biomarkers in stress related diseases/disorders: Diagnostic, prognostic, and therapeutic values. *Front Mol Biosci*, *6*, 91.

Djorovic, J., Markovic, J.M.D., Stepanic, V., et al. (2014). Influence of different free radicals on scavenging potency of gallic acid. *J Mol Model*, *20*, E2345.

Doroski, A., Klaus, A., Kozarski, M., et al. (2021). The influence of grape pomace substrate on quality characterization of *Pleurotusostreatus*-total quality index approach. *J Food Process Preserv*, *45*, e15096.

Echavarri-Erasun, C., & Johnson, E. A. (2002). Fungal carotenoids. *Applied Mycology and Biotechnology*, *2*, 45–85.

El Amrani, F. B. A., Perello, L., Real, J. A., et al. (2006). Oxidative DNA cleavage induced by an iron(III) flavonoid complex: Synthesis, crystal structure and characterization of chlorobis (flavonolato) (methanol) iron(III) complex. *J InorgBiochem*, *100*, 1208–1218.

Elsayed, A. E., Enshasy, H. E., Wadaan, M. A. M., & Aziz, R. (2014). Mushrooms: A potential natural source of anti-inflammatory compounds for medical applications. *MediatorsInflamm*, *2014*, ID 805841.

Falandysz, J. (2008). Selenium in edible mushrooms. *J Environ Sci Health*, *C26*, 256–299.

Falandysz, J., & Borovicka, B. (2013). Macro and trace mineral constituents and radionuclides in mushrooms: Health benefits and risks. *Appl Microbiol Biotechnol*, *97*, 477–501.

Ferreira, I. C. F. R., Barros, L., & Abreu, R. M. V. (2009). Antioxidants in wild mushrooms. *Curr Med Chem*, *16*, 1543–1560.

Fiedor, J., & Burda, K. (2014). Potential role of carotenoids as antioxidants in human health and disease. *Nutrients*, *6*, 466–488.

Fogarasi, M., Diaconeasa, Z. M., Pop, C. R., et al. (2020). Elemental composition, antioxidant and antibacterial properties of some wild edible mushrooms from Romania. *Agronomy*, *10*, 1972.

Foti, M.C. (2007). Antioxidant properties of phenols. *J Pharm Pharmacol*, *59*, 1673–1685.

Galano, A. (2011). On the direct scavenging activity of melatonin towards hydroxyl and a seriesof peroxyl radicals. *Phys Chem Chem Phys*, *13*, 7178–7188.

Gargano, M. L., Van Griensven, L. J. L. D., Isikhuemhen, O. S., Lindequist, U., Venturella, G., Wasser, S. P., et al. (2017). Medicinal mushrooms: Valuable biological resources of high exploitation potential. *Plant Biosyst*, *151*, 548–565.

Giustarini, D., Dalle-Donne, I., Tsikas, D., & Rossi, R. (2009). Oxidative stressand human diseases: Origin, link, measurement, mechanisms, and biomarkers. *Crit Rev Clin Lab Sci*, *46*, 241–281.

Gonzalez-Paramasa, A. M., Ayuda-Durana, B., Martineza, S., Gonzalez-Manzano, S., & Santos-Buelga, C. (2018). The mechanisms behind the biological activity of flavonoids. *CurrMedChem*, *25*, 1–10.

Hahn, D. Shin, S. H., & Bae, J. S. (2020). Natural antioxidant and anti-inflammatory compounds in foodstuff or medicinal herbs inducing heme oxygenase-1 expression. *Antioxidants*, *9*, 1191.

Harvie, M. (2014). Nutritional supplements and cancer: Potential benefits and proven harms. *Am Soc Clin Oncol EducBook*, *34*, e478–e486.

Hazell, L. J., & Stocker, R. (1997). Alpha-tocopherol does not inhibit hypochlorite-induced oxidation of apolipoprotein B-100 of low-density lipoprotein. *FEBS Lett*, *414*, 541–544.

He, L., He, T., Farrar, S., Ji, L., Liu, T., & Ma, X. (2017). Antioxidants maintain cellular redox homeostasis by elimination of reactive oxygen species. *Cell PhysiolBiochem*, *44*, 532–553.

Herman, F., Westfall, S., Brathwaite, J., & Pasinetti, G. M. (2018). Suppression of presymptomatic oxidative stress and inflammation in neuro degeneration by grape-derived polyphenols. *Front Pharmacol*, *9*, 867.

Hseu, Y. C., Gowrisankar, Y. V., Chen, X. Z., Yang, Y. C., & Yang., H. L. (2020). The antiaging activity of ergothioneine in uva-irradiated human dermal fibroblasts via the inhibition of the AP-1 pathway and the activation of Nrf2-mediated antioxidant genes. *Oxid Med Cell Longev*, *2020*, 2576823.

Hseu, Y. C., Lo, H. W., Korivi, M., Tsai, Y. C., Tang, M. J., & Yang, H. L. (2015). Dermato-protective properties of ergothioneine through induction of Nrf2/ARE-mediated antioxidant genes in UVA-irradiated human keratinocytes. *Free Radic Biol Med*, *86*, 102–117.

Hunyadi, A. (2019). The mechanism(s) of action of antioxidants: Fromscavenging reactive oxygen/nitrogen species toredox signaling and the generation of bioactive secondary metabolites. *Med Res Rev*, *39*, 2505–2533.

Hyde, K. D., Xu, J., Rapior, S., et al. (2019). The amazing potential of fungi: 50 ways we can exploit fungi industrially. *Fungal Diversity*, *97*, 1–136.

Ibs, K. H., & Rink, L. (2003). Zinc-altered immune function. *J Nutr*, *133*, 1452s–1456ss.

Joshi, R., Gangabhagirathi, R., Venu, S., Adhikari, S., & Mukherjee, T. (2012). Antioxidant activity and free radical scavenging reactions of gentisic acid: in-vitro and pulse radiolysis studies. *Free Radic Res, 46,* 11–20.

Justino, G. C., & Vieira, A. J. S. C. (2009). Antioxidant mechanisms of quercetin and myricetin in the gas phase and in solution – A comparison and validation of semi-empirical methods. *J Mol Model, 16,* 863–876.

Kasprzak, M. M., Erxleben, A., & Ochockia, J. (2015). Properties and applications of flavonoid metalcomplexes. *RSC Adv, 5,* 45853.

Khatua, S., Paul, S., & Acharya, K. (2013). Mushroom as the potential source of new generation of antioxidant: A review. *Research J Pharm and Tech, 6,* 496–505.

Kim, J. D., Liu, L., Guo, W., & Meydan, M. (2006). Chemical structure of flavonols in relation to modulationof angiogenesis and immune-endothelial cell adhesion. *J NutrBiochem, 17,* 165–176.

Kishk, Y. F. M., & Al-Sayed, H. M. A. (2007). Free-radical scavenging and antioxidative activities of somepolysaccharides in emulsions. *LWT Food Sci Technol, 40,* 270–277.

Klaus, A., Kozarski, M., Vunduk, J., et al. (2015). Biological potential of extracts of the wild edible Basidiomycete mushroom *Grifolafrondosa. Food Res In, 67,* 272–283.

Kozarski, M., Klaus, A., Jakovljevic, D., et al. (2014a). Dietary polysaccharide extracts of *Agaricus brasiliensis* fruiting bodies: Chemical characterization and bioactivities at different levels of purification. *Food Res In, 64,* 53–64.

Kozarski, M., Klaus, A., Jakovljevic, D., et al. (2015a). Antioxidants of edible mushrooms. *Molecules, 20,* 19489–19525.

Kozarski, M., Klaus, A., Jakovljevic, D., Todorovic, N., Wan-Mohtar, W. A. A. Q. I., & Niksic, M. (2019). *Ganoderma lucidum* as a cosmeceutical: Antiradical potential and inhibitory effect on hyper pigmentation and skin extracellular matrix degradation enzymes. *Arch Biol Sci, 71,* 253–264.

Kozarski, M., Klaus, A., Niksic, M., Jakovljevic, D., Helsper, J. P. F. G., & Van Griensven, L. J. L. D. (2011). Antioxidative and immuno modulating activities of polysaccharide extracts ofthe medicinal mushrooms *Agaricus bisporus, Agaricus brasiliensis, Ganoderma lucidum* and *Phellinus linteus. Food Chem, 129,* 1667–1675.

Kozarski, M. S., Klaus, A. S., Niksic, M. P., Van Griensven, L. J. L. D., Vrvic, M. M., & Jakovljevic, D. M. (2014b). Polysaccharides of higher fungi: Biological role, structure and antioxidative activity. *Hem Ind, 68,* 305–320.

Kozarski, M., Klaus, A., Vunduk, J., et al. (2015b). Nutraceutical properties of the methanolic extract of edible mushroom Cantharellus cibarius (Fries): Primary mechanisms. *Food Funct, 6,* 1875–1886.

Lagunes, I., & Trigos A. (2015). Photo-oxidation of ergosterol: Indirect detection of antioxidants photosensitizers or quenchers of singlet oxygen. *J PhotochemPhotobiolB, 145,* 30–34.

Laird, E., Rhodes, J., & Kenny, R. A. (2020). Vitamin D and inflammation-potential implications for severity of COVID-19. *Ir Med J, 113,* 1–3.

Lee, M. T., Lin, W. C., & Lee, T. T. (2019). Potential crosstalk of oxidative stress and immune response in poultry through phytochemicals. *Asian-Australas J Anim Sci, 32,* 309–319.

Lew, S. Y., Lim, S. H., Lim, L. W., & Wong, K. H. (2020). Neuroprotective effects of *Hericiumerinaceus* (Bull.: Fr.) Pers. against high-dosecorticosterone-induced oxidative stress inPC-12 cells. *BMC Complement Med Ther, 20,* 340.

Li, B., Lee, D. S., Kang, Y., Yao, N. Q., An, R. B., & Kim, Y. C. (2013). Protective effect of ganodermanondiol isolated from the Lingzhi mushroom against tert-butyl hydroperoxide-induced hepatotoxicity through Nrf2-mediated antioxidant enzymes. *Food Chem Toxicol, 53,* 317–324.

Li, S. P., Li, P., Dong, T. T. X., & Tsim, K. W. (2001). Anti-oxidation activity of different types of natural *Cordyceps sinensis* and cultured *Cordyceps* mycelia. *Phytomedicine, 8,* 207–212.

Litwinienko, G., & Ingold, K. U. (2004). Abnormal solvent effects on hydrogen atom abstraction. 2. Resolution of the curcumin antioxidant controversy. The role of sequential proton loss electron transfer. *J Org Chem, 69,* 5888–5896.

Litwinienko, G., & Ingold, K. U. (2005). Abnormal solvent effects on hydrogen atom abstraction. 3. Novel kinetics in sequential proton loss electron transfer chemistry. *J Org Chem, 70,* 8982–8990.

Litwinienko, G., & Ingold, K. U. (2007). Solvent effects on the rates and mechanisms of reaction of phenols with free radicals. *Acc Chem Res, 40,* 222–230.

Liu, Y., & Zhang, X. (2018). Heat shock protein reports on proteome stress. *Biotechnol J*, *13*, 1800039.

Liu, T., Zhou, J., Li, W., et al. (2020). Effects of sporoderm-broken spores of *Ganoderma lucidum* on growth performance, antioxidant function and immune response of broilers. *Anim Nutr*, *6*, 39–46.

Markovic, Z. (2016). Study of the mechanisms of antioxidative action of different antioxidants. *J Serbian Soc Comput Mech*, *10*, 135–150.

Markovic, Z. S., Dimitric Markovic, J. M., & Dolicanin, C. B. (2010). Mechanistic pathways for the reaction of quercetin with hydroperoxy radical. *Theor Chem Acc*, *127*, 69–80.

Markovic, Z., Dorovic, J., Dimitric Markovic, J. M., Zivic, M., & Amic, D. (2014). Investigation of theradical scavenging potency of hydroxybenzoic acids and their carboxylate anions. *Chem Mon*, *145*, 953–962.

Markovic, Z., Milenkovic, D., Dorovic, J., et al. (2012). Free radical scavenging activity of morin 20-O⁻ phenoxide anion. *Food Chem*, *135*, 2070–2077.

Meng, X., Li, Y., Li, S., et al. (2017). Dietary sources and bioactivities of melatonin. *Nutrients*, *9*, 367.

Meng, M., Zhang, R., Han, R., Kong, Y., Wang, R., & Hou, L. (2021). The polysaccharides from the *Grifolafrondosa* fruiting body prevent lipopolysaccharide/D-galactosamine-induced acute liver injury via the miR-122-Nrf2/ARE pathways. *Food Funct*, *12*, 973–982.

Mohanta, T. K. (2020). Fungi contain genes associated with flavonoid biosynthesis pathway. *J Funct Foods*, *68*, 103910.

Mustacich, D. J., Bruno, R. S., & Traber, M. G. (2007). Vitamin E. *Vitam Horm*, *76*, 1–21.

Nakayama, T., & Uno, B. (2015). Importance of proton-coupled electron transfer from natural phenolic compounds in superoxide scavenging. *Chem Pharm Bull*, *63*, 967–973.

Navegantes-Lima, K. C., Monteiro, V. V. S., de França Gaspar, S. L., et al. (2020). *Agaricus brasiliensis* mushroom protects against sepsis by alleviating oxidative and inflammatory response. *Front Immunol*, *11*, 1238.

Ng, F., Wijaya, L., & Tang, B. L. (2015). SIRT1 in the brain-connections with aging-associated disorders and lifespan. *Front Cell Neurosci*, *9*, 64.

Niki, E. (2000). Free radicals in the 1900's: From in vitro to in vivo. *Free Radic Res*, *33*, 693–704.

Nimse, S. B., & Pal, D. (2015). Free radicals, natural antioxidants, and their reaction mechanisms. *RSC Adv*, *5*, 27986.

Outila, T. A., Mattila, P. H., Piironen, V. I., & Lamberg-Allardt, C. J. (1999). Bioavailability of vitamin D from wild edible mushrooms (*Cantharellus tubaeformis*) as measured with a human. *Am J Clin Nutr*, *69*, 95–98.

Ozturk, B., Havsut, E., & Yildiz, K. (2021). Delaying the postharvest quality modifications of *Cantharellus cibarius* mushroom by applying citric acid and modified atmosphere packaging. *LWT-Food Sci Technol*, *138*, 110639.

Pekal, A., Biesaga, M., & Pyrzynska, K. (2011). Interaction of quercetin with copper ions: Complexation, oxidation and reactivity towards radicals. *BioMetals*, *24*, 41–49.

Phillips, K. M., Horst, R. L., Koszewski, N. J., & Simon, R. R. (2012). Vitamin D4 in mushrooms. *PLoS ONE*, *7*, e40702.

Pillai, R., Redmond, M., & Roding, J. (2005). Anti-wrinkle therapy: Significant new findings in the non-invasive cosmetic treatment of skin wrinkles with beta-glucan. *IFSCC Magazine*, *8*, 1–6.

Pilz, D., Norvell, L., Danell, E., & Molina, R. (2003). Ecology and management of commercially harvested chanterelle mushrooms. In: *General Technical Report PNW-GTR-576*. Portland: United States Department of Agriculture, Forest Service, Pacific Northwest Research Station, pp. 83.

Podkowa, A., Kryczyk-Poprawa, A., Opoka, W., & Muszynska, B. (2021). Culinary–medicinal mushrooms: A review of organic compounds and bioelements with antioxidant activity. *Eur Food ResTechnol*, *247*, 513–533.

Puttaraju, N. G., Venkateshaiah, S. U., Dharmesh, S. M., Urs, S. M., & Somasundaram, R. (2006). Antioxidant activity of indigenous edible mushrooms. *J Agric Food Chem*, *54*, 9764–9772.

Rao, A. V., & Rao, L. G. (2007). Carotenoids and human health. *Pharmacol Res*, *55*, 207–216.

Raphael, I., Joern, R. R., & Forsthuber, T. G. (2020). Memory CD4+ T cells in immunity and autoimmune diseases. *Cells*, *9*, 531.

Reiter, R. J., Tan, D. X., Mayo, J. C., Sainz, R. M., Leon, J., & Czarnocki, Z. (2003). Melatonin as an antioxidant: Biochemical mechanisms and pathophysiological implications in humans. *Acta Biochim Pol*, *50*, 1129–1146.

Roleira, F. M. F., Siquet, C., Orru, E., et al. (2010). Lipophilic phenolic antioxidants: Correlation between antioxidant profile, partition coefficients and redox properties. *Bioorg Med Chem*, *18*, 5816–5825.

Rahal, A., Kumar, A., Singh, V., et al. (2014). Oxidative stress, prooxidants, and antioxidants: The interplay. *Biomed Res Int*, *2014*, ID 761264.

Sabino Ferrari, A. B., de Oliveira, G. A., Mannochio Russo, H., et al. (2021). *Pleurotusostreatus*and *Agaricus subrufescens*: Investigation of chemical composition and antioxidant properties of these mushrooms cultivated with different handmade and commercial supplements. *Int. J. FoodSci*, *56*, 452–460.

Saltarelli, R., Ceccaroli, P., Buffalini, M., et al. (2015). Biochemical characterization and antioxidant and antiproliferative activities of different *Ganoderma* collections. *J Mol Microbiol Biotechnol*, *25*, 16–25.

Sami, R., Elhakem, A., Alharbi, M., et al. (2021). Evaluation of antioxidant activities, oxidation enzymes, and quality of nano-coated button mushrooms (*Agaricus bisporus*) during storage. *Coatings*, *11*, 149.

Sanchez, C. (2017). Reactive oxygen species and antioxidant properties from mushrooms. *Synth Syst Biotechnol*, *2*, 13–22.

Santos-Sanchez, N. F., Salas-Coronado, R., Villanueva-Canongo, C., & Hernandez-Carlos, B. (2019). Antioxidant compounds and their antioxidant mechanism. In Shalaby, E., ed., *Antioxidants*. London: IntechOpen, pp. 1–28.

Sies, H. (1985). *Oxidative Stress*. London: Academic Press.

Singh, M., Tulsawani, R., Koganti, P., Chauhan, A., Manickam, M., & Misra, K. (2013). *Cordyceps sinensis* increases hypoxia tolerance by inducing heme oxygenase-1 and metallothionein via Nrf2 activation in human lung epithelial cells. *Biomed ResInt*, *2013*, ID 569206.

Smith, R. E., Tran, K., Smith, C. C., McDonald, M., Shejwalkar, P., & Hara, K. (2016). The role of the Nrf2/ARE antioxidant system in preventing cardiovascular diseases. *Diseases*, *4*, 34.

Solkoff, D., Cumming, W. T., Liang, B. ,Inserra, P., & Watson, R. R. (1998). The antioxidant propertiesof melatonin. In: *Melatonin in the Promotion of Health*, R. R. Watson, ed. London: CRC Press, pp. 41–58.

Song, X., Ren, Z., Wang, X., Jia, L., & Zhang, C. (2020). Antioxidant, anti-inflammatory and renoprotective effects of acidic-hydrolytic polysaccharides by spent mushroom compost (Lentinula edodes) on LPS-induced kidney injury. *Int J BiolMacromol*, *151*, 1267–1276.

Spim, S. R. V., de Oliveira, B. G. C. C., Leite, F. G., et al. (2017). Effects of *Lentinula edodes* consumption on biochemical, hematologic and oxidative stress parameters in rats receiving high-fat diet. *Eur J Nutr*, *56*, 2255–2264.

Stamets, P., & Zwickey, H. (2014). Medicinal mushrooms: Ancient remedies meet modern science. *Integr Med (Encinitas)*, *13*, 46–47.

Traber, M. G., & Stevens, J. F. (2011). Vitamins C and E: Beneficial effects from a mechanistic perspective. *Free Radic Biol Med*, *51*, 1000–1013.

Tsiapali, E., Whaley, S., Kalbfleisch, J., Ensley, H. E., Browder, I. W., & Williams, D. L. (2001). Glucansexhibit weak antioxidant activity, but stimulate macrophage free radical activity. *Free RadicBiol Med*, *30*, 393–402.

Urbaniak, A., Molski, M., & Szelag, M. (2012). Quantum-chemical calculations of the antioxidant properties of *trans-p*-coumaric acid and *trans*-sinapinic acid. *CMST*, *18*, 117–128.

Valu, M. V., Soare, L. C., Sutan, N. A., et al. (2020). Optimization of ultrasonic extraction to obtain erinacine A and polyphenols with antioxidant activity from the fungal biomass of *Hericiumerinaceus*. *Foods*, *9*, 1889.

Van den Ende, W., Peshev, D., & De Gara, L. (2011). Disease prevention by natural antioxidants and pre-biotics acting as ROS scavengers in the gastrointestinal tract. *Trends Food Sci Technol*, *22*, 689–697.

Van Griensven, L. J. L. D. (2009). Culinary-medicinal mushrooms: Must action be taken? *Int J Med Mushrooms*, *11*, 281–286.

Villalba, J. M., & Alcain, F. J. (2012). Sirtuin activators and inhibitors. *Biofactors*, *38*, 349–359.

Vunduk, J., Klaus, A., Kozarski, M., et al. (2015). Did the iceman know better? Screening of the medicinal properties of the birch polypore medicinal mushroom, *Piptoporusbetulinus* (Higher Basidiomycetes). *Int J Med Mushrooms*, *17*, 1113–1125.

Vunduk, J., Kozarski, M., Djekic, I., Tomasevic, I., & Klaus, A. (2021). Effect of modified atmosphere packaging on selected functional characteristics of *Agaricus bisporus*. *Eur Food Res Technol*, *247*, 829–838.

Wei, S., & Van Griensven, L. J. L. D. (2008). Pro- and antioxidative properties of medicinal mushroom extracts. *Int J Med Mushrooms*, *10*, 315–324.

Yamaguchi, Y., Kagota, S., Nakamura, K., Shinozuka, K., & Kunitomo, M. (2000). Antioxidant activity of the extracts fromfruiting bodies of cultured *Cordyceps sinensis*. *Phytother Res*, *14*, 647–649.

Yang, Y., Zhang, H., Zuo, J., et al. (2019). Advances in research on the activeconstituents and physiological effects ofGanoderma lucidum. *BMC Dermatol*, *3*, 6.

Yatoo, M. I., Dimri, U., Mashooq, M., Saxena, A.,Gopalakrishnan, A., & Bashir, S. T. (2019). Redox disequilibrium vis-a-vis inflammatory cascade mediation of lymphocyte dysfunction, apoptosis, cytokine expression and activation of NF-kB in subclinical diabetic goats. *Indian J Anim Sci*, *89*, 40–45.

Yu, J. H., Pan, H. J., Guo, L. Q., Lin, J. F., Liao, H. L., & Li, H. Y. (2020). Successful biosynthesis of natural antioxidant ergothioneine in *Saccharomycescerevisiae* required only two genes from *Grifolafrondosa*. *Microb Cell Fact*, *19*, 164.

Zhai, F. H., Chen, Y. F., Zhang, Y., Zhao, W. J., & Han, J. R. (2021).Phenolic compounds and antioxidant properties of wheat fermented with *Agaricus brasiliensis* and *Agaricus bisporus*. *FEMS Microbiol Lett*, *368*, fnaa213.

Zhu,B. Z. Mao, L., Fan, R. M., et al. (2011). Ergothioneine prevents copper-induced oxidative damage to DNA and proteinby forming a redox-inactive ergothioneine-copper complex. *Chem Res Toxicol*, *24*, 30–34.

CHAPTER 11

Qualitative and quantitative techniques of analysis for mushrooms

Chitra Sonkar[1], Neha Singh[2], and Rohit Biswas[3]
[1]Assistant Professor, Department of Food Process Engineering, Sam Higginbottom University of Agriculture, Technology and Sciences, Allahabad, Uttar Pradesh, India
[2]Research Scholar, Rajendra Mishra School of Engineering Entrepreneurship, Indian Institue of Technology, Kharagpur, West Bengal, India
[3]Research Scholar, Department of Agricultural and Food Engineering, Indian Institute of Technology, Kharagpur, West Bengal, India

CONTENTS

11.1 INTRODUCTION

Mushrooms for a long time were recognized as protein-rich vegetables. However, recently the scientific community has also developed an interest in the biologically active mushroom compounds of high medicinal value. According to the source of their names, mushrooms consumed by humans are classified within five categories and are named after tastes, anatomy, sprouting habit, appearance, and habitat (Oso, 1975).

The following are the examples in each category:

- **Tastes** – *Volvariella volvacea, Volvariella esculenta, Ogiriagbe, Termitomyces clypeatus*
- **Anatomy** – *Termitomyces manniformis, Termitomyces robustus Schizophyllum commue, Agrocybe broadwayi*
- **Sprouting habit** –*Termitomyces globulus, Termitomyces microcarpus, Pleurotus tuber-regium*
- **Appearance** – *Pleurotus squarrouslus, Psathyrella atroumbonata, Auricularia auricula*
- **Habitat** – *Francolimus bicalcaratus, Calvatia cyathiformis, Termitomyces* spp.

Additionally, the observation is made by native people as various fungi grow on distinctive kinds of deadwood, while every fungus is assigned a name after the wood on which it proliferates. Mushrooms are considered healthy food due to the low calories and higher amounts of minerals, vitamins, and dietary fiber. The mushroom's primary composition is composed of moisture, fat, protein, and a healthy level of amino acids (Nketia et al., 2020).

11.2 PHYSICAL COMPONENT ANALYSIS

Several mushroom characters describe the quality (structure, appearance, texture) and stability (water activity) linked to their physical properties. Some of the physical component evaluation has been documented below.

11.2.1 Dry density

According to the method explained in ISO 9427:2003, the density can be calculated by selecting the oven-dry mass ratio to volume.

11.2.2 Moisture content

The following standard ISO 16979: 2003 can be used to calculate the moisture content.

$$M = \frac{(W_1 - W_2) * 100}{W_2}$$

where:
M is moisture content present (%)
W_1 is the weight of the sample containing moisture content (g)
W_2 is the weight of the sample without moisture content or oven-dried weight (g)

11.2.3 Mechanical stress test

The mushroom's hardness can be evaluated according to ASTM D3501 using a load-up bench with a load capacity of 100 kN and a load cell of 10 kN at room temperature (25°C) and approximately 50% relative humidity. The 10 kN load cell can provide the most accurate results. The displacement of the tests is regulated using a scale of 5 mm/minute. Due to the uneven surface of the sample, the contact surface area is not absolute. When the sample reaches a stable strain, the test ends, and its range is inbetween 70% and 80%. By using the following formula, the curve for load displacement is converted to a curve for stress-strain to obtain the strain ε and compressive stress σ:

$$\sigma = \frac{F}{A}$$

and

$$\varepsilon = \frac{\Delta X}{X_o}$$

where:

F is a compressive force (N)
A is the sample's preliminary cross-sectional area (mm^2)
ΔX is the displacement of those loading surfaces (mm)
X_o is the preliminary height of the section being examined (mm)

11.2.4 Thermal conductivity of mushroom

According to ASTM D5334-00, transients (which require unstable conditions) can be used to calculate mushroom samples' thermal conductivity. The thermal needle probe (TNP) conforms to the specifications. Another needle having a reduced diameter compared to the TNP is used to drill an initial guiding hole (2 mm) in the center of the cylinder-shaped sample. Lower thermal conductivity can be passed with a faster increase in temperature when a high electric current is employed. This standard does not mean that the temperature will rise quickly, as this may cause inaccurate readings. To obtain precise and reliable results, take the average value between the higher and lower currents. The thermal conductivity (λ) is calculated using the following formula:

$$\lambda = \frac{Q}{4\pi \, (T_2 - T_1)} * \ln\left(\frac{t_2}{t_1}\right)$$

where:

Q is the steady current (W/m)
T_1 & T_2 are the linear portions of the first and last temperature (°C), respectively
t_1 & t_2 are the linear portions of the first and final time (s).

11.2.5 Water absorbtion rate

The method for estimating water absorption rate by hydraulic cement concrete ASTM C 1585 can be used for mycelial composites with the most frequently reduced examples. The tests are carried out at a temperature of 20°C and evaluated based on the weight difference to the initial weight. The subsequent expression is used to calculate the water absorption rate:

$$m_w = \frac{M_a}{a * d}$$

where:

m_w is the water absorption (mm)
M_a is the altered sample weight (g)
a is the revealed area (cross-sectional) (mm^2)
d is the water density (g/mm^3)

11.2.6 Color measurement

The color of the mushroom sample is measured with a spectrophotometer in three different regions outside of the estimated mean values of the mushroom. A computer software using color data software value for Hunter color Lab to estimate the value, namely L*, a*, and b* using illuminant C and 8 mm aperture openings. An equipment standardization is performed on standard white plates before analysis (Fernandes et al., 2013).

11.2.7 Yield

The dry sample is ground into powder, and the weight of this powder is measured, followed by an estimate of the yield using the following equation (Negi, 2003):

$$Yield\,(kg) = \left(\frac{W_p}{W_r}\right) * 100$$

where:

W$_p$ is the weight of the product
W$_r$ is the raw material weight

11.2.8 Browning index

Five g of a desiccated mushroom sample is used to measure the browning index by optical density (OD), using a wavelength of 440 nm with a digitally operated spectrophotometer (Srivastava & Kumar, 2000).

11.2.9 Water-holding capacity

To estimate a mushroom sample's water retention, a powdered mushroom sample (2g) is stored in 20 mm of distilled water for 12 hours. For 20 minutes, the water-soaked samples are exposed to centrifugation at 3,000 rpm. Under centrifugal force, the water separates from the sample mixture and is estimated to calculate water retention by Nollet's expression (2004):

$$Water\ retention\left(\frac{mL}{g}\right) = \frac{water\ retained}{weight\ of\ sample}$$

where:

water retained (mL) = volume of water added (20 mL) – volume of water obtained after centrifugation.

11.2.10 Swelling index

The mixture of the desiccated sample and 100 milliliters of distilled water is stored at 20 to 30°C. (overnight). Using the expression given by Nollet, (2004), the swelling index is estimated.

$$Swelling\ index\left(\frac{mL}{g}\right) = \frac{change\ in\ vol\,(mL)}{amount\ of\ sample\,(g)}$$

11.2.11 Seed germination test

The germination test can be carried out with 1:5 w/v. Water extracts are obtained from compost at various composting sampling times. The extracts are prepared according to the procedure illustrated by Zhang et al. (2013). A double layer of filter paper is placed in a 90 mm diameter sterilized Petri dish to which compost extract of about 1 mL is added. The seeds (20 in number) of pakchoi (*Brassica rapa L.*, Chinensis group) are placed in a Petri dish. After 48 hours at a temperature of 25°C (in the dark), the amount of germinating seeds and the root radicals' length were estimated for every dish. The following equation calculated the germination index (GI):

$$GI\,(\%) = 100 \, * \, \left(\frac{N_d \, * \, l_d}{N_c \, * \, L_c} \right)$$

where:

N_d is the mean amount of germinated seeds each dish
l_d is the mean length of root each dish
N_c is the mean amount of germinated seeds (control conditions)
L_c is the average length of root (control conditions)

11.3 CHEMICAL COMPONENT ANALYSIS

The button mushroom (*Agaricus bisporus*) occupies a significant share of the total production of mushrooms worldwide, i.e. roughly more than half of the entire world's yield of mushrooms. However, exotic mushrooms, such as oyster (*Pleurotus* spp.), straw (*Volvariella volvacea*), shiitake (*Lentinula edodes*), and enokitake (*Flammulina velutipes*), are acquiring prominence. These mushroom varieties possess average amounts of superior characteristic protein and are excellent reservoirs of dietetic fiber, ascorbic acid, vitamin B complex, and mineral deposits necessary for human growth. Mushroom exhibits low lipids levels, but the ratio for unsaturated to saturated fatty acids is prominent (around 2.0–4.5:1). It has been reported that some exotic species (e.g. shiitake) collect cadmium, selenium, and other heavy metals, and certain species might also possess toxins that are high temperature-sensitive, for instance, volvatoxin, a protein found in the straw mushroom that is cardiotoxic. Intense scientific analyses, specially undertaken in Japan, have shown that a numeral variety of mushrooms possess healing and restorative properties for oral treatment of various malignant tumors, hypercholesterolemia, virus-related diseases (respiratory tract infections, polio), high blood pressure, and platelet aggregation in the blood (Khatun et al., 2012; Zhang et al., 2018). Investigations undertaken have mainly concentrated on shiitake, *Pleurotus* spp., enokitake as well as the usual unedible *Ganoderma* spp. Isolation and identification of many effective constituents, including nucleic acid products (the hypocholesterolemic eritadenine), polysaccharides (e.g., β-glucans), peptides, lipids, glycoproteins, and proteins have been done (Ma et al., 2018).

11.3.1 Reversed phase-HPLC analysis of toxins

HPLC grade of AA, BA, GA, PHN, and PCN solvent is used to prepare the stock solution of 100 μg/mL of concentration; 10, 20, 100, 200, 1,000, and 2,000 ng/mL concentrations of toxin is diluted in extraction fluid for calibration standard. For each toxin, calibration curves are produced over a linear range of interest ($R^2 > 0.99$) (Kaya et al., 2015). The chromatographic parameters for the method used in this study are provided by Kay et al. Therefore, the author reports that atomoxetine and glytoxin can be separated well by RP-HPLC and UV detection at two varied wavelengths. The mushroom extract was analyzed by reversed phase-HPLC on the HPLC system. RP-HPLC analysis of the standard solutions AA, BA, GA, PHN, PCN, and mushroom extracts were performed on 150 × 4.6 mm, 5 mm particle, C18 column and wavelength of 303 nm (for amatoxins), and 291 nm (for phallotoxins) at UV detector was used. The flow rate of 1 mL/minute was used in an isocratic profile for the mobile phase. the content of the mobile phase was 0.05 M ammonium acetate (pH 5.5 with acetic acid)/Acetonitrile (90:10 v/v). The detection limits were set at 3 ng/g and 2.5 ng/g for amatoxins and phallotoxins, respectively.

11.4 QUALITATIVE ESTIMATION

11.4.1 Extraction of polysaccharides from mushrooms

Before testing, organic solvents (degreaser) and alcohol are pretreated before extraction to remove any reduced molecular weight contamination or impurities, resulting in a more efficient extraction process. Ethanol precipitation, deproteinization, decolorization, dialysis, and fractionation result in further purification (Ruthes et al., 2015; Szwengiel & Stachowiak, 2016). Recovery yield, monosaccharide composition, molecular weight, structure, and spatial configuration and, most importantly, the isolated polysaccharides' bioactivity are affected by the extraction process (Qin et al., 2017; Morales et al., 2019). Physical techniques such as heat treatment, ultrasound, and microwave application can be used to extract functional polysaccharides from mushroom samples as they are water-soluble (Liu, 2019). Some non-polar polysaccharides, such as some β-glucans, require more extended periods of extractions or higher temperatures/pressures than most of the polar polysaccharides, which are thoroughly soluble in water and alkaline solution (Smiderle et al., 2017). The great difficulty is encountered during extraction in polysaccharides with a considerable molecular weight, resulting in weak solubility and high viscosity (Akram et al., 2017; Chawla et al., 2019). Longer treatment time or higher temperature and pressure are required to extract chitinous materials insoluble in water. These processes result in the precipitation of chitinous fractions and the dispersion of other components (Morales et al., 2019). Some limiting factors should also be considered when extracting polysaccharides from a mushroom. More importantly, the mushroom cell structure's complexity prevents the liberation of polysaccharides out of the intracellular matrix, making it economically unfavorable (Parniakov et al., 2014; Xue & Farid, 2015). The previously mentioned conventional extraction strategies cause the denaturation of proteins, the modification of structural polysaccharides, the breakdown of heat-sensitive bioactive compounds, and the discharge of vast quantities of unwanted cell wall components, pectins, and cell debris thus bring the higher cost of purification.

11.4.2 Hot water extraction (HWE) of polysaccharides

HWE extraction is quite effective for separating water-soluble polysaccharides, which uses hot water at a temperature of 50°C–100°C for about 1.5–5 hours. Hence, it has low running costs and fundamental equipment and tool application (Parniakov et al., 2014; Xue & Farid, 2015).

11.4.3 Alkaline- or acid-extraction (AE)

The use of sodium hydroxide or potassium hydroxide and acids like hydrochloric acid or ammonium oxalate increase the polysaccharides' extraction process from mushrooms (Yi, Xu, Wang, Huang & Wang, 2020), which commonly follow hot water extraction to improve the separation of polysaccharides from mushrooms.

11.4.4 Ultrasonic-assisted extraction (UAE)

Ultrasound-assisted extraction (UAE) is an alternative extraction method for bioactive in-gredients with higher recovery efficiency. The UAE forms cavitation with the acoustic within the solvent for the breakdown of asymmetrical microbubble, releasing enormous energy creating mi-crojets and shock waves with high shear force. The cavitation help in the breakdown of cell walls and, due to pore formation, capillary effects help in increased extraction yield combined with reduced particle sizes (You, Yin, & Ji, 2014; Ke, 2015). Diffusion of polysaccharides is assisted by the localized energy generated due to the thermal effect during the treatment facilitating the diffusion of polysaccharides within the solvent for improved extraction efficiency (Alzorqi et al., 2017).

11.5 MOLISCH'S TEST FOR CARBOHYDRATES

In this test, a small amount of Molisch's reagent is mixed with each portion dissolved in distilled water, and then 1 mL of concentrated H_2SO_4 is added by pouring it through the test tube's sides. This mixture is placed still for 2 minutes and then diluted with 5 mL of distilled water. Red or dull violet at the junction of the bilayer indicates a positive result (Sofowora, 1993).

11.6 QUANTITATIVE ANALYSIS

11.6.1 Proximate analysis

11.6.1.1 Estimation of moisture content

The moisture content of mushrooms can be determined by the approach of Raghuramulu et al. (2003). The cultivated mushroom samples' moisture content is calculated as follows: The sample's wet weight is measured with an electronic microbalance until the sensitivity limit is 0.001 g. The sample was dried under reduced pressure in a confined space and then dried under reduced pressure at 110°C on a rotary evaporator. The dry weight of the contents is measured, and then the total moisture content is calculated using the following expression and expressed as percentage moisture content per gram of wet tissue:

$$Moisture\,(\%) = (Initial\;weight - Final\;weight) * \left(\frac{100}{weight\;of\;sample} \right)$$

11.6.1.2 Estimation of total proteins

Preparation of tissue homogenate: About 1 g of wet mushroom mass is emulsified with 10 mL of saline buffered with phosphate (pH 7.4) and purified utilizing centrifugal force in a centrifuge moving at 10,000 rpm for 10 minutes. The supernatant (defined as tissue homogenate) is trans-ferred to a fresh tube and kept at 20°C up until use. Protein estimation: The entire protein content can be assessed by Bradford's method (1976). 100 μl of the test sample (tissue homogenate in saline buffered with phosphate, pH 7.4) is made to a volume of 500 μl with distilled water and mixed with 4.5 mL of Bradford's reagent (Coomassie Brilliant blue, 10% (w/v) in methanol, 50% (v/v) and 88% orthophosphoric acid, 50% (v/v), adjusted the optical density of OD550 nm to 1.1). The contents are incubated and mixed well and incubated for 5 minutes at ambient temperature. The optical density is evaluated at 595 nm against a blank reagent (without tissue homogenate) in ultraviolet-visible (UV-Vis) spectrophotometry. A standard graph is prepared using BSA ranges from 20 to 100 µl/mL. The protein concentration is calculated by plotting the OD of the test sample against the standard graph. The protein quantity is expressed as mg/g dry weight.

- **For Dried Mushrooms**
  Five grams of pulverized mushroom are boiled in a water bath for 30 minutes with 50 mL of NaOH (0.1 N). The solution is then cooled at room temperature and utilizing a tabletop centrifuge; it is centrifuged at 1,000 × g. The floating substance is accumulated, and the total amount of protein is evaluated by the method given by Lowry et al. (1951).
- **For Fresh Mushrooms**
  Five grams of sample is mixed with 50 mL phosphate buffer and homogenized with a tissue homogenizer to evaluate protein content. Five mL of homogenized sample is taken with 50 mL of 0.1 N NaOH, and protein content is evaluated as described previously.

11.6.1.3 Lowry assay by Folin reaction

The "Lowry Assay: Protein by Folin Reaction" (Lowry et al., 1951) is by far the most actively utilized technique to analyze the number of proteins (present in solution or solvable in dilute alkali) in test samples of biological origin. Firstly, pre-treatment of protein is done using copper ions present in the alkali, reducing Folin's reagent phosphomolybdate-phosphotungstic acid due to aromatic amino acids. The blue color indicates the termination of the reaction. Read the absorbance (at 750 nm) of the final Folin reaction product on the graph along the standard curve of the defined standard protein mixture. The amount of proteins present in the defined sample can be analyzed by assessing the absorbance (at 750 mm) of the Folin reaction's final product on the graph along with the standard curve of the principle protein mixture (in some cases; Bovine Serum Albumin – BSA – solution).

11.6.1.4 Determination of total lipids

The total lipid count can be estimated by acquiring a slight modification in the method of Folch et al. (1957).

- **Dried Mushrooms**
  Five grams of grounded mushrooms are dissolved in a mixture of 50 mL of chloroform: methanol (2:1 v/v), and then give it a thorough mix and left undisturbed for 3 days. The still mixture is then strained and centrifuged at 1,000 × g through a table centrifuge. Using a Pasteur pipette, the top layer of methanol can be pulled off, and then chloroform is vaporized using heat. Then the substance left behind is a crude lipid.
- **Fresh Mushrooms**
  A 5-gram sample with 50 mL of phosphate buffer is taken and homogenized with a tissue homogenizer to evaluate fresh mushrooms' total lipid. Lipid content is determined using a solution of 5 mL homogenized mixture and 50 mL of 2:1 chloroform and methanol solution by the previous method.

11.6.1.5 Crude fiber analysis

For crude fiber determination in mushrooms, the sample must be free of fats and moisture. Ten g of such a sample is placed in a beaker with 200 mL of 0.255 N H_2SO_4 boiling temperature. The boiling is continued for half an hour, and then carefully reduce the volume by adding water after regular intervals. This mixture is then strained using a muslin cloth, along the filtrate is rinsed with slightly boiling water until it becomes acid-free. This mixture is moved to the same container, and then 200 mL of a boiling 0.313 N NaOH solution is added. After the volume has been kept constant in a beaker and boiled for half an hour, this mixture is then strained with a muslin cloth, and the filtrate is rinsed with slightly boiling water until the filtrate is alkali-free. The filtered residue is washed with alcohol and ether. Now it is dried overnight in a crucible at 80°C to 1,000°C. After drying is complete, the residue is weighed on an electric weighing balance (Wd). The crucible is heated in a muffle furnace to 600°C for approximately 5 to 6 hours, then cooled and reweighed (Wa). The variation in weights (Wd–Wa) is the weight of the crude fiber can be found as (Raghuramulu et al., 2003):

$$Crude\ fiber\left(\frac{g}{100g}\ of\ sample\right) = [100 - (moisture\ content + fat)] * (W_d - W_a)$$

$$/(weight\ of\ sample)$$

11.6.1.6 Determination of total ash

To calculate the mushroom sample fat content, take 1 g of a carefully weighed sample in the crucible. The crucible is first heated on a clay pipe triangle until the material is wholly burnt, then heated in a muffle furnace to 600°C for about 5 to 6 hours. The sample is then cooled down using a desiccator and then weighed. To ensure that the sample is wholly ashed, the crucible is heated again in a muffle furnace for 1 hour, then cooled and weighed. This process is repeated until the subsequent weights are identical and the burned ash is white or grey-white. The formula used for calculating total ash is as follows (Raghuramulu et al., 2003):

$$Ash\ content\left(\frac{g}{100g\ of\ sample}\right) = weight\ of\ Ash * \frac{100}{weight\ of\ sample\ used}$$

11.6.1.7 Total carbohydrate estimation

The subsequent equation can be used to estimate the available carbohydrate in a mushroom sample is expressed in g/100 g of sample (Raghuramulu et al., 2003):

$$Carbohydrate = 100 - \{(moisture + protein + fat + crude\ fibre + ash)$$

11.6.1.8 Mineral analysis

The total ash is used to analyze the mushroom sample's mineral content. Two mL of concentrated nitric acid is added to the ash, and it is thermally treated for 2 minutes. A dash of hydrogen peroxide is added to this formulation, and then the mixture is transferred to a volumetric beaker. Further, the final volume is increased to 50 mL by adding water, which can later evaluate the calcium, iron, manganese, magnesium, zinc, selenium, and arsenic with flame and graphite method coupled with atomic absorption spectrophotometer.

11.6.1.9 Estimation of total sugar

The total sugar in a mushroom sample can be estimated by the phenolic acid hydrolysis technique (Dubois et al., 1956; Ashwell, 1957; Krishnaveni et al., 1984). In short, 100 µl of tissue homogenate is mixed with 100 µl of phenol reagent (Phenol, 5% (w/v) in distilled water), mixed well carefully. Five mL of concentrated sulfuric acid is mixed by pouring it gradually through the tube's sides with periodic stirring in a vortex mixer. The contents are mixed well and kept at 37°C for 20 minutes in a closed water bath. The optical density of the green-colored hydroxymethyl furfural is analyzed through a UV-Vis spectrophotometer. Ten to 100 µg of glucose range is used to draw a standard graph. By plotting the optical density of the test sample against the standard graph, the total sugar concentration is obtained and expressed as mg/g dry weight.

11.6.1.10 Soluble sugar assay

To extract free sugar, 10 mL of ethanol at a concentration of 85% in a sample of 1.0 g dry weight is used for 24 hours. Free sugars are extracted from the sample using 10 mL of 85% ethanol on a dry weight basis for 24 hours. HPLC analysis is performed using a C18 column (4.6 × 250 mm) with a column temperature of 35°C with a mobile phase flow rate of 1.2 mL/min in 75% acetonitrile. The free sugar is quantified by comparing the standard table.

11.7 AMINO ACID ASSAY

The configuration of the amino acids of mushrooms can be determined from samples that have been hydrolyzed with 6 N HCl at 105°C for 24 hours. Then, the obtained amino acids are detected by liquid chromatography. The retention times and ranges with the original standard mixtures are used to identify amino acids.

11.8 FATTY ACID ASSAY

11.8.1 Fatty acid profile

According to Hamilton (1992), the total lipids' fatty acid composition is obtained from dried mushroom samples and analyzed in the form of fatty acid methyl esters (FAMEs) through gas chromatography. Ovens are set at 140°C for 5 minutes. The sample is introduced inside the GC by use of an auto-injector. The temperature is programmed to 140°C for 5 minutes, then raised to 240°C at 4°C/minute and kept for 15 minutes. A constant flow rate of 20 cm/s of helium is used in the form of carrier gas. The injection port and oven temperature of the flame ionization detector are set at 260°C. FAMEs are diagnosed through the way of means of evaluating retention times with a genuine standard mixture.

11.9 QUALITATIVE PHYTOCHEMICAL ANALYSIS

11.9.1 Barfoed's test for monosaccharides

Barfoed's reagent can be efficiently utilized to detect monosaccharides in the mushroom sample; 0.5 g of sample is required for the analysis. The sample is mixed with distilled water, and the resultant solution is passed through a filter paper. A 1 mL each of filtrate and Barfoed's reagent placed in the test tube is heated for about 2 minutes onto the water bath. Monosaccharide presence is indicated by the reddish precipitation formed by the cuprous oxide (Sofowora, 1993).

11.9.2 Fehling's test for free reducing sugar

Fehling's reagent can be efficiently utilized to detect free reducing sugars in the mushroom sample; 0.5 g of sample is mixed with distilled water, and the resultant solution is passed through a filter paper. Fehling's solution A and B mixed with filtrate, 5 mL each is heated. According to Sofowora (1993), reducing sugar is confirmed by the precipitation of reddish cuprous oxide.

11.9.3 Test for tannins

According to Trease and Evans (2002), a solution of a 0.5 g sample and distilled water is prepared, filtered, and mixed with 2% of $FeCl_3$. The blue-black or blue-green precipitate confirms the presence of tannin.

11.9.3.1 Borntrager's test for anthraquinones

Borntrager's test can be performed to analyze anthraquinone's presence. According to this test, a solution of 0.2 g of sample and 10 mL of benzene is prepared. The filtrate is mixed with 5 mL of 10% of ammonia solution. The presence of anthraquinones is confirmed by pink, red, or violet color in the ammonical phase (Sofowora, 1993).

11.9.3.2 Liebermann-Burchard test for steroids

Steroids can be confirmed by violet to blue or bluish-green color obtained by mixing of 0.2 g of sample with 2 mL of acetic acid, which is cooled well by ice and further mixed by concentrated sulphuric acid (Aglcone portion of a cardiac glycoside is the steroidal ring) (Sofowora, 1993).

11.9.4 Test for terpenoids

A test of terpenoids is performed by mixing little quantity of sample with ethanol. One mL of acetic anhydride acid is slowly mixed with the sample, followed by concentrated H_2SO_4. The appearance of pink to violet color shows the presence of terpenoids in mushrooms (Sofowora, 1993).

11.9.4.1 Saponins test

One gram of mushroom sample is boiled with 5 mL of distilled water and filtered. The filtrate is mixed with 3 mL of distilled water and shaken vigorously for 5 minutes. Saponins are confirmed by the persistent froth obtained at higher temperatures (Sofowora, 1993).

11.9.4.2 Flavonoids by Shinoda's test

Initially, a mixture of ethanol and 0.5 mL is prepared, warmed, and passed through filter paper. Magnesium chips (three in number) and some drops of concentrated hydrochloric acid are then mixed with the filtrate. Flavanoids are confirmed by the appearance of orange, red, and pink to purple color (Trease & Evans, 2002).

11.9.4.3 Flavonoids by ferric chloride test

Distilled water mixed with 0.5 mL of sample is boiled and filtered. Flavanoids are confirmed by the appearance of green-blue or violet color when ferric chloride is mixed with 2 mL of sample filtrate (Trease & Evans, 2002).

11.9.4.4 Flavonoids by lead ethanoate test

The sample is dissolved in distilled water and filtered. Five mL of the filtrate is mixed with 3 mL of ethanoate solution. Flavanoids are confirmed by the appearance of buff-colored precipitate (Trease & Evans, 2002).

11.9.4.5 Flavonoids by NaOH test

According to Trease and Evans (2002), a little quantity of the sample is dissolved in water and filtered; to this 2 ml of the 10%, aqueous sodium hydroxide is later added to produce a yellow coloration. A change in color from yellow to colorless on the addition of dilute hydrochloric acid is an indication of the presence of flavonoids.

11.9.5 Test for alkaloids

The sample is mixed with 5 mL of 1% aq. HCL and heated on water bath followed by filtration. Two test tubes are filled with 1 mL of the sample filtrate, and the first tube is mixed with few drops of Dragendorrf's reagent, the appearance of red precipitate indicated the positive sign. The second test tube is mixed with a few drops of Mayer's reagent, indicating the presence of alkaloids on buff-colored precipitation (Sofowora, 1993).

11.9.6 Test for soluble starch

According to the procedure given by Vishnoi (1979), a little amount of the sample is heated with 1 mL of 5% KOH, followed by cooling and acidifying with H_2SO_4. The appearance of yellow color is marked as the presence of soluble starch.

11.10 QUANTITATIVE PHYTOCHEMICAL ANALYSIS

11.10.1 Test for organic acids

A standardized procedure is given by Barros, Pereira, and Ferreira (2013) and can be utilized to evaluate the amount of organic acid present in a mushroom sample. A wavelength of 215 and 245 nm is the ideal wavelength for this detection, which is done using a PAD. Previously standardized curves of commercial standards of various organic compounds are matched with the area of curves generated using the mushroom samples for organic acids at a wavelength of 215 nm. Organic acid includes fumaric, oxalic citric, and quinic acids. Grams per 100 g of dry basis is used as a unit.

11.10.2 Determination of TPC

The TPC can be quantified using the procedure of Siddhuraju and Becker (2007). An extracted sample is taken in a test tube and made up to 1 mL using distilled water. The solution is mixed with 0.5 and 2.5 mL of Folin–Ciocalteu reagent and sodium carbonate solution, respectively. The solution is mixed on a vortex and stored for 40 minutes in a completely dark place. The rested solution is subject to spectrophotometric analysis at 725 nm with a blank reagent solution. The TPC is quantified with the standard curve of catechol. The experimentation is performed in a replicatation of three.

11.10.3 Total flavonoid determination

The aluminum chloride colorimetric method can be employed to estimate the total flavonoids as reported by Chouhan et al. (2019). The standard curve of quercetin of 10, 20, 40, 60, and 80 μg/mL in methanol is used. The sample was mixed with methanol for extraction in a concentration of mg/mL of methanol was obtained. One mL of the extract is taken and mixed with 4 mL of distilled water, and

300 µL of 5% of sodium nitrite is added and incubated for 5 minutes. To the incubated solution, 300 µL of 10% aluminum chloride is mixed and left to stand for another 6 minutes. Following the addition of 1 M sodium hydroxide, the solution was made 10 mL with double distilled water. The solution was left to stand for another 15 minutes, and then absorbance was recorded at 510 nm. The flavonoid content was calculated using a calibration curve with $r^2 = 0.994$,

$$y = 0.006x + 0.005$$

mg g^{-1} of dried extract QUE (quercitin equivalent) is used to express the flavonoid content.

11.10.4 Estimation of total flavonoid content

Another method to determine total flavonoid content with slight modification is given by Zhishen et al. (1999); 0.5 mL sample extract is mixed with 2 mL of distilled water. The solution is mixed with 0.15 mL of 5% $NaNO_3$ solution. The solution is left for 6 minutes, and 0.15 mL of 10% $AlCl_3$ solution is added, and the final volume is made 5 mL using distilled water. The solution is mixed thoroughly and left undisturbed for 15 minutes. Spectrophotometric analysis is carried out at 510 nm with catechol as a standard compound for reference.

11.10.5 Test for 5′-nucleotide assay

The procedure given by Pei et al. (2014) can be followed for the extraction of 5′-nucleotides from the samples. For this purpose, freeze-dried samples are dissolved and extracted with 50 mL of deionized water. Centrifugation is done for 30 minutes at 4,000 rpm on the boiled extract obtained after cooling. The extracted sample goes for HPLC analysis after being filtered through a 0.22 µm nylon filter. HPLC was performed using the C18 column on the isocratic mobile phase of 5% methanol and 95% of 0.05% phosphoric acid with a flow rate of 0.7 mL/minute, and detection is done at 254 nm. With the help of authentic standard retention time, 5′-nucleotides are identified and quantified by the respective calibration curve.

11.10.6 1LC-MS analysis of phenolics

LC-MS analysis can be done using a mass spectrometer with an ESI, i.e. electrospray ionization source (ESI), which is combined with a mass spectrophotometer (MS) according to the procedure given by Benabderrahim et al. (2019). C18 column (250 × 2.1, 2.6 µm) is utilized to investigate and separate compounds. Two mobile phases consisted of 0.1% formic acid in distilled water, and methanol is utilized. In a linear gradient elution for 0 to 45 minutes, 10% to 100% B; 45 to 55 minutes, 100% B. Five µL was used for injection volume, with 0.4 mL/minute of flow rate and 30°C of temperature was maintained, and at 280 nm detention was performed (Benabderrahim et al., 2019).

11.11 NUTRITIONAL, TOXIC STUDIES FOR HUMAN HEALTH HAZARDS

Though consumption of mushrooms has many health benefits, Numerous studies have confirmed mushroom bioactivities, including anti-inflammatory, cholesterol-lowering antioxidant, antibacterial, immunoregulatory, antifungal, anticancer, antiallergic, antihypertensive, antidiabetic, liver-protective, and antiviral activities (Moro et al., 2012; Kosanić et al., 2016; Muszyńska et al., 2016; Nowakowski et al., 2020). Apart from the stated benefits, the particular risk is associated with mushroom consumption mainly due to the poisoning of heavy metals such as Cd, Hg, As, and Pb. The heavy metal results from bioaccumulation resulting in biomagnification of these hazards in the food chain resulting in consumer risk significantly due to Hg, in methylated form (Širić et al., 2017). Seafood and fish are a significant target of bioaccumulation but can also be found in mushrooms (Adel et al., 2016; Adel et al., 2018; Cammilleri et al., 2018; Cammilleri et al., 2019; Signorelli et al., 2019; Cammilleri et al., 2020).

11.11.1 Determination of mercury content in mushroom

Mercury content can be measured using the analysis protocol of the manufacturer in a single-purpose atomic absorption spectrometer. Dried mushroom is powdered and kept into the cuvette, and placed in an analyzer chamber. The sample is dried, followed by combustion at 600°C in the presence of oxygen prevailing in the chamber. Gold amalgamator traps the Hg vapors. The quantification of Hg is done on the following condition of drying time, decomposition time, and waiting time of 60, 150, and 65 seconds, respectively. The calibration is performed with the help of a standard working solution of mercury. Mercury is measured (wet basis) depending upon the sample's water content, and the result is expressed in $\mu g\ kg^{-1}$ on a dry and wet basis (Nowakowski et al., 2021).

11.11.2 Determination of lead and cadmium content in mushroom

Pb and Cd's content can be quantified using electrothermal atomic absorption analytical analysis (ETAAS) by Zeman's subsequent correction. Pb and Cd concentration in mineral samples was made at wavelengths of 283.3 and 228.8 nm, respectively; 0.5% ammonium dihydrogen phosphate or one gL^{-1} palladium nitrate was used as a matrix modifier for Pb and Cd, respectively (Omeljaniuk et al., 2018). Calibration curves for Pb and Cd were created using a fixed standard solution concentration of 0, 5, 10, 20, and 40 μgL^{-1} for Pb and 0, 0.5, 1, 2, and 4 μgL^{-1} for Cd. The content of the element in the sample is dependent upon the sample's water content. The results are presented on a wet basis in mg/kg (Nowakowski et al., 2021).

11.11.3 Determination of arsenic content in mushrooms

Arsenic content is quantified using Inductively coupled plasma mass spectrometry (ICP-MS) and the discriminatory kinetic energy chamber (KED). In this adjustment, the polyatomic interference was corrected by collision and electromagnetic distortion. The conditions for ICP-MS were as follows: mass – 74.9216 amu, stay time for each amu – 50 ms, integration time – 1,000 ms, and measurement mode for both detectors (Jablonska et al., 2017). Calibration curves were created from standard active solutions of arsenic at 0, 1, 2, 5, and 10 μgL^{-1} concentrations, prepared based on the As standard solution. The arsenic content is expressed on a wet basis depending upon the sample's water content. The results are expressed as $mg\ kg^{-1}$ on a dry and wet basis (Nowakowski et al., 2021).

11.12 ESTIMATION OF HUMAN HEALTH RISKS

Different indicators have been utilized in the study of health risks conferred by the consumption of mushrooms on humans. Some of these indicators are PTWI (Provisional Tolerable Weekly Intake) for Hg, PTMI (Provisional Tolerable Monthly Intake) for Cd, and BMDL (Benchmark Dose Lower Confidence Limit) for As and Pb (WHO, 2011a, 2011b, 2013). Target Hazard Quotient (THQ), Hazard Index (HI), Carcinogenic Risk (CR), and Estimated Daily Intake (EDI) have been established for the elements (Dadar et al.,2017; Ahmed et al., 2019). Studies indicate that Polish consumers consume 100 g of fresh mushrooms daily, which deems it as their standard portion (Wojciechowska-Mazurek et al., 2011). The daily permissible value was set as 0.57 µg/kg/day for Hg, 0.5 µg/kg/day for Pb, 0.83 µg/kg/day for Cd, and 3 µg/kg/day for As. These values have been calculated following the index set by World Health Organization (WHO) – PTWI for Hg is 4 µg/kg/week, PTMI for Cd – 25 µg/kg/month, and BMDL for Pb and As is 0.63 (nephrotoxicity effect) and 3 µg/kg/day (lung cancer), respectively (WHO, 2011a, 2011b, 2013). Studies have been conducted to calculate the maximum amount of mushroom that a person can consume with a 70 kg bodyweight that would exceed the daily permissible. EDI is an index that assesses the flow of elements from the consumption of food to the human body. The EDI value for a particular element is calculated by multiplying the mean value of element content in a mushroom sample by the number of mushrooms consumed by a person of average body weight (70 kg) daily. The following formula is used to calculate EDI:

$$EDI = FIR \ x \ C$$

where:

FIR is the average daily consumption of mushrooms (Kalač & Svoboda, 2000)
C is the mean concentration of the elements in the mushroom sample.

Target hazard quotient value (THQ) was used to assess the non-carcinogenic and long-term health hazards from hazardous elements resulting from consuming mushrooms. This index combines various factors, including concentrations and toxicity of Pb, Cd, Hg, and As. It also includes the quality and amount of food ingested, and the consumer's average body mass which is assumed to be 70 kg.

A THQ value higher than 1 indicates the possible deleterious health effect of ingesting a specific food product. THQ can be calculated using the following formula:

$$THQ = 10 - 3x \ (Efr * EDtot * FIR * C)/(RfDo * BWa * ATn)$$

where:

Efr shows exposure frequency – 365 days per year
EDtot is the duration of exposure
The average consumption of mushrooms daily is termed a FIR value and has been roughly approximated to be 27 g/day (10 kg/year) (Kalač & Svoboda, 2000).
C represents the toxic element's average concentrations in mushroom
RfDo is the oral reference dose for toxic elements (Kalac, 2019)
Bwa and ATn are the average value of body weight (70 kg) and average exposure, respectively (365 days per year).

Pb, Cd, and As are carcinogens that cause cancer as a consequence of their exposure; CR value represents the likelihood of developing cancer in a lifetime due to exposure to such cancer-causing agents. A CR value exceeding 10–4 shows a higher risk of cancer-causing effect, and a CR value lower than 10–4 represents an acceptable risk (Hu et al., 2017; USEPA, 2000). The CR value is calculated with the following formula:

$$CR = 10 - 3x \ (Efr \ x \quad EDtot \ x \ EDI \ x \ CSf)/ATn$$

CSf represents the oral slope factor of carcinogen, and it is 1.5, 0.0085, and 6.3 mg/kg/day for As, Pb Cd, respectively (USEPA, 2000).

The HI index is calculated as a summation of THQs of toxic effects of elements that have been studied (Pb, Cd, and As) and found in the physical part of the mushroom, and the equation is

$$HI = \sum_{i=k}^{n} THQs$$

where:

THQs is the estimated target hazard quotients for elements

The HI is considered a substantial risk and potential danger to health when it reaches a value higher than 1 (Ahmed et al., 2019; Dadar et al., 2017).

11.13 CONCLUSION

Mushrooms have a variety of nutrients that have applicability in a variety of places. Characterization of various types of mushrooms, based on their morphological characteristics, several physical tests such as density, mechanical, and color are done to sort in basic order. Similarly, to check the availability of a particular chemical without quantification, tests such as flavonoids, reducing sugar tannins, steroids, terpenoids, and alkaloids can be done without quantification. Several changes occur during the processing of mushrooms, especially in the case of drying; the morphological characteristics are changed drastically for which specific physical tests such as moisture retention test, rehydration ratio, and yield are performed to check the efficiency of the processing that has been carried out. Wild mushrooms are classified as both toxic and edible, and to quantify the toxic components of the mushroom, a specific test such as HPLC is performed to estimate the levels of toxic components that could render the mushrooms inedible. Mushrooms growing in the wild, especially near watersides, can also be bioaccumulation sources of heavy metals, as discussed, which have serious health hazards that need to be determined before consumption. Several guidelines are issued regarding the safe limits of toxic elements, such as Hg, As, Pb, and Cd to render the fungus edibility.

REFERENCES

Adel, M., Conti, G. O., Dadar, M., Mahjoub, M., Copat, C., & Ferrante, M. (2016). Heavy metal concentrations in edible muscle of whitecheek shark, Carcharhinus dussumieri (elasmobranchii, chondrichthyes) from the Persian Gulf: A food safety issue. *Food and Chemical Toxicology*, *97*, 135–140.

Adel, M., Copat, C., Asl, M. R. S., Conti, G. O., Babazadeh, M., & Ferrante, M. (2018). Bioaccumulation of trace metals in banded Persian bamboo shark (Chiloscyllium arabicum) from the Persian Gulf: A food safety issue. *Food and Chemical Toxicology*, *113*, 198–203.

Ahmed, A. S., Sultana, S., Habib, A., Ullah, H., Musa, N., Hossain, M. B., ..., & Sarker, M. S. I. (2019). Bioaccumulation of heavy metals in some commercially important fishes from a tropical river estuary suggests higher potential health risk in children than adults. *Plos one*, *14*(10), e0219336.

Akram, K., Shahbaz, H. M., Kim, G. R., Farooq, U., & Kwon, J. H. (2017). Improved extraction and quality characterization of water-soluble polysaccharide from gamma-irradiated lentinus edodes. *Journal of Food Science*, *82*(2), 296–303.

Alzorqi, I., Sudheer, S., Lu, T. J., & Manickam, S. (2017). Ultrasonically extracted β-d-glucan from artificially cultivated mushroom, characteristic properties and antioxidant activity. *Ultrasonics Sonochemistry*, *35*, 531–540.

Ashwell, G. (1957). Colorimetric analysis of sugars. *Methods in Enzymology*. Academic Press, Volume 3, 73–105.

Barros, L., Pereira, C., & Ferreira, I. C. (2013). Optimized analysis of organic acids in edible mushrooms from Portugal by ultra fast liquid chromatography and photodiode array detection. *Food Analytical Methods*, 6(1), 309–316.

Benabderrahim, M. A., Yahia, Y., Bettaieb, I., Elfalleh, W., & Nagaz, K. (2019). Antioxidant activity and phenolic profile of a collection of medicinal plants from Tunisian arid and Saharan regions. *Industrial Crops and Products*, 138, 111427.

Bradford, M. M. (1976). A rapid and sensitive method for the quantitation of microgram quantities of protein utilizing the principle of protein-dye binding. *Analytical Biochemistry*, 72(1–2), 248–254.

Cammilleri, G., Galluzzo, F. G., Fazio, F., Pulvirenti, A., Vella, A., Lo Dico, G. M., ..., & Ferrantelli, V. (2019). Mercury detection in benthic and pelagic fish collected from Western Sicily (Southern Italy). *Animals*, 9(9), 594.

Cammilleri, G., Galluzzo, P., Pulvirenti, A., Giangrosso, I. E., Lo Dico, G. M., Montana, G., ..., & Ferrantelli, V. (2020). Toxic mineral elements in Mytilus galloprovincialis from Sicilian coasts (Southern Italy). *Natural Product Research*, 34(1), 177–182.

Cammilleri, G., Vazzana, M., Arizza, V., Giunta, F., Vella, A., Lo Dico, G., ..., & Ferrantelli, V. (2018). Mercury in fish products: What's the best for consumers between bluefin tuna and yellowfin tuna?. *Natural Product Research*, 32(4), 457–462.

Chawla, P., Kumar, N., Kaushik, R., & Dhull, S.B. (2019). Synthesis, characterization and cellular mineral absorption of gum arabic stabilized nanoemulsion of *Rhododendron arboreum* flower extract. *Journal of Food Science and Technology*, 56(12), 5194–5203.

Chouhan, K. B. S., Tandey, R., Sen, K. K., Mehta, R., & Mandal, V. (2019). Extraction of phenolic principles: Value addition through effective sample pretreatment and operational improvement. *Journal of Food Measurement and Characterization*, 13(1), 177–186.

Dadar, M., Adel, M., Nasrollahzadeh Saravi, H., & Fakhri, Y. (2017). Trace element concentration and its risk assessment in common kilka (Clupeonella cultriventris caspia Bordin, 1904) from southern basin of Caspian Sea. *Toxin Reviews*, 36(3), 222–227.

Dubois, M., Gilles, K. A., Hamilton, J. K., Rebers, P. A., & Smith, F. J. A. C. (1956). Phenol sulphuric acid method for total carbohydrate. *Food Analytical Chemistry*, 26, 350.

Fernandes, Â., Antonio, A. L., Barreira, J. C., Botelho, M. L., Oliveira, M. B. P., Martins, A., & Ferreira, I. C. (2013). Effects of gamma irradiation on the chemical composition and antioxidant activity of Lactarius deliciosus L. wild edible mushroom. *Food and Bioprocess Technology*, 6(10), 2895–2903.

Folch, J., Lees, M., & Stanley, G. S. (1957). A simple method for the isolation and purification of total lipides from animal tissues. *Journal of Biological Chemistry*, 226(1), 497–509.

Hamilton, S. (1992). Extraction of lipids and derivative formation. *Lipid Analysis: A Practical Approach*, 13–64.

Hu, B., Jia, X., Hu, J., Xu, D., Xia, F., & Li, Y. (2017). Assessment of heavy metal pollution and health risks in the soil-plant-human system in the Yangtze River Delta, China. *International Journal of Environmental Research and Public Health*, 14(9), 1042.

Jablonska, E., Socha, K., Reszka, E., Wieczorek, E., Skokowski, J., Kalinowski, L., ... & Wasowicz, W. (2017). Cadmium, arsenic, selenium and iron – implications for tumor progression in breast cancer. *Environmental Toxicology and Pharmacology*, 53, 151–157.

Kalac, P. (2019). *Mineral Composition and Radioactivity of Edible Mushrooms*. Academic Press.

Kalač, P., & Svoboda, L. (2000). A review of trace element concentrations in edible mushrooms. *Food Chemistry*, 69(3), 273–281.

Kaya, E., Karahan, S., Bayram, R., Yaykasli, K. O., Colakoglu, S., & Saritas, A. (2015). Amatoxin and phallotoxin concentration in Amanita phalloides spores and tissues. *Toxicology and Industrial Health*, 31(12), 1172–1177.

Ke, L. Q. (2015). Optimization of ultrasonic extraction of polysaccharides from lentinus edodes based on enzymatic treatment. *Journal of Food Processing and Preservation*, 39(3), 254–259.

Khatun, S., Islam, A., Cakilcioglu, U., & Chatterjee, N. C. (2012). Research on mushroom as a potential source of nutraceuticals: A review on Indian perspective. *American Journal of Experimental Agriculture*, 2(1), 47.

Kosanić, M., Ranković, B., Rančić, A., & Stanojković, T. (2016). Evaluation of metal concentration and antioxidant, antimicrobial, and anticancer potentials of two edible mushrooms Lactarius deliciosus and Macrolepiota procera. *Journal of Food and Drug Analysis*, *24*(3), 477–484.

Krishnaveni, S., Balasubramanian, T., & Sadasivam, S. (1984). Sugar distribution in sweet stalk sorghum. *Food Chemistry*, *15*(3), 229–232.

Liu, C. (2019). Extraction, separation and purification of acidic polysaccharide from Morchella esculenta by high voltage pulsed electric field. *International Journal Bioautomation*, *23*(2), 193.

Lowry, O. H., Rosebrough, N. J., Farr, A. L., & Randall, R. J. (1951). Protein measurement with the Folin phenol reagent. *Journal of Biological Chemistry*, *193*, 265–275.

Ma, G., Yang, W., Zhao, L., Pei, F., Fang, D., & Hu, Q. (2018). A critical review on the health promoting effects of mushrooms nutraceuticals. *Food Science and Human Wellness*, *7*(2), 125–133.

Morales, D., Smiderle, F. R., Villalva, M., Abreu, H., Rico, C., Santoyo, S., ..., & Soler-Rivas, C. (2019). Testing the effect of combining innovative extraction technologies on the biological activities of obtained β-glucan-enriched fractions from Lentinula edodes. *Journal of Functional Foods*, *60*, 103446.

Moro, C., Palacios, I., Lozano, M., D'Arrigo, M., Guillamón, E., Villares, A., ..., & García-Lafuente, A. (2012). Anti-inflammatory activity of methanolic extracts from edible mushrooms in LPS activated RAW 264.7 macrophages. *Food Chemistry*, *130*(2), 350–355.

Muszyńska, B., Kała, K., Firlej, A., & Sułkowska-Ziaja, K. (2016). "Cantharellus cibarius": Culinary-medicinal mushroom content and biological activity. *Acta Poloniae Pharmaceutica Drug Research*, *73*(3).

Negi, A. (2003). Development and incorporation of fruit powders for value addition of cereal based food products. Doctoral dissertation, Chaudhary Charan Singh Haryana Agricultural University, Hisar.

Nketia, S., Buckman, E. S., Dzomeku, M., & Akonor, P. T. (2020). Effect of processing and storage on physical and texture qualities of oyster mushrooms canned in different media. *Scientific African*, *9*, e00501.

Nollet, L. M. (Ed.). (2004). *Handbook of Food Analysis: Volume 2: Residues and Other Food Component Analysis*. CRC Press.

Nowakowski, P., Markiewicz-Żukowska, R., Soroczyńska, J., Puścion-Jakubik, A., Mielcarek, K., Borawska, M. H., & Socha, K. (2021). Evaluation of toxic element content and health risk assessment of edible wild mushrooms. *Journal of Food Composition and Analysis*, *96*, 103698.

Nowakowski, P., Naliwajko, S. K., Markiewicz-Żukowska, R., Borawska, M. H., & Socha, K. (2020). The two faces of Coprinus comatus—Functional properties and potential hazards. *Phytotherapy Research*, *34*(11), 2932–2944.

Omeljaniuk, W. J., Socha, K., Soroczynska, J., Charkiewicz, A. E., Laudanski, T., Kulikowski, M., ..., & Borawska, M. H. (2018). Cadmium and lead in women who miscarried. *Clinical Laboratory*, *64*, 59–67.

Oso, B. A. (1975). Mushrooms and the Yoruba people of Nigeria. *Mycologia*, *67*(2), 311–319.

Parniakov, O., Lebovka, N. I., Van Hecke, E., & Vorobiev, E. (2014). Pulsed electric field assisted pressure extraction and solvent extraction from mushroom (Agaricus bisporus). *Food and Bioprocess Technology*, *7*(1), 174–183.

Pei, F., Shi, Y., Gao, X., Wu, F., Mariga, A. M., Yang, W., ..., & Hu, Q. (2014). Changes in non-volatile taste components of button mushroom (Agaricus bisporus) during different stages of freeze drying and freeze drying combined with microwave vacuum drying. *Food Chemistry*, *165*, 547–554.

Qin, Y., Zhang, Z., Song, T., & Lv, G. (2017). Optimization of enzyme-assisted extraction of antitumor polysaccharides from hericium erinaceus mycelia. *Food Science and Technology Research*, *23*(1), 31–39.

Raghuramulu, N., Nair, K. M., & Kalyanasundaram, S. (2003). *A Manual of Laboratory Techniques*. Hyderabad National Institute of Nutrition.

Ruthes, A. C., Smiderle, F. R., & Iacomini, M. (2015). D-Glucans from edible mushrooms: A review on the extraction, purification and chemical characterization approaches. *Carbohydrate Polymers*, *117*, 753–761.

Santo Signorelli, S., Conti, G. O., Zanobetti, A., Baccarelli, A., Fiore, M., & Ferrante, M. (2019). Effect of particulate matter-bound metals exposure on prothrombotic biomarkers: A systematic review. *Environmental Research*, *177*, 108573.

Siddhuraju, P., & Becker, K. (2007). The antioxidant and free radical scavenging activities of processed cowpea (Vigna unguiculata (L.) Walp.) seed extracts. *Food Chemistry*, *101*(1), 10–19.

Širić, I., Kasap, A., Bedeković, D., & Falandysz, J. (2017). Lead, cadmium and mercury contents and bioaccumulation potential of wild edible saprophytic and ectomycorrhizal mushrooms, Croatia. *Journal of Environmental Science and Health, Part B*, *52*(3), 156–165.

Smiderle, F. R., Morales, D., Gil-Ramírez, A., de Jesus, L. I., Gilbert-López, B., Iacomini, M., & Soler-Rivas, C. (2017). Evaluation of microwave-assisted and pressurized liquid extractions to obtain β-d-glucans from mushrooms. *Carbohydrate Polymers*, *156*, 165–174.

Sofowora, A. (1993). *Medicinal Plants and Traditional Medicine in Africa*. Spectrum Books, 191–289.

Srivastava, R. P., & Kumar, S. (2000). *Fruit and Vegetable Preservation Principles and Practices*. International Book Distributing Company, pp. 43–56.

Szwengiel, A., & Stachowiak, B. (2016). Deproteinization of water-soluble ß-glucan during acid extraction from fruiting bodies of Pleurotus ostreatus mushrooms. *Carbohydrate Polymers*, *146*, 310–319.

Trease, G. E., & Evans, W. C. (2002). *Pharmacognosy*. Saunders Publishers, 42–44.

USEPA, U. (2000). *Risk-Based Concentration Table*.

Vishnoi, N. R. (1979). *Advanced Practical Chemistry*. Yikas Publication House, pp. 447–449.

WHO. (2011a). *Evaluation of Certain Contaminants in Food*. World Health Organ Tech Rep Ser (959), 1–105 back cover.

WHO. (2011b). *Evaluation of Certain Food Additive and Contaminants*. World Health Organ Tech Rep Ser (960), 1–226 back cover.

WHO. (2013). *Evaluation of Certain Food Additives and Contaminants*. World Health Organ Tech Rep Ser (983), 1–75 back cover.

Wojciechowska-Mazurek, M., Mania, M., Starska, K., Rebeniak, M., & Karłowski, K. (2011). Pierwiastki szkodliwe dla zdrowia w grzybach jadalnych w Polsce. *Bromatologia i Chemia Toksykologiczna*, *44*(2), 143–149.

Xue, D., & Farid, M. M. (2015). Pulsed electric field extraction of valuable compounds from white button mushroom (Agaricus bisporus). *Innovative Food Science & Emerging Technologies*, *29*, 178–186.

Yi, Y., Xu, W., Wang, H. X., Huang, F., & Wang, L. M. (2020). Natural polysaccharides experience physio-chemical and functional changes during preparation: A review. *Carbohydrate Polymers*, *234*, 115896.

You, Q., Yin, X., & Ji, C. (2014). Pulsed counter-current ultrasound-assisted extraction and characterization of polysaccharides from Boletus edulis. *Carbohydrate Polymers*, *101*, 379–385.

Zhang, K., Pu, Y. Y., & Sun, D. W. (2018). Recent advances in quality preservation of postharvest mushrooms (Agaricus bisporus): A review. *Trends in Food Science & Technology*, *78*, 72–82.

Zhang, L., Sun, X., Tian, Y., & Gong, X. (2013). Effects of brown sugar and calcium superphosphate on the secondary fermentation of green waste. *Bioresource Technology*, *131*, 68–75.

Zhishen, J., Mengcheng, T., & Jianming, W. (1999). The determination of flavonoid contents in mulberry and their scavenging effects on superoxide radicals. *Food Chemistry*, *64*(4), 555–559.

CHAPTER 12

Toxic components and toxicology of wild mushrooms

Predrag Petrović
Innovation Center of the Faculty of Technology and Metallurgy, University of Belgrade Karnegijeva 4, Belgrade, Serbia

CONTENTS

DOI: 10.1201/9781003152583-15

12.1 INTRODUCTION

The earliest evidence for human mushroom consumption comes from human dental fossils ("the Red Lady") found in El Miron cave in Spain, dating from the Magdalenian phase of the upper Paleolithic period, some 18,700 years ago; spores of a bolete and an agaric mushroom were found in dental plaque of this fossil's teeth (Power et al., 2015; Straus et al., 2015), suggesting that wild mushrooms were part of people's diet for thousands of years. Mushrooms are regarded as functional food, low in total calories and fats, and high in fiber content, proteins, minerals, and certain vitamins. Many species are praised due to their organoleptic properties, and many are used because of their alleged medicinal properties (Kalač, 2009; Govorushko et al., 2019; Petrović et al., 2019c; Li et al., 2021). Unfortunately, a great number of these mushrooms cannot be cultivated, so the only way to obtain them is to collect them from the wild (Li et al., 2021). The wild mushroom foraging is, however, a potentially dangerous activity, as numerous poisonous species can be easily mistaken for edible mushrooms due to their macroscopic similarities (Unluoglu & Tayfur, 2003; Jo et al., 2014). There is a great variety of toxins produced by fungi, and mushroom poisonings can range from simple gastrointestinal disorders to potentially fatal multi-organ failures. Deaths due to mushroom poisonings have been reported annually. The list of poisonous mushrooms is growing continuously, as well as previously unknown mushroom poisoning syndromes (Beuhler, 2017). However, responsible toxins and/or mechanisms of their action are still not known for a great number of poisonous mushrooms. Mushroom poisonings can be classified in different ways, but they are usually classified by the clinical presentation of poisoning (Beuhler, 2017; White et al., 2019a). This chapter deals with the most important mushroom poisoning syndromes and sums up the relevant and newest information, with a critical approach.

12.2 HEPATOTOXIC MUSHROOMS – AMATOXIN POISONING

Amatoxin containing mushrooms, especially those of the section *Phalloidae* of the genus *Amanita*, are responsible for 90% of the lethal mushroom poisonings around the globe (Enjalbert et al., 2002), and one species is particularly infamous – *Amanita phalloides* or the death cap, which gives the name to the whole section of "lethal Amanitas". Although it originates from Europe, *A. phalloides* has been introduced to North America, Australia, New Zealand, and South Africa. Other notable members of the genus that contain amatoxins are *A. verna from* Europe, *A. virosa* from Europe and northeastern Asia, *A. fulliginea*, and *A. exitialis* from east Asia , *A. bisporigera*, and *A. ocreata* from North America. A recent genetic study showed that there is a great diversity of lethal species of the genus *Amanita* worldwide (Cai et al., 2014), but especially in East Asia, with several new species being described (Cai et al., 2016), with a total of 28 valid species recognized worldwide (Cai et al., 2014). *A. fulliginea* was reported as the most common cause of amatoxin poisonings in southern China, accounting for more than 40% of both mushroom poisonings and deaths due to mushroom poisonings (Chen et al., 2013). Several species in the genus *Galerina* and *Lepiota* also contain amatoxins and were implicated in lethal poisonings, most notably *G. marginata*, *L. helveola*, and *L. brunneoincarnata*, as well as *Conocybe filaris* (Akata et al., 2020, Sun et al., 2019, Ramirez et al., 1993, Haines et al., 1986), although poisoning with these species is far more rarely reported. A study conducted in 1976 reported that traces of amatoxins can be found in many species of edible mushrooms, including *Agaricus sylvaticus*, *Boletus edulis*, and *Cantharellus cibarius* (Faulstich & Cochet-Meilhac, 1976), although the sensitivity of the methods used must be questioned since newer studies failed to find amatoxins even in the closest relatives or most basal members of the *Amanita* sect: *Phalloidae*, *A. areolata*, *A. hesleri*, and *A. zangii* (Cai et al., 2014).

Photograph 12.1. *Amanita phalloides*, the most feared poisonous mushroom. Crni vrh, Serbia, 2014. Photo: P. Petrović

Amatoxins are bicyclic octapeptides and include at least nine related compounds, the best studied of them being α- and β-amanitin (Garcia et al., 2015b). They represent RiPPs: ribosomally synthesized and posttranslationally modified peptides; they are produced by proteolytic cleavage of proproteins of 35 aminoacids, together with modifications of amino acid residues, which includes the formation of an unusual 2′-indole thioether bond between the Cys and Trp residues, resulting in a unique moiety referred to as tryptathionine that forms an inner ring structure (Bills & Gloer, 2017). The structure of α-amanitin is given in Figure 12.1.

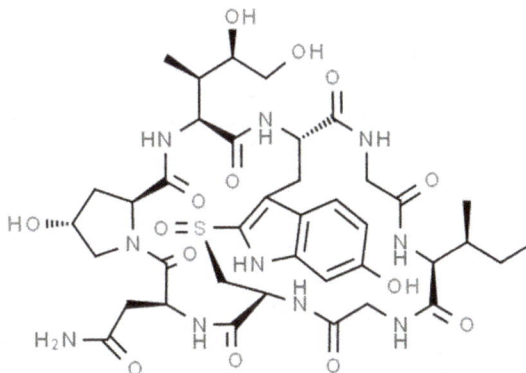

Figure 12.1 α-amanitin chemical structure.

Source: http://www.chemspider.com.

All parts of death cap's fruiting body were found to contain amatoxins: cap, stipe, volva, and even spores. The highest concentration was found in mushroom caps, while the lowest amatoxin content was found to be in volva and spores (Kaya et al., 2013; Garcia et al., 2015c). Zhou et al. (2017) had similar results concerning *A. fulliginea*; they found that the amatoxins in the cap were the most concentrated in lamellae, while spores contained the lowest amatoxin content. Amatoxins are thermostable but water-soluble so the concentration of the toxin in the fruiting bodies can be lowered significantly by parboiling; α-amanitin concentration was showed to be reduced about five times and β-amanitin about eight times (Sharma et al., 2021). They are also resistant to low stomach pH, as well as proteolytic enzymes (Garcia et al., 2015b). Amatoxins are readily absorbed from GIT and quickly eliminated from blood as they do not bind to albumin. The liver is the main target of their accumulation and toxicity (Garcia et al., 2015b) as they undergo the active uptake by hepatocytes via OATP1B3 ("Organic anion transporting polypeptide 1B3") and NTCP transporter (Rodrigues et al., 2020); kidneys are also exposed to the toxin as amatoxins are mainly excreted via urine. The part of the toxins that are excreted via feces can be reabsorbed from GIT, and thus undergo enterohepatic circulation (Garcia et al., 2015b). The LD_{50} in humans for orally administered α-amanitin is estimated at 0.1 mg/kg. The average fully developed fruiting body of *A. phalloides* contains 5–8 mg of amatoxins, which means that ingestion of a single fruiting body may be lethal (Sharma et al., 2021).

Apart from amatoxins, the deadly poisonous species in the *Amanita* genus contain structurally very similar phallotoxins and virotoxins. Phallotoxins also have bicyclic structure but are heptapeptides and they include at least seven different compounds, while virotoxins are monocyclic heptapeptides, consisting of at least five different compounds. Both phallotoxins and virotoxins are shown to interfere with cytoskeletal components *in vitro*, and cause death in mice after intraperitoneal injection after 2–5 hours, due to liver damage. LD_{50} dose of phallotoxins and virotoxins in mice varies from 1–20 mg/kg, depending on the compound. However, their contribution to the overall toxicity of deadly poisonous mushrooms in the *Amanita* genus is thought to be negligible, as they are poorly absorbed from GIT (Garcia et al., 2015b). A study showed that both amatoxins and phallotoxins may be stable in dry conditions for nearly two decades; they were found to be detectable in 17-year-old dried herbarium specimens of *A. phalloides* and 10-year-old specimens of *A. marmorata* (as *A. reidii*), despite some reports stating that phallotoxins are easily degraded upon drying (Hallen et al., 2002).

The best-understood amatoxin action in living cells is the inhibition of RNA polymerase II, which consequently blocks protein synthesis, thus interferes with normal cell functions, leading to cell death (Wieland, 1983). Although amatoxins bind non-covalently to the enzyme, the inhibition may be irreversible due to consequent degradation of RNA polymerase II largest subunit, RPB1, as suggested by Nguyen et al. (1996). However, other mechanisms of amatoxin action were also identified, questioning the precise role of protein depletion in early hepatic dysfunction stages. Several *in vitro* studies on hepatocytes found that α-amanitin causes induction of p53-mediated apoptosis (Chen et al., 2020; Garcia et al., 2015); it is suggested that this effect, seen in other RNA polymerase inhibitors as well, may be correlated with the transcription inhibition, although it is not clear how (Ljungman et al., 1999). When induced, p53 forms complexes with mitochondrial protective proteins, which leads to alteration of mitochondrial membrane permeability and release of cytochrome-C into the cytosol, causing activation of the mitochondrial apoptotic pathway. It was also found that p53-deficient cells were less sensitive to death induced by α-amanitin, confirming the importance of p53-mediated toxicity (Garcia et al., 2015b).

A recent *in vivo* metabolomic study in mice suggested that mitochondrial dysfunction plays an important role in the early phase of amatoxin poisoning; α-amanitin was shown to cause progressive energy cell disorders, affecting sugar, arginine, CoA, and TCA cycle metabolism; adaptive processes that tend to bring these pathways "back to the center" are soon blocked by inhibition of protein synthesis, which leads to cell dysfunction, death, and consequent liver failure (Chen et al., 2021). Oxidative stress is also thought to be an important part of the amatoxin mechanism of action; it was shown that α-amanitin increases the concentration of total and mitochondrial ROS in hepatocytes, and that activation of apoptotic mitochondrial pathways is associated with oxidative damage (Chen et al., 2020). The role of TNFα in the α-amanitin toxicity mechanism was also studied and there is evidence that it might exacerbate its toxicity (Garcia et al., 2015b). Combining proteomic, genomic, and metabolomic studies in the future may finally elucidate the precise mechanism of amatoxin actions and connect all the puzzle pieces, thus provide new and more effective therapeutical procedures.

Amatoxin poisoning has three distinct phases. During the first, gastrointestinal phase, symptoms of poisoning are non-specific and include nausea, vomiting, abdominal pain, and diarrhea, followed by general weakness due to loss of water and electrolytes. These symptoms take place shortly after amatoxin-containing mushroom ingestion, usually after 3–6 hours, but they may be delayed for up to 24 hours. It is believed that phallotoxins, and not amatoxins are responsible for transitory gastrointestinal dysfunction; phallotoxins, unlike amatoxins, do not undergo systemic absorption. In the next, latent phase, which typically starts 24 hours upon ingestion, gastrointestinal symptoms resolve, often misleading that the patient has recovered. With the progress of the second phase, the impairment of both hepatic and renal function, caused by amatoxins worsens but is still compensated so the symptoms associated with this impairment take place in the third and final, hepato-renal phase, 3–5 days after ingestion (Allen et al., 2012; Garcia et al., 2015b). Thrombocytopenia was also reported from a case series that involved *A. fuliginea* poisonings in China (Wang et al., 2020). If not treated, hepatic and renal failure may occur, leading to multiorgan failure and death. Early diagnosis is crucial for clinical outcomes. If amatoxin-containing mushroom poisoning is suspected during the early, gastrointestinal phase, gastric lavage, and/or administration of activated charcoal are recommended (Allen et al., 2012; Ye & Liu, 2018). Drug treatment includes several chemically different compounds, although none of them can prevent binding of amatoxin to RNA polymerase II and consequent degradation of ribosomal subunits; in other words – none acts as an antidote (Rodrigues et al., 2020). These drugs include *N*-acetylcysteine (NAC), benzylpenicillin, lipoic acid, sylibin, but also vitamins with antioxidant activity (vitamins C and E) and certain hormones (steroids and insulin combined with growth hormone or glucagon) (Enjalbert et al., 2002; Poucheret et al., 2010). As depletion of glutathione, an important intracellular antioxidant, is characteristic biochemical change during the amatoxin poisoning, its precursor, *N*-acetylcysteine has been used to boost cells' capacities to fight reactive oxygen species produced in hepatocytes; *N*-acetylcysteine alone also acts as a free radical scavenger and may alter amatoxin tryptathione bridge, which is essential for its activity. Although its efficacy has been questioned by some, the largest cohort study undertaken so far, which analyzed the documented results of various protocols used for the treatment of amatoxin poisonings found that *N*-acetylcysteine was correlated with positive outcomes. Other compounds with purely antioxidant activity did not show the same efficacy (Enjalbert et al., 2002; Poucheret et al., 2010; Allen et al., 2012; Ye & Liu, 2018). Benzylpenicillin is the most frequently used drug in amatoxin poisoning treatment. It acts as an amatoxin uptake inhibitor, competitively binding to the same receptor – OATP1B3 transporter. However, its clinical efficacy was seriously questioned nowadays, as it has been shown that benzylpenicillin does not have any influence on overall mortality. This might be because relatively high doses of benzylpenicillin are needed for efficient uptake-inhibitory effect, doses that might not be safe to use, so it is now recommended that benzylpenicillin should be administered only if other therapeutic options are not available. The time point of administration may be crucial for benzylpenicillin efficacy, considering its mechanism of action. Sylibin, the main constituent of sylimarine – a lignoflavonoid complex isolated from milk thistle (*Sylibum marianum*) may be the closest to an antidote when it comes to amatoxin poisoning treatment. Its clinical efficacy to prevent serious liver damage and death is well established, so it became a drug of the first choice in the treatment of amatoxin poisoning. It acts both as an uptake inhibitor, as benzylpenicillin, but also as a very potent antioxidant, scavenging ROS (Enjalbert et al., 2002; Poucheret et al., 2010; Allen et al., 2012; Ye & Liu, 2018). Polymyxin B was recently found to bind to the same site on RNA Polymerase II as α-amanitin, thus preventing amatoxin binding, and was marked as a potential "breakthrough drug" in the treatment of amatoxin poisoning. *In vivo* studies showed that it could significantly reduce hepatic and renal injury and increase survival in α-amanitin treated animals (Garcia et al., 2015a). *In vitro* studies, however, didn't confirm the proposed polymyxin's mode of action, as it had much less affinity to bind to RNA Polymerase II than α-amanitin and could not prevent its toxicity, but neither could benzylpenicillin, *N*-acetyl cysteine, or sylibin, suggesting the complexity of their *in vivo* actions (Rodrigues et al., 2020). *Ganoderma lucidum* decoction (referred to as "glossy ganoderma decoction"), a traditional Chinese remedy was recently shown, in a limited clinical study to be potentially beneficial as an adjuvant in the treatment of amatoxin poisoning, with reduced mortality among the poisoned, due to the alleged hepatoprotective properties (Xiao et al., 2007; Ye & Liu, 2018). In severe cases of liver failure, transplantation may be the only treatment option (Allen et al., 2012).

12.3 NEPHROTOXIC MUSHROOMS

12.3.1 Amanita renal failure poisoning syndrome

While species of genus *Amanita* in section *Phalloideae* cause predominantly hepatic failure, species in the section *Lepidella* and *Amidella* can cause severe renal failure (Fu et al., 2017). This syndrome was relatively recently described from North American *Amanita smithiana* (Pelizzari et al., 1994), and later from European *A. proxima* (De Haro et al., 1997) and several Asian species of the same section, such as *A. pseudoporphyria* (Iwafuchi et al., 2003), *A. oberwinklerana* (Fu et al., 2017), and *A. neoovoidea*. The previously reported primarily nephrotoxic poisoning caused by species of *Amanita* mushrooms might have been associated with *A. phalloides* (West et al., 2009). The symptoms of poisoning caused by ingestion of these mushrooms include transient gastrointestinal disorder which presents after 5–6 hours on average (20 minutes–12 hours) and include nausea, vomiting, abdominal pain, and sometimes diarrhea, similar to the early phase of amatoxin containing mushroom poisoning. The renal phase of poisoning can be seen on the very same day of mushroom ingestion or it may be delayed, and present after 3–6 days. A biopsy of a patient's kidney indeed showed all signs of tubular necrosis. A mild liver injury that resolves quickly can be seen in some cases, and it seems that it is associated with early renal failure (Bains and Chawla, 2020). The treatment is supportive and it includes hemodialysis, which may be required for up to seven weeks; however, renal failure is reversible and renal function seems to return to normal so patients do not require kidney transplantation (West et al., 2009; Warden & Benjamin, 1998). The non-proteinogenic amino acid, 2(S)-amino-4,5-hexadienoic acid ("allenic norleucine"; Figure 12.2), isolated in the 1970s, was suggested to be the main toxic principle responsible for this syndrome (West et al., 2009). Apart from *A. smithiana*, the presence of allenic norelucine was

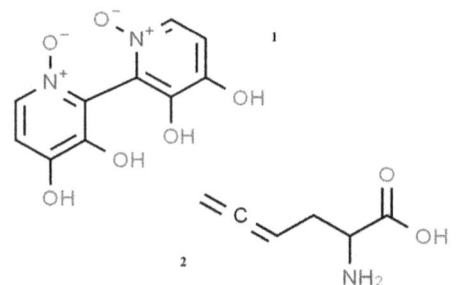

Figure 12.2   Chemical structures of orellanine (1) and 2(S)-amino-4,5-hexadienoic acid (2).

Source: http://www.chemspider.com.

confirmed in *A. smithiana*, *A. pseudoporphyria*, *A. neoovoidea*, and *A. abrupta* (Fu et al., 2017), the first three of which are known to have caused nephrotoxicity in humans. The *in vivo* studies in guinea pigs showed that administration of amino hexadienoic acid causes renal damage as seen in *A. smithiana* poisoning (West et al., 2009), but it had a relatively high lethal dose of 100 mg/kg, and the toxic effects seen at doses of 33 mg/kg (Pelizzari et al., 1994). On the other hand, allenic norleucine content in *A. smithiana* was found to be 0.1% dry weight, which can be characterized as relatively low, prompting the question if an unusual amino acid is the (main) responsible toxin in these mushrooms (Kirchmair et al., 2011). Pelizzari et al. (1994) partially purified a compound from *A. smithiana* by TLC, which showed toxicity towards two epithelial renal cell lines and referred to it as a possible amino sugar, consisting of allenic norleucine and a sugar moiety. The incubation time *in vitro* was found to be 12 hours, which is again consistent with *in vivo* poisoning progression (West et al., 2009). Kirchmair et al. (2011) used the same TLC technique and detected the identical compound in *A. gracilior*, *A. boudieri,* and *A. echinocephala*. They did not detect it in *A. strobiliformis*, an edible member of the section *Lepidella*, but neither in *A. proxima*, nephrotoxic species from the section *Amidella*. They argued that the toxic compound was either 2(S)-amino-4,5-hexadienoic acid, or an amino sugar, but did not perform any further analytical analysis to strengthen their theory. They also concluded that *A. proxima* contains different toxic principle(s). The mechanism of toxic compounds is yet to be elucidated (West et al., 2009).

12.3.2 Orellanine poisoning syndrome

Webcaps (*Cortinarius*) represent one of the largest mushroom genus in the world, with an estimated 2,000–3,000 species worldwide (Danel et al., 2001). However, little was known about how dangerous some *Cortinarius* species are, until a series of kidney failure–related deaths in Poland were associated with webcap ingestion during the 1950s. A toxin responsible for kidney injury was characterized several years later and named orellanine (Figure 12.2) after *Cortinarius orellanus*, the species involved in poisonings in Poland (Rapior et al., 1987). The ingestion of orellanine containing mushrooms causes primarily delayed nephrotoxicity (White et al., 2019a), making it difficult to establish a link between the ingestion of mushrooms and the symptoms. Orellanine is known to be present in species of the subgenus *Cortinarius* (formerly *Leprocybe*), section *Orellanii*, which include European species *C. orellanus* and *C. rubellus* (syn. *C. speciosissimus*, *C. orellanoides*), the most frequent causes of human poisonings, and also North American species *C. orellanosus* and *C. rainierensis* (possibly another synonym of *C. rubellus*) and in lower concentration in *C. armillatus*. There are many species of webcaps and they are difficult to identify, so edible webcaps, like *C. caperatus* and *C. praestans* may easily be mistaken for poisonous species in the genus. As poisonous webcaps have relatively small fruiting bodies, with reddish tones, they have been also mistaken for chanterelles (Shao et al., 2016). Reports of confusion of psychedelic *Psilocybe* mushrooms with *Cortinarius* species also exist (Wörnle et al., 2004; Calviño et al., 1998). The great majority of reported poisonings had taken place in Europe and more than half of that in Poland (Michelot & Tebbett, 1990, Danel et al., 2001). Poisoning with webcaps is more rarely reported from North America (Brandenburg & Ward, 2018). Orellanine is an alkaloid with a bipyridine structure that resembles diquat and paraquat, widely used synthetic herbicides. Interestingly, orellanine produces the same effect in plants as bipyridines, interfering with the photosynthesis process and decreasing chlorophyll levels; it is thought to act through its free radical form which damages chloroplast membranes (Høiland, 1983).

Orellanine is a white crystalline substance that may constitute up to 1.5% of the dry mass of *C. orellanus* and 0.9% of *C. rubellus* (as *C. speciosissimus*, Prast et al., 1988). It is photosensitive, with photodegradation products being non-toxic orellinine and orellin, which fluoresce under UV light (Laatsch & Matthies, 1991). Mushrooms containing orellanine may thus be identified using this ability. Orellanine is, on the other hand, heat resistant (Bunel et al., 2018).

Similar to liver's sensitivity to amatoxins, renal tissue seems to actively accumulate orellanine, probably via cellular uptake. This fixation to renal tissue might take place fast upon absorption, as orellanine could not be detected in blood or urine of victims two days upon ingestion. Mechanism of orellanine nephrotoxic action is not yet uncovered; the evidence suggests it acts at the level of the tubular system, proximal tubular epithelium, to be precise, so the glomerular function in the presence of the toxin is preserved (Nilsson et al., 2008). The evidence also suggests that the mechanism of toxicity includes promotion of oxidative stress as well as protein synthesis inhibition (Richard et al., 1991; Nilsson et al., 2008; Bunel et al., 2018), particularly those involved in cellular defense against oxidative damage, superoxide dismutase and catalase; it was found that these enzymes were down-regulated in renal tissue of rats, at both mRNA and protein level. The most dramatic effect is the down-regulation of glutathione peroxidase 3 (GPX3) at mRNA level in mice, to less than 20%, although the function of this circulating enzyme which is synthesized in kidneys is still unclear (Nilsson et al., 2008). *In vitro* studies showed that orellanine may act through the free radical generation and it is highly indicated that this happens *in vivo*, too. Orellanine was shown to easily go through the redox cycle *in vitro*, which may also occur under physiological conditions. Orellanine is oxidized to the orthosmeiquinone anion radical, which can then be reduced by glutathione or ascorbate; this *circulus viciosus* could easily lead to depletion of cell antioxidant capacities, namely gluthatione, and consume oxygen, leading to cell dysfunction (Richard et al., 1995; Bunel et al., 2018). The precise connection between protein synthesis inhibition and oxidative stress is not fully elucidated.

After speculations that there were toxins of peptidic nature present in poisonous *Cortinarius* species, Tebbett and Caddy (1984) reported the presence of cyclopeptide compounds from *Cortinarius* species, similar to amatoxins and phallotoxins. They allegedly elucidated the structures of three different types of these peptides, named cortinarin A, B, and C. They even performed *in vivo* toxicological studies, proving cortinarin A and B being allegedly nephrotoxic. They found that cortinarin A was practically ubiquitous within the genus, detecting it in more than 50 species. They also found that cortinarin B was present only in known toxic species, together with a relatively high content of cortinarin A. However subsequent studies could not confirm the validity of cortinarins' toxic nature, or even their existence. Prast et al. (1988) performed a thorough screening of *C. orellanus* and *C. rubellus* (as *C. speciosissimus*) for toxic compounds and the only toxic compound they identified in mushroom extracts was orellanine. They commented on the proposed toxic peptides and argued that taking into account the (scarce) data provided by Tebbett and Caddy, cortinarins must be present in much lower concentrations than orellanine in *C. rubellus* and contribute much less to nephrotoxicity, suggesting that could be a reason why they failed to detect them. Laatsch and Matthies (1991) questioned cortinarins' existence as they were not able to replicate the study and could not find any evidence of the existence of such compounds in *C. rubellus*. The fluorescent compound dubbed cortinarin A in the previous study was found to be a steroid compound, an ergosterol derivative. UV spectrum of supposed cortinarin A, provided by Tebbett and Caddy does resemble that of ergosterol. Despite the obvious lack of convincing data, cortinarins are still cited in the literature as toxic compounds of webcaps.

Clinical presentation of orellanine poisoning is best summed up by Danel (2001), who reviewed 245 reported cases. Two phases of poisoning can be distinguished, prerenal and renal phase. The prerenal phase often presents with delay, with first symptoms appearing 12 hours to 14 days (3 days median) after mushroom ingestion, which increases the risk of repeated ingestion of meals containing the mushrooms. The prerenal phase is characterized by non-specific symptoms, mainly of gastrointestinal nature. Vomiting was reported to be the most common GI-related symptoms, followed by nausea, abdominal pain, diarrhea, and anorexia. Besides this, burning sensation in the mouth and polydipsia was reported to occur in almost half of patients and polyuria in about one-fifth. Other symptoms include headache, asthenia, and myalgia. Liver injury was reported to occur in this early phase, however, data are inconsistent; if it does occur, it is likely transitory and resolves quickly. In cases where an impaired liver function was observed, it is possible that the identification of the consumed mushroom was incorrect. Mild poisonings may be limited only to the GI phase; however, about 70% of patients progress towards renal failure. A symptom-free period may precede the renal phase, which develops after 4–15 days after ingestion of orellanine containing mushrooms (Danel et al., 2001), and up to 20 days in some cases (Bouget et al., 1990). The renal phase is characterized by symptoms such are lumbar pain, anuria, oligo-, or polyuria and, biochemically, by leucocyturia, hematuria, and proteinuria, which may or may not develop. Kidney biopsy shows lesions due to tubulointerstitial nephritis with damaged tubular epithelium, which may be accompanied by interstitial edema with inflammatory infiltrates and early stadium of interstitial fibrosis. Treatment is supportive and there are no known antidotes or standardized therapeutic protocols. Haemodialysis is indicated in those with impaired kidney function. In about 50% of patients, acute renal failure progresses towards chronic renal failure; in severe cases kidney transplantation is necessary. Corticosteroids were reported to be inefficient in the treatment of orellanine poisoning (Bouget et al., 1990; Danel et al., 2001; Bunel et al., 2018). Antioxidant therapy using N-acetylcysteine was associated with favorable outcomes in a couple of cases (Wörnle et al., 2004), but the sample size is too small and inter-individual differences in response to toxin too great to have any definite conclusions. A study in mice using systemic administration of superoxide dismutase was also shown to be ineffective; it worsened kidney damage and increased mortality in tested animals (Nilsson et al., 2008). A great inter-individual response to orellanine was best seen in the case of the poisoning of 26 French soldiers who consumed webcaps during a survival exercise in 1987. They all consumed approximately the same portion of mushrooms, but only 12 of them developed acute renal failure and eight of them needed hemodialysis. Chronic renal failure that lasted for several months was consequently developed in four patients; one patient was maintained under chronic dialysis and one patient was successfully transplanted (Bouget et al., 1990). Genetic and other factors might be responsible for susceptibility to orellanine.

12.4 NEUROTOXIC MUSHROOM POISONING SYNDROMES

12.4.1 Muscarinic poisoning syndrome

Muscarine is probably the best-known mushroom toxin, giving its name to a type of acetylcholine receptors, the muscarinic receptors. Although isolated from *Amanita muscaria*, and named after it, the archetype poisonous mushroom contains only traces of muscarine and owes its toxicity primarily to isoxazole alkaloids (Michelot & Melendez-Howell, 2003). Muscarine is on the other hand the principal toxin of poisonous *Inocybe* and *Clitocybe* species, being found in variable content in these species and reaching up to 1.6% of dry mass (Lurie et al., 2009). Reported poisoning cases with mushrooms belonging to the *Inocybe* genus involved *I rimosa (as I. fastigiata), I. geophylla, I. patouillardii, I. tristis, I. serotine,* and others (Lurie et al., 2009; White et al., 2019a; Sai Latha et al., 2020; Xu et al., 2020), but there are dozens of other *Inocybe* species that are found to contain muscarine (Barceloux, 2008; Sai Latha et al., 2020), and the list is growing continuously. Some of the *Clitocybe* species that contain muscarine include *C. dealbata, C. rivulosa,* and *C. phyllophylla* (Genest et al., 1968, Barceloux, 2008). *Mycena pura* was also found to contain muscarine (Barceloux, 2008). Some species of the Boletales order were reported to be involved in muscarinic poisonings (Pauli & Foot, 2005), but there doesn't seem to be enough evidence for the claims.

Photograph 12.2. *Clitocybe dealbata*, one of the mushroom species containing high levels of muscarin. Bor district, Serbia, 2018. Photo: P. Petrović

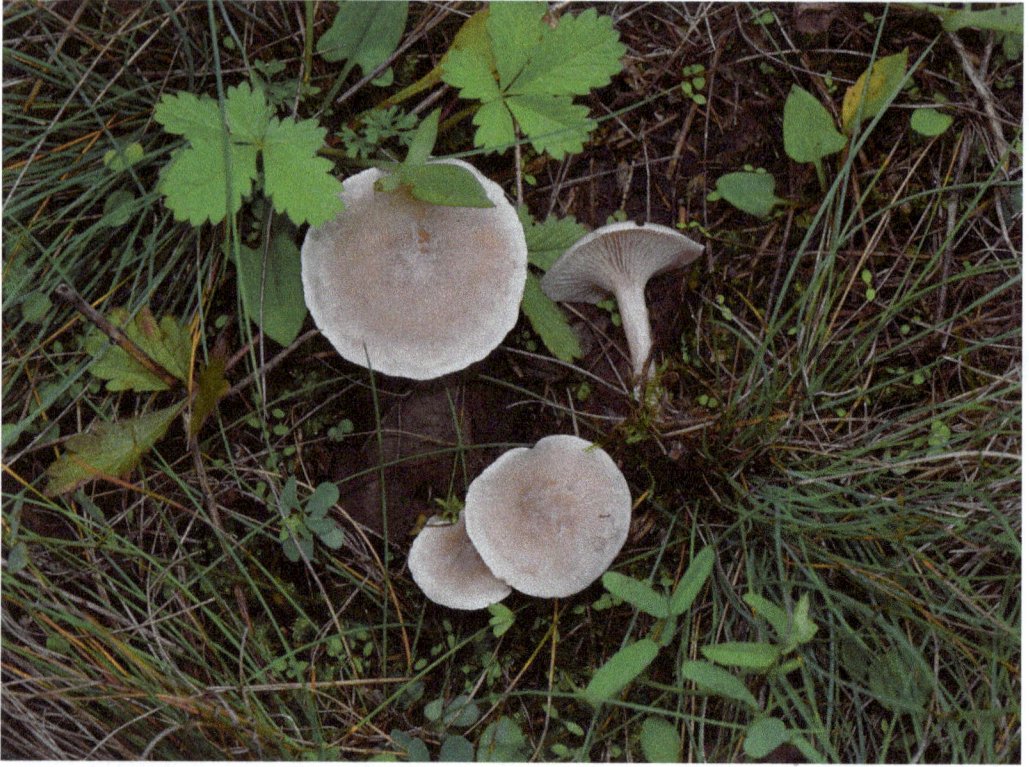

Figure 12.3 Chemical structure of muscarine.

Source: http://www.chemspider.com.

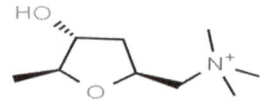

Muscarine (Figure 12.3) is water-soluble and heat resistant so heat treatment of mushrooms does not lower its content. There are four enantiomers of muscarine and only one is active, L-(+)-muscarine. Muscarine causes "cholinergic" or, more precisely "parasympathomimetic" poisoning as it acts as a non-selective agonist on muscarinic receptors, as it bears structural similarity with acetylcholine (Barceloux, 2008); muscarinic receptors are one of two types of acetylcholine receptors, the other being nicotinic receptors (Picciotto et al., 2012). There are five subtypes of muscarinic receptors, localized in different tissues; they are involved in regulating heartbeat, contraction of smooth muscles, glandular secretion, and various CNS functions (Kruse et al., 2014). Muscarin does not cross the blood-brain barrier so it exerts only peripheral effects (Peredy & Bradford, 2014). The symptom onset is relatively fast; they can occur within minutes to 2 hours (5 hours in one case) after poisonous mushroom ingestion, probably depending on the mushroom preparation method. The described symptoms of poisoning with muscarine containing mushrooms include increased secretion from endocrine and exocrine glands, as well as stomach, leading to profuse sweating, increased salivation and lacrimation; bradycardia and sometimes hypotension, bronchoconstriction and dyspnea, abdominal pain, and diarrhea (due to smooth muscle contraction), and miosis and blurred vision (Chew et al., 2008; George & Hegde, 2013; Peredy & Bradford, 2014). Unlike acetylcholine, it cannot be inactivated enzymatically by acetylcholine esterase, as it does not contain ester, but ether bond, so its activity is prolonged compared to acetylcholine. Muscarine-containing mushroom poisonings are usually mild, due to the low bioavailability of muscarine and the fact that most species do not contain large quantities of muscarine. Symptoms usually resolve within 6 hours, but in rare cases can be serious, leading to cardiovascular collapse (Barceloux, 2008), and sometimes even be fatal (George & Hegde, 2013). A lethal dose of muscarine in humans is estimated at 180–300 mg (Patočka et al., 2021). Death is the result of primarily cardiovascular collapse or lung failure. Besides supportive therapy, the specific treatment includes administration of atropine, which is a specific muscarine antidote (Barceloux, 2008), as it acts parasympatholytic, being a competitive muscarinic receptor antagonist (North & Kelly, 1987).

12.4.2 Mycoatropinic poisoning syndrome

"Mycoatropinic syndrome" is caused by several species belonging to the genus *Amanita*; the name refers to psychotropic effects that closely resemble those seen in poisoning with plants containing tropan-alkaloids such are *Atropa belladonna*, *Hyoscyamus niger,* or *Datura stramonium*. Instead of tropan-alkaloids, these mushrooms contain izoxazole derivatives (Satora et al., 2005). Fly agaric (*Amanita muscaria*) and its close relative, panther cap (*Amanita pantherina*) from the section *Amanita*, are the most well-known and the most important in this group of poisonous mushrooms. Both species are widely distributed across the Northern Hemisphere but can be found in South America, Africa, Australia, and New Zealand (Oda et al., 2004), where they have been introduced (Diaz, 2017). Other species from the same genus section also contain isoxazoles and may cause the same type of poisoning. *Amanita ibotengutake* was recently described in Japan (Oda et al., 2002). *A. gemmata* was reported to contain these substances as well, but in lower concentration than fly agaric and panther cap. Some of the other members of the genus that are reported to contain isoxazoles are *A. multisquamosa* (as *A. cothurnata*, sometimes considered as a variety of *A. pantherina*), *A. aprica*, and *A. regalis* (considered by some authors as a variety of *A. muscaria*) (Chilton & Ott, 1976; Moss & Hendrickson, 2018). Due to psychotropic effects, *A. muscaria* was once used for ritualistic/religious purposes in different parts of the world, particularly Siberia (Tsujikawa et al., 2006; Peredy & Bruce, 2014); however, more widespread use in Europe seems to be for killing flies (Lumpert & Kreft, 2016); hence the name (Peredy & Bruce, 2014). The compounds mainly responsible for the toxicity of *A. muscaria* and *A. pantherina* are isoxazole derivatives, ibotenic acid, and its decarboxylated derivative, muscimol (DeFeudis, 1980; Michelot & Melendez-Howell, 2003). Ibotenic acid (Figure 12.4) is structurally similar to glutamic acid, representing its "conformationally restricted analog" (Schwarcz et al., 1979), so it acts as a non-specific agonist of various glutamate receptors but has the greatest affinity for *N*-methyl-D-aspartic acid (NMDA) receptor, thus producing excitatory effects on neurons (Nielsen et al., 1985; Zinkand et al., 1992; Peredy & Bruce, 2014). As ibotenic acid has more affinity to glutamate receptors than glutamate itself, this can lead to neuronal death and the creation of brain lesions. This has led to the ibotenic acid's use as a model substance for inducing neurodegenerative processes in test animals (Schwarcz et al., 1979; Connor et al., 1991; Martínez-Torres et al., 2021). However, these effects are seen only when ibotenic acid is administered directly to certain brain regions, intracranially. It is generally thought that, when ingested, ibotenic acid acts through its metabolite, muscimol (Figure 12.4) and that most symptoms of *A. muscaria* poisoning correspond to the effects produced by muscimol when administered alone.

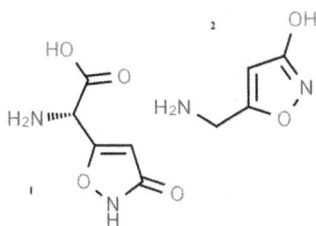

Figure 12.4   Chemical structures of ibotenic acid (1) and muscimol (2).
Source: http://www.chemspider.com.

Photograph 12.3. *Amanita muscaria*, the archetypal poisonous mushroom, Crni vrh, Serbia, 2014. Photo: P. Petrović

Ibotenic acid is relatively unstable and can easily undergo decarboxylation during fruiting body desiccation, heat treatment (cooking), or passage through acidic content of the stomach (Michelot & Melendez-Howell, 2003; Diaz, 2017). Decarboxylation was shown to take place spontaneously in physiological conditions and enzymatically, as ibotenic acid is a substrate for glutamate decarboxylase, and the enzymatic decarboxylation of the ibotenic acid happens at even faster rates than decarboxylation of glutamate, its physiological substrate (Nielsen et al., 1985). Despite this, ibotenic acid (as well as muscimol) can be found in urine and plasma even several hours after ingestion of *A. muscaria* or *A. pantherina*, and ibotenic acid was reported to be present at higher levels than muscimol in both plasma and urine (Stříbrný et al., 2011; Xu et al., 2020). Unlike glutamate and GABA, both ibotenic acid and muscimol can cross the blood-brain barrier, probably by active transport (Michelot & Melendez-Howell, 2003). Most studies suggest that, after oral administration, about 10–20% of ibotenic acid is converted to muscimol, a claim which is supported by comparing doses of ibotenic acid and muscimol needed to trigger psychotropic effects (Peredy & Bruce, 2014; Diaz, 2017; Voynova et al., 2020). Muscimol, on the other hand, bears structural similarity to the main inhibitory neurotransmitter in CNS, γ-aminobutyric acid (GABA), and acts as a non-selective agonist on GABA receptors. In humans, a dose of about 6–10 mg of muscimol is enough to produce psychotropic symptoms including dizziness, confusion, hallucinations, and (deep) sleep with vivid dreams, the same symptoms that are seen in most *A. muscaria*/*A. pantherina* poisoning cases. Higher doses however can cause nausea and vomiting (Voynova et al., 2020), which may explain gastrointestinal symptoms that are sometimes seen in *A. muscaria*/*A. pantherina* poisonings. Muscimol content in fresh *A. muscaria* is within the range of 0.03 up to 0.1% (DeFeudis, 1980), so one cap is enough to induce the psychotropic symptoms (Peredy & Bruce, 2014). The highest concentration of toxins is found in caps (Diaz, 2017), particularly in the cap cuticle and in the flesh right beneath the cuticle.

Both ibotenic acid and muscimol are readily soluble in water so it is reported that fly agaric may be safely consumed if the cuticle is peeled off and water in which mushrooms were cooked discarded (Michelot & Melendez-Howell, 2003). A study found that summer forms of these mushrooms have more psychotropic potential than autumn forms, with ten times more ibotenic acid content (Vendramin & Brvar, 2014). There are other related, biologically active compounds found in mushrooms of the section *Amanita*; muscazone is a dihydrogenated oxazole derivative, formed when ibotenic acid is subjected to UV light. It possesses similar psychotropic properties but to much less extent. Stizolobic and stizolobinic acids have been found only in *A. pantherina* and *A. gemmata* so farand are reported to act excitatory on isolated rat spinal cord. Muscarine, although named after *A. muscaria* from which it was originally isolated and characterized, is found in minute concentrations (0.0002–0.0003% fresh weight of *A. muscaria*) (Michelot & Melendez-Howell, 2003; Voynova et al., 2020) so muscarinic symptoms are very rare (Satora et al., 2005).

Poisoning symptoms start 30 minutes to 2 hours upon ingestion, peaking in 3 to 4 hours, and include confusion, agitation, euphoria, and hypersensitivity of visual and auditory perception, space and time distortion. Myoclonus and convulsions/seizures may appear in rare cases. After the excitatory phase, drowsiness (somnolence) takes place and victims usually fall into a long, deep coma-like sleep. The symptoms usually resolve within 12–24 hours. Gastrointestinal symptoms are reported, but inconsistently; vomiting is more often seen than diarrhea (Michelot & Melendez-Howell, 2003; Vendramin & Brvar, 2014; Diaz, 2017). One study that investigated cases of *A. muscaria* and *A. pantherina* poisoning found slight differences in the presentation of symptoms in these two species, with agitation and confusion more frequently experienced after *A. muscaria* ingestion and coma after *A. pantherina* ingestion. The slight difference in intoxication symptoms between these two mushrooms might be due to different ibotenic acid/muscimol content ratios some authors reported; *A. muscaria* was found to contain higher levels of ibotenic acid compared to muscimol, while *A. pantherina* contained more muscimol (Vendramin & Brvar, 2014); *A. pantherina* may thus act more quickly in respect to sedative effects, as ibotenic acid needs to be converted to muscimol first to activate GABA receptors. Most intoxications are mild and severe cases with death outcomes are very rare (Michelot & Melendez-Howell, 2003). In recent times, several deaths connected to poisonings with *A. muscaria* have been reported from North America; in most cases, ingestion of large quantities of *A. muscaria* caps was involved. In one fatal case, a patient suffered from cardiorespiratory arrest and consequent anoxic brain injury; in another, the fatal outcome was an indirect consequence of the induced comatose state, as a young camper froze to death after ingesting *A. muscaria* fruiting bodies (Watson et al., 2005; Beug et al., 2006; Beug, 2009). Severe cases in which prolonged psychosis or coma occurred and needed prolonged hospitalization were also reported from Slovenia, Poland, and Switzerland (Brvar et al., 2006; Mikaszewska-Sokolewicz et al., 2016; Rampolli et al., 2021). The largest cohort study on *A. muscaria* poisoning in Europe (Slovenia) found that in 90% of poisoning cases *A. muscaria* was mistaken for *A. caesarea*, a highly prized, edible member of the genus. *A. pantherina* is, on the other hand, most commonly confused with *A. rubescens* (Vendramin & Brvar, 2014), but also with *Macrolepiota procera* and *A. spissa* (Satora et al., 2006). Poisonings with other izoxasole containing mushrooms are rarely reported; a series of *A. regalis* poisonings were documented from Finland (Elonen et al., 1979) and one recent case of poisoning with *A. aprica* was reported from the USA (Moss & Hendrickson, 2018). The recreational use of *A. muscaria* seems to be increasing, particularly in younger people (Rampolli et al., 2021). In Japan, several products containing dried fly agaric and panther cap are reported to be circulating and can be even bought online (Tsujikawa et al., 2006).

12.4.3 Psilocybin-containing mushrooms

The greatest number of all known poisonous mushrooms probably belongs to a group of psychoactive mushrooms that contain psilocybin and psilocin as the main active compounds, tryptamine derivatives, and structural analogue of serotonin. These mushrooms are thus often referred to as psilocybin mushrooms, hallucinogenic mushrooms or "magic mushrooms" (Tylš et al., 2014; Dasgupta, 2019). There are more than 150 species known to produce psilocybin (Dasgupta, 2019), with the great majority of them belonging to the genus *Psilocybe* (Guzmán, 2008). Other genera with species that contain psilocybin are *Panaeolus, Stropharia, Conocybe, Gymnopilus, Pluteus,* and even *Inocybe* (Tylš et al., 2014), known primarily as muscarine-containing mushrooms. Species of the genus *Psilocybe* are however the most important from the toxicological point of view, as they have been implicated in most of the reported poisonings in the literature. Poisonings with psilocybin mushrooms are specific as they are almost exclusively due to intentional ingestion. Psilocybin mushrooms have been used for thousands of years for religious purposes; they are particularly well documented as being of great importance both in healing rituals and religious ceremonies of Pre-Columbian Mesoamerican cultures. Today, they are used recreationally (Badham, 1984; Carod-Artal, 2015), especially by young people (Dasgupta, 2019).

Psilocybin and its dephosphorylated analog psilocin (Figure 12.5) are the main active trypta-mines of these mushrooms (Dasgupta, 2019). Their content varies between 0.2 and 1% for most species (Tylš et al., 2014); the content of up to 2% of mushrooms' dry weight has been reported (Badham, 1984). able to cross the blood-brain. Other related compounds, found at lower con-centrations, include baeocystin and its dephosphorylated analogue, norpsilocyn, which doesn't seem to contribute to psychotropic effects. Aeruginascin is so far known only from *Inocybe aeruginascens* and its biological activity is still cross the blood-brainunknown. Finally, norbaeo-cystin is thought not to be able to cross the blood-brain barrier as being a primary amine, and thus being inactive *in vivo*. Psilocybin is rapidly dephosphorylated to psilocin, which represents its active metabolite. Psilocin crosses the blood-brain barrier and acts as an agonist on serotonin $5HT_2$ receptors, triggering psychedelic effects. Psilocin is relatively resistant to deamination by MAO, unlike some other short-acting psychedelic tryptamines. Although norpsilocin was shown to act as a $5HT_2$ agonist *in vitro*, with similar potency as psilocin, its precursor, baeocystin was not shown to be active when given orally to mice; it was suggested that these compounds, as secondary amines, were more prone to MAO, thus being quickly metabolized before reaching the target receptors in the brain (Sherwood et al., 2020). Effects of psilocybin mushrooms usually develop as soon as 30 minutes after the ingestion and the symptoms vary depending on the number of mushrooms and the content of psychoactive tryptamines in them, but there is also a great inter-individual response. The "average" dose of psilocybin mushrooms leads to an increase in body temperature, mydriasis, compulsive movement, muscle weakness, and drowsiness and altered mood often manifesting as euphoria and uncontrollable laughter. True hallucinations, like the imagination of abstract figures, have been reported rarely after ingestion of moderate doses, with the majority of people describing having visions of altered shapes and colors instead and the whole experience as a "dream-like" state (Badham, 1984; Amsterdam et al., 2011). The so-called "trip" lasts about 4–6 hours (Dasgupta, 2019). Apart from "enjoyable" sensory distortions, delusions, restlessness, anxiety and even panic attacks may occur. "Bad trips", the psychotic episodes with paranoid thoughts, which may lead to accidents, self-injury, or suicide attempts, may occur in

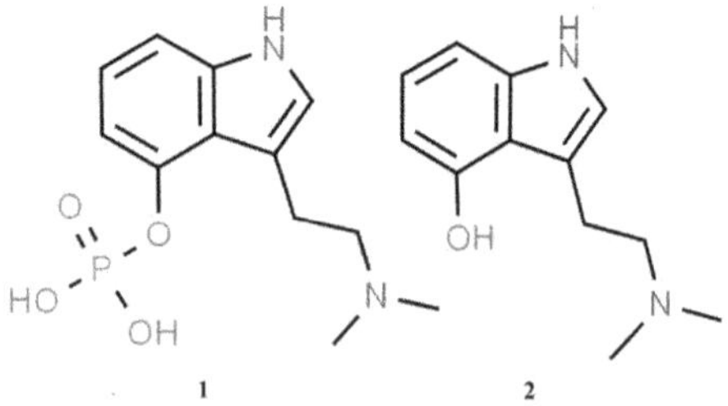

Figure 12.5 Chemical structures of psilocybin and psilocin.

Source: http://www.chemspider.com.

serious cases. The effects of psilocybin mushrooms may however be long-lasting in some in-dividuals, as recurring episodes of altered perceptions/hallucinations, referred to as hallucinogen persisting perception disorder or "flashbacks". Flashbacks can be experienced days, weeks, or even years after psilocybin use (Amsterdam et al., 2011). A dose of 10–20 mg of psilocybin is required to induce the psychedelic effects; lower doses cause only mild physical symptoms, while higher doses (>60 mg) are typically associated with negative experiences (Snook, 2016). Other sources state that as low as 4 mg of psilocybin can induce hallucinogenic effects (Beck et al., 1998), or about 10 g of fresh psilocybin mushrooms, although up to 50 g of fresh mushrooms are used for recreational purposes (Amsterdam et al., 2011). There are also atypical presentations of psilocybin mushroom poisoning described in the literature. De Sagun and Tabunar (2012) reported seizure and transient expressive aphasia in a patient that allegedly misidentified edible mushroom for *Psilocybe* sp. Austin et al. (2019) described acute renal failure associated with *P. cubensis* ingestion in a 15-year-old male. Intravenous poisonings with psilocybin mushrooms have also been reported; on one such occasion, a man with bipolar disorder injected himself with a decoction made from *P. cubensis* mushrooms and suffered from multiorgan failure and septic shock. Blood samples of the patient confirmed both bacteriemia and fungemia caused by *P. cubensis* that was ultimately cultured from the patient's blood (Giancola et al., 2021). Other known cases of intravenous mushroom poisonings had a milder clinical presentation, with headache, nausea, vomiting, diarrhea, myalgia, and arthralgia, and also hypoxemia, pyrexia, and mild methemoglobinemia (Curry & Rose, 1985). Deaths associated with psilocybin mushrooms are rare, and almost exclusively do not represent a direct consequence of poisoning (the estimated lethal dose of psilocybin mushrooms in humans is 17 kg), but altered state of mind; most deaths involved accidental self-harm and simultaneous use of other drugs, mostly alcohol (Amsterdam et al., 2011). Children may be more sensitive to psilocybin mushrooms toxicity; there is a report of a child that had eaten large quantities of *P. baeocystis* and died after developing severe pyrexia and convulsions (Badham, 1984). There is also a risk con-sidering foraging of wild-growing psilocybin mushrooms, as they can be mistaken for deadly poi-sonous species (Dasgupta, 2019); such a case was described by Franz et al. (1996) where a webcap mushroom (*Cortinarius* sp.) was mistaken for hallucinogenic mushrooms and resulted in acute renal failure in a young male patient.

Reports of so-called "mushroom madness", the alleged mushroom-consumption-associated behavioral changes seen in men and women of Kuma people in the Waghi Valley of Papua New Guinea, sparked several ethnomycological studies through the 1960s and 1970s, which were to reveal the psychotropic mushroom species responsible for this cultural phenomenon. These studies have been most recently summed up by Treu and Adamson (2006). According to the locals, several species identified to belong to the genus *Boletus* and *Russula* were held responsible to cause "mushroom madness" . In one of these species, named *Boletus manicus*, traces of indolic com-pounds were found although the precise nature of these compounds is unknown (an amino acid tryptophan has an indolic ring, for example). One of the main characteristics of the "mushroom madness" is that it affects men and women differently, with men being aggressive and women experiencing joyfulness, dancing and acting more openly; moreover, men and women are affected by different mushroom species. While boletes would allegedly affect everyone, species of *Russula* would affect only women. This led scientists to argue that "mushroom madness" is more of a ritual, that probably has origins in psychedelic mushroom consumption/poisoning events (one of the species found and described in the area is, presumably, psychoactive *Psilocybe kumaenorum*); Kuma people have been gathering they have been preparing them many mushroom species for food, and since they have been preparing them mixed together, the species that might have caused psychotropic effects could have easily been wrongly identified. "Mushroom madness" nowadays slowly vanishes even from the memory of the locals, like many other traditions. Despite the lack of evidence, *B. manicus* is still referred to as a psychedelic bolete in the literature.

12.4.4 Erythromelalgia-like syndrome

The syndrome of mushroom poisoning that resembles a rare peripheral pain disorder – erythromelalgia (also known as acromelalgia) – was known from Japan since the late 19th century (Bessard et al., 2004), and in 1918 a species responsible for the syndrome was described and named *Clitocybe acromelalga* (Ichimura, 1918); it was later placed in a separate genus *Paralepistopsis* (Çolak et al., 2017). The species is native to Japan and South Korea, so for a long time, such poisonings were confined to Far East (Saviuc & Danel, 2006). The same syndrome was, however, described in France in 2001, as a result of poisoning with *Paralepistopsis amoenolens* (syn. *Clitocybe amoenolens*) (Saviuc et al., 2001), a thermophile mushroom species of Mediterranean distribution, known from Morocco, France, Italy, and Turkey (Saviuc et al., 2001; Saviuc & Danel, 2006; Çolak et al., 2017). The poisoning event that took place in 1996 included two families who consumed the mushrooms mistaken for edible *Paralepista flaccida* (syn. *Lepista inversa*); a case that occurred in the same area, with a similar clinical presentation that involved mushroom consumption was dated back to 1979 (Saviuc et al., 2001). The syndrome was later recognized to had occurred in Italy as well (Saviuc & Danel, 2006). One of the main features of poisoning is the somewhat delayed onset of symptoms, 24 hours to several days, without any reported GIT disturbance, allowing repeated ingestion of mushrooms. The symptoms resemble those seen in erythromelalgia or diabetic peripheral neuropathy. They start with numbness of the limbs, followed by heat sensations and burning pain, which has episodic nature, lasting 2–3 hours; the pain is reluctant to most analgesics and is relieved only by applying cold water to the affected area. The symptoms can last up to 5 weeks, although in some cases up to 4 months. Redness and swelling of fingers and toes accompany sensations of pain. Paroxysmal pain episodes are more frequent at night; victims have troubled sleep, leading to general weakness. Deaths have been reported as a consequence of prolonged exhaustion and secondary infection (Saviuc et al., 2001; Saviuc & Danel, 2006). Treatment is symptomatic and involves the administration of analgesics, although most are ineffective. Treatment with acetylsalicylic acid, morphine, and clomipramine was found to be effective to some degree (Saviuc & Danel, 2006). Although the detailed mechanism of poisoning remains unknown, the agents responsible for the erythromelalgia-like symptoms are thought to be primarily specific amino acids, named acromelic acids, structurally similar to a powerful neurotoxin, kainic acid (Konno et al., 1988; Ishida & Shinozaki, 1988). These compounds represent non-proteinogenic amino acids with a characteristic pyrrolidine dicarboxylic acid moiety. So far, five types are characterized (A, B, C, D, and E) from *P. acromelalga* (Fushiya et al., 1993) and the presence of acromelic acid A was confirmed in *P. amoenolens* (Bessard et al., 2004). Acromelic acids are, like other kainoids, glutamate receptor agonists, causing powerful excitatory action. Acromelic acid A (Figure 12.6) is the most active and the most studied of the five related compounds; it was found to be one of the most powerful excitatory natural aminoacids in general, in both vertebrates and invertebrates (Ishida & Shinozaki, 1988).

Studies in mice showed that either systemic (peritoneal) or intrathecal administration of acromelic acid A leads to selective damage of interneurons in the lower spinal cord and prolonged rigid-spastic paraparesis (partial limb paralysis) (Shinozaki et al., 1989). At lower concentrations – extremely lower concentrations to be precise, allodynia – hypersensitivity to stimuli that aren't normally causing pain, was reproduced in mice after intrathecal injection of acromelic acid A and B. The allodynic effect of acromelic acid A was superior and observed at only 50 ag/kg; the dose-

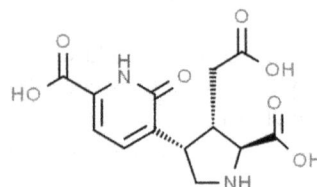

Figure 12.6 Chemical structure of acromelic acid A.

Source: http://www.chemspider.com.

dependent effect showed a bell-shaped pattern, with allodynia presenting at concentrations between 50 ag/kg and 50 ng/kg with the maximal effect observed at 50 fg/kg (1 fg/mouse). Peripheral pain hypersensitivity was noticed 5 minutes after the injection and the effect lasted for 4 weeks after a single dose (Minami et al., 2004). The actions of acromelic acid seen in testing animals were confusing at first and it was hypothesized that it acts through a specific, unknown subtype of glutamate receptors (Taguchi et al., 2009). A recent study showed that the acromelic acid A causes induction of the early phase of allodynia through activation of NMDA receptors, while late phase allodynia is most likely linked to consequent activation of spinal glia, specifically astrocytes, which have been recently found to have one of the crucial roles in neuropathic pain (Omoto et al., 2015). Thus, acromelic acid has so far shown helpful in establishing *in vivo* models of chronic neuropathic pain, one of the greatest therapeutic challenges of modern medicine, leading to the discovery of a synthetic analogue of acromeic acid A with anti-allodynic activity (Soen et al., 2007). Acromelic acids A, B, and C were found to account for 0.0014% of fresh *C. acromelalga* fruiting bodies, with acromelic acid A making about 50% of the toxins content. Intraperitoneal lethal doses in mice were found to be 7, 8, and 10 mg/kg for acromelic acid A, B, and C, respectively (Fushiya et al., 1993), although one study reported a "minimum" lethal dose of acromelic acid A in mice to be 4 mg/kg (Shinozaki et al., 1989). An intrathecal lethal dose of the acromelic acid A was found to be much lower, 500 ng/kg (Minami et al., 2004).

12.4.5 Polyporic acid – poisoning syndrome

Although polypore mushrooms are well known for their medicinal properties (Vunduk et al., 2015; Klaus et al., 2020), *Hapalopilus rutilans* (syn. *H. nidulans*) is a rare example of a poisonous polypore. Poisonings are very rare and the first case in modern times was reported in 1986 from Germany (then East Germany) (Kraft et al., 1998). There were only six reported cases in total in the literature as of 2013 (Villa et al., 2013). Symptom onset starts relatively late, after more than 12 hours after ingestion. Symptoms include gastrointestinal disturbance (nausea and vomiting), headache, weakness, and liver and kidney impairment (elevated blood levels of aminotransferases, creatinine, and urea, proteinuria, leukocyturia). The purple color of urine is the unique feature of poisoning with *H. rutilans*. There are also signs of peripheral neurotoxicity, manifesting as visual disturbances (blurred vision, diplopia), nystagmus (Kraft et al., 1998; Saviuc & Danel, 2006; Villa et al., 2013), but in one case (a child) there was also notable central neurotoxicity, with neurological features that included vertigo, somnolence, but also electroencephalographic abnormalities affecting all cerebral areas with signs of cerebral oedema. The central neurotoxic symptoms regressed after one week and the child fully recovered (Kraft et al., 1998). In a case report from 2013, an adult patient also had visual hallucinations and balance disorders. This patient had purple-colored urine for a week (Villa et al., 2013), suggesting ingestion of a relatively high content of mushroom. The mushroom was mistaken for edible *Fistulina hepatica* in all cases (Saviuc & Danel, 2006; Villa et al., 2013).

The responsible toxin is polyporic acid (Kraft et al., 1998), a well-known compound that is found in several other polypore species (Gill & Steglich, 1987), but what makes *H. rutilans* stand out is the extraordinary high content of this terphenylquinone derivate, which may account for 40% of fungal dry weight (Figure 12.7). Alkaline solutions of polyporic acid are violet, which explains the change of urine color in poisoned patients.

Polyporic acid acts as an inhibitor of dihydroorotate dehydrogenase, an enzyme involved in the synthesis of pyrimidine ring, located in mitochondria, which connects pyrimidine synthesis and mitochondrial oxidative reactions. Polyporic acid exhibited the same effects on rats as seen in human victims, which included reduced locomotor activity, vision impairment, and impaired maneuver abilities, as well as hepato-renal failure. *In vitro* studies showed that polyporic acid has a high affinity for neuronal cells, hence explaining the neurotoxicity (Kraft et al., 1998; Saviuc &

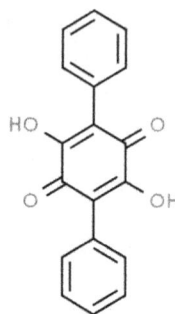

Figure 12.7   Chemical structure of polyporic acid.

Source: http://www.chemspider.com.

Danel, 2006). The exact connection between the pyrimidine synthesis pathway inhibition and the poisoning symptoms has not been well established yet.

Another polypore mushroom was involved in a case of poisoning with the neurological syndrome in 1987 in Canada; a 6-year-old child allegedly ingested a piece of raw "chicken of the wood" (*Laetiporus sulphureus*) and experienced movement and speech impairment, muscle weakness, and developed visual hallucinations of "lines and shapes of different colors", but also of "yellow and orange monster". The child's urine was negative for any known hallucinogens. *L. sulfureus* is considered edible, although it is known to cause GI disorders, and occasionally central nervous system-related symptoms like dizziness and disorientation in susceptible individuals. It is, however, never consumed raw. A combination of factors (age, susceptibility, ingestion of raw mushroom) may explain this unique presentation of poisoning with an edible mushroom (Appleton et al., 1988).

12.4.6 Morel neurological syndrome

True morels (*Morchella* spp.) are known to cause GI disturbance when poorly cooked, but there has been growing evidence in Europe that they can induce neurological symptoms as well; "morel toxic neurological syndrome" was described recently and proposed as a distinct type of mushroom poisoning (Saviuc et al., 2010). The syndrome is still poorly known and not very well defined, presenting with GI and various, inconsistent CNS-related symptoms (Saviuc et al., 2010; White et al., 2019a). The circumstances under which morels exhibit neurological disorders are unknown but are in most cases linked to consumption of large quantities of the mushroom. No links were established between the toxicity and the way of mushroom preparation or storage, co-ingestion with alcohol or drugs. Allergic reaction is also excluded as a possibility, so the presence of the unknown toxin is considered the most probable explanation. The onset of symptoms begins approximately 12 hours after ingestion; the symptoms are generally mild and include dizziness, tremor, ataxia, vision disorders (most frequently reported were diplopia and blurred vision), headache, paresthesia, muscle spasms, and confusion. Several authors proposed that the cerebellum is the main target of toxic compounds, as some of the symptoms imply cerebellum function impairment; however, the hypothesis was discredited. More than half of the patients with the neurological syndrome also had gastrointestinal symptoms (nausea, vomiting, and diarrhea). The symptoms resolve within 24 hours in most cases (Saviuc et al., 2010). Cases of morel poisonings with neurological symptoms are known from North America as well; Beug et al. (2006) reported several mushroom poisoning cases associated with morels, which presented with various central and peripheral symptoms like disorientation, sound sensitivity, salivation, and sweating.

Photograph 12.4. *Morchella esculenta*, a delicacy, a poisonous mushroom or both? Bor district, Serbia, 2015. Photo: P. Petrović

12.5 GYROMITRIN POISONING SYNDROME

"False morels" represent several mushroom species (Puschner, 2013) morphologically similar to morels, or "true morels" (Brooks & Graeme, 2016), well known as commercially important wild edible ethylidene gyromitrin (Braun et al., 1979), mushrooms (Arłukowicz-Grabowska et al., 2019). The name false morel however most often refers to *Gyromitra esculenta*, a spring-fruiting species that is widely distributed in Europe and North America (Michelot, 1991). There is a tradition of consuming this species in some parts of Europe, especially Eastern Europe and Scandinavia (Finland), where it has been considered a delicacy, even though it has also been known as - potentially fatal (Michelot & Toth, 1991; Benjamin, 2020). *G. esculenta* is highly poisonous when consumed raw, but drying and/or heat treatment may lower its toxicity and make it safe for human consumption (Pyysalo &Niskanen, 1977; Michelot & Toth, 1991; Brooks & Graeme, 2016). This is due to its unique, volatile toxic components, hydrazone derivatives, named gyromitrins. Ten different hydrazones, named "gyromitrins" were isolated from *G. esculenta*, which differ by *N*-linked alkenyl moiety, i.e. aldehyde which forms hydrazone. The most abandoned of these hydrazones is ethylidene gyromitrin (Braun et al., 1979), to which the name "gyromitrin" is applied today (Figure 12.8). Although volatile, these toxins may persist in mushrooms after cooking so consumption of false morels can never be considered safe (Brooks & Graeme, 2016).

The toxin content in *G. esculenta* varies from 50–300 mg/kg fresh weight, expressed as monomethylhydrazine, a metabolite of gyromitrin (Michelot & Toth, 1991); other members of the genus that are reported to contain gyromitrin include *G. gigas* and *G. fastigiata*. North American taxons *G.*

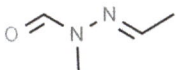

Figure 12.8 Chemical structure of gyromitrin.

Source: http://www.chemspider.com.

montana and *G. korfii* are sometimes treated as *G. gigas* or at least as very closely related species and are assumed to contain gyromitrin as well, although no studies were performed to confirm that (Benjamin, 2020). Gyromitrins are not confined only to the *Gyromitra* genus, or Helvellaceae family, and can be found in various species of the order Pezizales; *Helvella crispa*, *H. lacunosa*, *Leotia lubrica*, *Spathularia flavida*, *Otidea onotica*, *Cyathipodia macropus*, and *Leptopodia elastica* are all reported to contain gyromitrin, although in only minute concentrations. A significant amount of gyromitrin, however, comparable to that from *G. esculenta* was found in *Cudonia circinans* of the Geoglossaceae family (150 mg of monomethylhydrazin per kg of fresh weight) (Andary et al., 1985). *Sarcosphaera coronaria*, another ascomycete mushroom once considered edible, was occasionally reported to cause poisonings, with at least one casualty in Switzerland, back in 1920. It was believed *S. coronaria* could contain the same hydrazine toxins, but this was proved not to be the case; however, *S. coronaria* was found to be an incredible bioaccumulator of arsenic, which may account for up to 0.9% of the mushroom's dry weight. It was recently proposed that organic arsenic compounds may be responsible for the toxicity of this mushroom (Braeuer et al., 2020).

The specificity of gyromitrin poisoning is that it can affect different organs through different mechanisms of action and may cause GI disorders, neurological disorders, or exert hepatotoxicity leading to liver failure and death. Poisonings are overall associated with the consumption of raw and poorly cooked mushrooms, as well as repeated consumption, although there is a great inter-individual difference in respect to the manifestation of poisoning (Michelot & Toth, 1991). Most poisonings with gyromitrin-containing mushrooms are mild and are limited to GI dysfunction, with unspecific symptoms like nausea, vomiting, and diarrhea that resolve without further complications. These symptoms usually appear 5–12 hours after a meal but may develop even after more than 2 days upon ingestion. In severe cases, neurotoxicity is observed with symptoms ranging from nervousness to delirium, convulsions/seizures, and coma. Neurological disorders usually appear after GI dysfunction is resolved, sometimes with a symptom-free period between. Hepatotoxicity, if it develops, occurs within 2 days of mushroom ingestion and manifests as cytolytic hepatitis that leads to hepatic failure, characterized by elevated levels of blood transaminases, jaundice, and coagulopathy. The toxic effects on blood include hemolysis that sometimes accompanies hepatitis, and methemoglobinemia. Renal injury is infrequently reported and is thought to be the consequence of hemolysis and other disorders. The overall mortality rate is put at around 10% in those who develop symptoms of poisoning (Michelot & Toth, 1991; Brooks & Graeme, 2016), mainly due to liver failure. A fatal case of suspected gyromitrin poisoning with multi-organ failure accompanied by encephalopathy was described by Arłukowicz-Grabowska et al. (2019).

Several potential molecular mechanisms of action of hydrazones were proposed that could explain the overall toxicity of gyromitrin. Gyromitrin acts through its metabolites, *N*-methyl-*N*-formylhydrazine (MFH) and *N*-methylhydrazine or monomethylhydrazine (MMH). Gyromitrine is unstable and easily oxidizes and hydrolyzes at room temperature to acetaldehyde and MFH. This reaction probably takes place in acidic stomach conditions. MFH, although more stable than gyromitrin, can further hydrolyze to formic acid and MMH (Braun et al., 1979; Brooks & Graeme, 2016). Neurological symptoms that are often seen in Gyromitra poisonings are thought to be due to a decrease of GABA concentration in CNS caused by MMH; GABA is an inhibitory neurotransmitter and is synthesized from the excitatory neurotransmitter glutamate, via glutamic acid decarboxylase. The enzymatic reaction requires pyridoxal phosphate as a cofactor, which is the active form of pyridoxine (vitamin B6). On the other hand, the activation (phosphorylation) of pyridoxine is dependent on pyridoxine phosphokinase. MMH acts as an inhibitor of both glutamic acid decarboxylase and pyridoxine phosphokinase, plus it can directly inactivate phosphorylated pyridoxine by binding to it and forming a complex, which is excreted by urine (Brooks & Graeme, 2016). Hydrazines are metabolized in the liver by acetylation to acetylhydrazides, which is thought to prevent neurotoxicity; however, there is a great inter-individual variation in acetylation ability, which differentiates "slow" and "fast acetylators", explaining the differences seen in the clinical presentation of gyromitrin poisoning. Oxidation of the monosubstituted hydrazines can, however, lead to the formation of highly reactive intermediates, free methyl radical and methyl cation, both of which can alkylate various biological molecules. These reactive species can bind to amino acid residues in enzymes, and also lower antioxidant capacities by reacting with glutathione, leading to cell dysfunction and death, which is thought to be the main mechanism of hepatotoxicity (Brooks & Graeme, 2016; Michelot & Toth, 1991). Lethal doses of MMH in children and adults are estimated at 1.6–4.8 and 4.8–8 mg/kg, respectively (Andary et al., 1985). Apart from symptomatic treatment, administration of pyridoxine (vitamin B6) is indicated to treat neurological disorders, as well as methylene blue in cases of methemoglobinemia. Benzodiazepines, barbiturates, or propofol may be administered in cases of seizures (Brooks & Graeme, 2016; Michelot & Toth, 1991).

Apart from acute toxicity, gyromitrin also possesses carcinogenic ability; *in vivo* studies showed that both gyromitrin and its hydrazine derivatives can induce tumors in rodents, after repeated administration. The mechanism of genotoxicity probably involves the methylation of DNA by reactive alkylating metabolites and particularly the creation of O^6-methylguanine. The overall carcinogenic risk of false morel consumption in humans is however thought to be relatively low if results obtained in animal models are translated to humans. Those with low acetylation capacity of hydrazines may be at higher risk (Bergman & Hellenäs, 1992).

Agaritine, a phenylhydrazine derivative present in species of genus *Agaricus*, including widely consumed white button mushroom (*A. bisporus*), was also a subject of intense research due to its proposed genotoxic potential (Shephard & Schlatter, 1998; Roupas et al., 2010; Koge et al., 2011). Some studies reported that it can induce tumors in experimental animals, but these studies were criticized for using exceptionally large doses of synthetic hydrazines in comparison to the levels found in mushrooms. Studies that involved feeding with mushrooms have not been associated with any toxic effects of *A. bisporus* (Roupas et al., 2010); to the contrary, mushroom consumption is associated with a lower risk of cancer development (Ba et al., 2021). The available, relevant data thus suggests that there is no risk of cancer development due to consumption of *A. bisporus* (Roupas et al., 2010).

12.6 DISULFIRAM-LIKE MUSHROOM POISONING SYNDROME

Several mushroom species, mainly from the genus *Coprinus* sensu lato ("inky caps"), cause poisonings only when combined with alcohol. *Coprinopsis atramentaria* (previously known as *Coprinus atramentarius*) is the most commonly associated with this so-called "disulfiram-like" poisoning syndrome (Berger & Guss, 2005; White et al., 2019a). This is due to the clinical manifestation of poisonings, which resemble the side effects of a drug used in the treatment of alcoholism, disulfiram (Michelot, 1992). Disulfiram acts as an inhibitor of aldehyde dehydrogenase (ALDH), an enzyme involved in alcohol metabolism (Swift & Leggio, 2009); ethanol is metabolized in two steps; in the first step it is oxidized to acetaldehyde by alcohol dehydrogenase (ADH) and in the second step acetaldehyde is further oxidized to acetic acid by ALDH (Zakhari, 2006).

An amino acid, named coprine (Figure 12.9), which chemically represents N^5-(1-hydroxycyclopropyl)–L-glutamine was isolated from *Coprinopsis atramentaria* and is thought to be the main responsible toxic compound (Tottmar & Lindberg, 1977). Other related species were also found to contain the toxin (Govorushko et al., 2019). As *in vitro* studies indicate, coprine itself is not toxic but acts through its metabolite, 1-aminocyclopropanol, which inhibits ALDH irreversibly, thus interferes with ethanol metabolism just like disulfiram (Tottmar & Lindberg, 1977; Beuhler, 2017). This leads to the accumulation of acetaldehyde, which causes a very unpleasant reaction, resembling a hangover. The syndrome can occur if alcohol is ingested up to 72 h upon mushroom consumption and the severity of the reaction depends on the amount of ingested mushrooms and alcohol (Beuhler, 2017). The therapy includes only supportive measures. Specific symptoms like tachycardia may be treated with β-blockers like propranolol. The symptoms are usually mild, so they resolve on their own, typically within few hours, but sometimes may persist up to 2 days (Berger & Guss, 2005).

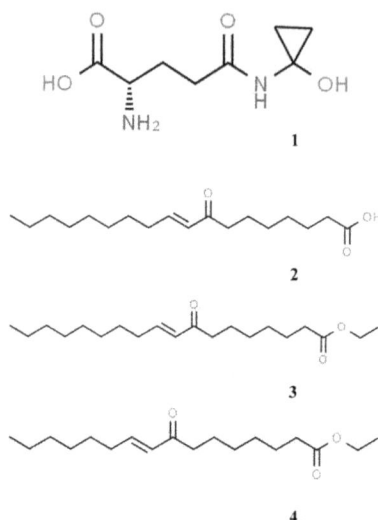

Figure 12.9 Chemical structures of coprine (1), (9E)-8-Oxo-9-octadecenoic acid (2), ethyl (9E)-8-oxo-9-octadecenoate (3) and ethyl (9E)-8-oxo-9-hexadecenoate (4).

Source: http://www.chemspider.com.

Ampulloclitocybe clavipes (formerly known as *Clitocybe clavipes*) were reported to trigger the same disulfiram-like reaction when consumed before alcohol ingestion, both from North America and Japan (Cochran & Cochran, 1978). Kawagishi et al. (2002) isolated several fatty acid derivatives from the mushroom that were shown to exhibit ALDH-inhibitory activity *in vitro* (Figure 12.9). All these compounds have an enone functional group (α,β – unsaturated ketone), which the authors speculated can be involved in forming a Michael addition adduct with the thiol group of the enzyme. The syndrome was also reported to occur after consumption of *Lepiota aspera* (Haberl et al., 2011).

12.7 RHABDOMYOLYSIS MUSHROOM POISONING SYNDROME

Rhabdomyolysis is a serious, life-threatening condition characterized by disruption of muscle fibers and release of their content into the bloodstream, which may lead to renal function impairment and failure. It is a rare disorder and usually results from trauma or intoxication with alcohol, pesticides, or certain drugs (Melli et al., 2005; Janković et al., 2013). It is only recently connected with the consumption of certain mushroom species. *Tricholoma equestre*, the yellow knight, which had a reputation of popular edible mushroom in western and northern Europe, was implicated in a series of rhabdomyolysis cases between 1992 and 2016 in France, Poland, Germany, and Lithuania, some of which had a fatal ending (Bedry et al., 2001; Anand et al., 2009; Laubner & Mikulevičienė, 2016; Rzymski & Klimaszyk, 2018). Since then it has been debated if this mushroom is a cause of rhabdomyolysis, with contradicted evidence from different studies (Rzymski & Klimaszyk, 2018). It also sparked research concerning the risk of consumption of other supposedly edible mushrooms, regarding their potential to cause such serious conditions (Nieminen et al., 2009; Nieminen & Mustonen, 2020). The symptoms of poisoning start with fatigue and weakness, which is followed by myalgia, particularly in the legs. Dark urine occurs due to the excretion of released myoglobin. Profuse sweating without fever is also a characteristic of the syndrome. Blood analysis shows a dramatic increase of creatine kinase (CK), an enzyme confined to muscle fibers, as well as alanine aminotransferase (ALT) and aspartate aminotransferase (AST). The treatment is only supportive (Bedry et al., 2001; Laubner & Mikulevičienė, 2016). Recovery takes about 15 days but in serious cases, there is a progression to renal failure or cardiovascular impairment and death (White et al., 2019a).

Photograph 12.5. *Tricholoma equestre s.l.*, a mushroom species connected with serious rhabdomyolysis poisoning syndrome. Crni vrh, Serbia, 2015. Photo: P. Petrović

Characteristic of all *T. equestre*–associated poisonings is the consumption of consecutive meals containing the mushroom, in some cases for days, suggesting cumulative toxicity. It is also suggested that there may be a genetic susceptibility to the toxins contained within the mushroom (Bedry et al., 2001; Laubner & Mikulevičienė, 2016; White et al., 2019a; White et al., 2019b).

Several studies performed using animal models showed increased CK in mice that were fed with *T. equestre* mushrooms or their extracts (Bedry et al., 2001; Nieminen et al., 2005, Nieminen et al., 2008), although the methodology of these studies has been criticized for allegedly using high, non-relevant doses of mushrooms or their product (Rzymski & Klimaszyk, 2018). The toxic principles were not characterized, nor isolated in these studies. Several, although limited clinical studies did not show any toxic reactions in humans who were put on a diet that included repeated daily *T. equestre* meals. One of these studies included patients with dyslipidemias on a statin and/or fibrate therapy, which may alone cause rhabdomyolysis as a rare and serious side effect, implying that even individuals with a higher risk of myotoxic events could safely consume *T. equestre* in reasonable amounts (Klimaszyk & Rzymski, 2018; Rzymski & Klimaszyk, 2018). Besides the methodology used in studies on animals, case reports that marked *T. equestre* as the cause of rhabdomyolysis were also questioned for proper identification of the species. Most of these reports did not involve confirmation by mycologists or molecular techniques and *T. equestre* is only one in the series of similar yellow species in the genus (Rzymski & Klimaszyk, 2018).

Russula subnigircans was on the other hand unambiguously marked as the cause of rhabdomyolysis in series of poisonings with fatalities in Taiwan, Japan, and China in the period from 1998–2013, although in Japan there were reports of such events that dated back to 1956. The course of poisoning is however slightly different than that associated with *T. equestre*; it starts with GIT issues (nausea, diarrhea) as quickly as 30 minutes after ingestion. Pain in the back and

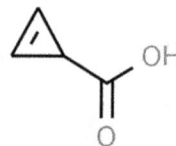

Figure 12.10   Chemical structure of cycloprop-2-ene carboxylic acid.

Source: http://www.chemspider.com.

shoulder stiffness is accompanied by speech impairment and pupil contractions. Myoglobinuria, as well as a marked increase in serum CK (greater than 90,000 U/L), occur as a result of muscle fiber destruction. Unconsciousness and heart weakening is seen in severe cases which often end fatally. Early recognition of the syndrome and intensive care are key factors to prevent death (Lee et al., 2001; Matsuura et al., 2009; Lin et al., 2015).

Highly strained, low-weight carboxylic acid, cycloprop-2-ene carboxylic acid was found to be responsible for the toxicity (Figure 12.10). Studies in mice confirmed that, when administered orally, it causes a dramatic elevation of serum CK and LD_{50} of this compound was found to be 2.4 mg/kg; the content of the toxin in the mushrooms was found to be very high, about 720 mg/kg. The mechanism of action is unknown, but the authors suggested that it does not cause direct damage to muscle fibers, but acts by triggering some other biochemical reaction (Matsuura et al., 2009).

Yin et al. (2014) dubbed *Tricholoma terreum*, one of the most popular edible mushrooms of the *Tricholoma* genus in Europe, as potentially dangerous rhabdomyolysis-causing species in a paper named "Chemical and toxicological investigations of a previously unknown poisonous European mushroom *Tricholoma terreum*"; the authors isolated several new triterpenoids from *T. terreum* fruiting bodies and found that two of them, saponaceolide B and M caused CK elevation in mice. LD_{50} values of orally administered compounds in mice were determined to be 88.3 and 63.7 mg/kg for saponaceolide B and M, respectively, and it was found that these saponaceolides were present in the mushrooms at levels of 414 and 184 mg/kg-1 dry weight. They concluded that *T. terreum* is in fact a poisonous species and that it might have been mistaken for *T. equestre* and was the actual cause of rhabdomyolysis. The study was heavily criticized for the interpretation of the results and conclusions drawn from them; heat and low pH resistance of the compounds were not tested. The reported LD_{50} values in mice were relatively high and toxin levels in mushrooms were low. The critics of the study also remarked that using stricter, body-surface area (BSA) scaling to translate the toxicity of the compounds to humans would end up in consumption of approximately 12–20 kg of fresh mushrooms to exert the LD_{50} values for these triterpenoids (Davoli et al., 2016).

Well-known toxic mushrooms that cause other poisoning syndromes were also occasionally reported to cause raise in CK blood levels; in a retrospective study of rhabdomyolysis causes during one year period in Serbia, it was reported that 5/28 patients with clinical presentation of hepatotoxic mushroom poisoning (presumably *Amanita phalloides*) had mild to moderate elevation of CK. As *A. phalloides* is not known to induce rhabdomyolysis, the authors suggested the possibility of co-ingestion of other poisonous mushroom species or another cause of CK elevation (Janković et al., 2013). Another case of suspected amatoxin poisoning with evident rhabdomyolysis, confirmed by both blood analysis and muscle biopsy had been reported earlier, in 1996, but again, no definitive identification of the species involved could be performed (Gonzalez et al., 1996). In some cases of poisoning with *Gyromitra esculenta*, rhabdomyolysis can be seen accompanying hemolysis and convulsions (Michelot & Toth, 1991; Brooks & Graeme, 2016).

Other, well-established edible mushrooms were also implicated in rhabdomyolysis cases; in a case report from Turkey, rhabdomyolysis that occurred in a married couple was claimed to be due to the consumption of *Agaricus bisporus*, bought in the local marketplace. No evidence was however provided for the claim, and the identification of the species was based on the patients' anamnesis (Akilli et al., 2014). This remains the only rhabdomyolysis report that was connected to *A. bisporus*. A rhabdomyolysis case in Poland was associated with repeated consumption of large amounts of edible species of *Boletus* and *Leccinum* genus (Chwaluk, 2013), although this report also lacks any details on mushroom identification other than patient's claims. To further complicate the subject, Nieminen et al. performed series of studies in mice to see whether the potential to cause myotoxicity was inherent to all mushrooms or restricted to a few species. Their find was surprising as they showed that all mushrooms tested when ingested several days in a row are capable of elevating serum CK levels (Nieminen et al., 2005, Nieminen et al., 2008). Again, these studies received criticism regarding dosing issues (Rzymski & Klimaszyk, 2018). More research will be needed to determine whether this applies to humans, what are the mechanisms of toxicity and, if this is true, what is the safe amount of mushrooms that can be eaten and how often should someone engage in consuming mushrooms. It is important to note that a great number of studies of mushroom diet in humans showed only beneficial effects on various physiological functions and health conditions.

12.8 PANCYTOPENIA – TRICHOTECENE-CONTAINING MUSHROOMS

Trichothecenes are one of the major groups of mycotoxins with a great impact on society, both from a health and economic aspect. They are primarily known as secondary metabolites of molds that affect industrial crops and include species of genera *Fusarium*, *Myrothecium*, *Spicellum*, *Stachybotrys*, *Cephalosporium*, *Trichoderma*, and *Trichothecium*, all being members of the order Hypocreales (McCormick et al., 2011). They are very potent cytotoxins and are toxic to animals, but plants and protists as well, practically all eukaryotic organisms (Ueno & Matsumoto, 1975). Trichothecenes are protein synthesis inhibitors that target the 60 S ribosomal subunit, causing impairment of the peptidyl transferase center, thus preventing peptide bond formation. They also act as protein synthesis inhibitors at the mitochondrial level (McCormick et al., 2011). Trichothecenes were also showed to be capable of *in vitro* inactivation of several SH-containing enzymes, creatine phosphokinase, lactate dehydrogenase, and alcohol dehydrogenase, presumably by the addition of SH groups to epoxide moiety of trichothecenes (Ueno & Matsumoto, 1975). They eventually lead to accumulation of reactive oxygen species and cause oxidative stress in the target cells (McCormick et al., 2011) and lead to apoptosis or necrosis. Trichothecens are amphiphilic molecules that are easily absorbed from GIT and they can passively move through cell membranes; the most susceptible are tissues and organs where active proliferation takes place (McCormick et al., 2011), causing the greatest impact on hematopoietic, lymphoid and digestive tract tissues (Park et al., 2016).

A series of poisonings with *Podostroma cornu-damae*, the rare "fire coral" mushroom native to East Asian countries and Oceania, has been documented since 1999 from Japan and South Korea, some of which were fatal (Saikawa et al., 2001; Kim et al., 2016; Ohta et al., 2020). The fact that trichothecenes are found to be principal toxins of this mushroom (Saikawa et al., 2001) is not surprising, as *P. cornu-damae* (syn. *Hypocrea cornu-damae*) is a member of the Hypocreales order and a teleomorph (sexual form) of a *Trichoderma* sp.; hence, the newly proposed/accepted name: *Trichoderma cornu-damae* (Bissett et al., 2015; Ohta et al., 2020). In most cases of poisoning, the victims confused fire coral with young fruiting bodies of either edible mushrooms from the genus *Clavulinopsis* or mushrooms used in traditional medicine, such as *Ganoderma lucidum*, *Ophiocordyceps sobolifera*, or *Cordyceps militaris* (Park et al., 2016; Choe et al., 2018; Ohta et al., 2020).

Trichothecenes of *T. cornu-damae* are of the macrocyclic type and include satratoxin H (Figure 12.11) and its mono- and diacetylated derivatives; satratoxin G; roridin D, E, and Q; verrucarin J; and verrucarol (Saikawa et al., 2001; Choe et al., 2018). All of these compounds, except for the verrucarin J, exhibited a lethal effect in mice at a single dose of 0.5 mg, within 24 hours (Choe et al., 2018). An early phase of poisoning is characterized by gastrointestinal symptoms (nausea, vomiting, diarrhea, and consequent dehydration). Hypotension and tachycardia, oliguria, weakness, headache, and consciousness impairment follow. These symptoms resolve within few days. However, severe pancytopenia soon develops, with a high risk of life-threatening sepsis, which had occurred in some cases (Ahn et al., 2013; Park et al., 2016).

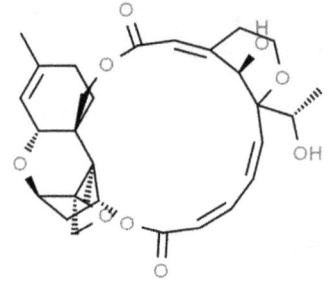

Figure 12.11 Chemical structure of satratoxin H.
Source: http://www.chemspider.com.

Leukopenia trend was well documented in four cases and it was observed that leukocyte numbers reached the lowest point after 17–20 days after *T. cornu-damae* ingestion (Park et al., 2016). Neutropenic fever consequently occurs. Desquamation of palms, soles, scalp, including hair loss, is also a highly characteristic feature of *T. cornu-damae* poisoning (Ahn et al., 2013; Park et al., 2016). However, the GI phase, as well as desquamation may not occur, and the course of symptoms seems to depend on the amount of ingested toxins. Victims usually die due to multiple organ failures, including renal failure, hepatic necrosis, and disseminated intravascular coagulation (Ahn et al., 2013). Fatal hemorrhagic lung necrosis was connected to *T. cornu-damae* poisoning in one case (Jang et al., 2013). The treatment of leucopenia and avoiding bacterial infections is crucial for a patient's survival. Granulocyte colony-stimulating factor may be beneficial, but there are mixed results. It often takes several weeks for full recovery (Park et al., 2016).

Ganoderma species, although considered safe and well known for their medicinal properties were implicated in cases of pancytopenia on two occasions; Kyeon et al. (1991) reported aplastic anemia in a husband and a wife who have consumed uncooked *G. lucidum* (as *G. japonicum*). Yoon et al. (2011) described two cases of poisoning with the same clinical presentation, again, in a married couple, which developed several days after repeated intake of decoctions prepared from *G. neojaponicum*. Due to the lack of any other possible cause, they assumed that *G. neojaponicum* is responsible for the pancytopenia. The specimens gathered in the wild by the couple were retrieved and described as "being rotten" at the time of examination as they had been left at room temperature. Photographs of the specimens were provided, showing greenish mold growing on the mushroom surface. Although it cannot be ruled out that certain individuals may have predispositions for rare toxic effects of species of *Ganoderma*, it is highly unlikely for this to occur in two unrelated individuals at the same time, on two different occasions. One explanation could be that *Ganoderma* specimens were contaminated by trichothecene-producing fungi as pancytopenia may be the only major symptom of trichothecene poisoning (Ahn et al., 2013); green mold that affects *G. sichuanense* (syn *G. lingzhi*), a mushroom disease of concern in China, is recently found to be caused by *Trichoderma atroviride* (Yan et al., 2019) and *T. hengshanicum* (Cai et al., 2020).

12.9 ENCEPHLOPATHY POISONING SYNDROME – *PLEUROCIBELLA PORRIGENS*

The angel's wing mushroom (*Pleurocybella porrigens*) is a widespread species and has been eaten particularly in Japan. However, it caused a series of serious poisonings in Japan in 2004 when 17 people in total died. The cause of death was acute encephalopathy, which followed the ingestion of the mushroom. Symptoms occurred 1–31 days after the first ingestion of mushrooms, with many patients having meals containing the mushroom for several days; they included disturbed consciousness, convulsions, and myoclonus. All victims had underlying chronic kidney diseases and were undergoing hemodialysis (Gejyo et al., 2005; Wakimoto et al., 2010). Several compounds were proposed to be responsible for toxicity, including steroids/vitamin D analogues, fatty acids and oligosaccharides, amino acids, and even cyanides (Gonmori et al., 2011); although

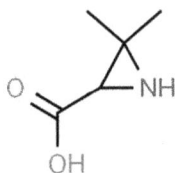

Figure 12.12 Chemical structure of pleurocybellaziridine.
Source: http://www.chemspider.com.

the mystery remains unsolved, recent studies support the latter two as more likely to be responsible for the toxicity of *P. porrigens*.

Several unusual amino acids were isolated from the mushroom after the poisoning event occurred, some of which showed toxicity towards mouse cerebrum glial cells (Kawaguchi et al., 2010). It had already been known that *P. porrigens* produces β-hydroxyvaline, a non-proteinogenic amino acid (Aoyagi & Sugahara, 1988), and the series of newly isolated amino acids had one thing in common – a β-hydroxyvaline moiety, attached to various endogenous compounds (Wakimoto et al., 2010). These unusual structures led researchers to speculate about the existence of a common precursor – a highly strained, unstable aziridine amino acid (Figure 12.12). As highly reactive, it could not be isolated, but its existence in the mushrooms was proven indirectly.

Pleurocybellariziridine, as it was named, showed *in vitro* toxicity to rat oligodendrocytes and was more active than its derivatives. Wakimoto et al. (2010) further speculated that demyelination, caused by oligodendrocyte damage could be a trigger for encephalopathy. On the other hand, Gonmori et al. (2011) hypothesized that hydrogen cyanide (HCN) was a causative toxin from the mushroom. There are several mushrooms capable of producing HCN and *P. porrigens* is one of them. It was showed that heat treatment lowers HCN content in mushrooms (the level of decrease depends on the type of treatment) but does not eliminate it. They speculated that environmental factors, specifically unusually wet weather conditions during the summer and early autumn in 2004 may have contributed to some changes in the biochemistry of *P. porrigens*, making the mushroom produce more HCN than usual. The mushroom was reported to fruit earlier that year, showing "abnormal massive genesis". The victims, who were all diagnosed with renal failure would be specifically prone to toxic effects of HCN, as a low-protein diet of kidney patients may lead to a deficiency of sulfur-containing amino acids and consequently to deficiency of rhodenase activity, leading to the formation of toxic cyanate OCN^- as a product of CN^- metabolism, instead of thiocyanate. OCN is a known cause of neurodegenerative diseases. Akiyama et al. (2015) further investigated this theory using an adenine-induced rat model of chronic renal damage; the rats received cyanogen glycoside fraction isolated from the mushrooms orally. They showed that impaired renal function affected cyanide metabolism, leading to accumulation of cyanide in hemolyzed blood of the rats 8 hours after single cyanogen glucoside administration and thiocyanate in both hemolyzed blood and brain 24 hours after the administration, which they explained by the incapability of thiocyanate excretion by kidneys. The clinical signs of poisoning in animals were however noticed only in one of three test animals that received a higher dose of cyanogen glucoside; the animal appeared weak, laying down in a prone position. The authors also quantified the cyanide levels in the *P. porrigens* samples from 2004, which were found to be 0–114 μg/g, and suggested that cyanides could truly be responsible for the series of acute encephalopathy associated with *P. porrigens* consumption.

12.10 YUNNAN SUDDEN UNEXPLAINED DEATH SYNDROME – *TROGIA VENENATA*?

Trogia venenata was scientifically described as recently as 2012 and was already suspected to be one of the most dangerous mushroom species, responsible for the deaths of over 400 people in Yunnan province in southwestern China since 1978. Mysterious deaths occurred in time-space clusters during the summer months, often among members of the same family. Previously healthy individuals would suddenly fall unconscious during diurnal activities (Shi et al., 2012a), and pass away within a day (2 hours median) and such events were dubbed as "sudden unexplained death" syndrome (SUD). The majority of victims did not show signs of any health issues up to 2 days before death. Cardiac-related symptoms such as dizziness, weakness, palpitation, and unconsciousness were seen in more than 60% of victims within 2 days before the onset of fatal events (Shi et al., 2006). Fieldwork initiated in 2005 as a part of a special SUD surveillance system soon found a connection between the incidence of SUDs and summer occurring, previously unknown "little white mushrooms" which grew in affected areas and were consumed by villagers (Shi et al., 2012a).

Soon after the implication of *T. venenata* in SUD, Zhou et al. (2012) isolated unusual nonproteinogenic without hypothetical predisposing risk factors amino acids from the fruiting bodies of this fungus, which were confirmed to be toxic in mice, as well as previously known toxin, γ-guanidinobutyric acid. Newly isolated 2R-amino-5-hexynoic acid and 2R-amino-4S-hydroxy-5-hexynoic acid (Figure 12.13) were shown to cause a rise in CK in mice and the action is dose dependant; however, the CK increase level is not as high as it is seen in rhabdomyolysis. The estimated LD_{50} value for these amino acids in mice is 84 and 71 mg/kg body weight, respectively, which can be regarded as relatively high; 2R-amino-4S-hydroxy-5-hexynoic acid was found in the heart blood sample of a SUD victim. The total combined content of these compounds in the fruiting bodies was found to be 2 g/kg dry weight; Zhou et al. estimated that the lethal dose for 60 kg human would be about 400 g of dry mushrooms, without taking into account hypothetical predisposing risk factors. The potential ability of the mushroom to cause hypoglycemia was also considered as a possible mechanism of mushroom's alleged toxicity; Shi et al. (2012b) tested the toxicity of *T. venenata* ethanol extracts in mice. Mice that received the lowest dose of 500 mg/kg did not develop any symptoms, while those that received doses from 1,500–3,500 mg/kg developed signs of illness, characterized primarily by the decrease of spontaneous activity; most of them died in less than 5 hours. Biochemical blood analysis showed that the extract dose of 2,000 mg/kg caused profound hypoglycemia, which was considered as the cause of death in mice. The authors estimated that the minimum fatal dose of fresh mushrooms in humans could be put between 140–430 g. The control group of mice that received 3,500 mg/kg ethanol extract of *Laccaria vinaceoavellanea*, a common edible mushroom found in the same area, survived through the 7-day observation period, without any pathological symptoms.

An autopsy performed on four victims of SUD with previous exposure to *T. venenata* showed that they all had damaged heart muscle fibers with focal lymphocytic infiltrations, however, there is an evident lack of clinical data for SUD victims as death takes place soon after symptom onset. A study that involved family members of SUD victims who consumed *T. venenata* found that 63% of them developed illness/unusual symptoms 16 hours–15 days after the last *T. venenata* meal. The symptoms included palpitations, dizziness, chest distress, shortness of breath, abdominal pain, headache, syncope, etc. The study also revealed that their levels of creatine kinase and its cardiac, MB isoenzyme, as well as myoglobin and AST, were 1.6 to 3.8 times elevated comparing to control villagers, who did not consume the mushroom, though the difference was statistically

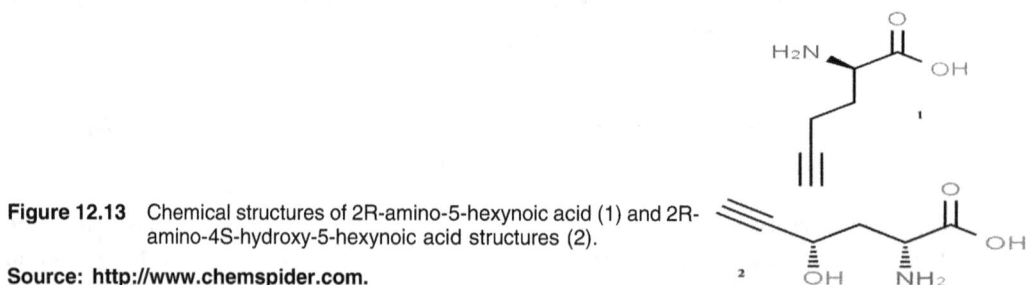

Figure 12.13 Chemical structures of 2R-amino-5-hexynoic acid (1) and 2R-amino-4S-hydroxy-5-hexynoic acid structures (2).

Source: http://www.chemspider.com.

significant only for AST. Among family members of SUD victims who did not have a history of *T. venenata* consumption, 17% developed an illness during the same period of time. There were also SUD cases that could not be connected to *T. venenata* consumption; however, Shi et al.(2012a) marked *T. venenata* as a strong suspect for SUD. After a special surveillance program that included education of local people and a warning not to eat unfamiliar mushrooms, the SUD cases dropped to nearly zero after 2008 (Shi et al., 2012a), raising hope that the Yunnan SUD mystery is solved, at least for the part of the scientific community. However, cases of SUD soon reappeared; three documented cases occurred in 2015. All members of the victims' family consumed several mushroom species but claimed not to have consumed any little white species. The genome of two victims and their living relatives was sequenced and revealed mutations in several genes, including certain ion channel genes, as well as genes that could be associated with glucose metabolism deficiency. Conditions that could result from such mutations (e.g. channelopathies) may remain unnoticed for a long time and maybe completely asymptomatic until a triggering event occurs. This study rejected the hypothesis that poisoning with *T. venenata is* the cause of unexpected deaths, as in some cases researchers failed to connect *T. venenata* consumption with SUDs. However, the authors did acknowledge that common exposure to external factors must be taken into account since there is clear time-space clustering of SUDs (Li et al., 2020). Whether *T. venenata* is the key common exposure factor is yet to be uncovered. Since barium is known to cause cardiac arrest and sudden death in humans, it was suggested at one point that *T. venenata* contains high levels of this metal. This theory was later rejected as it was found that barium content in mushrooms gathered from different sites was low and comparable to other foods (Zhang et al., 2012). Yunnan's sudden unexplained deaths remain a mystery, without scientific consensus regarding the cause.

12.11 GASTROINTESTINAL MUSHROOM POISONING SYNDROME

While the majority of all poisonous mushrooms cause gastrointestinal (GI) disorders, there are a great number of mushrooms that cause exclusively (or almost exclusively) gastrointestinal upset, which leads to the development of non-specific symptoms such as nausea, vomiting, and diarrhea, accompanied by cramps and abdominal pain. The disorder is self-limited in most cases, and the therapy is only supportive. Also, there is a great inter-individual difference in susceptibility to mushroom gastrointestinal irritants (Köppel, 1993; Berger & Guss, 2005;Beuhler, 2017; Cervellin et al., 2017; White et al., 2019a). GI syndrome is the most frequently reported clinical manifestation of mushroom poisoning (Beug et al., 2006; Keller et al., 2018). There is a great variety of species that cause GI syndrome, and the great majority of toxins are not identified yet (Köppel, 1993). Some of the best-known representatives of mushrooms that cause primarily GI disorders are given below, with emphasis on species for which suspected toxins were reported or with well-described clinical presentation of poisoning.

Among mushrooms of the genus *Tricholoma*, *T. ustale* is recognized as one of the three most common causes of overall mushroom poisoning in Japan; a bisphenylacetic acid derivative, named ustalic acid (Figure 12.14) was isolated from its fruiting bodies as well as several of its analogues, all of which were found to be inhibitors of Na^+,K^+-ATPase. Suppressing of a Na^+,K^+ pump causes inability of water absorption and resulting with diarrhea. Mice that were force-fed with 2, 5, or 10 mg of ustalic acid suffered from tremor and abdominal contractions, causing them to sit in a crouched position, with little movement. Symptom onset was 3 hours after ingestion, continuing up to 2 days afterward resulting in death in some cases; the dose of 10 mg caused death in all mice tested, although no information about the number of test animals was given. The authors reported the isolation of 191.4 mg of ustalic acid from 30.3 kg of fresh mushrooms, the toxin concentration being 6.3 mg/kg. The analogues, which also showed Na^+,K^+-ATPase inhibitory activity were found in much lower concentrations (Sano et al., 2002).

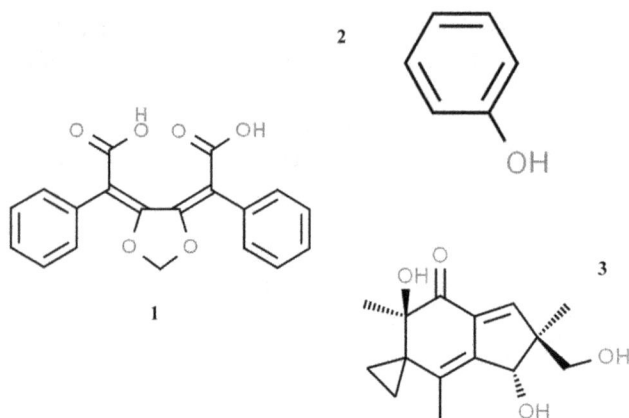

Figure 12.14 Chemical structures of ustalic acid, phenol and illudin S.
Source: http://www.chemspider.com.

Agaricus xanthodermus ("yellow stainer") is a rare example of a poisonous member of the genus *Agaricus* (Gill & Strauch, 1984), which includes the widely consumed white button mushroom (*A. bisporus*) (Djekic et al., 2016). It is often mistaken for other, edible, wild-growing species of the genus such as *A. campestris*, causing vomiting and diarrhea in susceptible individuals. It is found that *A. xanthodermus* contains phenol (Figure 12.14) in the concentration of up to 0.1% fresh weight, which is sufficient to cause acute toxicity. The symptoms of *A. xanthodermus* poisoning are consistent with those caused by the ingestion of phenol. p-Quinol, which gives the mushroom the characteristic odor, is also known to be toxic (Gill & Strauch, 1984).

Omphalotus olearius can be found in Europe where it is sometimes mistaken for edible chanterelles (*Cantharellus cibarius*) and is known to cause GI disorders; it is also reported to be sold as a chanterelle in the markets, presumably by mistake. Two reports of poisoning with this mushroom in Croatia during the 1960s and 1970s included 29 people in total, showing a variety of symptoms, ranging from mild gastrointestinal to muscarinic-like syndromes, with symptoms like salivation, bradycardia, and even diplopia. Transient liver injury was also noticed in few cases. Misidentification of toxic mushrooms in some cases cannot be ruled out, although there were other reports from France which questioned whether *O. olearius* contains muscarine (Maretić, 1967; Maretić et al., 1975; Vanden Hoek et al., 1991). Poisonings with North American *O. illudens*, for a long time considered the same species as *O. olearius*, cause similar GI symptoms, primarily vomiting, although the poisonings are rarely reported. Elevation in liver enzymes was reported in two cases and hypokalemia in one (Vanden Hoek et al., 1991). Severe poisonings with eastern Asian *O. japonicas* (as *Lampteromyces japonicus*) were reported to be accompanied by delayed intestinal edema, after initial GI disorder symptoms which include nausea, vomiting and diarrhea (Kobayashi et al., 2017). Illudin S (Figure 12.14), a potent cytotoxic sesquiterpene, is often referred to as the toxic principle of *O. japonicus* (Tanaka et al., 1994; Beuhler, 2017); however, data that clearly connect oral administration of this compound to GI symptoms seen in human poisonings by this mushroom are insufficient.

Rubroboletus satanas (formerly known as *Boletus satanas*), or devil's bolete, is known to cause serious gastroenteritis, with severe bloody diarrhea, although there are no reported deaths due to poisoning with this, rather rare European mushroom (Kretz et al., 1989; Patočka, 2018). Kretz et al. (1989) isolated a toxic protein (in later publications referred to as glycoprotein) named bolesatin which was shown to inhibit protein synthesis in cell cultures. Bolesatine is said to be "relatively thermostable" at 70°C for 60 minutes as well as resistant to proteolytic enzymes.

Detailed studies on bolesatine's toxicokinetics were performed using ^{14}C labeled protein. It was found that most of the bolesatine is excreted in feces (63%, during 24 hours) and the rest is absorbed; after 3–6 hours the organ distribution is stable, with intestine tissue containing the highest proportion of the toxin (13.9%), followed by liver (2.5%) and kidneys (1.5%). The absorbed bolesatine is excreted in urine (17.4%, during 24 hours); after 31 hours, bolesatine is almost eliminated/excreted from the body. Oral LD_{50} in mice was determined to be 3.3 mg/kg and postmortem autopsy showed signs of hepatotoxicity. Although no data was given whether autopsy showed signs of GI damage (Kretz et al., 1991b), in consequent papers bolesatine was referred to as causing hepatotoxicity and "slight gastro-enteritis" in mice (Kretz et al., 1991a). Bolesatin was also shown to have lectinic properties and to agglutinate both human erythrocytes and platelets (Gachet et al., 1996). Ricin, a well-known toxic lectin that acts as protein synthesis inhibitors cause GI disorders when ingested orally, the GI symptoms start after several hours (Audi et al., 2005); GI symptoms, however, start relatively quickly upon *R. satanas* ingestion, within 2 hours (Merlet et al., 2011). Nevertheless, Kretz et al. proposed that bolesatine is at least partly responsible for the toxicity of *R. satanas*.

Before isolation and characterization of bolesatine, a toxic protein named bolaffinine was isolated from *Xanthoconium affine* (as *Boletus affinis*), which was implicated in deadly poisonings of cattle in Madagascar. Bolaffinine was found to be thermolabile, which explained why the mushroom was safe for human consumption (Razanamparany et al., 1986). Bolevenine, another toxic protein isolated from *Neoboletus venenatus* (as *B. venenatus*) that known to cause GI disorders grows in Japan and China and is known to cause GI disorders, was also found to be thermolabile (Matsuura et al., 2007). More detailed information on thermostability however lacks for bolesatine.

Interestingly, it was recently found that *R. satanas* may cause significant elevation of serum procalcitonin (hyperprocalcitonemia), as well as c-reactive protein (CRP) levels. The underlying mechanism is unknown and *R. satanas* is the only known mushroom that can cause this distinct poisoning symptom (Merlet et al., 2011).

Chlorophyllum molybdites is one of the leading causes of mushroom poisoning in North America (Lehmann & Khazan, 1992), although poisonings due to consumption of this parasol mushroom species are known from India, Australia, and Brazil as well (Young, 1989; Natarajan & Kaviyarasan, 1991; Meijer et al., 2007; Bijeesh et al., 2017). Eilers and Nelson (1974) isolated a high molecular weight thermolabile protein (~400 kDa) from the *C. molybdites* fruiting bodies, which was shown to be toxic to mice and chicks when administered systemically; although there was no toxic reaction when protein was force-fed to animals it was still suggested that it might be the main toxic principle of the mushroom; Lehmann and Khazan (1992) later expressed doubts about the claim.

Poisonings with *Entoloma sinuatum* (syn. *E. lividum*) and *Tricholoma pardinum* (syn. *T. tigrinus*) have been reported quite frequently in Europe (Alder, 1961; Schmutz et al., 2016; Cervellin et al., 2017; Clericuzio et al., 2020; Wennig et al., 2020), although there is a lack of specific reports describing such cases in recent times. Other species known to produce GI upset include *Russula emetica*, *Lactarius torminosus*, *L. helvus*, *Tylopilus felleus*, *Ramaria pallida*, *R. formosa*, *Scleroderma citrinum*, *Clitocybe nebularis*, *Hebeloma crustuliniforme*, *Hypholoma fasciculare*, etc. (Köppel, 1993; Schenk-Jaeger et al., 2012; Wennig et al., 2020.

A fact that should be always highlighted is that edible mushrooms are responsible for a great number of gastrointestinal mushroom poisonings, as a result of inappropriate preparation or storage (Gawlikowski et al., 2014); mushrooms are products with a relatively short shelf-life at ambient temperatures (Djekic et al., 2017). Some may cause GI disorders due to consumption of large quantities and low digestibility (White et al., 2019a). Inability to hydrolyze mushroom disaccharide, trehalose, due to trehalase deficiency, may also lead to GI discomfort (Gawlikowski et al., 2014). In a retrospective study conducted in Switzerland, it was found that in most mushroom poisoning cases an edible species were involved (Schenk-Jaeger et al., 2012). Among the top six species involved in poisonings, four were edible, commercial species: *Boletus edulis*, *Agaricus bisporus*, *Cantharellus cibarius,* and

Morchella esculenta. B. edulis was also found to be one of the main causes of mushroom poisoning in Italy, despite being one of the most popular mushroom delicacies (Cervellin et al., 2017). A study in Poland also found that more than 87% of mushroom poisonings with GI symptoms were due to ingestion of edible mushrooms (Gawlikowski et al., 2014).

Photograph 12.6. *Entoloma lividum*, one of many mushroom species that cause gastro-intestinal upset. Bor district, Serbia, 2014. Photo: P. Petrović

12.12 SHIITAKE DERMATITIS

Shiitake (*Lentinula edodes*) is a very popular edible commercial mushroom species, particularly in the Far East, but under certain circumstances it can cause characteristic flagellate dermatitis, which is mostly connected with consumption of raw or undercooked mushrooms, suggesting a thermolabile toxic compound (Nakamura, 1992; Boels et al., 2014). First signs of skin reaction arise 12 hours to 5 days after consumption (Boels et al., 2014). It was proposed that a lentinan, a β-glucan isolated from shiitake that is used as an adjuvant in cancer therapy in Japan, or a polysaccharide like lentinan is responsible for the reaction (Nakamura, 1992; Boels et al., 2014), although there is no direct evidence for this claim. It is, however, known that one of the rare side effects of intravenously administered lentinan is the development of a streaky rash (Nakamura, 1992). Most cases are reported from Asia where shiitake is widely produced, but in recent times cases are reported from Europe and America due to the popularization of the mushroom. The condition is treated with corticosteroids or and antihistaminics, leading to full recovery (Boels et al., 2014). Although current evidence suggests that shiitake flagellate dermatitis is a result of toxicity and not hypersensitivity, shiitake growers can also develop various occupational immunological conditions like contact dermatitis/protein contact dermatitis, as well as conjunctivitis and respiratory allergic reactions (Aalto-Korte et al., 2005; Boels et al., 2014).

12.13 MUSHROOMS THAT CAUSE HYPERSENSITIVITY REACTIONS

12.13.1 Immunohaemolytic anemia–inducing mushrooms – *Paxillus* syndrome

Paxillus involutus was once thought to be a good edible species, although it was known to cause GI discomfort associated with consumption of undercooked mushrooms, due to thermolabile toxins. However, there were cases of poisonings related to *P. involutus* that had symptoms of systemic poisoning and some ended fatally; all of them involved people who had consumed *P. involutus* repeatedly. It was found that certain strains of this mushroom produce a compound with antigenic properties, capable of provoking an immune response in humans. Repeated consumption of *P. involutus* can thus lead to sensibillization in susceptible people and the production of IgG antibodies. The reaction between the antigen from the mushroom and antibodies leads to the formation of the antigen-antibody complexes. These immune complexes target specifically erythrocytes and bind to their surface; erythrocytes agglutinate which leads to their destruction and consequently to hemolytic anemia and all the systemic disorders that are connected with it. The symptoms, which occur 30 mininutes to 3 hours after ingestion of the mushroom, apart from hemolysis, include abdominal pain, vomiting, jaundice, anuria, circulatory shock, and pulmonary failure (Winkelmann et al., 1986; Pohle, 1995; Stöver et al., 2019). Deaths are reported due to multiorgan failure and disseminated intravascular coagulation (Stöver et al., 2019). The treatment is supportive and includes the elimination of immune complexes by plasma exchange; acute renal failure is treated with hemodialysis (Winkelmann et al., 1986).

Even though today *P. involutus* is treated as a poisonous species, it is still consumed in some regions (Stöver et al., 2019).

12.13.2 Lycoperdonosis

Puffballs are a group of fungi that produce globous fruiting bodies; they are edible when young but upon maturation, they go through a process of autolysis in which their inside, gleba, is transformed into a powdery mass of spores (Petrović et al., 2016). Gleba of mature puffballs has been used in traditional medicine, as hemostyptic. Although mature puffballs contain compounds that may be beneficial in skin disorders (Petrović et al., 2019a; Petrović et al., 2019b), their spores can cause dangerous "toxic" acute reaction if inhaled. The condition was described as "lycoperdonosys", after a puffball genus *Lycoperdon* (Strand et al., 1967). Lycoperdonosys is a type of inflammatory, hypersensitivity reaction that occurs primarily after inhalation of puffball spores. The symptoms include nausea and vomiting, fever, and pneumonia with bilateral reticulonodular infiltrate, also accompanied by myalgia and fatigue. It is still not clear if lycoperdonosis is connected with some predisposing factors, such as asthma. Treatment is supportive and includes corticosteroids (Diaz, 2018).

ACKNOWLEDGMENT

This work was supported by the Ministry of Education, Science and Technological Development of the Republic of Serbia (Contract No. 451-03-68/2022-14/200135 & Contract No. 451-03-68/2022-14/200051)

REFERENCES

Aalto-Korte, K., Susitaival, P., Kaminska, R., & Makinen-Kiljunen, S. (2005). Occupational protein contact dermatitis from shiitake mushroom and demonstration of shiitake-specific immunoglobulin E. *Contact Dermatitis, 53*(4), 211–213.

Ahn, J. Y., Seok, S. J., Song, J. E., Choi, J. H., Han, S. H., Choi, J. Y., Kim, C. O., Song, Y. G., & Kim, J. M. (2013). Two cases of mushroom poisoning by Podostroma cornu-damae. *Yonsei Medical Journal*, *54*(1), 265.

Akata, I., Yilmaz, I., Kaya, E., Coskun, N. C., & Donmez, M. (2020). Toxin components and toxicological importance of Galerina marginata from Turkey. *Toxicon*, *187*, 29–34.

Akilli, N. B., Dundar, Z. D., Koylu, R., Gunaydin, Y. K., & Cander, B. (2014). Rhabdomyolysis induced by Agaricus bisporus. *Journal of Academic Emergency Medicine*, *13*(4), 212–213.

Akiyama, H., Matsuoka, H., Okuyama, T., Higashi, K., Toida, T., Komatsu, H., Sugita-Konishi, Y., Kobori, S., Kodama, Y., Yoshida, M., & Endou, H. (2015). The acute encephalopathy induced by intake of Sugihiratake mushroom in the patients with renal damage might be associated with the intoxication of cyanide and thiocyanate. *Food Safety*, *3*(1), 16–29.

Alder, A. E. (1961). Erkennung und Behandlung Der Pilzvergiftungen. *DMW - Deutsche Medizinische Wochenschrift*, *86*(23), 1121–1127.

Allen, B., Desai, B., & N. Lesenbee (2012). Amatoxin: A review. *Emergency Medicine: Open Access*, *02*(04), 1–4.

Amsterdam, J. V., Opperhuizen, A., & Brink, W. V. (2011). Harm potential of magic mushroom use: A review. *Regulatory Toxicology and Pharmacology*, *59*(3), 423–429.

Anand, J. S., Chwaluk, P., & Sut, M. (2009). Acute poisoning with Tricholoma equestre. *Przegled Lekarski*, *66*(6), 339–340.

Andary, C., Privat, G., & Bourrier, M. (1985). Variations of Monomethylhydrazine content in Gyromitra esculenta. *Mycologia*, *77*(2), 259.

Antkowiak, W. Z., & Gessner, W. P. (1979). The structures of orellanine and orelline. *Tetrahedron Letters*, *20*(21), 1931–1934.

Aoyagi, Y., & Sugahara, T. (1988). β-hydroxy-l-valine from Pleurocybella porrigens. *Phytochemistry*, *27*(10), 3306–3307.

Appleton, R. E., Jan, J. E., & Kroeger, P. D. (1988). Laetiporus sulphureus causing visual hallucinations and ataxia in a child. *Canadian Medical Association Journal*, *139*, 48–49.

Arłukowicz-Grabowska, M., Wójcicki, M., Raszeja-Wyszomirska, J., Szydłowska-Jakimiuk, M., Piotuch, B., & Milkiewicz, P. (2019). Acute liver injury, acute liver failure and acute on chronic liver failure: A clinical spectrum of poisoning due to Gyromitra esculenta. *Annals of Hepatology*, *18*(3), 514–516.

Audi, J., Belson, M., Patel, M., Schier, J., & Osterloh, J. (2005). Ricin poisoning. *JAMA*, *294*(18), 2342.

Austin, E., Myron, H. S., Summerbell, R. K., & Mackenzie, C. A. (2019). Acute renal injury cause by confirmed Psilocybe cubensis mushroom ingestion. *Medical Mycology Case Reports*, *23*, 55–57.

Ba, D. M., Ssentongo, P., Beelman, R. B., Muscat, J., Gao, X., & Richie, J. P. (2021). Higher mushroom consumption is associated with lower risk of cancer: A systematic review and meta-analysis of observational studies. *Advances in Nutrition*, *12*(5), 1691–1704.

Badham, E. R. (1984). Ethnobotany of psilocybin mushrooms, especially Psilocybe cubensis. *Journal of Ethnopharmacology*, *10*(2), 249–254.

Bains, A., & Chawla, P. (2020). In vitro bioactivity, antimicrobial and anti-inflammatory efficacy of modified solvent evaporation assisted Trametes versicolor extract. *3 Biotech*, *10*(9), 1–11.

Barceloux, D. G. (2008).Muscarine - Containing mushrooms and muscarine toxicity (Clitocybe and inocybe species). In *Medical Toxicology of Natural Substances*. John Wiley & Sons, pp. 303–306.

Beck, O., Helander, A., Karlson-Stiber, C., & Stephansson, N. (1998). Presence of Phenylethylamine in hallucinogenic Psilocybe mushroom: Possible role in adverse reactions. *Journal of Analytical Toxicology*, *22*(1), 45–49.

Bedry, R., Baudrimont, I., Deffieux, G., Creppy, E. E., Pomies, J. P., Ragnaud, J. M., Dupon, M., Neau, D., Gabinski, C., De Witte, S., Chapalain, J. C., Beylot, J., & Godeau, P. (2001). Wild-mushroom intoxication as a cause of Rhabdomyolysis. *New England Journal of Medicine*, *345*(11), 798–802.

Benjamin, D. R. (2020). Gyromitrin poisoning: More questions than answers. *Fungi*, *13*(1), 36–39.

Berger, K. J., & Guss, D. A. (2005). Mycotoxins revisited: Part II. *The Journal of Emergency Medicine*, *28*(2), 175–183.

Bergman, K., & Hellenäs, K. (1992). Methylation of rat and mouse DNA by the mushroom poison gyromitrin and its metabolite monomethylhydrazine. *Cancer Letters*, *61*(2), 165–170.

Bessard, J., Saviuc, P., Chane-Yene, Y., Monnet, S., & Bessard, G. (2004). Mass spectrometric determination of acromelic acid a from a new poisonous mushroom: Clitocybe amoenolens. *Journal of Chromatography A*, *1055*(1–2), 99–107.

Beug, M. W. (2009). NAMA toxicology committee report for 2007: Recent mushroom poisonings in North America. *McIlvainea*, *18*(1), 40–44.

Beug, M. W., Shaw, K. M., & Cochran, K. W. (2006). Thirty-plus years of mushroom poisoning: Summary of the approximately 2,000 reports in the NAMA case registry. *McIlvainea*, *16*(2), 47–68.

Beuhler, M. C. (2017). Overview of mushroom poisoning. In *Critical care toxicology* (pp. 2105–2128). Cham: Springer.

Bijeesh, C., Vrinda, K. B., & Pradeep, C. K. (2017). Mushroom poisoning by Chlorophyllum molybdites in Kerala. *Journal of Mycopathological Research*, *54*(4), 477–483.

Bills, G. F., & Gloer, J. B. (2017). Biologically active secondary metabolites from the fungi. *The Fungal Kingdom*, 1087–1119. Washington: ASM Press.

Bissett, J., Gams, W., Jaklitsch, W., & Samuels, G. J. (2015). Accepted Trichoderma names in the year 2015. *IMA Fungus*, *6*(2), 263–295.

Boels, D., Landreau, A., Bruneau, C., Garnier, R., Pulce, C., Labadie, M., De Haro, L., & Harry, P. (2014). Shiitake dermatitis recorded by French poison control centers – new case series with clinical observations. *Clinical Toxicology*, *52*(6), 625–628.

Bouget, J., Bousser, J., Pats, B., Ramee, M., Chevet, D., Rifle, G., Giudicelli, C., & Thomas, R. (1990). Acute renal failure following collective intoxication byCortinarius orellanus. *Intensive Care Medicine*, *16*(8), 506–510.

Braeuer, S., Borovička, J., Kameník, J., Prall, E., Stijve, T., & Goessler, W. (2020). Is arsenic responsible for the toxicity of the hyperaccumulating mushroom Sarcosphaera coronaria? *Science of TheTotal Environment*, *736*, 139524.

Brandenburg, W. E., & Ward, K. J. (2018). Mushroom poisoning epidemiology in the United States. *Mycologia*, *110*(4), 637–641.

Braun, R., Greeff, U., & Netter, K. (1979). Liver injury by the Morel poison gyromitrin. *Toxicology*, *12*(2), 155–163.

Brooks, D. E., & Graeme, K. A. (2016). Gyromitra mushrooms. *Critical Care Toxicology*, 1–12. Cham: Springer.

Brvar, M., Možina, M., & Bunc, M. (2006). Prolonged psychosis after amanita muscaria ingestion. *Wiener klinische Wochenschrift*, *118*(9–10), 294–297.

Bunel, V., Souard, F., Antoine, M., Stévigny, C., & Nortier, J. (2018). Nephrotoxicity of natural products: Aristolochic acid and fungal toxins. *Comprehensive Toxicology* (Third Edition), Vol. *14*, 340–379. Amsterdam: Elsevier.

Cai, M., Idrees, M., Zhou, Y., Zhang, C., & Xu, J. (2020). First report of green mold disease caused by Trichoderma hengshanicum on Ganoderma lingzhi. *Mycobiology*, *48*(5), 427–430.

Cai, Q., Cui, Y., & Yang, Z. L. (2016). Lethal amanita species in China. *Mycologia*, *108*(5), 993–1009.

Cai, Q., Tulloss, R. E., Tang, L. P., Tolgor, B., Zhang, P., Chen, Z. H., & Yang, Z. L. (2014). Multi-locus phylogeny of lethal amanitas: Implications for species diversity and historical biogeography. *BMC Evolutionary Biology*, *14*(1), 143.

Calviño, J., Romero, R., Pintos, E., Novoa, D., Güimil, D., Cordal, T., Mardaras, J., Arcocha, V., Lens, X., & Sanchez-Guisande, D. (1998). Voluntary ingestion of Cortinarius mushrooms leading to chronic interstitial nephritis. *American Journal of Nephrology*, *18*(6), 565–569.

Carod-Artal, F. (2015). Hallucinogenic drugs in pre-Columbian mesoamerican cultures. *Neurología (English Edition)*, *30*(1), 42–49.

Cervellin, G., Comelli, I., Rastelli, G., Sanchis-Gomar, F., Negri, F., De Luca, C., & Lippi, G. (2017). Epidemiology and clinics of mushroom poisoning in northern Italy: A 21-year retrospective analysis. *Human & Experimental Toxicology*, *37*(7), 697–703.

Chen, X., Shao, B., Yu, C., Yao, Q., Ma, P., Li, H., Cai, W., Fu, H., Li, B., & Sun, C. (2020). The cyclopeptide -amatoxin induced hepatic injury via the mitochondrial apoptotic pathway associated with oxidative stress. *Peptides*, *129*, 170314.

Chen, X., Shao, B., Yu, C., Yao, Q., Ma, P., Li, H., Li, B., & Sun, C. (2021). Energy disorders caused by mitochondrial dysfunction contribute to α-amatoxin-induced liver function damage and liver failure. *Toxicology Letters*, *336*, 68–79.

Chen, Z., Zhang, P., & Zhang, Z. (2013). Investigation and analysis of 102 mushroom poisoning cases in southern China from 1994 to 2012. *Fungal Diversity*, *64*(1), 123–131.

Chew, K. S., Mohidin, M. A., Ahmad, M. Z., Tuan Kamauzaman, T. H., & Mohamad, N. (2008). Early onset muscarinic manifestations after wild mushroom ingestion. *International Journal of Emergency Medicine*, *1*(3), 205–208.

Chilton, W. S., & Ott, J. (1976). Toxic metabolites of Amanita pantherina, A. cothurnata, A. muscaria and other Amanita species. *Lloydia*, *39*(2–3), 150–157.

Choe, S., In, S., Jeon, Y., Choi, H., & Kim, S. (2018). Identification of trichothecene-type mycotoxins in toxic mushroom Podostroma cornu-damae and biological specimens from a fatal case by LC–QTOF/MS. *Forensic Science International*, *291*, 234–244.

Chwaluk, P. (2013). Rhabdomyolysis as an unspecyfic symptom of mushroom poisoning–a case report. *Przegląd Lekarski*, *70*, 684–686.

Clericuzio, M., Hussain, F. H., Amin, H. I., Salis, A., Damonte, G., Pavela, R., & Vidari, G. (2020). New acetylenic metabolites from the toxic mushroom Tricholoma pardinum. *Natural Product Research*, *35*(23), 5081–5088.

Cochran, K. W., & Cochran, M. W. (1978). Clitocybe clavipes: Antabuse-like reaction to alcohol. *Mycologia*, *70*(5), 1124.

Connor, D. J., Langlais, P. J., & Thal, L. J. (1991). Behavioral impairments after lesions of the nucleus basalis by ibotenic acid and quisqualic acid. *Brain Research*, *555*, 84–90.

Curry, S. C., & Rose, M. C. (1985). Intravenous mushroom poisoning. *Annals of Emergency Medicine*, *14*(9), 900–902.

Danel, V., Saviuc, P., & Garon, D. (2001). Main features of Cortinarius spp. poisoning: A literature review. *Toxicon*, *39*(7), 1053–1060.

Dasgupta, A. (2019). Abuse of magic mushroom, peyote cactus, LSD, khat, and volatiles. In *Critical Issues in Alcohol and Drugs of Abuse Testing*. Academic Press, pp. 477–494

Davoli, P., Floriani, M., Assisi, F., Kob, K., & Sitta, N. (2016). Comment on "Chemical and toxicological investigations of a previously unknown poisonous European MushroomTricholoma terreum". *Chemistry – A European Journal*, *22*(16), 5786–5788.

De Haro, L., Arditti, J., David, J., Rascol, J., & Jouglard, J. (1997). Acute renal failure due to amanita Proxima: Experience of the Marseilles poison centre. *Toxicon*, *35*(6), 809.

De Sagun, S., & Tabunar, S. (2012). Seizure and transient expressive aphasia in hallucinogenic mushroom (Psilocybe) poisoning: A case report. *The Journal of Emergency Medicine*, *43*(5), 932.

DeFeudis, F. V. (1980). Physiological and behavioral studies with muscimol. *Neurochemical Research*, *5*(10), 1047–1068.

Diaz, J. H. (2017). Colorful mushroom ingestion. *Wilderness & Environmental Medicine*, *28*(4), 362–364.

Diaz, J. H. (2018). A puff of spores. *Wilderness & Environmental Medicine*, *29*(1), 119–122.

Djekic, I., Vunduk, J., Tomašević, I., Kozarski, M., Petrovic, P., Niksic, M., Pudja, P., & Klaus, A. (2016). Total quality index of Agaricus bisporus mushrooms packed in modified atmosphere. *Journal of the Science of Food and Agriculture*, *97*(9), 3013–3021.

Djekic, I., Vunduk, J., Tomašević, I., Kozarski, M., Petrovic, P., Niksic, M., Pudja, P., & Klaus, A. (2017). Application of quality function deployment on shelf-life analysis of Agaricus bisporus Portobello. *LWT*, *78*, 82–89.

Eilers, F. I., & Nelson, L. R. (1974). Characterization and partial purification of the toxin of Lepiota morganii. *Toxicon*, *12*(6), 557–563.

Elonen, E., Tarssanen, L., & Härkönen, M. (1979). Poisoning with Brown fly agaric, amanita Regalis. *Acta Medica Scandinavica*, *205*(1–6), 121–123.

Enjalbert, F., Rapior, S., Nouguier-Soulé, J., Guillon, S., Amouroux, N., & Cabot, C. (2002). Treatment of Amatoxin poisoning: 20-Year retrospective analysis. *Journal of Toxicology: Clinical Toxicology*, *40*(6), 715–757.

Faulstich, H., & Cochet-Meilhac, M. (1976). Amatoxins in edible mushrooms. *FEBS Letters*, *64*(1), 73–75.

Franz, M., Regele, H., Kirchmair, M., Kletzmayr, J., Sunder-Plassmann, G., Horl, W. H., & Pohanka, E. (1996). Magic mushrooms: Hope for a 'cheap high' resulting in end-stage renal failure. *Nephrology Dialysis Transplantation*, *11*(11), 2324–2327.

Fu, X., Fu, B., He, Z., Gong, M., Li, Z., & Chen, Z. (2017). Acute renal failure caused by amanita oberwinklerana poisoning. *Mycoscience*, *58*(2), 121–127.

Fushiya, S., Sato, S., Kusano, G., & Nozoe, S. (1993). β-cyano-l-alanine and N-(γ-l-glutamyl)-β-cyano-l-alanine, neurotoxic constituents of Clitocybe acromelalga. *Phytochemistry*, *33*(1), 53–55.

Gachet, C., Ennamany, R., Kretz, O., Ohlmann, P., Krause, C., Creppy, E., Dirheimer, G., & Cazenave, J. (1996). Bolesatine induces agglutination of rat platelets and human erythrocytes and platelets in vitro. *Human & Experimental Toxicology*, *15*(1), 26–29.

Garcia, J., Costa, V. M., Carvalho, A. T., Silvestre, R., Duarte, J. A., Dourado, D. F., Arbo, M. D., Baltazar, T., Dinis-Oliveira, R. J., Baptista, P., De Lourdes Bastos, M., & Carvalho, F. (2015a). A breakthrough on amanita phalloides poisoning: An effective antidotal effect by polymyxin B. *Archives of Toxicology*, *89*(12), 2305–2323.

Garcia, J., Costa, V. M., Carvalho, A., Baptista, P., De Pinho, P. G., De Lourdes Bastos, M., & Carvalho, F. (2015b). Amanita phalloides poisoning: Mechanisms of toxicity and treatment. *Food and Chemical Toxicology*, *86*, 41–55.

Garcia, J., Oliveira, A., De Pinho, P. G., Freitas, V., Carvalho, A., Baptista, P., Pereira, E., De Lourdes Bastos, M., & Carvalho, F. (2015c). Determination of amatoxins and phallotoxins in amanita phalloides mushrooms from northeastern Portugal by HPLC-DAD-MS. *Mycologia*, *107*(4), 679–687.

Gawlikowski, T., Romek, M., & Satora, L. (2014). Edible mushroom-related poisoning: A study on circumstances of mushroom collection, transport, and storage. *Human & Experimental Toxicology*, *34*(7), 718–724.

Gejyo, F., Homma, N., Higuchi, N., Ataka, K., Teramura, T., Alchi, B., Suzuki, Y., Nishi, S., & Narita, I. (2005). A novel type of encephalopathy associated with mushroom Sugihiratake ingestion in patients with chronic kidney diseases. *Kidney International*, *68*(1), 188–192.

Genest, K., Hughes, D., & Rice, W. (1968). Muscarine in Clitocybe species. *Journal of Pharmaceutical Sciences*, *57*(2), 331–333.

George, P., & Hegde, N. (2013). Muscarinic toxicity among family members after consumption of mushrooms. *Toxicology International*, *20*(1), 113.

Giancola, N. B., Korson, C. J., Caplan, J. P., & McKnight, C. A. (2021). A "Trip" to the intensive care unit: An intravenous injection of psilocybin. *Journal of the Academy of Consultation-Liaison Psychiatry*, *62*(3), 370–371.

Gill, M., & Steglich, W. (1987). Pigments of fungi (Macromycetes). *Progress in the Chemistry of Organic Natural Products*, *51*, 1–286.

Gill, M., & Strauch, R. J. (1984). Constituents of Agaricus xanthodermus Genevier: The first naturally endogenous azo compound and toxic phenolic metabolites. *Zeitschrift für Naturforschung C*, *39*(11–12), 1027–1029.

Gonmori, K., Fujita, H., Yokoyama, K., Watanabe, K., & Suzuki, O. (2011). Mushroom toxins: A forensic toxicological review. *Forensic Toxicology*, *29*(2), 85–94.

Gonzalez, J., Lacomis, D., & Kramer, D. (1996). Mushroom myopathy. *Muscle & Nerve*, *19*, 790–792.

Govorushko, S., Rezaee, R., Dumanov, J., & Tsatsakis, A. (2019). Poisoning associated with the use of mushrooms: A review of the global pattern and main characteristics. *Food and Chemical Toxicology*, *128*, 267–279.

Guzmán, G. (2008). Hallucinogenic mushrooms in Mexico: An overview. *Economic Botany*, *62*(3), 404–412.

Haberl, B., Pfab, R., Berndt, S., Greifenhagen, C., & Zilker, T. (2011). Case series: Alcohol intolerance with coprine-like syndrome after consumption of the mushroomLepiota aspera(Pers.:Fr.) Quél., 1886 (Freckled dapperling). *Clinical Toxicology*, *49*(2), 113–114.

Haines, J. H., Lichstein, E., & Glickerman, D. (1986). A fatal poisoning from an amatoxin containing Lepiota. *Mycopathologia*, *93*(1), 15–17.

Hallen, H., Adams, G., & Eicker, A. (2002). Amatoxins and phallotoxins in Indigenous and introduced South African amanita species. *South African Journal of Botany*, *68*(3), 322–326.

Høiland, K. (1983). Extracts of Cortinarius speciosissimus affecting the photosynthetic apparatus of lemna minor. *Transactions of the British Mycological Society*, *81*(3), 633–635.

Ichimura, T. (1918). A new poisonous mushroom. *Botanical Gazette*, *65* (1), 109–111.

Ishida, M., & Shinozaki, H. (1988). Acromelic acid is a much more potent excitant than kainic acid or domoic acid in the isolated rat spinal cord. *Brain Research*, *474*(2), 386–389.

Iwafuchi, Y., Morita, T., Kobayashi, H., Kasuga, K., Ito, K., Nakagawa, O., Kunisada, K., Miyazaki, S., & Kamimura, A. (2003). Delayed onset acute renal failure associated with amanita pseudoporphyria Hongo ingestion. *Internal Medicine*, *42*(1), 78–81.

Jang, J., Kim, C., Yoo, J. J., Kim, M. K., Lee, J. E., Lim, A. L., Choi, J., Hyun, I. G., Shim, J. W., Shin, H., Han, J., & Seok, S. J. (2013). An elderly man with fatal respiratory failure after eating a poisonous MushroomPodostroma cornu-damae. *Tuberculosis and Respiratory Diseases*, *75*(6), 264.

Janković, S., Jović-Stosić, J., Vucinić, S., Perković-Vukcević, N., & Vuković-Ercegović, G. (2013). Causes of rhabdomyolysis in acute poisonings. *Vojnosanitetski Pregled*, *70*(11), 1039–1045.

Jo, W., Hossain, M. A., & Park, S. (2014). Toxicological profiles of poisonous, edible, and medicinal mushrooms. *Mycobiology*, *42*(3), 215–220.

Kalač, P. (2009). Chemical composition and nutritional value of European species of wild growing mushrooms: A review. *Food Chemistry*, *113*(1), 9–16.

Kawagishi, H., Miyazawa, T., Kume, H., Arimoto, Y., & Inakuma, T. (2002). Aldehyde Dehydrogenase Inhibitors from the Mushroom Clitocybe clavipes. *Journal of Natural Products*, *65*, 1712–1714.

Kawaguchi, T., Suzuki, T., Kobayashi, Y., Kodani, S., Hirai, H., Nagai, K., & Kawagishi, H. (2010). Unusual amino acid derivatives from the mushroom Pleurocybella porrigens. *Tetrahedron*, *66*(2), 504–507.

Kaya, E., Yilmaz, I., Sinirlioglu, Z. A., Karahan, S., Bayram, R., Yaykasli, K. O., Colakoglu, S., Saritas, A., & Severoglu, Z. (2013). Amanitin and phallotoxin concentration in amanita phalloides Var. alba mushroom. *Toxicon*, *76*, 225–233.

Keller, S., Klukowska-Rötzler, J., Schenk-Jaeger, K., Kupferschmidt, H., Exadaktylos, A., Lehmann, B., & Liakoni, E. (2018). Mushroom poisoning—A 17 year retrospective study at a level I University emergency department in Switzerland. *International Journal of Environmental Research and Public Health*, *15*(12), 2855.

Kim, H. N., Do, H. H., Seo, J. S., & Kim, H. Y. (2016). Two cases of incidental Podostroma cornu-damae poisoning. *Clinical and Experimental Emergency Medicine*, *3*(3), 186–189.

Kirchmair, M., Carrilho, P., Pfab, R., Haberl, B., Felgueiras, J., Carvalho, F., Cardoso, J., Melo, I., Vinhas, J., & Neuhauser, S. (2011). Amanita poisonings resulting in acute, reversible renal failure: New cases, new toxic amanita mushrooms. *Nephrology Dialysis Transplantation*, *27*(4), 1380–1386.

Klaus, A., Petrović, P., Vunduk, J., Pavlović, V., & Van Griensven, L. J. (2020). The antimicrobial activities of silver nanoparticles synthesized from medicinal mushrooms. *International Journal of Medicinal Mushrooms*, *22*(9), 869–883.

Klimaszyk, P., & Rzymski, P. (2018). The yellow knight fights back: Toxicological, epidemiological, and survey studies defend edibility of Tricholoma equestre. *Toxins*, *10*(11), 468.

Kobayashi, F., Karasawa, T., Matsushita, T., Komatsu, O., & Adachi, W. (2017). Tsukiyotake (Lampteromyces japonicus) poisoning: Summary of 6 cases. *Journal Of The Japanese Association Of Rural Medicine*, *66*(4), 499–503.

Koge, T., Komatsu, W., & Sorimachi, K. (2011). Heat stability of agaritine in water extracts from Agaricus blazei and other edible fungi, and removal of agaritine by ethanol fractionation. *Food Chemistry*, *126*(3), 1172–1177.

Konno, K., Hashimoto, K., Ohfune, Y., Shirahama, H., & Matsumoto, T. (1988). Acromelic acids a and B. Potent neuroexcitatory amino acids isolated from Clitocybe acromelalga. *Journal of the American Chemical Society*, *110*(14), 4807–4815.

Kraft, J., Bauer, S., Keilhoff, G., Miersch, J., Wend, D., Riemann, D., Hirschelmann, R., Holzhausen, H., & Langner, J. (1998). Biological effects of the dihydroorotate dehydrogenase inhibitor polyporic acid, a toxic constituent of the mushroom Hapalopilus rutilans, in rats and humans. *Archives of Toxicology*, *72*(11), 711–721.

Kretz, O., Creppy, E., & Dirheimer, G. (1991a). Characterization of bolesatine, a toxic protein from the mushroom boletus Satanas Lenz and its effects on kidney cells. *Toxicology*, *66*(2), 213–224.

Kretz, O., Creppy, E. E., Boulanger, Y., & Dirheimer, G. (1989). Purification and some properties of Bolesatine, a protein inhibiting in vitro protein synthesis, from the mushroom boletus Satanas Lenz (Boletaceae). *Archives of Toxicology*, *13*, 422–427.

Kretz, O., Creppy, E. E., & Dirheimer, G. (1991b). Disposition of the toxic protein, bolesatine, in rats: Its resistance to proteolytic enzymes. *Xenobiotica*, *21*(1), 65–73.

Kruse, A. C., Kobilka, B. K., Gautam, D., Sexton, P. M., Christopoulos, A., & Wess, J. (2014). Muscarinic acetylcholine receptors: Novel opportunities for drug development. *Nature Reviews Drug Discovery*, *13*(7), 549–560.

Kyeon, H. M., Song, W. T., In, K. H., & Kim, J. S. (1991). A case of reversible hypoplastic pancytopenia due to Ganoderma japonicum. *The Korean Journal of Hematology*, *26*, 129–133.

Köppel, C. (1993). Clinical symptomatology and management of mushroom poisoning. *Toxicon*, *31*(12), 1513–1540.

Laatsch, H., & Matthies, L. (1991). Fluorescent compounds in Cortinarius speciosissimus: Investigation for the presence of Cortinarins. *Mycologia*, *83*(4), 492.

Laubner, G., & Mikulevičienė, G. (2016). A series of cases of rhabdomyolysis after ingestion of Tricholoma equestre. *Acta Medica Lituanica*, *23*(3), 193–197.

Lee, P., Wu, M., Tsai, W., Ger, J., Deng, J., & Chung, H. (2001). Rhabdomyolysis: An unusual feature with mushroom poisoning. *American Journal of Kidney Diseases*, *38*(4), e17.1–e17.5.

Lehmann, P. F., & Khazan, U. (1992). Mushroom poisoning by Chlorophyllum molybdites in the Midwest United States. *Mycopathologia*, *118*(1), 3–13.

Li, H., Tian, Y., Menolli Jr, N., Ye, L., Karunarathna, S., Perez, J., Rahman, M. M., Rashid, M. H., Phengsintham, P., Rizal, L., Kasuya, T., Lim, Y. W., Dutta, A. K., Khalid, A. N., Huyen, L. T., Balolong, M. P., Baruah, G., Madawala, S., Thongklang, N., Hyde, K. D., Kirk, P. M., Xu, J., Sheng, J., Boa, E., & Mortimer, P. (2021). Reviewing the World's Edible Mushroom Species: A New Evidence- Based Classification System. *Comprehensive Reviews In Food Science And Food Safety*, *20*, 1982–2014.

Li, L., Wang, Y., Qu, P., Ma, L., Liu, K., Yang, L., Nie, S., Xi, Y., Jia, P., Tang, X., Sun, Z., Huang, W., Li, Y., Dong, Y., & Lei, P. (2020). Genetic analysis of Yunnan sudden unexplained death by whole genome sequencing in Southwest of China. *Journal of Forensic and Legal Medicine*, *70*, 101896.

Lin, S., Mu, M., Yang, F., & Yang, C. (2015). Russula subnigricans poisoning: From gastrointestinal symptoms to Rhabdomyolysis. *Wilderness & Environmental Medicine*, *26*(3), 380–383.

Ljungman, M., Zhang, F., Chen, F., Rainbow, A. J., & McKay, B. C. (1999). Inhibition of RNA polymerase II as a trigger for the p53 response. *Oncogene*, *18*(3), 583–592.

Lumpert, M., & Kreft, S. (2016). Catching flies with amanita muscaria: Traditional recipes from Slovenia and their efficacy in the extraction of ibotenic acid. *Journal of Ethnopharmacology*, *187*, 1–8.

Lurie, Y., Wasser, S. P., Taha, M., Shehade, H., Nijim, J., Hoffmann, Y., Basis, F., Vardi, M., Lavon, O., Suaed, S., Bisharat, B., & Bentur, Y. (2009). Mushroom poisoning from species of genusInocybe(fiber head mushroom): A case series with exact species identification. *Clinical Toxicology*, *47*(6), 562–565.

Maretić, Z. (1967). Poisoning by the mushroom Clitocybe olearia Maire. *Toxicon*, *4*(4), 263–267.

Maretić, Z., Russell, F., & Golobić, V. (1975). Twenty-five cases of poisoning by the mushroom Pleurotus olearius. *Toxicon*, *13*(5), 379–381.

Martínez-Torres, N. I., Vázquez-Hernández, N., Martín-Amaya-Barajas, F. L., Flores-Soto, M., & González-Burgos, I. (2021). Ibotenic acid induced lesions impair the modulation of dendritic spine plasticity in the prefrontal cortex and amygdala, a phenomenon that underlies working memory and social behavior. *European Journal of Pharmacology*, *896*, 173883.

Matsuura, M., Saikawa, Y., Inui, K., Nakae, K., Igarashi, M., Hashimoto, K., & Nakata, M. (2009). Identification of the toxic trigger in mushroom poisoning. *Nature Chemical Biology*, *5*(7), 465–467.

Matsuura, M., Yamada, M., Saikawa, Y., Miyairi, K., Okuno, T., Konno, K., Uenishi, J., Hashimoto, K., & Nakata, M. (2007). Bolevenine, a toxic protein from the Japanese toadstool boletus venenatus. *Phytochemistry*, *68*(6), 893–898.

McCormick, S. P., Stanley, A. M., Stover, N. A., & Alexander, N. J. (2011). Trichothecenes: From simple to complex mycotoxins. *Toxins*, *3*(7), 802–814.

Meijer, A. A., Amazonas, M. A., Rubio, G. B., & Curial, R. M. (2007). Incidences of poisonings due to Chlorophyllum molybdites in the state of parana, Brazil. *Brazilian Archives of Biology and Technology*, *50*(3), 479–488.

Melli, G., Chaudhry, V., & Cornblath, D. R. (2005). Rhabdomyolysis: an evaluation of 475 hospitalized patients. *Medicine (Baltimore)*, *84*(6), 377–385.

Merlet, A., Dauchy, F., & Dupon, M. (2011). Hyperprocalcitonemia due to mushroom poisoning. *Clinical Infectious Diseases*, *54*(2), 307–308.

Michelot, D. (1992). Poisoning by Coprinus atramentarius. *Natural Toxins*, *1*, 73–80.

Michelot, D., & Melendez-Howell, L. M. (2003). Amanita muscaria: Chemistry, biology, toxicology, and ethnomycology. *Mycological Research*, *107*(2), 131–146.

Michelot, D., & Tebbett, I. (1990). Poisoning by members of the genus Cortinarius — A review. *Mycological Research*, *94*(3), 289–298.

Michelot, D., & Toth, B. (1991). Poisoning by Gyromitra esculenta – a Review. *Journal Of Applied Toxicology*, *11*(4), 235–243.

Mikaszewska-Sokolewicz, M. A., Pankowska, S., Janiak, M., Pruszczyk, P., Łazowski, T., & Jankowski, K. (2016). Coma in the course of severe poisoning after consumption of red fly agaric (Amanita muscaria). *Acta Biochimica Polonica*, *63*(1).

Minami, T., Matsumura, S., Nishizawa, M., Sasaguri, Y., Hamanaka, N., & Ito, S. (2004). Acute and late effects on induction of allodynia by acromelic acid, a mushroom poison related structurally to kainic acid. *British Journal of Pharmacology*, *142*(4), 679–688.

Moss, M. J., & Hendrickson, R. G. (2018). Toxicity of muscimol and ibotenic acid containing mushrooms reported to a regional poison control center from 2002–2016. *Clinical Toxicology*, *57*(2), 99–103.

Nakamura, T. (1992). Shiitake (Lentinus edodes) dermatitis. *Contact Dermatitis*, *27*(2), 65–70.

Natarajan, K., & Kaviyarasan, V. (1991). Chlorophyllum molybdites poisoning in India — A case study. *Mycologist*, *5*(2), 70.

Nguyen, V. T., Giannoni, F., Dubois, M., Seo, S., Vigneron, M., Kédinger, C., & Bensaude, O. (1996). In vivo degradation of RNA polymerase II largest subunit triggered by Alpha-amanitin. *Nucleic Acids Research*, *24*(15), 2924–2929.

Nielsen, E. Ø., Schousboe, A., Hansen, S. H., & Krogsgaard-Larsen, P. (1985). Excitatory amino acids: Studies on the biochemical and chemical stability of Ibotenic acid and related compounds. *Journal of Neurochemistry*, *45*(3), 725–731.

Nieminen, P., Kärjä, V., & Mustonen, A. (2008). Indications of hepatic and cardiac toxicity caused by subchronic Tricholoma flavovirens consumption. *Food and Chemical Toxicology*, *46*(2), 781–786.

Nieminen, P., Kärjä, V., & Mustonen, A. (2009). Myo- and hepatotoxic effects of cultivated mushrooms in mice. *Food and Chemical Toxicology*, *47*(1), 70–74.

Nieminen, P., Mustonen, A., & Kirsi, M. (2005). Increased plasma creatine kinase activities triggered by edible wild mushrooms. *Food and Chemical Toxicology*, *43*(1), 133–138.

Nieminen, P., & Mustonen, A. (2020). Toxic potential of traditionally consumed mushroom species—A controversial continuum with many unanswered questions. *Toxins*, *12*(10), 639.

Nilsson, U. A., Nyström, J., Buvall, L., Ebefors, K., Björnson-Granqvist, A., Holmdahl, J., & Haraldsson, B. (2008). The fungal nephrotoxin orellanine simultaneously increases oxidative stress and down-regulates cellular defenses. *Free Radical Biology and Medicine*, *44*(8), 1562–1569.

North, R. V., & Kelly, M. E. (1987). A review of the uses and adverse effects of topical administration of atropine. *Ophthalmic and Physiological Optics*, *7*(2), 109–114.

Oda, T., Tanaka, C., & Tsuda, M. (2004). Molecular phylogeny and biogeography of the widely distributed amanita species, A. muscaria and A. pant henna. *Mycological Research*, *108*(8), 885–896.

Oda, T., Yamazaki, T., Tanaka, C., Terashita, T., Taniguchi, N., & Tsuda, M. (2002). Amanita ibotengutake Sp. Nov., a poisonous fungus from Japan. *Mycological Progress*, *1*(4), 355–365.

Ohta, H., Watanabe, D., Nomura, C., Saito, D., Inoue, K., Miyaguchi, H., Harada, S., & Aita, Y. (2020). Toxicological analysis of satratoxins, the main toxins in the mushroom Trichoderma cornu-damae, in human serum and mushroom samples by liquid chromatography–tandem mass spectrometry. *Forensic Toxicology*, *39*(1), 101–113.

Omoto, H., Matsumura, S., Kitano, M., Miyazaki, S., Minami, T., & Ito, S. (2015). Comparison of mechanisms of allodynia induced by acromelic acid a between early and late phases. *European Journal of Pharmacology*, *760*, 42–48.

Park, J., Min, J., Kim, H., Lee, S., Kang, J., &An, J. (2016). Four cases of successful treatment after Podostroma cornu-damae intoxication. *Hong Kong Journal of Emergency Medicine*, *23*(1), 55–59.

Patočka, J., Wu, R., Nepovimova, E., Valis, M., Wu, W., & Kuca, K. (2021). Chemistry and toxicology of major Bioactive substances in Inocybe mushrooms. *International Journal of Molecular Sciences*, *22*(4), 2218.

Patočka, J. (2018). Bolesatine, a toxic protein from the mushroom rubroboletus Satanas. *Military Medical Science Letters*, *87*(1), 14–20.

Pauli, J. L., & Foot, C. L. (2005). Fatal muscarinic syndrome after eating wild mushrooms. *Medical Journal of Australia*, *182*(6), 294–295.

Pelizzari, V., Feifel, E., Rohrmoser, M. M., Gstraunthaler, G., & Moser, M. (1994). Partial purification and characterization of a toxic component of amanita smithiana. *Mycologia*, *86*(4), 555.

Peredy, T., & Bradford, H. (2014). Mushrooms, muscarine. *Encyclopedia of Toxicology* (Third Edition), 416–417. Cambridge: Academic Press.

Peredy, T., & Bruce, R. (2014). Mushrooms, Ibotenic acid. *Encyclopedia of Toxicology* (Third Edition), 412–413. Cambridge: Academic Press.

Petrović, P., Ivanović, K., Jovanović, A., Simović, M., Milutinović, V., Kozarski, M., Petković, M., Cvetković, A., Klaus, A., & Bugarski, B. (2019b). The impact of puffball autolysis on selected chemical and biological properties: Puffball extracts as potential ingredients of skin-care products. *Archives of Biological Sciences*, *71*(4), 721–733.

Petrović, P., Ivanović, K., Octrue, C., Tumara, M., Jovanović, A., Vunduk, J., Nikšić, M., Pjanović, R., Bugarski, B., & Klaus, A. (2019c). Immobilization of Chaga extract in alginate beads for modified release: Simplicity meets efficiency. *Chemical Industry*, *73*(5), 325–335.

Petrović, P., Vunduk, J., Klaus, A., Kozarski, M., Nikšić, M., Žižak, Ž., Vuković, N., Šekularac, G., Drmanić, S., & Bugarski, B. (2016). Biological potential of puffballs: A comparative analysis. *Journal of Functional Foods*, *21*, 36–49.

Petrović, P., Vunduk, J., Klaus, A., Carević, M., Petković, M., Vuković, N., Cvetković, A., Žižak, Ž.,& Bugarski, B. (2019a). From mycelium to spores: A whole circle of biological potency of mosaic puffball. *South African Journal of Botany*, *123*, 152–160.

Picciotto, M., Higley, M., & Mineur, Y. (2012). Acetylcholine as a neuromodulator: Cholinergic signaling shapes nervous system function and behavior. *Neuron*, *76*(1), 116–129.

Pohle, W. (1995). Paxillus involutus - a dangerous mushroom? *Czech Mycology*, *48*(1), 31–38.

Poucheret, P., Fons, F., Doré, J. C., Michelot, D., & Rapior, S. (2010). Amatoxin poisoning treatment decision-making: Pharmaco-therapeutic clinical strategy assessment using multidimensional multi-variate statistic analysis. *Toxicon*, *55*(7), 1338–1345.

Power, R. C., Salazar-García, D. C., Straus, L. G., González Morales, M. R., & Henry, A. G. (2015). Microremains from el Miron cave human dental calculus suggest a mixed plant–animal subsistence economy during the magdalenian in northern Iberia. *Journal of Archaeological Science*, *60*, 39–46.

Prast, H., Werner, E. R., Pfaller, W., & Moser, M. (1988). Toxic properties of the mushroom Cortinarius orellanus. *Archives of Toxicology*, *62*(1), 81–88.

Puschner, B. (2013). Mushrooms. In *Small Animal Toxicology* (pp. 659–676). St. Louis: Saunders.

Pyysalo, H., & Niskanen, A. (1977). Occurrence of N-methyl-N-formylhydrazones in fresh and processed false Morel, *Gyromitra esculenta.Journal of Agricultural and Food Chemistry*, *25*(3), 644–647.

Ramirez, P., Parrilla, P., Bueno, F. S., Robles, R., Pons, J. A., Bixquert, V., Nicolas, S., Nuñez, R., Alegria, M. S., Miras, M., & Rodriguez, J. M. (1993). Fulminant hepatic failure after Lepiota mushroom poisoning. *Journal of Hepatology*, *19*(1), 51–54.

Rampolli, F. I., Kamler, P., Carlino, C. C., & Bedussi, F. (2021). The Deceptive Mushroom: Accidental Amanita muscaria Poisoning. *European Journal of Case Reports in Internal Medicine*, *8*. 10.12890/2021_002212

Rapior, S., Andary, C., & Mousain, D. (1987). Cortinarius section orellani: Isolation and culture of Cortinarius orellanus. *Transactions of the British Mycological Society*, *89*(1), 41–44.

Razanamparany, J. L., Creppy, E. E., Perreau-Bertrand, J., Boulanger, Y., & Dirheimer, G. (1986). Purification and characterization of bolaffinine, a toxic protein from boletus affinis peck (Boletaceae). *Biochimie*, *68*(10–11), 1217–1223.

Richard, J., Cantin-Esnault, D., & Jeunet, A. (1995). First electron spin resonance evidence for the production of semiquinone and oxygen free radicals from orellanine, a mushroom nephrotoxin. *Free Radical Biology and Medicine*, *19*(4), 417–429.

Richard, J., Ekue Creppy, E., Benoit-Guyod, J., & Dirheimer, G. (1991). Orellanine inhibits protein synthesis in Madin-Darby canine kidney cells, in rat liver mitochondria, and in vitro: Indication for its activation prior to in vitro inhibition. *Toxicology*, *67*(1), 53–62.

Rodrigues, D. F., Pires Das Neves, R., Carvalho, A. T., Lourdes Bastos, M., Costa, V. M., & Carvalho, F. (2020). In vitro mechanistic studies on α-amanitin and its putative antidotes. *Archives of Toxicology*, *94*(6), 2061–2078.

Roupas, P., Keogh, J., Noakes, M., Margetts, C., & Taylor, P. (2010). Mushrooms and agaritine: A mini-review. *Journal of Functional Foods*, *2*(2), 91–98.

Rzymski, P., & Klimaszyk, P. (2018). Is the yellow knight mushroom edible or not? A systematic review and critical viewpoints on the toxicity of Tricholoma equestre. *Comprehensive Reviews in Food Science and Food Safety*, *17*(5), 1309–1324.

Sai Latha, S., Shivanna, N., Naika, M., Anilakumar, K. R., Kaul, A., & Mittal, G. (2020). Toxic metabolite profiling of Inocybe virosa. *Scientific Reports*, *10*(1).

Saikawa, Y., Okamoto, H., Inui, T., Makabe, M., Okuno, T., Suda, T., Hashimoto, K., & Nakata, M. (2001). Toxic principles of a poisonous mushroom Podostroma cornu-damae. *Tetrahedron*, *57*(39), 8277–8281.

Sano, Y., Sayama, K., Arimoto, Y., Inakuma, T., Kobayashi, K., Koshino, H., & Kawagishi, H. (2002). Ustalic acid as a toxin and related compounds from the mushroom Tricholoma ustale. *Chemical Communications*, (13), 1384–1385.

Sato, Y., Tomonari, H., Kaneko, Y., & Yo, K. (2019). Mushroom poisoning with scleroderma albidum: A case report with review of the literature. *Acute Medicine & Surgery*, *7*(1).

Satora, L., Pach, D., Butryn, B., Hydzik, P., & Balicka-Ślusarczyk, B. (2005). Fly agaric (Amanita muscaria) poisoning, case report and review. *Toxicon*, *45*(7), 941–943.

Satora, L., Pach, D., Ciszowski, K., & Winnik, L. (2006). Panther cap amanita pantherina poisoning case report and review. *Toxicon*, *47*(5), 605–607.

Saviuc, P., & Danel, V. (2006). New syndromes in mushroom poisoning. *Toxicological Reviews*, *25*(3), 199–209.

Saviuc, P. F., Danel, V. C., Moreau, P. M., Guez, D. R., Claustre, A. M., Carpentier, P. H., Mallaret, M. P., & Ducluzeau, R. (2001). Erythromelalgia and mushroom poisoning. *Journal of Toxicology: Clinical Toxicology*, *39*(4), 403–407.

Saviuc, P., Harry, P., Pulce, C., Garnier, R., & Cochet, A. (2010). Can morels (Morchellasp.) induce a toxic neurological syndrome? *Clinical Toxicology*, *48*(4), 365–372.

Schenk-Jaeger, K. M., Rauber-Lüthy, C., Bodmer, M., Kupferschmidt, H., Kullak-Ublick, G. A., & Ceschi, A. (2012). Mushroom poisoning: A study on circumstances of exposure and patterns of toxicity. *European Journal of Internal Medicine*, *23*(4), e85–e91.

Schmutz, M., Carron, P., Yersin, B., & Trueb, L. (2016). Mushroom poisoning: A retrospective study concerning 11-years of admissions in a Swiss emergency department. *Internal and Emergency Medicine*, *13*(1), 59–67.

Schwarcz, R., Hökfelt, T., Fuxe, K., Jonsson, G., Goldstein, M., & Terenius, L. (1979). Ibotenic acid-induced neuronal degeneration: A morphological and neurochemical study. *Experimental Brain Research*, *37*(2).

Shao, D., Tang, S., Healy, R. A., Imerman, P. M., Schrunk, D. E., & Rumbeiha, W. K. (2016). A novel orellanine containing mushroom Cortinarius armillatus. *Toxicon*, *114*, 65–74.

Sharma, S., Aydin, M., Bansal, G., Kaya, E., & Singh, R. (2021). Determination of amatoxin concentration in heat-treated samples of amanita phalloides by high-performance liquid chromatography: A forensic approach. *Journal of Forensic and Legal Medicine*, *78*, 102111.

Shephard, S., & Schlatter, C. (1998). Covalent binding of Agaritine to DNA in vivo. *Food and Chemical Toxicology*, *36*(11), 971–974.

Sherwood, A. M., Halberstadt, A. L., Klein, A. K., McCorvy, J. D., Kaylo, K. W., Kargbo, R. B., & Meisenheimer, P. (2020). Synthesis and biological evaluation of Tryptamines found in hallucinogenic mushrooms: Norbaeocystin, Baeocystin, Norpsilocin, and Aeruginascin. *Journal of Natural Products*, *83*(2), 461–467.

Shi, G., Huang, W., Zhang, J., Zhao, H., Shen, T., Fontaine, R. E., Yang, L., Zhao, S., Lu, B., Wang, Y., Ma, L., Li, Z., Gao, Y., Yang, Z., & Zeng, G. (2012a). Clusters of sudden unexplained death associated with the mushroom, Trogia venenata, in rural Yunnan province, China. *PLoS ONE*, *7*(5), e35894.

Shi, G., He, J., Shen, T., Fontaine, R. E., Yang, L., Zhou, Z., Gao, H., Xu, Y., Qin, C., Yang, Z., Liu, J., Huang, W., & Zeng, G. (2012b). Hypoglycemia and death in mice following experimental exposure to an extract of Trogia venenata mushrooms. *PLoS ONE*, *7*(6), e38712.

Shi, G. Q., Zhang, J., Huang, W. L., Yang, T., Ye, S. D., Sun, X. D., Li, Z. X., Xie, X. H., Li, F. R., Wang, Y. B., Ren, J. M., Fontaine, R. E., & Zeng, G. (2006). Retrospective study on 116 unexpected sudden cardiac deaths in Yunnan. *China.Chinese journal of Epidemiology*, *27*(2), 96–101.

Shinozaki, H., Ishida, M., Gotoh, Y., & Kwak, S. (1989). Specific lesions of rat spinal interneurons induced by systemic administration of acromelic acid, a new potent kainate analogue. *Brain Research, 503*(2), 330–333.

Snook, C. P. (2016). Indole hallucinogens. *Critical Care Toxicology,* 1–22. Cham: Springer.

Soen, M., Minami, T., Tatsumi, S., Mabuchi, T., Furuta, K., Maeda, M., Suzuki, M., & Ito, S. (2007). A synthetic kainoid, (2s,3r,4r)-3-carboxymethyl-4-(phenylthio)pyrrolidine-2-carboxylic acid (PSPA-1) serves as a novel anti-allodynic agent for neuropathic pain. *European Journal of Pharmacology, 575*(1–3), 75–81.

Strand, R. D., Neuhauser, E. B., & Sornberger, C. F. (1967). Lycoperdonosis. *New England Journal of Medicine, 277*(2), 89–91.

Straus, L. G., González Morales, M. R., Carretero, J. M., & Marín-Arroyo, A. B. (2015). "The red lady of el Miron". Lower magdalenian life and death in oldest Dryas Cantabrian Spain: An overview. *Journal of Archaeological Science, 60,* 134–137.

Stöver, A., Haberl, B., Helmreich, C., Müller, W., Musshoff, F., Fels, H., Graw, M., & Groth, O. (2019). Fatal Immunohaemolysis after the consumption of the poison Pax mushroom: A focus on the diagnosis of the Paxillus syndrome with the aid of two case reports. *Diagnostics, 9*(4), 130.

Stříbrný, J., Sokol, M., Merová, B., & Ondra, P. (2011). GC/MS determination of ibotenic acid and muscimol in the urine of patients intoxicated with amanita pantherina. *International Journal of Legal Medicine, 126*(4), 519–524.

Sun, J., Zhang, H., Li, H., Zhang, Y., He, Q., Lu, J., Yin, Y., & Sun, C. (2019). A case study of Lepiota brunneoincarnata poisoning with endoscopic nasobiliary drainage in Shandong. *China. Toxicon, 161,* 12–16.

Swift, R., & Leggio, L. (2009). Adjunctive Pharmacotherapy in the Treatment of Alcohol and Drug Dependence. In *Evidence-Based Addiction Treatment* (pp. 287–310). Cambridge: Academic Press.

Taguchi, T., Tomotoshi, K., & Mizumura, K. (2009). Excitatory actions of mushroom poison (acromelic acid) on unmyelinated muscular afferents in the rat. *Neuroscience Letters, 456*(2), 69–73.

Tanaka, K., Inoue, T., Kanai, M., & Kikuchi, T. (1994). Metabolism of illudin S, a toxic substance ofLampteromyces japonicus.IV.Urinary excretion of an illudin S metabolite in rat. *Xenobiotica, 24*(12), 1237–1243.

Tebbett, I. R., & Caddy, B. (1984). Mushroom toxins of the genus Cortinarius. *Experientia, 40,* 441–446.

Tottmar, O., & Lindberg, P. (1977). Effects on rat liver acetaldehyde Dehydrogenases in vitro and in vivo by Coprine, the disulfiram-like constituent of Coprinus atramentarius. *Acta Pharmacologica et Toxicologica, 40*(4), 476–481.

Treu, R., & Adamson, W. (2006). Ethnomycological Notes from Papua, New Guinea. *McIlvainea, 16*(2), 3–10.

Tsujikawa, K., Mohri, H., Kuwayama, K., Miyaguchi, H., Iwata, Y., Gohda, A., Fukushima, S., Inoue, H., & Kishi, T. (2006). Analysis of hallucinogenic constituents in amanita mushrooms circulated in Japan. *Forensic Science International, 164*(2–3), 172–178.

Tylš, F., Páleníček, T., & Horáček, J. (2014). Psilocybin – Summary of knowledge and new perspectives. *European Neuropsychopharmacology, 24*(3), 342–356.

Ueno, Y., & Matsumoto, H. (1975). Inactivation of some thiol-enzymes by trichothecene mycotoxins from Fusarium species. *Chemical and Pharmaceutical Bulletin, 23*(10), 2439–2442.

Unluoglu, I., & Tayfur, M. (2003). Mushroom poisoning: An analysis of the data between 1996 and 2000. *European Journal of Emergency Medicine, 10*(1), 23–26.

Vanden Hoek, T. L., Erickson, T., Hryhorczuk, D., & Narasimhan, K. (1991). Jack O'Lantern Mushroom Poisoning. *Annals of Emergency Medicine, 20*(5), 559–561.

Vendramin, A., & Brvar, M. (2014). Amanita muscaria and amanita pantherina poisoning: Two syndromes. *Toxicon, 90,* 269–272.

Villa, A. F., Saviuc, P., Langrand, J., Favre, G., Chataignerl, D., & Garnier, R. (2013). Tender nesting Polypore (Hapalopilus rutilans) poisoning: Report of two cases. *Clinical Toxicology, 51*(8), 798–800.

Voynova, M., Shkondrov, A., Kondeva-Burdina, M., & Krasteva, I. (2020). Toxicological and pharmacological profile of amanita muscaria (L.) Lam. – a new rising opportunity for biomedicine. *Pharmacia, 67*(4), 317–323.

Vunduk, J., Klaus, A., Kozarski, M., Petrovic, P., Zizak, Z., Niksic, M., & Van Griensven, L. J. (2015). Did the iceman know better? Screening of the medicinal properties of the birch Polypore medicinal mushroom, Piptoporus betulinus (Higher basidiomycetes). *International Journal of Medicinal Mushrooms, 17*(12), 1113–1125.

Wakimoto, T., Asakawa, T., Akahoshi, S., Suzuki, T., Nagai, K., Kawagishi, H., & Kan, T. (2010). Proof of the existence of an unstable amino acid: Pleurocybellaziridine in Pleurocybella porrigens. *Angewandte Chemie*, *123*(5), 1200–1202.

Wang, Q., Sun, M., Lv, H., Lu, P., Ma, C., Liu, Y., Liu, S., Tong, H., Hu, Z., & Gao, Y. (2020). Amanita fuliginea poisoning with thrombocytopenia: A case series. *Toxicon*, *174*, 43–47.

Warden, C. R., & Benjamin, D. R. (1998). Acute renal failure associated with suspected amanita smithiana mushroom ingestions: A case series. *Academic Emergency Medicine*, *5*(8), 808–812.

Watson, W. A., Litovitz, T. L., Rodgers, G. C., Klein-Schwartz, W., Reid, N., Youniss, J., Flanagan, A., & Wruk, K. M. (2005). 2004 annual report of the American Association of poison control centers toxic exposure surveillance system. *The American Journal of Emergency Medicine*, *23*(5), 589–666.

Wennig, R., Eyer, F., Schaper, A., Zilker, T., & Andresen-Streichert, H. (2020). Mushroom poisoning. *Deutsches Ärzteblatt International*, *117*(42), 701–708.

West, P. L., Lindgren, J., & Horowitz, B. Z. (2009). Amanita smithiana mushroom ingestion: A case of delayed renal failure and literature review. *Journal of Medical Toxicology*, *5*(1), 32–38.

White, J., Weinstein, S. A., De Haro, L., Bédry, R., Schaper, A., Rumack, B. H., & Zilker, T. (2019a). Mushroom poisoning: A proposed new clinical classification. *Toxicon*, *157*, 53–65.

White, J., Weinstein, S. A., De Haro, L., Bédry, R., Schaper, A., Rumack, B. H., & Zilker, T. (2019b). Reply to Rzymski and Klimaszyk regarding comment on "Mushroom poisoning: A proposed new clinical classification". *Toxicon*, *160*, 59.

Wieland, T. (1983). The toxic peptides from Amanita mushrooms. *International Journal of Peptide and Protein Research*, *22*, 251–276.

Winkelmann, M., Stangel, W., Schedel, I., & Grabensee, B. (1986). Severe hemolysis caused by antibodies against the mushroomPaxillus involutus and its therapy by plasma exchange. *Klinische Wochenschrift*, *64*(19), 935–938.

Wörnle, M., Angstwurm, M. W., & Sitter, T. (2004). Treatment of intoxication with Cortinarius speciosissimus using an antioxidant therapy. *American Journal of Kidney Diseases*, *43*(4), e16.1–e16.4.

Xiao, G., Zhang, C., Liu, F., Chen, Z., & Hu, S. (2007). Clinical experience in treatment of amanita mushroom poisoning with glossy Ganoderma decoction (灵芝煎剂) and routine western medicines. *Chinese Journal of Integrative Medicine*, *13*(2), 145–147.

Xu, F., Zhang, Y., Zhang, Y., Guan, G., Zhang, K., Li, H., & Wang, J. (2020). Mushroom poisoning from Inocybe serotina: A case report from Ningxia, northwest China with exact species identification and muscarine detection. *Toxicon*, *179*, 72–75.

Xu, X., Zhang, J., Huang, B., Han, J., & Chen, Q. (2020). Determination of ibotenic acid and muscimol in plasma by liquid chromatography-triple quadrupole mass spectrometry with bimolecular dansylation. *Journal of Chromatography B*, *1146*, 122128.

Yan, Y., Zhang, C., Moodley, O., Zhang, L., & Xu, J. (2019). Green mold caused by Trichoderma atroviride on the Lingzhi medicinal mushroom, Ganoderma lingzhi (Agaricomycetes). *International Journal of Medicinal Mushrooms*, *21*(5), 515–521.

Ye, Y., & Liu, Z. (2018). Management of amanita phalloides poisoning: A literature review and update. *Journal of Critical Care*, *46*, 17–22.

Yin, X., Feng, T., Shang, J., Zhao, Y., Wang, F., Li, Z., Dong, Z., Luo, X., & Liu, J. (2014). Chemical and toxicological investigations of a previously unknown poisonous European mushroom Tricholoma terreum. *Chemistry - A European Journal*, *20*, 7001–7009.

Yoon, Y., Choi, S., Cho, H., Moon, S., Kim, J., & Lee, S. (2011). Reversible pancytopenia following the consumption of decoction ofGanoderma neojaponicumImazeki. *Clinical Toxicology*, *49*(2), 115–117.

Young, T. (1989). Poisonings by Chlorophyllum molybdites in Australia. *Mycologist*, *3*(1), 11–12.

Zakhari, S. (2006). Overview: How Is Alcohol Metabolized by the Body? *Alcohol Research & Health*, *29*(4), 245–254.

Zhang, Y., Li, Y., Wu, G., Feng, B., Yoell, S., Yu, Z., Zhang, K., & Xu, J. (2012). Evidence against barium in the mushroom Trogia venenata as a cause of sudden unexpected deaths in Yunnan, China. *Applied and Environmental Microbiology*, *78*(24), 8834–8835.

Zhou, Q., Tang, S., He, Z., Luo, T., Chen, Z., & Zhang, P. (2017). Amatoxin and phallotoxin concentrations in amanita fuliginea: Influence of tissues, developmental stages and collection sites. *Mycoscience*, *58*(4), 267–273.

Zhou, Z., Shi, G., Fontaine, R., Wei, K., Feng, T., Wang, F., Wang, G., Qu, Y., Li, Z., Dong, Z., Zhu, H., Yang, Z., Zeng, G., & Liu, J. (2012).Evidence for the natural toxins from the mushroom Trogia ve-nenata as a cause of sudden unexpected death in Yunnan province, China. *Angewandte Chemie International Edition, 51*(10), 2368–2370.

Zinkand, W. C., Moore, W. C., Thompson, C., Salama, A. I., & Patel, J. (1992). Ibotenic acid mediates neurotoxicity and phosphoinositide hydrolysis by independent receptor mechanisms. *Molecular and Chemical Neuropathology, 16*(1–2), 1–10.

CHAPTER 13

Influence of food-processing conditions on bioactivity and nutritional components of edible mushroom

Jyoti Singh[1], Jaspreet Kaur[1], Vishesh Bhadariya[2], Priyanka Kundu[1], Sapna Jarial[3], Kartik Sharma[4], and Simran Gogna[1]

Department of Food Technology and Nutrition, School of Agriculture, Lovely Professional University, Punjab, India

Department of Chemical and Petroleum Engineering, School of Chemical Engineering and Physical Sciences, Lovely Professional University, Phagwara, Punjab, India

Department of Agricultural Economics and Extensions, School of Agriculture, Lovely Professional University, Punjab, India

Department of Biotechnology, Council of Scientific and Industrial Research- Institute of Himalayan Bioresource Technology (CSIR-IHBT), Palampur, Himachal Pradesh, India

CONTENTS

DOI: 10.1201/9781003152583-16

13.1 INTRODUCTION

Mushrooms are considered among the most popular foods due to their unique flavor as well as textural properties. Mushrooms have been extensively used as part of the human diet for centuries as they are not only nutritionally rich but also possess various biological activities such as anti-inflammatory, hypotensive, antioxidant activity, anti-cancerous, anti-allergic, anti-cholesterolemic, antidiabetic, etc. (Moon & Lo, 2014). On average, a consumer consumes around 5 kg mushrooms per year. As the awareness regarding the health benefits of mushrooms is increasing among the population, this rate is expected to increase further. Mushrooms have gained popularity in the field of processed foods, where they are being used in the form of powder in various food products such as muffins, patties, pasta, snacks, etc. to increase their nutritional value. There is a hike of 3.34 million metric tons in global production of mushrooms in a decade (2008–2017), with China (77% of the world's production of mushrooms), the United States, the Netherlands, Poland, and Spain among the top producers. *Agaricus bisporus*, also known as common mushroom, is the most cultivated variety of edible mushroom, followed by *Lentinus elodes* or Shiitake, *Pleurotus* spp., and *Flammulinavelutipes*. Mushrooms, being low in fats and calories, contain a sufficient amount of proteins, ergothioneine, vitamins such as vitamin B and D, and minerals such as phosphorus, copper, selenium, magnesium, potassium, etc. along with dietary fibers in them and are therefore considered healthy to be part of the diet. Consumption of mushrooms enhances the immune system of the body, acts as antibacterial, and lowers the cholesterol level in the body. It helps in lowering the risk of various diseases such as strokes, hypertension, cancer, Alzheimer's, Parkinson's disease, etc. (Valverde et al., 2015; Ho et al., 2020). Mushrooms generally undergo different processing conditions using oil or water as a medium along with a wide range of temperature treatments before consumption and are rarely eaten in raw form. The type of processing depends upon the storage and usage of mushrooms with the end product. The preliminary processing includes washing, blanching, soaking, vacuum moistening in a solution containing table salt, ascorbic acid, citric acid, sodium erythorbate, versenic acid or EDTA, hydrogen peroxide, cysteine hydrochloride, ethyl-alcohol vapors, methyl jasmonate vapors, calcium chloride, metabisulfites, etc. Metabisulfites are frequently used to inhibit the activity of lactase, polyphenol oxidase, lipoxygenase, peroxidase, and other such enzymes responsible for alteration in color. Washing using simple water as preliminary processing decreases the overall quality of mushroom products due to the darkening of pilei, but still, it is being performed as it is important from a hygienic point of view. In such cases, washing is usually preferred using solutions containing sodium metabisulfite, which prevents the formation of undesirable color on the pile and thus has a preventive effect on the whiteness of pilei (Bernas et al., 2006).

Therefore, this temperature treatment, along with various processing conditions, plays a crucial role as these are responsible for causing variation in chemical composition which ultimately alters the final composition of the product and thereby its health and nutritional properties too (Tsai & Chen, 2019). Majorly, two treatments are recommended that are most widely used by various researchers before blanching: soaking and vacuum moistening. Moistening of mushrooms using a vacuum helps in retention of the natural color of pilei in blanched mushrooms. Along with this, it also helps in the reduction of weight loss during blanching, which is reduced almost half when compared with mushrooms that have not undergone vacuum-moistening as pre-processing (Czapski, 1994). Soaking of fresh mushrooms in solutions of ascorbic acid and citric acid before blanching reduces the weight loss by 10% associated with blanching; however, it led to a reduction of water-soluble compounds present in mushrooms (Bernas et al., 2006). Various treatments have been standardized and optimized by various researchers for the increased yield production with the enhanced nutritional value of mushrooms. This chapter will focus on the effect of various processing techniques such as fermentation, pasteurization, soaking, blanching, irradiation, low temperature, vacuum moistening, drying, salting, pickling, and various heat treatments on the nutritional composition of edible mushrooms.

13.2 EFFECT OF FERMENTATION ON EDIBLE MUSHROOMS

Fermentation is a food preservation process that has been used since ancient times to extend the shelf life and increase the nutritional profile of food products (Hutkins, 2018). This technique is also used for mushrooms to make them more valued and delicious. Solid-state fermentation and lactic acid fermentation are the common fermentation processes that are used to increase the yield of mushrooms and for waste product utilization also (Barros, Cruz, Baptista, Estevinho, & Ferreira, 2008).

As all the fermentation techniques have different effects on mushrooms, it is important to analyze the effect of these processes on the nutritional value and sensory characteristics of the final product. The most effective technique should be chosen depending on the analysis. In Table 13.1, a few studies are summarized, including the fermentation methods used and their effect on various edible mushrooms.

13.3 EFFECT OF PASTEURIZATION ON EDIBLE MUSHROOMS

Mushrooms are very perishable, due to moisture deficiency, color shift, and ineffective preservation processes. Improved shelf life and quality characteristics of mushrooms would improve marketability and add value to the agricultural supply chain. To ensure food security, farmers must learn modern storage approaches that are both economical and easily available to them (Moda et al., 2005). Mushrooms are an essential component of natural waste management ad can be strengthened by cultivation. Pasteurization is a common practice among mushroom growers to avoid the growth of competitor molds and to improve the substrate's suitability for mushroom mycelial growth (Ziombra, 2000).

Table 13.1 Research findings of fermentation techniques and their effects on edible mushrooms

Technique used	Substrate used	Findings	References
Solid-state fermentation	Wood flour	Anaerobic solid-state fermentation was used to test the extraction of biohydrogen from mushroom cultivation waste. Polypropylene bag stuffed with wood flour is used as mushroom cultivation waste. The finding indicates that anaerobic mixed microflora using solid-state fermentation can produce hydrogen from mushroom cultivation waste.	Lin *et al.* (2016)
Lactic acid fermentation	Starter culture of Lactic acid	To achieve higher yield and higher quality end product, a starter culture of lactic acid bacillus is used which enables the process to be repeated. Owing to the presence of viable LAB cells, lactic acid fermentation resulted in an improvement in the health beneficial value of the mushroom due to the presence of viable cells of lactic acid fermentation.	Skapska et al. (2008)
Lactic acid fermentation	Blanched, salted, and raw *Termitomyces robustus*	The effect of fermentation on the mineral content of the edible mushroom *Termitomyces robustus* (blanched, raw, and salted) was investigated. The findings revealed that salted and fermented mushrooms had a higher mineral content (potassium, calcium, magnesium, and iron).	Bello and Akinyele (2007)

Pasteurization appears to be the best overall concept for the processes of heating mushroom substrate to eliminate weeds, viruses, and pests. If we talk about substrates for pasteurization, the cotton hull is almost universally used as a substrate, although it is not readily available to most farmers and is relatively expensive in comparison to other agricultural wastes (de Siqueira et al., 2012). So, for effective mushroom production, identifying different substrates from various types of crops, as well as pasteurization methods, is crucial (Zied et al., 2011).

13.3.1 Techniques of pasteurization

Several preservation techniques have been employed like hot water treatment, steam pasteurization, and self-heating pasteurization (Table 13.2). In hot-water pasteurization, the straw fragments are filled in gunny bags and placed in hot water at 80–90°C for 40 minutes for proper pasteurization. This method is often used by small growers (Oseni et al., 2012). On the other hand, in steam pasteurization, for 5–6 hours, continuous steam is transferred under pressure into a room containing the wet substrate, and placed in trays at a temperature of 65°C. The temperature of the air in the room is maintained at a steady level. As self-heating pasteurization is an aerobic treatment that takes the very least time to complete, this treatment aims to raise the temperature enough to destroy most contaminating species. This treatment conserves water and allows for accurate monitoring of the moisture content of the substrate. However, this method is a successful option since it saves resources (Morales & Sánchez, 2017).

13.4 EFFECT OF SOAKING AND BLANCHING ON EDIBLE MUSHROOMS

Soaking and blanching are the economical and simple methods of preservation of fruits and vegetables in which the fruit or vegetable slices will be immersed in water alone or combined with some chemicals at different temperatures (Table 13.3). These methods can also be used for the shelf-life extension of mushrooms. By steeping the mushrooms in various chemical solutions, the shelf life of the mushroom can be extended for a short period (Chawla et al., 2019). Salts and acids can be used for the steeping solutions. The uses of various chemicals in steeping or blanching solutions in permissible limits can be proven the effective pretreatments before going for further processing like drying, osmotic dehydration, canning, and freezing (Jabłońska-Ryś, Sławińska, Radzki, & Gustaw 2016).

13.5 EFFECT OF IRRADIATION ON EDIBLE MUSHROOMS

The applications of irradiation, when combined with other food preservation techniques, were found more effective when compared with individual preserving methods. Ultraviolet (UV-C) irradiation is more beneficial in causing chemical sterilization and microbial reduction in food products and can also be used as a food disinfectant. A positive effect was shown by ultraviolet (UV-C) irradiation on mushrooms (*L. edodes*) at low temperatures, for 15 days. The wavelength of UV-C radiation is 200–280 nm. The effect of UV-C irradiation on *L. edodes* showed increased antioxidant capacity (Jiang et al., 2010). The wavelength of ultraviolet A (UV-A) radiation is 315–400 nm. Vitamin D2 content increased using the UV-A light irradiation on species (*A. bisporus, A. portabella, P. ostreatus, B. edulis, C. cibarius, C. tubaeformis, L. edodes*) (Teichmann et al., 2007). Ergosterol content was also increased using LED irradiation on mushroom species *Lyophyllumulmarium* (Kim, Son, Martin Lo, Lee, & Moon 2014). Ergothioneine (2-mercaptohistidine trimethylbetaine) is a naturally occurring amino acid that can be synthesized in some bacteria and fungi. The pulsed light irradiation treatment did not affect the ergothioneine

Table 13.2   Research findings of pasteurization techniques and their effects on edible mushrooms

Pasteurization techniques/temp	Substrate used	Findings	References
Cold-water and hot-water pasteurization	Saw-dust, maize stalk, wheat straw, Teff straw, WTF straw, Fababean stalk, saw, and Fababean	This analysis aimed to see how various organic substrates and pasteurization methods affected the yield and quality of oyster mushrooms. Overall, the findings of this study revealed that hot-water pasteurization of the substrates was successful in reducing competent microorganisms. And if we talk about substrates, Fababean and maize stalk had favorable association with mushroom yield.	Ejigu and Kebede (2015)
Stem pasteurization and hot-water pasteurization	Cotton waste	In this study, various pasteurization techniques were used on cotton waste to determine the best way for mycelial growth of three species (*Pleurotus florida, pleurotus pulmonarius, pleurotus ostreatus*) of oyster mushroom. The results showed that steam pasteurization produces the most mycelial growth in the shortest period.	Ali et al. (2007)
Self-heat pasteurization	Dry grass + 2% $Ca(OH)_2$	In this study, the productivity of four species (*Pleurotus ostreatus, Pleurotus eryngii, Pleurotus citrinopileatus,* and *Pleurotus djamor*) of oyster mushroom was analyzed by using self-heat pasteurization and alkaline immersion disinfection techniques. It was revealed from the results that the self-heating pasteurization process is effective in the highest production of mushrooms.	Avendaño-Hernandez and Sánchez (2013)
Stem pasteurization	Grass and Straw mixers	Two experiments were conducted to examine the effect of grass and straw mixture on the growth of three different types of mushrooms i.e. *Pleurotus ostreatus, Pleurotus pulmonarius, and Pleurotus eryngii.* According to the findings, steam pasteurization is a very good method for the sustainable cultivation of mushrooms.	de Siqueira et al. (2012)
Hot-water pasteurization	Rye and wheat straw	The yield of *Pleurotus cornucopiae* grown on wheat and rye straw substrates pasteurized for 24, 48, 72, and 96 hours was checked. Results showed that rye straw pasteurized for 72 hours and wheat straw that had been pasteurized for 48 hours were responsible for the highest yield.	Ziombra (2000)
Hot-water and chemical pasteurization	Wheat straw substrate	A study has been conducted to know the effect of hot water and chemical pasteurization (with formaldehyde and carbendazim) on *Pleurotus* species (PL-17-7, PL-17-10, PL-17-11, PL-17-12) yield with wheat straw as substrate was investigated. Results revealed that the Chemical and hot-water pasteurization methods were found to be the most effective for the pasteurization of wheat straw. Based on economic yield and other yield attributes of oyster mushroom on chemical and hot-water pasteurization method was better compared to lime and plane water treatment of wheat straw.	Kerketta et al. (2019)

Table 13.3 Research findings of blanching and boiling techniques and their effects on edible mushrooms

Technique used	Solution/ Time period	Findings	References
Soaking and boiling	15 min, 30 min, 45 min, 60 min, 75 min, 90 min 24-hour soaking + 30 min boiling	*Pleurotus sajor- caju singer*, an oyster mushroom, was grown in polythene bags containing 500 g of fresh, dry, and chopped leaves of sugarcane. The spawning was conducted at 50 g/bag, then the substrate was soaked and boiled. It was depicted that soaking and boiling the substrate for 75 minutes results in a higher yield percentage. Similar results were also obtained by other researchers.	Pathan et al. (2009)
Steam blanching and soaking	Steam blanching (3 min) and combination of steam blanching (3 min) + soaking for 30 min in 0.5% citric acid	Effect of pretreatments (blanching and combination of blanching with soaking) followed by drying on nutritional profile and physicochemical factors of oyster mushrooms were investigated and findings revealed that drying without pretreatments was found the best method for oyster mushrooms. This suggests that blanching and soaking have no positive correlation with mushroom drying.	Maray et al. (2018)
Blanching and cooking	Blanching (5 min with citric acid) and boiling (15 min)	The impact of blanching and boiling on phenolic content and antioxidant activity of three edible mushroom species (*Agaricus bisporus, Pleurotus ostreatus, Lentinula edodes*) was investigated. Observation led us to the conclusion that in general hydrothermal processing could lead to the decrease in total phenolics and antioxidant content of mushrooms. Boiling for 15 min had a negative impact. On the other hand, blanching in the citric acid solution for 5 min resulted in an increase of alcohol-soluble phenolics in *P. ostreatus*.	Radzki et al. (2016)
Water blanching and steam blanching	Water and steam blanching with potassium bisulfite, lemon juice, and vinegar	The nutritional value of oyster mushroom after no blanching, water blanching, and steam blanching accompanied by chemical pretreatments was investigated. Blanching followed by chemical pretreatments resulted in higher content of phenolic and flavonoid content than lemon juice. It can be inferred that appropriate chemical treatment would result in an improved quality processed product.	Mutukwa (2014)
Soaking	Soaking water enriched with wheat bran, corn flour, soy bean flour, and rice bran	Impact of additional nutrient-rich ingredients (wheat bran, corn flour, soy bean flour, rice bran) into soaking water treatment on edible mushroom shiitake production and other properties. According to findings, rice and wheat bran–enriched soaking treatments were found most effective in increasing mushroom production.	Ranjbar et al. (2017)

content in the mushrooms when the average dosage of 11.50 kJ m^2/pulse was given (Tsai & Chen, 2019). Ionizing treatments were applied on the edible variety of mushrooms (*Agaricus bisporus*) at 0.5 kilograys (kGy), 1.5 kGy, and 2.5 kGy, to check the effect on phenylalanine ammonia-lyase activity, total phenolic content, and respiration. It was observed that at 1.5 kGy and 2.5 kGy, the rate of respiration was decreased. Phenylalanine ammonia-lyase activity and total phenolic content were increased between 1–4 days but decreased after that. There was an increase in the activity of polyphenol oxidase (PPO) treated differently does of kGy. Loss of the white color of mushrooms during storage was observed (Benoit et al., 2000).

Mushrooms generally have a shorter shelf life (1–2 days) due to browning, stalk elongation, and cap opening, but this shelf life can be prolonged by exposing the mushrooms to a particular dosage level of kilogray. Gamma irradiation (upto 10 kGy) could prevent the enzymatic browning of mushrooms and extend their shelf life by inactivating the enzyme polyphenol oxidase (PPO) in mushrooms. Also, when the doses were as low as 4.5 kGy/h, it provided a more valuable effect in extending the shelf life when compared with doses of 32 kGy/h. A shelf life of 5 days can be achieved by exposing *Agaricus bisporus* mushrooms to the dose of 1 kGy, whereas a dose of 2 kGy extended the shelf life for 3–4 days only (Akram & Kwon, 2010). The effect of gamma irradiation on the major aromatic compounds agaritine, GHB (gamma-glutaminyl-4-hydroxybenzene) as well as its effect on the total phenolic content and antioxidant capacity was observed. There was no effect on agaritine, until the dosage of 3 kGy. But at doses of 5 kGy, there was a reduction in the dry weight of mushrooms (from 1.54 to 1.35 g/kg), and GHB levels by 31%, while there was no effect on the total phenolic content and antioxidant capacity of mushrooms (Sommer et al., 2009).

Gamma and electron beam irradiation extends the postharvest shelf life of fresh mushrooms (*Agaricus bisporus, Lentinus edodes,* and *Pleurotusostreatus*), destroying the microorganisms or insects present in the food. It also inhibited cap opening and browning, stalk elongation, reduced the level of microbial contamination without affecting the taste qualities of mushrooms. It has also been observed that when gamma irradiation is combined with refrigeration, it extends the shelf life by reducing the loss of moisture and improving the color and appearance. Electron beam irradiation also reduces the harmful bacteria in foodstuffs such as fruits and vegetables and preserves the fresh taste, aroma, texture, and nutritional composition of mushrooms (Fernandes et al., 2012). A delay in browning, as its effect on color, is also observed when these treatments are given. The dose of 1.2 kGy delayed fruit body softening and splitting (by 6–9 days) on the mushroom (*P. nebrodensis*) (Xiong et al., 2009). Electron beam irradiation of 0.5, 1, 3.1 kGy provided firmness to the samples, while the dose of 5.2 kGy kept the sample soft and did not provide the firmness desired (Koorapati et al., 2004).

Gamma irradiation of *H. marmoreus* or electron beam irradiation of *A. bisporus* at low dosage showed a smaller rate of decrease of reducing sugars during the storage period (Duan et al., 2010). But the dose of 1.0 kGy showed an increase in total sugar content of variety (*L. edodes*) (Jiang et al. 2010). The content of ergosterol in mushroom varieties *H. marzuolus* and *A. bisporus* is the highest (6.81–6.42 mg/g), followed by *B. edulis, L. edodes, C. gambosa,* and *P. ostreatus* (4–3.31 mg/g). On exposure of ergosterol to UV light at wavelength 280–320 nm, it undergoes photolysis and produces photo-irradiation products such as vitamin D2, tachysterol, and lumisterol (Mattila et al., 2002). Oxidation of ergosterol can be seen when mushrooms (*C. cornucopioides*) are exposed to UV irradiation for 10 minutes, while the varieties *A. bisporus* and *H. marzuolus* had a lower extent of oxidation (Villares, Mateo-Vivaracho, García-Lafuente, & Guillamón 2014).

When ultraviolet B (UV-B) radiation treatment at 280–360 nm was given to the mushroom species for 2 hours at 25°C, the vitamin D2 content of the fruiting bodies increased from 0–3.93 mg/g to 15.06–208.65 mg/g. The highest content was found in golden oyster mushroom (from 2.72 to 81.67 µg/g) followed by oyster (from 5.93 to 81.71 µg/g) and pink oyster mushrooms (0.28 to 66.03 µg/g) (Huang, Lin, & Tsai 2015). This drastic increase in vitamin D2 content in mushrooms is due to photosynthetic or thermal processes occurring when ergosterol is exposed to UV light. There was a decrease in polysaccharide content after the exposure to UV-B rays for 2 hours. The EC50 values of irradiated fruiting bodies were lower than those of the non-irradiated controls (6.76, 5.58, and 3.48 mg/g for reducing power, scavenging, and chelating abilities, respectively) (Huang, Lin, & Tsai 2015).

13.6 EFFECT OF LOW TEMPERATURE (FREEZING) ON EDIBLE MUSHROOMS

Freezing is regarded as the best preservative method for maintaining the natural taste of mushrooms. To maintain the texture and color of mushrooms, a temperature of 3.5–20°C was deemed effective (Mohapatra et al., 2010). When the mushrooms were stored at 4°C for 3–15 days, there was a decrease in the ergothioneine concentration of the *P. citrinopileatus* with the increase of storage time. The freeze-drying method showed higher ergothioneine content than cold-air drying and hot-air drying, with 255.27 µg/g dw in *F. velutipes* and 2,643.20 µg/g dw in *P. citrinopileatus* (Tsai & Chen, 2019). Before freezing the mushrooms, they were washed in the solution of metabisulphites to prevent the adverse changes in their color and flavor, and preserve their texture even when stored for 3 months in refrigeration. The best freezing temperatures of –25°C to –30°C were the most common temperature used. It was also found that blanching in 1% citric acid solution and 2% salt solution before freezing mushrooms, had shown poor taste, aroma, and consistency than the unblanched ones (Bernas et al., 2006).

The deep-frozen (at –20°C) and lyophilized mushrooms contained more concentration of silver (Ag) and barium (Ba), but less chromium (Cr), which is possible due to leaching (Drewnowska et al., 2017). The ice crystal size is the most important parameter during the freezing process, which is directly proportional to the freezing rate (Tu et al., 2015). To preserve the structural quality of the food and reduce the microbial and enzymatic activities, a quick-freezing rate is more acceptable, which leads to the uniform destruction of small ice crystals (Xue, Hao, Yu, & Kou 2017). No effect on dry matter, ash, and pH level in *Boletus edulis* mushrooms was observed, but on the contrary, a decrease in *B. edulis* caps and stipes hardness, chewiness, and gumminess were observed. Freezing the mushrooms also caused a reduction in free amino acids by 39.8% and in essential amino acids by 39% (Liu et al., 2014). A decrease in the concentrations of thiamine (B1), riboflavin (B2), and vitamin C and loss of phenolic compounds in *Boletus edulis* mushrooms was reported (Jaworska & Bernaś, 2009).

The effect of freezing on bioactivity showed high antioxidant-reducing power (0.5 absorbance at 1.27 mg/mL) due to disruption of the cell walls, causing the extraction of intracellular compounds in the frozen samples (Fernandes et al., 2017). When frozen at –18°C for 15 days, the 2-thiol-l-histidinebetaine (l-ergothioneine, ESH), total phenols (TP) content, and DPPH radical scavenging ability remained unchanged (Nguyen et al., 2012). When mannitol solution was added to cultures of *V. volvacea* fruiting bodies, it was noted that intracellular mannitol was catabolized as an energy storage material and the expression of genes encoding enzymes was also inhibited. Also, organs of *V. volvacea* began to shrink and collapse and no significant elongation was found during storage at 4°C. But the sensory characteristics of mannitol treated variety were much better than the control (Zhao et al., 2019). A reduction in the diversity of the bacterial community was found when the *V. volvacea* was stored at 15°C (Wang et al., 2019).

Frozen storage had caused a decrease in hardness (by 88%), springiness (by 30%), and chewiness (by 85%) on the texture of unblanched mushroom caps when compared to caps directly after freezing. But when mushrooms had undergone a pretreatment such as blanching, then no major changes occurred as a result of freezing on the texture. The texture of *Boletus edulis* was observed when frozen for 12 months in frozen storage, and it was found that there was a decrease in hardness, chewiness, and gumminess (77–100%) and an increase in cohesiveness (121–521%) and wateriness (1.8 –4.0 points) after frozen storage when compared to the raw material (Jaworska & Bernaś, 2010). The effect of freeze drying on button mushrooms was checked at temperatures –2°C, –5°C, and –8°C, and it was noticed that freeze-drying temperatures affected only the ascorbic acid content, while there was no effect on protein and antioxidant (Tarafdar et al., 2017).

When the mushrooms were exposed to temperatures as low as 3°C, the highest moisture content of about 85.3% and shelf life of 11.92 days, the lowest dry matter content of 14.97%, and weight loss of 4% was recorded. So, it can be concluded that at 3°C, mushrooms had the best quality in the category of weight loss, disease severity, color, firmness, freshness, appearance, flavor, texture, and dry matter content, leading to prolonged shelf life when compared to those exposed to ambient temperature (Rahman et al., 2020). After freezing the species *Pleurotuseryngii* and comparing the two technologies that are at −20°C by natural air convection freezing (NF) or by individual quick freezing (IQF) at −62.5°C (with speed 8.23 m/s), it was observed that IQF samples of mini and diced mushroom had higher quality when compared to NF samples in all the aspects including color, flavor, hardness, and shelf life. It showed that IQF technology can be used as a new preservation method of mushrooms (Li et al., 2018).

13.7 EFFECT OF VACUUM MOISTENING ON EDIBLE MUSHROOMS

Besides soaking, the other most frequently used method before blanching is vacuum moistening. It plays a major role in reducing the weight loss of mushrooms due to blanching and also helps to retain the natural color of blanched food products (Bernas et al., 2006). Research studies conducted by Czapski (1994) and Beelman et al. (1973) revealed that vacuum moistening has a positive effect on the quality of mushrooms and color losses were also observed to be less for the mushrooms which have undergone this processing method as compared to the ones in which vacuum moistening was not done. Thus, vacuum moistening is a pre-processing method that is performed before blanching to prevent significant weight loss in mushrooms (Jaworska & Bernaś, 2009).

13.8 EFFECT OF DRYING ON EDIBLE MUSHROOMS

Drying mushrooms is the oldest and simplest way to process and is an essential role in mushroom preservation. Drying means removing water from food to lengthen their life span (Moon & Lo, 2014). There are various methods of drying mushrooms which are as follows.

13.8.1 Solar drying of mushrooms

Owing to energy-saving, low-cost, and simple operation, sunlight drying is used widely around the world. The most popular method is to thinly spread the mushrooms on a tray and expose them to sunlight and wind; nevertheless, due to contamination, prolonged drying, microbial development, and drying cause loss. Also, as a rule, sun drying depends greatly on weather conditions. Solar equipment like hybrid solar dryers (Reyes et al., 2014) are designed to address such issues, and mushrooms are placed in the chamber to keep them away from spoiling. Hot air drying (HAD) is widely used for preserving mushrooms because of the low cost and easy operation, although it frequently results in undesired quality degradation. The best drying method for energy consumption is the microwave dryer (Motevali et al., 2011). Novel technologies like air-impingement jet drying (AID) are promising drying technology for obtaining high sensory quality mushrooms. In shiitake mushrooms, AID improved the characteristic flavors (onion-like odor and umami) of dried mushrooms by partially inhibiting enzymatic and Maillard reactions (Luo et al., 2021). The solar drying method is cheap and better in quality than the hot air dryer (60°C) (Jo et al., 2009). The shelf life of mushrooms is enhanced by thermal drying, but their flavors are affected. Thermal drying with ultraviolet light transformed ergosterol to vitamin D. Thermal drying enhanced the nutritional value of all edible mushrooms (Jiang et al., 2020). High protein content, catalase, and peroxidase levels were found in sun-dried mushrooms. (Arumuganathan et al., 2010).

13.8.2 Hot air drying

The most accessible and most commonly used dryer is hot air drying (Argyropoulos, Heindl, & Müller 2011). A standard method of hot air drying is to place mushrooms on a tray and push them into a cabinet or tunnel dryer. Heating to a temperature of 50–70°C (Argyropoulos et al., 2011), with air convection, the hot wind also removes the mushroom's moisture. Besides, due to improper drying, a smoky flavor can also occur (Argyropoulos et al., 2011).

Furthermore, microwave drying and infrared techniques have been developed to circumvent these problems. These drying methods have been implemented separately or combined with hot air drying. While traditional hot air drying is thought to be a relatively easy, cost-effective, and reliable method of extending mushroom shelf life, fluidized bed drying is faster and produces a better quality product (Kulshreshtha et al., 2009; Arumuganathan et al., 2009). Protein content and residual enzyme activity are more significant in sun-dried mushrooms while they were lower in fluidized bed-dried mushrooms (Arumuganathan et al., 2010).

13.8.3 Freeze drying

Freeze drying research is conducted in some mushroom species, such as button mushrooms (Pei et al., 2014) and shiitake mushrooms (Zhao et al., 2016). For obtaining high quality (higher ergothioneine content), the freeze-drying method can be used (Tsai & Chen, 2019). The texture of oyster mushrooms dried using the freeze-drying method is the least firm in texture. In oyster mushrooms, osmo-air drying results in hard and tough dried mushrooms (Arumuganathan et al., 2010). Freeze drying is the most expensive process for producing dehydrated products and also demands expensive equipment (Donsì et al., 2001; Irzyniec et al., 1995). Consequently, the freeze-drying technology is limited. Many attempts have focused on some assistive technologies such as microwave, infrared, and ultrasonic drying are increasingly used to improve drying efficiency and reduce energy consumption.

13.8.4 Microwave drying

Mushrooms, as a heat-sensitive product, can be dried using microwave or microwave-assisted drying than conventional thermal drying. Nevertheless, microwave energy improper use causes irreversible changes in product quality, such as charred edge, excessive oxidation, etc. Continuous exposure to microwaves can also cause local overheating (Das & Arora, 2018). Microwaves alone are not ideal for mushroom drying, although the drying time is short (Walde et al., 2006). Combined drying (hot air and microwave vacuum drying) can make mushroom structure fluffy and crispy, and this technique can develop novel snacks (Argyropoulos et al., 2011a). Mushrooms can be dried rapidly by alternating microwave and convective hot air, which enabled the mushrooms to reach the commercial quality standard in a relatively short time of 72 minutes (Das & Arora, 2018).

13.8.5 Far-infrared drying

Most food components absorb far-infrared radiation energy among different infrared technologies. Infrared drying can be used to ascertain the nutritional, structural, and sensory properties of *Lentinus edodes*. The results showed that far-infrared drying speed is fast compared with freeze drying and can save up to 66.25% time. Among the five drying methods, far-infrared dried *Lentinus edodes* had better overall quality (Zhao et al., 2019).

13.8.6 Ultrasound-assisted drying

Ultrasound-assisted drying or frying of mushrooms improves the quality of products (Devi et al., 2018). Low-temperature drying of white button mushrooms with ultrasound leads to drying time reduction (Vallespir et al., 2019). The mushroom's drying rate is increased with the combination of electro-plasmolysis and ultrasound pretreatments by 37.10% (Çakmak et al., 2016). Mushrooms subjected to ultrasonic pretreatment dried faster with better rehydration performance (Jambrak et al., 2007).

13.9 EFFECT OF SALTING AND PICKLING ON EDIBLE MUSHROOMS

Salting is the process of two-way mass transfer in which the migration of water takes place from the product to the brine and the migration of solute takes place from brine to product. This type of processing is mostly used for fish and meat and less frequently used for edible mushrooms. The parameters used for the process include a concentration of brine, use of agitation, and the sample. The principle of drying includes the concept of higher osmotic pressure is achieved by using salt and fermentation by microorganisms (Xue et al., 2017).

The study conducted by Liu et al. (2014) has reported the effect of salting on *A. bisporus* which was nutritionally different from fresh one and the differences included the significant increase in the dry matter content from 7.29 to 23.23% and reduction in protein content from 26.27 to 16.5%. The change was observed due to the water loss process during salting and the precipitation of water and salt soluble protein. The process of salting has also reduced the non-volatile taste compounds like essential amino acids and free amino acids by 90% and 92.8% in salted samples and the amount of MSG-like amino acids and 5-nucleotides in salted *A. bisporus* was lower than canned and frozen products due to the Maillard reaction in which amino acids reaction substrates were irreversible transformation. The extraction rate of crude polysaccharides of *Agaricus bisporus* has shown a significant difference after salting; however, infrared ray (IR) assays have shown that the process of salting does not change the structure of *A. bisporus* polysaccharides (Yang et al., 2007). The antioxidant capacity of pure and crude polysaccharides of unsalted *Agaricus bisporus* was higher than the salted one, which is due to the leaching out of water-soluble proteins, vitamins, and minerals during the liquid exuding process in salting. The reduction in the amount of protein, vitamins, and minerals reduces biological activities.

Auricularia auricula mushroom pickle was prepared and the effect of processing techniques was studied by Khaskheli et al. (2015). The pickle was prepared by adding mustard oil, vinegar, and salt (MOVS) and another sample by adding soft water, vinegar, and pickle (SWVS). The sensory analysis result showed the highest acceptance for MOVS formulated pickle and the microbiological studies showed that total viable counts were high in SWVS pickle than MOVS pickle. The results for storage activity showed that pickles have a good shelf life of 90 days at ambient temperature (26 ± 4°C) without any significant change in the quality attributes of the MOVS pickle.

The traditional fermented oyster mushrooms were used for the osmotic and pickling method by Temesgen & Workneh (2015) to study the effect on its nutritional quality and acceptance. The pretreatment effects of ascorbate and osmotic solution on oven-dried and pickled mushrooms were evaluated and the result showed that ascorbate concentration increased the ash, fat, and protein content of mushrooms and osmotic solution pretreated mushrooms showed the highest rehydration capacity. They concluded that pickling done by pretreating it with ascorbate is advantageous for improving the quality of mushrooms as compared to oven drying and osmotic pretreatments.

Pickling is one of the ancient methods for the preservation of food where the food is either subjected to lactic acid fermentation or marinated by adding acetic acid (Jabłońska-Ryś et al., 2019).

Pickling in mushrooms involves the lactic bacteria which is beneficial for human organisms and gives a pleasant taste and aroma to the pickled food *Pleurotusostreatus* Kumm (Bernas et al., 2006). Species of mushrooms are edible for pickling and the good results were reported by Kreb and Lelley (1991) of pickling it with shredded cabbage for 10 days at 21°C. The obtained product has a mild and pleasant taste similar to pickled cabbage and can be stored for up to 6 months.

13.10 EFFECT OF HEAT TREATMENT ON EDIBLE MUSHROOMS

Mushrooms are usually cooked by different cooking methods including steaming, boiling, frying, pressure cooking, stir frying, and microwaving, which alter the total antioxidant values and degrade its nutritional value (Ng & Tan, 2017). The heating process makes the food less healthy compared to fresh one due to the loss of heat-labile nutrients. However, many studies have stated that thermally processed foods including vegetables and fruits enhance biological activities due to the chemical changes during heat treatment (Choi et al., 2006). However, mushrooms area highly perishable food commodity henceforth demand the extension of shelf life is necessary by using different handling process like sterilizing, drying, cooling, and freezing (Saenmuang et al., 2017).

In this regard, Ng and Tan, (2017) have investigated the effect of four different cooking methods with different time intervals on five edible varieties of mushrooms namely *Agaricus bisporus, Flammulinavelutipes, Lentinula edodes, Pleurotusostreatus,* and *Pleurotuseryngii.* The antioxidant activity, total flavonoid, phenolic, and ascorbic acid were evaluated and *A. bisporus* showed the highest values among all the varieties and short duration heating i.e. for 3 minutes has shown the increased flavonoid content and ascorbic acid and on the other hand, pressure cooking for 15 minutes has decreased the water-soluble phenolic content in all the varieties of mushrooms. The optimized cooking method reported was pressure cooking to increase the antioxidant values in edible mushrooms.

The effect of heat treatments on free radical scavenging activity and phenolic compounds were studied by Saenmuang et al. (2017). *Tylopilusalboater,* a wild mushroom was evaluated after boiling, steaming, and at the fresh state and the result showed that after boiling for 15 and 30 minutes the total polyphenolic content and flavonoid content significantly reduced whereas, in the steamed sample, the sight reduction was observed. The DPPH scavenging activity of heated mushrooms was more than the fresh one and the boiled samples showed the highest scavenging activity. They reported that the boiling method for 30 minutes gives better results in the case of bitter mushrooms. The different heat treatments for mushroom cultivation in Egypt were evaluated by Hassan et al. (2012). Three different varieties namely *Pleurotusostreatus, Flammulina velutipes,* and *Hericiumerinaceus* were grown on pasteurized, unpasteurized, and sterilized rice straw and the yield was studied. They reported that the sterilized medium increased the yield for *H. erinaceus,* pasteurized conditions increased the yield of *F. velutips* and unpasteurized medium increased the yield of *P. ostreatus.* They concluded that alternative media preparation could be time-saving and reduce the money and effort for mushroom cultivation.

13.11 CONCLUSION

Many studies have explored the emerging processing and preservation technologies or have developed new technologies to increase the shelf life and nutritional quality of edible fungi during the commercialization and transportation process. As mushroom is highly perishable and has a low shelf life, various methods like sterilization, margination, and fermentation by lactic acid bacteria could be a great alternative for preserving the mushrooms. The appropriate pretreatments and mentioned food processing methods like drying, salting, pickling, etc. solve the perishability problem of mushrooms.

These methods improve the nutritional value as well as bioactive components in the mushrooms; however, the retention of bioactive compounds are dependent on the variety of mushrooms other than the cooking method.

REFERENCES

Akram, K., & Kwon, J.-H. (2010). Food irradiation for mushrooms: A review. *Journal of the Korean Society for Applied Biological Chemistry*, 53(3), 257–265.

Ali, M. A., Mehmood, M. I., Rab Nawaz, M., Hanif, A., & Wasim, R. (2007). Influence of substrate pasteurization methods on the yield of oyster mushroom (Pleurotus species). *Pakistan Journal of Agricultural*, *44*, 300–303.

Anju, B., Anisa, M., & Naseer, A. Effect of steeping solution on the quality of button mushrooms (A. bisporus) preserved under ambient conditions, *10*(1), 1–6.

Argyropoulos, D., Heindl, A., & Müller, J. (2011a). Assessment of convection, hot-air combined with microwave-vacuum and freeze-drying methods for mushrooms with regard to product quality. *International Journal of Food Science & Technology*, 46(2), 333–342.

Argyropoulos, D., Khan, M. T., & Müller, J. (2011b). Effect of air temperature and pre-treatment on color changes and texture of dried Boletus edulis mushroom. *Drying Technology*, 29(16), 1890–1900.

Arumuganathan, T., Manikantan, M. R., Indurani, C., Rai, R. D., & Kamal, S. (2010). Texture and quality parameters of oyster mushroom as influenced by drying methods. *International Agrophysics*, 24(4), 339–342.

Arumuganathan, T., Manikantan, M. R., Rai, R. D., Anandakumar, S., & Khare, V. (2009). Mathematical modelling of drying kinetics of milky mushroom in a bed dryer. *International Agrophysics*, 23(1), 1–7.

Avendaño-Hernandez, R. J., & Sánchez, J. E..(2013). Self-pasteurised substrate for growing oyster mushrooms (Pleurotus spp.). *African Journal of Microbiology Research.*, 7(3), 220–226.

Benoit, M. A., D'Aprano, G., & Lacroix, M. (2000). Effect of γ-irradiation on phenylalanine ammonia-lyase activity, total phenolic content, and respiration of mushrooms (Agaricus bisporus). *Journal of Agricultural and Food Chemistry*, 48(12), 6312–6316.

Barros, L., Cruz, T., Baptista, P., Estevinho, L. M., & Ferreira, I. C. F. R. (2008). Wild and commercial mushrooms as source of nutrients and nutraceuticals. *Food and Chemical Toxicology*, 46(8), 2742–2747.

Beelman, R. B., Kuhn, G. D., & McArdle, J. (1973). Influence of post-harvest storage and soaking treatments on the yield and quality of canned mushrooms. *Journal of Food Science*, 38(6), 951–953.

Bello, B. K., & Akinyele, B. J. (2007). Effect of fermentation on the microbiology and mineral composition of an edible mushroom Termitomyces robustus (Fries). *International Journal of Biological Chemistry*, *1*(4), 237–243.

Bernas, E., Jaworska, G., & Lisiewska, Z. (2006). Edible mushrooms as a source of valuable nutritive constituents. *Acta Scientiarum Polonorum Technologia Alimentaria*, 5, 5–20.

Çakmak, R. Ş., Tekeoğlu, O., Bozkır, H., Ergün, A. R., & Baysal, T. (2016). Effects of electrical and sonication pretreatments on the drying rate and quality of mushrooms. *LWT-Food Science and Technology*, 69, 197–202.

Chawla, P., Kumar, N., Kaushik, R., Dhull, S. B. (2019). Synthesis, characterization and cellular mineral absorption of gum arabic stabilized nanoemulsion of *Rhododendron arboreum* flower extract. *Journal of Food Science and Technology*, 56(12), 5194–5203.

Choi, Y., Lee, S. M., Chun, J., Lee, H. B., & Lee, J. (2006). Influence of heat treatment on the antioxidant activities and polyphenolic compounds of Shiitake (Lentinus edodes) mushroom. *Food Chemistry*, 99(2), 381–387.

Czapski, J. (1994). Effect of some technological treatments on the yield and quality of blanched mushrooms stored in brine. *Bulletin of Vegetable Crops Research Work, 2,* 101–119.

Das, I., & Arora, A. (2018). Alternate microwave and convective hot air application for rapid mushroom drying. *Journal of Food Engineering*, 223, 208–219.

de Siqueira, F. G., Maciel, W. P., Martos, E. T., Duarte, G. C., Miller, R. N. G., da Silva, R., & Dias, E. S. (2012). Cultivation of Pleurotus mushrooms in substrates obtained by short composting and steam pasteurization. *African Journal of Biotechnology, 11*(53), 11630–11635.

Devi, S., Zhang, M., & Law, C. L. (2018). Effect of ultrasound and microwave assisted vacuum frying on mushroom (Agaricus bisporus) chips quality. *Food Bioscience, 25*, 111–117.

Donsì, G., Ferrari, G., & Matteo, D. I. (2001). Utilization of combined processes in freeze-drying of shrimps. *Food and Bioproducts Processing, 79*(3), 152–159.

Drewnowska, M., Falandysz, J., Chudzińska, M., Hanć, A., Saba, M., & Barałkiewicz, D. (2017). Leaching of arsenic and sixteen metallic elements from Amanita fulva mushrooms after food processing. *LWT, 84*, 861–866.

Duan, Z., Xing, Z., Shao, Y., & Zhao, X. (2010). Effect of electron beam irradiation on postharvest quality and selected enzyme activities of the white button mushroom, Agaricus bisporus. *Journal of Agricultural and Food Chemistry, 58*(17), 9617–9621.

Ejigu, D., & Kebede, T. (2015). Effect of Different Organic Substrates and their Pasteurization Methods on Growth Performance, Yield and Nutritional values of Oyster Mushroom (Pleurotusostreatus). *Global Journal of Science Frontier Research, 15*(6), 27–29.

Fernandes, Â., Antonio, A. L., Beatriz, M., Oliveira, P. P., Martins, A., & Ferreira, I. C. F. R. (2012). Effect of gamma and electron beam irradiation on the physico-chemical and nutritional properties of mushrooms: A review. *Food Chemistry, 135*(2), 641–650.

Fernandes, Â., Barreira, J. C. M., Günaydi, T., Alkan, H., Antonio, A. L., Beatriz, M. , Oliveira, P. P., Martins, A., & Ferreira, I. C. F. R. (2017). Effect of gamma irradiation and extended storage on selected chemical constituents and antioxidant activities of sliced mushroom. *Food Control, 72*, 328–337.

Hassan, F. R. H., Medany, G. M., & Hussein, S. D. A. (June 2012) Evaluation of different heat treatments for mushroom cultivation medium in Egypt. *Journal of Applied Sciences Research, 8*(6), 3012–3018.

Ho, L.-H., Zulkifli, N., & Tan, T.-C. (2020). Edible mushroom: nutritional properties, potential nutraceutical values, and its utilisation in food product development. *An Introduction to Mushroom.* 10.5772/intechopen.91827

Huang, S.-J., Lin, C.-P., & Tsai, S.-Y. (2015). Vitamin D2 content and antioxidant properties of fruit body and mycelia of edible mushrooms by UV-B irradiation. *Journal of Food Composition and Analysis, 42*, 38–45.

Hutkins, R. W. (2018). *Microbiology and Technology of Fermented Foods.* Hoboken, NJ.

Irzyniec, Z., Klimczak, J., & Michalowski, S. (1995). Freeze-drying of the black currant juice. *Drying Technology, 13*(1-2), 417–424.

Jabłońska-Ryś, E., Skrzypczak, K., Sławińska, A., Radzki, W., & Gustaw, W. (2019). Lactic acid fermentation of edible mushrooms: Tradition, technology, current state of research: A review. *Comprehensive Reviews in Food Science and Food Safety, 18*(3), 655–669.

Jabłońska-Ryś, E., Sławińska, A., Radzki, W., & Gustaw, W. (2016). Evaluation of the potential use of probiotic strain Lactobacillus plantarum 299v in lactic fermentation of button mushroom fruiting bodies. *Acta ScientiarumPolonorumTechnologia Alimentaria, 15*(4).

Jambrak, A. R., Mason, T. J., Paniwnyk, L., & Lelas, V. (2007). Accelerated drying of button mushrooms, Brussels sprouts and cauliflower by applying power ultrasound and its rehydration properties. *Journal of Food Engineering, 81*(1), 88–97.

Jaworska, G., & Lisiewska, Z. (2006). Edible mushrooms as a source of valuable nutritive constituents. Acta Scientiarum Polonorum Technologia Alimentaria, 5, 5–20.

Jaworska, G., & Bernaś, E. (2010). Effects of pre-treatment, freezing and frozen storage on the texture of Boletus edulis (Bull: Fr.) mushrooms. *International Journal of Refrigeration, 33*(4), 877–885.

Jaworska, G., & Bernaś, E. (2009). The effect of preliminary processing and period of storage on the quality of frozen Boletus edulis (Bull: Fr.) mushrooms. *Food Chemistry, 113*(4), 936–943.

Jiang, Q., Zhang, M., & Mujumdar, A. S. (2020). UV induced conversion during drying of ergosterol to vitamin D in various mushrooms: Effect of different drying conditions. *Trends in Food Science & Technology, 105*(2020), 200–210.

Jiang, T., Jahangir, M. M., Jiang, Z., Lu, X., & Ying, T. (2010). Influence of UV-C treatment on antioxidant capacity, antioxidant enzyme activity and texture of postharvest shiitake (Lentinus edodes) mushrooms during storage. *Postharvest Biology and Technology, 56*(3), 209–215.

Jiang, T., Luo, S., Chen, Q., Shen, L., & Ying, T. (2010). Effect of integrated application of gamma irradiation and modified atmosphere packaging on physicochemical and microbiological properties of shiitake mushroom (Lentinus edodes). *Food Chemistry, 122*(3), 761–767.

Jo, W.-S., Park, S.-D., Park, S.-C., Chang, Z.-Q., Seo, G.-S., Uhm, J.-Y., & Jung, H.-Y. (2009). Changes in quality of Phellinus gilvus mushroom by different drying methods. *Mycoscience, 50*(1), 70–73.

Kerketta, V., Shukla, C. S., Singh, H. K., & Kerketta, A. (2019). Effect of grain substrates on spawn development and their impact on yield and yield attributing characters of Pleurotus spp. *Journal of Pharmacognosy and Phytochemistry, 8*(6), 722–725.

Khaskheli, S. G., Zheng, W., sheikh, S. A., Khaskheli, A. A., Liu, Y., Wang, Y., & Huang, W. (2015). Effect of processing techniques on the quality and acceptability of Auricularia auricula mushroom pickle. *Journal of Food and Nutrition Research, 3*(1), 46–51.

Kim, S. Y., Son, J., Martin Lo, Y., Lee, C., & Moon, B. (2014). Effects of Light-Emitting Diode (LED) Lighting on the Ergosterol Content of Beech Mushrooms (L yophyllumulmarium). *Journal of Food Processing and Preservation, 38*(4), 1926–1931.

Koorapati, A., Foley, D., Pilling, R., & Prakash, A. (2004). Electron-beam irradiation preserves the quality of white button mushroom (Agaricus bisporus) slices. *Journal of Food Science, 69*(1), SNQ25–SNQ29.

Krishnamurthy, K., Khurana, H. K., Soojin, J., Irudayaraj, J., & Demirci, A. (2008). Infrared heating in food processing: an overview. *Comprehensive reviews in Food Science and Food Safety, 7*(1), 2–13.

Kreb, M., & Lelly, J. (1991). Preservation of Oyester mushrooms by lactic acid fermentation. In Science and cultivation of edible fungi. M.J Maher Conference. Dublin, Republic of Ireland, 1–6.

Kulshreshtha, M., Singh, A., & Vipul, D. (2009). Effect of drying conditions on mushroom quality. *Journal of Engineering Science and Technology, 4*(1), 90–98.

Kurtzman Jr, Ralph H. (2010). Pasteurization of mushroom substrate and other solids. *African Journal of Environmental Science and Technology, 4*(13), 936–941.

Li, B., Liu, C., Fang, D., Yuan, B., Hu, Q., & Zhao, L. (2019). Effect of boiling time on the contents of flavor and taste in Lentinus edodes. *Flavour and Fragrance Journal, 34*(6), 506–513.

Li, T., Lee, J.-W., Luo, L., Kim, J., & Moon, B. (2018). Evaluation of the effects of different freezing and thawing methods on the quality preservation of Pleurotuseryngii. *Applied Biological Chemistry, 61*(3), 257–265.

Lin, C. Y., Lay, C. H., Chen, C. C., Sen, B., & Sung, I. Y. (2016). Biohydrogen Production from Mushroom Cultivation Waste by Anaerobic Solid-state Fermentation. *Journal of the Chinese Chemical Society, 63*(2), 199–204.

Liu, Y., Huang, F., Yang, H., Ibrahim, S. A., Wang, Y.-f., & Huang, W. (2014). Effects of preservation methods on amino acids and 5′-nucleotides of Agaricus bisporus mushrooms. *Food chemistry, 149*, 221–225.

Luo, D., Wu, J., Ma, Z., Tang, P., Liao, X., & Lao, F. (2021). Production of high sensory quality Shiitake mushroom (Lentinus edodes) by pulsed air-impingement jet drying (AID) technique. *Food chemistry, 341*, 128290.

Maray, A. R. M., Mostafa, M. K., & El-Fakhrany, Alaa El-Din MA. (2018). Effect of pretreatments and drying methods on physico-chemical, sensory characteristics and nutritional value of oyster mushroom. *Journal of Food Processing and Preservation, 42*(1), e13352.

Mattila, P., Lampi, A.-M., Ronkainen, R., Toivo, J., & Piironen, V. (2002). Sterol and vitamin D2 contents in some wild and cultivated mushrooms. *Food Chemistry, 76*(3), 293–298.

Moda, E. M., Horii, J., & Spoto, M. H. F. (2005). Edible mushroom Pleurotussajor-caju production on washed and supplemented sugarcane bagasse. *Scientia Agricola, 62*(2), 127–132.

Mohapatra, D., Bira, Z. M., Kerry, J. P., Frías, J. M., & Rodrigues, F. A. (2010). Postharvest hardness and color evolution of white button mushrooms (Agaricus bisporus). *Journal of Food Science, 75*(3), E146–E152.

Moon, B., & Lo, Y. M. (2014). Conventional and novel applications of edible mushrooms in today's food industry. *Journal of Food Processing and Preservation, 38*(5), 2146–2153.

Morales, V., & Sánchez, J. E. (2017). Self-heating pasteurization of substrates for culinary-medicinal mushrooms cultivation in Mexico. *International Journal of Medicinal Mushrooms, 19*(5), 477–484.

Motevali, A., Minaei, S., Khoshtaghaza, M. H., & Amirnejat, H. (2011). Comparison of energy consumption and specific energy requirements of different methods for drying mushroom slices. *Energy, 36*(11), 6433–6441.

Ng, Z. X., & Tan, W. C. (2017). Impact of optimised cooking on the antioxidant activity in edible mushrooms. *Journal of Food Science and Technology*, *54*(12), 4100–4111.

Nguyen, T. H., Nagasaka, R., & Ohshima, T. (2012). Effects of extraction solvents, cooking procedures and storage conditions on the contents of ergothioneine and phenolic compounds and antioxidative capacity of the cultivated mushroom Flammulinavelutipes. *International Journal of Food Science & Technology*, *47*(6), 1193–1205.

Nölle, N., Argyropoulos, D., Ambacher, S., Müller, J., & Biesalski, H. K. (2017). Vitamin D2 enrichment in mushrooms by natural or artificial UV-light during drying. *LWT-Food Science and Technology*, *85*, 400–404.

Oseni, T. O., Dlamini, S. O., Earnshaw, D. M., & Masarirambi, M. T. (2012). Effect of substrate pretreatment methods on oyster mushroom (Pleurotusostreatus) production. *International Journal of Agriculture and Biology*, *14*(2), 251–255.

Pathan, A. A., Jiskani, M. M., Pathan, M. A., Wagan, K. H., & Nizamani, Z. A. (2009). Effect of soaking and boiling of substrate on the growth and productivity of oyster mushroom. *Pakistan Journal of Phytopathology*, *21*(1), 1–5.

Phillips, K. M., & Rasor, A. S. (2013). A nutritionally meaningful increase in vitamin D in retail mushrooms is attainable by exposure to sunlight prior to consumption. Journal of Nutrition & Food Sciences, 3, 1.

Pei, F., Yang, W.-j., Shi, Y., Sun, Y., Mariga, A. M., Zhao, L.-y., Fang, Y., Ma, N., An, X.-x., & Hu, Q.-h. (2014). Comparison of freeze-drying with three different combinations of drying methods and their influence on colour, texture, microstructure and nutrient retention of button mushroom (Agaricus bisporus) slices. *Food and Bioprocess Technology*, *7*(3), 702–710.

Qi, L.-L., Zhang, M., Mujumdar, A. S., Meng, X.-Y., & Chen, H.-Z. (2014). Comparison of drying characteristics and quality of shiitake mushrooms (Lentinus edodes) using different drying methods. *Drying Technology*, *32*(15), 1751–1761.

Radzki, W., Slawinska, A., Jablonska-Rys, E., & Michalak-Majewska, M. (2016). Effect of blanching and cooking on antioxidant capacity of cultivated edible mushrooms: A comparative study. *International Food Research Journal*, *23*(2), 599.

Rahman, M., Hassan, M., & Talukder, F. U. (2020). Effect of low temperature on postharvest behaviors of oyster mushroom (Pleurotus spp.). *International Journal of Horticultural Science and Technology*, *7*(3), 213–225.

Ranjbar, M. E., Olfati, J. A., & Amani, M. (2017). Influence of enriched soaking water on shiitake (Lentinus edodes (Berk.) Singer) mushroom yield and properties. *Acta AgriculturaeSlovenica*, *109*(3), 555–560.

Reyes, A., Mahn, A., & Vásquez, F. (2014). Mushrooms dehydration in a hybrid-solar dryer, using a phase change material. *Energy Conversion and management*, *83*, 241–248.

Saenmuang, S., Sirijariyawat, A., & Aunsri, N. (2017). The Effect of Moisture Content, Temperature and Variety on Specific Heat of Edible-Wild Mushrooms: Model Construction and Analysis. *Engineering Letters*, *25*(4), 446–454.

Skapska, S., Owczarek, L., Jasinska, U., Halasinska, A., Danielczuk, J., Sokolowska, B., & InstytutBiotechnologiiPrzemysluRolno-Spozywczego. (2008). Changes in the antioxidant capacity of edible mushrooms during lactic acid fermentation. *Zywnosc Nauka TechnologiaJakosc (Poland)*, *15*(4), 243–250.

Sommer, I., Schwartz, H., Solar, S., & Sontag, G. (2009). Effect of γ-irradiation on agaritine, γ-glutaminyl-4-hydroxybenzene (GHB), antioxidant capacity, and total phenolic content of mushrooms (Agaricus bisporus). *Journal of Agricultural and Food Chemistry*, *57*(13), 5790–5794.

Tarafdar, A., Shahi, N. C., Singh, A., & Sirohi, R. (2017). Optimization of freeze-drying process parameters for qualitative evaluation of button mushroom (Agaricus bisporus) using response surface methodology. *Journal of Food Quality*, *2017*.

Teichmann, A., Dutta, P. C., Staffas, A., & Jägerstad, M. (2007). Sterol and vitamin D2 concentrations in cultivated and wild grown mushrooms: Effects of UV irradiation. *LWT-Food Science and Technology*, *40*(5), 815–822.

Temesgen, M., & Workneh, T. S. (2015). Effect of Osmotic and Pickling Pretreatments on Nutritional Quality and Acceptance of Traditional Fermented Oyster Mushrooms. *Food Science and Quality Management*, *37*, 64–73.

Tsai, S.-Y., & Chen, Z.-Y. (2019). Influence of Cold Storage and Processing of Edible Mushroom on Ergothioneine Concentration. *Inflammation*, *10*, 12.

Tu, J., Zhang, M., Xu, B., & Liu, H. (2015). Effects of different freezing methods on the quality and microstructure of lotus (Nelumbo nucifera) root. *International Journal of Refrigeration*, *52*, 59–65.

Vallespir, F., Crescenzo, L., Rodríguez, Ó., Marra, F., & Simal, S. (2019). Intensification of low-temperature drying of mushroom by means of power ultrasound: effects on drying kinetics and quality parameters. *Food and Bioprocess Technology*, *12*(5), 839–851.

Valverde, M. E., Hernández-Pérez, T., & Paredes-López, O. (2015). Edible mushrooms: improving human health and promoting quality life. *International Journal of Microbiology*, *2015*.

Villares, A., Mateo-Vivaracho, L., García-Lafuente, A., & Guillamón, E. (2014). Storage temperature and UV-irradiation influence on the ergosterol content in edible mushrooms. *Food Chemistry*, *147*, 252–256.

Walde, S. G., Velu, V., Jyothirmayi, T., & Math, R. G. (2006). Effects of pretreatments and drying methods on dehydration of mushroom. *Journal of Food Engineering*, *74*(1), 108–115.

Wang, W. H. (2015). Effect of processing techniques on the quality and acceptability of Auricularia auricula mushroom pickle. *Journal of Food and Nutrition Research*, 3(1), 46–51.

Wang, X., Liu, S., Chen, M., Yu, C., Zhao, Y., Yang, H., Zha, L., & Li, Z. (2019). Low Temperature (15°C) Reduces Bacterial Diversity and Prolongs the Preservation Time of Volvariellavolvacea. *Microorganisms*, *7*(10), 475.

Xiong, Q.-l., Xing, Z.-t., Feng, Z., Tan, Q., & Bian, Y.-b. (2009). Effect of 60Co γ-irradiation on postharvest quality and selected enzyme activities of Pleurotusnebrodensis. *LWT-Food Science and Technology*, *42*(1), 157–161.

Xue, Z., Hao, J., Yu, W., & Kou, X. (2017) Effects of processing and storage preservation technologies on nutritional quality and biological activities of edible fungi: A review. *Journal of Food Process Engineering*, *40*(3), e12437.

Yang, L.-h., Liu, L.-d., Wang, X.-j., Qu, H.-g., Ma, Q.-h., & Cai, D.-h. (2007). Effects of Salting on Extraction of Agaricus bisporus Polysaccharides [J]. *Food Science*, *10*.

Zhao, J.-H., Ding, Y., Nie, Y., Xiao, H.-W., Zhang, Y., Zhu, Z., & Tang, X.-M. (2016). Glass transition and state diagram for freeze-dried Lentinus edodes mushroom *Thermochimica Acta*, *637*, 82–89.

Zhao, Y., Bi, J., Yi, J., Jin, X., Wu, X., & Zhou, M. (2019). Evaluation of sensory, textural, and nutritional attributes of shiitake mushrooms (Lentinula edodes) as prepared by five types of drying methods. *Journal of Food Process Engineering*, *42*(4), e13029.

Zied, D. C., Savoie, J.-M., & Pardo-Giménez, A. (2011). Soybean the main nitrogen source in cultivation substrates of edible and medicinal mushrooms *Soybean and nutrition*, *22*, 433–452.

Ziombra, M. (2000). Influence of substrate and pasteurization on yield of Pleurotuscornucopiae (Paul.: Pers.) roll. *Journal of Vegetable Crop Production*, *6*(2), 69–73.

PART IV

Specific applications of wild mushrooms

CHAPTER **14**

Extracellular enzymes of wild mushrooms

Kanika Dulta[1], Arti Thakur[1], Somvir Singh[1], Parveen Chauhan[1], Kumari Manorma[2], Vinod Kumar[3], and P. K. Chauhan[1]

[1]Faculty of Applied Sciences and Biotechnology, Shoolini University, Solan, Himachal Pradesh, India
[2]Dr YSP & UHF, Nauni, Solan, Himachal Pradesh
[3]Peoples' Friendship University of Russia (RUDN University), Moscow, Russian Federation

CONTENTS

DOI: 10.1201/9781003152583-18

14.1 INTRODUCTION

Fungi is an essential part of diverse ecosystems in the forests; it is concerned with processes like nutrient cycling and breakdown of organic substances (Herrera and Ulloa 1990). The white-rot fungi can degrade all components of plants, including lignin, cellulose, and hemicellulose (Martínez et al., 2005). The word mushroom is widely used throughout the world to express the different species of fungi, which belong to the order of Basidomyecets or Ascomycetes. Mushrooms can occur all over the place in soils, which are rich in organic substance and humus, moist wood, and waste of animals following heavy rainfall or an unexpected change in temperature. Mushrooms are well recognized for their dietary and culinary values and are well-known as a necessary element of the human physical condition and nourishment as a substitute to plant as well as animal-derived products (Rathee et al., 2012; Oboh and Shodehinde, 2009). Traditionally, mushrooms have been widely used throughout the world as a basis of foodstuff and medicines (Wasser and Weis, 1999). Globally, it is estimated that 1069 species of mushrooms are reported as being utilized for purposes of food (Boa, 2004). For example, on the African continent, most tribal or rural communities consumed wild mushrooms as foodstuff; nearly 300 species have been reported that they are used by human beings to a small extent (Rammeloo and Walleyn, 1993). Miles and Chang, 2004 and Cowan 2001 and other researchers reported 27,000 fungal species in India. Wild mushrooms are considered as one of the most essential forest stuff and eaten by approximately 3,000 species globally (Garibay-Orijel et al., 2009). A mushroom is a saprophytic fungus that grows on a moist part of the wood trunk of trees, decomposing organic substances in moist soil. During the mycelium growth of mushrooms and the progress to mature fruiting body, biochemical changes take place. And the secretion of extracellular enzymes corrupts the insoluble molecules into soluble material, i.e. substrates which are used subsequently by intracellular enzymes within the mushroom. Therefore, enzymes take an essential role in the growth of mushrooms which affect the nutrients of food, taste, and the life span of fungi (Oei, 1991). Some proteins among fascinating natural behavior are expanded through the mushrooms and these proteins comprise potentially appropriate actions. Mushrooms are a wealth of numerous enzymes with industrial significance and purpose. Over 500 products are prepared by industrial enzymes; in 2009 the industrial advertising of these biomolecules reached $5100 billion, which was categorized into the subsequent area of applications: food 45% (out of which starch processing represents 11%), detergents 34%, textiles 11%, leather 3%, pulp, and paper 1.2% (Boa 2004; Sanchez and Demain, 2011). Such enzymes include cellulases, laccases, lipases, amylases, and proteases. Cellulose is the mainly verdant biomass on Earth. It is often commonly degraded by an enzyme called cellulase (Immanuel et al., 2006). The extra and multi-enzyme complexes provide multiple functions and occupy a broad range of ecological and industrial habitats (Sadhu and Maiti, 2013). Cellulases are used in various industries such as the textile industry, for softening of cotton, and in laundry detergents regarding the care of colors (Ramanathan et al., 2010). Laccases are copper-containing enzymes that degrade lignin and catalyze the phenolic compound oxidation and pungent amines through molecular oxygen utilized as electron acceptors (Baldrian, 2006). Laccases obtained from mushrooms can be used in the bleaching of paper-pulp (Annunziatini et al., 2005), in textile dyes and synthetic dyes decolorization (Nagai et al., 2002), for bioremediation (Jaouani et al., 2005), and in chemical synthesis (Karamyshev et al., 2003). An excellent example of an enzyme with a lot of space for expansion is lipase. This enzyme has the unusual ability to hydrolyze the ester bonds at the substrate-water interface for water-insoluble substrates. This could alter the reaction in both aqueous and non-aqueous mediums (Laachari et al., 2015; Lee et al., 2015). Due to the catalytic activity of lipase over a variety of chemical reactions (Chowdary et al., 2000; Rao and Divakar, 2001), it serves an extensive application in chemical, food industries, detergents, and for pharmaceuticals, etc., (Ananthi et al., 2014; Thakur, Tewari and Sharma, 2014). Amylases are a combination of starch degrading enzymes with significance in the biotechnology industries. Amylases and cellulases exhibit various utilities in foodstuff, textile, detergents, drinks, production of paper, animal feed, and various processes related to fermentation (Pandey et al., 2000). Similarly, proteases have lots of utility in the processing of leather, detergents, silver recovery, medicinal purposes, feeds, food processing, chemical industry, as well as treatment of waste (Ma et al., 2007). Proteases also take part in physiological processes like turnover of protein, sporulation and conidial release, germination, enzyme alteration, regulation and nourishment of gene expression, etc. (Rao et al., 1998). Because of their widespread applicability, these enzymes have gained significant significance in a variety of fields (Pliego et al., 2015). The United States and Europe have the largest share of the enzyme market in the manufacturing sector; however, Asia Pacific has seen the highest growth, with an annual growth rate of more than 8%. In comparison to Japan, the United States, and Canada, where the economy is fully grown, the industrial enzyme is in higher demand in Africa, Asia-Pacific, and the Middle East (Norus, 2006). Because of the high demand for enzymes, 31% are used for food enzymes, 6% for feed enzymes, and the rest are used for technical purposes. As a result, the focus of this chapter is on an extracellular enzyme found in wild mushrooms and their various applications.

14.2 MUSHROOMS

Crous et al. (2006), studied the fungal biodiversity throughout the world and 1.5 million species have been reported (Hawksworth, 2004), and 50% of them have been characterized (Manoharachary et al., 2005). According to various researchers, it is estimated that 70,000 species of fungi are present all over the world, out of which 2,000 species of mushrooms under 31 genera are primarily consumed. Approximately, 10% of species are toxic which is comparatively small and considered as dangerous (Teferi et al., 2013). The edibility capability of mushrooms, toxic nature, and associations with mycorrhizal and parasites with the trees make these fungi efficiently important and for interest to study. Various scientists have studied wild mushrooms and documented more than 2,000 species of edible mushrooms worldwide, out of which 283 edible species are reported from India; among these some are cultivated (Adhikari, 2000; Purkayastha and Chandra, 1985). For thousands of years, non-cultivated species of edible mushrooms were eaten by the inhabitants. These are a delicious source of food all over the world. Wild mushroom have a rich amount of protein; vitamins like B, C, D; and minerals; as a result, the increase in the consumption of wild mushrooms is increasing still in the developed countries. Compared to vegetables and fruit, the nutritional value of non-cultivated mushrooms is higher nearly two times. Along with its use as foodstuff, satisfactory evidence recommends that lots of species have substances that may prevent or reduce the chances of contracting cancer, heart diseases, diabetes, and viral infections (Oei, 1991). Various types of bioactive compounds obtained from the mushrooms are linked to the cell wall of the mycelial that assist in increasing resistance against carcinogens (Ramesh and Pathar, 2010). Nutraceuticals are food and part of food that provides nutrition, as well as health benefits, which leads to the mushroom, and can act as nutraceuticals. A mushroom is known to have a potential antioxidant activity (Jayakumar et al., 2006; Nitha et al., 2010). The non-cultivated mushrooms are accepted foodstuff and therapeutic sources in several areas of India (Purkayastha and Chandra, 1985). Ethno-mycological studies were done in different areas of the world including India. Most usually, edible species that are consumed are *Cordyceps, Amanita, Astraeus, Termitomyces, Cantharellus, Pleurotus, Morchella, Russula,* and *Lactarius.* Local inhabitants consumed wild mushrooms as food and it is also a source of income by selling them in markets.

14.2.1 Significance of mushrooms

Consumption of mushrooms plays an essential role in human body health. A wild edible mushroom eaten in our daily diet provides us strong resistance (immunity) against various infections. Mushrooms are naturally without gluten, similarly in fruit and green vegetables and they formulate a tasty and nourishing diet. Beta-glucans occurred in a lot of species of mushrooms and have the immunity-stimulating property, which provides resistance against allergies and takes part in physiological processes associated with the metabolism of sugars and fats in the individual body. The edible species of mushroom-like shiitake, oyster, and split gill mushrooms contain beta-glucans, which are considered to be generally efficient (Duyff, 2012). Mushrooms are a superior component of vitamin B, such as riboflavin, niacin, and pantothenic acid, which assist to supply energy through the breakdown of carbohydrates, proteins, and fats (U.S. Department of Agriculture, Agricultural Research Service, USDA Nutrient Data Laboratory, 2009). Mushrooms have lower calories, are cholesterol-free, fat-free, and have a very low content of sodium; they provide several nutrients that are present in animal food (U.S. Food and Drug Administration, 1994). Shiitake is a well-liked mushroom consumed as a vegetable and in many dishes all over the world. It contains a variety of health-promoting agents, for example, lentinan. It has antitumor properties that lead to the treatment of cancers (Sasidharan et al., 2010). Shiitake has also possesses the antiviral capacity (including against hepatitis, HIV, and the common cold), antibacterial, and antifungal activity; helps in the blood-sugar stabilization; reduces aggregation of platelets; and overcomes atherosclerosis (Yamada et al., 2002). Mushrooms are a rich source of natural antibiotics (Chawla et al., 2019). The presence of glucans in the cell wall of mushrooms is recognized for their immunomodulatory activities and the secondary metabolites found to be efficient against microorganisms such as bacteria and viruses (Kupka, Anke, Oberwinkler, Schramm and Steglich, 1979; Suzuki et al., 1990). Exudates obtained from the mycelia of mushrooms are active against protozoans like the malaria parasite (*Plasmodium falciparum*) (Isaka et al., 2001). Fungi and humans have common microbial antagonists, for instance *Escherichia coli*, *Pseudomonas aeruginosa*, and *Staphylococcus aureus*, so humans can benefit from natural defense strategies of fungi. It is well known that polypores provide a protective immunological shield against a variety of infectious diseases.

14.3 EXTRACELLULAR ENZYMES OF WILD MUSHROOMS

Enzymes are mainly vital products obtained for human needs through microbial sources. Various industrial processes in the field of food and environmental biotechnology use enzymes at different stages. Present development in biotechnology is providing a novel utility for enzymes (Pandey et al., 1999). In the present techno-economic era, procurance of energy is the main problem that is faced by humanity (Figure 14.1).

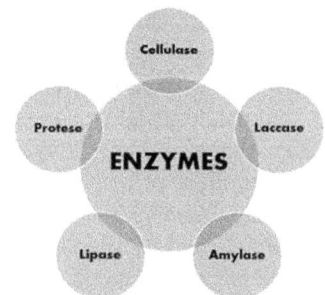

Figure 14.1   Extracellular enzymes of wild mushrooms.

14.3.1 Cellulases

Cellulases are the enzymes that hydrolyze β-1,4 linkages in cellulose chains. Cellulase has catalytic modules that are categorized into various families depending on their sequences of amino acid and crystal structures (Henrissat, 1991). It contains non-catalytic carbohydrate-binding modules (CBMs) and other functions. Both known or unknown modules may be situated at the N- or C-terminus of a catalytic site. It speeds up the hydrolysis of cellulose and oligosaccharide derivatives, which is considered as a probable instrument for saccharification in industries if there is cellulosic biomass (Berry and Paterson, 1990), and a cost-effective method which that is considered to be important for the successful use of cellulosic resources (Nwodo-Chinedu et al., 2007). Cellulases have been used and studied for the 20th century and are the most commercially important of all the enzyme families. The enzyme activities were increased about 30–80% when produced by SSF in comparison with conventional SmF enzyme production. The cost of cellulase production is brought downwards by several approaches, which include the utilization of low-cost lignocellulosic substrates and by cost-efficient fermentation methods such as solid-state fermentation (Sukumaran et al., 2009). The enzyme production is good by most of the fungi like *Aspergillus* and *Trichoderma* species, which break down and metabolize cellulose. The microorganisms must secrete the cellulases that are free or bonded to a surface. Cellulases are also used for a variety of industrial purposes such as in the pulp, textile, paper, and food industry, as well as a preservative in detergents and increase the digestibility of animal feed. Now, cellulases account for a significant split of the world's trade enzyme market. The rising concerns about the reduction of crude oil and the emissions of greenhouse gases have stimulated the construction of bioethanol from lignocellulose, especially by enzymatic hydrolysis of lignocelluloses materials-sugar platform (Bayer et al., 2007). However, the expenses of cellulase for hydrolysis of pretreated lignocellulosic resources need to be reduced, and their catalytic competence should be extra enhanced to formulate the procedure economically possible (Sheehan & Himmel, 1999).

14.3.1.1 Classification of cellulases

In natural conditions, hydrolysis of cellulose is mediated by a mixture of three types of cellulases: (1) endoglucanases (EC 3.2.1.4); (2) exoglucanases, including cellobiohydrolases (CBHs) (EC 3.2.1.91); and (3) β -glucosidase (BG) (EC 3.2.1.21).

14.3.1.1.1 Endoglucanase

Endoglucanase (β-1,4-D-glucanase, endo-β-1,4-D-glucan-4- glucano-hydrolase) – also known as CMCase – hydrolyzes carboxymethyl cellulose (CMC) or swollen cellulose randomly. Accordingly, the duration of the polymer decreases, resulting in the augment of reducing sugar concentration. Endoglucanase also acts on cellodextrins. It is the middle product of hydrolysis of cellulose and converts it into glucose and cellobiose (Begum and Absar, 2009).

14.3.1.1.2 Exoglucanase

Exoglucanase (β-1,4-D glucanase, cellobiohydrolase) degrades cellulose by breaking down the cellobiose units from the non-reducing part of the chain. It is also active against partly degraded amorphous substrates and cellodextrins but does not hydrolyze soluble derivatives of cellulose like carboxymethyl cellulose and hydroxyethyl cellulose. Several cellulase systems also contain glucohydrolase (Exo-1,4-D-glucan-4-glucohydrolase) as a small part (Joshi and Pandey, 1999).

14.3.1.1.3 β- glucosidase

β-glucosidase completes the course of hydrolysis of cellulose by breaking down cellobiose and removing glucose from the non-reducing part (i.e. with a free hydroxyl group at C-4) of oligosaccharides. The enzyme also hydrolyzes aryl β-glucosides and alkyl (Kubicek et al., 1993).

14.3.1.2 Source of cellulases

Microorganisms (cellulolytic) are isolated from environments such as manure, soil compost, municipal solid waste, distillery sludge, wood material, sewerage sludge, hot springs, marine sediments, alkaline environments, and geysers. Among the bacterial isolates explored, *Bacillus amyloliquefaciens, Bacillus megaterium, Bacillus subtilis, Acidothermus cellulyticus, Pseudomonas cellulose, Ruminococcus albus, Micrococcus* species, etc. are found to produce cellulase enzyme (Moorthy et al., 2018). Most industrial planning of cellulase is of fungal source taking place on *Trichoderma reesei.* Cellulase enzymes are also produced from several studied fungi such as *Candida, Streptomyces, Aspergillus, Actinomadura, Trichoderma, Neurospora, Saccharomonospora, Penicillum, Piromyces,* and *Rhizopus* strains (Oksanen et al., 2000).

14.3.1.3 Mechanism of cellulases

Cellulases hydrolyze β-1, 4-D-glucan linkages in cellulose and create glucose, cellobiose, and cello- oligosaccharides. Production of cellulases is possible by several microorganisms and comprises numerous different enzyme classifications. Three major types of cellulase enzymes i.e. cellobiohydrolase (CBH), endo-β-1, 4- glucanase (EG), and β-glucosidase (BGL)] are playing important roles in the hydrolysis of cellulose. There are many enzymes inside these classifications. For instance, the mainly studied fungus for the production of cellulase is *Trichoderma reesei,* which produces 2 CBH, 8 EG, and 7 BGL (Aro et al., 2005). EGs make incisions in cellulose polymer exposing reducing and non-reducing ends, CBH acts upon these reducing and non-reducing ends to release cello oligosaccharides and cellobiose units, and BGL cleaves cellobiose to release glucose by completing hydrolysis (Figure 14.2). The whole cellulase system comprises BGL, EG, and CBH mechanisms which consequently act synergistically to change crystalline cellulose to glucose (Béguin and Aubert, 1994). The majority of cellulases take a distinctive two-domain structure with a catalytic domain in addition to a cellulose-binding domain (CBD; also known as a carbohydrate-binding module (CBM)) linked through a linker peptide. Core domain or catalytic domain contains catalytic position while CBDs assist in the binding of the enzyme to cellulose. Deprivation of native cellulase requires diverse levels of support among cellulases. Such synergisms occur between endo- and exo-glucanases (Sakka et al., 2000). In the initial type, EC action produces unrestricted ends, on which exo-glucanases act, and in the second one, exo-glucanases collaborate by acting on reducing and non-reducing ends to bring about efficient cellulose degradation. Individual enzymes are not capable to corrupt cellulose totally, while a combination of enzymes increases the effectiveness of saccharification. Supplementation of heterogeneous BGL is held to improve the hydrolytic potential of FLs synergistically, while there are some conflicting reports (Massadeh et al., 2001).

14.3.2 Laccases

Enzymes have gained immense significance in industries; laccases are one of them, which are broadly present in nature. Laccases are the generally studied and oldest enzymatic systems since the end of the 19th century (Williamson, 1994). These enzymes comprise 15 to 30% carbohydrates and, in addition, have a molecular mass of 60 – 90 kDa. Laccase (benzenediol: oxygen oxidoreductase, EC 1.10.3.2) is an element of a broad cluster of enzymes termed *polyphenol oxidases* that comprise

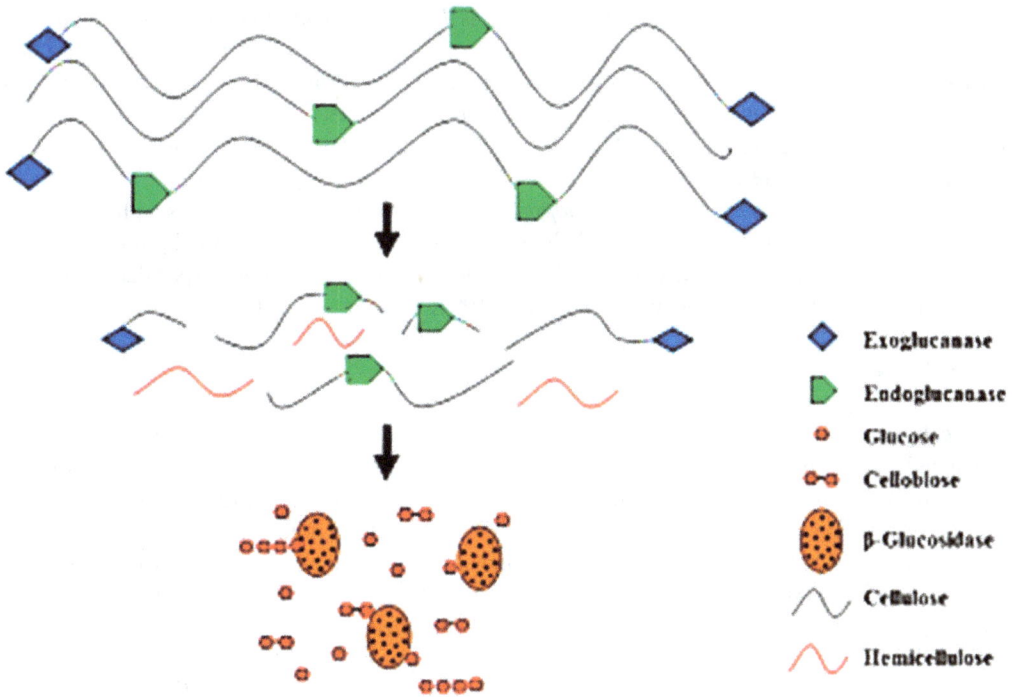

Figure 14.2 Mode of cellulase action.

copper atoms in the catalytic center; in addition, are frequently known as multicopper oxidases. Laccases comprise three categories of copper atoms, one of which is responsible for their featured blue color. The enzymes missing a blue copper atom are known as yellow or white laccases. Typically, laccase mediates catalysis occurs through the reduction of oxygen to water conveyed by the oxidation of the substrate. Laccases are consequently oxidases that oxidize polyphenols, aromatic diamines, and methoxy-substituted phenols, in addition to a variety of other compounds (Baldrian, 2006). These enzymes are polymeric and usually hold each type 1, 2, and 3 copper center/subunit, where the type 2 and type 3 are close together, forming a trinuclear copper cluster. Yoshida first described laccase in 1883, once he extracted it from the exudates of the Japanese lacquer tree, *Rhus vernicifera*. In 1896, laccase was validated to be present in fungi for the primary period by both Bertrand and Laborde (Alcalde, 2007; Beloqui et al., 2006). Since then, laccases have been established in Ascomycetes, Deuteromycetes, and Basidiomycetes; chiefly plentiful in a lot of white-rot fungi that are concerned with lignin metabolism (Leontievsky et al., 1997). Fungal laccases have superior redox potential compared to bacterial or plant laccases (up to +800 mV), and their action seems to be appropriate in natural world judgment with significant applications in biotechnology. Thus, fungal laccases are concerned with the deprivation of lignin or in the elimination of potentially poisonous phenols that arise during the degradation of lignin. Fungal laccases are hypothesized to play a role in the production of dihydroxy naphthalene melanins, darkly pigmented polymers that organisms make against environmental trauma or in fungal morphogenesis by catalyzing the construction of extracellular pigments (Henson et al., 1999).

14.3.2.1 Sources of laccases

Laccases are generally present in higher plants and fungi, but are also found in some bacteria such as *S. lavendulae*, *S. cyaneus*, and *Marinomonas mediterranea*. In fungi, laccases come out more than the higher plants. Basidiomycetes such as *Phanerochaete chrysosporium*, *Theiophora terrestris*, and *Lenzites*, *betulina*, and white-rot fungus, such as *Phlebia radiate*, *Pleurotus ostreatus*, and *Trametes versicolour* also produce laccases. Many *Trichoderma* species like *T. atroviride*, *T. harzianum*, and *T. longibrachiatum* are the sources of laccases (Velázquez-Cedeño et al., 2004; Shekher 2011). Laccase from *Monocillium indicum* was the first enzyme to be considered from Ascomycetes, which shows peroxidase activity. *Pycnoporuscinn abarinus* produces laccase as a ligninolytic enzyme while *Pycnoporuss anguineus* produces laccase as a phenoloxidase (Eggert et al., 1996; Pointing and Vrijmoed, 2000). In plants, laccases take part in lignification while in fungi they have been concerned with delignification, pigment production, sporulation, fruiting body formation, and plant pathogenesis (Thurston, 1994; Yaver et al., 2001).

14.3.2.2 *Mechanisms of laccases*

The laccase catalysis was found due to the reduction of one oxygen molecule to water accompanied by the oxidation of one electron with a wide variety of perfumed compounds, which include polyphenol (Bourbonnais and Paice, 1990) and methoxy-substituted monophenols, in addition to aromatic amines (Bourbonnais et al., 1995). Laccases hold four copper atoms termed Cu T1 (reducing substrate binds this site) and trinuclear copper bunch T2/T3 (electrons move from type 1 to type 2 and type 3 trinuclear clusters/ reduction of oxygen to water at the trinuclear cluster) (Gianfreda et al., 1999). These four copper ions are classified into three categories: Type 1 (T1), Type 2 (T2), and Type 3 (T3). These groups are distinguished by the use of UV/visible and electronic paramagnetic resonance (EPR) spectroscopy. When oxidized, Type 1 Cu provides a blue color to the protein at an absorbance of 610 nm, which is EPR observable. Type 2 Cu does not provide color but is EPR noticeable, and Type 3 Cu contains a pair of atoms in a binuclear conformation that provide a weak absorbance at the near UV area but is not detected by EPR signals (Thurston, 1994). The Type 2 copper and Type 3 copper form a trinuclear center, which are concerned with the enzyme catalytic mechanism. The O_2 molecule linked to the trinuclear bunch for asymmetric activation, and it is postulated that the O_2 binding section appears to restrict the access of oxidizing agents. During a stable state, laccase catalysis revealed that O_2 reduction occurred (Gianfreda et al., 1999). Laccase operates as a battery and stores electrons from a person's oxidation reaction to decrease molecular oxygen. Hence, the oxidation of four reducing substrate molecules is essential for the whole reduction of molecular O_2 to H_2O. Production of free radicals by the laccase oxidizes substrate. The lignin deprivation proceeded by phenoxy radical leads to oxidation at α-carbon or breakage of the bond between α-carbon and β-carbon. This oxidation results in an oxygen-centered free radical, which can be changed into a second enzyme-catalyzed reaction to a quinone. The free radicals and quinone can cause polymerization (Thurston, 1994) (Figure 14.3).

The association of the sites of copper in laccase is explained by the spectroscopic studies (Quintanar et al., 2005), which exposed that Type 2 copper coordinates two His-N and one oxygen atom as -OH, while each copper of Type 3 coordinate 3 His residues. Further, both T2 and T3 copper sites have open synchronization positions near the center of the trinuclear cluster with the negative protein segment (Zoppellaro, 2000). The laccase-mediated catalysis can be comprehensive to nonphenolic substrates by the inclusion of mediators. Organic compounds which that have a low molecular weight are oxidized by a laccase known as a mediator. Oxidation of nonphenolic compounds by the active cation radicals is not possible by the laccase only; 1-hydroxy benzotriazole (HOBT), N-hydroxyphthalimide (NHPI), 2,2-azinobis-(3-ethylbenzthiazoline-6-sulfonate) (ABTS), and 3-hydroxyanthranilic acids are the utmost frequently used synthetic mediators (Gochev and Krastanoy, 2007). Oxygen uptake by laccase remains rapid than the HOBT in the existence of ABTS.

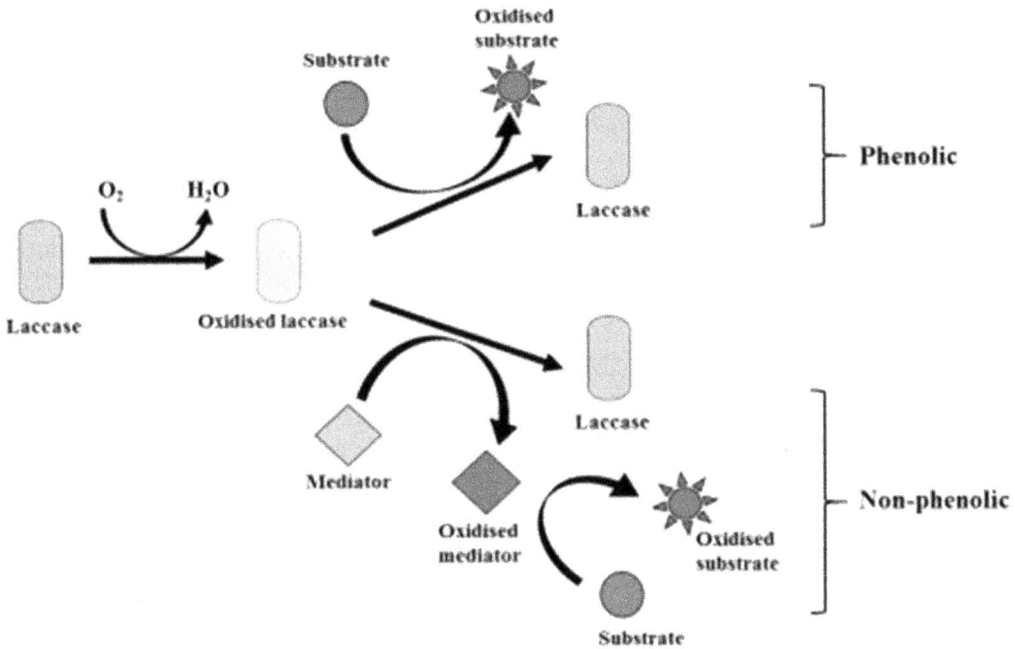

Figure 14.3 Mode of action of laccase on both substrates of phenolic and nonphenolic.

14.3.3 Amylases

Amylases are a class of enzymes that are capable of digesting glycosidic linkages present in starch. It can be obtained from a variety of sources. They exist in all living organisms; nevertheless, the enzymes differ in activity and specificity, and in requirements from species to species and even from tissue to tissue in a similar organism. The amylase-catalyzed hydrolysis of starch is among the most significant industrially applied enzyme reaction (Gupta et al., 2015). The vulnerability of starch to amylase attack based on the properties of the specific starch, such as degree of gelatinization, and the characteristics of the specific amylase. A lot of studies deal with the amylolysis of native starch and focus on the effects of substrate characteristics, like granule size, shape, structure, and amylose content. In addition, amylases showed extensive, diverse performance on different kinds of solubilized starches (Mukerjea et al., 2006). Amylases are often defined by their starch corrupting quality and fewer by their action on the individual starch polymers. It is frequently conventional that amylases randomly hydrolyze amylose and amylopectin. Sprouted seeds of pulses such as *Cicer arietinium, Cecineri,* and *Pisum sativum* were established to have a rich source of amylase (Rani and Chauhan, 2014). Hence, the most extensively used thermostable enzymes are the amylases, which have a significant role in the industry of starch (Sarikaya et al., 2000).

14.3.3.1 Classification of amylases

Amylases are glycoside hydrolases (GHs), which act upon the bonds among the glucose units of the starch polymers. GHs are considered by the number EC 3.2.1.x, where the x expresses the substrate specificity or in some cases the molecular method or type of linkage (e.g. EC 3.2.1.1, α-amylase; EC 3.2.1.2, β-amylase; EC 3.2.1.3, glucoamylase). Complications occur with this substrate specificity-based classification when an enzyme hydrolyses many substrates.

14.3.3.1.1 α-amylase

(EC 3.2.1.1) (1,4-α-D-glucan glucanohydrolase; glycogenase) amylase is calcium me-talloenzyme and acts at random sites of the starch chainthat lead to the collapse of long-chain carbohydrates into maltotriose and maltose from amylose, or maltose and glucose and limit dextrin from amylopectin. α-amylase tends to be faster acting than β-amylase and, in animals, it is a major digestive enzyme with an optimum pH is 6.7 to 7.0. In human physiology, both the salivary and pancreatic amylases are α-amylases and also occur in plants (barley), pulses, fungi (ascomycetes and basidiomycetes), and bacteria (*Bacillus*) (Rani, 2012b; Rani, 2012d; Rani, 2012e).

14.3.3.1.2 β-amylase

(EC 3.2.1.2) (alternate names: 1,4-α-D-glucanmaltohydrolase; glycogenase; saccharogen amylase) synthesis of β-amylase by fungi, bacteria, pulses, seeds, and plants catalyzes the hydrolysis of the second α-1,4glycosidic bond and breakage of two glucose units (maltose) at a time. During the ripening of fruit, β-amylase breaks down starch into sugar that leads to the sweetening of ripe fruit. β-amylase occurs before germination, whereas α-amylase and proteases occurred occur when the germination starts. Animal tissues do not contain β-amylase, though it may be present in microorganisms present within the digestive tract (Rani, 2012b, Rani, 2012c and Rani, 2012d).

14.3.3.1.3 γ-amylase

(EC 3.2.1.3) (alternative names: Glucan 1,4-α-glucosidase; amyloglucosidase; Exo-1,4-α-glucosidase; glucoamylase; lysosomal α-glucosidase; 1,4-α-D-glucan glucohydrolase), γ-amylase cleaves last α (1-4) glycosidic linkage at the non-reducing end of amylose and amylopectin, yielding glucose along with α (1-6) glycosidic linkages. Unlike the other kind of amylase, γ-amylase is most capable in an acidic environment and has an optimum pH of 3 (Rani, 2012c).

14.3.3.2 Sources of amylases

Several amylase-producing microorganisms have been isolated and characterized for many decades. Bacteria and fungi exude amylases from their cells to perform extracellular digestion. *Aspergillus niger, Aspergillus oryzae, Thermomyces lanuginosus,* and *Penicillium expansum* are among the mold species that produce high amounts of amylase (Arnesen et al., 1998), in ad-dition to various species of the genus *Mucor* (Gams and Anderson, 1980). Amylolytic yeasts diverge strongly concerning amylase discharge and the extent of starch hydrolysis. Strains of *Filobasidiuim capsuligenum* can hydrolyze the starch. In context to bacteria, *Bacillus* spp. and the linked genera make a large variety of extracellular enzymes, of which amylases are of exacting implication to the trade e.g. *B. cereus, B. circulans, B. subtilis, B. licheniformis,* and *Clostridium thermosulfurogenes* (El-Banna et al., 2008; El-Banna et al., 2007). Bacteria belong chiefly to the genus *Bacillus* which has been widely used for the commercial manufacture of thermostable α-amylase (Tonkova, 2006). Many alkaline amylases have been found in cultures of *Bacillus* sp. These alkaline amylases are all of the saccharifying types, excepting the enzymes from *Bacillus* sp. strain 707 and *B. licheniformis* TCRDC-B13. Thermostable β-amylases have been cut off from *Bacillus* species and *Lactobacillus plantarum* strain A6 has been selected for their capability to manufacture large amounts of extracellular α-amylase (Giraud et al., 1991; El-Fallal et al., 2012).

14.3.3.3 Mode of action

It is believed that α-amylases are endo-acting amylases that hydrolyze α-(1-4) glycosidic bonds of the starch polymers inside. Several models for the amylase action pattern have been proposed, such as the unsystematic action and the several attack actions. The random act has also been referred to as a single attack or multi-chain attack action (Azhari and Lotan, 1991). In the previous, the polymer molecule is hydrolyzed completely before dissociation of the enzyme-substrate complex, whereas in the latter only one bond is hydrolyzed per efficient encounter.

The attack action is an intermediate between the single-chain and the multi-chain action (Bijttebier et al., 2007a), where the enzyme cleaves numerous glycosidic bonds sequentially after the first (random) hydrolytic attack previous to degrading from the substrate. In short, it can be seen that several attack actions are generally the accepted idea to explain the difference in the action model of amylases (Kramhøft et al., 2005; Svensson et al., 2002). However, mainly the endo-amylases take a low to very low level of many attack actions. While only a few reports deal with the power of pH and temperature on the action pattern of amylases, this influence was complete. Bijttebier et al. (2007b) showed that the level of many attacks of several endo-amylases were amplified with temperature to a degree depending on the amylase itself (Figure 14.4).

14.3.4 Lipases

The requirement of an enzyme in global demand is by 12 main producers and 400 small suppliers. Approximately 60% of the world's industrial enzymes are produced in Europe. Hydrolysis occurs in at least 75% of all industrial enzymes (including lipases). Lipases (EC 3.1.1.3) are called triacylglycerol acyl hydrolase, which acts on carboxylic ester bonds is the division of the hydrolases family (Kapoor and Gupta, 2012). They do not need any cofactor and belong to the group of serine hydrolases (Beisson et al., 2000). Using lipases, triglycerides are hydrolyzed into diglycerides, monoglycerides, fatty acids, and glycerol (Figure 14.5). The esterases hydrolyzed carboxylic ester bonds by adding up to the lipases (Pascoal et al., 2018; Almeida et al., 2019). Under natural conditions, lipases catalyze the hydrolysis of ester bonds at the interface i.e. between an unknown phase of the substrate and an aqueous phase, where the enzymes remain liquefied. Mainly lipases have an optimal variety of action and constancy for pH

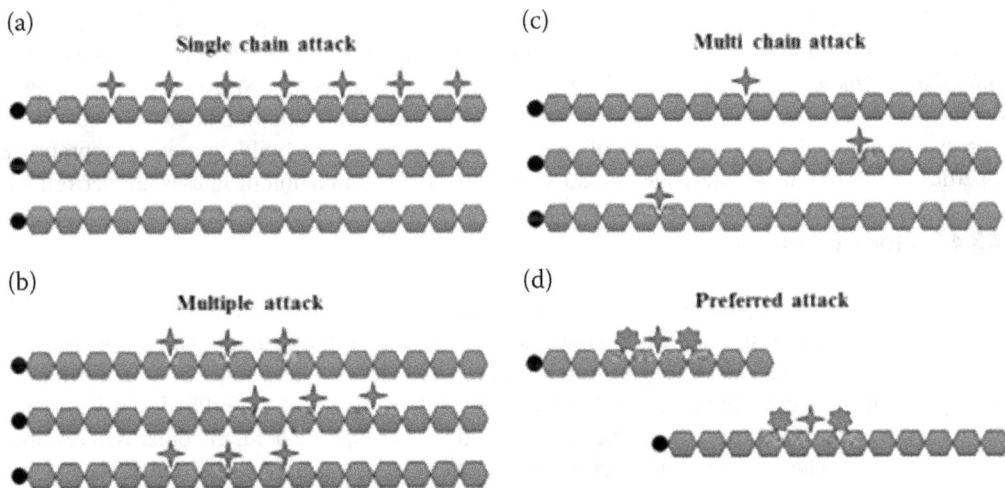

Figure 14.4 Mode of action of amylase action patterns on amylose and amylose fragments.

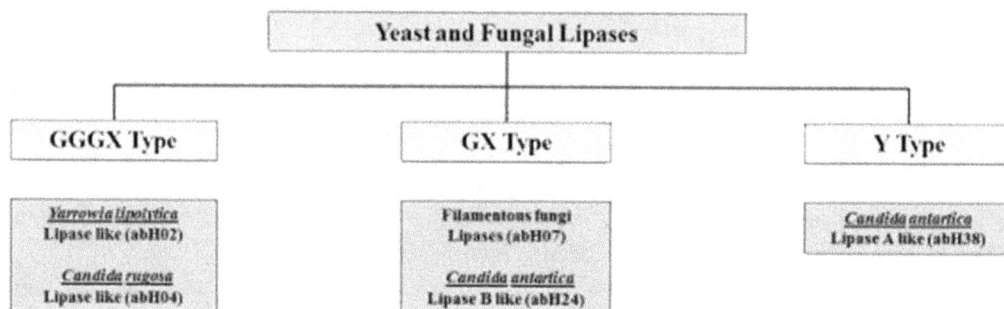

Figure 14.5 Classification of lipases based on lipase engineering database.

values between 6.0–8.0 and temperature between 30–40°C. However, *Pseudomonas aeruginosa*, *Candida anatarctica*, and *Burkholderia glumae* exhibit a lid but did not demonstrate interfacial activation (Stergiou et al., 2013). Esterification, interesterification, transesterification, acidolysis, alcoholysis, and aminolysis translation reactions are catalyzed by lipases (Borchert et al., 2017; Jiao et al., 2018). The occurrence of a lid and the interfacial activation are not the appropriate conditions to classify a true lipase (Lei et al., 2016).

14.3.4.1 Classification of lipases

In the Lipase Engineering Database (LED) a new categorization was recently reported (http://www.led.uni-stuttgart.de), which today includes not only the bacteria but also yeast, fungi, and mammalian lipases. This distributes the lipases into three classes based on the oxyanion hole: GX, GGGX, and Y (Fischer & Pleiss, 2003; Gupta et al., 2015). Based on this categorization and of the amino-acid series similarities, yeasts and fungal lipases have been divided into five diverse sub-classes, two in the GX class and GGGX class, and one in the Y class (Figure 14.5).

14.3.4.2 Sources for lipases

These enzymes are usually found in organisms like plants, microbes, and animals. Several important lipases generating bacterial groups are *Burkholderia*, *Bacillus* and *Pseudomonas*, *Staphylococcus,* and *Chromobacterium*. The main genera of fungi are *Rhizopus, Aspergillus, Penicillium, Mucor, Ashbya, Geotrichum, Beauveria, Humicola, Rhizomucor, Fusarium, Acremonium, Alternaria, Eurotrium,* and *Ophiostoma* (Chandra et al., 2020). Due to the requirement of precise features, there is a scarcity of lipases. In recent days, to build lipases of appropriate properties and to use on a trade scale, biotransformation and combination of lipases are performed.

14.3.4.3 Mode of action

The method of the lipase to catalyze ester hydrolysis is like carboxylesterases and serine proteases, and involve a first nucleophilic attack of the serine on the carbonyl carbon of the ester bond, yielding a covalent acyl-enzyme intermediate and releasing alcohol, i.e. after forming a hydroxyl group in a triacylglycerol molecule, a diacylglycerol would be released (Adlercreutz, 2013) (Figure 14.6). The process is stabilized by histidine and aspartic acid, the other two residues of the lively site. The acyl-enzyme intermediate is then hydrolyzed by water and a second nucleophilic harass occurs, thereby producing a carboxylic acid. Many different compounds can serve as acyldonors and many nucleophilic compounds in addition to water can perform the same function and sever the acyl-enzyme intermediate (Jung et al., 2013). Because of their strong

(a)

(b)

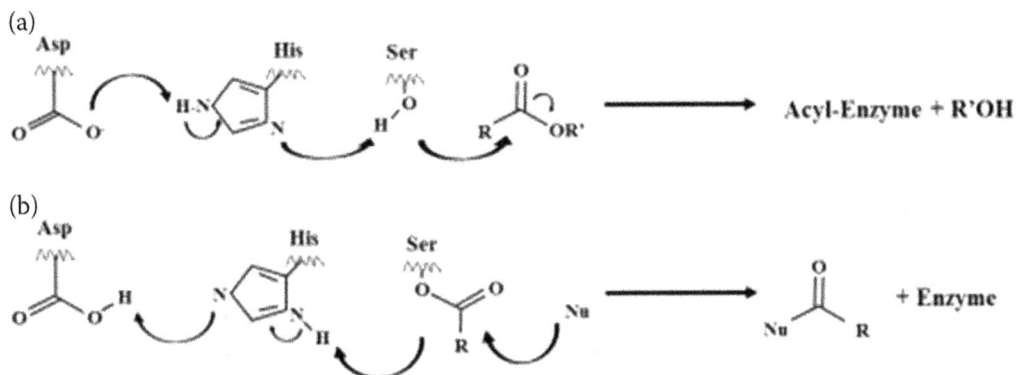

Figure 14.6 Mode of action of lipases.

substrate specificity, lipases can conduct a variety of reactions other than acylglycerol hydrolyses, such as trans-esterification, esterification, interesterification, and acidolysis. Lipases catalytic possibilities, on the other hand, have expanded to a variety of other synthetic or non-conventional substrates and forms of reactions, ranging from using amines as nucleophilic compounds to conducting aldol trappings (Branneby et al., 2004; Vongvilai et al., 2011).

14.3.5 Proteases

The history of proteolytic enzymes or peptidases can be traced back at least to the late 18th century. However, in recent years, the work in this field has expanded significantly, fueled by various functional applications in biotechnology, and the recognition that they are among the main beneficial targets. A protease is an enzyme that hydrolyzes peptide bonds in proteins. Classification of proteases is hierarchical, based on the principles of catalytic form, clan, family, and peptidase activity. This breaks the long-chained molecules of proteins into shorter parts. Analytical advances have shown that proteases perform extremely complex and selective protein modifications such as activation of zymogenic types of enzymes through minimal proteolysis, blood clotting, fibrin clot lysis, processing, and transfer of secretory proteins through membranes. The current projected value of the global commercial enzyme revenues is $1 billion. Proteases are one of the three main classes of industrial enzymes, which account for about 60% of overall global enzymes sales and 75% of hydrolytic enzymes (Godfrey and West, 1996).

14.3.5.1 Classification of proteases

Proteases are divided into two main groups based on their mode of action: exopeptidases and endopeptidases. Exopeptidases cleave peptide bonds close to the amino or carboxyl termini of the substrate, while endopeptidases cleave peptide bonds far from the substrate termini.

14.3.5.1.1 Exopeptidases

Exopeptidases are enzymes that cleave peptide bonds at the end of polypeptide chains. Aminopeptidases are exopeptidases that cleave the bond from the N-terminus to create a single amino acid residue or a di- or tripeptide. The carboxypeptidases are exopeptidases that operate on the C-terminus and contain either a single amino acid residue or dipeptides.

14.3.5.1.2 Endopeptidases

Endopeptidases are enzymes that cleave the bonds between the termini. Proteolytic enzymes are divided into four groups depending upon the mode of action, which is determined by the catalytic amino acid residue and that involved: serine, aspartate, cysteine, and metalloproteases (Madala et al., 2010). Serine proteases, which are active at neutral and alkaline pH, may be exo- or endo-peptidase depending on their mode of action, e.g. subtilisin, an essential component of detergents, is classified as a serine protease (Rao et al., 1998). Aspartic proteases are exo-peptidases that are active at acidic pH, e.g. chymosin is used in the production of cheese. Cysteine proteases are typically active at neutral pH and often involve the presence of a reducing agent. Papain is a well-known cysteine protease. Metalloproteases are distinguished by their need for a divalent cation activity; e.g. collagenase is a metalloprotease. They can be active in alkaline or acidic pH conditions. MEROPS is a database in which proteolytic enzymes are divided into groups based on amino acid sequence similarity, and homologous families are grouped into clans.

14.3.5.2 Source of proteases

Proteolytic enzymes are formed by all living things, including plants, animals, and microorganisms. Proteolytic enzymes derived from these sources have been used at some stage in human evolution. However, to satisfy the massive demand of factories, microorganisms have become the primary source of proteolytic enzymes. Microorganisms, which grow quickly and need less space than plant or animal sources, maybe cultured to provide a high yield of proteolytic enzymes. Bacterial proteolytic enzymes are generally active in neutral or alkaline pH. *Bacillus* and *Streptomyces* are the most widely seen at the manufacturing level. Fungi, on the other hand, are capable of producing acidic, neutral, and alkaline protease enzymes. A diverse variety of fungi, including the genera *Aspergillus, Mucor*, and *Rhizopus*, as well as bacteria, including the genera *Clostridium, Bacillus*, and *Pseudomonas* are potential sources of proteases (Kuberan et al., 2010; Ghasemi et al., 2011).

14.3.5.3 Mode of action

Proteases have specific subsites on one or both sides of the catalytic site, each of which can accept the side chain of a single amino acid residue from the substrate. These sites are numbered from the catalytic site S1 to Sn toward the structure N terminus and S19 to Sn9 toward the structure C terminus. The residues that accommodate the substrate are labeled Pl through Pn and P19 through Pn9, respectively (Figure 14.7).

14.3.5.3.1 Serine proteases

Serine proteases usually use a two-step hydrolysis reaction in which a covalently bound enzyme-peptide intermediate is formed by the loss of the amino acid or peptide fragment. This acylation step is followed by a deacylation process which involves a nucleophilic attack on the intermediate by water, resulting in peptide bond hydrolysis. Serine endopeptidases are divided into three types depending on their primary substrate preference: (i) trypsin-like, which cleave after positively charged residues; (ii) chymotrypsin-like, which cleave after large hydrophobic residues; and (iii) elastase-like, which cleave after small hydrophobic residues. The Pl residue determines the site of peptide bond cleavage. Only the Pl residues influence the primary specificity; the residues in other positions influence the rate of cleavage. The subsite associations are restricted to various amino acids surrounding the Pl residue to a distinct group of sequences on the enzyme (Masaki et al., 1978; Yoshimoto, 1980). From *Achromobacter* spp. some serine peptidases are

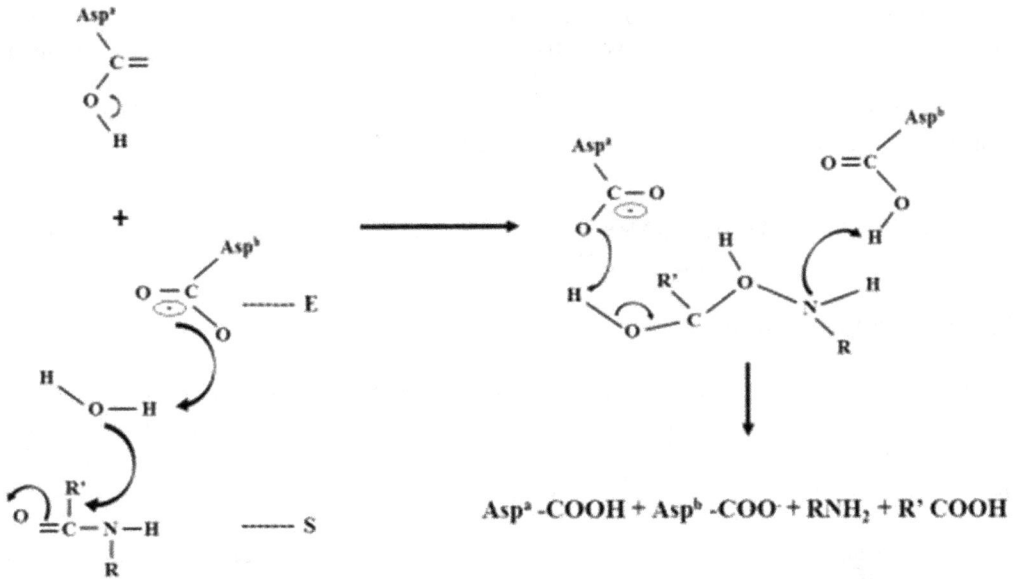

Figure 14.7 Mode of action of proteases.

lysine-specific enzymes, while from *Clostridium* spp. they are arginine specific (clostripain) and from *Flavobacterium* spp. they are post-proline specific. Endopeptidases are specific to aspartic acid and glutamic acid residues and have also been discovered in *B. licheniformis* and *S. aureus*. However, peptidase A from *E. coli* and the repressor LexA have distinct mechanisms of action that do not include the classic Ser-His-Asp triad. Few glycine residues are conserved near the catalytic serine residue, but their precise positions vary. Almost all chymotrypsin-like enzymes are found in mammals, except trypsin-like enzymes found in actinomycetes and *Saccharopolyspora* spp., as well as the fungus *Fusarium oxysporum*. Few subtilisin family serine proteases have a catalytic triad consisting of the same residues as the chymotrypsin family; however, the residues appear in a different order (Asp-His-Ser). Any subtilisin family members from the yeasts *Tritirachium* and *Metarhizium* spp. require thiol for their activity. Cys173 near the active site of histidine is responsible for the thiol dependency. Among the serine-dependent enzymes, carboxypeptidases are unique in that they are most active at acidic pH. These enzymes have a Glu residue before the catalytic Ser, which is thought to be responsible for their high acidic pH. Even though the catalytic domain is present in most of the serine proteases, Glu-specific proteases have a strong preference for Glu-Xaa bonds over Asp-Xaa bonds (Drapeau et al., 1972; Jany et al., 1986; Rao et al., 1998).

14.3.5.3.2 Aspartic proteases

Aspartic endopeptidases depend on residues of aspartic acid for their catalytic activity. A wide-ranging base catalytic mechanism has been suggested for the hydrolysis of proteins by aspartic proteases such as penicillopepsin and endothiapepsin. Crystallographic studies have revealed that the enzymes of the pepsin family are bilobed molecules with the active-site cleft located among the lobes in addition, each lobe subsidizing one of the pair of aspartic acid residues that is vital for the catalytic action. The lobes are homologous to one another, having ascended by gene duplication. The retro pepsin molecule has only one lobe, which carries only one aspartic residue; in addition, the activity needs the formation of a noncovalent homodimer. In the majority of pepsin-family enzymes, the catalytic Asp residues are confined in an Asp-Thr-Gly-Xaa motif in both the N- and

C-terminal lobes, where Xaa is Ser or Thr, whose side chains will bind to the hydrogen of Asp. However, Xaa is Ala in the furthermost of the retro pepsins. Noticeable conservation of cysteine residue is also evident in aspartic proteases. The pepsins, as well as most of other members of the family, show specificity for the cleavage of bonds in peptides with at least six residues through hydrophobic amino acids in both the Pl and Pl9 sites (Rao et al., 1998).

14.4 APPLICATIONS OF EXTRACELLULAR ENZYMES OF WILD MUSHROOMS (MCKELVEY AND MURPHY, 2011)

14.4.1 Food processing

One of the major or utilizations of these enzymes is done by the food industry, which includes the role of catalases and lipases in cheese ripening and production, α-amylases in liquefaction, and proteases in meat tenderization. Lipases have applications in the production of leaner fish, refining rice flavor, and modifying soybean milk. Proteases are essential in meat tenderization because they can hydrolyze connective tissue as well as muscle fiber proteins. Currently, cellulase, protease, and amylase are being used as dough improvers. Their use improves the texture, flavor, freshness, and volume as well as improved dough machinability.

14.4.2 Tanning industry

Historically, lime and sodium sulfide mixtures were used in the leather industry to dehair the skins and hides. This approach pollutes the environment and is irritating. However, with the advancement of biotechnology, cleaner, environmentally sustainable methods of handling animal skins have emerged. From fungal strains such as *Aspergillus,* proteases and lipases are extracted and used for cocktails. The mixture of the enzyme causes the swelling of hair roots, which allows the hair to be removed easily.

14.4.3 Animal feed

With an annual supply of more than 950 million tons of feed worth more than the US$50 billion, the animal feed industry is a vital component of the world's agro-industrial operations. In the past few years, this industry has gone through many changes as consumers and the industry itself have looked more closely than ever before into how compound animal feeds are produced,

Table 14.1 Major enzymes and their applications

Enzymes	Type of Industry	Uses
Amylases, proteases	Food processing	Degradation of starches, proteins, and meat softening
Amylases, proteases, lipases, and cellulases	Detergent	Removing insoluble starch in dish washing, removing protein after staining, to remove fats and oils and to enhance the efficiecny of detergents
Amylases, lipases, and cellulases	Paper and pulp	Degrading the starches to lower viscosity, aiding sizing, deinking, and paper coating, cellulases and lipases reduce pitch and enzymes remove lignin to soften paper
Amylases, proteases, and cellulases	Textiles	The reduce starch size, to remove glue and waxes between fiber core
Laccases and lipases	Biopolymer and plastic	To form cross-links in biopolymers to produce materials in situ by means of polymerization

how the animals are reared, and how the systems of animal husbandry in use today affect the environment. The integration of enzymes such as cellulase, amylase, and protease into feedstuffs has been shown to improve production efficiency.

14.4.4 Waste treatment

The use of enzymes for waste disposal is extensive, and a variety of enzymes are used in the removal of harmful contaminants. Industrial effluents and household waste contain a variety of chemical commodities that are hazardous or poisonous to living beings and the environment. Amylases, cellulases, lipases, and proteases are some of the enzymes that function in waste management. Oxidative binding detoxification of toxic organic compounds is mediated by oxidoreductases. Laccase, for example, is an enzyme that catalyzes the removal of chlorinated phenolic compounds from industrial effluents. Microbial enzymes are often used to recycle waste for reuse, such as recovering additional oil from oil crops, converting starch to sugar, and converting whey to a variety of useful items.

14.5 CONCLUSION

Mushrooms are well known for various pharmaceutical, nutraceutical, and extracellular enzyme productions. They are an important cultural patrimony, used since time immemorial as food and medicines according to traditional ecological knowledge. The chemical and biological characteristics of wild mushrooms are of interest because they are a natural source of great importance to produce compounds with potential biotechnological applications. Fungi can produce a variety of enzymes to help them survive in a variety of environments. Fungal enzymes are used in a wide range of commercial applications. Their remarkable high stability is one of the examples, as they have instinctively evolved to function in a comparatively hostile extracellular environment.

REFERENCES

Adhikari, M. K. (2000). *Mushrooms of Nepal*. Kathmandu, Nepal: P.U. Printers.

Adlercreutz, P. (2013). Immobilisation and application of lipases in organic media. *Chemical Society Reviews*, *42*(15), 6406–6436.

Aikawa, J., Nishiyama, M., & Beppu, T. (1992). Protein engineering of the milk-clotting aspartic proteinases. *Scandinavian Journal of Clinical and Laboratory Investigation*, *52*, 51–58.

Alcalde, M. (2007). Laccases: biological functions, molecular structure and industrial applications. *In Industrial enzymes*. Springer, pp. 461–476

Almeida, J. M., Martini, V. P., Iulek, J., Alnoch, R. C., Moure, V. R., Müller-Santos, M., & Krieger, N. (2019). Biochemical characterization and application of a new lipase and its cognate foldase obtained from a metagenomic library derived from fat-contaminated soil. *International Journal of Biological Macromolecules*, *137*, 442–454.

Ananthi, S. U., Ramasubburayan, R. A., Palavesam, A., & Immanuel, G. (2014). Optimization and purification of lipase through solid state fermentation by *bacillus cereus* MSU as isolated from the gut of a marine fish *Sardinella longiceps*. *International Journal of Pharmacy and Pharmaceutical Sciences*, *6*(5), 291–298.

Anbu, P., Gopinath, S. C., Cihan, A. C., & Chaulagain, B. P. (2013). Microbial enzymes and their applications in industries and medicine.

Annunziatini, C., Baiocco, P., Gerini, M. F., Lanzalunga, O., & Sjögren, B. (2005). Aryl substituted N-hydroxyphthalimides as mediators in the laccase-catalysed oxidation of lignin model compounds and delignification of wood pulp. *Journal of Molecular Catalysis B: Enzymatic*, *32*(3), 89–96.

Arnesen, S., Eriksen, S. H., Ørgen Olsen, J., & Jensen, B. (1998). Increased production of α-amylase from *Thermomyces lanuginosus* by the addition of Tween 80. *Enzyme and Microbial Technology*, 23(3–4), 249–252.

Aro, N., Pakula, T., & Penttilä, M. (2005). Transcriptional regulation of plant cell wall degradation by filamentous fungi. *FEMS Microbiology Reviews*, 29(4), 719–739.

Azhari, R., & Lotan, N. (1991). Enzymic hydrolysis of biopolymers via single-scission attack pathways: a unified kinetic model. *Journal of Materials Science: Materials in Medicine*, 2(1), 9–18.

Baldrian, P. (2006). Fungal laccases–occurrence and properties. *FEMS Microbiology Reviews*, 30(2), 215–242.

Bayer, E. A., Lamed, R., & Himmel, M. E. (2007). The potential of cellulases and cellulosomes for cellulosic waste management. *Current Opinion in Biotechnology*, 18(3), 237–245.

Béguin, P., & Aubert, J. P. (1994). The biological degradation of cellulose. *FEMS Microbiology Reviews*, 13(1), 25–58.

Begum, M. F., & Absar, N. (2009). Purification and characterization of intracellular cellulase from *Aspergillus oryzae* ITCC-4857.01. *Mycobiology*, 37(2), 121–127.

Beisson, F., Tiss, A., Rivière, C., & Verger, R. (2000). Methods for lipase detection and assay: A critical review. *European Journal of Lipid Science and Technology*, 102(2), 133–153.

Beloqui, A., Pita, M., Polaina, J., Martínez-Arias, A., Golyshina, O. V., Zumárraga, M., Yakimov, M. M., García-Arellano, H., Alcalde, M.,Fernández, V. M., & Elborough, K. (2006). Novel polyphenol oxidase mined from a metagenome expression library of bovine rumen: Biochemical properties, structural analysis, and phylogenetic relationships. *Journal of Biological Chemistry*, 281(32), 22933–22942.

Berry, D. R., & Paterson, A. (1990). Enzymes in food industry: In enzyme chemistry, impact and applications. 2nd edn. Suckling C. J., ed., 306–351.

Bijttebier, A., Goesaert, H., & Delcour, J. A. (2007a). Temperature impacts the multiple attack action of amylases. *Biomacromolecules*, 8(3), 765–772.

Bijttebier, A., Goesaert, H., & Delcour, J. A. (2007b). Action pattern of different amylases on amylose and amylopectin. In *3rd Symposium on the Alpha-Amylase Family, Programme and Abstracts* (p. 52).

Boa, E. R. (2004). Wild edible fungi: A global overview of their use and importance to people.

Borchert, E., Selvin, J., Kiran, S. G., Jackson, S. A., O'Gara, F., & Dobson, A. D. (2017). A novel cold active esterase from a deep sea sponge *Stelletta normani* metagenomic library. *Frontiers in Marine Science*, 4, 287.

Bourbonnais, R., & Paice, M. G. (1990). Oxidation of non-phenolic substrates: an expanded role for laccase in lignin biodegradation. *FEBS Letters*, 267(1), 99–102.

Bourbonnais, R., Paice, M. G., Reid, I. D., Lanthier, P., & Yaguchi, M. (1995). Lignin oxidation by laccase isozymes from Trametes versicolor and role of the mediator 2,2'-azinobis (3-ethylbenzthiazoline-6-sulfonate) in kraft lignin depolymerization. *Applied and Environmental Microbiology*, 61(5), 1876–1880.

Branneby, C., Carlqvist, P., Hult, K., Brinck, T., & Berglund, P. (2004). Aldol additions with mutant lipase: analysis by experiments and theoretical calculations. *Journal of Molecular Catalysis B: Enzymatic*, 31(4–6), 123–128.

Chandra, P., Singh, R., & Arora, P. K. (2020). Microbial lipases and their industrial applications: A comprehensive review. *Microbial Cell Factories*, 19(1), 1–42.

Chawla, P., Kumar, N., Kaushik, R., & Dhull, S. B. (2019). Synthesis, characterization and cellular mineral absorption of gum arabic stabilized nanoemulsion of *Rhododendron arboreum* flower extract. *Journal of Food Science and Technology*, 56(12), 5194–5203.

Chinedu, S. N., Okochi, V. I., Smith, H. A., Okafor, U. A. Onyegeme-Okerenta, B. M., & Omidiji, O. (2007). Effect of carbon sources on cellulase (EC 3. 2. 1. 4) production by *Penicillium chrysogenum* PCL501. *African Journal of Biochemistry Research*, 1(1), 006–010.

Chowdary, G. V., Ramesh, M. N., & Prapulla, S. G. (2000). Enzymic synthesis of isoamyl isovalerate using immobilized lipase from *Rhizomucor miehei:* A multivariate analysis. *Process Biochemistry*, 36(4), 331–339.

Cowan, A. (2001). Fungi-life support for ecosystems. *Essential ARB*, 4, 1–5.

Crous, P. W., Rong, I. H., Wood, A., Lee, S., Glen, H., Botha, W., & Hawksworth, D. L. (2006). How many species of fungi are there at the tip of Africa? *Studies in mycology*, 55, 13–33.

Doyle, E., Noone, A., Kelly, C., Quigley, T., & Fogarty, W. (1998). Mechanisms of action of the maltogenic α-amylase of Byssochlamys fulva. *Enzyme and Microbial Technology*, 22(7), 612–616.

Drapeau, G. R., Boily, Y., & Houmard, J. (1972). Purification and properties of an extracellular protease of *Staphylococcus aureus*. *Journal of Biological Chemistry*, *247*(20), 6720–6726.

Duyff, R. L. (2012). American dietetic association complete food and nutrition guide *Houghton Mifflin Harcourt* Wiley & Sons, 3rd Edition.

Eggert, C., Temp, U., & Eriksson, K. E. (1996). The ligninolytic system of the white rot fungus *Pycnoporus cinnabarinus:* purification and characterization of the laccase. *Applied and Environmental Microbiology*, *62*(4), 1151–1158.

El-Banna, T., Abd-Aziz, A., Abou-Dobara, M., & Ibrahim, R. (2008). Optimization and Immobilization of α-amylase from *Bacillus licheniformis*. *Catrina: The International Journal of Environmental Sciences*, *3*(1), 45–53.

El-Banna, T. E., Abd-Aziz, A. A., Abou-Dobara, M. I., & Ibrahim, R. I. (2007). Production and immobilization of alpha-amylase from *Bacillus subtilis*. *Pakistan Journal of Biological Sciences: PJBS*, *10*(12), 2039–2047.

El-Fallal, A., Dobara, M. A., El-Sayed, A., & Omar, N. (2012). Starch and microbial α-amylases: from concepts to biotechnological applications. *Carbohydrates–Comprehensive Studies on Glycobiology and Glycotechnology*, 459–488.

Fischer, M., & Pleiss, J. (2003). The Lipase Engineering Database: a navigation and analysis tool for protein families. *Nucleic ~~Acid Research~~*, *31*(1), 319–321.

Gams, W., & Anderson, T. H. (1980). *Compendium of Soil Fungi*. Academic press.

Garibay-Orijel, R., Martínez-Ramos, M., & Cifuentes, J. (2009). Disponibilidad de esporomas de hongos comestibles en los bosques de pino-encino de Ixtlán de Juárez, Oaxaca. *Revista Mexicana de Biodiversidad*, *80*(2), 521–534.

Ghasemi, Y., Rsoul, A. S., Ebrahiminezhad, A., Kazemi, A., Shahbazi, M., & Talebnia, N. (2011). Screening and isolation of extracellular protease producing bacteria from the Maharloo Salt Lake. *Iranian Journal of Pharmceutical Science*, *7*, 175–180.

Gianfreda, L., Xu, F., & Bollag, J. M. (1999). Laccases: A useful group of oxidoreductive enzymes. *Bioremediation Journal*, *3*(1), 1–26.

Giraud, E., Brauman, A., Keleke, S., Lelong, B., & Raimbault, M. (1991). Isolation and physiological study of an amylolytic strain of *Lactobacillus plantarum*. *Applied Microbiology and Biotechnology*, *36*(3), 379–383.

Gochev, V. K., & Krastanov, A. I. (2007). Fungal laccases. *Bulgarian Journal of Agricultural Science*, *13*(1), 75.

Godfrey, T., & West, S. (1996). *Industrial Enzymology*, 2nd ed., New York, N.Y: Macmillan Publishers Inc., pp. 3.

Gupta, R., Kumari, A., Syal, P., & Singh, Y. (2015). Molecular and functional diversity of yeast and fungal lipases: their role in biotechnology and cellular physiology. *Progress in Lipid Research*, *57*, 40–54.

Hawksworth, D. L. (2004). Fungal diversity and its implications for genetic resource collections. *Studies in Mycology*, *50*(1), 9–17.

Henrissat, B. (1991). A classification of glycosyl hydrolases based on amino acid sequence similarities. *Biochemical Journal*, *280*(2), 309–316.

Henson, J. M., Butler, M. J., & Day, A. W. (1999). The dark side of the mycelium: melanins of phyto-pathogenic fungi. *Annual Review of Phytopathology*, *37*(1), 447–471.

Herrera, T., & Ulloa, M. (1990). El reino de los hongos. *Universidad Nacional Autónoma de México*.

Immanuel, G., Dhanusha, R., Prema, P., & Palavesam, A. (2006). Effect of different growth parameters on endoglucanase enzyme activity by bacteria isolated from coir retting effluents of estuarine environment. *International Journal of Environmental Science and Technology*, *3*(1), 25–34.

Isaka, M., Tanticharoen, M., Kongsaeree, P., & Thebtaranonth, Y. (2001). Structures of Cordypyridones A–D, Antimalarial N-Hydroxy-and N-Methoxy-2-pyridones from the Insect Pathogenic Fungus Cordyceps n ipponica. *The Journal of Organic Chemistry*, *66*(14), 4803–4808.

Jany, K. D., Lederer, G., & Mayer, B. (1986). Amino acid sequence of proteinase K from the mold *Tritirachium album* Limber: Proteinase K—a subtilisin-related enzyme with disulfide bonds. *FEBS Letters*, *199*(2), 139–144.

Jaouani, A., Guillén, F., Penninckx, M. J., Martínez, A. T., & Martínez, M. J. (2005). Role of *Pycnoporus coccineus* laccase in the degradation of aromatic compounds in olive oil mill wastewater. *Enzyme and Microbial Technology*, *36*(4), 478–486.

Jayakumar, T., Ramesh, E., & Geraldine, P. (2006). Antioxidant activity of the oyster mushroom, Pleurotus ostreatus, on CCl4-induced liver injury in rats. *Food and Chemical Toxicology, 44*(12), 1989–1996.

Jiao, Y., Tang, J., Wang, Y., & Koral, T. L. (2018). Radio-frequency applications for food processing and safety. *Annual Review of Food Science and Technology, 9,* 105–127.

Joshi, V. K., & Pandey, A. (1999). Food fermentation. *Biotechnology.* Asiotech Publishers Inc.

Jung, S., Kim, J., & Park, S. (2013). Rational design for enhancing promiscuous activity of Candida antarctica lipase B: a clue for the molecular basis of dissimilar activities between lipase and serine-protease. *RSC Advances, 3*(8), 2590–2594.

Kapoor, M., & Gupta, M. N. (2012). Lipase promiscuity and its biochemical applications. *Process Biochemistry, 47*(4), 555–569.

Karamyshev, A. V., Shleev, S. V., Koroleva, O. V., Yaropolov, A. I., & Sakharov, I. Y. (2003). Laccase-catalyzed synthesis of conducting polyaniline. *Enzyme and Microbial Technology, 33*(5), 556–564.

Kramhøft, B., Bak-Jensen, K. S., Mori, H., Juge, N., Nøhr, J., & Svensson, B. (2005). Involvement of individual subsites and secondary substrate binding sites in multiple attack on amylose by barley α-amylase. *Biochemistry, 44*(6), 1824–1832.

Kuberan, T., Sangaralingam, S., & Thirumalai Arasu, V. (2010). Isolation and optimization of protease producing bacteria from halophilic soil. *Journal of Biosciences, 1*(3), 163–174.

Kubicek, C. P., Messner, R., Gruber, F., Mach, R. L., & Kubicek-Pranz, E. M. (1993). The Trichoderma cellulase regulatory puzzle: from the interior life of a secretory fungus. *Enzyme and Microbial Technology, 15*(2), 90–99.

Kupka, J., Anke, T., Oberwinkler, F., Schramm, G., & Steglich, W. (1979). Antibiotics from basidiomycetes. VII. Crinipellin, a new antibiotic from the basidiomycetous fungus *Crinipellis stipitaria* (Fr.) Pat. *The Journal of Antibiotics, 32*(2), 130–135.

Laachari, F. El, Bergad, F., Sadiki, M., Sayari, A., Bahafid, W., Elabed, S., & Ibnsouda, S. K. (2015). Higher tolerance of a novel lipase from Aspergillus flavus to the presence of free fatty acids at lipid/water interface. *African Journal of Biochemistry Research, 9*(1), 9–17.

Lee, L. P., Karbul, H. M., Citartan, M., Gopinath, S. C., Lakshmipriya, T., & Tang, T. H. (2015). Lipase-secreting *Bacillus* species in an oil-contaminated habitat: Promising strains to alleviate oil pollution. *BioMed Research International, 2015.*

Lei, C., Guo, B., Cheng, Z., & Goda, K. (2016). Optical time-stretch imaging: Principles and applications. *Applied Physics Reviews, 3*(1), 011102.

Leontievsky, A., Myasoedova, N., Pozdnyakova, N., & Golovleva, L. (1997). Yellow'laccase of Panus ti-grinus oxidizes non-phenolic substrates without electron-transfer mediators. *FEBS Letters. 413*(3), 446–448.

Levine, W. G. (1965). Laccase, a review, In: *The Biochemistry of Copper.* New York: Academic Press Inc., pp. 71–385.

Ma, C., Ni, X., Chi, Z., Ma, L., & Gao, L. (2007). Purification and characterization of an alkaline protease from the marine yeast *Aureobasidium pullulans* for bioactive peptide production from different sources. *Marine Biotechnology, 9*(3), 343–351.

Madala, P. K., Tyndall, J. D., Nall, T., & Fairlie, D. P. (2010). Update 1 of: proteases universally recognize beta strands in their active sites. *Chemical Reviews, 110,* 1–31.

Manoharachary, C., Sridhar, K., Singh, R., Adholeya, A., Suryanarayanan, T. S., Rawat, S., & Johri, B. N. (2005). Fungal biodiversity: distribution, conservation and prospecting of fungi from India. *Current Science, 89,* 58–71.

Martínez, Á. T., Speranza, M. Ruiz-Dueñas, F. J., Ferreira, P. Camarero, S., Guillén, F., & Río Andrade, J. C. D. (2005). Biodegradation of lignocellulosics: microbial, chemical, and enzymatic aspects of the fungal attack of lignin.

Masaki, T., Nakamura, K., Isono, M., & Soejima, M. (1978). A new proteolytic enzyme from *Achromobacter lyticus* M497-1. *Agricultural and Biological Chemistry, 42*(7), 1443–1445.

Massadeh, M. I., Yusoff, W. M. W., Omar, O., & Kader, J. (2001). Synergism of cellulase enzymes in mixed culture solid substrate fermentation. *Biotechnology Letters, 23*(21), 1771–1774.

McKelvey, S. M., & Murphy, R. A. (2011). Biotechnological use of fungal enzymes. *Biology and Applications, 179.*

Miles, P. G., & Chang, S. T. (2004). *Mushrooms: Cultivation, Nutritional Value, Medicinal Effect, and Environmental Impact.* CRC Press.

Moorthy, M., Anbalagan, S., & Sankareswaran, M. (2018). Microbial cellulases enzyme: Characteristics, sources, and their industrial applications. *Journal of Pharmaceutical Investigation.*

Mukerjea, R., Slocum, G., Mukerjea, R., & Robyt, J. F. (2006). Significant differences in the activities of α-amylases in the absence and presence of polyethylene glycol assayed on eight starches solubilized by two methods. *Carbohydrate Research, 341*(12), 2049–2054.

Nagai, M., Sato, T., Watanabe, H., Saito, K., Kawata, M., & Enei, H. (2002). Purification and characterization of an extracellular laccase from the edible mushroom Lentinula edodes, and decolorization of chemically different dyes. *Applied Microbiology and Biotechnology, 60*(3), 327–335.

Nitha, B., De, S., Adhikari, S. K., Devasagayam, T. P. A., & Janardhanan, K. K. (2010). Evaluation of free radical scavenging activity of morel mushroom, *Morchella esculenta* mycelia: A potential source of therapeutically useful antioxidants. *Pharmaceutical Biology, 48*(4), 453–460.

Norus, J. (2006). Building sustainable competitive advantage from knowledge in the region: The industrial enzymes industry. *European Planning Studies, 14*(5), 681–696.

Oboh, G., & Shodehinde, S. A. (2009). Distribution of nutrients, polyphenols and antioxidant activities in the pilei and stipes of some commonly consumed edible mushrooms in Nigeria. *Bulletin of the Chemical Society of Ethiopia, 23*(3).

Oei, P. (1991). *Manual on Mushroom Cultivation: Techniques, Species and Opportunities for Commercial Application in Developing Countries.* Tool. pp. 21–26.

Oei, P. ed. (2016). *Mushroom Cultivation IV: Appropriate Technology for Mushroom Growers.* ECO Consult Foundation.

Oksanen, T., Pere, J., Paavilainen, L., Buchert, J., & Viikari, L. (2000). Treatment of recycled kraft pulps with *Trichoderma reesei* hemicellullases and cellulases. *Journal of Biotechnology, 78*, 39–48.

Pandey, A., Nigam, P., Soccol, C. R., Soccol, V. T., Singh, D., & Mohan, R. (2000). Advances in microbial amylases. *Biotechnology and Applied Biochemistry, 31*(2), 135–152.

Pandey, A., Selvakumar, P., Soccol, C. R., & Nigam, P. (1999). Solid state fermentation for the production of industrial enzymes. *Current Science, 77*, 149–162.

Pascoal, A., Estevinho, L. M., Martins, I. M., & Choupina, A. B. (2018). Novel sources and functions of microbial lipases and their role in the infection mechanisms. *Physiological and Molecular Plant Pathology, 104*, 119–126.

Pliego, J., Mateos, J. C., Rodriguez, J., Valero, F., Baeza, M., Femat, R., & Herrera-López, E. J. (2015). Monitoring lipase/esterase activity by stopped flow in a sequential injection analysis system using p-nitrophenyl butyrate. *Sensors, 15*(2), 2798–2811.

Pointing, S. B., & Vrijmoed, L. L. P. (2000). Decolorization of azo and triphenylmethane dyes by *Pycnoporus sanguineus* producing laccase as the sole phenoloxidase. *World Journal of Microbiology and Biotechnology, 16*(3), 317–318.

Purkayastha, R. P. & Chandra, A. (1985). *Manual of Indian Edible Mushrooms.* New Delhi, India: Jagendra Book Agency.

Quintanar, L., Yoon, J., Aznar, C. P., Palmer, A. E., Andersson, K. K., Britt, R. D., & Solomon, E. I. (2005). Spectroscopic and electronic structure studies of the trinuclear Cu cluster active site of the multicopper oxidase laccase: nature of its coordination unsaturation. *Journal of the American Chemical Society, 127*(40), 13832–13845.

Ramanathan, G., Banupriya, S., & Abirami, D. (2010). Production and optimization of cellulase from *Fusarium oxysporum* by submerged fermentation.

Ramesh, C. H., & Pattar, M. G. (2010). Antimicrobial properties, antioxidant activity and bioactive compounds from six wild edible mushrooms of western ghats of Karnataka, India. *Pharmacognosy Research, 2*(2), 107.

Rammeloo, J., & Walleyn, R. (1993). *Edible fungi of Africa south of the Sahara.* National Botanic Garden of Belgium. *Scripta Botanica Belgica, 5*, 1–62.

Rani, K. (2012b). Aqueous two phase purification of pulses amylases and study its applications in desizing of fabrics. *Asian Journal of Biochemical and Pharmaceutical Research, 2*(3), 215–221.

Rani, K. (2012c). Extraction and study of kinetic parameters of variety of sprouted pulses β-amylases. *International Journal of Pharmaceutical and Life Sciences, 3*(8), 1893–1896.

Rani, K. (2012d). *Production of Amylase and Alkaline Phosphatase.* Germany: Lap Lambert Academic Publ, pp. 1–56.

Rani, K. (2012e). Comparative study of kinetic parameters of bacterial and fungal amylases. *Journal of Bio Innovation, 3*, 48–57.

Rani, K., & Chauhan, C. (2014). Biodegradation of *Cicer Arietinum* Amylase loaded Coconut oil driven Emulsified Bovine Serum Albumin Nanoparticles and their application in Washing Detergents as Eco-Friendly Bio-Active Addictive. *World Journal of Pharmacy and Pharmaceutical Sciences, 3*(12), 924–936.

Rao, M. B., Tanksale, A. M., Ghatge, M. S., & Deshpande, V. V. (1998). Molecular and biotechnological aspects of microbial proteases. *Microbiology and Molecular Biology Rreviews, 62*(3), 597–635.

Rao, P., & Divakar, S. (2001). Lipase catalyzed esterification of α-terpineol with various organic acids: Application of the Plackett–Burman design. *Process Biochemistry, 36*(11), 1125–1128.

Rathee, S., Rathee, D., Rathee, D., Kumar, V., & Rathee, P. (2012). Mushrooms as therapeutic agents. *Revista Brasileira de Farmacognosia, 22*(2), 459–474.

Sadhu, S., & Maiti, T. K. (2013). Cellulase production by bacteria: a review. *Microbiology Research Journal International, 3*, 235–258.

Sakka, K., Kimura, T., Karita, S., & Ohmiya, K. (2000). Molecular breeding of cellulolytic microbes, plants, and animals for biomass utilization. *Journal of Bioscience and Bioengineering, 90*(3), 227–233.

Sanchez, S., & Demain, A. L. (2011). Enzymes and bioconversions of industrial, pharmaceutical and bio-technological significance. *Organic Process Research and Development, 15*, 224–230.

Sarikaya, E., Higasa, T., Adachi, M., & Mikami, B. (2000). Comparison of degradation abilities of α-and β-amylases on raw starch granules. *Process Biochemistry, 35*(7), 711–715.

Sasidharan, S., Aravindran, S., Latha, L. Y., Vijenthi, R., Saravanan, D., & Amutha, S. (2010). In vitro antioxidant activity and hepatoprotective effects of *Lentinula edodes* against paracetamol-induced hepatotoxicity. *Molecules, 15*(6), 4478–4489.

Sheehan, J., & Himmel, M. (1999). Enzymes, energy, and the environment: a strategic perspective on the US Department of Energy's research and development activities for bioethanol. *Biotechnology Progress, 15*(5), 817–827.

Shekher, R., Sehgal, S., Kamthania, M., & Kumar, A. (2011). Laccase: microbial sources, production, purification, and potential biotechnological applications. *Enzyme Research, 2011.*

Stergiou, P. Y., Foukis, A., Filippou, M., Koukouritaki, M., Parapouli, M., Theodorou, L. G., & Papamichael, E. M. (2013). Advances in lipase-catalyzed esterification reactions. *Biotechnology Advances, 31*(8), 1846–1859.

Sukumaran, R. K., Singhania, R. R., Mathew, G. M., & Pandey, A. (2009). Cellulase production using biomass feed stock and its application in lignocellulose saccharification for bio-ethanol production. *Renewable Energy, 34*(2), 421–424.

Suzuki, H., Iiyama, K., Yoshida, O., Yamazaki, S., Yamamoto, N., & Toda, S. (1990). Structural characterization of the immunoactive and antiviral water-solubilized lignin in an extract of the culture medium of *Lentinus edodes* mycelia (LEM). *Agricultural and Biological Chemistry, 54*(2), 479–487.

Svensson, B. T., Jensen, M., Mori, H., Bak-Jensen, K. S., Bonsager, B., Nielsen, P. K., & Driguez, H. (2002). Fascinating facets of function and structure of amylolytic enzymes of glycoside hydrolase family 13. *Biologia-Bratislava-,, 57*, 5–20.

Teferi, Y., Muleta, D., & Woyessa, D. (2013). Mushroom consumption habits of Wacha Kebele residents, southwestern Ethiopia. *Journal of Agricultural, Biological and Environmental Statistics, 4*(1), 6–16.

Thakur, V., Tewari, R., & Sharma, R. (2014). Evaluation of production parameters for maximum lipase production by P. stutzeri MTCC 5618 and scale-up in bioreactor. *Chinese Journal of Biology, 2014*, 1–14.

Thurston, C. F. (1994). The structure and function of fungal laccases. *Microbiology, 140*(1), 19–26.

Tonkova, A. (2006). Microbial starch converting enzymes of the α-amylase family. *Microbial Biotechnology in Horticulture, 1*, 421–472.

Toshinungla, A. O., Deb, C. R., & Neilazonuo, K. (2016). Wild edible mushrooms of Nagaland, *India: a potential food resource. Journal of Experimental Biology and Agricultural Sciences, 4*(1), 59–65.

U.S. Department of Agriculture, Agricultural Research Service, USDA Nutrient Data Laboratory. (2009). USDA National Nutrient Database for Standard Reference, Release 22. Available at www.ars.usda.gov/nutrientdata

U.S. Food and Drug Administration (1994). Center for Food Safety & Applied Nutrition. A Food Labeling Guide. http://www.cfsan.fda.gov/~dms/flg-toc.html

Velázquez-Cedeño, M. A., Farnet, A. M., Ferre, E., & Savoie, J. M. (2004). Variations of lignocellulosic activities in dual cultures of *Pleurotus ostreatus* and *Trichoderma longibrachiatum* on unsterilized wheat straw. *Mycologia, 96*(4), 712–719.

Vongvilai, P., Linder, M., Sakulsombat, M., Svedendahl Humble, M., Berglund, P., Brinck, T., & Ramström, O. (2011). Racemase activity of *B. cepacia* lipase leads to dual-function asymmetric dynamic kinetic resolution of α-aminonitriles. *Angewandte Chemie International Edition, 50*(29), 6592–6595.

Wang, H., & Ng, T. B. (2001). Pleureryn, a novel protease from fresh fruiting bodies of the edible mushroom *Pleurotus eryngii. Biochemical and Biophysical Research Communications, 289*(3), 750–755.

Wasser, S. P., & Weis, A. L. (1999). Therapeutic effects of substances occurring in higher Basidiomycetes mushrooms: a modern perspective. *Critical Reviews™ in Immunology, 19*(1).

Williamson, P. R. (1994). Biochemical and molecular characterization of the diphenol oxidase of *Cryptococcus neoformans*: identification as a laccase. *Journal of Bacteriology, 176*(3), 656–664.

Yamada, T., Oinuma, T., Niihashi, M., Mitsumata, M., Fujioka, T., Hasegawa, K., & Itakura, H. (2002). Effects of Lentinus edodes mycelia on dietary-induced atherosclerotic involvement in rabbit aorta. *Journal of Atherosclerosis and Thrombosis, 9*(3), 149–156.

Yaver, D. S., Berka, R. M., Brown, S. H., & Xu, F. (2001). *The Presymposium on recent advances in lignin biodegradation and biosynthesis.* Finland: Vikki Biocentre, University of Helsinki, pp. 40.

Yoshimoto, T. (1980). Proline-specific endopeptidase from *Flavobacterium. Purification and Properties, 255*(10), 4786–4792.

Zoppellaro, G. (2000). Kinetic studies on the fully reduced laccase with dioxygen. *Inorganic Reaction Mechanisms, 2*, 79–84.

CHAPTER 15

Application of mushrooms in the degradation of xenobiotic components and the reduction of pesticides

Karishma Joshi[1], Anamika Das[2], Gaurav Joshi[3], and Bibekananda Sarkar[4]

[1]Department of Botany, Banaras Hindu University, Varanasi, Uttar Pradesh, India
[2]Department of Paramedical Sciences, Guru Kashi University, Bhatinda, India
[3]School of Pharmacy, Graphic Era Hill University, Dehradun, Uttarakhand, India
[4]Department of Zoology, B.S.S. College, Supaul, Bihar, India

CONTENTS

15.1 INTRODUCTION

Xenobiotic is a broad term referring to foreign substances for human life (Rieger et al., 2002). This includes food, plant-derived constituents, pesticides, drugs, industrial chemicals, flavoring agents, cosmetics, additives, and pollutants, to name a few (Abdelsalam, 2020). This xenobiotic invades our body via our lifestyle choices, diet, drinking water, air etc. Once entered, numerous detoxication processes are catalyzed by our body to eliminate these xenobiotic components. However, incomplete detoxication may be followed by adversities which may include most fatal disease conditions, including cancer (Nakov & Velikova, 2020; Croom, 2012). Among known xenobiotics, pesticides are one of the significant xenobiotics that affects both flora and fauna of the biodiversity (Maurya & Malik, 2016). Thus, keeping the check on their levels along with other significant xenobiotics inversely affecting the health and environment is vital and a major need of the hour. The process of keeping this check is called remediation (Liu et al., 2018). It is a broad term that is concerned with partial or

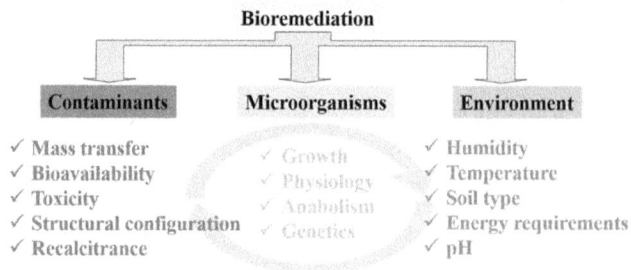

Figure 15.1 Illustration of underlying factors in bioremediation.

complete removal of contaminants or harmful xenobiotics from the environment that directly or in-directly affects human health. This is achieved by utilizing physical, chemical, or biological means. Among these, biological remediation, also called bioremediation, is widely accepted. This is because it is an eco-friendly and cost-effective means of removing toxicity (Vidali, 2001). Bioremediation is achieved by the use of biological agents, including microbes, plants, fungi, or any other living things (Juwarkar et al., 2010). The bioremediation process is thus primarily relying on microorganisms, types of contaminants, and prevailing environment factors, as outlined in Figure 15.1.

Among various biological agents involved in bioremediation, the mushroom is one the most emerging agents. Mushrooms belong to the class of fungi, basidiomycetes, and are one of the major sources of proteins. They are also in possession of key enzymatic machinery, including laccase, lignin peroxidase, peroxidase, and oxidases associated with bioremediation (Rhodes, 2014; Kulshreshtha et al., 2014). The bioremediation by mushrooms is called "mycoremediation" and was first coined by Paul Stamets in the year 2005 in his book *Mycelium Running* (Stamets, 2005). The mycoremediation also possesses various advantages in comparison to traditional bioremediation. The importance in-cludes low cost, simple cultivation, low maintenance, and more public acceptance as by-products are reusable. This is also important since other remediation processes include insertion of high-cost factors, inadequate applications, and are deprived of soil heath improving assets (Koul et al., 2021); not only mycoremediation, but mushrooms are also used for their medicinal attributes (*Ganoderma Schizophyllan commune, Ganoderma lucidum, Pleurotus, Agaricus*); as antioxidants (*Phellinus ri-mosus, Ganoderma lucidum,* and *Pleurotus pulmonaris*), as antimutagenic or antigenotoxic assets and food (*Agaricus, Pleurotus*) (Sharifi-Rad et al., 2020). This chapter therefore is kept forth to explain different applications of mushrooms for remediation of different xenobiotic compounds.

15.2 ESSENTIAL CHEMICALS AND ENZYMES FROM MUSHROOMS INVOLVED IN MYCOREMEDIATION

Mushroom enzymatic substrates possess a high structural homology with various xenobiotic chemicals with pharmacophoric features of polycyclic aromatic hydrocarbons, nitrotoluenes pentachlorophenol, pesticides, dioxins, dyes, lignin, and cellulose derivatives. The critical enzymes involved in remediation include peroxidases, oxidases, ligninase, pectinases, cellulases, and xy-lanases (Phan & Sabaratnam, 2012; Gulzar et al., 2020). These enzymes catalyze the degradation and breakdown of functional moieties like ester, amides, ether aromatic ring, or aliphatic side chains of the mentioned xenobiotics. Apart from these, xenobiotics are also utilized as a source of energy, nitrogen, sulphur, and carbon for their own growth and development. There exists evidence that ligninolytic enzymes catalyze the degradation of anthracene (polycyclic aromatic hydro-carbons) to anthraquinone and subsequently to phthalic acid and carbon dioxide. Further, evidence of non-ligninolytic degradation of 2,4,6-trinitrotoluene, atrazine, and terbuthylazine is also

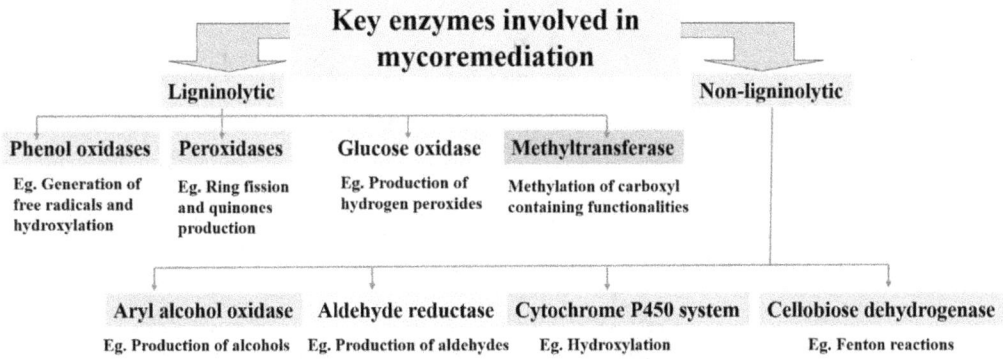

Figure 15.2   Illustration highlighting critical enzymes involved in mycoremediation.

reported (Gulzar et al., 2020; Mathur & Gehlot, 2021). The brief outline of crucial enzymes present in the mushroom are depicted in Figure 15.2.

15.3 NEED FOR DEGRADATION OF XENOBIOTICS AND REDUCTION OF PESTICIDES

Today, effective agricultural management practices are the need of the hour, for which we need the efficient management of biotic factors such as insects, pests, various diseases, weeds, etc.

To achieve this target, farmers have started to use a substantial amount of various chemical fertilizers, insecticides, pesticides, and weedicides in their fields, thus releasing xenobiotic compounds into the environments (Anode & Onguso, 2021). According to a study, the use of pesticides in India considerably increased after 2009–2010 and has been doubled in 2014–2015 with as much as 0.29 kg/ha. The main problem is that only a handful of these chemicals are target specific and, thus, a majority of the chemicals get released into the soil and water. From there, these harmful xenobiotics through biomagnification ultimately reach the food chain. Besides agriculture, other anthropogenic activities such as industrialization and landfill are significant contributors to the release of xenobiotics, such as phenolic compounds, dyes, PAHs, heavy metals, etc., into the environment. Contamination of soil and water with these harmful compounds leads to a number of ailments. Synthetic dyes and heavy metals are one of the significant carcinogenic sources. The well-known Minamata disease caused by water contaminated with mercury is one of the many examples in this regard. Diclofenac is an example of the adverse effect of xenobiotics in animals. Diclofenac, a pain killer drug, when through the food chain entered the system of Old World Gyps vultures, evoked visceral gout, renal necrosis, and mortality within a few days of exposure. Enzyme inhibition is one of the many adverse effects of xenobiotics. Carbon monoxide poisoning is a prevalent example of atmospheric xenobiotic. Further, bioremediation could assist in the conversion of toxic xenobiotics and pesticide waste into inexpensive organic by-products (Jasmin et al., 2020; Ihsanullah et al., 2020). In a nutshell, bioremediation based upon the use of mushrooms could assist in degrading the complex waste derived from the use of heavy metals, pesticides, pharmaceutical drugs, herbicides, and others, as illustrated in Figure 15.3.

15.4 MECHANISM OF DEGRADATION BY MUSHROOMS

The mycoremediation of xenobiotics or pesticides usually falls under three broad categories of mechanisms. These include enzymatic biodegradation, biosorption, and bioconversion (Figure 15.4)

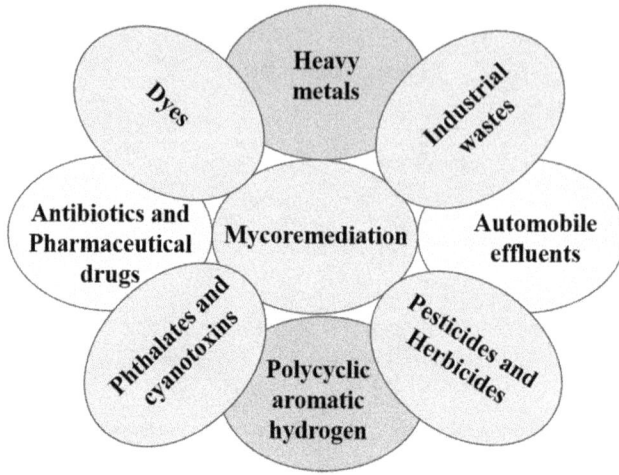

Figure 15.3 Illustration suggesting the utility of mycoremediation in allied areas involved in the generation of potentially toxic xenobiotics.

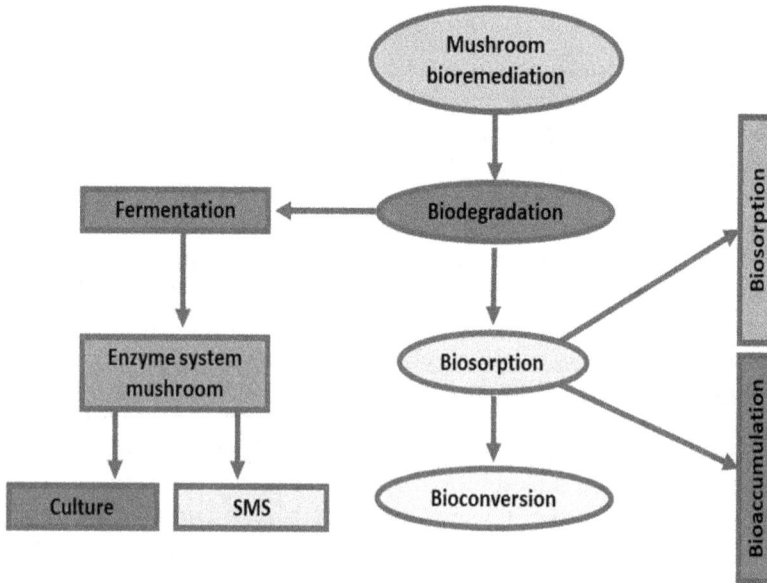

Figure 15.4 The critical mechanism involved in mycoremediation for degradation of xenobiotics and pesticides.

(Kulshreshtha et al., 2014). Enzymatic biodegradation is concerned with the degradation of xenobiotics or pesticides by inherent mushroom enzymes. It is a complex process in which organic molecules are degraded into simpler units like CO_2, H_2O, and other inorganic compounds by living organisms by enzymatic actions (Barh et al., 2019). Various reports have suggested the extracellular enzymes produced by mushroom species play a vital role in pesticide degradation, polymeric and non-polymeric degradation, along with dye decolorization. The significant enzymes include lignin peroxidases (liginases), lignocellulolytic enzymes, peroxidase, laccase, manganese-dependent peroxidase, lignin peroxidase, manganese peroxidase, and many more. The critical mechanism(s) involves the generation of reactive free radicals (via H_2O_2), initiating cascades of redox reaction in bringing out degradation (Rhodes, 2014).

A plethora of evidence is available for enzymatic degradation by mushroom-derived enzymes. A few representative examples are compiled in Table 15.1.

The second underlying mechanism in mycoremediation is biosorption. This method involves the sorption of metals, pollutants or xenobiotics present in waste or effluent. These biosorbents are reported to be developed involving mushroom species. The key mechanism involves bioaccumulation followed by biosorption. Bioaccumulation is an active metabolism dependent process that foresees transport of waste or effluent into the cellular component of mushroom and partitioning thereafter into the individual compartment, which further allows binding of biomass via adsorption, ion exchange or covalent binding allowing biosorption. The polar amino acids from proteins of mushroom polysaccharides are known to play a vital role in biosorption (Haile et al., 2021). The

Table 15.1 Examples of a few important mushroom species, their key enzymes, and involvement in xenobiotic degradation

Mushroom species	Enzyme involved	Type of xenobiotic	Ref
Pleurotus ostreatus	Ligno-cellulolytic enzymes	Plastics	(da Luz et al., 2015)
	Lignin peroxidase, laccase, and manganese peroxidase	Anthracene	(Zebulun et al., 2011)
	Laccase oxidoreductases	Green polyethene and P.C.B.s	(da Luz et al., 2015)
Lentinula edodes	Ligninolytic enzyme-derived vanillin	2,4-Dichlorophenol	(Tsujiyama et al., 2013)
Pleurotus pulmonarius	Ligninolytic enzymes	Radioactive cellulosic-based waste	(Eskander et al., 2012)
	Peroxidase	Crude oil	(Olusola, & Ejiro, 2011)
Coriolus versicolor	Laccase, manganese-dependent peroxidase, and lignin peroxidase	PAHs	(Jang et al., 2009)
Pleurotus palmonarius, Pleurotus tuberregium, Lentinus squarrosulus	Ligninolytic enzymes	Crude oil	(Adedokun, & Ataga, 2014)
Pleurotus tuber-regium	Ligninolytic enzymes	Crude oil	(Isikhuemhen et al., 2003)
Bjerkandera adusta	Lignin-degrading enzyme	PAHs, P.C.B.s	(Bumpus et al., 1985)
Irpex lacteus	Laccase, lignin peroxidase, manganese peroxidase, versatile peroxidase	PAHs, T.N.T., bisphenol, dimethyl, phthalate	(Novotný et al., 2000)
	Oxidoreductases	PCBs	(Stella et al., 2017)
Phanerochaete chrysosporium	Lip, MnP	D.D.T., P.H.A.s, P.C.B.s	(Singh, 2006)
	Peroxidases (LiP, MnP)	PAHs	(Spadaro et al., 1992)
	Peroxidases	Styrene	(Braun-Lüllemann et al., 1997)
Schizophyllum commune, Polyporus sp.	Ligninolytic enzymes	Malachite green dye	(Yogita et al., 2011)
Trametes versicolor	Ligninolytic enzymes	Lignin, polycyclic aromatic hydrocarbons, polychlorinated biphenyl mixtures, and a number of synthetic dyes	(Novotný et al., 2000)

Table 15.2 Key mushroom species and their involvement in biosorption

Mushroom species	Xenobiotic undergoing biosorption	Ref
Agaricus bisporus	Cadmium ions	(Yildirim & Hilal, 2020; Eliescu et al., 2020; Shamim, 2018; Vimala, & Das, 2009; Das et al., 2008; Das, 2005)
Lactarius piperatus	Cadmium ions	
Fomes fasciatus	Copper ions	
Pleurotus platypus	Zinc, iron, cadmium, lead, nickel	
Agaricus bisporus	Zinc, iron, cadmium, lead, nickel	
Calocybe indica	Zinc, iron, cadmium, lead, nickel	
Flammulina velutipes	Copper	
Pleurotus tuber-regium	Heavy metals	
Pleurotus ostreatus	Cadmium	
Pleurotus sajor-caju	Heavy metals like zinc	

Table 15.3 Examples of various mushroom species in bioconversion (Kulshreshtha et al., 2014)

Mushroom species	Type of xenobiotic undergoing bioconversion
Pleurotus ostreatus	Extracts derived from sawdust
P. ostreatus var. florida	Solid sludges and effluents of the paper industry, industrial cardboard waste
Pleurotus tuber-regium	Waste derived as agricultural waste from cotton, rice straw, sawdust, fruit peels
Volvariella volvacea	Agro-industrial waste, particularly banana leaves
Lentinula tigrinus	Wheat and rice straw
Lentinula edodes	Agricultural waste derived from wheat, rice, and barley straw

biosorption capacity of mushroom is reported to be higher for dead mushrooms. The dead mushrooms are much more effective than live mushrooms since they can maintain their remedial power irrespective of adversities that may include varying pH, nutrient supply, temperature, metal ion concentration, or other inherent factors (Kulshreshtha et al., 2014). Further, the dead mushroom may be chiefly available as a fermentation by-product of industrial waste. This method is exceptionally advantageous due to its low cost of implementation and high uptake capacity. The uptake capacity is further reported to get enhanced by treating the fermented by-product (mushroom) with strong acid and bases and elevating the temperatures, which drastically improves biosorption property by many folds (Thakur, 2019). Various evidence has been reported for biosorption by mushrooms as a mean of mycoremediation. Some noteworthy examples are compiled in Table 15.2.

Further, another mechanism involved in mycoremediation is bioconversion. The process involves the conversion of harmful xenobiotics or wastes into some usable or commercial form. The primary bioconversion product produced is a mushroom, which is generated as a fermented by-product from lignocellulosic waste (Sotthisawad et al., 2017). Different product waste gives rise to different mushroom species. A brief compilation is made in Table 15.3.

15.5 UNDERLYING FACTORS INFLUENCING BIOREMEDIATION BY MUSHROOM

The majority of intrinsic (nutrients, pH, moisture content, electrical conductivity and oxygen availability) and extrinsic factors (temperature, humidity, daylight, CO_2 level, O_2 exchange) are

Table 15.4 Numerous intrinsic and extrinsic factors affecting mycoremediation

	Optimum value	Implication	Ref
Intrinsic Factor(s)			
Nutrient carbon-nitrogen (C:N) ratio	C:N ratio of 5:2	Improves mycelial growth	(Zhang, & Elser, 2017)
pH	4.0–7.0	Assist in mycelium growth	(Kulshreshtha et al., 2014)
	6.5–7.0	Assist in basidiocarp development	
Moisture content	50–75% RH	Assist in transferring nutrients from mycelium to fruiting body, allowing the fruit formation	(Endeshaw et al., 2017)
Particle size and oxygen availability	High	High aeration improves particle size of mycelium and spores, leading to better growth	(Endeshaw et al., 2017)
Electrical conductivity	—	Soil with high electrical conductivity promotes growth	(Guo et al., 2001)
Extrinsic Factor(s)			
Temperature	10 to 16°C (cold regions)	Temperature is mushroom species-specific parameters and also depends upon a geographical location. A temperature below optimal range devoid pinhead and delay fruit body development and may induce contamination	(Bellettini et al., 2019)
	17 to 23°C (warmer regions)		
	24 to 30°C (hot regions)		
Humidity	2,000–5,000 ppm	Assist in fruiting body development. The high concentration was fruit deformity	
O_2 exchange	Optimal	Optimal adjust the humidity level and maintains CO_2 level allowing optimal growth	

known to affect mycoremediation. The extrinsic factors affect the growing and fruiting of the mushroom, while intrinsic factor will affect the development of the mycelium and primordia formation in addition to fruiting initiation. Both of these underlying factors are associated with mushroom survival and growth. The key factors are briefly compiled in Table 15.4.

15.6 LIMITATIONS OF MYCOREMEDIATION

Although mycoremediation finds its immense use not only in agricultural or biological sciences but also in allied fields including biochemistry, genetics, chemical engineering, chemistry, and other subject areas as depicted in Figure 15.5. The data was retrieved from the Scopus database using keywords "Mushroom" AND "Bioremediation" on April 1, 2021.

Further, the data also suggested a total of 346 articles have been published using mentioned keywords. The majority of articles have been published by China (76), followed by India (36), Spain (25), and the United States (24), as illustrated in Figure 15.6, revealing the importance of bioremediation in these densely populated countries.

Although mycoremediation is very popular, at the same time it has numerous drawbacks, which is a setback to their popularity (Kulshreshtha et al., 2014; Sasek & Cajthaml, 2005). The important ones include, **i.** rate of degradation by mushroom is slow; **ii.** often lead to the generation of toxic secondary metabolites; **iii.** diversity among mushroom species makes them non-selective for a particular xenobiotic; **iv.** mushroom species may undergo evolutionary changes and may confer resistance to biodegradation; mushroom dependency on extrinsic factors, including soil and climate, further affects the remediation process; **v.** genomic makeup to allow synergism of mushrooms with bacteria is a tedious task.

Total Publications across various disciplines

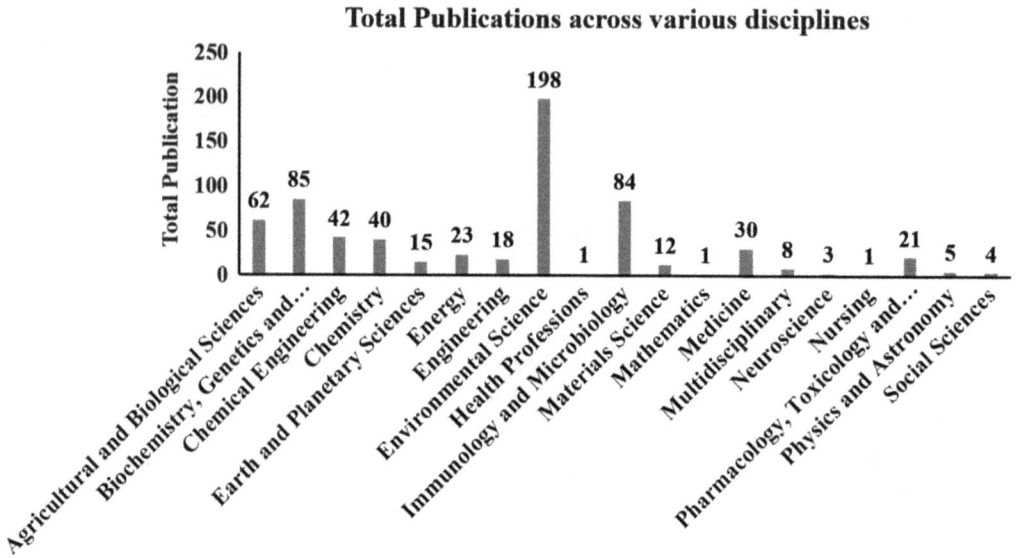

Figure 15.5 Bar graph representation suggesting popularity of mycoremediation across various subject disciplines.

Publications by countries

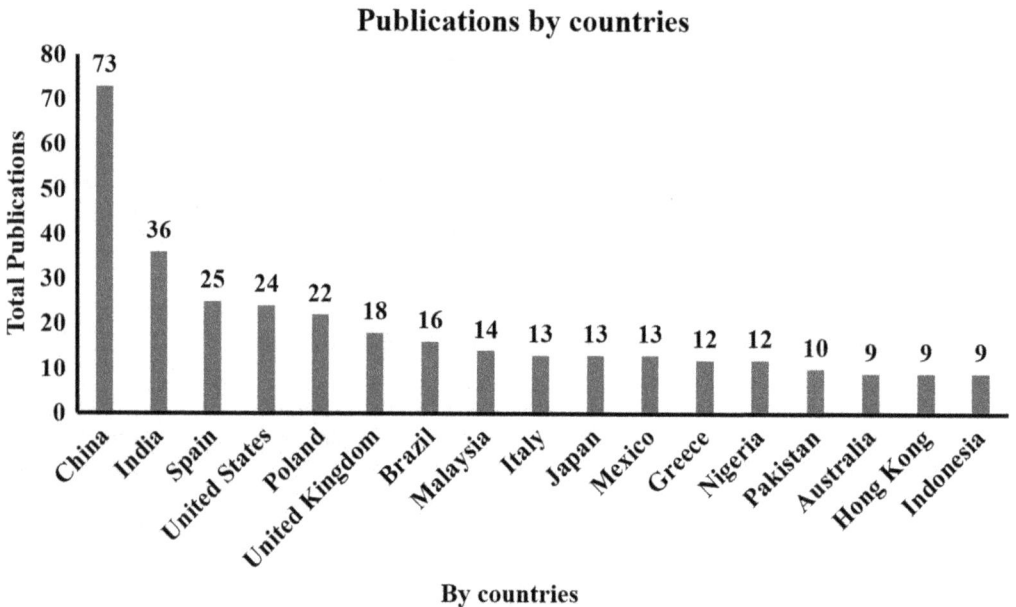

By countries

Figure 15.6 Bar graph suggesting publications by countries on mycoremediation. For clarity, only those countries that published more than nine papers have been included

15.7 CONCLUSION AND FUTURE AVENUES

Mushrooms are seen as one of the limited options of mycoremediation to address real-world solution for the degradation of pesticides and xenobiotics. The mycoremediation using mushrooms could be a game-changer in reducing waste accumulation and, at the same time, could serve as a nutritious food source for an ever-expanding population. This process may not be used in full scale due to its limitation but with the help of nanotechnology we can use this in a larger prospect.

FUNDING SOURCE

This research did not receive any specific grant from funding agencies in the public, commercial, or not-for-profit sectors.

AUTHOR CONTRIBUTIONS

The manuscript has been prepared and drafted with the understanding and contributions of all the authors.

The authors declare that they have no competing financial interests.

ACKNOWLEDGMENTS

All the authors thank their respective affiliated working institute and universities for providing the necessary infrastructure and support to carry out the current work.

REFERENCES

Abdelsalam, N. A., Ramadan, A. T., ElRakaiby, M. T., & Aziz, R. K. (2020). Toxicomicrobiomics: The human microbiome vs. pharmaceutical, dietary, and environmental xenobiotics. *Frontiers in Pharmacology*, *11*, 390. 10.3389/fphar.2020.00390

Adedokun, O. M., & Ataga, A. E. (2014). Oil spills remediation using native mushroom-A viable option. *Research Journal of Environmental Sciences*, *8*(1), 57.

Anode, S., & Onguso, J. (2021). Cuxrrent methods of enhancing bacterial bioremediation of pesticide residues in agricultural farmlands. *Microbial Rejuvenation of Polluted Environment*, Springer, pp. 167–187.

Barh, A., Kumari, B., Sharma, S., Annepu, S. K., Kumar, A., Kamal, S., & Sharma, V. P. (2019). Mushroom mycoremediation: Kinetics and mechanism. *Smart Bioremediation Technologies*, Elsevier, pp. 1–22.

Bellettini, M. B., Fiorda, F. A., Maieves, H. A., Teixeira, G. L., Ávila, S., Hornung, P. S., Júnior, A. M., & Ribani, R. H. (2019). Factors affecting mushroom Pleurotus spp, *Saudi Journal of Biological Sciences*, *26*(4), 633–646. doi:https://doi.org/10.1016/j.sjbs.2016.12.005

Braun-Lüllemann, A., Majcherczyk, A., & Hüttermann, A. (1997). Degradation of styrene by white-rot fungi. *Applied Microbiology Biotechnology Letters*, *47*(2), 150–155.

Bumpus, J. A., Tien, M., Wright, D., & Aust, S. D. (1985). Oxidation of persistent environmental pollutants by a white rot fungus. *Science*, *228*(4706), 1434–1436.

Croom, E. (2012). Metabolism of xenobiotics of human environments. *Progress in Molecular Biology Translational Science*, *112*, 31–88.

da Luz, J. M. R., Paes, S. A., Ribeiro, K. V. G., Mendes, I. R., & Kasuya, M. C. M. (2015). Degradation of green polyethylene by Pleurotus ostreatus. *Plos One*, *10*(6), e0126047.

Das, N. (2005). Heavy metals biosorption by mushrooms. *Indian Journal of Natural Products and Resources (IJNPR)*, *4*(6), 454–459. http://nopr.niscair.res.in/handle/123456789/8140?mode=full

Das, N., Vimala, R., & Karthika, P. (2008). Biosorption of heavy metals – An overview. *Indian Journal of Biotechnology*, *7*, 159–169.

Eliescu, A., Georgescu, A. A., Nicolescu, C. M., Bumbac, M., Cioateră, N., Mureşeanu, M., & Buruleanu, L. C. (2020). Biosorption of Pb (II) from aqueous solution using mushroom (Pleurotus ostreatus) biomass and spent mushroom substrate. *Analytical Letters*, *53*(14), 2292–2319.

Endeshaw, A., Birhanu, G., Zerihun, T., & Misganaw, W. (2017). Application of microorganisms in bioremediation-review. *Environmental Microbiology*, *1*(1), 2–9.

Eskander, S., El-Aziz, A., El-Sayaad, H., & Saleh, H. (2012). Cementation of bioproducts generated from biodegradation of radioactive cellulosic-based waste simulates by mushroom. *International Scholarly Research Notices*, *2012*.

Guo, M., Chorover, J., Rosario, R., & Fox, R. H. (2001). Leachate chemistry of field-weathered spent mushroom substrate, *Journal of Environmental Quality*, *30*(5), 1699–1709.

Gulzar, A. B. M., Vandana, U. K., Paul, P., & Mazumder, P. B. (2020). *The Role of Mushrooms in Biodegradation and Decolorization of Dyes, An Introduction to Mushroom*, IntechOpen.

Haile, S., Alemu, D., Tesfaye, T., & Kamaraj, M. (2021). Mycoremediation: fungal-based technology for biosorption of heavy metals–A review. *Strategies Tools for Pollutant Mitigation: Avenues to a Cleaner Environment 355*.

Ihsanullah, I., Jamal, A., Ilyas, M., Zubair, M., Khan, G., & Atieh, M. A. (2020). Bioremediation of dyes: Current status and prospects. *Journal of Water Process Engineering*, *38*, 101680.

Isikhuemhen, O. S., Anoliefo, G. O., & Oghale, O. I. (2003). Bioremediation of crude oil polluted soil by the white rot fungus. *Pleurotus tuberregium (Fr.) Sing, Environmental Science Pollution Research*, *10*(2), 108–112.

Jang, K.-Y., Cho, S.-M., Scok, S.-J., Kong, W.-S., Kim, G.-H., & Sung, J.-M. (2009). Screening of biodegradable function of indigenous ligno-degrading mushroom using dyes. *Mycobiology*, *37*(1), 53–61.

Jasmin, M., Syukri, F., Kamarudin, M., & Karim, M. (2020). Potential of bioremediation in treating aquaculture sludge. *Aquaculture*, *519*, 734905.

Juwarkar, A. A., Singh, S. K., & Mudhoo, A. (2010). A comprehensive overview of elements in bioremediation. *Reviews in Environmental Science and Bio/Technology*, *9*(3), 215–288.

Koul, B., Ahmad, W., & Singh, J. (2021). *Mycoremediation: A Novel Approach for Sustainable Development, Microbe Mediated Remediation of Environmental Contaminants*, Elsevier, pp. 409–420.

Kulshreshtha, S., Mathur, N., & Bhatnagar, P. (2014). Mushroom as a product and their role in mycoremediation, *A.M.B. Express*, *4*(1), 1–7.

Liu, L., Li, W., Song, W., & Guo, M. (2018). Remediation techniques for heavy metal-contaminated soils: Principles and applicability. *Science of the Total Environment*, *633*, 206–219.

Mathur, M., & Gehlot, P. (2021). Mechanistic evaluation of bioremediation properties of fungi, *New and Future Developments in Microbial Biotechnology and Bioengineering*, Elsevier, pp. 267–286.

Maurya, P. K., & Malik, D.S. (2016). Bioaccumulation of xenobiotics compound of pesticides in riverine system and its control technique: A critical review. *Journal of Industrial Pollution Control*, *32*(2) .

Nakov, R., & Velikova, T. (2020). Chemical metabolism of xenobiotics by Gut microbiota, *Current Drug Metabolism*, *21*(4), 260–269.

Novotný, Č., Erbanova, P., Cajthaml, T., Rothschild, N., Dosoretz, C., & Šašek, V. (2000). Irpex lacteus, a white rot fungus applicable to water and soil bioremediation. *Applied Microbiology Biotechnology Letters*, *54*(6), 850–853.

Olusola, S., & Ejiro, A. (2011.) Bioremediation of a crude oil-polluted soil with Pleurotus pulmonarius and Glomus mosseae, 10th African Crop Science Conference Proceedings, Maputo, Mozambique, 10–13 October 2011, African Crop Science Society, pp. 269–271.

Phan, C.-W., & Sabaratnam, V. (2012). Potential uses of spent mushroom substrate and its associated lignocellulosic enzymes, *Applied Microbiology Biotechnology*, *96*(4), 863–873.

Rhodes, C. (2014). Mycoremediation (bioremediation with fungi)–growing mushrooms to clean the earth. *Chemical Speciation Bioavailability*, *26*(3), 196–198.

Rieger, P.-G., Meier, H.-M., Gerle, M., Vogt, U., Groth, T., & Knackmuss, H.-J. (2002). Xenobiotics in the environment: Present and future strategies to obviate the problem of biological persistence. *Journal of Biotechnology*, *94*(1), 101–123.

Sasek, V., & Cajthaml, T. (2005). Mycoremediation: Current state and perspectives, *International Journal of Medicinal Mushrooms*, *7*(3).

Sharifi-Rad, J., Butnariu, M., Ezzat, S. M., Adetunji, C. O., Imran, M., Sobhani, S. R., Tufail, T., Hosseinabadi, T., Ramírez-Alarcón, K., & Martorell, M. (2020). Mushrooms-rich preparations on wound healing: From nutritional to medicinal attributes, *Frontiers in Pharmacology*, *11*, 1452. 10.3389/fphar.2020.567518

Shamim, S. (2018). Biosorption of heavy metals, *Biosorption*, *2*, 21–49.

Singh, H. (2006). *Mycoremediation: Fungal Bioremediation*, John Wiley, & Sons.

Spadaro, J. T., Gold, M. H., & Renganathan, V. (1992). Degradation of azo dyes by the lignin-degrading fungus Phanerochaete chrysosporium. *Applied environmental microbiology*, *58*(8), 2397–2401.

Sotthisawad, K., Mahakhan, P., Vichitphan, K., Vichitphan, S., & Sawaengkaew, J. (2017). Bioconversion of mushroom cultivation waste materials into cellulolytic enzymes and bioethanol. *Arabian Journal for Science Engineering*, *42*(6), 2261–2271.

Stamets, P. (2005). *Mycelium Running: How Mushrooms can Help Save the World*, Random House Digital, Inc.

Stella, T., Covino, S., Čvančarová, M., Filipová, A., Petruccioli, M., D'Annibale, A., & Cajthaml, T. (2017). Bioremediation of long-term PCB-contaminated soil by white-rot fungi. *Journal of Hazardous Materials*, *324*, 701–710.

Tsujiyama, S., Muraoka, T., & Takada, N. (2013). Biodegradation of 2,4-dichlorophenol by shiitake mushroom (Lentinula edodes) using vanillin as an activator. *Biotechnology letters*, *35*(7), 1079–1083.

Thakur, M. (2019). *Mushrooms as a Biological Tool in Mycoremediation of Polluted Soils, Emerging Issues in Ecology and Environmental Science*. Springer, pp. 27–42.

Vidali, M. (2001). Bioremediation. An overview. *Pure Applied Chemistry*, *73*(7), 1163–1172.

Vimala, R., & Das, N. (2009). Biosorption of cadmium (II) and lead (II) from aqueous solutions using mushrooms: a comparative study. *Journal of hazardous materials*, *168*(1), 376–382.

Yogita, R., Simanta, S., Aparna, S., & Kamlesh, S. (2011). Biodegradation of malachite green by wild mushroom of Chhatisgrah. *Journal of Experimental Sciences*, *2*(10), 69–72.

Yildirim, A., & Hilal, A. (2020). Biosorption studies of mushrooms for two typical dyes. *Journal of the Turkish Chemical Society Section A: Chemistry*, *7*(1), 295–306.

Zebulun, H. O., Isikhuemhen, O. S., & Inyang, H. (2011). Decontamination of anthracene-polluted soil through white rot fungus-induced biodegradation. *The Environmentalist*, *31*(1), 11–19.

Zhang, J., & Elser, J. J. (2017). Carbon: Nitrogen: Phosphorus Stoichiometry in fungi: A meta-analysis. *Frontiers in Microbiology*, *8*, 1281–1281. doi:10.3389/fmicb.2017.01281

CHAPTER **16**

Wild Mushrooms: Characteristics, Nutrition, and Processing

**Liliana Aguilar-Marcelino[1], Laith Khalil Tawfeeq Al-Ani[2A,2B],
Filippe Elias de Freitas Soares[3], Fabio Ribeiro Braga[4], AnaVictoria Valdivia Padilla[5], and
Ashutosh Sharma[6]**

[1]Centro Nacional de Investigación Disciplinaria en Salud Animal e Inocuidad, INIFAP, Morelos, México
[2A]Department of plant protection, College of Agriculture, University of Baghdad, Baghdad, Iraq
[2B]School of Biology Science, UniversitiSains Malaysia, Minden, Pulau Pinang, Malaysia
[3]Department of Chemistry, Universida de Federal de Lavras, Minas Gerais, Brazil
[4]Universidade Vila Velha, Brazil
[5]Tecnologico de Monterrey, School of Engineering and Sciences, Campus Queretaro, Av. Epigmenio Gonzalez San Pablo, Querétaro, Mexico
[6]Tecnologico de Monterrey, School of Engineering and Sciences, Campus Queretaro, Av. Epigmenio Gonzalez San Pablo, Querétaro, Mexico

CONTENTS

16.1 INTRODUCTION

Wild mushroom are macrofungi that grow in forests on dead parts of hardwood, such as fallen branches and trunks. One of the major environmental problems nowadays in our world is the contamination of soil, water, and air with synthetic organic compounds coming from the waste

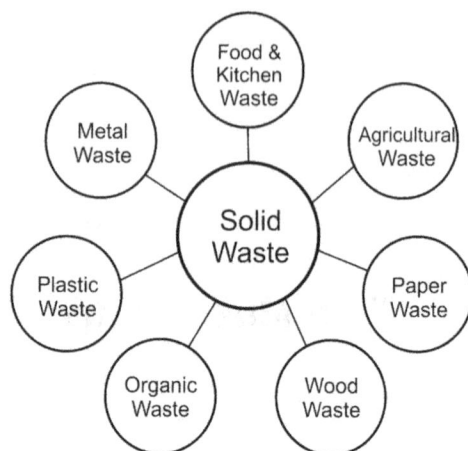

Figure 16.1 Different sources of solid waste.

produced by different activities like agriculture, municipal solid waste, and different industrial activity like pharmaceuticals, textile industries, food processing industries, etc. Most of these toxic compounds inserted into nature are called xenobiotics, which are materials that do not occur naturally in the biosphere and are probably not easily degraded by the indigenous microflora and fauna. Annually massive amounts of waste are being released into the environment, these wastes can be founded in three different forms: solid, liquid, and semi-solid waste and are being dumped or burned without proper recycling, thus contributing to human health hazards, environmental pollution, and global warming (Woldemariam, 2019). Solid waste can be defined as "any unwanted or useless solid materials generated from combined residential, industrial and commercial activities". Due to the increased pollution, industrialization, and urbanization, a trend of a significant increase in solid waste generation has been recorded worldwide. Waste generation has been observed to increase annually in proportion to the rise in population and urbanization (Jebapriya et al., 2013). According to data of the World Bank, 2.01 billion tons of municipal solid waste are produced annually in the world, with at least 33% of that not managed in an environmentally safe manner, and it is expected to grow to 3.40 billion tons by 2050, with the continuous trend in population growth (Kaza et al., 2018). The different sources of solid waste are illustrated in Figure 16.1.

Furthermore, the rapid industrialization and modernization of the recent years have brought, as a consequence, the release of industrial effluents and the accumulation of toxic substances into the biosphere, destroying the environment by the interaction with various components of the natural ecosystem. These wastes contained various synthetic dyes, toxic heavy metals, polycyclic aromatic hydrocarbons, and other chemicals that come in contact directly or indirectly with water and soil and destroy their natural properties like pH, total organic carbon (TOC), biological oxygen demand (BOD), and chemical oxygen demand (COD) (Gulzar et al. 2020; Kaushik et al., 2017).

Nowadays, incineration is the most common and effective remediation practice to reduce waste mass, but it is extremely costly in terms of money and energy used, and even though the ash that remains from this process is extremely small in quantity, compared to the original mass waste, it contains various toxic compounds, heavy metals, and induces the formation of other recalcitrant derivates that require further treatment and, if not disposed of them correctly, they can cause serious harm to the population's health and the environment. The concentration of carcinogenic heavy metals like As, Cd, Co, Cr, Cu, Fe, Hg, Mn, Ni, among others, is relatively high in untreated industrial wastes (Gulzar et al. 2020; Kaushik et al., 2018). Also, soils contaminated with heavy metals and other organic pollutants are generally left abandoned for several years because they may not be safe for agricultural production. Different classes of chemicals have been targeted by the United States Environmental Agency (USEPA) as priority pollutants due to their toxic effects on the environment and human health. These chemicals include polycyclic aromatic hydrocarbons, pentachlorophenols, polychlorinated biphenyls, 1,1,1- trichloro – 2,2-bis (4-chlorophenyl) ethane, benzene, toluene, ethylbenzene xylene, and trinitrotoluene. For all these reasons, a solution for the elimination of the wide ranges of pollutants and wastes from the soil and water is an absolute requirement to promote sustainable development of our society with low environmental impact (Woldemariam, 2019; Chawla et al., 2018). Due to the magnitude of this problem and the lack of a rapid cost-effective and ecologically responsible method, it is necessary to develop technology that helps minimize the impact that these wastes can have on the planet's ecosystems. In recent years, various physical and chemical remediation technologies have been developed to achieve the complete or partial removal of contaminants from the polluted sites and therefore provide a sustainable environment, but its high cost, limited application with limited opportunities, and the inability to improve the intrinsic health of the soil have made this technology almost obsolete and abandoned (Gulzar et al., 2020). Environmental biotechnology has been concentrating on the development of new "clean technologies" with a biological approach that focuses on reducing waste generation as well as its treatment and conversion in some useful form (Kulshreshtha et al., 2014). Bioremediation refers to the use of biological agents such as microbes, plants, fungi, or any other living things that help to either restore or clean up contaminated sites (Vishwakarma et al., 2020). The general approaches to bioremediation are to enhance natural biodegradation by natural organisms. This technology can be applied *in situ* when the contaminated materials are treated at the site, and *ex situ* when it is necessary to remove the contaminated material to be treated elsewhere (C. O. 2012). Generally, these treatments involved the use of microorganisms and their enzymes for the degradation and transformation of pollutants into another form that is less toxic for the environment. Various species of archaea, bacteria, algae, and fungi have shown bioremediation activity; also, plants and their associated bacteria play a significant role in the degradation of toxic compounds present, mainly in the soil and air (Sharma et al., 2018). The bioremediation rate of success depends on many factors, including the chemical nature and concentration of pollutants, the physicochemical characteristics of the environment, and their availability to microorganisms (Abatenh et al., 2017).

Fungi are one of the most widely used organisms of this type of technology since they are known as one of the most outstanding biological decomposed and they also play a crucial role in converting these wastes into valuable products. They exist in a variety of habitats due to their versatile physiological nature, thus found in acidic pH, temperature, oxygen concentrations, salinity, and heavy metal concentrations (Woldemariam, 2019). Mycoremediation based on detoxification of contaminated soil with the use of fungi is defined as a process of sequestration of contaminated soil or water by using fungi to reduce contaminants (Barkat Md Gulzar et al., 2020). The fungi are unique among the living organisms and are omnipresent in the biosphere. They are eukaryotic, spore-bearing, achlorophyllous microorganisms, and entirely heterotrophic. These organisms had the biochemical and ecological function to degrade organic chemicals, by producing extracellular enzymes which help in the assimilation of complex carbohydrates without prior hydrolysis. This leads them to have the ability to degrade a wide range of pollutants by chemical modification or by influencing chemical bioavailability (Vishwakarma et al., 2020).

Leaf litter can be found naturally on the forest floor. It cannot be digested directly by plants and used as a substrate to grow because the fallen leaves are too tough to be broken down. Fungus is a unique organism that can decompose leaf litter, mainly its mycelium, the vegetative part of the fungus, which we often see as fine white threads that grow from dead wood and leaves. Fungi are the only organisms that can decompose or break down wood. Their mycelium exudes powerful extracellular enzymes and acids that can decompose lignin and cellulose, the two essential components of plant fiber. A complex and rich material called humus is formed when fungi decompose wood and leaves. A variety of organisms of different species and kingdoms take advantage of these different substrates that are present in humus. The efficiency of this degradation becomes maximum when a good supply of nutrients are present in the soil, mainly elements such as N, P, K, and other essential inorganic elements (Rhodes, 2014). Another key consideration for successful mycoremediation is the correct selection of the fungal species. This should be based on the target pollutant to be removed. The most used fungi in bioremediation are wood-rot Basidiomycetes capable of degrading lignin (ligninolytic fungi). Most of these fungi cause white rot of wood, and so they are often called white-rot fungi (WRF). White-rot fungi has the capacity to degrade lignin due to a complex of extracellular enzymes—namely, lignin peroxidase, manganese dependent peroxidase, hydrogen peroxide generating oxidases, and phenol oxidases such as laccase (Sasek & Cajthaml, 2005).

16.2 KINDS OF WILD MUSHROOMS

A mushroom is a general term utilized mostly for the fruiting body of the macrofungi, belongs to Ascomycota and Basidiomycota families, and represents only a short reproductive stage in their life cycle (Krishna et al., 2015). The fruiting body of mushrooms consists of a steam (stipe) with a bearing cap (pileus). A spore-forming part (sporophore) emerges from an extensive underground network of threadlike strands (mycelium) (Woldemariam, 2019). A mushroom, or toadstool, is the fleshy, spore-bearing fruiting body of a fungus, typically produced aboveground on soil or its food source. A mushroom can be a variety of gilled fungi, with or without stems; therefore, the term is used to describe the fleshy fruiting bodies of some Ascomycota. Microscopic spores are produced by these gills, which help the fungus spread through the soil (Garziano, 2018).

Over 300 genera of mushrooms and related basidiomycetes have been identified; only a very few species of these fungi are cultivated commercially. This is since the majority of these species can only grow in very specific conditions in the wild (Woldemariam, 2019). It is estimated that there are approximately 1.5 million species of fungi on the planet. Of this number, it is believed that 140,000 species produce fruiting bodies of sufficient size and suitable structure to be considered macrofungi, which can be called mushrooms; and only 10% of these species have been identified and classified (Cheung, 2009). Because the use of the term "mushroom" has been so widely used, there are different ways in which they can be classified, as it is shown in Figure 16.2. Two different classifications present the first one, depending on their ecological role in nature, and the second one, according to some of their most notable functional characteristics.

16.2.1 Ecological classification

Mushrooms can be ecologically classified into three categories: parasites, saprophytes, and mycorrhizae (Cheung, 2009). Parasite mushrooms attack living organisms, penetrate their outer defenses, invade them, and obtain nourishment from living cytoplasm, thereby causing disease and sometimes death of the host. Most of these mushrooms are parasites of plants. There are only a few known parasitic mushrooms. One species with this characteristic is the honey fungus (*Armillaria mellea),* shown in Figure 16.3; it mainly attacks forest and fruit trees and causes them serious root rot.

Another type of mushroom in this classification is the mycorrhiza; for example, Perigold black truffle (*Tuber melanosporum*) and Matsutake mushroom (*Tricholoma matsutake*) show in Figure 16.4. These are considered some of the most expensive edible mushrooms in the world due to their scarcity. Unlike other mushrooms, cultivating them artificially is very difficult and are mostly found exclusively in the wild.

The symbiotic relationships that form between fungi and plants are called mycorrhizae (Chawla et al., 2019). When fungi occupy the root system of a plant, they provide greater

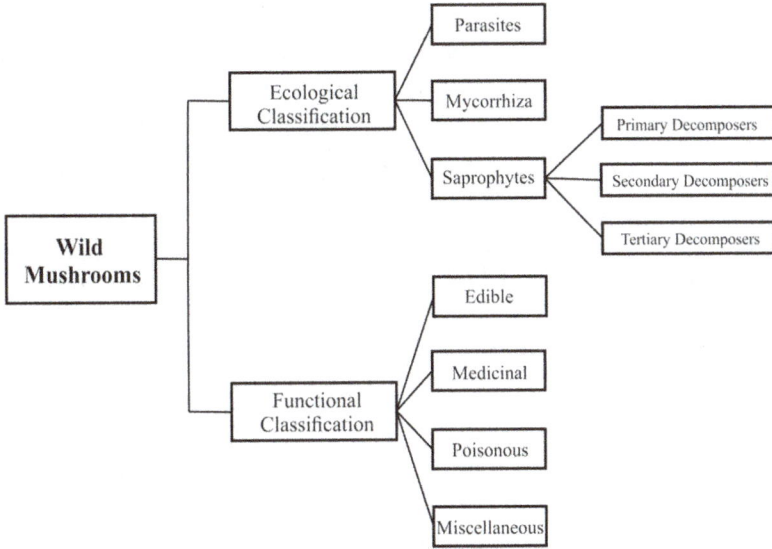

Figure 16.2 Schematic representation of wild mushroom classification according to ecological role and functional properties.

Figure 16.3 Parasite mushroom *Armillaria mellea* or honey fungus, growing in tree roots.

(a)   (b)

Figure 16.4 Examples of mycorrhizae mushrooms. (a) Perigold black truffle (*Tuber melanosporum*), (b) Matsutake mushroom (*Tricholoma matsutake*).

absorption of water and nutrients from the soil, while the plant supplies the fungus with carbo-hydrates that it produces during photosynthesis. Mycorrhizae can also increase the protection of the host plant against some pathogens (Writers of The New York Botanical Garden, 2003).

The last type of mushroom in this classification is the saprophyte. These correspond to the vast majority of wild mushrooms and are known as the premier recyclers on the planet. These mushrooms obtain nutrients from dead organic materials. The filamentous mycelial network is designed to weave through the plant cell wall, due to the secretion of enzymes and acids that can degrade complex molecules into simpler compounds. All ecosystems depend upon fungi´s ability to decompose organic plant matter and the return of carbon, hydrogen, nitrogen, and minerals back into the ecosystem in forms usable to other organisms (Stamets, 2001). As decomposers, they can be separated into three key groups, depending on their nutritional and ecological requirements (Chawla et al., 2021).

Primary Decomposers. These mushrooms can grow on fresh or almost fresh wood residues. They are typically fast growers with a mycelium that can quickly attach and decompose plant tissue. These fungi are the first to start the wood degradation process. They have the enzymatic machinery to degrade lignin-cellulose materials and absorb nutrients from the decaying material and prefer environments with little to no competition from other organisms. Some examples of these mushrooms are shiitake *(Lentinula edodes)*, oyster mushrooms *(Pleurotus ostreatus)*, enoki *(Flammulina velutipes)*, cloud ear *(Auriculariapolytricha)*, golden pholiota *(Pholiota aurivella)*,

Figure 16.5 Species of saprophytes mushrooms, primary decomposers growing in wood. (a) Reishi *(Ganoderma lucidum)*; (b) oyster mushrooms *(Pleurotus ostreatus)*; (c) enoki *(Flammulina velutipes)*; (d) spring agaric *(Agrocybe praecox)*.

golden jelly fungus (*Tremella mesenterica*), spring agaric (*Agrocybe praecox*), and reishi (*Ganoderma lucidum*), shown in Figure 16.5.

Secondary Decomposers. These mushrooms rely on the previous activity of primary decomposers to partially break down a substrate to a state they can take advantage of and get nutrients to grow (Bains et al., 2021). Secondary decomposers normally grow from composted material. This environment is accompanied by other organisms like insects, bacteria, actinomycetes, yeast, etc.; all of these organisms degraded the plant material and reduced mass, structure, and composition of the compost, and proportionately available nitrogen is increased (Stamets, 2001). Some commonly secondary decomposers are the paddy straw mushroom (*Volvariella volvaceae*) and inkcaps (*Coprinus atramentarius*), shown in Figure 16.6, and the fungi questionable stropharia *(Stropharia ambigua).*

Tertiary Decomposers. These are the fungi found toward the end of the decomposition process. They thrive in habitats created by primary and secondary decomposers are amorphous and difficult to categorize groups and rely upon highly complex microbial environments. The differentiation between secondary and tertiary decomposers is often complicated, so mycologists often refer to these mushrooms as "soil dwellers". Fungus existing in these reduced subtracts is remarkable in that the habitat appears inhospitable for most other mushrooms (Stamets, 2001). Some examples of this category are the orange peel mushroom (*Aleuria aurantia*) shown in Figure 16.7; and the

Figure 16.6 Illustration of the mushroom inkcap *(Coprinus atramentarius),* growing in forest soil.

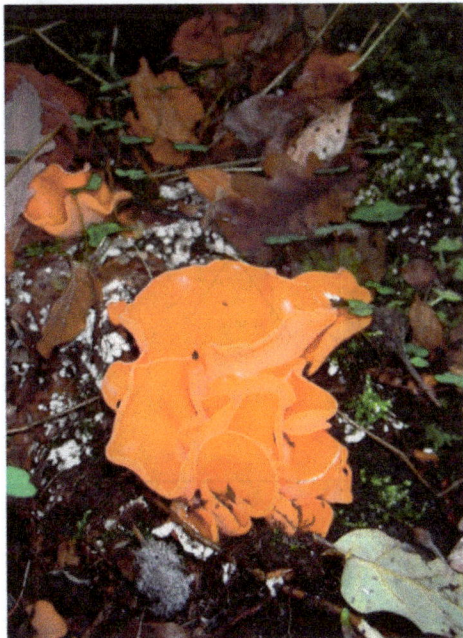

Figure 16.7 An orange peel mushroom (*Aleuria aurantia*) growing in composted soil.

banded mottle gill *(Panaeolus cinctulus).* Some other species are *Conocybe, Agrocybe, Pluteus,* and *Agaricus,* known as tertiary decomposes.

It is important to mention that some mushrooms can share two of these categories; for example, some *Ganoderma* spp., including the lingzhi mushroom (*Ganoderma lucidum*) that is a common saprophyte but can be pathogenic too; also, the matsutake mushroom (*Tricholoma matsutake*), while initially appearing to be mycorrhizal on young roots, soon becomes pathogenic and finally exhibits some saprophytic ability (Cheung, 2009). Figure 16.8 shows the triangle model for mushroom's ecological classification and examples of mushrooms that can belong to more than one group.

16.2.2 Functional classification

Another way to categorize fungi depends on the important characteristics or properties that they present. Mushrooms then can be roughly divided into four categories: (1) edible mushrooms, (2) medicinal, (3) poisonous or toxic, and (4) miscellaneous (Chang & Miles, 2004). Certainly, this approach of classifying mushrooms is not absolute as many kinds of mushrooms can belong to two or more categories, some mushrooms for example are not only edible but also possess tonic and medicinal qualities.

Edible Mushrooms. Wild edible mushrooms have been part of the human diet for thousands of years due to their nutritional, organoleptic characteristics, and medicinal properties. They are a source of fiber, protein, vitamins, minerals, and are poor in calories and fat. Moreover, edible mushrooms provide a nutritionally significant content of vitamins B1, B2, B12, C, D, and E; they have been considered as a very useful complement to the vegetarian diet since mushrooms can provide all the essential amino acids for adults requirements, have more protein than most ve-getables, and are the only non-animal natural source of vitamin D (Sun et al., 2020).

The most common edible part of mushrooms is their fruiting bodies, although the sclerotia of some mushrooms can also be consumed (Das et al., 2021). Mycelia are less commonly utilized as a human food despite their shorter production time and comparable nutritional value to fruiting bodies (Cheung, 2009).

Less than 25 species of edible mushrooms are broadly cultivated and accepted as food of economic importance. Worldwide, the most cultivated mushroom is *Agaricus bisporus*, followed by *Lentinus edodes*, *Pleurotus* spp., and *Flammulina velutipes*. However, wild mushrooms are becoming more important for their nutritional, sensory, and especially pharmacological char-acteristics (Valverde et al., 2017). Because of their high nutritional value and unique flavor, edible mushrooms are known and consumed as a delicacy. Mushrooms are widely distributed across the

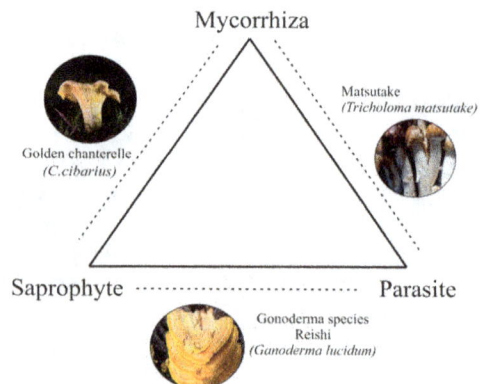

Figure 16.8 Triangular model for the ecological classifica-tion of mushrooms. Modified from (Cheung, 2009).

earth, with approximately 20,000 species; among them, more than 3,000 are edible, and 200 of those are wild species (Sun et al., 2020).

China ranks high in the early and later historical record of wild edible mushrooms. In this country, many species of mushrooms have been considered valuable for several years, not only for their taste and nutritional worth but also for their healing properties. Other countries that have a long and notable tradition of wild edible fungi are Turkey, Mexico, and major areas of central and southern Africa (Boa, 2008). Figure 16.9 shows some of the most famous consumed wild mushroom species, including the chanterelle mushroom *(Cantharellus cibarius)*, morchella *(Morchella esculenta)*, lion's mane mushroom *(Hericium erinaceus)*, maitake mushroom *(Grifola frondosa),* and oyster mushroom *(Pleurotus ostreatus).*

Medicinal Mushrooms. A large number of species of mushrooms have been used to treat diseases over the years as a tradition in different cultures through their immunomodulatory and antineoplastic properties. Nowadays, the study of the medicinal properties of mushrooms has increased rapidly, and it has been suggested that many mushrooms are like mini-pharmaceutical factories producing compounds with miraculous biological properties (Kaushik et al., 2013). Some examples of bioactive molecules synthesized by mushrooms are polysaccharides, proteins, fats, minerals, glycosides, alkaloids, volatile oils, terpenoids, tocopherols, phenolics, flavonoids, carotenoids, folates, lectins, enzymes, ascorbic, and organic acids, in general (Valverde et al., 2017).

More than a hundred medicinal functions have been proven to be produced by fungi and mushrooms. Some of their medicinal studied actions included antitumor, immunomodulating, antioxidant, radical scavenging, cardiovascular, anti-hypercholesterolemia, antiviral, antibacterial, antiparasitic, antifungal, detoxification, hepatoprotective, and antidiabetic effects (Wasser, 2011). Some of the most important drugs in recent years like antibiotics (penicillin, tetracycline, and erythromycin), antiparasitic (avermectin), antimalarials (quinine, artemisinin), lipid control agents (lovastatin and analogs), and anticancer drugs (taxol, doxorubicin) were discovered from components found in fungi. Of the 14,000 to 15,000 species of mushrooms in the world, around 700 have known medicinal properties. However, it has been estimated that there are about 1,800 species of mushrooms that have potential medicinal attributes (Chang and Miles, 2004).

Ganoderma species, the "mushroom of immortality," are commonly known as reishi, and are an example of a wild medicinal mushroom. Studies of *Ganoderma lucidum* have shown that this fungus contains edible compounds in high concentrations like organic germanium, polysaccharides, and triterpenes. These active components have been proven for strength, regulate the immune system, and eliminate allergic reactions such as asthma, rheumatoid, arthritis, and lupus. *G. lucidum* can also

(a)   (b)

Figure 16.9 Common edible wild mushrooms species. (a) Morchella mushroom *(Morchella esculeta);* (b) chanterelle mushroom *(Cantharellus cibarius).*

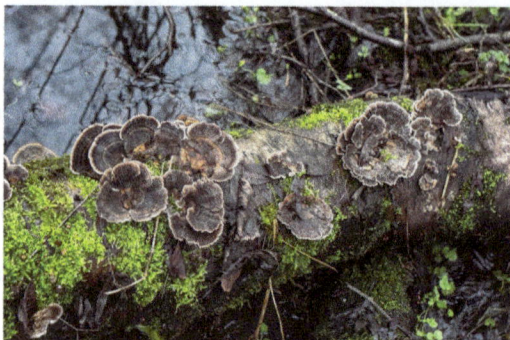

Figure 16.10 Medicinal mushroom Turkey tail has shown to have chemopreventive potential and the inhibition of several human cancer cell lines.

stimulate the production of interleukin-II due to the presence of ganoderic acid, which is active against liver cancer (Olusola Ogidi, Olusegun Oyetayo, and Juliet Akinyele 2020). Some other species of mushrooms that also possess health benefits are the Turkey tail mushroom *(Trametes versicolor)*, shown in Figure 16.10, that promote chemopreventive potential. It inhibits the growth of several human cancer cell lines and act as an adjuvant in breast cancer prevention. Also, maitake mushroom *(Grifola frondosa)* is promoted as an anticancer agent, particularly on human gastric carcinoma. Such an effect results from the induction of cell apoptosis and could significantly accelerate the anticancer activity (Valverde et al., 2017).

Poisonous or Toxic Mushrooms. This category of mushrooms represents less than 1% of the world's identified mushrooms. This species contained toxins with different chemical compositions, and thus the effect of poisoning differs considerably according to the species involved (Chang & Miles, 2004). Poisonous mushrooms vary in their effects of the mild stomach and digestive upsets to more serious problems like liver damage and death. There are no general guidelines for distinguishing between poisonous and edible species (Dhull et al., 2020). The only means by which a non-specialist can determine the edibility or toxicity of a given mushroom is to carry out an accurate identification of the specimen. Such identification may be obtained by consulting the relevant literature, preferably with illustrations, or experts in the subject (Boa, 2008).

Several species of *Amanita* (Figure 16.11) are extremely poisonous, but obvious symptoms do not appear until 8–12 hours after ingestion. The dead cap fungi *(Amanita phalloides)* group causes the majority of recorded deaths from mushrooms. The general symptoms of this type of poisoning are severe abdominal pains, nausea, violent vomiting, diarrhea, cold sweats, and excessive thirst

(a)   (b)

Figure 16.11 *Amanita* species with poisonous activity. (a) The dead cap fungi *(Amanita phalloides)*; (b) the fly agaric or fly amanita *(Amanita muscaria).*

(Chang & Miles, 2004). The psychotropic or hallucinogenic activity can also be an effect of poisonous mushrooms. Hallucinogenic compounds contained in fungi of *Psilocybea* and *Amanita* genera like psilocin, ibotenic acid, and muscimol act on the central nervous system, producing distortions in vision and tactile sensations, and a similar neurotoxic reaction as LSD (*d*-lysergic acid diethylamide). Chronic exposure to these substances can induce serious disorders like cardiovascular complications and neurodegenerative diseases. Shrooms or "magic mushrooms" *(Psilocybe cubensis)* are the more known and common sources of psilocybin (Stebelska, 2013).

Miscellaneous Mushrooms. This category includes those mushrooms whose properties remain less well defined and can group the mushrooms used in other applications not mentioned before, like the production of enzymes, pigments, bioremediation of xenobiotics, and lignocellulosic waste degradation. The use of fungi as a remediation tool is also called "mycoremediation". Mycoremediation refers to the use of mushrooms and their enzymatic ability to degrade a wide variety of environmentally persistent pollutants and transform industrial and agro-industrial wastes into products (Kulshreshtha et al., 2014). Mushrooms are capable of converting lignocellulosic waste materials into food, feed, and fertilizers. They can be cultivated in containers, providing artificially controlled conditions, are relatively fast-growing organisms, and can also act as effective bio absorbents of toxic metals like Cd, Cu, Pb, Cr, and Mn (Woldemariam, 2019). The enzymes produced by mushrooms, which are lignin peroxidase, manganese peroxidase, and laccase penetrate, break, and digest, or mineralizes harmful substances in waste. Fungi such as the golden oyster mushroom *(Pleurotus citrinopileatus)* can be cultivated with the use of handmade paper sludge and industrial cardboard waste as a substrate to grow, turning these organic wastes into compost (humus), which has very high agricultural value. It is used as fertilizer, and it is nonodorous and free of pathogens (Jebapriya et al., 2013).

16.3 DISTRIBUTION OF WILD MUSHROOMS

Fungi are the most diverse organisms on Earth since they have a crucial role in the ecology of the planet (Tedersoo et al., 2014). Wild edible mushrooms have been collected and consumed by people for millions of years. Wild edible mushrooms have been found in fossilized wood with a time of approximately 300 million years (Ram et al., 2010). As an example of these fungi, some genera have been described such as *Volvarias, Polipore,* and tuber fungi that used ethnobotanical food by the different tribal of forest regions of India and Nepal (Alexopolous et al., 1996). Nowadays, about 1,200 species of fungi belonging to the orders Agaricales, Russulales, and Boletales are described in comparison to about 14,000 species of mushrooms reported worldwide that contribute 10% of the global mushroom flora. So far, about 1,105 to 1,208 species of mushrooms belonging to 128–130 genera have been documented and, among these, 300–315 species belonging to 75–80 genera are considered edible (Thiribhuvanamala et al., 2011). The distribution of wild edible mushrooms is divided into forests, deserts, and mountains. The distribution of the aforementioned wild edible mushrooms is described below.

16.3.1 Forests

The forest is a refuge from the daily routine of work, and the proliferation of fungi contributes significantly to improve their quality of life for people. For mushroom pickers living in rural areas, the occurrence and availability of mushroom picking sites are often even more important as it contributes to their diet or even provides an occasional additional income. In this context, a study by Olah et al. (2020) studied the ecological preferences of mainly ectomycorrhizal fungi and applied them as parameters to model the potential that forests represent in mushroom cultivation in central Slovakia. Subsequently, they analyzed the theoretical aspects of the demand for wild mushrooms as a subsistence provisioning service for the local population with a special focus on the socially disadvantaged inhabitants. The results showed that there is an overlapping space of forest stands with a high potential for mushroom cultivation and the districts with the highest proportion of unemployed or inhabitants receiving social benefits, and the best mushroom forest the stands are located within walking distance of the settlements. This supports the initial assumption that wild mushrooms can contribute to improving the lives of disadvantaged local communities. Wild edible mushrooms in the forest have become challenging, as some species of these mushrooms can be cultivated. Mycorrhization practices offer a promising advance to produce currently uncultivable wild edible mushrooms. The persistence of the production of edible species, whether cultivated or wild, depends both on several factors, among which the tree stands out, as well as on ecology and the environment (fungal communities, climate, soil, tree development) (Savoie & Largeteaue, 2011).

16.3.2 Deserts

Within the wild edible fungi are the mushrooms that grow in extreme environments such as the desert. A very interesting fungus that grows in these environments is *Podaxis pistillaris*. The fungus *P. pistillaris* is commonly found in most desert areas worldwide. The oval-shaped peridium, the rigid woody stipe, plus a 10–15 μm spore size stand out among its morphological features. Even though this fungus is used for human consumption and several traditional remedies, a lack of knowledge regarding its fibrinolytic enzymatic system still prevails. This fungus was collected from the central region of the Mexico Is Sonoran Desert (29° 07.23′ 97" LN and 110° 53.58′ 02" LW, 238 masl).

Mexico has an important fungal diversity. In the north of the country, specifically Sonora, the vegetation is made up of desert scrub, microphyllous desert scrub, tropical thorn forest, tropical deciduous forest, oak, and oak-pine forest. Many of these fungi tolerate long dry periods, which characterize arid and semi-arid regions (200–350 mm/yr) and play a decisive role in the conservation of ecosystems, especially in the recycling of organic matter. In these extreme environments, fungi were isolated. These mushrooms of agaricoid and gastroid habits, traditionally classified in the artificial group Gasteromycetes. Nine species of mushrooms, which are little known worldwide, or which occur in a very small area, were identified from samples collected during the last decade in Sonora. These include *Endoptychum arizonicum, Araneosa columellate,* and *Calvatia pygmaea* (Moreno et al., 2007).

16.3.3 Mountains

Hawksworth (2001) reported that only about 6.7% of the estimated 1.5 million species of fungi in the world have been described and most of them are in temperate regions. The tropical region that undoubtedly harbors the highest mycodiversity has been inadequately sampled and my oflora poorly documented (Douanla-Meli et al., 2007; Hawksworth, 2004). In 2010, in the mountains of the Peruvian Andes, edible mushrooms were found at high altitudes of 3,800–4,000 masl, a characteristic mushroom that they found was *Calvatia cynthiformis;* this edible mushroom contains 55% protein and is used in the Andes as meat or cheese when it is young (Trutmann and Luque, 2011).

16.4 TYPES OF WASTE MATERIAL

The FAOSTAT (The Food and Agriculture Organization Statistical Database) reported in 2018 that China is the country with the highest mushroom production, followed by the United States and the Netherlands with 8.99 million tons. On the other hand, on the American continent, a production of 666,000 tons was estimated (Sánchez et al., 2018). Derived from the production of wild edible mushrooms, currently the generation of agro-industrial waste worldwide is considerable and the elimination and burning of this agro-industrial waste have generated various environmental impact problems around the world. On the other hand, agricultural waste is composed of various inputs as detailed: banana leaves (Tarrés et al., 2017), coffee husk, corn bran (Graminha et al., 2008), grasses (Sun & Chen, 2002), oat bran (Graminha et al., 2008), rice straw (El-Tayeb et al., 2012), pine sawdust (Buzala et al., 2015), wheat bran (Graminha et al., 2008), among others. However, agro-industrial waste can be turned into different high value-added products such as biofuels, fine chemicals, and cheap energy sources for microbial fermentation and enzyme production. Agro-industrial waste for disposal can represent a source of energy and carbon. Additionally, this form of waste is a source of the necessary nutrients for the growth of edible and lignocellulolytic fungi and production of enzymes by solid-state fermentation (Kumla et al., 2020). The fungal species employed as pre-treatments of substrates used for biogas and methane production are *Pleurotus* spp. (wheat straw) (Albornoz et al., 2018), *Lentinula edodes* (shiitake) (wood chips) (Lin et al., 2015), and *Flammulina velutipes* (*Agropyron elongatum*) (Lalak et al., 2016).

16.5 BIOLOGY OF WILD MUSHROOMS AND ENZYMES PRODUCED

Enzymes are macromolecules with the biological function of catalyzing chemical reactions. This is because they can lower the energy barrier needed for the reaction to happen, thereby increasing their speed. Except for a small group of catalytic RNA molecules, called ribozymes, all enzymes are proteins. Sometimes the same enzyme has two or more names, or two enzymes have the same name. Due to this ambiguity and also to the increasing number of enzymes that are discovered, biochemists, through an international agreement, adopted an enzyme nomenclature and classification system. This system divides the enzymes into six classes, each with subclasses, based on the types of reactions that catalyze: oxidoreductases, transferases, hydrolases, lyases, isomerases, and ligases. A four-part classification number and a systematic name, which identifies the catalyzed reaction, are specific to each enzyme (Nelson & Cox, 2018).

Microorganisms are the main agents used for the production of enzymes (Sanchez & Demain, 2017). Among these microorganisms, wild mushrooms stand out. The mushrooms belong to the fungi kingdom and can be classified as edible, poisonous, and hallucinogenic (Castañeda-Ramírez et al., 2020). During the process of mycelium colonization and in the production of mushrooms, enzymes play a fundamental role (Chang & Wasser, 2018). Mushrooms produce extracellular enzymes that are induced according to the medium in which these fungi grow. In particular, mushrooms produce hydrolytic and oxidative enzymes that are involved in the degradation of material rich in lignin, such as wood (Tan & Wahab, 1997). Wood-degrading fungi can be classified into white, brown, and soft rot fungi based on their growth substrate preferences and wood-decaying patterns. Certain mushrooms are classified as white and brown rot fungi (Kameshwar & Qin, 2016; Peralta et al., 2017).

Brown rot fungi produce bracket-shaped fruiting bodies, although their nomenclature derivates from the brown decaying wood produced by these fungi. The color brown comes from the brownish residue that breaks into cubical fragments. This residue is observed because brown rot fungi attack cellulose but do not significantly degrade lignin. On the other hand, most of the cellulose and hemicelluloses are degraded, leaving the lignin more or less intact (Krah et al., 2018; Peralta et al., 2017). Brown-rot fungi can metabolize cellulose and hemicellulose; however, they have no lignin-degrading enzymes except small molecule reactive species to depolymerize lignin (Andlar et al., 2018). In these mushrooms, a chelator-mediated Fenton (CMF) system has evolved to replace at least some of the cellulolytic enzymatic machinery (Zhu et al., 2020). CMF system works based on oxygen radical reactions, trigging the formation of hydroxyl radicals ($\cdot$OH) that attack the wood cell wall-associated at sites where iron-binding occurs. It allows nonenzymatic deconstruction of the cellulose (Andlar et al., 2018; Umezawa et al., 2020). Brown rot fungi are generally found in gymnosperms, considered specialists in substrates derived from this group of plants (Umezawa et al., 2020). Their prevalence in nature among wood-rotting basidiomycetes species is approximately 7% (Hatakka & Hammel, 2011).

White rot fungi account for over 90% of all wood-rotting mushrooms (Gilbertson, 1980). They are classified into the Agaricomycetes class, with about 10,000 species, and subdivided into six families: Phanerochaetaceae, Polyporaceae, Marasmiaceae, Pleurotaceae, Hymenochaetaceae, Ganodermataceae, and Meruliaceae (Peralta et al., 2017). However, only a few dozen have been properly studied. These fungi are more commonly found on angiosperm than on gymnosperm wood species in nature (Hatakka & Hammel, 2011). These mushrooms have this nomenclature (white rot) due to the white aspect that the wood affected by these fungi develops (Peralta et al., 2017). White-rot fungi can decompose lignin, cellulose, and hemicellulose. In comparison with brown rot and soft rot fungi, their degradation of lignin is much more efficient, because they possess a unique ability to complete mineralization to CO_2 due to their arsenal of enzymes (Couturier & Berrin, 2013; Andlar et al., 2018). White rot fungi have a high copy number of genes encoding different carbohydrate-active enzymes (CAZymes), specially ligninolytic enzymes (Kameshwar & Qin, 2016). Kameshwar and Qin (2016) suggest that white rot fungi possess higher lignin and soft rot fungi potentially possess higher cellulolytic, hemicellulolytic, and pectinolytic abilities, based on a comparative genome study. Later, we will discuss more of the ligninolytic enzymes produced by wild mushrooms. It is necessary to emphasize that countless other enzymes are produced by wild mushrooms and that they have fundamental importance in their metabolism and nutrition. One class of these enzymes, for example, are proteases, which play a key role in the process of nematode consumption by these fungi (Ferreira et al., 2019; Soares et al., 2019).

16.6 THE ROLE OF WILD MUSHROOMS IN BIOREMEDIATION

Currently, various activities such as the burning of agricultural inputs, wood, fossil fuels, coal mining, oil extraction, and release recalcitrant pollutants such as aromatic polycyclic hydrocarbons are easy to eliminate once they have been released into the environment (Rhodes, 2013; Singh, 2006). All these activities represent a threat to the sustainability and sustainability of the environment. Another serious problem is the disposal of pesticides and anthelmintic in aquifers. They do not carry an adequate pre-treatment process for effective elimination. In this context, it is necessary and urgent to propose and use sustainable alternatives for the treatment of these polluting wastes of the environment (Hyde et al., 2019).

In general, fungi are the most effective decomposers in nature, the process of bioremediation with fungi is also known as "mycoremediation" and this process is the use of live fungi to clean up contamination through mineralization using various enzymes or by absorption. The use of mycoremediation has been applied to clean the air, particularly with *Agaricus bisporus* spent substrate has been evaluated as a mix with other materials for removal of H_2S (Shojaosadati & Siamak, 1999) or volatile organic compounds (Mohseni & Allen, 1999).

Regarding the cleaning of contaminated water, the spent substrate of *Pleurotus* spp. has been investigated for the removal of copper (Tay et al., 2010) and nickel-contaminated water (Tay et al., 2011), reduction of phenol content, toxicity in olive mill waste (Martirani et al., 1996), pesticides in effluents from the fruit packing industry (Karas et al., 2015), antibiotics in swine wastewater (Chang et al., 2014), and textile dyes (Singh et al., 2011). Regarding the cleaning of contaminated soil, two edible fungi have also been used, such as *Agaricus bisporus* and *Pleurotus* spp. *A. bisporus* spent substrate-enzymes has been studied for their effect on the distribution of zinc, cadmium, and lead among soil fractions; the amelioration of zinc toxicity; and the degradation of chlorophenols, polycyclic aromatic hydrocarbons, or aromatic monomers (Rinker, 2017). On the other hand, *Pleurotus* spp. spent substrate has been explored for degradation of polycyclic aromatic hydrocarbons, removal or degradation petroleum, and remediation of mining contaminated soils (Rinker, 2017). Finally, the species of fungi that have been used to degrade pesticides are *Aspergillus niger, A. terreus, Cladosporium oxysporum, Fusarium ventricosum, Rhizopus oryzae, Trichoderma harzianum*, and *Phanerochaetechrysosporium* (Bhalerao and Puranik 2007; León-Santiesteban et al., 2016).

16.7 THE IMPORTANCE OF LIGNOCELLULOLYTIC ENZYME-PRODUCING BY WILD MUSHROOMS

The main constituents of lignocellulose are cellulose (40–50%), followed by hemicellulose (25–30%) and lignin (15–25%) (Stech et al., 2014). The presence of these polysaccharides makes the cell wall of the plant recalcitrant. Cellulose is a linear homopolymer composed of linear chains of D-glucose linked by β-1,4-glycosidic bonds. Cellobiose is the repeating disaccharide unit of cellulose (Parisutham et al., 2017). Hemicellulose, the second most abundant polysaccharide of plants, is mainly composed of xylan (β (1→4) D-xylose units), xyloglucan, β-glucans (β (1→3) (1→4) D-glucose), and mannan (β 1→4 D-mannose) (Kumar et al., 2008; Sjostrom, 2013). Lignin is a complex aromatic heteropolymer derived from three cinnamyl alcohol monomers (p-coumaryl alcohol, coniferyl alcohol, and sinapyl alcohol). There are three basic structural monomers in lignin: p-phenyl monomer (H type) derived from coumaryl alcohol, guaiacyl monomer (G type) derived from coniferyl alcohol, and syringyl monomer (S type) derived from sinapyl alcohol (Ahuja & Roy, 2020).

Lignocellulose complete degradation requires the synergistic action of a large number of oxidative, hydrolytic, and nonhydrolytic enzymes (Peralta et al., 2017). In nature, the degradation of lignin is carried out mostly by basidiomycetes, especially white-rot fungi. As seen above, brown rot fungi attack cellulose but do not significantly degrade lignin(Krah et al., 2018; Peralta et al., 2017). Carbohydrate-active enzymes (CAZy) are associated with lignocellulose degradation catalysis. They have been classified into six classes: glycoside hydrolase (GH), glycosyltransferase (GT), auxiliary activity (AA), carbohydrate esterase (CE), polysaccharide lyases (PL), and carbohydrate-binding domains (Lombard et al., 2013). For complete catalysis of cellulose degradation, several glycoside hydrolases act in synergism, including endocellulases (EC 3.2.1.4), exocellulases (cellobiohydrolases, CBH, EC 3.2.1.91), glucano-hydrolases (EC 3.2.1.74), and beta-glucosidases (EC 3.2.1.21). Cellulose hydrolysis can be summarized by (i) the adsorption of cellulases via carbohydrate-binding modules, (ii) complexation of cellulose with the catalytic domain, (iii) hydrolysis of the glycosidic bonds, and (iii) desorption of cellulases (Jeoh et al., 2017; Houfani et al., 2020). Also, enzymes called lytic polysaccharide monooxygenases act synergistically with cellulases during cellulose hydrolysis catalysis (Arfi et al., 2014; Houfani et al., 2020).

Hemicellulose hydrolysis catalysis involves the synergic action of several enzymes, at different points of the hemicellulolytic matrix, including endo-1,4-β-xylanase (EC 3.2.1.8), β-xylosidase (EC 3.2.1.37), β-mannanase (EC 3.2.1.78), β-mannosidase, (EC 3.2.1.25), α-glucuronidase (EC 3.2.1.1), α-l-arabinofuranosidase (EC 3.2.1.55), acetyl xylan esterase (EC 3.1.1.72), p-coumaric and ferulic acid esterases (EC 3.1.1.1), and feruloyl esterases (EC 3.1.1.73). This cooperative action promotes the hydrolysis of glycosidic bonds, ester bonds, and removes the substituent chains or side chains (Peralta et al., 2017; Andlar et al., 2018). Xylan is the main hemicellulose polysaccharide in hardwoods and herbaceous biomass. On the other hand, mannan is the major component of hemicellulose in softwood (Hu et al., 2020; Andlar et al., 2018). Thus, the nature of the material to be degraded by the fungus must be taken into account.

Cellulases and hemicellulases have several important applications in biotechnology, for example, in the conversion of biomass to biofuels. Mushrooms are the main source of production for these enzymes, with much research and investment being made available for the development of increasingly efficient and cheap enzymatic cocktails. However, it is worth noting that the amount of mushrooms already studied is small, close to the total available in nature. Thus, the bioprospecting of wild mushrooms aiming at the production of cellulases is very important.

In nature, white rot fungi are the main actors in lignin degradation, performing it more rapidly and extensively than other microorganisms (Woiciechowski et al., 2013). Some white-rot fungi species produce enzymes that preferentially catalyzes lignin degradation than hemicellulose and cellulose degradation, e.g. *Ceriporiopsis subvermispora*, *Phellinus pini*, *Phlebia* sp., *Pleurotus* sp., *Phanerochaete chrysosporium*, *Trametes versicolor*, *Heterobasidion annosum*, and *Irpex lacteus* (Andlar et al., 2018). White rot fungi can mineralize lignin into carbon dioxide and water by their unique H_2O_2 production and extracellular enzyme system, which includes the H_2O_2 production system, with glucose 1-oxidase (EC1.1.3.4, CAZy AA3), glyoxal oxidase (EC 1.2.3.5), and cellobiose dehydrogenase (EC 1.1.99.18; CAZy AA3) and the lignin oxidase system, which includes laccases (EC 1.10.3.2; CAZy AA1), manganese peroxidases (EC 1.11.1.13; CAZy AA2), and lignin peroxidases (EC1.11.1.14; CAZy AA2) (Kersten & Cullen, 2007; Kracher & Ludwig, 2016).

Laccases are multicopper enzymes able to oxidize phenolic compounds and generate water as a by-product using oxygen as an electron acceptor. These "blue oxidases" contain four copper atoms per mol of the enzyme, arranged into three metallocentres between three structural domains formed by a single polypeptide of about 500 amino acids in length. They can be monomeric, dimeric, or tetrameric glycoproteins (Tramontina et al., 2020; Andlar et al., 2018; Hatakka & Hammel, 2011). Moreover, laccases are considered eco-friendly catalysts, once they do not generate toxic by-products. Thus, laccases can be used in various fields such as biofuels and food production, pulp, and paper treatments, textile industry, nanobiotechnology, soil bioremediation, synthetic chemistry, and cosmetics (Moreno et al., 2020).

Lignin peroxidases catalyze the depolymerization of lignin in the presence of H_2O_2. They can oxidize phenolic aromatic substrates and also a variety of non-phenolic lignin model compounds as well as a range of organic compounds (Wong, 2009). On the other hand, manganese-dependent peroxidases catalyze the oxidation of $Mn2+$ ions to $Mn3+$ in the presence of hydrogen peroxide. White rot fungi secrete several types of manganese-dependent peroxidases, with a molecular weight ranging from 38 to 62.5 kDa, ~ 350 amino acid residues (Martin, 2002; Andlar et al., 2018). Both classes of oxidases have research on biocatalysts for lignin and lignocellulose conversion, dye compound degradation, activation of aromatic compounds, and biofuel production (Lundell et al., 2017). Classification and mechanism of

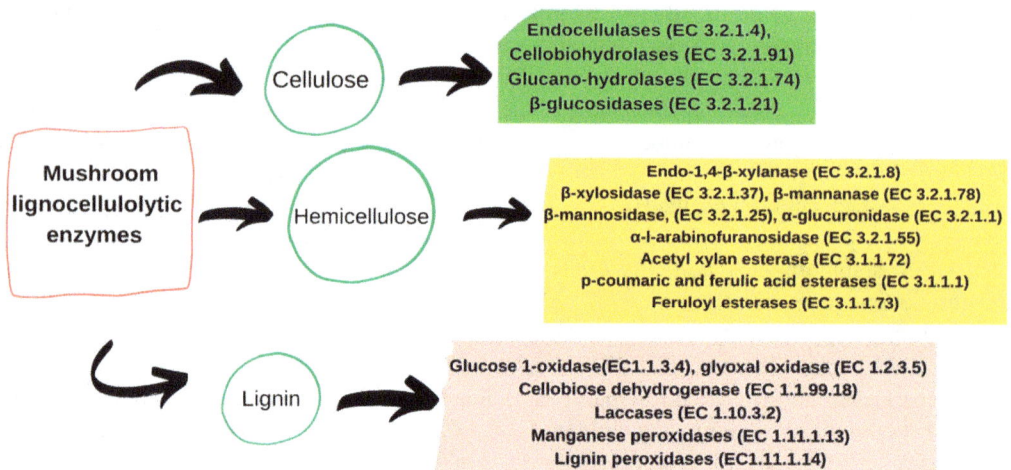

Figure 16.12 Classification and mechanism of action of mushroom lignollulolytic enzymes.

action of mushroom lignocellulolytic enzymes are shown in Figure 16.12.

Research on enzyme applications has shifted from focusing on a single enzyme preparation to enzyme cocktails (Wong, 2009). Despite all the progress, further research and method developments are needed for the industrial utilization of those enzymes. The tremendous biotechnological potential of wild mushrooms is far from being fully discovered and used.

16.8 CONCLUSION

Wild mushrooms are the more important macrofungi in the world. The capability of wild mushrooms in degrading the waste is very clear. The accumulation of waste results from an increase in solid waste. The increase in levels of industrial production, agriculture production, and others without thinking of the increase in the level of reduction of these wastes represents a highly significant danger to the ecosystem. Wild mushrooms can play an important role in removing many types of solid waste. Wild mushrooms secrete different enzymes that are more helpful in biodegradation and bioremediation for solid waste without any effect on the ecosystem to curb global warming.

REFERENCES

Abatenh, E., Gizaw, B., Tsegaye, Z., & Wassie, M. (2017). The Role of microorganisms in bioremediation-A review. *Open Journal of Environmental Biology*, 2(1), 38–46. doi:10.17352/ojeb.000007

Adenipekun, C. O. (2012). Uses of mushrooms in bioremediation: A Review. *Biotechnology and Molecular Biology Reviews*, 7(3). doi: 10.5897/bmbr12.006

Ahuja, V., & Roy, R. (2020). Lignin Synthesis and Degradation. *In Lignin*. Cham: Springer, (pp. 77–113).

Albornoz, S., Wyman, V., Palma, C., & Carvajal, A. (2018). Understanding of the contribution of the fungal treatment conditions in a wheat straw biorefinery that produces enzymes and biogas. *Biochemical Engineering Journal*, 140, 140–147. 10.1016/j.bej.2018.09.011

Alexopolous, C. J., Mims, C. W., & Blackwell M. (1996). Introductory mycology and Sons Inc., *Schultesfil. (Liliaceae). Sistematik Botanik Dergisi*, 15(2),115–124.

Andlar, M., Rezić, T., Marđetko, N., Kracher, D., Ludwig, R., & Šantek, B. (2018). Lignocellulose degradation: an overview of fungi and fungal enzymes involved in lignocellulose degradation. *Engineering in life sciences*, 18(11), 768–778.

Arfi, Y., Shamshoum, M., Rogachev, I., Peleg, Y., & Bayer, E. A. (2014). Integration of bacterial lytic polysaccharide monooxygenases into designer cellulosomes promotes enhanced cellulose degradation. *Proceedings of the National Academy of Sciences*, 111(25), 9109–9114.

Bains, A., Kaushik, R., Dhull, S. B., & Chawla, P. (2021). Synthesis and characterization of gum Arabic stabilized red rice extract nano-emulsion: in vitro antimicrobial and anti-inflammatory application. *Food Chemistry*, Accepted.

Barkat Md Gulzar, A., Vandana, U. K., Paul, P., & Mazumder, P. B. (2020). The Role of Mushrooms in Biodegradation and Decolorization of Dyes. *An Introduction to Mushroom*, 1, 1–23. doi:10.5772/intechopen.90737

Bhalerao, T. S., & Puranik, P. (2007). Biodegradation of organochlorine pesticide, endosulfan, by a fungal soil isolate, Aspergillus niger. *Int Biodeterior Biodegrad*, 59, 315–321

Boa, E. (2008). Wild edible fungi a global overview of their use and importance to people. *Biology*.

Buzala, K., Przybysz, P., Rosicka-Kaczmarek, J., & Kalinowska, H. (2015). Comparison of digestibility of wood pulps produced by the sulfate and TMP methods and woodchips of various botanical origins and sizes. *Cellulose*, 22, 2737–2747.

Castañeda-Ramírez, G. S., Torres-Acosta, J. F. D. J., Sánchez, J. E., Mendoza-de-Gives, P., González-Cortázar, M., Zamilpa, A., & Aguilar-Marcelino, L. (2020). The possible biotechnological use of edible mushroom bioproducts for controlling plant and animal parasitic nematodes. *BioMed Research International.* doi: 10.1155/2020/6078917

Chang, B. V., Hsu, F. Y., & Liao, H. Y. (2014). Biodegradation of three tetracyclines in swine wastewater. *Journal of Environmental Science and Health Part B, 49*(6), 449–455.

Chang, S. T., & Wasser, S. P. (2018). Current and future research trends in agricultural and biomedical applications of medicinal mushrooms and mushroom products. *International Journal of Medicinal Mushrooms, 20*(12), 95–134.

Chang, S. T., & Miles, P. G. (2004). Mushrooms: Cultivation, Nutritional Value, Medicinal Effect, and Environmental Impact: Second Edition. *Mushrooms: Cultivation, Nutritional Value, Medicinal Effect, and Environmental Impact, 1*, 1–480. doi:10.1663/0013-0001

Chawla, P., Kaushik, R., Shiva-Swaraj, V. J., & Kumar, N. (2018). Organophosphorus pesticides residues in food and their colorimetric detection. *Environmental Nanotechnology Monitoring and Management, 10*(2018), 292–307.

Chawla, P., Kumar, N., Kaushik, R. & Dhull, S. B. (2019). Synthesis, characterization and cellular mineral absorption of gum arabic stabilized nanoemulsion of *Rhododendron arboreum* flower extract. *Journal of Food Science and Technology, 56*(12), 5194–5203.

Chawla, P., Najda, A., Bains, A., Nurzy´nska-Wierdak, R., Kaushik, R., & Tosif, M. M. (2021). Potential of Gum Arabic Functionalized Iron Hydroxide Nanoparticles Embedded Cellulose Paper for Packaging of Paneer. *Nanomaterials, 11*, 1308. 10.3390/nano11051308

Cheung, P. C. K. (2009). Mushrooms as Functional Foods. *Mushrooms as Functional Foods*, 1st edition, *1*, 1–296. doi:10.1002/9780470367285

Couturier, M., & Berrin, J.-G. (2013). The saccharification step: Themain enzymatic components, In: Faraco V. ed., *Lignocellulose Conversion: Enzymatic and Microbial Tools for BioethanolProduction,* Berlin Heidelberg: Springer Verlag.

Das, A. K., Nanda, P. K., Dandapat, P., Bandyopadhyay, S., Gullón, P., Sivaraman, G. K., McClements, D. J., Gullón, B., & Lorenzo, J. M. (2021). Edible mushrooms as functional ingredients for development of healthier and more sustainable muscle foods: A flexitarian approach. *Molecules, 26*, 2463. doi: 10.3390/molecules26092463

Dhull, S. B., Chawla, P., & Kaushik, R. (2020). *Nanotechnological Approaches in Food Microbiology.* CRC Press/Taylor & Francis, ISBN: 9780367359447.

Douanla-Meli, C., Ryvarden, L., & Langer, E. (2007). Studies of tropical African pore fungi (Basidiomycota, Aphyllophorales): Three new species from Cameroon. *Nova Hedwigia, 84*(3–4), 409–420.

El-Tayeb, T. S., Abdelhafez, A. A., Ali, S. H., & Ramadan, E. M. (2012). Effect of acid hydrolysis and fungal biotreatment on agro-industrial wastes for obtainment of free sugars for bioethanol production. *Brazilian Journal of Microbiology, 43*, 1523–1535.

FAOSTAT. Food and Agriculture Data. Available online: http://www.fao.org/faostat/en/#home

Ferreira, J. M., Carreira, D. N., Braga, F. R., & de Freitas Soares, F. E. (2019). First report of the nematicidal activity of *Flammulina velutipes*, its spent mushroom compost and metabolites. *3 Biotech, 9*(11), 1–6.

Garziano, G. (2018). Mushrooms Classification – Part 1: Visualizing Data in R. *Data Science.* https://datascienceplus.com/mushrooms-classification-part-1/

Gilbertson, R. L. (1980). Wood-rotting fungi of North America. *Mycologia, 72*(1), 1–49.

Gouvea, B. M., Torres, C., Franca, A. S., Oliveira, L. S., & Oliveira, E. S. (2009). Feasibility of ethanol production from coffee husks. *Biotechnology Letter, 31*, 1315–1319.

Graminha, E. B. N., Gonçalvez, A. Z. L., Pirota, R. D. P. B., Balsalobre, M. A. A., da Silva, R., & Gomes, E. (2008). Enzyme production by solid-state fermentation: Application to animal nutrition. *Animal Feed Science and Technology, 144*, 1–22.

Gulzar, A.B.Md., Vandana, K.U., Paul, P., & Mazumder, P.B. (2020). The role of mushrooms in biodegration and decolorization of dyes. In Passari, K.A., & Sánchez, S., eds, *An Introduction to Mushroom.* CDMX, Mexico: IntechOpen, 1–22 pp.

Hatakka, A., & Hammel, K. E. (2011). Fungal biodegradation of lignocelluloses. *In Industrial applications.* Berlin, Heidelberg: Springer, pp. 319–340.

Hawksworth, L. D. (2001). The magnitude of fungal diversity: the 1.5 million species estimate revisited. *Mycological Research, 105*, 1422–1432.

Hawksworth, L. D. (2004). Fungal diversity and its implications for genetic resource collections. *Studies in Mycological, 50,* 9–18.

Houfani, A. A., Anders, N., Spiess, A. C., Baldrian, P., & Benallaoua, S. (2020). Insights from enzymatic degradation of cellulose and hemicellulose to fermentable sugars–a review. *Biomass and Bioenergy, 134,* 105481.

Hu, B., Xie, W. L., Li, H., Li, K., Lu, Q., & Yang, Y. P. (2020). On the mechanism of xylan pyrolysis by combined experimental and computational approaches. *Proceedings of the Combustion Institute, 38*(3), 4215–4223.

Hyde, D. et al. (2019). The amazing potential of fungi: 50 ways we can exploit fungi industrially. *Fungal Diversity, 97,* 1–36. 10.1007/s13225-019-00430-9

Jebapriya, G., & Roseline, V. D. V. G., & Gnanadoss, J. J. (2013). APPLICATION OF MUSHROOM FUNGI IN SOLID WASTE MANAGEMENT Material & Packaging Suppliers Manufacturers Distributors Waste Collection Consumers Wholesalers & Retailers Waste Sorting Recycling Energy Recovery Landfill, no. October, 279–285.

Jeoh, T., Cardona, M. J., Karuna, N., Mudinoor, A. R., & Nill, J. (2017). Mechanistic kinetic models of enzymatic cellulose hydrolysis—A review. *Biotechnology and Bioengineering, 114*(7), 1369–1385.

Kameshwar A. K. S., & Qin W. 2016. Lignin degrading fungalenzymes. In: *Production of Biofuels and Chemicals From Lignin.* Springer, pp. 81–130.

Karas, P., Metsoviti, A., Zisis, V., Ehaliotis, C., Omirou, M., & Papadopoulou, E. S. (2015). Dissipation, metabolism and sorption of pesticides used in fruit-packaging plants: Towards an optimized depuration of their pesticide-contaminated agro-industrial effluents. *Science of the Total Environment. 531,* 129–139.

Kaushik, R., Chawla, P., Kumar, N., & Kolish, M. (2017). Effect of pre-milling treatments on wheat flour quality. *The Annals of the University Dunarea de Jos of Galati – Food Technology, 41*(2), 141–152.

Kaushik, R., Chawla, P., Kumar, N., Jhangu, S., & Lohan, A. (2018). Effect of pre-milling treatments on wheat gluten extraction and noodle quality. *Food Science and Technology International, 24*(7), 627–636.

Kaushik, R., Sharma, N., Swami, N., Sihag, M., Goyal, A., Chawla, P., Kumar, A., & Pawar, A. (2013). Physico-chemical properties, extraction and characterization of Gluten from different Indian wheat cultivars. *Research & Reviews: A Journal of Crop Science and Technology, 2*(2), 37–42.

Kaza, S., Lisa, C. Y., Bhada-Tata, P., & Woerden, F. V. (2018). *What a Waste 2.0: A Global Snapshot of Solid Waste Management to 2050. Proceedings of the Institution of Civil Engineers: Municipal Engineer. Vol. 139.* Washington, DC: World Bank. doi:10.1596/978-1-4648-1329-0

Kersten, P., & Cullen, D. (2007). Extracellular oxidative systems of thelignin-degrading Basidiomycete *Phanerochaete chrysosporium. Forest. Genet. Biol., 44,* 77–87.

Kracher, D., & Ludwig, R. (2016). Cellobiose dehydrogenase: An essentialenzymefor lignocellulose degradation in nature—Areview. *Journal of Land Management, Food and Environment, 67,* 145–163.

Krah, F. S., Bässler, C., Heibl, C., Soghigian, J., Schaefer, H., & Hibbett, D. S. (2018). Evolutionary dynamics of host specialization in wood-decay fungi. *BMC Evolutionary Biology, 18*(1), 1–13.

Krishna, G., Samatha, B., Hima Bindu Nidadavolu, S. V. S. S. S. L., Prasad, M. R., Rajitha, B., & Charaya, M. A. S. (2015). Macrofungi in Some Forests of Telangana State, India. *Journal of Mycology, 2015*(April), 1–7. doi:10.1155/2015/382476

Kulshreshtha, S., Mathur, N., & Bhatnagar, P. (2014). Mushroom as a product and their role in mycoremediation. *AMB Express, 4*(1), 1–7. doi:10.1186/s13568-014-0029-8

Kumar, R., Singh, S., & Singh, O. V. (2008). Bioconversion of lignocellulosic biomass: biochemical and molecular perspectives. *Journal of Industrial Microbiology and Biotechnology, 35*(5), 377–391.

Kumla, J., Suwannarach, N., Sujarit, K., Penkhrue, W., Kakumyan, P., Jatuwong, K., Vadthanarat, S., & Lumyong, S. (2020). Cultivation of mushrooms and their lignocellulolytic enzyme production through the utilization of agro-industrial waste. *Molecules, 25,* 2811: doi:10.3390/molecules25122811

Lalak, J., Kasprzycka, A., Martyniak, D., & Tys, J. (2016). Effect of biological pre-treatment of Agropyron elongatum 'BAMAR' on biogas production by anaerobic digestion, *200,* 194–200. 10.1016/j.biortech. 2015.10.022

León-Santiesteban, H. H., Wrobel, K., Revah, S., & Tomasini A. (2016). Pentachlorophenol removal by Rhizopus oryzae CDBB-H-1877 using sorption and degradation mechanisms. *Journal of Chemical Technology & Biotechnology, 91*, 65–71.

Lin, Y., Ge, X., Liu, Z., & Li, Y. (2015). Integration of shiitake cultivation and solid-state anaerobic digestion for utilization of woody biomass. *Bioresource Technology, 182*, 128–135. 10.1016/j.biortech.2015.01.102

Lombard, V., Ramulu, H. G., Drula, E., Coutinho, P. M., & Henrissat, B. (2013). The carbohydrate-active enzymes database (CAZy) in 2013. *Nucleic Acids Research, 42*(1). doi: 10.1093/nar/gkt1178

Lundell, T., Bentley, E., Hildén, K., Rytioja, J., Kuuskeri, J., F Ufot, U., ... & T Smith, A. (2017). Engineering towards catalytic use of fungal class-II peroxidases for dye-decolorizing and conversion of lignin model compounds. *Current Biotechnology, 6*(2), 116–127.

Martin H. (2002). Review: Lignin conversion by manganese peroxidase (MnP). *Enzyme Microb Technol, 30*, 454–466.

Martirani L., Giardina P., Marzullo L., & Sannia G. (1996). Reduction of phenol content and toxicity in olive oil mill waste waters with the ligniolytic fungus Pleurotusostreatus. *Water Research, 30*(8), 1914–1918.

Mohseni, M., & Allen, D. G. (1999). Transient performance of biofilters treating mixtures of hydrophilic and hydrophobic volatile organic compounds. *Journal of the Air and Waste Management Association, 49*, 1434–1441.

Momeni, M. H., Bollella, P., Ortiz, R., Thormann, E., Gorton, L., & Abou Hachem, M. (2019). A novel starch-binding laccase from the wheat pathogen *Zymoseptoria tritici* highlights the functional diversity of ascomycete laccases. *BMC Biotechnology, 19*(1), 1–12.

Moreno, A. D., Ibarra, D., Eugenio, M. E., & Tomás-Pejó, E. (2020). Laccases as versatile enzymes: from industrial uses to novel applications. *Journal of Chemical Technology & Biotechnology, 95*(3), 481–494.

Moreno, G., Esqueda, M., Pérez-Silva, E., Herrera, T., & Altés, A. (2007). Some interesting Gasteroid and Secotioid fungi from Sonora, Mexico. *Persoonia, 19*(2), 265–280.

Nelson, D. L., & Cox, M. M. (2018). Princípios de Bioquímica de Lehninger-7. Artmed Editora.

Olah, B., Kunka, V., & Gallay, I. (2020). Assessing the potential of forest stands for ectomycorrhizal mushrooms as a subsistence ecosystem service for socially disadvantaged people: A case study from Central Slovakia. *Forest, 11*, 282. doi:10.3390/f11030282

Olusola Ogidi, C., Oyetayo, V. O., & Akinyele, B. J. (2020). Wild medicinal mushrooms: Potential applications in phytomedicine and functional foods. *An Introduction to Mushroom, 1*, 1–148. doi:10.5772/intechopen.90291

Parisutham, V., Chandran, S. P., Mukhopadhyay, A., Lee, S. K., & Keasling, J. D. (2017). Intracellular cellobiose metabolism and its applications in lignocellulose-based biorefineries. *Bioresource Technology, 239*, 496–506.

Peralta, R. M., da Silva, B. P., Côrrea, R. C. G., Kato, C. G., Seixas, F. A. V., & Bracht, A. (2017). Enzymes from basidiomycetes—peculiar and efficient tools for biotechnology. *In Biotechnology of Microbial Enzymes*. Academic Press, pp. 119–149.

Ram, R. C., Pandey, V. N., & Singh, H. B. (2010). Morphological characterization of edible fleshy fungi from different forest regions. *Indian Journal Scientific Research, 1*(2), 33–35.

Rascón-Chu, A., Contreras-Vergara, C. A., Figueroa-Soto, C. G., González-Soto, T. E., Esqueda-Valle, M., & Sánchez-Villegas, J. A. (2019). Fibrolytic activity of Podaxis pistillaris fungus in submerged culture. *Biotecnia, XXII*(1). https://doi.org/10.18633/biotecnia.v21i1.874

Rhodes, C. J. (2013). Applications of bioremediation and phytoremediation. *Science Progress, 96*, 417–427.

Rhodes, C. J. (2014). Mycoremediation (Bioremediation with Fungi) - growing mushrooms to clean the earth. *Chemical Speciation and Bioavailability, 26*(3), 196–198. doi:10.3184/095422914×14047407349335

Rinker, D. L. (2017). Bioremediation. In: *Edible and Medicinal Mushrooms. Technology and Applications*. Zied Diego Cunha and Pardo-Giménez Arturo, eds., West Sussex, UK: John Wiley & Sons Ltd, pp 429–430.

Rodríguez-Couto, S. (2017). Industrial and environmental applications of White-Rot fungi. *Mycosphere, 8*(3), 456–466. doi:10.5943/mycosphere/8/3/7

Sánchez, J. E., Zied, D., & Albertó, E. (2018). Edible mushroom production in the Americas Abstracts of the 9th international conference on mushroom biology and mushroom products: 2–11. Shanghai, China. 12-15 November of 2018.

Sanchez, S., & Demain, A. L. (2017). Useful microbial enzymes—an introduction. *In Biotechnology of Microbial Enzymes.* Academic Press, pp. 1–11.

Sasek, V., & Cajthaml, T. (2005). Mycoremediation: Current State and Perspectives. *International Journal of Medicinal Mushrooms, 7*(3), 360–361. doi:10.1615/intjmedmushr.v7.i3.200

Savoie, J. M., & Largeteaue, M. L. (2011). Production of edible mushrooms in forests: trends in development of a mycosilviculture. *Applied Microbiology and Biotechnology, 89,* 971–979. doi: 10.1007/s00253-01 0-3022-4

Sharma, B., Dangi, A. K., & Shukla, P. (2018). Contemporary enzyme based technologies for bioremediation: A review. *Journal of Environmental Management, 210,* 10–22. doi:10.1016/j.jenvman.2017.12.075 .Elsevier Ltd

Shojaosadati S. A., & Siamak E. (1999). Removal of hydrogen sulfide by the compost biofilter with sludge of leather industry. *Resources, Conservation and Recycling, 27*(1–2), 139–144.

Singh, A. D., Vikineswary, S., Abdullah, N., & Sekaran, M. (2011). Enzymes from spent mushroomsubstrate of Pleurotus sajor-caju for the decolourisation and detoxification of textile dyes. *World Journey of Microbiology and Biotechnolog, 27,* 535–545.

Singh, H. (2006). *Mycoremediation: Fungal Bioremediation.* Hoboken: Wiley.

Sista Kameshwar, A. K., & Qin, W. (2018). Comparative study of genome-wide plant biomass-degrading CAZymes in white rot, brown rot and soft rot fungi. *Mycology, 9*(2), 93–105.

Sjostrom E. (2013). *Wood Chemistry: Fundamentals and Applications.* Elsevier.

Soares, F. E. F., Nakajima, V. M., Sufiate, B. L., Satiro, L. A. S., Gomes, E. H., Fróes, F. V., …, & de Queiroz, J. H.(2019). Proteolytic and nematicidal potential of the compost colonized by *Hypsizygus marmoreus. Experimental Parasitology, 197,* 16–19.

Stamets, P. (2001). The role of mushrooms in nature. In *The Overstory Book: Cultivating Connections with Trees,* Elevitch CR, ed. Holualoa, USA.

Stebelska, K. (2013). Fungal Hallucinogens Psilocin, ibotenic acid, and Muscimol: Analytical methods and biologic activities. *Therapeutic Drug Monitoring, 35*(4), 420–442. doi:10.1097/FTD.0b013e31828741a5

Stech, M., Hust, M., & Schulze, C., Dubel, S. and Kubick, S. (2014). Cellfreeeukaryotic systems for the production, engineering, andmodification of scFv antibody fragments. *Engineering in Life Sciences, 14,* 387–398.

Sun, L. B., Zhang, Z., Xin, G., Sun, B. X., Bao, X., Wei, Y. Y., Zhao, X., & Xu, H. R. (2020). Advances in umami taste and aroma of edible mushrooms. *Trends in Food Science and Technology, 96* (October 2019), 176–187. doi:10.1016/j.tifs.2019.12.018

Sun, Y., & Chen, J. (2002). Hydrolysis of lignocellulosic material for ethanol production: A review, *Bioresour. Technol, 83,* 1–11.

Tan, Y. H., & Wahab, M. N. (1997). Extracellular enzyme production during anamorphic growth in the edible mushroom, *Pleurotus sajor-caju. World Journal of Microbiology and Biotechnology, 13*(6), 613–617.

Tarrés, Q., Espinosa, E., Domínguez-Robles, J., Rodríguez, A., Mutjé, P., & Aguilar, M. D. (2017). The suitability of banana leaf residue as raw material for the production of high lignin content micro/nano fibers: From residue to value-added products. *Industrial Crops and Products, 99,* 27–33.

Tay, C.-C., Redzwan, G., Liew, H.-H., Yong, S.-K., Surif, S., & Abdul-Talib, S. (2010). Copper (II) biosorption characteristic of Pleurotus spent mushroom compost. Proceedings: 2010 International Conference on Science and Social Research (CSSR 2010), December 5–7, 2010, Kuala Lumpur, Malaysia, pp 6–10.

Tay, C. C., Liew, H. H., Redzwan, G., Yong, S. K., Surif, S., & Abdul-Talib, S. (2011). Pleurotusostreatus spent mushroom compost as green biosorbent for nickel (II) biosorption. *Water Science and Technology, 64*(12), 2425–2432.

Tedersoo, L., Bahram, M., Põlme, S., Kõljalg, U., Yorou, N. S., Wijesundera, R., Ruiz, L. V., Vasco-Palacios, A. M., Thu, P. Q., Suija, A. et al. (2014). Global diversity and geography of soil fungi. *Science, 346,* 1078–1088.

Thiribhuvanamala, G., Prakasam, V., Chandraseker, G., Sakthivel, K., Veeralakshmi, S., Velazhahan R., & Kalaiselvi, G. (2011). Biodiversity, conservation and utilization of mushroom flora from the Western Ghats region of India. Proceedings of the 7th International Conference on Mushroom Biology and Mushroom Products (ICMBMP7). 155–164.

Tramontina, R., Brenelli, L. B., Sodré, V., Cairo, J. P. F., Travália, B. M., Egawa, V. Y., …, & Squina, F. M. (2020). Enzymatic removal of inhibitory compounds from lignocellulosic hydrolysates for biomass to bioproducts applications. *World Journal of Microbiology and Biotechnology*, *36*(11), 1–11.

Trutmann, P., & Luque, A. (2011). Fungi and mushrooms: indicators of human and environmental wellbeing In mountains, and beacons of opportunity in Peru. *Annual Report 2011 Global Mountain Action*. www.globalmountainaction.org

Umezawa, K., Niikura, M., Kojima, Y., Goodell, B., & Yoshida, M. (2020). Transcriptome analysis of the brown rot fungus *Gloeophyllum trabeum* during lignocellulose degradation. *PloS One*, *15*(12), e0243984.

Valverde, M. E., Hernández-Pérez, T., & Paredes-López, O. (2017). Edible Mushrooms: Improving Human Health and Promoting Quality Life. *Phytochemistry Letters* 20 (Table 1): IFC. http://linkinghub.elsevier.com/retrieve/pii/S1874390017303634.

Vishwakarma, G. S., Bhattacharjee, G., Gohil, N., & Singh, V. (2020). *Current Status, Challenges and Future of Bioremediation. Bioremediation of Pollutants*. INC. doi:10.1016/b978-0-12-819025-8.00020-x

Wasser, S. P. (2011). Current findings, future trends, and unsolved problems in studies of medicinal mushrooms. *Applied Microbiology and Biotechnology*, *89*(5), 1323–1332. doi:10.1007/s00253-010-3067-4

Woiciechowski, A. L., Porto de Souza Vandenberghe, L., Karp, S. G. et al. 2013. The pretreatment step in lignocellulosic biomassconversion: Current systems and new biological systems. In: Faraco V. ed., *Lignocellulose Conversion: Enzymatic and MicrobialTools for Bioethanol Production* Verlag Berlin Heidelberg: Springer.

Woldemariam, W. G. (2019). Mushrooms in the Bio-Remediation of Wastes from Soil. *Advances in Life Science and Technology*, *76*, 41–47. doi:10.7176/alst/76-04

Wong, D. W. (2009). Structure and action mechanism of ligninolytic enzymes. *Applied Biochemistry and Biotechnology*, *157*(2), 174–209.

Writers of The New York Botanical Garden. (2003). NYBG.Org: Hidden Partners: Mycorrhizal Fungi and Plants. *The New York Botanical Garden*. http://sciweb.nybg.org/Science2/hcol/mycorrhizae.asp.html

Zhu, Y., Plaza, N., Kojima, Y., Yoshida, M., Zhang, J., Jellison, J., …, & Goodell, B. (2020). Nanostructural analysis of enzymatic and non-enzymatic brown rot fungal deconstruction of the lignocellulose cell wall. *Frontiers in Microbiology*, *11*.

CHAPTER 17

Cultivation of wild mushrooms using lignocellulosic biomass-based residue as a substrate

Pawan Kumar Rose[1], Sanju Bala Dhull[2], and Mohd. Kashif Kidwai[1]
[1]Department of Energy and Environmental Sciences, Chaudhary Devi Lal University, Sirsa, Haryana, India
[2]Department of Food Science and Technology, Chaudhary Devi Lal University, Sirsa, Haryana, India

CONTENTS

17.1 INTRODUCTION

The existence of mushrooms on planet Earth is evidence from the palaeontological fossil record of around 408–438 million years ago (Silurian period) (Reddy, 2015). In ancient times, European, Asian, and American countries acknowledged mushrooms as a palatable food (Ma et al., 2018; Borthakur & Joshi, 2019), whereas in India, it entered from the Northwest through Afghanistan and registered their presence during the Indus Valley civilization (Reddy, 2015). Only 10% of mushroom species on Earth have been nomenclatured among the 150,000 mushroom-forming species (Niksic et al., 2016). Among 2,300 known species of edible and medicinal mushrooms, around 80 species are accepted as food and out of which 20 species are commercially cultivated (Silva et al., 2007; Philippoussis, 2009; Valverde et al., 2015). The majority of commercial edible mushrooms belong to the genera *Agaricus, Agrocybe, Auricularia, Flammulina, Ganoderma, Hericium, Lentinula, Lentinus, Pleurotus, Tremella*, and *Volvariella* (Kumla et al., 2020). In 2017, Royse et al.

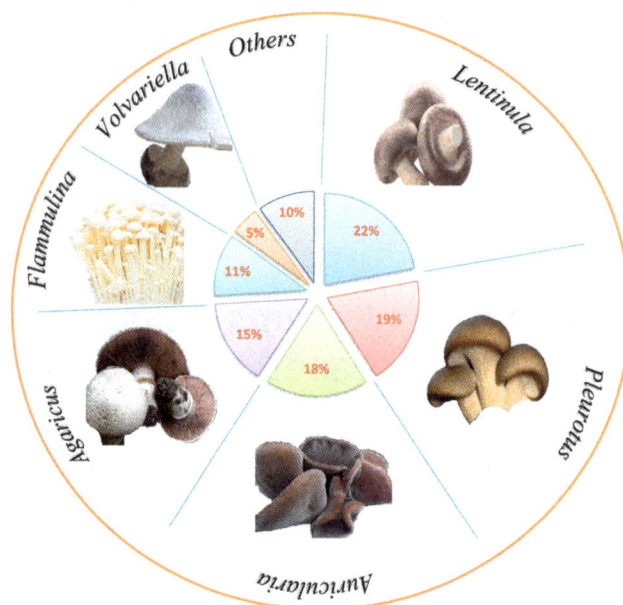

Figure 17.1 World mushroom production among various genera.

distinguished world mushroom production among various genera (Figure 17.1). The world's top four most cultivated edible mushrooms embrace the genera: *Lentinula* (shiitake and relatives), *Pleurotus* (oyster mushroom), *Auricularia* (wood ear mushroom), and *Agaricus* (button mushroom and relatives) (Royse et al., 2017; Kumla et al., 2020). Among the US$38 billion mushroom market in 2018, cultivated edible mushroom shares the highest portion followed by medicinal mushroom and wild mushroom. Asian countries clamped the first position in mushroom production with a 76% share followed by Europe (17.2%) and the United States (5.9%) (Sande et al., 2019; Mahari et al., 2020). In recent years, mushroom production has increased significantly, and almost doubled from 2015 to 2017 (Ritota & Manzi, 2019), and is anticipated to cross more than $59.48 billion in 2021 (Zionmarketresearch, 2021). Mushrooms are high in nutritional value, popularly recognized as nutraceutical foods, and considerable attention was drawn due to organoleptic merit, medicinal properties, and economic significance causing an increase in mushroom production and consumption throughout the world (Chang & Miles, 2008; GünçErgönül et al., 2013; Valverde et al., 2015; Ma et al., 2018).

17.2 WILD MUSHROOM CULTIVATION: NUTRITIONAL ASPECTS

On a dry weight basis, mushrooms contain a high proportion of carbohydrates (63%) followed by protein (25%), fat (4%), and the rest (8%) is contributed by minerals (Rajarathnam & Shashirekha, 2003). Mushrooms contain glycogen as polysaccharides instead of starch as in plants. Monosaccharides are mainly available in form of mannitol, trehalose, and glucose. Trehalose and glucose are found in lower concentrations in comparison to mannitol. Further, the mannitol percentage varies among basidiomycetes like *Cantharellus cibarius* (13.9%), *Lactariusdeliciosus* (13.7%), *Ramaria botrytis* (11.7%), *Agaricus arvensis* (6.5%), *Tricholoma portentosum* (1.0%), and *Lycoperdon perlatum* (0.2%) (Barros et al., 2007; Kalac, 2009; Borthakur & Joshi, 2019).

Mushrooms comprises a considerable amount of dietary fiber and proven superiority over cereals in terms of their chemical structure, which exhibits immunostimulatory and anticancer activity (Cheung, 2013). On a dry weight basis, *Boletus* spp. contain 4.2%–9.2% of soluble fiber and 22.4%–31.2% insoluble fibers (Cheung, 2013). Edible mushrooms are a highly nutritious food with an appreciable amount of protein (20–25 g/100 g of dry matter) and comprise a noteworthy amount of amino acids like leucine, valine, glutamine, glutamic, aspartic acids, etc. (Valverde et al., 2015). Moreover, the protein content in wild mushrooms is higher than commercial varieties and reported to be a better substitute for the intake of dietary protein of animal and plant origin (Borthakur & Joshi, 2019). Wild mushrooms contain protein content up to 65% (dry weight) and exhibit dominancy among other protein sources such as soybean (39.1%), milk (25.2%), pork meat (20%), wheat (13.2%), rice (7.3%), etc. (Chang, 1990; Atri et al., 2012; Borthakur & Joshi, 2019). The protein of mushrooms also has an edge over cereals in terms of digestibility and presence of the essential amino acids like arginine, histidine, lysine, threonine, and tryptophan (Rajarathnam & Shashirekha, 2003; Borthakur & Joshi, 2019).

The ash content in edible mushrooms lies in the range from 8–12 g/100 g (dry matter) including potassium, phosphorus, calcium, magnesium, copper, iron, zinc, etc. (Valverde et al., 2015). Potassium is reported to be the most abundant mineral in mushrooms followed by phosphorus. However, the potassium concentration varies with body parts of growing mushrooms, i.e. highest in cap and lowest in spores (Borthakur & Joshi, 2019). Inter-variety differences in mushroom species offer a diverse amount of potassium and phosphorus (mg/100 g dry matter), respectively, such as *Agaricus campestris* (4,762 and 1,429), *Volvariella diplasia* (3,333 and 1,042), *Flammulina velutipes* (2,981 and 278), *Pholiota nameko* (2,083 and 771), and *Lentinus edodes* (1,246 and 650) (Rajarathnam & Shashirekha, 2003). The phosphorus in mushroom fruiting bodies generally ranges from 0.5 to 1.0 g/100 g of dry matter (Kalac, 2009). Several mushroom species also contain selenium, which exhibits excellent antioxidant functionality and considers necessary for the selenoenzymes biosynthesis that ranges from 5 µg/g (*Lycoperdon* spp.) to 200 µg/g (*Albatrellus pes-caprae*) (Borthakur & Joshi, 2019). Mushrooms are documented to accumulate trace minerals like cadmium, mercury, lead, copper, and antimony from their growth medium in various quantities (Faik et al., 2011). *Lactarius* spp. is best known for the accumulation of zinc, whereas *Clitocybe alexandri* for copper accumulation (Borthakur & Joshi, 2019). On the contrary, they become dangerous for consumption if the accumulation of toxic minerals or elements takes place and may affect human health (Rose & Devi, 2018). Mleczek et al. (2021) reported a worrying level of heavy metals e.g. aluminium, nickel, and lead in cultivated wild species of mushrooms such as *Pleurotus* ostreatus, *Flammulina velutipes*, *Lentinus edodes* etc., and mercury in wild-growing mushrooms such as *Boletus edulis Bull*. Mushrooms are an excellent source of vitamins such as niacin, B1, B2, B9, B12, C, D, and E. *Pleurotus* species are acknowledged for their B-complex vitamins and folic acid (Rajarathnam & Shashirekha, 2003). A mushroom is the only vegetarian food with vitamin D, and especially D2 when the wild mushroom grows in darkness with UV-B light (Valverde et al., 2015). The amount (mg/100 g dry weight) of niacin, thiamine, and riboflavin, respectively, present in the different mushroom species like *Agaricus bisporus* (56.19, 1.14, and 4.95), *Volvariella volvacea* (59.5, 0.32, and 2.73), and *Pholiota nameko* (72.9, 18.8, and 14.6) (Rajarathnam & Shashirekha, 2003). Mushrooms accommodate a lower amount of fat (1.3–8% of dry matter); therefore, they complement a low-calorie food. Moreover, the presence of palmitic (C16:0), oleic (C18:1), and linoleic (C18:2), enabling mushrooms a healthier source of nutrition for the human diet (Valverde et al., 2015). In addition to the nutritional components, the mushroom varieties also contain bioactive compounds like secondary metabolites, glycoproteins, and polysaccharides, primarily β-glucans in high quantity (Valverde et al., 2015; Ma et al., 2018). Mushrooms retain B-complex vitamins in their body, even utilizing lignocellulosic biomass (Rajarathnam & Shashirekha, 2003).

17.3 WILD MUSHROOM CULTIVATION: ENVIRONMENTAL ASPECTS

Edible mushrooms are saprophytic fungi that naturally grow on the diverse lignocellulosic biomass-based substrate and draw nutrients for their growth via degrading lignocellulosic components using lignocellulolytic enzymes produced by them. Mushroom species degrade lignocellulosic biomass either in their raw or composted form. The lignocellulosic biomass is degraded by two different extracellular enzymatic systems. Firstly, the hydrolytic system, which degrades cellulose and hemicellulose, and the second, an exclusive oxidative ligninolytic system for the lignin degradation (Baldrian, 2006; Philippoussis, 2009; Rosero-Delgado et al., 2021). Lignocellulosic biomass degradation is accomplished through the symbiotic functions of hydrolytic and oxidative enzymes. The cellulose degradation is achieved mainly through three enzymes: endoglucanase (break β-1, 4-glucosidic linkages and produced several reducing ends), exoglucanase (act on reducing or non-reducing end and release cellobiose), and glucosidase (convert cellobiose into glucose) (Kumla et al., 2020). The main hemicellulose degrading enzymes are endoxylanase (convert xylan into xylose), endomannanase (convert mannan to mannose), and arabinanases (convert arabinan to L-arabinose) (Sørensen et al., 2013). Lignin degradation has a primordial and necessary phase in lignocellulosic biomass degradation because it provides accessibility for hydrolytic action. Ligninolytic enzymes primarily include laccases, lignin peroxidases, versatile peroxidases, manganese peroxidases, and dye decolorizing peroxidases (Familoni et al., 2018). The majority of cellulose and hemicellulose degrading enzymes are contained by mushroom species like *Pleurotus ostreatus*, *Lentinula edodes*, *Agaricus bisporus*, and *Volvariella volvacea* (Philippoussis, 2009). However, the nature and amount of the enzymes produced by a mushroom species during the vegetative growth phase are allied with the type and composition of the lignocellulosic biomass (Elisashvili et al., 2008; Philippoussis, 2009). White rot, a wood-degrading mushroom-forming fungus, like *Pleurotus* spp., *Lentinula edodes*, *Ganoderma* spp., etc., contains enzymes required for concurrent degradation of all the three lignocellulosic components, i.e. cellulose, hemicellulose, and lignin (Philippoussis, 2009). *Ganoderma lucidum* is acknowledged as an efficient wood biomass degrader (Rashad et al., 2019). However, brown rot mushroom-forming fungus, i.e. *Agaricus bisporus and Volvariella volvacea* are deficient in lignin-degrading enzymes; that's why they are commercially cultivated on the composted lignocellulosic biomass (Ritota & Manzi, 2019). *Agaricus* spp., is known as a litter-decomposing fungi (LDFs), due to low lignin degradation rate in comparison to white rot fungi (Ritota & Manzi, 2019). *Agaricus bisporus* degrade lignin through laccase and manganese peroxidase enzymes and exhibit a lower degradation rate than white rot fungi (Steffen et al., 2007; Philippoussis, 2009). *Pleurotus* spp. cultivation on the lignocellulosic biomass has shown superiority over *Agaricus* spp. (world's most cultivated mushroom) as they include all the enzymes and simply chopped, water-soaked, and non-composted straw are sufficient to achieve desirable degradation of lignocellulosic biomass (Cohen et al., 2002; Philippoussis, 2009; Ritota & Manzi, 2019). Furthermore, *Agaricus bisporous* and *Lentinula edodes* require logs or moisturized bag sawdust cultivation techniques with a supplement such as cereal bran (Philippoussis, 2009). Mushrooms have a better ligninolytic enzyme system support than hydrolytic action of cellulase and hemicellulose enzymes. Sardar et al. (2017) reported a remarkable decrease in cellulose, hemicellulose, and lignin content by the *Pleurotus eryngii*. The highest degradation of cellulose (40%), hemicellulose (43%), and lignin (29%) were reported in cotton waste and the lowest for cellulose (22%), hemicellulose (39%), and lignin (17%) in sawdust by *Pleurotus eryngii*. Moreover, *Lentinula edodes* (white rot fungus) proved their high selectivity for lignin degradation (Philippoussis, 2009; Ritota & Manzi, 2019). The cellulose to lignin ratio demonstrated a positive correlation between mycelial growth rate and mushroom yield. The highest yield and biological efficiency were reported with the substrate having lower cellulose to lignin ratio and vice versa (Atila, 2019). Atila (2019) documented lower cellulose to lignin ratio for sunflower head residue (2:1) and chickpea straw (3.1) and associated maximum total yield (g/kg) (233.7 and 228.1, respectively) and biological efficiency (51.0% and 50.6%, respectively) for *Lentinula edodes*. However, the corn stalk substrate with high cellulose to lignin ratio (8.1) resulted in a lower yield (87.9 g/kg) and biological efficiency (20.1%). Biological efficiency validates the ability of a particular mushroom species to utilize substrate productively. Biological efficiency is a function of the genotype of mushrooms, nature, and concentration of nutrients and minerals in the substrate (Sardar et al., 2017; Chawla et al., 2019). Better fruiting growth and yield were documented with the cellulose-rich substrate (Ashraf et al., 2013). Cellulose, hemicellulose, and nitrogen are the key components of any agriculture and agro-industrial residue, and their degradation depends on their respective amount in the residue and mushroom species (Chukwurah et al., 2013). The maximum value of biological efficiency labels the quantifiable degradation of lignocellulosic components of the agriculture and agro-industrial residue by lignocellulolytic enzymes (Shashirekha et al., 2002; Sardar et al., 2017). But, high lignin content (≥15%) causes slow decomposition as well as nitrogen immobilization that end in the poor nutrient profile of growing mushrooms (Chanakya et al., 2015). The high lignin content of sawdust (30.5%) and sugarcane bagasse (21.22%) show a low lignin degradation rate of 17% and 13%, respectively, and protein content of 19% and 20%, respectively with white rot fungi, i.e. *Pleurotus eryngii*. On the contrary, the low lignin content of lignocellulosic biomass such as cotton waste (4.82%) and corncob (7.46%) revealed improved lignin degradation (29% in cotton waste and 22% in corncob) and enriched protein content (25.36% in cotton waste and 23.72% in corncob) (Sardar et al., 2017). The protein content of mushrooms is strongly influenced by numerous factors, i.e. nature and nutrient profile of the substrate, varieties of mushroom, development stages, and analysis time after harvest (Gothwal et al., 2012; Sardar et al., 2017). Moreover, substrate rich in nitrogen content results in high protein content in the fruiting body (Hoa, et al., 2015; Ritota & Manzi, 2019). Koutrotsios et al. (2014) documented a negative correlation between the hemi-cellulose content of substrate and crude protein content in growing mushrooms. Accordingly, a substrate with a low C:N ratio favors a high-protein yield in the mushroom (Wang et al., 2013; Hoa et al., 2015; Ritota & Manzi, 2019). During the production phase, the breakdown of organic matter in substrate causes a reduction in carbon content and increment in nitrogen content that ultimately decline the C:N ratio (Atila, 2019). The higher nitrogen content in the lignocellulosic biomass during the production cycle help in the development of high-protein content. Hoa et al. (2015) reported high protein content in *Pleurotus ostreatus* and *Pleurotus cystidiosus* with corncob substrate (C:N = 34.57) and low protein value with sawdust substrate (C:N = 51.7). Koutrotsios et al. (2014), Sardar et al. (2017), and Atila (2019) made similar observations using different varieties of lignocellulosic biomass.

Each mushroom species has its optimum C:N ratio that ensures maximum yield in minimum time (Zied et al., 2011; Atila, 2019). A low value of the optimum C:N ratio was reported for *Agaricus* spp. like *Agaricus bisporus*, (19/1), *Agaricus brasiliensis* (27/1), and moderate C:N ratio for *Pleurotus* spp. like *Pleurotus eryngii* (45–55/1), *Pleurotus ostreatus* (45–60/1). *Ganoderma lucidum* exhibits a high value of the optimum C:N ratio, i.e. 70–80/1. The optimum C:N ratio for some common wild mushroom species are as *Lentinula edodes* (30–35/1), *Flammulina velutipes* (30/1), *Lentinus sajor-caju* (45–55/1), and *Volvariella volvacea* (40–60/1) (Atila, 2019; Kumla et al., 2020). Numerous works of literature demonstrate the ability of wild mushrooms to utilize diverse ranges of lignocellulosic biomass in individual or combination as a substrate such as wheat straw, rice straw, sawdust, beech sawdust, sugarcane bagasse, sugarcane leaves, cotton waste, cottonseed hull, corncob, soybean straw, oat straw, bean straw, oil palm whole stalk, cassava peels, potato peels, etc. (Tables 17.2 and 17.3). Generally, cultivated edible mushrooms can degrade lignocellulosic biomass through lignocellulolytic enzymes and release material used as nutrients for their growth which making their application in solid-state fermentation for the bio-transformation of lignocellulosic biomass having value addition, economic, and ecological importance for their utilization (Philippoussis, 2009; Cherubin et al., 2018).

17.4 LIGNOCELLULOSIC BIOMASS COMPOSITION AND SUITABILITY AS A SUBSTRATE FOR WILD MUSHROOM CULTIVATION

Generally, agriculture and agro-industrial residues are water-insoluble materials reported to be rich in the lignocellulosic components that include cellulose, hemicellulose, and lignin, collectively referred to as "lignocellulosic materials", a most common natural polymer on Earth. Typically, 85–90% (dry weight) of the lignocellulosic materials comprise cellulose, hemicellulose, and lignin, whereas the rest are balanced by ash and other minor components. However, the quantitative and qualitative aspects of these components are defined by the nature of the source of the residues, harvesting techniques, processing techniques, storage conditions, etc. (Mulkey et al., 2006; Pasangulapati et al., 2012). On a dry weight basis, softwood comprises 33–42% of cellulose, 22–40% of hemicellulose, and 27–32% of lignin (Nhuchhen et al., 2014; Tarasov et al., 2018). Hardwood contains 38–51% of cellulose, 17–38% of hemicellulose, and 21–31% of lignin (Menon & Rao, 2012; Tarasov et al., 2018). Herbaceous crops contain 25–95% of cellulose, 20–50% of hemicellulose, and 0–40% of lignin (Smit & Huijgen, 2017; Tarasov et al., 2018). Therefore, based on quantity, cellulose is one of the prominent ubiquitous organic polymers on Earth in lignocellulosic biomass residues like plant wood, cotton, sugarcane, cereals, etc. (Havstad, 2020).

Cellulose is a linear glucan chain of more than 10,000 anhydrous D-glucose units united together via β-1,4-glycosidic bonds and cellobiose residues function as repeating units at various degrees of polymerization. This ailment helps in the development of a strong microfibrils structure through hydrogen bonding at the intracellular molecular level and van der Waals forces at the intermolecular level that subsequently provides high strength to cellulose and other significant mechanical properties (Bai et al., 2019). Microfibrils are usually fixed on hemicellulose and lignin-containing matrix (Pasangulapati et al., 2012). Hemicellulose is a heteropolymer that contain various sugar molecules divided into three major groups, i.e. xylans, mannans, and galactans (Philippoussis, 2009; Pasangulapati et al., 2012). The number of sugar units, length, branching, etc., of the chain is the key factor of hemicellulose structure (Kumla et al., 2020).

Moreover, microfibrils formation is absent in hemicellulose, but they do from hydrogen bonds with cellulose and lignin, henceforth known as "cross-linking glucans" (Pasangulapati et al., 2012). After cellulose, lignin is the most abundant organic polymer on Earth. Lignin is insoluble in water and long-chain three-dimensional polymer mainly includes phenylpropane units linked by ether bonds (Saha et al., 2009) and is one of the chief components of straw (Xu, 2010). The complexity in lignin structure arises due to the presence of varieties of functional groups (Kumla et al., 2020).

Lignin acts like a cementing material, which provides elasticity and mechanical strength to the wood (Pasangulapati et al., 2012). In addition to these three structural components, crop residues also contain (dry weight, approximately) nitrogen (0.5–1.5%), phosphorus (0.15–0.2%), potassium (1.0%), calcium (1.0%), magnesium (0.5%), manganese (30.0 mg/kg), iron (100.0 mg/kg), zinc (30.0 mg/kg), copper (5.0 mg/kg), boron (20.0 mg/kg), etc. (Philippoussis, 2009; Cherubin et al., 2018). However, the proportion of these minor components in lignocellulosic biomass depends on various parameters, i.e. crop type, part of the plant, season of growth, and other environmental factors during their growth (Kumla et al., 2020). Both, crop-based and processing-based lignocellulosic biomass residues are used in mushroom cultivation (Tables 17.1).

17.5 SOURCES, COMPOSITION, AND AVAILABILITY OF LIGNOCELLULOSIC BIOMASS

According to an estimate, 50% of the world biomass is composed of lignocellulosic materials (10–50 billion tons per annum) (Tsoutsos, 2010), and a major portion is shared by agriculture and agro-industrial residues. In the year 2019, the world top ten produced agricultural commodities

Table 17.1 Chemical properties (dry weight basis) of lignocellulosic biomass used as substrates for wild mushroom cultivation

Lignocellulosic biomass	Cellulose (%)	Hemicellulose (%)	Lignin (%)	Cellulose/lignin ratio	N (%)	C/N ratio	References
Crop-based residue							
Rice straw	32.0–47.0	19.0–27.0	5.0–24.0	2.20–3.24	1.0–3.0	33–60	Dotaniya et al., 2016; Khir & Pan, 2019; Millati et al., 2019
Wheat straw	31.0–38.5	27.2–37.6	5.6–11.6	3.60–4.47	0.4–0.8	80–125	Philippoussis, 2009; Koutrotsios et al., 2014
Corn stover	24.0–40.0	20.0–29.0	12.0–23.0	1.37–2.28	0.6–0.9	55–57	Philippoussis, 2009; NRCS, 2011; Dotaniya et al., 2016; Ruan et al., 2019
Corn stalk	38.7	30.4	4.8	8.06	0.8	50–93.2	NRCS, 2011; Atila, 2019
Processing-based residue							
Corncob	32.0–40.5	35.4–41.6	2.5–9.0	5.56–7.04	1.0	50–71.6	Philippoussis, 2009; NRCS, 2011; Koutrotsios et al., 2014; Sardar et al., 2017: Millati et al., 2019
Sugarcane bagasse	34.2–47.0	24.5–31	14.0–21.22	1.94–2.67	0.2–0.8	116–190	Philippoussis, 2009; Dotaniya et al., 2016; Sardar et al., 2017; Hernández et al., 2019
Rice husk	31.12–43.3	22.48–28.6	22.0–22.34	1.40–1.95	0.3–0.4	83–136	Thiyageshwari et al., 2018; Khir & Pan, 2019; Millati et al., 2019
Softwood sawdust	33.0–42.0	22.0–40.0	27.0–32.0	1.11–1.42	0.1	310–520	Philippoussis, 2009; Nhuchhen et al., 2014; Tarasov et al., 2018
Hardwood sawdust	38.0–51.0	17.0–38.0	21.0–31.0	1.46–1.96	0.1–0.2	150–450	Philippoussis, 2009; Menon & Rao, 2012; Tarasov et al., 2018

Table 17.2 Biological efficiency and chemical composition of some wild mushrooms cultivated on the different lignocellulosic biomass-based substrates

Lignocellulosic substrate	Mushroom species	Biological efficiency (%)	Chemical composition (%, dry weight)					References
			Crude protein	Carbohydrate	Fat	Crude fiber	Ash	
Wheat straw	Pleurotus ostreatus	77.26	15.22	73.75	2.54	18.99	8.49	Koutrotsios et al., 2019
	Pleurotus sajor-caju	44.7–74.86	22.90–29.36	32.16–56.00	2.07–2.55	7.10	6.65–8.05	Gupta et al., 2013; Patil, 2013
	Agaricus bisporus	47.1–51.0	21.3–27.0	38.3–48.9	2.53–3.92	17.7–23.3	7.77–11.0	Tsai et al., 2007
	Agrocybe cylindracea	61.40	1.50	89.60	0.30	40.4	8.60	Koutrotsios et al., 2014
	Hericium erinaceus	43.50	26.80	58.92	3.73	ND	10.55	Hassan, 2007
	Lentinula edodes	66.0–93.1	14.4–15.4	63.7–65.6	1.1–1.5	ND	3.8–4.4	Gaitán-Hernández & Norberto Cortés, 2014
	Pleurotus citrinopileatus	98.32	25.28	64.00	2.67	ND	8.05	Medany, 2014
Rice straw	Pleurotus eryngii	48.29	21.10	56.10	2.40	13.35	7.62	Sardar et al., 2017
	Hericium erinaceus	33.9	24.07	60.50	4.16	ND	11.27	Hassan, 2007
	Pleurotus sajor-caju	78.33	23.40	55.00	2.40	7.90	6.85	Patil, 2013
	Pleurotus citrinopileatus	77.52	22.84	64.93	3.17	ND	9.06	Medany, 2014
	Pleurotus eryngii	45.99	20.87	52.45	1.87	9.15	8.57	Sardar et al., 2017
	Lentinula edodes	48.68	16.30	78.22	5.80	ND	3.32	Gao et al., 2020
	Pleurotus ostreatus	10.67	14.85	47.87	6.33	12.39	18.56	Emiru et al., 2016
	Volvariella volvacea	11.75	13.12–20.00	ND	ND	ND	ND	Biswas & Layak, 2014
Sawdust	Pleurotus ostreatus	7.00–46.44	12.72–19.52	43.80–51.26	1.32–6.13	19.47–22.00	5.90–17.89	Hoa et al., 2015; Emiru et al., 2016
	Hericium erinaceus	50.30	24.83	60.95	3.59	ND	10.63	Hassan, 2007
	Pleurotus citrinopileatus	51.58	24.07	65.58	2.59	ND	7.76	Medany, 2014
	Pleurotus eryngii	35.47	18.93	52.53	2.42	6.78	7.51	Sardar et al., 2017
	Auricularia polytricha	44.63–99.49	10.22	78.45	0.85	ND	4.15	Liang et al., 2019
Beech sawdust	Ganoderma lucidum	61.24	16.84	77.86	2.21	47.93	3.10	Koutrotsios et al., 2019
	Agrocybe cylindracea	38.3	ND	ND	ND	ND	ND	Koutrotsios et al., 2014
	Pleurotus ostreatus	33.5	0.30	97.7	0.30	50.9	1.80	Koutrotsios et al., 2014
Eucalyptus sawdust	Ganoderma lucidum	0.30	9.6	ND	ND	41.2	1.10	de Carvalho et al., 2015

Table 17.2 (Continued) Biological efficiency and chemical composition of some wild mushrooms cultivated on the different lignocellulosic biomass-based substrates

Lignocellulosic substrate	Mushroom species	Biological efficiency (%)	Chemical composition (%, dry weight)					References
			Crude protein	Carbohydrate	Fat	Crude fiber	Ash	
Sugarcane bagasse	*Pleurotus ostreatus*	65.65	27.13	34.94	1.32	22.00	6.68	Hoa et al., 2015
	Pleurotus eryngii	41.31	19.98	49.53	3.10	6.59	7.78	Sardar et al., 2017
	Lentinula edodes	130.2	13.80	73.0	1.0	ND	6.2	Salmones et al., 1999
Depithed bagasse	*Pleurotus ostreatus*	66.64	12.83	44.78	1.48	12.83	6.28	Aguilar-Rivera & De Jesús-Merales, 2010
Bagasse from mill	*Pleurotus ostreatus*	70.09	15.81	45.14	1.94	12.89	5.81	Aguilar-Rivera & De Jesús-Merales, 2010
Sugarcane leaves	*Lentinula edodes*	82.70	14.40	78.20	0.90	ND	6.50	Salmones et al., 1999
Cotton stalk	*Pleurotus ostreatus*	20.60	22.80	58.40	2.10	10.50	6.20	Sardar et al., 2020
	Pleurotus sajor-caju	41.42	31.40	42.70	0.95	ND	ND	Ragunathan & Swaminathan, 2003
Cotton waste	*Pleurotus eryngii*	71.56	25.36	59.69	5.18	9.10	8.12	Sardar et al., 2017
	Volvariella volvacea	ND	1.41–34.17	1.02	0.10	2.83–11.9	0.65–10.8	Ul Haq et al., 2011; Adedokun & Akuma, 2013
Cottonseed hull	*Pleurotus ostreatus*	8.97–61.00	14.27–17.5	49.54–65.9	1.2–6.10	10.2–26.53	3.56–5.2	Emiru et al., 2016; Sardar et al., 2020
Corncob	*Pleurotus ostreatus*	18.67–66.08	18.64–29.70	30.78–34.08	2.67–6.40	25.66–29.75	7.10–15.22	Hoa et al., 2015; Emiru et al., 2016
	Agrocybe cylindracea	46.80	2.70	93.30	0.60	35.9	3.40	Koutrotsios et al., 2014
	Pleurotus eryngii	51.80	23.72	54.69	1.88	9.68	6.97	Sardar et al., 2017
Soybean straw	*Pleurotus sajor-caju*	83.00	25.80	52.20	2.82	6.70	7.30	Patil, 2013
Jowar straw	*Pleurotus sajor-caju*	70.83	23.10	58.50	2.62	7.30	7.10	Patil, 2013
Coast-cross straw	*Agaricus bisporus*	44.1–58.3	29.80–37.88	ND	1.90–2.36	5.64–7.56	10.31–11.39	Andrade et al., 2008
Tifton straw	*Agaricus bisporus*	24.4–53.4	31.37–35.21	ND	2.05–2.34	6.07–7.60	10.60–11.77	Andrade et al., 2008
Oat straw	*Agaricus bisporus*	47.2–52.9	26.78–36.28	ND	2.34–3.06	6.57–10.31	9.83–11.36	Andrade et al., 2008
Bean straw	*Ganoderma lucidum*	2.0	9.9	ND	ND	56.2	1.0	de Carvalho et al., 2015
	Ganoderma lucidum	4.7	12.3	ND	ND	57.3	1.4	de Carvalho et al., 2015
Brachiaria grass straw	*Ganoderma lucidum*	3.50	11.60	ND	ND	48.80	1.3	de Carvalho et al., 2015

(Continued)

Table 17.2 (Continued)   Biological efficiency and chemical composition of some wild mushrooms cultivated on the different lignocellulosic biomass-based substrates

Lignocellulosic substrate	Mushroom species	Biological efficiency (%)	Chemical composition (%, dry weight)					References
			Crude protein	Carbohydrate	Fat	Crude fiber	Ash	
Sunflower stalk	Pleurotus sajor-caju	63.13	21.00	50.70	2.75	7.65	6.90	Patil, 2013
Pigeaon stalk	Pleurotus sajor-caju	75.43	24.20	48.20	2.45	7.78	6.80	Patil, 2013
Oil palm whole stalk	Volvariella volvacea	5.46	29.67	49.26	3.52	8.46	9.09	Triyono et al., 2019
Sorghum stover	Pleurotus sajor-caju	36.84	36.2	43.2	1.41	ND	ND	Ragunathan & Swaminathan, 2003
Date palm tree leave	Pleurotus ostreatus	55.70	16.13	72.77	3.41	19.89	7.83	Koutrotsios et al., 2014
	Agrocybe cylindracea	52.4	6.20	82.8	0.10	32.9	10.9	Koutrotsios et al., 2014
Plantain leave	Volvariella volvacea	ND	1.01	1.37	0.10	2.77	0.95	Adedokun and Akuma, 2013
Cassava peels	Pleurotus ostreatus	26.00	10.48	73.07	2.18	8.88	7.69	Kortei et al., 2014
	Volvariella volvacea	1.47	14.24	51.42	2.44	0.4	6.16	Apetorgbor & Apetorgbor, 2015
Yam peel	Volvariella volvacea	4.49	14.63	50.31	2.23	0.59	3.85	Apetorgbor & Apetorgbor, 2015
Potato peel	Volvariella volvacea	1.05	14.49	48.32	2.31	0.56	6.84	Apetorgbor & Apetorgbor, 2015
Pine needles	Pleurotus ostreatus	24.74	22.74	75.88	2.44	13.68	7.50	Koutrotsios et al., 2014
	Agrocybe cylindracea	93.50	9.90	63.30	1.60	26.40	25.2	Koutrotsios et al., 2014
Oil palm stem	Volvariella volvacea	1.90	38.27	40.56	4.14	7.19	9.83	Triyono et al., 2019
Oil palm hump	Volvariella volvacea	3.48	32.47	47.15	3.45	7.57	9.35	Triyono et al., 2019
Tifton grass	Ganoderma lucidum	5.50	12.50	ND	ND	48.2	1.6	de Carvalho et al., 2015
Bracts of pineapple crown	Lentinula edodes	37.50	14.0	78.8	0.50	ND	6.7	Salmones et al., 1999
Extrated olive press cake	Agrocybe cylindracea	29.20	7.80	87.3	1.30	47.6	3.6	Koutrotsios et al., 2014
Coir fiber	Pleurotus sajor-caju	27.33	44.3	45.2	1.22	ND	ND	Ragunathan & Swaminathan, 2003

ND : Not Determined

Table 17.3 Biological efficiency and chemical composition of some wild mushrooms cultivated on the different lignocellulosic biomass-based substrates in various combinations

Mushroom species	Lignocellulosic substrate	Biological efficiency (BE) (%)	Chemical Composition (%, Dry Weight)					Change in BE% (Compared with control)	Change in protein% (Compared with control)	References
			Crude protein	Carbohydrate	Fat	Crude Fiber	Ash			
Agrocybe cylindracea	Wheat straw + Raw two-phase olive mill waste (20%)	100.67	ND	ND	ND	ND	ND	45.53% ↑	ND	Zervakis et al., 2013
Auricularia polytricha	Sawdust + Panicum repens stalk (1:1 w/w)	124.70	10.37	78.49	0.87	ND	3.49	25.33% ↑	1.46% ↑	Liang et al., 2019
Auricularia polytricha	Sawdust + Pennisetum purpureum stalk (1:1 w/w)	129.54	9.13	79.65	0.74	ND	3.94	30.20% ↑	10.66% ↓	Liang et al., 2019
Auricularia polytricha	Sawdust + Zea mays stalk (1:1 w/w)	144.15	10.12	77.74	1.06	ND	4.35	44.88% ↑	0.97% ↓	Liang et al., 2019
Ganoderma lucidum	Beech sawdust + Olive pruning residues (1:1 w/w)	20.52	15.28	79.28	2.03	43.80	3.41	66.49% ↓	9.26% ↓	Koutrotsios et al., 2019
Ganoderma lucidum	Beech sawdust + Olive-mill wastes (1:1 w/w)	20.52	22.21	72.44	1.10	49.34	4.26	66.49% ↓	31.88% ↑	Koutrotsios et al., 2019
Lentinula edodes	Sawdust + Rice straw (80+20%)	36.09	15.80	78.22	5.75	ND	3.52	25.86% ↓	3.06% ↓	Gao et al., 2020
Pleurotus citrinopileatus	Rice straw + Wheat straw (1:1 w/w)	111.88	23.60	64.28	3.23	ND	8.89	44.22% ↑	3.32% ↑	Medany, 2014
Pleurotus citrinopileatus	Sawdust + Rice straw (1:1 w/w)	56.02	22.91	65.33	2.92	ND	8.84	8.60% ↑	4.81% ↑	Medany, 2014
Pleurotus citrinopileatus	Sawdust + Wheat straw (1:1 w/w)	51.83	26.01	63.77	2.54	ND	7.68	0.48% ↑	8.05% ↑	Medany, 2014
Pleurotus cystidiosus	Sawdust + Sugarcane bagasse (1:1 w/w)	44.11	18.66	46.86	3.28	24.5	6.70	21.61% ↑	10.00% ↑	Hoa et al., 2015
Pleurotus cystidiosus	Sawdust + Corncob (1:1 w/w)	43.57	21.47	44.85	2.50	23.58	7.30	20.12% ↑	36.92% ↑	Hoa et al., 2015
Pleurotus eryngii	Wheat straw + Raw two-phase olive mill waste (20%)	96.12	ND	ND	ND	ND	ND	9.82% ↑	ND	Zervakis et al., 2013

(Continued)

Table 17.3 (Continued) Biological efficiency and chemical composition of some wild mushrooms cultivated on the different lignocellulosic biomass-based substrates in various combinations

Mushroom species	Lignocellulosic substrate	Biological efficiency (BE) (%)	Chemical Composition (%, Dry Weight)						Change in BE% (Compared with control)	Change in protein% (Compared with control)	References
			Crude protein	Carbohydrate	Fat	Crude Fiber	Ash				
Pleurotus eryngii	Wheat straw + Grape Marc (1:1 w/w)	87.24	ND	ND	ND	ND	ND	52.17% ↑	ND	Koutrotsios et al., 2018
Pleurotus ostreatus	Saw dust + Rice straw (1:1 w/w)	13.67	15.29	41.96	5.40	27.01	10.33	95.28% ↑	20.20% ↑	Emiru et al., 2016
Pleurotus ostreatus	Saw dust + Cottonseed hull (1:1 w/w)	27.00	15.00	42.08	3.50	30.59	8.83	285% ↑	17.92% ↑	Emiru et al., 2016
Pleurotus ostreatus	Saw dust + Maize cob (1:1 w/w)	8.67	17.04	29.96	8.40	32.60	12.00	23.85% ↑	33.96% ↑	Emiru et al., 2016
Pleurotus ostreatus	Rice straw + Cottonseed hull (1:1 w/w)	6.00	14.12	51.44	4.00	27.10	3.33	40% ↓	4.91% ↓	Emiru et al., 2016
Pleurotus ostreatus	Rice straw + Maize cob (1:1 w/w)	8.33	15.43	43.75	4.35	23.80	12.67	16.7% ↓	3.90% ↑	Emiru et al., 2016
Pleurotus ostreatus	Cottonseed hull + Maize cob (1:1 w/w)	33.33	16.31	48.16	5.50	24.70	5.33	185.60% ↑	14.29% ↑	Emiru et al., 2016
Pleurotus ostreatus	Paddy straw + Soybean straw (1:1 w/w)	81.69	23.00	50.50	2.70	7.68	6.42	3.39% ↓	1.70% ↓	Patil et al., 2010
Pleurotus ostreatus	Wheat straw + Soybean straw (1:1 w/w)	77.66	21.10	52.00	2.56	7.40	6.15	7.77% ↑	4.76% ↑	Patil et al., 2010
Pleurotus ostreatus	Wheat straw + Paddy straw (1:1 w/w)	71.76	20.33	56.20	2.58	7.50	5.90	0.41% ↓	3.19% ↓	Patil et al., 2010
Pleurotus ostreatus	Cotton stalk + Cottonseed hell (1:1 w/w)	20.2	22.8	58.0	2.9	10.8	5.5	1.94% ↓	0%	Sardar et al., 2020
Pleurotus ostreatus	Wheat straw + Olive pruning residues (1:1 w/w)	56.79	19.88	71.76	1.87	16.54	6.49	26.49% ↓	30.61% ↓	Koutrotsios et al., 2019
Pleurotus ostreatus	Wheat straw + Olive-mill wastes (1:1 w/w)	71.33	19.32	68.56	2.70	12.97	9.42	7.67% ↓	26.93% ↑	Koutrotsios et al., 2019

Table 17.3 (Continued) Biological efficiency and chemical composition of some wild mushrooms cultivated on the different lignocellulosic biomass-based substrates in various combinations

Mushroom species	Lignocellulosic substrate	Biological efficiency (BE) (%)	Chemical Composition (%, Dry Weight)					Change in BE% (Compared with control)	Change in protein% (Compared with control)	References
			Crude protein	Carbohydrate	Fat	Crude Fiber	Ash			
Pleurotus ostreatus	Cassava peels + Corncobs (1:1 w/w)	33.7	10.66	74.80	2.16	8.79	7.65	29.11% ↑	0.47% ↑	Kortei et al., 2014
Pleurotus ostreatus	Sawdust + Sugarcane bagasse (1:1 w/w)	58.94	24.17	37.88	2.50	28.75	6.70	26.91% ↑	23.82% ↑	Hoa et al., 2015
Pleurotus ostreatus	Sawdust + Corncobs (1:1 w/w)	58.82	25.65	37.50	1.80	28.35	6.80	26.65% ↑	31.40% ↑	Hoa et al., 2015
Pleurotus ostreatus	Wheat straw + Grape Marc (1:1 w/w)	84.14	ND	ND	ND	ND	ND	5.62% ↑	ND	Koutrotsios et al., 2018
Pleurotus sajor-caju	Sugarcane bagasse + Elephant grass (1:1 w/w)	71.90	26.35	34.50	ND	26.55	6.95	355% ↑	37.45% ↓	Bernardi et al., 2019
Pleurotus sajor-caju	Rice straw + Elephant grass (1:1 w/w)	79.64	23.92	58.66	ND	8.69	6.78	25.29% ↓	14.60% ↓	Bernardi et al., 2019
Pleurotus sajor-caju	Sugarcane bagasse + Rice straw (1:1 w/w)	95.48	30.92	33.85	ND	24.51	6.95	504% ↑	26.60% ↓	Bernardi et al., 2019
Pleurotus sajor caju	Wheat straw + Mahua cake (10%)	ND	33.00	31.86	1.76	ND	8.19	ND	12.39% ↑	Gupta et al., 2013
Volvariella volvacea	Rice straw + Banana pseudo stem (1:1 w/w)	14.90	ND	ND	ND	ND	ND	34.23% ↑	ND	Biswas and Layak, 2014
Volvariella volvacea	Rice straw + Water hyacinth (1:1 w/w)	12.05	ND	ND	ND	ND	ND	8.55% ↑	ND	Biswas & Layak, 2014

ND: Not Determined; ↑: Increase; ↓: Decrease

were cereals (2.97 billion tons), sugar crops primary (2.22 billion tons), sugarcane (1.94 billion tons), maize (1.14 billion tons), vegetable primary (1.13 billion tons), oil crops (1.10 billion tons), fruits primary (0.883 billion tons), roots and tubers (0.861 billion tons), wheat (0.765 billion tons), and rice (0.755 billion tons) (FAOSTAT, 2021). Agricultural practices are significantly associated with essential agro-industrial sectors, like food, feed, fiber, etc. (Kapoor et al., 2016) which produce an enormous (1.6 billion tons per year) quantity of agriculture and agro-industrial residues/waste/by-product. Agriculture waste/residue is defined as the plant materials obtained from agriculture activities, especially post-harvest, and the agro-industrial residue is the materials result of agriculture or animal products processing (Kumla et al., 2020). Mande (2005) sub-grouped agriculture residues into crop-based residues and processing-based residues. Crop-based residues are generated on-site consisting of plant materials left after the harvesting of the main crop. These materials are morphologically distinct and consist of various plant parts such as roots, stem, stalk stick, branches, leaves, fibrous materials, empty seed pods, straw, etc. The major sources of these materials are oilseed crops, horticulture crops, cereal crops, etc. The processed-based residue is the by-product generated during industrial processing of the crop as post-harvest processes that include husks, peels, pulp, and shells, etc. Agro-industrial residues include bran, peels, bagasse, etc., generated from the industrial processing of horticultural crops (Philippoussis, 2009; Sadh et al., 2018; Kumla et al., 2020). Food industries produce an enormous quantity of processed residues such as bagasse, husks, hulls, corncobs, etc. Further, wood residues like bark, chips, sawdust, etc., are also included in processed-based residues (Figure 17.2) (Philippoussis, 2009).

The conventional use of agriculture and agro-industrial-based residues include cattle feeding, animal bedding, soil mulching, organic manure, thatching for rural homes, fuel for domestic purposes, etc. Nowadays, with available non-conventional strategies, these residues are converted into different high-value products such as biofuel, enzymes, chemicals, etc. (Da Silva, 2016; Kumla et al., 2020). After being used in various applications, nearly 30% of these residues remain available as surplus, which are disposed of by farmers and industries in an inappropriate manner (Devi et al., 2017). The most common practices for crop-based and processed-based residue disposal are onsite burnings and open dumping. Burning is the most common on-site practice for rice, wheat, and maize crop residues that not only release shoot particles (causing health problems), and generate greenhouse gases but also cause loss of important crop nutrients like nitrogen, phosphorus, potassium, selenium, etc. (Domínguez-Escribá & Porcar, 2010; Kapoor et al., 2016). The unattended open dumping of residues like cereal straw causes rotting and their associated environmental issues (Kapoor et al., 2016). The processed-based residue like cereal bran, is a good source of various important nutrients used in supplemented feed for ruminants and poultry (Muazu & Stegemann, 2015). These practices of crop residues (field or processed) handling cause various environmental problems and wasting valuable resources of energy and carbon (Momayez et al., 2019). These residues act as a pool of nutrients that are suitable for mushroom cultivation using the solid-state fermentation method (Sánchez, 2009; Kumla et al., 2020).

In terms of production and yield, sugarcane, maize, rice, and wheat are the top crops (Millati et al., 2019) and, except sugarcane, the remaining three fulfilled more than 42% of calorie demand of the human population (Ricepedia, 2021). The utility of these kinds of materials as a substrate for mushroom cultivation has been discussed further in the chapter.

17.5.1 Rice

Rice (*Oryza sativa*) holds the tenth place in the world of most produced commodities with 755 million tons of production in the year 2019, and continent leader was Asia (89.6%) followed by Africa (5.1%), America (4.7%), Europe (0.5%), and Oceania (0.1%) (FAOSTAT, 2021). Among the countries, China shares a 27.74% portion followed by India (23.51%), and both together share more than half of the total global rice production (FAOSTAT, 2021). Rice cultivation produces

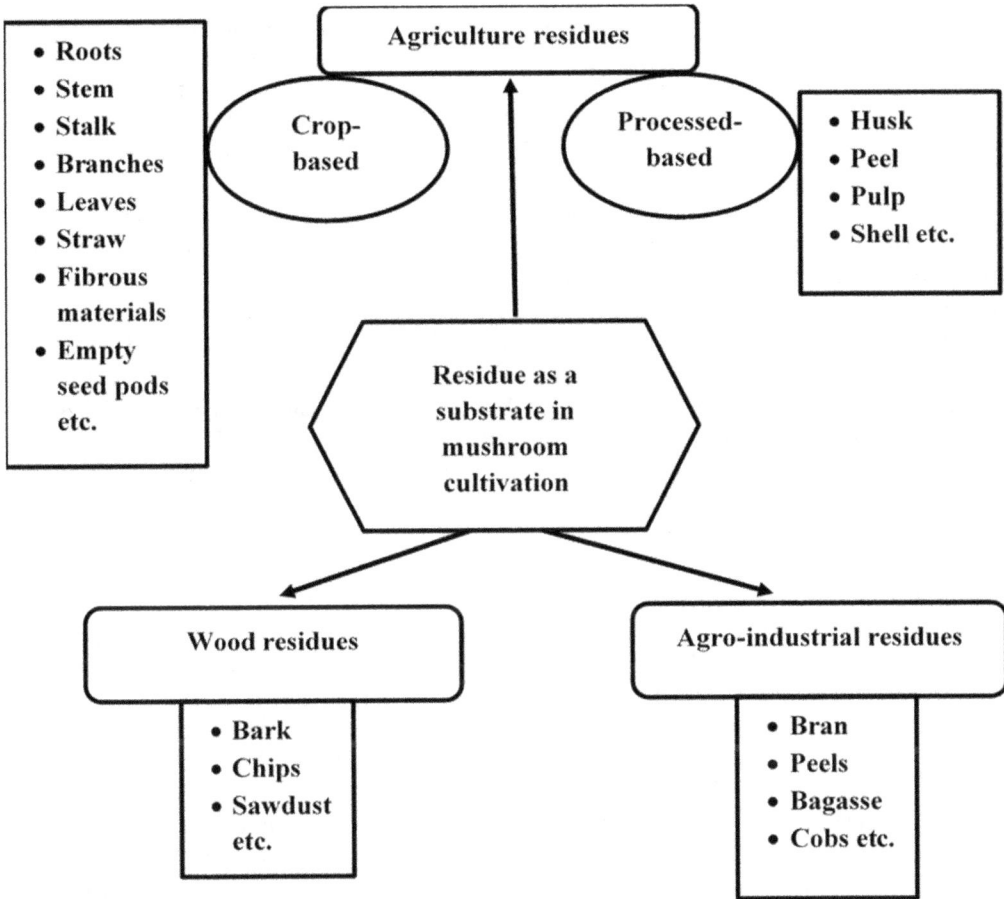

Figure 17.2 Type of agriculture and agro-industrial residues.

two major rice residues, i.e. straw and husk. Rice husks (known as rice hulls) contain mainly silica and lignin and roughly 1 kg processed white rice generates 0.28 kg of the husk (Rice Knowledge Bank, 2021; Millati et al., 2019). Therefore, in accordance with the aforementioned ratio, 140.5 million tons of rice husks were produced globally in the year 2020 (USDA, 2020). A typical composition of rice husk includes cellulose (28.6–43.30%), hemicellulose (22.0–29.7%), and lignin (19.2–24.4%) (Mirmohamadsadeghi & Karimi, 2020). Rice straw is another major residue of rice cultivation, and the amount is dependent on the cutting height (Van Hung et al., 2020). Moreover, various environmental and genetic factors also define the straw-to-grain ratios such as the variety of crop, seasonal fluctuation, plant height at maturity, soil texture and fertility, plant density, water availability, fertilizer quality, and quantity, nature of weeds, and their control, crop harvesting methods, and stages. Typically, the straw-to-grain ratio range between 0.7–1.5 (Lal, 2005; Kapoor et al., 2016) and, considering an average value (1.0), the rice straw production was more than 755 million tons in the year 2019. The rice straw utilization is restricted to around 20% of total production in applications like biofuel production, animal feed, paper, fertilizer production, etc. The rest of the portion is either burned at the site or incorporated in soil (Hanafi et al., 2012; Goodman, 2020;Van Hung et al., 2020).

Rice straw is a lignocellulosic material that consists of 32–38.6% of cellulose, 19.7–35.7% of hemicellulose, and 13.5–22.3% of lignin (Goodman, 2020). The rice straw acts as the main source of carbon and nitrogen and is efficiently utilized for mushroom cultivation in their raw or com-posed form. However, the high carbon-to-nitrogen ratio indicates a poor source of nitrogen supply for the desired growth of mushrooms. Nitrogen deficiency of the substrates for mushroom culti-vation can be compensated by adding various organic natural materials like cereal bran, shell, meal, manure, etc. However, the superiority of inorganic materials like ammonium, chloride, and urea over natural materials is the timing and efficiency (Ragunathan & Swaminathan, 2003; Cueva et al., 2017; Kumla et al., 2020). The low bulk density and better water-holding capacity make rice residue a suitable substrate for mycelium formation and better growth of the fruiting body of mushrooms (Khir & Pan, 2019). Moreover, the high carbon-to-nitrogen (C:N) ratio of rice husk and rice straw is responsible for their poor degradation by mushrooms but can be overcome by combining with other lignocellulosic biomass and adding supplements (Table 17.3). White rot fungi such as *Pleurotus ostreatus* degrade 41% of lignin, 52% of hemicellulose, and 44% of cellulose (Taniguchi et al., 2005; Khir & Pan, 2019). Another mushroom species is a rice-straw mushroom, *Volvariella volvacea*, which grows better on rice straw and a yield of 0.8 kg per 10 kg of dried rice straw can be achieved after the third week of inoculation (Van Hung et al., 2020).

17.5.2 Wheat

Wheat crop (*Triticum aestivum*) is the fourth most popular crop, reported to meet 21% of the world food demand with annual global production of more than 765 million tons in the year 2019 (Momayez et al., 2019; FAOSTAT, 2021). Wheat straw, a harvesting residue of wheat grains mainly consists of internodes, nodes, leaves, chaffs, and rachis (Kapoor et al., 2016). Wheat straw is a low commercial value product generated in abundant quantity depending on climate and agronomic parameters. A typical average range of straw to grain ratio of 1.3–1.4 was reported from numerous areas (Saha et al., 2005; Lal, 2005; Momayez et al., 2019). According to the mentioned ratio, the global annual production of wheat straw was more than 995 million tons in the year 2019 and shared in the following proportions, i.e. Asia (44.1%), Europe (34.8%), America (15.3%), Africa (3.5%), and Oceania (2.4%) (FAOSTAT, 2021). The leader nations in wheat straw pro-duction are China and India, i.e. 31% together. Such an amount raising the concern of sustainable management of straw. Several conventional uses of wheat straw include feed sources for animals (especially dairy animals), for thatching (craft of building roof), and other tradable commodities. However, on-site crop residue burning is still one of the most popular methods for straw man-agement, which ruins valuable resources (Momayez et al., 2019; Duncan et al., 2020). Wheat straw is a lignocellulosic biomass that represents a valuable source of cellulose (34–40%), hemicellulose (20–25%), lignin (20%), and other organic and inorganic compounds (Rodriguez-Gomez et al., 2012; Momayez et al., 2019). Wheat bran is a by-product obtained after the milling process of wheat grain that generally contributes 14–19% (w/w) portion (Cui et al., 2013). Wheat bran contains 15% protein and other nutrients, which make it a better supplement to compensate ni-trogen deficiency of lignocellulosic biomass for the commercial cultivation of mushrooms (Tiefenbacher, 2017). Similar to rice straw, the high C:N ratio of wheat straw also restricts the adequate utilization of lignocellulosic biomass by mushrooms for their growth but either com-bining with other low C:N ratios lignocellulosic biomass or supplement can remove hurdles for commercial cultivation of mushrooms (Table 17.3).

17.5.3 Maize

Maize (*Zea mays*) is the most popular commercial agriculture crop, where global production increased about 40% in the last decade, and the production was about 1148 million tons in the year 2019 (FAOSTAT, 2021). The top three producers of maize were the United States (30.21%), China (22.70%), and Brazil (8.80%) with a total share of 61.71% of the world's entire maize production

(FAOSTAT, 2021). Cornstalk, corn leaves, and corn husks (together known as corn stover) are the main residues obtained after harvesting the ear of maize and corncob residue produced after the milling process of maize grain. Typically, 1.01 kg (expressed in dry matter) of corn stover is obtained from 1.0 kg of harvested maize grain (Mazurkiewicz et al., 2019; Ruan et al., 2019) and, according to this ratio, the world corn stover generation was over 1 billion tons in the year 2019. Corn stover is a lignocellulosic biomass consists (w/w) of cellulose (about 35%), hemicellulose (about 20%), and lignin (about 12%) (Ruan et al., 2019). Corncob, the central core of the maize ear, is the most commonly used substrate in mushroom cultivation because of its high cellulose content (40%) with comparatively low lignin content (8%) (Table 17.1) (Millati et al., 2019). The processing of 1.0 kg of ear corn produces 0.22 kg of corncob (Basalan et al., 1995; Millati et al., 2019). Following the aforementioned ratio, the global production of corncob was approximately 253 million tons in the year 2019. Due to the low lignin content of maize crop residues, white rot fungi (like *Pleurotus* spp., *Lentinula edodes*, *Ganoderma* spp.) efficiently utilize the corn residue as a substrate through the action of lignocellulosic enzymes. Moreover, for litter-decomposing fungi (LDFs), like *Agaricus* spp. corncobs serve as the best substrate as they are deficit in lignin enzymes (Tables 17.2 and 17.3). The amount and composition of corncob residue make its efficient utility in mushroom cultivation with less effort due to the better utilization of substrates.

17.5.4 Sugarcane

Sugarcane (*Saccharum officinarum*) holds the third position in terms of the world's largest produced commodities with 1.9 billion tons production in the year 2019 (FAOSTAT, 2021). Brazil (0.75 billion tons) is the largest producer of sugarcane, followed by India (0.40 billion tons) and both together share around 60% of the global total production. The sugarcane processing industries produce two types of residues, i.e. bagasse (fibrous fraction as a result of the cleaning, preparation, and extraction of sugarcane juice) and straw (harvest residue). The residue ratio (unit weight of residue to product) for sugarcane bagasse was reported to be about 0.25–0.30 (Hernández et al., 2019). According to the mentioned ratio (considering average value, 0.275), about 536 million tons of sugarcane bagasse were produced globally in the year 2019. Sugarcane bagasse and straw consist of cellulose (34–47% and 33–45%), hemicellulose (24–31% and 18–30%), and lignin (14–22% and 17–41%), respectively (Costa et al., 2013; Kapoor et al., 2016).

17.5.5 Sawdust

Sawdust is obtained from two different types of wood, i.e. hardwood and softwood. Hardwood belongs to deciduous trees and softwood belongs to coniferous and evergreen trees. Wood wastes produced from both primary and secondary manufacturing processes include sawdust, chips, coarse residues, etc. (Millati et al., 2019). Sawdust comprises 11%–15% of the total percentage in wood (Millati et al., 2019). In the year 2019, the industrial round wood production was about 2 billion m^3 (FAOSTAT, 2021) and, according to the aforesaid proportion (13%, average value considered), the sawdust produced was around 263 million m^3. Sawdust consists of cellulose, hemicellulose, and lignin as lignocellulosic components that vary with the types and parts of the tree. The hardwood sawdust consists of 38–51% of cellulose, 17–38% of hemicellulose, and 21–31% of lignin (Table 17.1). Softwood sawdust consists of 33–42% cellulose, 22–40% hemicellulose, and 27–32% lignin (Table 17.1). Sawdust produced from wood or forest waste is a plentiful source of lignocellulosic biomass that is proficiently utilized for mushroom cultivation. *Ganoderma lucidum* is one of the most efficient wood biomass degrader mushrooms (Rashad et al., 2019).

17.6 RAW LIGNOCELLULOSIC BIOMASS AS A SUPPLEMENT FOR WILD MUSHROOM GROWTH

The heterotrophic nature of mushroom means its dependence on external nutrients for growth. Carbon, nitrogen, phosphorus, potassium, and magnesium are the major macronutrients, and iron, selenium, zinc, manganese, copper, and molibidium as trace elements are mandatory for better growth of mushroom. Along with nutrients, suitable oxygen availability and specific pH range are also required for normal metabolism, which enhances mushroom growth (Carrasco et al., 2018). The vegetative mycelium draws nutrients from the growing substrate, mainly of either two origins, i.e. composted materials acquired via fermentation and pasteurization process (Vos et al., 2017) or non-composted materials (a mixture of various lignocellulosic biomass) after steam sterilization (Yamanaka, 2017). Whatever the substrate, both are procured from agriculture and agro-industrial residues (Carrasco et al., 2018). Supplementation of substrates for nutrient enrichment stimulates the mycelium growth, promotes fructification, and enrich nutrient profile in mushrooms. Supplements are of a variety of origin depending on the country like defatted vegetable meal and dry nuts, cereal bran, fruits peels, grape pomace, cottonseed meal, chicken manure, urea, super-phosphate, ammonium sulfate, etc. (Pardo-Giménez et al., 2018). Cereal straw (wheat, rice) and sawdust are usually considered the base material for substrates; however, their nitrogen supply is exhausted in the mycelium formation, which results in the decline of yield and protein content (Pardo-Giménez, Carrasco, Roncero, Álvarez-Ortí, Zied, & Pardo-González 2018; Pardo-Giménez et al., 2020). The nitrogen deficiency can be compensated by animal litter (chicken or turkey manure) in the compost along with gypsum, which conserves nitrogen and balances the desired pH throughout mushroom development (Pardo-Giménez et al., 2020).

The low-cost lignocellulosic biomass-based residues as supplements are a promising approach that balances the nutrient availability problems for better mushroom cultivation as well as offers better management of lignocellulosic biomass as waste through mushroom cultivation. Processed-based residues like wheat bran, rice bran, mustard cake, cottonseed cake, soya cake powder, carrot pulp, molasses, etc., are low cost and are easily available supplements that compensate nutrient deficiency of base substrates (Jafarpour et al., 2010; Salama et al., 2019). The C:N ratio, pH, moisture content, compaction, oxygen, and carbon dioxide concentrations, and temperature of the substrates also influence mushroom growth; however, they are strongly influenced by the type of base substrate and supportive supplement percentage along with precise timing and proper method of application (Salama et al., 2019). The nutritional aspects of *Pleurotus ostreatus* cultivated with various types of sawdust (eucalyptus tree, mahogany tree, rain tree, ipil ipil tree, fig tree), as base substrates in individual and in combination were evaluated with 30% wheat bran as a supplement (Telang et al., 2010). The results revealed the highest protein content (27.30%) with ipil ipil tree sawdust, and fig tree sawdust showed the highest amount of dry matter (10.53%), lipid (4.46%), nitrogen (4.52%), iron (42.55 mg/100 g), zinc (27.65 mg/100 g), and selenium (6.77 mg/100 g). Moreover, mahogany tree sawdust exhibited the highest amount of carbohydrates (42.36%), calcium (31.98 mg/100 g), and magnesium (19.85 mg/100 g). Pardo-Giménez et al., (2018) estimated the protein content of *Agaricus bisporus* and *Pleurotus ostreatus* cultivated on wheat straw as substrates supplemented with defatted pistachio meal (5 g/kg dose) and reported a high content of protein (207 g/kg for *Agaricus bisporus* and 147 g/kg for *Pleurotus ostreatus*) in comparison to the control.

In another study, Pardo-Giménez et al., (2018) used a mixture of compost (chicken manure), and wheat straw as the substrate supplemented with a 5 g/kg dose of defatted pistachio meal and reported protein content of 201 g/kg and 153 g/kg (dry weight) in *Agaricus bisporus* and *Pleurotus ostreatus*, respectively. In both studies, supplemented substrate produced high protein content in mushrooms than the non-supplemented substrate. Jeznabadi et al. (2016) evaluated various combinations of agriculture residue as substrate and supplements for protein content (g/100 g edible weight) in *Pleurotus eryngii*. The results revealed the highest value of protein content of 13.66% for wheat straw, wheat bran, and soybean powder combination, followed by a wheat straw, soybean powder, and rice bran combination (13.53%) and barley straw and soybean powder combination (11.82%). The barley straw and wheat bran combination produced the least protein content (4.64%) in *Pleurotus eryngii*. Moreover, *Pleurotus eryngii* showed high-protein content in the fruit body with wheat bran supplement (7.70%) in comparison to rice bran (6.74%).

Moonmoon et al. (2011) calculated the biological yield of *Lentinus edodes* grown on the sawdust substrate supplemented with wheat bran and rice bran and observed the higher biological yield (153.3/500 g packet) with 25% wheat bran in comparison to rice bran (105.3/500 g packet). Furthermore, a better-quality mushroom was achieved with 40% wheat bran supplement and sawdust-based substrate. Lignocellulosic biomass as a basic substrate and bran as the supplement has shown improved characteristics for nutrient enhancement in growing mushrooms due to efficient uptake of naturally available nitrogen of organic origin such as wheat bran by mushroom instead of synthesizing the molecules (Salama et al., 2019). The wheat bran presence improved the water retention in the substrate as well as the high content of protein of wheat bran (15.6%) over rice bran (13.4%) makes wheat bran a better supplement that enhanced mushroom yield and quality by providing certain amino acids available in wheat bran (Tiefenbacher, 2017; Salama et al., 2019).

A mixture of cotton stalk, rice straw, sugarcane bagasse, and wheat bran improved the protein content (16.69%), polysaccharides (3.61%), and minerals (3,433 mg/100 g) (Rashad et al., 2019). The nutritional aspect of *Pleurotus ostreatus* grown on conifer sawdust (CS) and bamboo (*Phyllostachys pubescencs*) sawdust (BS) as a base substrate and sweet potato schochu lees (SPSL) (dry) as a supplement was explored (Yamauchi et al., 2019). BS+SPSL combination expressed superiority over CS+SPSL in terms of protein content (43.8 and 41.1 g/100 g, dry weight), phosphorous value (1,573 and 1,474 mg/100 g), leucine amino acid value (453 and 252 mg/100 mg), isoleucine amino acid value (301 and 135 mg/100 mg), and valine amino acid value (355 and 180 mg/100 mg), respectively. The results show an enriched nutrient profile in *Pleurotus ostreatus* with BS in comparison to CS substrate. Rice straw with wheat bran and rice bran revealed a higher potassium content of 0.97 g/100 gm and 0.81 g/100 g, respectively, than the control (0.33 g/100 g) in *Pleurotus ostreatus* (Salama et al., 2019). *Schizophyllum commune* grown on 94% grape residue supplemented with 5% wheat bran displayed high protein content (16.59%) compared to the control (9.96%) (Basso et al., 2020). Wheat bran (25%) supplemented waste paper (75%) substrate indicated improvement in biological efficiency from 179% to 522% with *Pleurotus ostreatus* (Tesfay et al., 2020).

17.7 MODIFIED LIGNOCELLULOSIC BIOMASS AS A SUPPLEMENT FOR THE GROWTH OF WILD MUSHROOMS

Biochar is a pyrogenous carbon-rich solid derived from lignocellulosic biomass, commonly produced via the pyrolysis process, offers dual benefits. Biochar production serves as a sustainable option for lignocellulosic biomass-based waste management as well as it is an environmentally friendly low-cost material with various environmental applications (Tomczyk et al., 2020). Biochar exhibits high-carbon content, resistivity to hydrolysis and chemical reaction (Nam et al., 2018), high porosity, and large surface area (200 m^2/g–400 m^2/g) (Li et al., 2016). The adsorptive porous structure of biochar improves the retention and translocation of nutrients and water (Nam et al., 2018; Mahari et al., 2020). These promising features make biochar a stable supplement in mushroom cultivation. The slow pyrolysis (known as bio-carbonization) process produces a greater amount of biochar yield (35%, dry weight) in comparison to fast pyrolysis (12%, dry weight) (Mahari et al., 2020). Microwave vacuum pyrolysis (MVP) is a recently developed, promising, and innovative technique for biochar production (Lam et al., 2016; Nam et al., 2018; Liew et al., 2018;

Kong et al., 2019; Mahari et al., 2020; Foong et al., 2020). Nam et al. (2018) evaluated the growth of *Pleurotus ostreatus* on a mixture of sawdust (883 g), rice bran (85 g), and supplemented with biochar (20 g) which was produced by microwave vacuum pyrolysis technique from the palm kernel shell. The biochar presence reduced the mycelium colonization period to 21 days in comparison to conventional mushroom cultivation methods, i.e. 32–49 days. Furthermore, biochar significantly improved the mushroom yield (400–600 g/month) compared to the control (100–300 g/month). Spent mushroom substrate (SMS) transformed biochar via microwave vacuum pyrolysis enhanced the water retention up to 59%; total mycelium growth volume of 317 cm^3 and *Pleurotus* species yield of 200 g/month (Lam et al., 2019).

Mahari et al. (2020) used biochar produced by microwave vacuum pyrolysis (750 W) of palm kernel shell (yield of 28 wt%) with a mixture of rice bran and sawdust as a growth medium for *Pleurotus ostreatus* cultivation. The results revealed that a 1-month period with 150 g of biochar was sufficient for complete colonization of the mycelium (Mahari et al., 2020). High temperatures (750 W) resulted in a low biochar yield (28 wt%) and vice versa (550 W, 36 wt%) because high temperatures intensified the process of biomass decomposition, which reduced the carbon content, active surface area, and porosity (Nam et al., 2018; Mahari et al., 2020). The adsorptive porous surface of the biochar provides various sites and porous networks that absorb, retain, and translocate the necessary nutrients from rice bran for faster mycelium growth (Zhu et al., 2017; Nam et al., 2018). The addition of 10% biochar in a substrate consisting of a mixture of cotton stalks (27%), rice straw (27%), sugarcane bagasse (27%), wheat bran (18%), and $CaCO_3$ (1%) responded to the highest yields (238.40 g/kg), biological efficiencies (23.84%), protein content (19.58%), and minerals (4,092 mg/100 g) (Rashad et al., 2019). Better retention of water, nutrients, and neutralizing potential made biochar a potential additive in mushroom cultivation (Peiris et al., 2019; Mahari et al., 2020).

17.8 CONCLUSION

The cultivation of different mushroom species is established as a popular agro-commercial activity recognized for high nutritional and medical values. Besides being rich in protein content and other nutritional components, mushrooms also produce various bioactive compounds that are used in prevention and treatment for various chronic diseases like cancer, asthma, hepatitis, cardiovascular problems, etc. *Ganoderma lucidum* and *Lentinula edodes* have been widely explored for their antitumor activity. Moreover, some of the researchers also reported the use of different mushroom species for animal feed, making them a valuable resource for animal husbandry. The cultivation of mushrooms is primarily associated with nutritional aspects but also exhibits both ecological and economical gain. Besides nutritional and health aspects, some of the edible wild and cultivated mushrooms are reported to be associated with toxicity issues and may be consumed judiciously.

Mushroom cultivation is one of the important commercial activities associated with the sustainable use of different lignocellulosic biomass-based residues as substrates. Mushroom cultivation on a lignocellulosic biomass-based substrate not only enhances the protein content but also transforms them into efficient animal feed. The lignocellulosic biomass residue like rice straw, sugarcane bagasse, etc. can easily be degraded by white rot mushroom-forming fungi and so obtain a fermented product used as ruminant feed supplement due to enhanced digestibility. Wild mushrooms draw nutrients from their growing substrate via solid-state fermentation of the lignocellulosic biomass with the help of their enzymatic system. This clears a way for eco-friendly management of different agriculture and agro-industrial residue (lignocellulosic biomass-based), resulting in the production of value-added products for different human needs such as enzymes, etc. *Pleurotus* spp. and *Lentinula edodes* are widely explored for enzyme production of industrial applications as well as efficient bioremediation and biodegradation potential for toxic substances.

Apart from all the positive aspects of the usage of different lignocellulosic biomass-based residues as substrates for the cultivation of mushroom species, some risks are also associated with the consumption of edible mushrooms as they tend to accumulate various heavy metals, which may pose hazardous effects on human health and the food chain, e.g., *Pleurotus* ostreatus, *Flammulina velutipes*, *Lentinus edodes*, *Boletus edulis Bull*, *Lactarius* spp., *Clitocybe alexandri*, etc.

Various lignocellulosic biomass-based residues utilized as substrates play an important role in supplying the nutritional requirement for the growth of different mushroom varieties. However, growth conditions also play an important role in mushroom cultivation as different factors or variables influence the vegetative growth of mushroom species such as the C:N ratio, temperature, moisture, competition with other organisms, microbial contamination, etc. The major issue related to the use of lignocellulosic biomass-based resides as substrates for mushroom cultivation is the inadequate supply of nitrogen until the maturity of the fruiting body that reduced the protein content and other nutritional qualities of mushrooms. However, mixing with other lignocellulosic biomasses, the addition of agriculture and agro-industrial based products like bran and modified lignocellulosic biomasses like biochar overcome this major problem.

Intensive studies involving biochemical, nutrigenomics, and molecular studies should be conducted to explore novel mechanisms and characterization of biologically important metabolites and the farming community may be made aware regarding the production of this value-added product for the sustainable use of resources with good economic gain. Mushroom cultivation is a small commercial activity involving the agricultural community and technical experts, which fulfills three sustainable development goals, i.e., nutrition (proteinaceous food), good health and well-being (medicinal value), and lignocellulosic biomass-based waste management, making Earth a better place for human survival with less environmental and nutritional challenges.

REFERENCES

Adedokun, O. M., & Akuma, A. H. (2013). Maximizing agricultural residues: Nutritional properties of straw mushroom on maize husk, waste cotton, and plantain leaves. *Natural Resources*, *4*(8), 534–537

Aguilar-Rivera, N., & De Jesús-Merales, J. (2010). Edible mushroom *Pleurotusostreatus* production on cellulosic biomass of sugar cane. *Sugar Tech*, *12*(2), 176–178.

Ahmed, M., Abdullah, N., Ahmed, K. U., & Bhuyan, M. H. M. (2013). Yield and nutritional composition of oyster mushroom strains newly introduced in Bangladesh. *Pesquisa Agropecuária Brasileira*, *48*(2), 197–202

Andrade, M. C. N. D., Zied, D. C., Minhoni, M. T. D. A., & Kopytowski Filho, J. (2008). The yield of four *Agaricus bisporus* strains in three compost formulations and chemical composition analyses of the mushrooms. *Brazilian Journal of Microbiology*, *39*(3), 593–598.

Apetorgbor, M. M., & Apetorgbor, A. K. (2015). Comparative studies on yield of *Volvariellavolvacea* using root and tuber peels for improved livelihood of communities. *Journal of Ghana Science Association*, *16*(1), 34–43.

Ashraf, J., Ali, M. A., Ahmad, W., Ayyub, C. M., & Shafi, J. (2013). Effect of different substrate supplements on oyster mushroom (*Pleurotus* spp.) production. *Food Science and Technology*, *1*(3), 44–51.

Atila, F. (2017). Evaluation of suitability of various agro-wastes for productivity of *Pleurotusdjamor*, *Pleurotuscitrinopileatus* and *Pleurotuseryngii* mushrooms. *Journal of Experimental Agriculture International*, *17*(5), 1–11.

Atila, F. (2019). Compositional changes in lignocellulosic content of some agro-wastes during the production cycle of shiitake mushroom. *Scientia Horticulturae*, *245*, 263–268.

Atri, N. S., Sharma, S. K., Joshi, R., Gulati, A., & Gulati, A. (2012). Amino acid composition of five wild *Pleurotus* species chosen from North West India. *EuropeanJournal of Biological Sciences*, *4*(1), 31–34.

Bai, F. W., Yang, S., & Ho, N. W. Y. (2019). Fuel Ethanol Production From Lignocellulosic Biomass. In Murray Moo-Young (ed), *Comprehensive Biotechnology* (Third Edition). New York: Elsevier Inc, pp. 49–65.

Baldrian, P. (2006). Fungal laccases-occurrence and properties. *FEMS Microbiology Reviews, 30*(2): 215–242.

Barros, L., Baptista, P., Correia, D. M., Sá Morais, J., & Ferreira, I. C. (2007). Effects of conservation treatment and cooking on the chemical composition and antioxidant activity of Portuguese wild edible mushrooms. *Journal of Agricultural and Food Chemistry, 55*(12), 4781–4788.

Basalan, M., Bayhan, R., Secrist, D. S., Hill, J., Owens, F. N., Witt, M., & Kreikemeier K. (1995). Corn maturation: changes in the grain and cob. *Animal Science Research Report, 1*, 92–98.

Basso, V., Schiavenin, C., Mendonça, S., de Siqueira, F. G., Salvador, M., & Camassola, M. (2020). Chemical features and antioxidant profile by *Schizophyllum commune* produced on different agroindustrial wastes and byproducts of biodiesel production. *Food Chemistry, 329*, 127089.

Bernardi, E., Volcão, L. M., de Melo, L. G., & do Nascimento, J. S. (2019). Productivity, biological efficiency and bromatological composition of *Pleurotussajor-caju*growth on different substrates in Brazil. *Agriculture and Natural Resources, 53*(2), 99–105.

Biswas, M. K., & Layak, M. (2014). Techniques for increasing the biological efficiency of paddy straw mushroom (*Volvariellavolvacea*) in eastern India. *Food Science and Technology, 2*(4), 52–57.

Borthakur, M., & Joshi, S. R. (2019). Wild Mushrooms as Functional Foods: The Significance of Inherent Perilous Metabolites. In Gupta V. K., & Pandey A., eds, *New and Future Developments in Microbial Biotechnology and Bioengineering*, New York: Elsevier Inc, pp. 1–12

Carrasco, J., Zied, D. C., Pardo, J. E., Preston, G. M., & Pardo-Giménez, A. (2018). Supplementation in mushroom crops and its impact on yield and quality. *AMB Express, 8*(1), 1–9.

Chanakya, H. N., Malayil, S., & Vijayalakshmi, C. (2015). Cultivation of *Pleurotus* spp. on a combination of anaerobically digested plant material and various agro-residues. *Energy for Sustainable Development, 27*, 84–92.

Chang, S. T. (1990). Future trends in cultivation of alternative mushrooms. *Mushroom Journal, 215*, 422–423.

Chang, S. T., & Miles, P. G. (2008). *Mushrooms: Cultivation, Nutritional Value, Medicinal Effect, and Environmental Impact* (Second Edition). Boca Raton, Fla, USA: CRC Press.

Cherubin, M. R., Oliveira, D. M. D. S., Feigl, B. J., Pimentel, L. G., Lisboa, I. P., Gmach, M. R., Varanda, L. L., Morais, M. C., Satiro, L. S., Popin, G. V., & Paiva, S. R. D. (2018). Crop residue harvest for bioenergy production and its implications on soil functioning and plant growth: A review. *Scientia Agricola, 75*(3), 255–272.

Cheung, P. C. (2013). Mini-review on edible mushrooms as source of dietary fiber: Preparation and health benefits. *Food Science and Human Wellness, 2*(3–4), 162–166.

Chukwurah, N. F., Eze, S. C., Chiejina, N. V., Onyeonagu, C. C., Okezie, C. E. A., Ugwuoke, K. I., Ugwu, F. S. O., Aruah, C. B., Akobueze, E. U., & Nkwonta, C. G. (2013). Correlation of stipe length, pileus width and stipe girth of oyster mushroom (*Pleurotusostreatus*) grown in different farm substrates. *Journal of Agricultural Biotechnology and Sustainable Development, 5*(3), 54–60.

Chawla, P., Kumar, N., Kaushik, R., & Dhull, S. B. (2019). Synthesis, characterization and cellular mineral absorption of gum arabic stabilized nanoemulsion of *Rhododendron arboreum* flower extract. *Journal of Food Science and Technology, 56*(12), 5194–5203.

Cohen, R., Persky, L., & Hadar, Y. (2002). Biotechnological applications and potential of wood-degrading mushrooms of the genus *Pleurotus. Applied Microbiology and Biotechnology, 58*(5), 582–594.

Costa, S. M., Mazzola, P. G., Silva, J. C., Pahl, R., Pessoa Jr., A., & Costa, S. A. (2013). Use of sugarcane straw as a source of cellulose for textile fiber production. *Industrial Crops and Products, 42*, 189–194.

Cueva, M. B. R., Hernáadez, A., & Nino-Ruiz, Z. (2017). Influence of C/N ratio on productivity and the protein contents of *Pleurotusostreatus* grown in different residue mixtures. *Revista de la Facultad de CienciasAgrariasUNCuyo, 49*(2), 331–334.

Cui, S. W., Wu, Y., & Ding, H. (2013). The range of dietary fibre ingredients and a comparison of their technical functionality. In Delcour J. A., & Poutanen K., eds, *Fibre-Rich and Wholegrain Foods: Improving Quality*. New York: Elsevier Inc, pp. 96–119.

Da Silva, L. L. (2016). Adding value to agro-industrial wastes. *Industrial Chemistry, 2*(2), e103.

de Carvalho, C. S. M., Sales-Campos, C., de Carvalho, L. P., de Almeida Minhoni, M. T., Saad, A. L. M., Alquati, G. P., & de Andrade, M. C. N. (2015). Cultivation and bromatological analysis of the medicinal mushroom *Ganoderma lucidum* (Curt.: Fr.) P. Karst cultivated in agricultural waste. *African Journal of Biotechnology, 14*(5), 412–418.

Devi, S., Gupta, C., Jat, S. L., & Parmar, M. S. (2017). Crop residue recycling for economic and environmental sustainability: The case of India. *Open Agriculture, 2*(1), 486–494.

Domínguez-Escribá, L., & Porcar, M. (2010). Rice straw management: the big waste. *Biofuels, Bioproducts and Biorefining, 4*(2), 154–159.

Dotaniya, M. L., Datta, S. C., Biswas, D. R., Dotaniya, C. K., Meena, B. L., Rajendiran, S., Regar, K. L., & Lata, M. (2016). Use of sugarcane industrial by-products for improving sugarcane productivity and soil health. *International Journal of Recycling of Organic Waste in Agriculture, 5*(3), 185–194.

Duncan, A. J., Samaddar, A., & Blümmel, M. (2020). Rice and wheat straw fodder trading in India: Possible lessons for rice and wheat improvement. *Field Crops Research, 246*, 107680.

Elisashvili, V., Penninckx, M., Kachlishvili, E., Tsiklauri, N., Metreveli, E., Kharziani, T., & Kvesitadze, G. (2008). *Lentinus edodes* and *Pleurotus species* lignocellulolytic enzymes activity in submerged and solid-state fermentation of lignocellulosicwastes of different composition. *Bioresource Technology, 99*(3), 457–462

Emiru, B., Zenebech, K., & Kebede, F. (2016). Effect of substrates on the yield, yield attribute and dietary values of oyster mushroom (*Pleurotusostreatus*) in the pastoral regions of northern Ethiopia. *African Journal of Food, Agriculture, Nutrition and Development, 16*(4), 11198–11218.

Faik, A., Hülya, A., Ahmet, T., Ertugrul, C., Mark, S., M., & Robert H. G. (2011). Macro and microelement contents of fruiting bodies of wild-edible mushrooms growing in the East Black Sea region of Turkey. *Food and Nutrition Sciences, 2*(2), 53–59.

Familoni, T. V., Ogidi, C. O., Akinyele, B. J., & Onifade, A. K. (2018). Evaluation of yield, biological efficiency and proximate composition of *Pleurotus* species cultivated on different wood dusts. *Czech Mycology, 70*, 33–45.

FAOSTAT. (2021). FAO-Food and Agriculture Organization of the United Nations. Availablefrom *FAOSTAT Statistics Database-Agriculture*. Rome, Italy: Food and AgricultureOrganization of the United Nations. Available online: http://www.fao.org/faostat/en/ (accessedon 15 February 2021).

Foong, S. Y., Latiff, N. S. A., Liew, R. K., Yek, P. N. Y., & Lam, S. S. (2020). Production of biochar for potential catalytic and energy applications via microwave vacuum pyrolysis conversion of cassava stem. *Materials Science for Energy Technologies, 3*, 728–733.

Gaitán-Hernández, R., & Norberto Cortés, G. M. (2014). Improvement of yield of the edible and medicinal mushroom *Lentinula edodes* on wheat straw by use of supplemented spawn. *Brazilian Journal of Microbiology, 45*(2), 467–474.

Gao, S., Huang, Z., Feng, X., Bian, Y., Huang, W., & Liu, Y. (2020). Bioconversion of rice straw agroresidues by *Lentinula edodes* and evaluation of non-volatile taste compounds in mushrooms. *Scientific Reports, 10*(1), 1–8.

Goodman, B. A. (2020). Utilization of waste straw and husks from rice production: A review. *Journal of Bioresources and Bioproducts, 5*, 143–162.

Gothwal, R., Gupta, A., Kumar, A., Sharma, S., & Alappat, B. J. (2012). Feasibility of dairy wastewater (DWW) and distillery spent wash (DSW) effluents in increasing the yield potential of *Pleurotusflabellatus* (PF 1832) and *Pleurotussajor-caju* (PS 1610) on bagasse. *3Biotech, 2*, 249–257.

GünçErgönül, P., Akata, I., Kalyoncu, F., & Ergönül, B. (2013). Fatty acid compositions of six wild edible mushroom species. *The Scientific World Journal, 163964*.

Gupta, A., Sharma, S., Saha, S., & Walia, S. (2013). Yield and nutritional content of *Pleurotussajor-caju* on wheat straw supplemented with raw and detoxified mahua cake. *Food Chemistry, 141*(4), 4231–4239.

Hanafi, E. M., El Khadrawy, H. H., Ahmed, W. M., & Zaabal, M. M. (2012). Some observations on rice straw with emphasis on updates of its management. *World Applied Sciences Journal, 16*(3), 354–361.

Hassan, F. R. H. (2007). Cultivation of the monkey head mushroom (*Hericiumerinaceus*) in Egypt. *Journal of Applied Sciences Research, 3*(10), 1229–1233.

Havstad, M. R. (2020). Biodegradable plastics. In Letcher T. M., ed., *Plastic Waste and Recycling 97-129*. New York: Elsevier Inc.

Hernández, C., Escamilla-Alvarado, C., Sánchez, A., Alarcón, E., Ziarelli, F., Musule, R., & Valdez-Vazquez, I. (2019). Wheat straw, corn stover, sugarcane, and Agave biomasses: chemical properties, availability, and cellulosic-bioethanol production potential in Mexico. *Biofuels, Bioproducts and Biorefining, 13*(5), 1143–1159.

Hoa, H. T., Wang, C. L., & Wang, C. H. (2015). The effects of different substrates on the growth, yield, and nutritional composition of two oyster mushrooms (*Pleurotusostreatus* and *Pleurotuscystidiosus*). *Mycobiology, 43*(4), 423–434.

Jafarpour, M., Jalali, A. J., Dehdashtizadeh, B., & Eghbalsaied, S. (2010). Evaluation of agricultural wastes and food supplements usage on growth characteristics of *Pleurotusostreatus*. *African Journal of Agricultural Research 5*(23), 3291–3296.

Jeznabadi, E. K., Jafarpour, M., & Eghbalsaied, S. (2016). King oyster mushroom production using various sources of agricultural wastes in Iran. *International Journal of Recycling of Organic Waste in Agriculture, 5*(1), 17–24.

Kalac, P. (2009). Chemical composition and nutritional value of European species of wild growing mushrooms: a review. *Food Chemistry, 113*(1), 9–16.

Kapoor, M., Panwar, D., & Kaira, G. S. (2016). Bioprocesses for enzyme production using agro-industrial wastes: technical challenges and commercialization potential. In Dhillon G. S., & Kaur S., eds, *Agro-Industrial Wastes as Feedstock for Enzyme Production.* New York: Elsevier Inc, pp. 61–93

Khir, R., & Pan, Z. (2019). Rice. In Pan Z., Zhang R., & Zicari S., eds, *Integrated Processing Technologies for Food and Agricultural By-Products.* New York: Elsevier Inc, pp. 21–58.

Kong, S. H., Lam, S. S., Yek, P. N. Y., Liew, R. K., Ma, N. L., Osman, M. S., & Wong, C. C. (2019). Self-purging microwave pyrolysis: an innovative approach to convert oil palm shell into carbon-rich biochar for methylene blue adsorption. *Journal of Chemical Technology & Biotechnology, 94*(5), 1397–1405.

Kortei, N. K., Dzogbefia, V. P., & Obodai, M. (2014). Assessing the effect of composting cassava peel based substrates on the yield, nutritional quality, and physical characteristics of *Pleurotusostreatus* (Jacq. ex Fr.) Kummer. *Biotechnology Research International, 571520.*

Koutrotsios, G., Kalogeropoulos, N., Kaliora, A. C., & Zervakis, G. I. (2018). Toward an increased functionality in oyster (*Pleurotus*) mushrooms produced on grape marc or olive mill wastes serving as sources of bioactive compounds. *Journal of Agricultural and Food Chemistry, 66*(24), 5971–5983.

Koutrotsios, G., Mountzouris, K. C., Chatzipavlidis, I., &Zervakis, G. I. (2014). Bioconversion of lignocellulosic residues by *Agrocybecylindracea* and *Pleurotusostreatus* mushroom fungi-Assessment of their effect on the final product and spent substrate properties. *Food Chemistry, 161,* 127–135.

Koutrotsios, G., Patsou, M., Mitsou, E. K., Bekiaris, G., Kotsou, M., Tarantilis, P. A., Pletsa, V., Kyriacou, A. & Zervakis, G. I. (2019). Valorization of olive by-products as substrates for the cultivation of *Ganoderma lucidum* and *Pleurotusostreatus* mushrooms with enhanced functional and prebiotic properties. *Catalysts, 9*(6), 537.

Kumla, J., Suwannarach, N., Sujarit, K., Penkhrue, W., Kakumyan, P., Jatuwong, K., Vadthanarat, S., & Lumyong, S. (2020). Cultivation of mushrooms and their lignocellulolytic enzyme production through the utilization of agro-industrial waste. *Molecules, 25*(12), 2811.

Lal, R. (2005). World crop residues production and implications of its use as a biofuel. *Environment International, 31*(4), 575–584.

Lam, S. S., Lee, X. Y., Nam, W. L., Phang, X. Y., Liew, R. K., Yek, P. N., Ho, Y. L., Ma, N. L., & Rosli, M. H. (2019). Microwave vacuum pyrolysis conversion of waste mushroom substrate into biochar for use as growth medium in mushroom cultivation. *Journal of Chemical Technology & Biotechnology, 94*(5), 1406–1415.

Lam, S. S., Liew, R. K., Jusoh, A., Chong, C. T., Ani, F. N., & Chase, H. A. (2016). Progress in waste oil to sustainable energy, with emphasis on pyrolysis techniques. *Renewable and Sustainable Energy Reviews, 53,* 741–753.

Li, J., Dai, J., Liu, G., Zhang, H., Gao, Z., Fu, J., He, Y., & Huang, Y. (2016). Biochar from microwave pyrolysis of biomass: A review. *Biomass and Bioenergy, 94,* 228–244.

Liang, C. H., Wu, C. Y., Lu, P. L., Kuo, Y. C., & Liang, Z. C. (2019). Biological efficiency and nutritional value of the culinary-medicinal mushroom *Auricularia* cultivated on a sawdust basal substrate supplement with different proportions of grass plants. *Saudi Journal of Biological Sciences, 26*(2), 263–269.

Liew, R. K., Nam, W. L., Chong, M. Y., Phang, X. Y., Su, M. H., Yek, P. N. Y., Ma, N. L., Cheng, C. K., Chong, C. T., & Lam, S. S. (2018). Oil palm waste: An abundant and promising feedstock for microwave pyrolysis conversion into good quality biochar with potential multi-applications. *Process Safety and Environmental Protection, 115*, 57–69.

Ma, G., Yang, W., Zhao, L., Pei, F., Fang, D., & Hu, Q. (2018). A critical review on the health promoting effects of mushrooms nutraceuticals. *Food Science and Human Wellness, 7*(2), 125–133.

Mahari, W. A. W., Nam, W. L., Sonne, C., Peng, W., Phang, X. Y., Liew, R. K., Yek, P. N. Y., Lee, X. Y., Wen, O. W., Show, P. L., & Chen, W. H. (2020). Applying microwave vacuum pyrolysis to design moisture retention and pH neutralizing palm kernel shell biochar for mushroom production. *Bioresource Technology, 312*, 123572.

Mahari, W. A. W., Peng, W., Nam, W. L., Yang, H., Lee, X. Y., Lee, Y. K., Liew, R. K., Ma, N. L., Mohammad, A., Sonne, C., & Van Le, Q. (2020). A review on valorization of oyster mushroom and waste generated in the mushroom cultivation industry. *Journal of Hazardous Materials, 400*, 123156.

Mande, S. (2005). Biomass gasifier-based power plants: potential, problems, and research needs for decentralized rural electrification. In Lal B., & Reddy M. R. V. P., eds, *Wealth from Waste: Trends and Technologies*. New Delhi: TERI press, pp. 1–28.

Mazurkiewicz, J., Marczuk, A., Pochwatka, P., & Kujawa, S. (2019). Maize straw as a valuable energetic material for biogas plant feeding. *Materials, 12*(23), 3848.

Medany, G. M. (2014). Cultivation possibility of golden oyster mushroom (*Pleurotuscitrinopileatus*) under the Egyptian conditions. *Egyptian. Journal of Agriculture Research, 92*(2), 749–761.

Menon, V., & Rao, M. (2012). Trends in bioconversion of lignocellulose: biofuels, platform chemicals & biorefinery concept. *Progress in Energy and Combustion Science, 38*(4), 522–550.

Millati, R., Cahyono, R. B., Ariyanto, T., Azzahrani, I. N., Putri, R. U., & Taherzadeh, M. J. (2019). Agricultural, industrial, municipal, and forest wastes: An Overview. In Taherzadeh M. J., Bolton K., Wong J., & Pandey A., eds, *Sustainable Resource Recovery and Zero Waste Approaches*. New York: Elsevier Inc, pp. 1–22.

Mirmohamadsadeghi, S., & Karimi, K. (2020). Recovery of silica from rice straw and husk. In Varjani S., Pandey A., Gnansounou E., Khanal S. K., & Raveendran S., eds, *Current Developments in Biotechnology and Bioengineering*. New York: Elsevier Inc, pp. 411–433

Mleczek, M., Budka, A., Siwulski, M., Mleczek, P., Budzyńska, S., Proch, J., Gąsecka, M., Niedzielski, P., & Rzymski, P. (2021). A comparison of toxic and essential elements in edible wild and cultivated mushroom species. *European Food Research and Technology, 247*(5), 1249–1262.

Momayez, F., Karimi, K., & Horváth, I. S. (2019). Sustainable and efficient sugar production from wheat straw by pretreatment with biogas digestate. *RSC advances, 9*(47), 27692–27701.

Moonmoon, M., Shelly, N. J., Khan, M. A., Uddin, M. N., Hossain, K., Tania, M., & Ahmed, S. (2011). Effects of different levels of wheat bran, rice bran and maize powder supplementation with saw dust on the production of shiitake mushroom (*Lentinus edodes (Berk.) Singer*). *Saudi Journal of Biological Sciences, 18*(4), 323–328.

Muazu, R. I., & Stegemann, J. A. (2015). Effects of operating variables on durability of fuel briquettes from rice husks and corncobs. *Fuel Processing Technology, 133*, 137–145.

Mulkey, V. R., Owens, V. N., & Lee, D. (2006). Management of switchgrass-dominated conservation reserve program lands for biomass production in South Dakota. *Crop Science, 46*(2), 712–720.

Nam, W. L., Phang, X. Y., Su, M. H., Liew, R. K., Ma, N. L., Rosli, M. H. N. B., & Lam, S. S. (2018). Production of bio-fertilizer from microwave vacuum pyrolysis of palm kernel shell for cultivation of Oyster mushroom (*Pleurotusostreatus*). *Science of the Total Environment, 624*, 9–16.

Nhuchhen, D. R., Basu, P., & Acharya, B. (2014). A comprehensive review on biomass torrefaction. *International Journal of Renewable Energy & Biofuels, 506376*.

Niksic, M., Klaus, A., & Argyropoulos, D. (2016). Safety of foods based on mushrooms. In Prakash V., Martín-Belloso O., Keener L., Astley S., Braun S., McMahon H., & Lelieveld H., eds, *Regulating Safety of Traditional and Ethnic Foods*. New York: Elsevier Inc, pp. 421–439.

Nrcs, U. (2011). Carbon to nitrogen ratios in cropping systems. *Fact Sheet, 2*. https://www.nrcs.usda.gov (accessed on 28-01-2021).

Pardo-Giménez, A., Carrasco, J., Roncero, J. M., Álvarez-Ortí, M., Zied, D. C., & Pardo-González, J. E. (2018). Recycling of the biomass waste defatted almond meal as a novel nutritional supplementation for cultivated edible mushrooms. *Acta Scientiarum. Agronomy, 40,* e39341.

Pardo-Giménez, A., Catalán, L., Carrasco, J., Álvarez-Ortí, M., Zied, D., & Pardo, J. (2016). Effect of supplementing crop substrate with defatted pistachio meal on *Agaricus bisporus* and *Pleurotusostreatus* production. *Journal of the Science of Food and Agriculture, 96*(11), 3838–3845.

Pardo-Giménez, A., Pardo, J. E., Dias, E. S., Rinker, D. L., Caitano, C. E. C., & Zied, D. C. (2020). Optimization of cultivation techniques improves the agronomic behavior of *Agaricus subrufescens*. *Scientific Reports, 10*(1), 1–9.

Pasangulapati, V., Ramachandriya, K. D., Kumar, A., Wilkins, M. R., Jones, C. L., & Huhnke, R. L. (2012). Effects of cellulose, hemicellulose and lignin on thermochemical conversion characteristics of the selected biomass. *Bioresource Technology, 114,* 663–669.

Pathmashini, L., Arulnandhy, V., & Wijeratnam, R. S. W. (2008). Cultivation of oyster mushroom (*Pleurotusostreatus*) on sawdust. *Ceylon Journal of Science (Biological Science), 37*(2), 177–182.

Patil, S. S. (2013). Productivity and proximate content of *Pleurotussajor-caju*. *Bioscience Discovery, 4*(2), 169–172.

Patil, S. S., Ahmed, S. A., Telang, S. M., & Baig, M. M. V. (2010). The nutritional value of *Pleurotusostreatus* (Jacq.: Fr.) kumm cultivated on different lignocellulosic agrowastes. *Innovative Romanian Food Biotechnology, 7,* 66–76.

Peiris, C., Gunatilake, S. R., Wewalwela, J. J., & Vithanage, M. (2019). Biochar for sustainable agriculture: Nutrient dynamics, soil enzymes, and crop growth. In Ok Y. S., Tsang D. C. W., Bolan N., & Novak J. M., eds, *Biochar from Biomass and Waste*. New York: Elsevier Inc, pp. 211–224.

Philippoussis, A. (2009). Production of Mushrooms Using Agro-Industrial Residues as Substrates. In Singh nee' Nigam P., & Pandey A., eds, *Biotechnology for Agro-Industrial Residues Utilisation: Utilisation of Agro-Residues*. Dordrecht: Springer, pp. 163–196.

Philippoussis, A., Zervakis, G., & Diamantopoulou, P. (2001a). Bioconversion of lignocellulosic wastes through the cultivation of the edible mushrooms *Agrocybeaegerita, Volvariellavolvacea* and *Pleurotus*spp. *World Journal of Microbiology and Biotechnology, 17*(2), 191–200.

Ragunathan, R., & Swaminathan, K. (2003). Nutritional status of *Pleurotus* spp. grown on various agrowastes. *Food Chemistry, 80*(3), 371–375.

Rashad, F. M., El Kattan, M. H., Fathy, H. M., Abd El-Fattah, D. A., El Tohamy, M., & Farahat, A. A. (2019). Recycling of agro-wastes for *Ganoderma lucidum* mushroom production and *Ganoderma* post mushroom substrate as soil amendment. *Waste Management, 88,* 147–159.

Reddy, S. M. (2015). Diversity and applications of mushrooms. In Bahadur B., Rajam M. V., Sahijram L., & Krishnamurthy K. V., eds, *Plant Biology and Biotechnology*. Switzerland: Springer Nature, pp. 231–261

Rice Knowledge Bank (2021). http://www.knowledgebank.irri.org/step-by-step-production/postharvest/rice-by-products/rice-husk (accessed on 10-02-2021)

Ricepedia. (2021). http://ricepedia.org/rice-as-food/the-global-staple-rice-consumers (accessed on 12-02-2021).

Ritota, M., & Manzi, P. (2019). *Pleurotus* spp. cultivation on different agri-food by-products: Example of biotechnological application. *Sustainability, 11*(18), 5049.

Rodriguez-Gomez, D., Lehmann, L., Schultz-Jensen, N., Bjerre, A. B., & Hobley, T. J. (2012). Examining the potential of plasma-assisted pretreated wheat straw for enzyme production by *Trichoderma reesei*. *Applied Biochemistry and Biotechnology, 166*(8), 2051–2063.

Rose, P. K., & Devi, R. (2018). Heavy metal tolerance and adaptability assessment of indigenous filamentous fungi isolated from industrial wastewater and sludge samples. *Beni-Suef University Journal of Basic and Applied Sciences, 7*(4), 688–694.

Rosero-Delgado, E. A., Zambrano-Arcentales, M. A., Gómez-Salcedo, Y., Baquerizo-Crespo, R. J., & Dustet-Mendoza, J. C. (2021). Biotechnology Applied to Treatments of Agro-industrial Wastes. In Maddela N. R., Cruzatty L. C. G., Chakraborty S., eds, *Advances in the Domain of Environmental Biotechnology*. Switzerland: Springer Nature, pp. 277–311.

Royse, D. J., Baars, J., & Tan, Q. (2017). Current overview of mushroom production in the world. In Diego C. Z., & Pardo-Giménez A., eds, *Edible and Medicinal Mushrooms: Technology and Applications*. England: Wiley-Blackwell, pp. 5–13.

Ruan, Z., Wang, X., Liu, Y., & Liao, W. (2019). Corn. In Pan Z., Zhang R., & Zicari S., eds, *Integrated Processing Technologies for Food and Agricultural By-Products*. New York: Elsevier Inc, pp. 59–72.

Rajarathnam, S., & Shashirekha, M. N. (2003). Mushrooms and truffles | Use of wild mushrooms. In. Caballero, B. ed, *Encyclopedia of Food Sciences and Nutrition* (Second Edition). New York: Elsevier Inc, pp. 4048–4054.

Sadh, P. K., Duhan, S., & Duhan, J. S. (2018). Agro-industrial wastes and their utilization using solid state fermentation: a review. *Bioresources and Bioprocessing*, *5*(1), 1–15.

Saha, B. C., Iten, L. B., Cotta, M. A., & Wu, Y. V. (2005). Dilute acid pretreatment, enzymatic saccharification and fermentation of wheat straw to ethanol. *Process Biochemistry*, *40*(12), 3693–3700.

Saha, B. C., Jordan, D. B., & Bothast, R. J. (2009). Enzymes, Industrial (overview). In. Schaechter, M. ed, *Encyclopedia of Microbiology* (Third Edition). New York: Elsevier Inc, pp. 281–294.

Salama, A. N., Abdou, A. A., Helaly, A. A., & Salem, E. A. (2019). Effect of different nutritional supplements on the productivity and quality of oyster mushroom (*Pleurotusostreatus*). *Al-Azhar Journal of Agricultural Research*, *44*(2), 12–23.

Salmones, D., Mata, G., Ramos, L. M., & Waliszewski, K. N. (1999). Cultivation of shiitake mushroom, *Lentinula edodes*, in several lignocellulosic materials originating from the subtropics. *Agronomie*, *19*(1), 13–19.

Sánchez, C. (2009). Lignocellulosic residues: biodegradation and bioconversion by fungi. *Biotechnology Advances 27*(2), 185–194.

Sande, D., de Oliveira, G. P., eMoura, M. A. F., de Almeida Martins, B., Lima, M. T. N. S., & Takahashi, J. A. (2019). Edible mushrooms as a ubiquitous source of essential fatty acids. *Food Research International*, *125*, 108524.

Sardar, A., Satankar, V., Jagajanantha, P., & Mageshwaran, V. (2020). Effect of substrates (cotton stalks and cotton seed hulls) on growth, yield and nutritional composition of two oyster mushrooms (Pleurotusostreatus and Pleurotusflorida). *Journal of Cotton Research and Development*, *34*(1), 135–145.

Sardar, H., Ali, M. A., Anjum, M. A., Nawaz, F., Hussain, S., Naz, S., & Karimi, S. M. (2017). Agro-industrial residues influence mineral elements accumulation and nutritional composition of king oyster mushroom (*Pleurotuseryngii*). *Scientia Horticulturae*, *225*, 327–334.

Sarker, N. C., Ahmed, S., Hossain, K., Jahan, A., & Quddus, N. M. M. (2009). Performance of different strains of shiitake mushroom (*Lentinus edodes*) on saw dust. *Bangladesh Journal of Mushroom*, *3*(1), 1–7.

Shashirekha, M. N., Rajarathnam, S., & Bano, Z. (2002). Enhancement of bioconversion efficiency and chemistry of the mushroom, *Pleurotussajor-caju* (Berk and Br.) Sacc. produced on spent rice straw substrate supplemented with oil seed cakes. *Food Chemistry*, *76*(1), 27–31.

Silva, E. S., Cavallazzi, J. R. P., Muller, G., & Souza, J. V. B. (2007). Biotechnological applications of *Lentinus edodes*. *International Journal of Food, Agriculture and Environment*, *5*(3,4), 403–407.

Smit, A., & Huijgen, W. (2017). Effective fractionation of lignocellulose in herbaceous biomass and hardwood using a mild acetone organosolv process. *Green Chemistry*, *19*(22), 5505–5514.

Sørensen, A., Lübeck, M., Lübeck, P. S., & Ahring, B. K. (2013). Fungal beta-glucosidases: A bottleneck in industrial use of lignocellulosic materials. *Biomolecules*, *3*, 612–631.

Steffen, K. T., Cajthaml, T., Snajdr, J., & Baldrian, P. (2007). Differential degradation of oak (*Quercus petraea*) leaf litter by litter-decomposing basidiomycetes. *Research in Microbiology*, *158*(5), 447–455.

Taniguchi, M., Suzuki, H., Watanabe, D., Sakai, K., Hoshino, K., & Tanaka, T. (2005). Evaluation of pretreatment with *Pleurotusostreatus* for enzymatic hydrolysis of rice straw. *Journal of Bioscience and Bioengineering*, *100*(6), 637–643.

Tarasov, D., Leitch, M., & Fatehi, P. (2018). Lignin-carbohydrate complexes: properties, applications, analyses, and methods of extraction: a review. *Biotechnology for Biofuels*, *11*(1), 1–28.

Telang, S. M., Patil, S. S., & Baig, M. M. V. (2010). Biological efficiency and nutritional value of *Pleurotus sapidus* cultivated on different substrates. *Food Science Research Journal*, *1*(2), 127–129.

Tesfay, T., Godifey, T., Mesfin, R., & Kalayu, G. (2020). Evaluation of waste paper for cultivation of oyster mushroom (*Pleurotusostreatus*) with some added *supplementary materials*. *AMB Express*, *10*(1), 1–8.

Tiefenbacher, K. F. (2017). Technology of Main Ingredients-Water and Flours. In Tiefenbacher K. F. ed, *Wafer and Waffle*. New York: Elsevier Inc, pp. 15–121

Thiyageshwari, S., Gayathri, P., Krishnamoorthy, R., Anandham, R., & Paul, D. (2018). Exploration of rice husk compost as an alternate organic manure to enhance the productivity of blackgram in typic haplustalf and typic rhodustalf. *International Journal of Environmental Research and Public Health*, *15*(2), 358.

Tomczyk, A., Sokołowska, Z., & Boguta, P. (2020). Biochar physicochemical properties: pyrolysis temperature and feedstock kind effects. *Reviews in Environmental Science and BioTechnology*, *19*(1), 191–215.

Triyono, S., Haryanto, A., Telaumbanua, M., Lumbanraja, J., & To, F. (2019). Cultivation of straw mushroom (*Volvariellavolvacea*) on oil palm empty fruit bunch growth medium. *International Journal of Recycling of Organic Waste in Agriculture*, *8*(4), 381–392.

Tsai, S. Y., Wu, T. P., Huang, S. J., & Mau, J. L. (2007). Non-volatile taste components of *Agaricus bisporus*harvested at different stages of maturity. *Food Chemistry*, *103*(4), 1457–1464.

Tsoutsos, T. (2010). Modelling hydrolysis and fermentation processes in lignocelluloses-to-bioalcohol production. In Waldron K., ed., *Bioalcohol Production*. New York: Elsevier Inc, pp. 340–362.

Ul Haq, I., Khan, M. A., Khan, S. A., & Ahmad, M. (2011). Biochemical analysis of fruiting bodies of *Volvariellavolvacea* strain Vv pk, grown on six different substrates. *Soil & Environment*, *30*(2), 146–150.

USDA (2020). Grain: world market and trade rice. https://www.fas.usda.gov/data/grain-world-markets-and-trade (accesses on 28-12-2020).

Valverde, M. E., Hernández-Pérez, T., & Paredes-López, O. (2015). Edible mushrooms: improving human health and promoting quality life. *International Journal of Microbiology*, *376387*.

Van Hung, N., Maguyon-Detras, M. C., Migo, M. V., Quilloy, R., Balingbing, C., Chivenge, P., & Gummert, M. (2020). Rice straw overview: availability, properties, and management practices. In Gummert M., Hung N. V., Chivenge P., & Douthwaite B., eds, *Sustainable Rice Straw Management*. Switzerland: Springer nature, pp. 1–13.

Vos, A. M., Jurak, E., Pelkmans, J. F., Herman, K., Pels, G., Baars, J. J. Hendrix, E., Kabel, M. A., Lugones, L. G. & Wösten, H. A. (2017). H₂O₂ as a candidate bottleneck for MnP activity during cultivation of *Agaricus bisporus* in compost. *Amb Express*, *7*(1), 1–9.

Wang, J. T., Wang, Q., & Han, J. R. (2013). Yield, polysaccharides content and antioxidant properties of the mushroom *Agaricus subrufescens* produced on different substrates based on selected agricultural wastes. *Scientia Horticulturae*, *157*: 84–89.

Xu, F. (2010). Structure, Ultrastructure, and Chemical Composition. In Run-Cang Sun (ed), *Cereal Straw as a Resource for Sustainable Biomaterials and Biofuels*. New York: Elsevier Inc, pp. 9–47.

Yamanaka, K. (2017). Cultivation of mushrooms in plastic bottles and small bags. In Diego C. Z., & Pardo-Giménez A., eds, *Edible and Medicinal Mushrooms: Technology and Applications*. England: Wiley-Blackwell, pp. 309–338.

Yamauchi, M., Sakamoto, M., Yamada, M., Hara, H., Taib, S. M., Rezania, S., Fadhil, M. D. M., & Hanafi, F. H. M. (2019). Cultivation of oyster mushroom (*Pleurotusostreatus*) on fermented moso bamboo sawdust. *JournalF of King Saud University-Science*, *31*(4), 490–494.

Zervakis, G. I., Koutrotsios, G., & Katsaris, P. (2013). Composted versus raw olive mill waste as substrates for the production of medicinal mushrooms: an assessment of selected cultivation and quality parameters. *BioMed Research International*, *546830*.

Zhu, X., Chen, B., Zhu, L., & Xing, B. (2017). Effects and mechanisms of biochar-microbe interactions in soil improvement and pollution remediation: a review. *Environmental Pollution*, *227*, 98–115.

Zionmarketsearch, (2021). Global Mushroom Market worth USD 59.48 billion in 2021. www.zionmarketresearch.com (accessedon 01 February 2021).

Zied, D. C., Savoie, J. M., & Pardo-Giménez, A., (2011). Soybean the main nitrogen source cultivation substrates of edible and medicinal mushrooms. In El-Shamy H., ed., *Soybean and Nutrition*. Croatia: Intechopen, pp. 434–452.

Index

cultivation, 70–71
  casing, 71
  compost preparations, 70
  harvesting, 71
  spawning, 70–71
drying, 73–78
  freeze drying, 76–77
  hot air drying, 74–75
  microwave drying, 75–76
  osmotic dehydration, 77–78
  traditional drying, 73–74
electrolyzed water, 83
freezing, 78–79
Ganoderma species, 69
Grifolafrondosa, 70
irradiation, 82
Lentinus species, 69
ozone, 83
packaging, 82–83
pickling, 81–82
Pleurotusostreatus, 70
pulsed electric field, 83
Rigidoporus species, 70
species, 69–70
substrate for cultivation, 71–72
ultrasound, 83
primary antioxidants, synergism-regeneration of, 333–334
pro-oxidative effect of antioxidants, 340–342
profile, mineral, factors affecting, 159–160
protease classification, 447–448
  endopeptidases, 448
  exopeptidases, 447
proteases, 447–450
  aspartic proteases, 449–450
  mode of action, 448–450
  serine proteases, 448–449
  source of, 448
protein digestibility, edible mushrooms, 173–174
protein sources, mushrooms, 169–192
proteins, 195–196
  bioactive function of molecules from mushrooms, 195–196
  estimation of, 356–357
  from mushrooms, 171–172
proton coupled electron transfer mechanism, hydrogen atom transfer mechanism, 324–326
psilocybin-containing mushrooms, 381–383
psilocybin for treatment of major depression, 238–240
pulsed electric field, 83

qualitative estimation, 355–356
  alkaline/acid-extraction (AE), 355
  extraction of polysaccharides from mushrooms, 355
  hot water extraction (HWE) of polysaccharides, 355
  ultrasonic-assisted extraction (UAE), 356
qualitative phytochemical analysis, 359
  alkaloids, test for, 361
  Barfoed's test for monosaccharides, 359
  Fehling's test for free reducing sugar, 359
  soluble starch, test for, 361
  tannins, test for, 360

terpenoids, test for, 360–361
qualitative phytochemical analysis qualitative phytochemical analysis, test for terpenoids
  flavonoids by ferric chloride test, 360
  flavonoids by lead ethanoate test, 360
  flavonoids by NaOH test, 361
  flavonoids by Shinoda's test, 360
  saponins test, 360
qualitative/quantitative analysis techniques, 349–368
  amino acid assay, 359
  chemical component analysis, 354–355
    reversed phase-HPLC analysis of toxins, 354–355
  estimation of human health risks, 363–365
  fatty acid assay, 359
    fatty acid profile, 359
  Molisch's test for carbohydrates, 356
  nutritional, toxic studies for human health hazards, 362–363
    determination of arsenic content in mushrooms, 363
    determination of lead/cadmium content in mushroom, 363
    determination of mercury content in mushroom, 363
  physical component analysis, 351–354
    browning index, 353
    color measurement, 353
    dry density, 351
    mechanical stress test, 351–352
    moisture content, 351
    seed germination test, 354
    swelling index, 353
    thermal conductivity of mushroom, 352
    water absortion rate, 352
    water-holding capacity, 353
    yield, 353
qualitative estimation, 355–356
  alkaline- or acid-extraction (AE), 355
  extraction of polysaccharides from mushrooms, 355
  hot water extraction (HWE) of polysaccharides, 355
  ultrasonic-assisted extraction (UAE), 356
qualitative phytochemical analysis, 359
  Barfoed's test for monosaccharides, 359
  Fehling's test for free reducing sugar, 359
  test for alkaloids, 361
  test for soluble starch, 361
  test for tannins, 360
  test for terpenoids, 360–361
quantitative analysis, 356–359
  crude fiber analysis, 357–358
  determination of total ash, 358
  determination of total lipids, 357
  estimation of moisture control, 356
  estimation of total proteins, 356–357
  estimation of total sugar, 358–359
  Lowry assay by Folin reaction, 357
  mineral analysis, 358
  proximate analysis, 356–359
  soluble sugar assay, 359
  total carbohydrate estimation, 358
quantitative phytochemical analysis, 361–362
  determination of TPC, 361
  estimation of total flavonoid content, 362

For Product Safety Concerns and Information please contact our EU
representative GPSR@taylorandfrancis.com
Taylor & Francis Verlag GmbH, Kaufingerstraße 24, 80331 München, Germany

www.ingramcontent.com/pod-product-compliance
Lightning Source LLC
Chambersburg PA
CBHW060952210326
41598CB00031B/4801